Evolutionary Analysis

Scott Freeman
University of Washington

Jon C. Herron
University of Washington

PRENTICE HALL
Upper Saddle River, New Jersey 07458

Library of Congress Cataloging-in-Publication Data

Freeman, Scott,
    Evolutionary analysis / Scott Freeman, Jon C. Herron.
      p. cm.
    Includes bibliographical references and index.
    ISBN 0-13-568023-9
    1. Evolution (Biology) 2. Evolution (Biology)—Research.
  I. Herron, Jon C. II. Title.
  QH366.2.F73   1998
  576.8—dc21
                                         97-36380
                                         CIP

Executive Editor: Sheri L. Snavely
Production Editor: Debra A. Wechsler
Editor in Chief: Paul F. Corey
Editorial Director: Tim Bozik
Assistant Vice President of Production & Manufacturing: David W. Riccardi
Executive Managing Editor, Production: Kathleen Schiaparelli
Director of Marketing: Kelly McDonald
Marketing Manager: Jennifer Welchans
Creative Director: Paula Maylahn
Art Director: Heather Scott
Art Manager: Gus Vibal
Art Editor: Karen Branson
Interior Designer: Judith A. Matz-Coniglio
Cover Designer: Heather Scott
Cover Illustrations: Jon C. Herron
Copy Editor: Christianne Thillen
Manufacturing Manager: Trudy Pisciotti
Manufacturing Buyer: Ben Smith
Photo Research: Stuart Kenter Associates
Photo Editor: Lorinda Morris-Nantz/Melinda Reo
Illustrators: Laszlo Meszoly; Douglas & Gayle; Jon C. Herron
Cover Research: Karen Sanatar
Media Coordinator: Andrew T. Stull

Cover Art: (jumping spider and tephritid fly, photo and datagraphic) Reprinted with permission from *Science* 236:310–312. © 1987 American Association for the Advancement of Science. Courtesy of Dr. Erick Greene, University of Montana, Division of Biological Sciences; (alpine skypilot photo) © 1982 Richard Parker/Photo Researchers, Inc.; (datagraphic) Candace Galen, University of Missouri, Division of Biological Sciences; *(Archaeopteryx)* Sinclair Stammers/Science Photo Library/Photo Researchers, Inc.; (datagraphic) Reprinted with permission from *Nature* 378:349–354. © 1995 Macmillan Magazines Limited; (marine iguana, photo and datagraphic) Courtesy of Dr. Martin Wikelski, University of Washington, Dept. of Zoology

© 1998 by Scott Freeman and Jon C. Herron
Published by Prentice-Hall, Inc.
Simon & Schuster/A Viacom Company
Upper Saddle River, New Jersey 07458

Printed in the United States of America
10 9 8 7 6 5 4 3 2 1

ISBN 0-13-568023-9

Prentice-Hall International (UK) Limited, *London*
Prentice-Hall of Australia Pty. Limited, *Sydney*
Prentice-Hall Canada Inc., *Toronto*
Prentice-Hall Hispanoamericana, S.A., *Mexico*
Prentice-Hall of India Private Limited, *New Delhi*
Prentice-Hall of Japan, Inc., *Tokyo*
Simon & Schuster Asia Pte, Ltd., *Singapore*
Editora Prentice-Hall do Brasil, Ltda., *Rio de Janeiro*

Brief Contents

Contents

P A R T III   The History of Life   361

Preface

Evolutionary Analysis is a text for undergraduates majoring in biology and related fields. We assume that readers have completed their introductory coursework and are studying evolution, among other topics, to prepare for careers in human and animal healthcare, environmental management and conservation, primary and secondary education, science journalism, and biological and medical research. We hope this book will be a valued addition to your course. Our goal is to help students learn to:

- Ask interesting questions about evolution;
- Design experiments and plan observations that would answer those questions;
- Read and critically evaluate papers from the primary literature; and
- Contribute to an informed conversation about evolution between specialists and the lay public.

In short, we hope to help readers learn to think like evolutionary biologists.

This approach is inspired by the movement to reform undergraduate science education and its theme of teaching science as a process of discovery. Presenting science as science is done, instead of as a body of facts, is laudable and exciting. But what does this concept actually look like in practice? Can a textbook cover the foundations of a scientific field—its conceptual framework, classical results, and current research topics—in the context of exploring questions? Can instructors motivate students of biology with the same spirit of curiosity and investigation that motivates practicing scientists?

We wrote *Evolutionary Analysis* because we believe that the answer to these questions is yes. Instead of surveying facts, the book introduces each major organizing theme in evolution through a question. How has HIV become drug-resistant? Why did the dinosaurs, after dominating the land vertebrates for 150 million years, suddenly go extinct? Are humans more closely related to gorillas or to chimpanzees? Next we develop the strategies that evolutionary biologists use for finding an answer. Often a theoretical treatment helps to focus the question, frame hypotheses, and make predictions. Then we consider observations and experiments that test the predictions made by competing hypotheses, and discuss how the data are interpreted. Frequently, the conclusions prompt a new round of questions. Throughout the exposition our objective is to present evolutionary biology as a dynamic field of inquiry.

This investigative framework, and the theme of thinking like evolutionary biologists, unifies the book's four parts. These chapter groupings are intended to support several different approaches to organizing a course on evolution.

- Part I, Darwinism and the Fact of Evolution, demonstrates the relevance of evolution to real-life problems, establishes the fact of evolution, presents natural selection as an observable process, and introduces some of the major approaches and methods used by contemporary researchers.
- Part II, Mechanisms of Evolutionary Change, develops the theoretical under-

pinnings of the Modern Synthesis by exploring the mechanisms of evolution and speciation, the nature of adaptation, and molecular evolution.

- Part III, The History of Life, emphasizes phylogenetic analysis and contemporary inquiry, with chapters devoted to methods for reconstructing phylogenies, the origin of life and early evolution, the Cambrian explosion, and mass extinctions.
- Part IV, Current Research—A Sampler, provides in-depth coverage of other issues inspiring contemporary investigations, including human evolution, sexual selection, life histories, and conservation.

Most chapters include boxes that cover special topics or methods, provide more detailed analyses, or offer derivations of equations. All of the chapters end with a set of questions that encourage readers to review the material, apply their understanding to new issues, talk to their peers and instructors, and explore the current literature.

Studying evolution should be relevant to readers' personal and professional lives, and interesting to students new to the field. As a result, we include discussions of the evolution of drug resistance in pathogens, evolution and human health, the evolution of senescence, sexual selection, social behavior, eugenics, and biodiversity and conservation.

In partnership with the book, the text-specific website *Evolution in Action* provides a forum where professors and students can interact with their colleagues and peers at other campuses. The site includes research postings, virtual experiments and simulations, current hyperlinks, book updates, and additional questions to test students' knowledge. *Evolution in Action* is located at

http://www.prenhall.com/freeman

To access the site enter the username **beagle** and the password **galapagos.**

Like the science, this book and the website are a work in progress. If we have left things out that interrupt the flow of your course, if there are examples that would illustrate concepts better than the ones we use, if sections prove confusing, or if we have done something right, please let us know. We will do our utmost to improve future editions of the book based on your input. Please contact us through the website.

Acknowledgments

Evolutionary Analysis was vastly improved by the thoughtful, thorough, and critical reviews provided by the following people. We thank them for their insight.

Mary V. Ashley, *The University of Illinois at Chicago*
Steven N. Austad, *University of Idaho*
Leticia Avilés, *University of Arizona*
David R. Begun, *University of Toronto*
Timothy J. Bell, *Chicago State University*
David Berrigan, *University of Washington*
Edmund D. Brodie III, *Indiana University*
Jennifer A. Clack, *Cambridge University*

George A. Clark, Jr., *University of Connecticut*
Ross H. Crozier, *La Trobe University*
Lynda Delph, *Indiana University*
Charles D. Dieter, *South Dakota State University*
Thomas Dowling, *Arizona State University*
Ann E. Edwards, *University of Washington*
Stephen T. Emlen, *Cornell University*
Douglas H. Erwin, *National Museum of Natural History*

Paul W. Ewald, *Amherst College*
Candace Galen, *University of Missouri–Columbia*
Charles J. Goodnight, *The University of Vermont*
James L. Gould, *Princeton University*
Michael C. Grant, *University of Colorado, Boulder*
Erick Greene, *University of Montana*
Martha Groom, *North Carolina State University*
John Hartung, *State University of New York at Brooklyn*
Marshal C. Hedin, *University of Arizona*
Willem J. Hillenius, *University of California, Los Angeles*
Kathleen Hunt, *University of Washington*
David Jablonski, *University of Chicago*
Joel Kingsolver, *University of Washington*
Andrew H. Knoll, *Harvard University*
Wen-Hsiung Li, *University of Texas, Houston*
Charles A. Long, *University of Wisconsin, Stevens Point*
Charles Lydeard, *University of Alabama*
Robin S. Manasse, *National Research Council*
Gary Meffe, *Savannah River Ecology Lab*
Alexander F. Motten, *Duke University*
Timothy A. Mousseau, *University of South Carolina*

Daniel M. Pavuk, *Bowling Green State University*
Patricia A. Peroni, *Davidson College*
Grant Pogson, *University of California, Santa Cruz*
David Queller, *Rice University*
Mark D. Rausher, *Duke University*
Paul A. Roberts, *Oregon State University*
W. Bruce Saunders, *Bryn Mawr College*
Stephen W. Schaeffer, *Pennsylvania State University*
Daniel J. Schoen, *McGill University*
Paul Stimers, *University of Washington*
Ethan J. Temeles, *Amherst College*
A. Spencer Tomb, *Kansas State University*
Robin W. Tyser, *University of Wisconsin, La Crosse*
Shawn van Meter, *University of Puget Sound*
Sara Via, *Cornell University*
Martin Wikelski, *University of Washington*
Ed Wiley, *University of Kansas*
Peter Wimberger, *University of Puget Sound*
Carl R. Woese, *University of Illinois*
David S. Woodruff, *University of California, San Diego*

Additional special thanks go to Lynda Delph for reading the entire manuscript in proof and providing supportive and constructive feedback.

We thank Kathleen Hunt for drafting end-of-chapter questions, Sara Tilstra for assistance with library research, and the many researchers and publications who allowed us to use their data, photographs, and illustrations.

Our publisher's enthusiastic and talented team of editors, designers, and producers have been generous, tireless, patient, and good-humored while guiding this book through conception, creation, production, and launch. Editor in Chief of Prentice Hall Science Paul F. Corey has been rock-steady in his leadership and support. Executive Managing Editor Kathleen Schiaparelli kept the project on schedule. Senior Production Editor Debra Wechsler was tolerant with us and eagle-eyed with the manuscript, ensuring that the art and printed text are accurate and complete, and that a million different pieces came together in the end. Art Editors Karen Branson and Gus Vibal managed the art production and cheerfully gave emergency tutorials in electronic art preparation. Art Director Heather Scott coordinated the design of the book and the cover. Chris Thillen carefully copyedited the manuscript. Senior Marketing Manager Jennifer Welchans is providing tremendous energy and enthusiasm in marketing the book.

Last and most, we offer our deep admiration and gratitude to our editor and muse Sheri Snavely, who has been a constant source of inspiration, support, ideas, motivation, and wisdom.

Scott Freeman
Jon C. Herron
Seattle, Washington

PART I

DARWINISM AND THE FACT OF EVOLUTION

One of the most attractive aspects of evolutionary biology is that it is rooted in a single, general, organizing mechanism called natural selection. Gaining a thorough understanding of natural selection is thus the first challenge for anyone new to the science of evolution. Natural selection is simple in concept, but widely misunderstood. Understanding how the process works requires moving far beyond slogans like "survival of the fittest."

The primary goal of Part I (Chapters 1–3) is to introduce natural selection as the primary agent of evolutionary change. This presentation lays the groundwork for exploring the other mechanisms that cause change through time, in Part II.

A secondary objective of the first three chapters is to introduce the experimental and analytical methods used by researchers in evolutionary science. These methods are a prominent theme in Chapters 1 and 2 and are the primary focus of Chapter 3. Our goal is to help readers learn to ask questions, design experiments, analyze data, and critically review scientific papers, in addition to mastering facts. The detailed examples we present not only make the general concepts of evolutionary biology clear but also provide insight into how we know what we know.

A Case for Evolutionary Thinking: Understanding HIV

At the start of a course it can be useful to step back and ask two questions: What sort of content will be covered? How will this information help me in my professional and everyday life? To help answer these questions, we explore the evolution of human immunodeficiency viruses 1 and 2 (HIV-1 and HIV-2). These are the pathogens that cause the acquired immune deficiency syndrome (AIDS).

In this chapter we introduce the scope of evolutionary analysis through an in-depth look at a prominent contemporary issue. Our goals are to illustrate the kinds of questions that evolutionary biologists investigate, demonstrate how an evolutionary perspective can inform research throughout the biological sciences, and introduce concepts that will be explored in detail later in the text.

HIV is a compelling case study because it raises issues that are almost certain to influence the professional and personal lives of every one of our readers. As an emerging virus that rapidly evolves drug resistance, HIV exemplifies two of the most pressing public health issues of our time. AIDS may, in fact, become one of the most devastating epidemics ever experienced by our species.

Here are the questions we address:

- Why have promising AIDS treatments, like the drug AZT, proven ineffective in the long run?
- How does HIV win its battle against the immune system?
- Why does HIV kill people?
- Where did HIV come from?
- How can we best manage the AIDS epidemic?

These questions may not sound as if they have anything to do with evolutionary biology. But evolution is the science devoted to understanding two things:

As a case study, HIV will demonstrate how evolutionary biologists study adaptation and diversity.

(1) how populations change through time in response to modifications in their social, biological, and physical environments, and (2) how new species come into being. More formally, evolutionary biologists study adaptation and diversity. These are *exactly* the issues targeted by our questions about AIDS and HIV. But before we tackle them, we need to delve into some background biology.

1.1   The Natural History of the Epidemic, HIV, the Immune System, and AIDS

In late 1996 the World Health Organization estimated that almost 23 million people worldwide were infected with HIV (Holden 1996; Figure 1.1). Most of these infections resulted from two related but distinct epidemics that occurred during the 1980s and early 1990s: one among heterosexual men and women in urban areas of sub-Saharan Africa, and the other among homosexual men and intravenous drug users in the United States and Europe. These two waves of infection are distinguished by the number of people involved and by the disease's mode of transmission.

The nature and size of the AIDS epidemics in Africa, North America and Europe, and Asia contrast sharply.

In sub-Saharan Africa the number of AIDS cases is almost overwhelming in magnitude (see De Cock et al. 1990; Palca 1991; Paley 1994; Piot 1996). Epidemiologists estimate that over 14 million Africans are now infected with HIV-1, and project that one-third of the workforce in some countries may eventually die of AIDS. In some African cities AIDS is already the leading cause of adult death. In this epidemic HIV is transmitted primarily through heterosexual intercourse, and men and women are about equally likely to be infected.

In contrast, the AIDS epidemic in North America and Europe is characterized by transmission via homosexual intercourse and needle sharing among intravenous drug users. Rates of infection in the general population are much lower than in Africa (Figure 1.2), and men are much more likely to be infected with HIV than women. For example, about one in a hundred men and one in eight hundred women, for a total of over 835,000 people, are now thought to be infected in the United States (Levy et al. 1994). Although rates of

Figure 1.1 **Distribution of HIV infections** This map shows the geographic distribution of HIV-1 and HIV-2 infections. The most dramatic conclusion to be drawn from this figure is that AIDS is primarily a disease of developing nations. It is estimated that over 90% of HIV-infected people live in the poor countries of the southern hemisphere. Two-thirds of these are in sub-Saharan Africa alone.

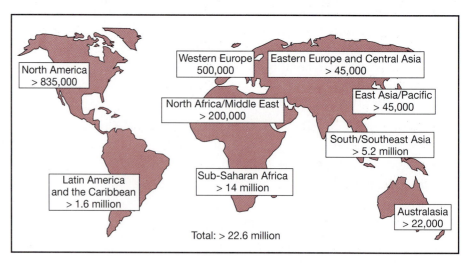

North America
> 835,000

Western Europe
500,000

Eastern Europe and Central Asia
> 45,000

North Africa/Middle East
> 200,000

East Asia/Pacific
> 45,000

South/Southeast Asia
> 5.2 million

Latin America and the Caribbean
> 1.6 million

Sub-Saharan Africa
> 14 million

Australasia
> 22,000

Total: > 22.6 million

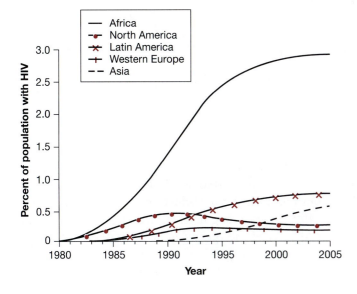

Figure 1.2   HIV infection rates by geographic area
This graph plots the documented and estimated rates of
HIV infection for five broad geographic areas, starting with
the first confirmed AIDS case in 1981 and ending with
projections beyond the turn of the century. Infection rates
are expected to drop or plateau in the wealthy countries of
North America and Europe, but increase steadily in the
poor nations of Africa, Latin America, and Asia.

heterosexual transmission continue to increase in the United States, efforts by
AIDS educators and researchers are paying off: The total annual number of new
AIDS cases in North America is leveling off, and will soon begin declining
(Figure 1.3).

Now a third wave of infections, transmitted via heterosexual intercourse as in
Africa, has begun in south and east Asia—particularly in Thailand and India
(Cohen 1995a, b). If rates of infection continue to increase among the large
populations of Asia, and if the African epidemic does not slow, at least 40 million people, and possibly as many as 100 million, will become infected with the
virus by the year 2000 (Levy et al. 1994; Tarantola et al. 1994). Most people infected with HIV-1 die within 15 years.

What is HIV? Some Background on Viruses and Retroviruses

Viruses are intracellular parasites, incapable of an independent life. As a group
they are remarkably diverse in the hosts they afflict. But individually, viruses are

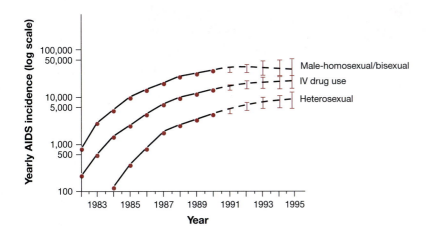

**Figure 1.3   Incidence of AIDS cases, by
risk group, in the United States**   This
graph shows the increase in the number of
new AIDS cases reported each year from the
early 1980s to mid-1990s. While illness in
male homosexuals and IV drug users
plateaued or began to decline by 1991, illness
in heterosexuals continues to increase. Note
that the y-axis has a logarithmic scale.

(a)

(b)

(c)

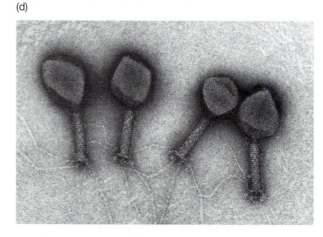

(d)

Figure 1.4 **Viral diversity** Virions can be shaped like rods, spheres, tetrahedrons, or rods with tetrahedral "heads." They consist of one or more molecules of a nucleic acid (DNA or RNA) and a protein coat. Some viruses also have a membrane containing lipid and glycoprotein. Note that viruses are not cells, and are usually referred to simply as "particles."   (a) Sendai virus (×227,220) is an RNA virus closely related to the measles virus. (K.G. Murti/Visuals Unlimited)   (b) Rotavirus (×528,000). (K.G. Murti/Visuals Unlimited)   (c) Adenovirus (×300,000). Experiments on this DNA virus confirmed that mRNAs undergo several processing steps prior to being translated. (Biophoto Associates/Science Source/Photo Researchers, Inc.)   (d) Bacteriophage T4 (×165,000). This DNA virus infects the bacterium *Escherichia coli* ("bacteriophage" or "phage" are general names for viruses that infect bacteria). Many of the details of DNA replication were worked out in experiments involving this virus. (Lee D. Simon/Photo Researchers, Inc.)

highly specific in the organisms they infect and in the cell types they parasitize. To the best of our knowledge there are viruses capable of parasitizing every organism known, and virtually every tissue and cell type within those organisms. The viruses featured in Figure 1.4 represent just a tiny sample of the diversity found in the size, shape, and target cells of viruses.

Figure 1.5 shows the life cycle of a "typical" virus, the bacteriophage T4. Notice that the life cycle alternates between an extracellular or infectious phase, when the virus can be transmitted from host to host, and an intracellular or parasitic phase, when the virus replicates. During the parasitic phase, the virus uses the host cell's own enzymatic machinery—its polymerases, ribosomes, and

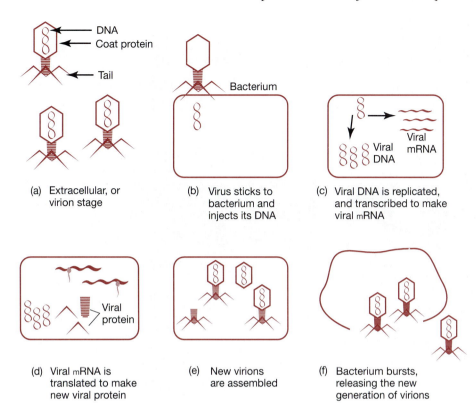

(a) Extracellular, or virion stage

(b) Virus sticks to bacterium and injects its DNA

(c) Viral DNA is replicated, and transcribed to make viral mRNA

(d) Viral mRNA is translated to make new viral protein

(e) New virions are assembled

(f) Bacterium bursts, releasing the new generation of virions

Figure 1.5 **The life cycle of bacteriophage T4** This virus infects the bacterium *Escherichia coli. E. coli* is abundant in the human gut.

tRNAs—to grow and reproduce. This is why viral diseases are notoriously difficult to treat: Drugs that interrupt the viral life cycle are almost certain to interfere with the host cell's enzymatic functions as well.

The genetic, or information-bearing, molecule that directs a virus's attack on a host cell can be either DNA or RNA, but not both. In DNA viruses the flow of genetic information follows the familiar route from DNA to mRNA to proteins. In some RNA viruses genetic information flows from RNA to proteins, with the same RNA molecule serving as both chromosome and mRNA. But in another group of RNA viruses, discovered by Howard Temin and colleagues in 1970, genetic information takes a different path. These "retroviruses" have an enzyme, called reverse transcriptase, that catalyzes the synthesis of DNA from an RNA template. In retroviruses information flow is from RNA to DNA to mRNA to proteins. It is that first step, with a backward flow of information, that inspired the "retro" in *retrovirus.*

HIV is a retrovirus. It was first isolated and named in 1984, in Luc Montagnier's lab at the Pasteur Institute in Paris and in Robert Gallo's lab at the National Institutes of Health in Bethesda, Maryland. HIV is fairly complex compared with other viruses. Many viruses have only three or four genes, but HIV has nine (Table 1.1). HIV is also slightly unusual because it has a diploid genome. There are two RNA strands in each virus particle, or virion (Figure 1.6).

Like most viruses, HIV parasitizes a specific cell type. HIV particles infect a category of lymphocytes ("white blood cells") called helper T cells. Specifically,

TABLE 1.1   HIV's nine genes and the functions of the proteins they code for

Note that by convention the names given to genes are short (often 3–4 letters) and italicized; they are often acronyms for words that describe their function. In contrast to viruses, humans are thought to have about 100,000 genes.

Gene	Function
gag	The protein produced by this gene splits into several "structural core" proteins that assemble into a shell-like structure around the virus's RNA.
env	("envelope") This gene produces a long protein that splits into the gp120 and gp41 glycoproteins. These form the outer capsule of the virion. They are also the components of the virus that the immune system "sees" and responds to (see Figure 1.8).
pol	("polymerase") This is the gene for reverse transcriptase. RT does three things: it makes a DNA complementary to the viral RNA (i.e., it functions as a polymerase), it digests the RNA strand of the subsequent RNA/DNA hybrid (i.e., it functions as a nuclease), and it then makes the remaining DNA strand double-stranded, so it can incorporate into the host genome.
tat	This gene makes a protein that increases viral replication.
rev	This gene codes for a protein product that prevents the host cell from splicing small mRNAs within the nucleus. As a result, long viral mRNAs can enter the cytoplasm and be translated.
nef	("negative factor") The function of this gene is still unclear. Experiments have shown that it can suppress or enhance HIV replication, depending on cell type.
vif	This gene seems to help in infectivity and replication.
vup	This gene seems to help in infectivity and replication.
vpr	This gene seems to help in infectivity and replication.

Sources: Levy et al. 1994; Brookmeyer and Gail 1994

HIV attacks helper T cells that have a protein called CD4 on their surface. The life cycle of HIV, from fusion with the cell surface through reverse transcription, replication, and release of a new generation of virions, is diagrammed in Figure 1.7. Because helper T cells with CD4, called CD4+ lymphocytes, are crucial to the functioning of the immune system, and because HIV infection

(a)

(b)

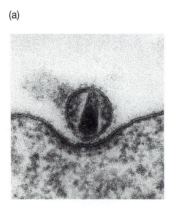
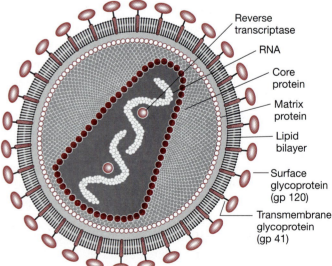

Reverse transcriptase

RNA

Core protein

Matrix protein

Lipid bilayer

Surface glycoprotein (gp 120)

Transmembrane glycoprotein (gp 41)

Figure 1.6 **HIV structure**   (a) Transmission electron micrograph of an HIV virion (×100,000). (Hans Gelderblom/Visuals Unlimited)   (b) Diagrammatic view of the HIV virion; see Table 1.1 to match HIV's genes with the proteins labeled on this figure.

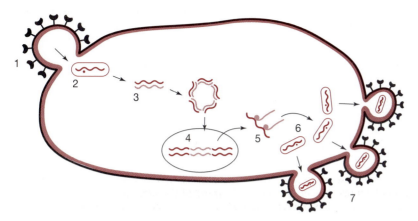

Figure 1.7   The life cycle of HIV   (1) The HIV virion attaches to the CD4 protein on the surface of a macrophage or helper T lymphocyte. The CD4 protein has been called the doorknob that virions "grasp" to enter the cell. (2) HIV's RNA, reverse transcriptase molecules, and regulator proteins enter the cell. (3) Reverse transcriptase synthesizes viral DNA in the host cell cytoplasm. (4) The viral DNA enters the nucleus, in circular form, and integrates into the host's genome at a transcriptionally active site. The host cell's DNA polymerase II then transcribes the proviral genome. (5) Proviral mRNAs are translated into proteins by the host cell's rRNA machinery. (6) Virions are assembled in the host cell cytoplasm. (7) The new generation of infectious virions bursts into the bloodstream (after Levy et al. 1994).

kills the affected cell, advanced HIV infections undermine the immune response. The collapse of the immune response leads to the condition known as AIDS. AIDS is characterized by opportunistic infections with bacterial and fungal pathogens that do not ordinarily cause problems for people with robust immune systems.

HIV attacks helper T cells that are crucial to the immune system's response to infection.

Cellular Defenses: The Immune Response to Infection

The human body responds to HIV infection in two ways: by destroying virions floating in the bloodstream, and by killing its own infected T cells before new virions are assembled and released.

The key to destroying free virions is for the immune system to recognize that they are foreign, or non-self. As Figure 1.8 illustrates, this recognition depends crucially on a role performed by helper T cells. Helper Ts circulate in the bloodstream, checking the surfaces of a class of white blood cells called macrophages. Macrophages are the immune system's vacuum cleaners. They engulf bacteria or virions and display fragments of the invaders' proteins, called epitopes, on their cell surfaces. When a helper T cell encounters a displayed epitope that it recognizes as foreign, the helper T cell triggers the immune system's humoral response by releasing hormones called lymphokines. The lymphokines help activate B cells that manufacture and release antibodies. The antibodies mark free virions for destruction (see figure).

Similarly, the key to destroying infected cells is to recognize that they are parasitized. Helper T cells are again crucial. The lymphokines they release, in combination with epitopes displayed by macrophages, activate killer T cells. The

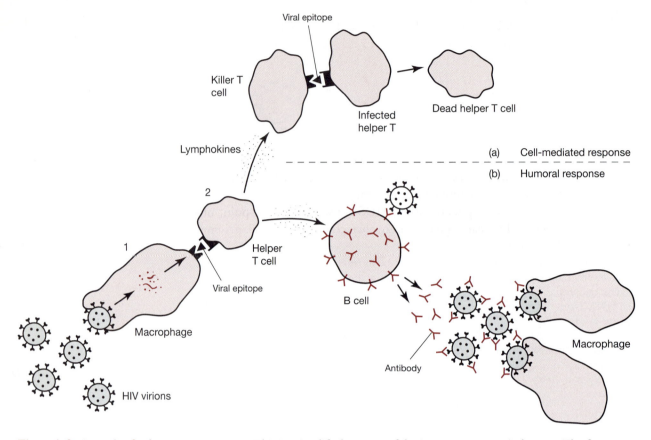

Figure 1.8 Steps in the immune response This is a simplified cartoon of the immune response in humans. The figure shows the immune reaction to HIV infection as an example.
(1) White blood cells called macrophages engulf HIV virions, break them up, and display pieces of the viral *env* proteins on their surface. These *env* protein fragments are called epitopes. (2) A few helper T cells have receptors that specifically recognize and bind to an epitope displayed by a macrophage. Once bound, the helper T cells divide and begin secreting hormones called lymphokines.   (a) Cell-mediated response: Killer T cells bind to the *env* epitope displayed by macrophages, and are stimulated by the lymphokines released by helper T cells. The stimulated killer T cells divide and begin destroying infected T cells, which they can identify because the infected cells display the *env* epitope.   (b) Humoral response: B cells bind to the *env* epitope on the surface of free virions, and are stimulated by the lymphokines secreted by helper T cells. The stimulated B cells divide and produce antibodies to the virus. The antibodies bind to free virions, marking them for more efficient destruction by macrophages.

activated killer T cells divide, then go about identifying infected host cells—by the epitopes they display—and destroying them. This is the immune system's cell-mediated response (Figure 1.8).

Each helper T cell in the body recognizes just a handful of epitopes out of the millions that exist (see Nowak et al. 1995; Nowak and McMichael 1995). Therefore only a tiny fraction of the helper T cells in each person's body are capable of recognizing HIV's epitopes, which are derived from glycoproteins in the virion's envelope (Figure 1.6).

Each helper T cell responds to a limited number of epitopes from invading pathogens.

To summarize, two points about immune system function are central to understanding the natural history of AIDS and the evolution of HIV:

1. Helper T cells are the key to both types of immune response. If the CD4+ helper T cells become infected by HIV and are destroyed, either by the virus itself or by killer cells, then the immune system's ability to respond to other types of invaders is compromised.

2. The recognition step is extremely specific. Each helper T, killer T, and B cell lineage binds and responds to less than 10 epitopes. If the epitope presented by the virions in an infection changes, the cell lineages that responded to the original epitope may not respond to the new one.

AIDS

HIV does not kill people directly. Rather, the destruction of CD4+ T cells weakens the immune system so thoroughly that rare diseases, including cancers like Kaposi's sarcoma and unusual strains of pneumonia like *Pneumocystis carinii,* can overwhelm the immune system and cause death.

HIV does not kill people quickly, either. AIDS is usually the end-point of a three-phase sequence. The early phase, often characterized by swollen lymph nodes, skin rashes, and fever, occurs within a few weeks of infection but in only about 30% of patients. The second, or asymptomatic, phase can last much longer: an average of 10 years among infected people in North America. Although few virions are circulating in the bloodstream during this time, the HIV population is anything but latent: the patient's lymph tissue is loaded with virions. As the infection progresses, the number of CD4+ T cells in the patient's bloodstream gradually but steadily declines from a normal level of about 1,000 cells/milliliter of blood to only 200 cells/mL. It is at this threshold that patients enter the third, or symptomatic, phase of infection. This is the phase called AIDS. Most patients with AIDS have died within 2 years.

With the preceding information as background, we are ready to explore our five questions about HIV evolution. The first question has frustrated everyone involved in fighting the epidemic: Why has it been so hard to develop an effective, lasting treatment? It is certainly not for lack of trying; governments and private companies have committed tens of millions of dollars to AIDS research and drug development. The story of AZT, one of the first anti–AIDS drugs, has turned out to be typical. Early on AZT looked promising, but ultimately it proved a disappointment. To explain why, we need to introduce the principle of natural selection.

1.2   Why Does AZT Work in the Short Run, but Fail in the Long Run?

Viral diseases are difficult to treat because they use the host cell's own enzymes in almost every step of their life cycle (see Figure 1.5 and Figure 1.7). A drug that inhibits transcription of viral DNA or blocks translation of viral proteins, for example, is likely to poison the host cell as well.

Fortunately, rational drug design can take advantage of the fact that HIV is a retrovirus (Connelly and Jenkins 1994). A drug that interrupts reverse transcription should effectively and specifically kill retroviruses, with minimal side effects. That is exactly the rationale behind azidothymidine, or AZT.

Note the *thymidine* in AZT's name: AZT is a nucleotide analog, similar in structure to normal thymidine, that "fools" reverse transcriptase. When AZT is present in a cell, reverse transcriptase mistakenly adds it to the growing DNA strand where a thymidine 5′-triphosphate should go. This misstep terminates

(a)   (b)   (c)

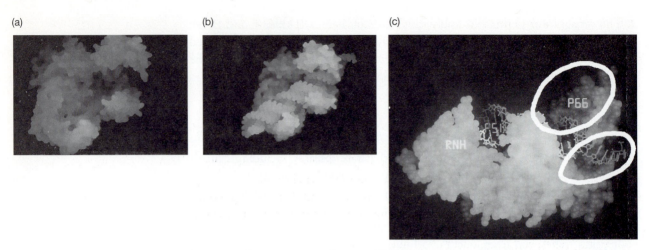

***Figure 1.9* Computer-generated images of reverse transcriptase**   (a) This image shows the large groove in the enzyme where the substrate (RNA) binds.   (b) This is a DNA/RNA hybrid molecule showing reverse transcriptase bound to DNA. (c) The dark spheres on this image inside the white circles, indicate the locations of amino acid substitutions correlated with resistance to AZT. Note that they are localized in the enzyme's groove, or active site. (Photos by Thomas A. Steitz, Yale University)

reverse transcription, because AZT will not accept the addition of the next nucleotide to the growing chain. AZT thus interrupts the pathway to new viral proteins and new virions, and stops the infection.

In early tests, AZT worked. It effectively halted the decline of CD4+ cell counts in AIDS patients. AZT produced serious side effects, because it also sometimes fools DNA polymerase and interrupts DNA synthesis in host cells. But it promised to at least halt the progression of the disease. By 1989, however, after only a few years of use, patients stopped responding to treatment, and their CD4+ cell counts again began to fall. Why?

Consider a thought experiment. If we wanted to genetically engineer an HIV virion capable of replicating in the presence of AZT, what would we do? The simplest answer might be to change the active site in the reverse transcriptase enzyme, to make it less likely to mistake AZT for the normal nucleotide. In practice, we could use a mutagenic chemical or ionizing radiation to generate strains of HIV with altered nucleotide sequences in their genomes, and thus altered amino acid sequences in their proteins. If we generated many mutants, at least a few would have a change in the part of the reverse transcriptase molecule that recognizes and binds to normal thymidine (Figure 1.9a). If one of these altered sequences were less likely to mistake AZT for the normal nucleotide, this mutant variant of HIV would be able to continue replicating in the presence of the drug. In populations of HIV virions grown in test tubes with AZT, strains unable to replicate in the presence of AZT would decline in numbers, and the new form would come to dominate the test tube HIV populations.

The steps involved in this thought experiment have actually occurred, inside the bodies of individual HIV patients. How do we know? Researchers took repeated samples of the HIV virions, throughout the course of treatment, from patients receiving AZT. In each sample of the virus, the researchers sequenced the reverse transcriptase gene. They found that viral strains present late in treatment were genetically different from viral strains present before treatment *in the same host individuals.* The viral populations had become resistant to AZT. The mutations associated with resistance were often the same from patient to patient (St. Clair et al. 1991; Mohri et al. 1993; Shirasaka et al. 1993), and were located in the active site of reverse transcriptase (Figure 1.9c). Researchers have observed the evolution of AZT resistance in dozens of AIDS patients. In each of these individuals, mutations in the HIV genome led to specific amino acid substitutions in the active site of reverse transcriptase. These genetic changes allowed the mutant strains of the virus to replicate in the presence of AZT. Unlike the situation in our thought experiment, however, no conscious genetic engineering took place. How, then, did the change in the viral strains occur?

Some mutations in the active site of reverse transcriptase make the enzyme less susceptible to AZT.

The keys are that reverse transcriptase is error-prone and that HIV's genome has no instructions for making error-correcting enzymes (study the list of genes in Table 1.1 again). (In this regard HIV is like most retroviruses, but unlike DNA-based cellular organisms such as ourselves.) As a result, over half of the DNA transcripts produced by reverse transcriptase contain at least one mistake (Carman et al. 1993). Because thousands of generations of HIV replication take place within each patient, a single strain of HIV can produce hundreds of different reverse transcriptase variants during the course of a single infection.

Simply because of their numbers, it is a virtual certainty that one or more of these variants contains an amino acid substitution that lessens the reverse transcriptase molecule's affinity for AZT. If the patient takes AZT, the replication of unaltered HIV variants is suppressed, but the resistant mutants will still be able to synthesize some DNA and produce new virions. As the resistant virions reproduce and the nonresistant virions fail, the fraction of the virions in the patient's body that are resistant to AZT increases over time. Furthermore, each new generation of virions will contain virions with additional new mutations. Some of these additional mutations may further enhance the ability of reverse transcriptase to function in the presence of AZT. Because they reproduce faster, the virions that carry these new mutations will also increase in frequency at the expense of their less resistant ancestors.

This process of change over time in the composition of the viral population is called evolution by natural selection. It has occurred so consistently in patients taking AZT that the use of AZT alone has been virtually abandoned as an AIDS therapy. In addition, when other nucleotide analogs like ddI or ddC have been used by themselves or in conjunction with AZT, HIV populations have evolved multiple-drug resistance (Larder et al. 1993; Shirasaka et al. 1993; Mohri et al. 1993; Ewald et al. 1994a; but see Holden 1995). And HIV is not unusual. The same type of response to drug therapies has been observed in populations of the flu and hepatitis B viruses (Carman et al. 1993), and in populations of the bacterium that causes tuberculosis (Bishai et al. 1996).

Changes in the genetic makeup of HIV populations over time have led to increased drug resistance. This is an example of evolution by natural selection.

Now consider a slightly different question. We have been following the fate of virions carrying different versions of the reverse transcriptase gene when AZT is present in the environment (that is, in host cells). Are the mutant HIV strains also more efficient at reproducing when host cells do *not* contain AZT? The answer is no: When AZT therapy has been started and then stopped, the proportion of AZT-resistant virions in the virus population has fallen over time back toward what it was before AZT treatment began. Back-mutations that restored reverse transcriptase's amino acid sequence to its original configuration were favored by natural selection (St. Clair et al. 1991). Notice how dynamic natural selection is: Without AZT present, natural selection favors nonmutant virions; with AZT present, natural selection favors mutant virions. Is evolution by natural selection unidirectional or irreversible? The answer, emphatically, is no.

Finally, HIV responds to natural selection involving forces in its environment other than antiviral drugs. For example, S. Itescu and colleagues (1994) have shown that HIV virions found in the lungs and blood of the same patient are highly divergent. Specifically, the *env* genes of each lineage are different. The *env* gene codes for the protein that is presented as an epitope by macrophages to T cells. The researchers suggest that the mutations they identified allow the virions that carry them to better elude the different macrophage types found in blood versus lungs. The research of Itescu and colleagues shows that HIV strains have diverged from the common ancestor that initiated the infection, and have become adapted to succeed in different host cell tissue types.

Note that the process we have just reviewed involves four steps:

1. Transcription errors, made by reverse transcriptase, led to mutations in the reverse transcriptase gene.
2. These mutations produced variability in enzyme function among virions.
3. Some virions were better able to survive and reproduce in an environment containing AZT than others, because of the functional properties of their mutant reverse transcriptase enzymes.
4. These mutations were passed on to the offspring of AZT-resistant virions.

The outcome is that new virion forms came to dominate HIV populations within hosts. The genetic composition of the HIV population at the end of the process was different from what it was at the start. This is evolution by natural selection.

Natural selection is pervasive and powerful, and is responsible for most of the adaptation and diversity observed among Earth's life forms. It is also responsible for HIV's ability to crush the immune system.

1.3   How Does HIV Defeat the Immune Response?

A hallmark of the immune response is its specificity. Each helper T-, B-, and killer T–cell lineage recognizes and responds to only a few epitopes. If each disease-causing agent produced a limited number of epitopes, the immune system could respond efficiently and prevent illness. But HIV, due to its high mutation rate, is constantly changing the epitopes it presents.

One hypothesis for how HIV defeats the immune response focuses on this point. The escape, or diversity threshold, hypothesis proposes that during the infection's latent phase, a pitched battle is going on in the lymph nodes (Holmes et al. 1992; Coffin 1995; Nowak and McMichael 1995; Nowak et al. 1995). The virus eventually wins because its population evolves by natural selection.

HIV has an extraordinarily high mutation rate . . .

HIV's weapon in its battle against the immune system is its mutation rate, which is among the highest ever recorded (Nowak 1990; Carman et al. 1993;

Ewald 1994a and b). Reverse transcriptase makes an error once in every 1,000 to 10,000 nucleotide bases it polymerizes. This error rate is a million times higher than that of DNA polymerase in humans, and a thousand times higher than that of influenza A's DNA polymerase (Li et al. 1988; Gardner 1994). Because HIV's genome contains just under 10,000 bases, a round of reverse transcription produces from 1 to 10 substituted bases per genome. Accepting J. Coffin's (1995) estimate that there are 300 viral replication cycles per year during the asymptomatic period (Coffin 1995), we can calculate that over the 10-year course of infection there are 3,000 viral generations. During this time sequence, diversity among HIV strains routinely increases from near 0% to 10% in each infected person (Wain-Hobson 1993a). Martin Nowak and Andrew McMichael (1995) point out that in the course of an infection in a single individual, HIV populations can undergo as much genetic change as humans have undergone in millions of years.

The escape hypothesis maintains that new helper T-cell lineages have to respond to epitope after epitope, produced through repeated rounds of replication and mutation by HIV. Note that from a virion's point of view, the immune system, like AZT, is an agent of selection. The immune response constantly selects against current epitopes. For any individual virus, there is a strong selective advantage to having a novel genotype. Virions with new mutations in their epitopes always have the highest survival and reproductive rate, because the immune system has not yet had time to mount a response against them. Over time, the diversity of virion types gets higher and higher, until there are so many different virion epitopes that the immune system cannot keep up with them all. At this threshold, an epitope is produced that is not recognized by the remaining helper T cells. The virions with this epitope escape the immune system and multiply without response, killing CD4+ cells as they go. As a result, the immune system collapses.

In fact, unless it is caused by a virus or bacterium that hides out in tissues not patrolled by the immune system, *any* disease with a long latency has to generate high epitope diversity to survive. Otherwise the immune system would eliminate it.

This logic suggests that HIV's high mutation rate is not accidental, but an adaptation: a genetically based response to environmental conditions. This hypothesis makes two predictions that we can test. First, in a comparison of HIV strains, some versions of reverse transcriptase should be more error-prone than others. That is, there should be genetically based variation in mutation rate. Although we do not yet have direct evidence for this in HIV, there is genetically based variation among strains of influenza A viruses in the mutation rate in *env* genes (Suárez et al. 1992). Second, HIV variants with high mutation rates should be more successful at replicating and sustaining an infection than variants with low mutation rates. If these predictions are tested and substantiated, it would suggest that HIV's mutation rate has indeed evolved to maximize the diversity of epitopes it presents.

. . . which may be an adaptation to produce new epitopes and escape detection by the immune system.

The escape hypothesis, however, offers only a partial explanation for what is going on in HIV evolution. The hypothesis addresses the "how" question of progression to AIDS, or what biologists call proximate causation. It explains how HIV defeats the immune system and causes death. But what about the "why" issue, or what we call ultimate causation?

1.4 Why is HIV Fatal?

One of the keys to becoming an evolutionary biologist is to learn how to think like an organism. This means adopting what is called "selection thinking." For example, from HIV's point of view the tendency to cause disease in a host, or what we call virulence, is largely a function of the virus's own growth rate. Illness in the host is a side effect of high growth rates. Extremely high growth rates can result in the death of the host. According to selection thinking, the key to understanding why HIV is fatal is to understand why it is advantageous for a virion to replicate as quickly as it does. We have just reviewed evidence that HIV can quickly adapt to differences among tissues and to changes in its environment brought about by drug therapies. If HIV can evolve so readily, why has it not evolved to have less of a negative impact on its host?

Sometimes the answer to questions like this is because it cannot. That is, perhaps competing HIV strains would infect more people if they grew more slowly and did not kill their hosts, but they are not able to because of some unchangeable property of reverse transcriptase, or because targeting CD4+ helper T cells inevitably makes infections fatal. As we will see in Chapters 2, 8, and 17, organisms are constrained in a variety of ways. Natural selection cannot optimize every aspect of adaptation.

The evidence suggests that constraints are not the cause of HIV-1's high virulence, however. There are, for example, nonlethal diseases that afflict CD4+ helper T cells. These include human herpes virus 6, which appears to cause only a rash similar to the mild childhood affliction called roseola (see Culliton 1990), and HIV-2, which may often be nonlethal (Ewald 1994a; Marlink et al. 1994). These observations suggest that diseases of CD4+ cells are not highly virulent by definition.

Another possibility is that evolution to a benign state has not occurred simply due to a lack of variation in the degree of virulence. If there are no mutations that alter the level of virulence, then virulence cannot evolve by natural selection. The data also argue against this hypothesis, however. We now have compelling evidence that HIV-1 strains vary in their degree of virulence. HIV-1 strains that dominate late in a given infection, when the patient is symptomatic, grow much faster in culture than strains present earlier in the infection (see Goldsmith 1990, Ewald 1994a). Base substitutions that are associated with virulence can be localized to the gp120 protein, which is coded for by the *env* gene (Goldsmith 1990; Groenink et al. 1993). This observation suggests a molecular mechanism for virulence: Rapidly growing strains are those with high mutation rates that maximize epitope diversity.

An alternative hypothesis is that natural selection has adjusted the virulence of HIV-1. This is the claim Paul Ewald makes with his transmission rate hypothesis (Ewald 1994a), illustrated in Figure 1.10. The transmission rate hypothesis works as follows: If transmission of sexually transmitted diseases from an existing host to new hosts is frequent, natural selection will favor increased virulence. But if transmission to new hosts is infrequent, selection will favor more benign strains. Note that we have now shifted from thinking about viral reproduction as the reproduction of individual virions inside single hosts to thinking about viral reproduction as infections hopping from host to host.

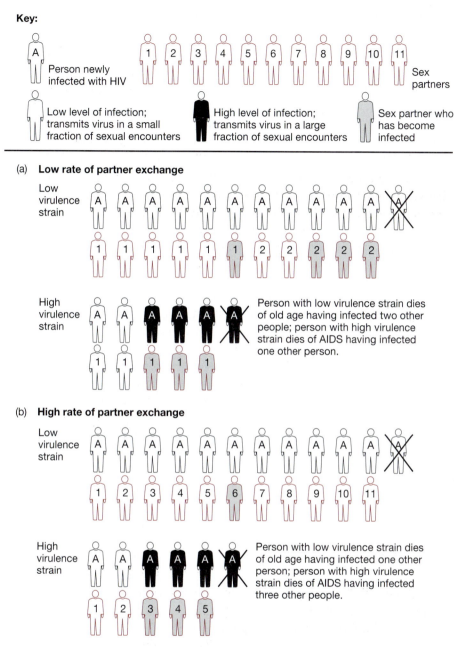

Key:

A — Person newly infected with HIV

1 2 3 4 5 6 7 8 9 10 11 — Sex partners

Low level of infection; transmits virus in a small fraction of sexual encounters

High level of infection; transmits virus in a large fraction of sexual encounters

Sex partner who has become infected

(a) Low rate of partner exchange

Low virulence strain

High virulence strain — Person with low virulence strain dies of old age having infected two other people; person with high virulence strain dies of AIDS having infected one other person.

(b) High rate of partner exchange

Low virulence strain

High virulence strain — Person with low virulence strain dies of old age having infected one other person; person with high virulence strain dies of AIDS having infected three other people.

Figure 1.10   **The transmission rate hypothesis**   The key to the diagram is at the top. Parts (a) and (b) illustrate the logic behind the transmission rate hypothesis. When rates of partner exchange are low, as in part (a), low virulence strains of virus are transmitted to the most new hosts. When rates of partner exchange are high, as in part (b), high virulence strains are transmitted to the most new hosts.

The key to the transmission rate hypothesis is the concept of trade-offs. Evolutionary biologists analyze trade-offs in terms of the costs and benefits of competing strategies. In this case, the strategies are growing slowly or growing rapidly. For a virion, the benefit of rapid growth is increasing its prevalence in the bloodstream of its host, and consequently increasing its chance of

Differences in virulence among HIV strains can be analyzed as competing strategies, each with costs and benefits. Which strategy succeeds depends on environmental conditions.

becoming transmitted during an episode of sexual intercourse. The cost of rapid growth is killing enough CD4+ cells to make the host ill and less likely to engage in intercourse.

The logic behind the transmission rate hypothesis now becomes clear: If people change sexual partners infrequently, then any single episode of intercourse is unlikely to involve a new partner. Thus transmission to a new host is unlikely during any single episode of intercourse (Figure 1.10). In the case of monogamy, highly virulent forms of the virus are almost certain to kill their host before transmission to a new host can occur. Highly virulent forms will thus die out. In this case, the slowly reproducing forms of HIV are present in fewer numbers in the host's bloodstream or semen at any time, but will transmit more efficiently over time and thus increase in the overall population. Conversely, if promiscuity is prevalent, then highly virulent strains will be transmitted to new hosts quickly, and much more frequently than slowly replicating forms. They will increase in frequency in the total viral population.

According to the transmission rate hypothesis, HIV first became deadly and spread because of changes in the sexual practices of heterosexual men and women in eastern and central Africa, and in homosexual men in the United States and Europe. Because of large-scale economic changes in the 1970s and 1980s, millions of men migrated from their homes in rural areas to work in the large cities of Zaire, Uganda, and Kenya. A large commercial sex industry sprang up in response, with prostitutes in some cities having up to 1,000 sexual partners per year (see Ewald 1994a). Likewise, rates of partner change among homosexual men in the United States and Europe in the late 1970s and early 1980s were high—up to 10 per six months (Koblin et al. 1992). The hypothesis contends that changes in transmission dynamics, caused by these behavioral changes in host populations, strongly favored the evolution of more virulent strains of the virus.

The transmission rate hypothesis predicts that different populations of people will harbor HIV strains with different degrees of virulence, depending on how often they change partners or share needles. Two natural experiments are now underway that may help us understand just how valid the hypothesis is.

- In the past several years, rates of partner change among homosexual men in the United States and Europe have decreased dramatically (Winkelstein et al. 1988; Sittirai et al. 1990; Adib et al. 1991), and rates of condom use have increased (Catania et al. 1992). In contrast, condom use and sexual practices in Africa seem to have changed little since the start of the epidemic (Editorial staff 1995). The transmission rate hypothesis predicts that HIV-1 will slowly become less deadly in North America and Europe, but continue to kill people, after relatively brief infections, in central and East Africa and perhaps in Southeast Asia as well.

- HIV-2, a close relative to HIV-1, is similar in life cycle and gene composition but is much more benign (De Cock et al. 1993). Interestingly, HIV-2's reverse transcriptase has a much lower mutation rate than HIV-1's reverse transcriptase (see Ewald 1994b). HIV-2 has, however, recently moved from its proposed historical center of incidence, West Africa (Figure 1.11), where sexual transmission rates due to partner changes are low, to India, where rates may

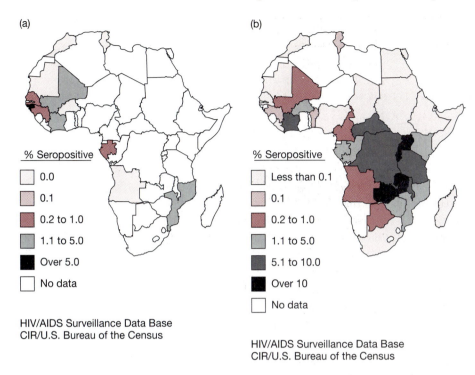

(a)

% Seropositive

- [ ] 0.0
- [ ] 0.1
- [ ] 0.2 to 1.0
- [ ] 1.1 to 5.0
- [ ] Over 5.0
- [ ] No data

HIV/AIDS Surveillance Data Base
CIR/U.S. Bureau of the Census

(b)

% Seropositive

- [ ] Less than 0.1
- [ ] 0.1
- [ ] 0.2 to 1.0
- [ ] 1.1 to 5.0
- [ ] 5.1 to 10.0
- [ ] Over 10
- [ ] No data

HIV/AIDS Surveillance Data Base
CIR/U.S. Bureau of the Census

Figure 1.11   **HIV in Africa**
(a) The distribution of HIV-2, showing highest rates of infection in the historical epicenter of West Africa.   (b) The distribution of HIV-1, showing highest rates of infection in central and East Africa. In both figures, incidence rates are measured as the percent of the general urban population that is seropositive, meaning that antibodies to HIV are present in the blood.

be much higher (Ewald 1994a). If research confirms that HIV-2 is being transmitted more rapidly in India than West Africa because of more promiscuous sexual practices, the transmission rate hypothesis predicts that we should soon witness the evolution of highly virulent Asian strains of HIV-2.

Is the transmission rate hypothesis correct? Only time, and data, will tell.

1.5   Where Did HIV Come From?

Phylogenetic systematics is a branch of evolutionary biology that we will visit again and again in this text. It is the study of evolutionary relationships, or who evolved from whom and when. A picture of these relationships is called a cladogram or phylogenetic tree. What is the phylogeny of HIV? And what, if anything, can it tell us about where HIV came from?

We can reconstruct an evolutionary tree for HIV and related viruses by comparing their nucleotide sequences. Although the methodology for reconstructing phylogenies is complex (we devote all of Chapter 10 to this topic), the basic logic of the process is simple: In general, closely related species should have more similar DNA or RNA sequences than distantly related forms. Figure 1.12 shows an evolutionary tree of HIV-1 and -2, some of the simian immunodeficiency viruses (SIVs), and several related retroviruses. This phylogeny of the immunodeficiency viruses was reconstructed by assessing similarities and differences in the sequences of their *gag* genes.

The tree shown in Figure 1.12 demonstrates something important: The pattern of viral evolution does not mimic the pattern of primate evolution. HIV strains occur on three separate branches of the tree. Human HIV-1 is more

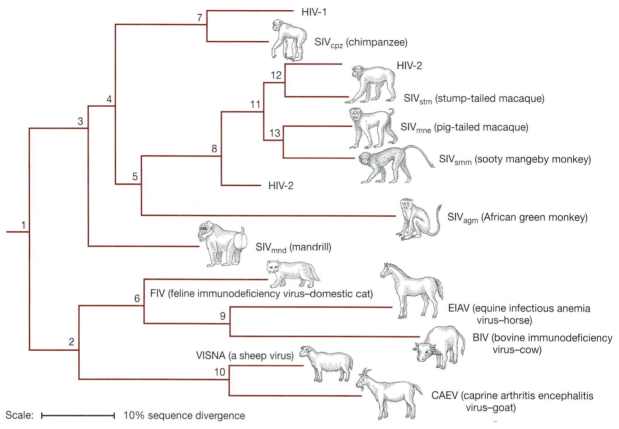

Figure 1.12   A phylogeny of some retroviruses   The evolutionary relationships diagrammed in this tree were reconstructed by comparing the nucleotide sequences of the *gag* genes in HIV-1, two divergent strains of HIV-2, a variety of simian immuno-deficiency viruses (SIVs), and immunodeficiency and other viruses from cats, cows, goats, and other mammals (Myers et al. 1992; see also Yokoyama 1988; Tsujimoto et al. 1989; Miura et al. 1990; Gao et al. 1992; Novembre et al. 1992; Tomonaga et al. 1993). This last group of viruses was included in the analysis to serve as an "outgroup." An outgroup is any taxonomic, or named, unit that is clearly not part of the ingroup. Outgroups share a common ancestor at the base of the tree with all the other taxa in the study. Locating this ancestral node is important because it identifies which branching events occurred first during the evolution of the group we are interested in (in this case, the HIVs and SIVs). In this way, placing the outgroup "roots the tree."

Evolutionary trees can be oriented vertically or horizontally, but are always read from the base, or root, to the tips. This one is horizontal and is read from left to right. The branching points, or nodes, identify common ancestors. These are populations that diverged into two or more descendant lineages.

Two additional points will aid in interpreting this figure:

• In this tree, the lengths of the branches are meaningful. The horizontal lines are drawn to represent the amount of divergence, in terms of base substitutions in *gag* genes, between each set of taxa.

• The vertical lines, however, are only a convenience. They are drawn to arbitrary lengths simply to make the tree more readable.

Most of the SIVs, by the way, do not cause serious disease in their natural hosts (Essex and Kanki 1988).

An analysis of the HIV phylogeny shows that the virus has moved between host species.

closely related to chimpanzee SIV than it is to human HIV-2, and one strain of human HIV-2 is more closely related to stump-tailed macaque SIV than it is to the other strain of human HIV-2. This observation means that viral evolution has not been host-dependent. If it had been, new viral species would have been formed only when host species split into new species. As a result, the viral and primate phylogenies would match exactly. Instead, it is clear that viruses have jumped from one species to another.

How could this process occur? People routinely hunt monkeys for food in parts of Africa. Chimps are also known to eat monkeys. So a virus could easily move from monkeys to either humans or chimpanzees. Humans sometimes eat chimpanzees, so a virus could move from chimpanzees to humans. The virus could also have moved in any direction from host to host in zoos or animal care facilities in Europe or North America, where humans and a variety of monkey and chimpanzee species are in close contact. How can we determine whether HIV jumped from monkeys to humans, or from humans to monkeys?

Because cross-species transmission of viruses is a rare event, we can use a principle called parsimony to answer this question. Invoking parsimony in evolutionary biology means that when drawing conclusions about what happened in the course of evolution, we favor simple explanations over complicated ones. That is, we prefer interpretations of data that minimize the amount of evolutionary change that has occurred. This is appropriate when we are studying events that occur infrequently.

Using parsimony, we can use the tree illustrated in Figure 1.12 to establish whether the HIVs are derived from viruses in monkeys that moved to humans, or whether a human virus began infecting monkeys. The "humans first" hypothesis requires that the virus represented at node 3 in Figure 1.12 infected humans, and implies that an HIV jumped to monkeys 5 times: to become SIV_{mnd} (at node 3), SIV_{cpz} (at node 7), SIV_{agm} (at node 5), SIV_{mne} and SIV_{smm} (at node 13), and SIV_{stm} (at node 12). The "monkeys first" hypothesis implies that an SIV jumped to humans three times: to become HIV-1 (at node 7), a strain of HIV-2 (at node 8), and a second strain of HIV-2 (at node 12). Based on our assumption of parsimony, the phylogeny favors the monkeys first hypothesis: HIVs arose when monkey viruses jumped to humans.

Although we would also like to know how long HIVs have been present in human populations, we do not yet have a definitive answer. The 1984 Montagnier–Gallo discovery of HIV came just three years after the first deaths due to AIDS in the United States were reported in gay men from San Francisco, but the epidemic is certainly older than that. For example, a blood sample taken in Zaire in 1959, then preserved and analyzed years later, tested positive for antibodies to HIV-1. But virtually no older blood from Africa, where SIVs from Old World monkeys apparently began infecting humans for the first time, is available to test (Ewald 1994a).

1.6   What, if Anything, Does Evolutionary Biology Have to Say About Ways to Stem the AIDS Epidemic?

Understanding evolutionary biology may help inform the fight against AIDS, if only in a minor way.

First, as our experience with cold and flu viruses suggests, and as many researchers in the field are beginning to conclude (for example, Wain-Hobson 1993b; Gardner 1994), the search for an AIDS vaccine may be futile. Vaccines consist of epitopes from killed or incomplete virions. Though no infection occurs after a vaccination, the immune system responds by activating and expanding the lineages of helper T, B, and killer T cells that recognize the epitopes presented. Then, should an authentic infection start later, the "pre-primed" immune system can respond quickly and effectively.

HIV's mutation rate is so high that in all likelihood no vaccine could effectively prepare the immune system for a challenge with live strains of HIV. The current consensus is that we simply could not design a vaccine with enough epitopes to match HIV strain diversity. As the evidence we have reviewed shows, and as the escape hypothesis predicts, HIV's reverse transcriptase has been under strong selection to be error prone.

Second, if the transmission rate hypothesis is correct, our best defense against HIV is unquestionably education and changes in behavior. Increased condom use, sexual fidelity, abstinence, and use of clean needles would not only decrease rates of transmission directly, but potentially help halt the epidemic indirectly by leading to the evolution of dramatically reduced virulence. The transmission rate hypothesis argues that changes in transmission dynamics will have a multiplicative effect in stemming the death rates, by lowering both the incidence *and* the severity of the disease.

Summary

In this chapter we focused on adaptation and diversification in a virus and introduced topics that will resonate throughout the text: mutation and variation, competition, natural selection, phylogeny reconstruction, lineage diversification, and applications of evolutionary theory to scientific and human problems. Our task now, in subsequent chapters, is to widen the fine focus of this introduction, and introduce the diversity of theory, experiments, and models that make up evolutionary biology.

This story begins, in Chapter 2, with the fact of evolution. Understanding that species have changed through time, and that the organisms alive today are descended from forms that lived in the past, is what motivated Darwin to seek an explanatory mechanism: a process that could create change through time. What, then, did Darwin propose? And just what *is* the evidence for the fact of evolution?

Questions

1. AIDS is primarily a disease of young adults. How does a disease like this affect the size, age distribution, and growth rate of human populations over time?

2. In the early 1990s researchers began to find AZT-resistant strains of HIV-1 in recently infected patients who had never received AZT. How can this be?

3. The text discusses some possible advantages to HIV of having a high mutation rate. But high mutation rates can have drawbacks as well. How might this relate to the common observation of HIV-infected cells that contain many incompletely assembled and dysfunctional virions?

4. At the start of an infection, why would sequence diversity among HIV strains within an individual be "near 0," as the text states? What does this observation im-

ply about the number of virions that are passed from one host to another during a transmission event?

5. In this chapter we have discussed two different types of selection: selection of different virus strains within one host, and selection of those virus strains that are able to transmit themselves from host to host. Suppose an HIV counselor is talking to a patient who is worried that he or she might have a virulent strain of HIV. "Don't worry," the counselor says. "Even if you have a virulent strain now, as long as you are monogamous the virulent strain will die out." How has the counselor misinterpreted the transmission rate hypothesis?

6. An alternative to the transmission rate hypothesis, traditionally championed by biomedical researchers, is that disease-causing agents "naturally" evolve into more benign forms as the immune systems of their hosts evolve more efficient responses to them. What predictions does this "coevolution hypothesis" make about evolution of HIV virulence in the United States vs. Africa vs. South Asia? What data would help you decide whether the transmission rate hypothesis or coevolution hypothesis is correct? Do the hypotheses suggest different ways to spend the limited budget for HIV research and education?

7. Respond to the following quote, from the character named Dr. Spock in "Star Trek": "A truly successful parasite is commensal, living in amity with its host, or even giving it positive advantages, as, for instance, the protozoans who live in the digestive system of your termites and digest for them the wood they eat. A parasite that regularly and inevitably kills its host cannot survive long, in the evolutionary sense, unless it multiplies with tremendous rapidity . . . it is not pro-survival."

8. Why is the use of parsimony justified only when we are studying infrequent events?

Exploring the Literature

9. The diversity threshold hypothesis has been controversial. Look up these papers for insights into how hypotheses are tested and how scientific debates are conducted.

Nowak, M.A., R.M. Anderson, M.C. Boerlijst, S. Bonhoeffer, R.M. May, A.J. McMichael, S.M. Wolinsky, K.J. Kunstmann, J.T. Safrit, R.A. Koup, A.U. Neumann, and B.T.M. Korber. 1996. HIV-1 evolution and disease progression. *Science* 274:1008−1011.

Weiss, R.A. 1996. HIV receptors and the pathogenesis of AIDS. *Science* 272:1885−1886.

Wolinsky, S.M., B.T.M. Korber, A.U. Neumann, M. Daniels, K.J. Kunstman, A.J. Whetsell, M.R. Furtado, Y. Cao, D.D. Ho, J.T. Safrit, and R.A. Koup. 1996. Adaptive evolution of human immunodeficiency virus-type 1 during the natural course of selection. *Science* 272:537−542.

10. For tests of the transmission rate hypothesis in other organisms, see the following papers:

Herre, E.A. 1993. Population structure and the evolution of virulence in nematode parasites of fig wasps. *Science* 259:1442−1445.

Lipsitch, M., S. Siller, and M.A. Nowak. 1996. The evolution of virulence in pathogens with vertical and horizontal transmission. *Evolution* 50:1729−1741.

Citations

Adib, S.M., J.G. Joseph, D.G. Ostrow, M. Tal, and S.A. Schwartz. 1991. Relapse in sexual behavior among homosexual men: A 2-year follow-up from the Chicago MACS/CCS. *AIDS* 5:757–760.

Bishai, W.R., N.M.H. Graham, S. Harrington, C. Page, K. Moore-Rice, N. Hooper, and R.E. Chaisson. 1996. Rifampin-resistant tuberculosis in a patient receiving rifabutin prophylaxis. *New England Journal of Medicine* 334:1573–1576.

Brookmeyer, R., and M.H. Gail. 1994. *AIDS Epidemiology: A Quantitative Approach* (New York: Oxford University Press).

Carman, W., H. Thomas, and E. Domingo. 1993. Viral genetic variation: Hepatitis B virus as a clinical example. *Lancet* 341:349–353.

Catania, J.A., T.J. Coates, R. Stall, H. Turner, J. Peterson, N. Hearst, M.M. Dolcini, E. Hudes, J. Gagnon, J. Wiley, and R. Groves. 1992. Prevalence of AIDS-related risk factors and condom use in the United States. *Science* 258:1101–1106.

Coffin, J.M. 1995. HIV population dynamics in vivo: Implications for genetic variation, pathogenesis, and therapy. *Science* 267:483–489.

Cohen, J. 1995a. Differences in HIV strains may underlie disease patterns. *Science* 270:30–31.

Cohen, J. 1995b. Thailand weighs AIDS vaccine test. *Science* 270: 904–907.

Connelly, L., and M. Jenkins. 1994. Strategies for antiviral therapy based on the retroviral life cycle. In Cohen, P.T., M.A. Sande, and P.A. Volberding, eds., *The AIDS Knowledge Base* (Boston: Little, Brown, and Company), 3.5-1–3.5-24.

Culliton, B. 1990. Emerging viruses, emerging threat. *Science* 247:279–280.

De Cock, K.M., B. Barrere, L. Diaby, M.-F. Lafontaine, E. Gnaore, A. Porter, D. Pantobe, G.C. Lafontant, A. Dago-Akribi, M. Ette, K. Odehouri, and W.L. Heyward. 1990. AIDS—The leading cause of adult death in the West African city of Abidjan, Ivory Coast. *Science* 249:793–796.

De Cock, K.M., G. Adjorlolo, E. Ekpini, T. Sibailly, J. Kouadio, M. Maran, K. Brattegaard, K.M. Vetter, R. Doorly, and H.D. Gayle. 1993. Epidemiology and transmission of HIV-2. *Journal of the American Medical Association* 270:2083–2086.

Editorial staff. 1995. Death by denial. *Lancet* 345: 1519–1520.

Essex, M., and P.J. Kanki. 1988. The origins of the AIDS virus. *Scientific American* 259:64–71.

Ewald, P.W. 1994a. *Evolution of Infectious Disease* (Oxford: Oxford University Press).

Ewald, P.W. 1994b. Evolution of mutation rate and virulence among human retroviruses. *Philosophical Transactions of the Royal Society of London,* Series B 346:333–343.

Gao, F., L. Yue, A.T. White, P.G. Pappas, J. Barchue, A.P. Hanson, B.M. Greene, P.M. Sharp, G.M. Shaw, and B.H. Hahn. 1992. Human infection by genetically diverse SIV_{sm}-related HIV-2 in West Africa. *Nature* 358:495–499.

Gardner, M.B. 1994. "HIV vaccine development." In Cohen, P.T., M.A. Sande, and P.A. Volberding, eds., *The AIDS Knowledge Base* (Boston: Little, Brown, and Company), 3.6-1–3.6-45.

Goldsmith, M.F. 1990. Science ponders whether HIV acts alone or has another microbe's aid. *Journal of the American Medical Association* 264:665–666.

Groenink, M., R.A.M. Fouchier, S. Broersen, C.H. Baker, M. Koot, A.B. van't Wout, H.G. Huisman, F. Miedema, M. Tersmette, and H. Schuitemaker. 1993. Relation of phenotype evolution of HIV-1 to envelope V2 configuration. *Science* 260:1513–1515.

Holden, C. 1995. Confirmation for combination AIDS therapy. *Science* 270:33.

Holden, C. 1996. Scourge of Africa. *Science* 274:923.

Holmes, E.C., L.Q. Zhang, P. Simmonds, C.A. Ludlam, and A.J.L. Brown. 1992. Convergent and divergent sequence evolution in the surface envelope glycoprotein of human immunodeficiency virus type 1 within a single infected patient. *Proceedings of the National Academy of Sciences, USA* 89:4835–4839.

Itescu, S., P.F. Simonelli, R.J. Winchester, and H.S. Ginsberg. 1994. Human immunodeficiency virus type 1 strains in the lungs of infected individuals evolve independently from those in peripheral blood and are highly conserved in the C-terminal region of the envelope V3 loop. *Proceedings of the National Academy of Sciences, USA* 91:11378–11382.

Koblin, B.A., J.M. Morrison, P.E. Taylor, R.L. Stoneburner, and C.E. Stevens. 1992. Mortality trends in a cohort of homosexual men in New York City, 1978–1988. *American Journal of Epidemiology* 136:646–656.

Larder, B.A., P. Kellam, and S.D. Kemp. 1993. Convergent combination therapy can select viable multidrug-resistant HIV-1 in vitro. *Nature* 365:451–453.

Levy, J.A., H. Fraenkel-Conrat, and R.A. Owens. 1994. *Virology.* (Englewood Cliffs, NJ: Prentice Hall).

Li, W.-H., M. Tanimura, and P.M Sharp. 1988. Rates and dates of divergence between AIDS virus nucleotide sequences. *Molecular Biology and Evolution* 5:313–330.

Marlink, R., P. Kanki, I. Thior, K. Travers, G. Eisen, T. Siby, I. Traore, C-C. Hsieh, M.C. Dia, E-H. Gueye, J. Hellinger, A. Gueye-Ndiaye, J-L. Sankalé, I. Ndoye, S.

Mboup, and M. Essex. 1994. Reduced rate of disease development after HIV-2 infection as compared to HIV-1. *Science* 265:1587–1590.

Miura, T., J. Sakuragi, M. Kawamura, M. Fukasawa, E.N. Moriyami, T. Gojobori, K. Ishikawa, J.A.A. Mingle, V.B.A. Nettey, H. Akari, M. Enami, H. Tsujimoto, and M. Hayami. 1990. Establishment of a phylogenetic survey system for AIDS-related lentiviruses and demonstration of a new HIV-2 subgroup. *AIDS* 4:1257–1261.

Mohri, H., M.K. Singh, W.T.W. Ching, and D.D. Ho. 1993. Quantitation of zidovudine-resistant human immunodeficiency virus type 1 in the blood of treated and untreated patients. *Proceedings of the National Academy of Sciences, USA* 90:25–29.

Morse, S.S. (ed.) 1993. *Emerging Viruses* (New York: Oxford University Press).

Myers, G., K. MacInnes, and B. Korber. 1992. The emergence of simian/human immunodeficiency viruses. *AIDS Research and Human Retroviruses* 8:373–386.

Novembre, F.J., V.M. Hirsch, H.M. McClure, P.N. Fultz, and P.R. Johnson. 1992. SIV from stump-tailed macaques: Molecular characterization of a highly transmissible primate lentivirus. *Virology* 186:783–787.

Nowak, M. 1990. HIV mutation rate. *Nature* 347:522.

Nowak, M.A., and A.J. McMichael. 1995. How HIV defeats the immune system. *Scientific American* August:58–65.

Nowak, M.A., R.M. May, R.E. Phillips, S. Rowland-Jones, D.G. Lalloo, S. McAdam, P. Klenerman, B. Köppe, K. Sigmund, C.R.M. Bangham, and A.J. McMichael. 1995. Antigenic oscillations and shifting immunodominance in HIV-1 infections. *Nature* 375:606–611.

Osmond, D.H. 1994. "Prevalence of infection and projections for the future." In Cohen, P.T., M.A. Sande, and P.A. Volberding, eds., *The AIDS Knowledge Base* (Boston: Little, Brown, and Company), 1.4-1–1.4-11.

Palca, J. 1991. The sobering geography of AIDS. *Science* 252:372–373.

Paley, C.L. 1994. "HIV disease in Africa." In Cohen, P.T., M.A. Sande, and P.A. Volberding, eds., *The AIDS Knowledge Base* (Boston: Little, Brown, and Company), 1.6-1–1.6-21.

Piot, P. 1996. AIDS: A global response. *Science* 272:1855.

St. Clair, M.H., J.L. Martin, G. Tudor-Williams, M.C. Bach, C.L. Vavro, D.M. King, P. Kellam, S.D. Kemp, and B.A. Larder. 1991. Resistance to ddI and sensitivity to AZT induced by a mutation in HIV-1 reverse transcriptase. *Science* 253:1557–1559.

Shirasaka, T., R. Yarchoan, M.C. O'Brien, R.N. Husson, B.D. Anderson, E. Kojima, T. Shimada, S. Broder, and H. Mitsuya. 1993. Changes in drug sensitivity of human immunodeficiency virus type 1 during therapy with azidothymidine, didieoxycytidine, and dideoxyinosine: An *in vitro* comparative study. *Proceedings of the National Academy of Sciences, USA* 90:562–566.

Sittirai, W., T. Brown, and J. Sterns. 1990. Opportunities for overcoming the continuing restraints to behavior change and HIV risk reduction. *AIDS* 4 (Suppl. 1):S269–276.

Suárez, P., J. Valcárcel, and J. Ortín. 1992. Heterogeneity of the mutation rates in influenza A viruses: Isolation of mutator mutants. *Journal of Virology* 66:2491–2494.

Tarantola, D., J. Mann, C. Mantel, and C. Cameron. 1994. Projecting the course of the HIV/AIDS pandemic and the cost of adult AIDS care in the world. In Kaplan, E.H., and M.L. Brandeau, eds., *Modeling the AIDS Epidemic: Planning, Policy, and Prediction* (New York: Raven Press, Ltd.), 3–27.

Tomonaga, K., J. Katahira, M. Fukasawa, M.A. Hassan, M. Kawamura, H. Akari, T. Miura, T. Goto, M. Nakai, M. Suleman, M. Isahakia, and M. Hayami. 1993. Isolation and characterization of simian immunodeficiency virus from African white-crowned mangabey monkeys (*Cercocedus torquatus lunulatus*). *Archives of Virolology* 129:77–92.

Tsujimoto, H., A. Hasegawa, N. Maki, M. Fukasawa, T. Miura, S. Speidel, R.W. Cooper, E.N. Moriyama, T. Gojobori, and M. Hayami. 1989. Sequence of a novel simian immunodeficiency virus from a wild-caught African mandrill. *Nature* 341:539–541.

Wain-Hobson, S. 1993a. Viral burden in AIDS. *Nature* 366: 22.

Wain-Hobson, S. 1993b. The fastest genome evolution ever described: HIV variation *in situ. Current Opinion in Genetics and Development* 3:878–883.

Winkelstein, W. Jr., J.A. Wiley, N.S. Padian, M. Samuel, S. Shiboski, M.S. Ascher, and J.A. Levy. 1988. The San Francisco Men's Health Study: Continued decline in HIV seroconversion rates among homosexual/bisexual men. *American Journal of Public Health* 78:1472–1474.

Yokoyama, S. 1988. Molecular evolution of the human and simian immunodeficiency viruses. *Molecular Biology and Evolution* 5:645–659.

Darwinism and the Fact of Evolution

Evolutionary biology addresses fundamental questions about the living world: Where do living things come from? Why are there so many different kinds of organisms? How have they come to be so well fitted to their environments?

Charles Darwin was an English naturalist who devoted his life to answering these questions, working to solve what he called the "mystery of mysteries." Darwin began to study biology seriously as a college student in the early 1820s. At that time, the leading explanation in Europe for the origin of species was the Theory of Special Creation. This theory held that all organisms were created by God during the six days of creation described in Genesis 1–2:4. The ideal types formed by this special process, including Adam and Eve, were the progenitors of all individuals living today. The theory stated that the types were unchanged since their creation, or immutable, and that variation within each type was strictly limited. The creation event was also believed to be recent: In 1664 Archbishop James Ussher of the Irish Protestant Church used Old Testament genealogies to calculate that the Earth was precisely 5668 years old. He wrote that "Heaven and Earth, Centre and substance were made in the same instant of time and clouds full of water and man were created by the Trinity on the 26th of October 4004 B.C. at 9:00 in the Morning."

By the time Darwin began working on the problem in the 1830s, however, dissatisfaction with the Theory of Special Creation had already begun to grow. Research in the biological and geological sciences was advancing rapidly, and the data clashed with creationism's central tenets and predictions.

2.1 The Fact of Evolution

Evidence from biology and geology was contradictory to creationism because it showed that species had changed through time and descended with modification from common ancestors, instead of each remaining unchanged since a special and independent origin. We review three types of data. Taken together they were convincing enough to shake many scientists' belief in the Theory of Special Creation.

Relatedness of Life Forms

Homology—literally, the study of likeness—was a key concept in the new field of comparative anatomy. For example, the naturalist Louis Agassiz was among many who observed that the embryos of vertebrates ranging from fish to humans are strikingly similar, especially early in development (Figure 2.1a). Richard Owen, Britain's leading anatomist, and Baron Georges Cuvier of Paris, the founder of comparative anatomy, described homologies among vertebrate skeletons. Darwin (1859) cited their work when he wrote, "What could be more curious than that the hand of a man, formed for grasping, that of a mole for digging, the leg of the horse, the paddle of the porpoise, and the wing of the bat, should all be constructed on the same pattern, and should include the same

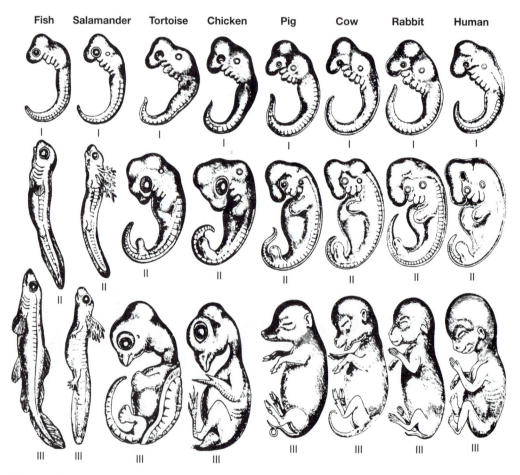

| Fish | Salamander | Tortoise | Chicken | Pig | Cow | Rabbit | Human |

Figure 2.1 **Structural homologies**   (a) Embryos from different vertebrate classes are similar, especially early in development. In this classical illustration, adapted from the work of embryologist Ernst Haeckel, embryo sizes have been scaled to be comparable. The I, II, and III identify the stage of development when each drawing was made. Note that all stage I and II embryos, including humans, have gill slits and tails. But see Richardson et al. (1997). *Anatomy and Embryology* 196: 91–106.

bones, in the same relative positions?" (Figure 2.1b). Darwin (1862) himself researched the anatomy of orchid flowers, and showed that they are actually constructed from the same suite of component pieces, though they are wildly diverse in shape.

What causes homology? Agassiz, Owen, Cuvier, and other early workers recognized that homologous structures in adults develop from the same groups of cells in embryos. As a consequence, homology was defined as similarity due to shared developmental pathways. But why should some organisms share developmental pathways? Darwin argued that descent from a common ancestor was the most logical and parsimonious explanation. He contended that the embryos in Figure 2.1a are similar because all vertebrates evolved from the same common ancestor, and because early developmental stages have remained unchanged as fish, mammals, amphibians, and reptiles diversified.

Structural homologies are a product of shared developmental pathways, which are a product of shared ancestry.

Darwin's recognition of relationship through shared descent was an authentic conceptual breakthrough, and extended to phenomena other than homology. Darwin's trip to the Galápagos Islands had a strong influence on his thinking. While he was aboard the H.M.S. Beagle during a five-year mapping and exploratory mission, Darwin collected and cataloged the flora and fauna encountered during the voyage. He was especially impressed by the mockingbirds during his work in the Galápagos, because each island had a distinct population. Although they were all similar in color, size, and shape and thus clearly related to one another, each mockingbird population seemed distinct enough to be classified as a separate species. This was confirmed, later, by Darwin's taxonomist colleagues back in England. Darwin and others followed up on this result with studies showing the same pattern in Galápagos tortoises and finches: The various islands hosted different, but closely related, species.

Like structural homologies, this radiation of related forms across island chains was a logical outcome of modification with descent. In contrast, both patterns were inconsistent with special creation, which predicted that organisms were

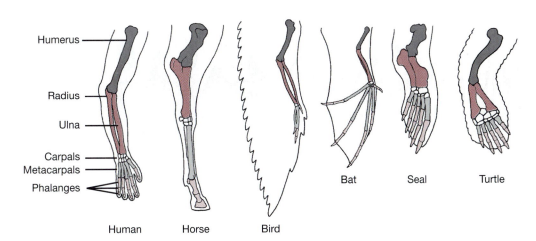

Figure 2.1   (b) These vertebrate forelimbs are used for different functions, but all have the same sequence and arrangement of bones. In this illustration, homologous bones are colored or shaded in the same way, and are labeled on the human arm.

created independently. Under special creation, no particular patterns are expected in the morphological or geographic relationships of organisms.

Change Through Time

One of the central tenets of the Theory of Special Creation was that species, once created, were immutable. Four lines of evidence challenged this claim.

An early eighteen-century paleontologist named William Clift was the first to publish an observation later confirmed and expanded upon by Darwin (Eisely 1958). Fossil and living mammals in the same geographic region are related to each other, and distinctly different from faunas of other areas. Clift worked on the extinct marsupial fauna of Australia and noted its close relationship to forms alive today; Darwin analyzed the armadillos of Argentina and their relationship to the fossil glyptodonts he excavated there (Figure 2.2). This general result, termed the law of succession, provided strong documentation for evolution.

Darwin added another piece of evidence for change through time: the presence of vestigial structures, or rudimentary organs, in a wide variety of organisms (Figure 2.3). Some blind, cave-dwelling fish have eye sockets but no eyes; flightless birds and insects have reduced wings; some snakes have tiny hips and rear legs; humans have a nonfunctional appendix and reduced tailbone. We also have muscles that make our body hair stand on end when we get cold or excited. Erectile fur is found in many mammals, including our close relatives the primates, and is important in signaling alarm or aggressive intent. In humans, it results in goose bumps.

Close relationships between fossil and existing species, vestigial structures, extinction, and obvious changes in landforms all argue that the earth and its species have changed through time.

All of the vestigial structures we have listed are useless homologs of functioning structures in closely related species. Again, Darwin argued that their existence is inexplicable under special creation. But they are readily interpretable as a result of modification with descent.

Baron Georges Cuvier's confirmation of the fact of extinction (Figure 2.4) delivered the same message: Earth's flora and fauna have changed through time.

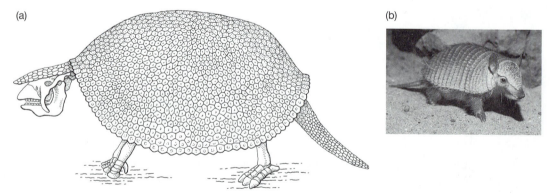

(a)   (b)

Figure 2.2   The "Law of Succession"   Comparative anatomists confirmed that the fossil glyptodont (a) and the contemporary pygmy armadillo *(Zaedyus pichiy)* (b) are close relatives. Close relationships between fossil and extant species from the same geographical area, and between fossil forms in adjacent rock strata, were observed so routinely in the early days of paleontology that the pattern became known as the law of succession. (Photo by Tom McHugh/Photo Researchers, Inc.)

(a)

(b)

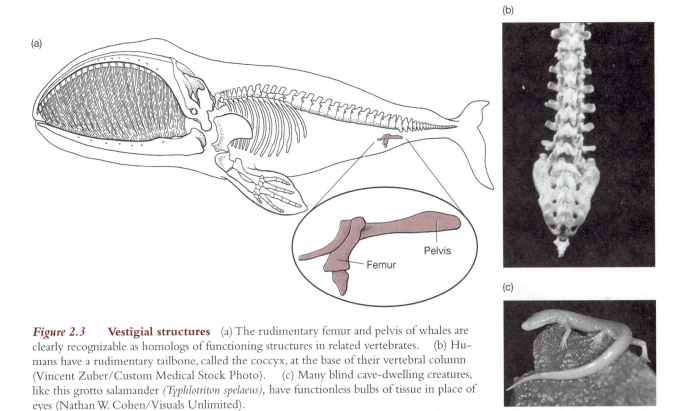

(c)

Figure 2.3 **Vestigial structures** (a) The rudimentary femur and pelvis of whales are clearly recognizable as homologs of functioning structures in related vertebrates. (b) Humans have a rudimentary tailbone, called the coccyx, at the base of their vertebral column (Vincent Zuber/Custom Medical Stock Photo). (c) Many blind cave-dwelling creatures, like this grotto salamander *(Typhlotriton spelaeus),* have functionless bulbs of tissue in place of eyes (Nathan W. Cohen/Visuals Unlimited).

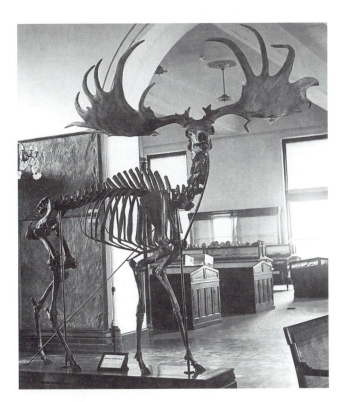

Figure 2.4 **Cuvier's analysis of the Irish Elk confirms the fact of extinction** In 1801 the comparative anatomist Baron Georges Cuvier published a list of 23 species that were no longer in existence. By 1812 his careful examination of fossils of the Irish Elk—the huge ice-age deer shown here, which was native to northern Europe and the British Isles—proved that it was not the same as any living species (Neg. No. 22581. Photo AMNH. Courtesy Dept. of Library Services. American Museum of Natural History).

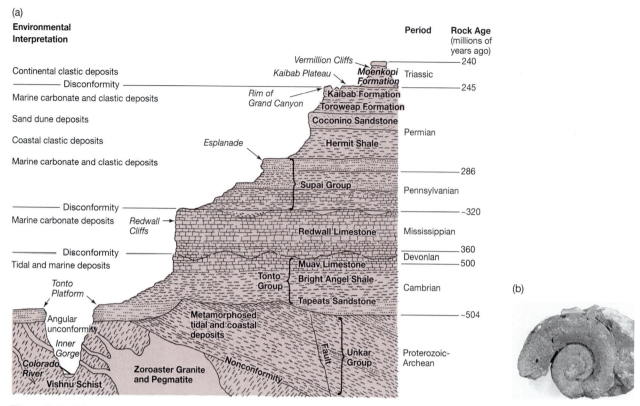

Figure 2.5 **The stratigraphy of the Grand Canyon** (a) This diagrammatic cross-section illustrates one of the most complete exposed rock sequences on Earth: the nearly 1.2 vertical miles of strata mapped in the Grand Canyon. These rocks represent almost 3 billion years of geologic history. There are two points to note: (1) The mudstones, shales, sandstones, and limestones indicate that the American Southwest, now a desert, was covered by shallow seas for most of its history; and (2) there are several "unconformities," or gaps, produced when rock layers were eroded away completely before additional deposits formed on top. Geologists can fill in these gaps by finding nearby areas where the missing material was not completely eroded away—for example, where Devonian rocks are still sandwiched between the Redwall and Muav Limestones. The completed, composite record is called the geologic column. Note that the geological periods and rock ages listed on the right can be matched to entries in the geological time scale of Figure 2.6. (b) This ammonite, a marine fossil, was found in the Kaibab limestone (Grand Canyon National Park #6952).

The recognition of marine fossils in contemporary nonmarine environments, like the Andes in South America and the Grand Canyon in America's arid southwest (Figure 2.5), argued that the Earth is not static, either. Habitats and landforms are being continuously generated and modified, just as species are.

The Age of the Earth

By the time Darwin first began working on "the species problem," data from the young science of geology had challenged a linchpin in the Theory of Special Creation: that the Earth was only about 6,000 years old. Evidence was mounting that the Earth was ancient.

James Hutton was the first to articulate a principle called uniformitarianism: Geological processes taking place now operated similarly in the past. This assumption led Hutton, and later Charles Lyell, to infer that the geological time scale was unimaginably long in human terms. This conclusion was driven by data: Given the measurements these early geologists made of ongoing rock-forming processes like deposition at beaches and river deltas and the accumulation of marine shells (the precursors of limestone), it was clear that vast stretches of time were required to produce the immense rock formations they were studying.

When Darwin began his work, Hutton and followers were already in the midst of a 50-year effort to put the major rock formations and fossil-bearing strata of Europe in a younger-to-older sequence. Their technique, called relative dating, was an exercise in logic based on the following assumptions:

- Younger rocks are deposited on top of older rocks (this is called the principle of superposition).

- Lavas and sedimentary rocks like sandstones, limestones, and mudstones were originally laid down in a horizontal position, so that any tipping or bending events occurred after deposition (principle of original horizontality).

- Rocks that intrude into seams or as dikes are younger than their host rocks (principle of cross-cutting relationships).

- Boulders, cobbles, or other fragments found in a body of rock are older than their host rock (principle of inclusions).

- Earlier fossil life forms are simpler than more recent forms, and more recent forms are most similar to existing forms (principle of faunal succession).

Using these rules, geologists established the chronology known as the geologic time scale (Figure 2.6) and created the concept of the geologic column. This is a history of the Earth based on a composite, older-to-younger sequence of rock strata. Taken together with the principle of uniformitarianism, the geologic time scale and geologic column furnished impressive evidence for an ancient Earth. Geologists began working in time scales of tens of millions of years, instead of a few thousand years, long before Darwin published his ideas on the origin of species.

Ironically, most of the data we have reviewed here, which Darwin used as evidence for the fact of evolution, were gathered by ardent creationists like Cuvier, Agassiz, and Owen. As persuasive as the evidence is, though, it is important to note that there was no one grand experiment that swept aside belief in the Theory of Special Creation and opened the way for a new theory. Observations like structural and developmental homologies did not falsify special creation outright. Rather, evolution was simply much more powerful in explaining the data. Instead of collapsing dramatically, the Theory of Special Creation began to creak and groan under the weight of the evidence, which accumulated steadily over the first half of the 19th century. Eventually it gave way.

2.2 Natural Selection: Darwin's Four Postulates

Because homologies, extinction, and the law of succession were widely recognized in Western Europe, the idea of evolution had been in the air long before Darwin began his work on the problem. Several writers, including Charles' grandfather Erasmus Darwin, had proposed that existing species are the

Eon	Era	Period	Epoch	Age Ma	Life Forms
Phanerozoic	Cenozoic	Quaternary	Holocene		
			Pleistocene		
		Tertiary / Neogene	Pliocene	1.8	Earliest *Homo*
			Miocene	5.2	
		Tertiary / Paleogene	Oligocene	23.8	First apes
			Eocene	33.5	First whales
			Paleocene	55.6	First horses
				65	Extinction of dinosaurs
	Mesozoic	Cretaceous	Late		First placental mammals
			Early	98.9	First flowering plants
		Jurassic	Late	144	First birds
			Middle	160	
			Early	180	
		Triassic	Late	206	First mammals
			Middle	228	First dinosaurs
			Scythian		
	Paleozoic	Permian		251	
		Carboniferous / Pennsylvanian		290	First mammal-like reptiles
		Carboniferous / Mississippian			First reptiles
		Devonian		353.7	First amphibians
					First insects
		Silurian		408.5	First land plants
					First fish with jaws
		Ordovician		439	First fish (no jaws)
		Cambrian		495	
				543	First shelled organisms
Proterozoic					First multicellular organisms
					First eukaryotes
				2500	
Archaean					First bacteria
				3600	Origin of life?
Hadean					Oldest rocks
				4600	Formation of the Earth

Figure 2.6 **The geologic time scale**   The sequence of eons, eras, periods, and epochs shown on the left part of this diagram was established through the techniques of relative dating. Each named interval of time is associated with a distinctive fossil flora and fauna. The absolute ages included here were added much later, when radiometric dating systems became available (see section 2.5). The abbreviation Ma stands for millions of years ago.

modified descendants of forms that existed previously. But no one had yet proposed a satisfactory mechanism for *how* a population of organisms could change through time. Understanding the mechanism that produces the phenomena we observe is the heart and soul of a scientific explanation. Put another way, the early evolutionists had discovered an important pattern. A growing body of facts indicated that fossil and living organisms had descended from common ancestors. But what process could produce this pattern?

Darwin's solution is the theory of evolution by natural selection. It can be stated concisely, as the logical outcome of the following four postulates:

Natural selection is a process that produces descent with modification, or evolution.

1. Individuals within species are variable.

2. Some of these variations are passed on to offspring.

3. In every generation, more offspring are produced than can survive.

4. Survival and reproduction are not random: The individuals that survive and go on to reproduce, or who reproduce the most, are those with the most favorable variations. They are naturally selected.

Darwin laid out these four postulates in his introduction to *On the Origin of Species by Means of Natural Selection,* published in 1859, and considered the rest of the book "one long argument" in their support.

Evolutionary change is an outcome of the process described by Darwin, which he called natural selection. The logic is clear: If there is variation among individuals in a population that can be passed on to offspring, and if there is differential success among those individuals in surviving and/or reproducing and thus in passing on those variations, then the characteristics of the population will change slightly with each succeeding generation. This is Darwinian evolution: gradual change in populations over time. Changes in populations are an inevitable product of natural selection on individuals.

Recall the HIV virions we described in Chapter 1. Individual virions within the same host varied in their ability to synthesize DNA in the presence of AZT, because of differences in the amino acid sequences of the reverse transcriptase active site. Virions with forms of reverse transcriptase that were less likely to bind AZT reproduced more than virions with forms that bound AZT readily. In the next generation, then, a higher percentage of virions had the modified form of reverse transcriptase than in the generation before. This is evolution by natural selection.

Darwin referred to the individuals who win this competition (it is rarely an actual head-to-head contest), and whose offspring make up a greater percentage of the population in the next generation, as more fit. In doing so he gave the everyday English words *fit* and *fitness* a new meaning. Darwinian fitness is the ability of an individual to survive and reproduce in its environment. An important aspect of fitness is its relative nature. It refers to how well an individual survives and how many offspring it produces compared to other individuals of its species. We use the word *adaptation* to refer to traits or characteristics of organisms, like a modified form of reverse transcriptase, that increase their fitness relative to individuals without the traits.

An adaptation is a characteristic that increases the fitness of an individual compared to individuals without the trait.

The same theory had, incidentally, been developed independently by a colleague of Darwin's named Alfred Russel Wallace. Though trained in England,

Wallace had been making his living in Malaysia by selling natural history specimens to private collectors. While recuperating from a bout with malaria in 1858, he wrote a manuscript explaining natural selection and sent it to Darwin. Darwin, who had written his first draft on the subject in 1842 but never published it, immediately realized that he and Wallace had formulated the same theory independently. The two agreed to have their papers read together before the Royal Society of London, and Darwin then rushed *On the Origin of Species* into publication (17 years after he had written the first draft). Today Darwin's name is more prominently associated with the theory of evolution by natural selection for two reasons: he had clearly thought of it first, and *On the Origin of Species* provided a full exposition of the idea, along with massive documentation.

One of the most attractive aspects of the Darwin–Wallace theory is that we can verify each of the four postulates and the logical consequence independently. That is, the theory is testable. There are no hidden assumptions, and nothing that we have to accept uncritically. In the next section we examine each of the four assertions by reviewing an ongoing study of finches in the Galápagos Islands off the coast of Ecuador. Can natural selection lead to evolutionary change in something other than viruses?

2.3   The Evolution of Beak Shape in Galápagos Finches

Peter Grant and Rosemary Grant and their colleagues have been studying several species of Galápagos finches continuously and on various islands in the Galápagos archipelago since 1973 (Grant 1986; Grant and Grant 1989). The 13 finch species found on the islands are similar in size and coloration, ranging from three to six inches in length and from brown to black in color. Two traits do show remarkable variation among species, however: the size and shape of the finches' beaks. The beak is the primary tool used by birds in feeding, and the enormous range of beak morphologies in the Galápagos finches (Figure 2.7) reflects the diversity of foods they eat. The warbler finch *(Cactospiza olivacea)* feeds on small arthropods and nectar; woodpecker and mangrove finches *(C. pallida and C. heliobates)* use twigs or cactus spines as tools to pry insect larvae or termites from dead wood; several ground finches in the genus *Geospiza* pluck ticks from iguanas and tortoises in addition to eating seeds; the vegetarian finch *(Platyspiza crassirostris)* eats leaves and fruit (Grant 1981a, 1986).

Because it is the most important trait used in acquiring food, the size and shape of a bird's beak has important consequences for its fitness.

To test the theory of evolution by natural selection, we will focus on data Grant and Grant and colleagues have gathered on the medium ground finch, *Geospiza fortis,* on Isla Daphne Major.

Daphne Major's size and location make it a superb natural laboratory. The island is tiny. It is just under 40 hectares (about 80 football fields) in extent, with a maximum elevation of 120 meters. Like all of the islands in the Galápagos, it is the top of a volcano. The climate is seasonal even though the location is equatorial (Figure 2.8). A warmer, wetter season from January through May alternates with a cooler, drier season from June through December. The vegetation consists of dry forest and scrub, with several species of cactus.

The *Geospiza fortis* on Daphne Major make an ideal study population because few finches migrate onto or off of the island (Boag and Grant 1981; Grant

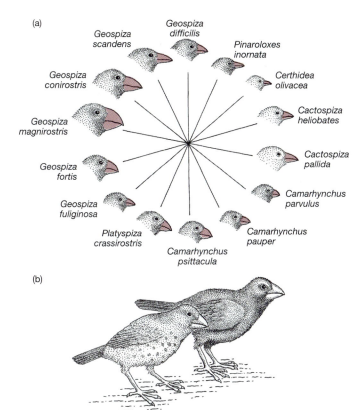

(a)

Geospiza scandens

Geospiza difficilis

Pinaroloxes inornata

Geospiza conirostris

Certhidea olivacea

Geospiza magnirostris

Cactospiza heliobates

Geospiza fortis

Cactospiza pallida

Geospiza fuliginosa

Camarhynchus parvulus

Platyspiza crassirostris

Camarhynchus pauper

Camarhynchus psittacula

(b)

Figure 2.7 **The Galápagos finches, and the medium ground finch *Geospiza fortis*** Many of the Galápagos and Cocos Island finches do not differ markedly in body size or coloration. The species do differ substantially, however, in beak size and shape. After Grant (1986).

1991), and the population is small enough to be studied exhaustively. In an average year there are about 1,200 individuals on the island. By 1977 Grant and Grant's team had captured and marked over half the individuals present; since 1980 virtually 100% of the population has been marked (Grant 1991).

The medium ground finch is primarily a seed eater. The birds crack seeds by grasping them at the base of the bill and then applying force. Grant and Grant and their colleagues have shown that both within and across finch species, beak size is correlated with the size of seeds harvested: In general, birds with bigger beaks eat larger seeds, and birds with smaller beaks eat smaller seeds. This is because birds with different beak sizes are able to handle different sizes of seeds more efficiently (Bowman 1961; Grant et al. 1976; Abbott et al. 1977; Grant 1981b).

Testing Postulate 1: Are Populations Variable?

Grant and Grant mark every finch they catch by placing colored aluminum bands around each of its legs. This allows the researchers to identify individual birds in the field. The scientists also weigh each finch and measure its wing length, tail length, beak width, beak depth, and beak length. All of the traits they have investigated are variable. When Grant and Grant plotted measurements of beak depth in the Isla Daphne Major population of *G. fortis,* for example, the data form a bell-shaped curve (Figure 2.9a). All of the finch characteristics they have measured clearly conform to Darwin's first postulate. As we will see in Chapter 4, variation among individuals is virtually universal.

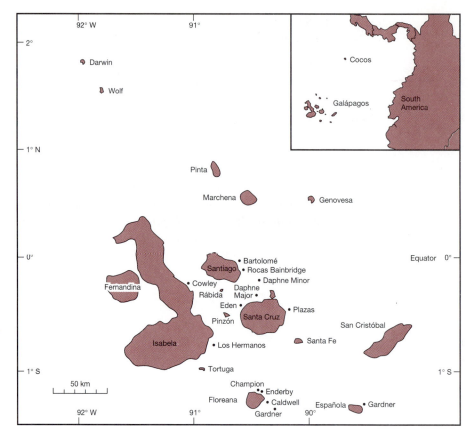

Figure 2.8   The Galápagos archipelago, and the island of Daphne Major   Thirteen species of finch inhabit the Galapágos, and one finch resides on Cocos. On this map, Isla Daphne Major is a tiny speck between Santa Cruz and Santiago.

Figure 2.9   Beak depth in medium ground finches
(a) This histogram shows the distribution of beak depth in medium ground finches in 1976, at the start of the Grant study. A few birds have shallow beaks, less than 8 mm deep. Most birds have medium beaks, 8 to 11 mm deep. A few birds have deep beaks, more than 11 mm deep. Overall, variation in beak depth exceeds 100%.   (b) This is the distribution of beak depth in the same population of finches, after the drought of 1977 (Grant 1986). (The birds were measured in March 1978.)

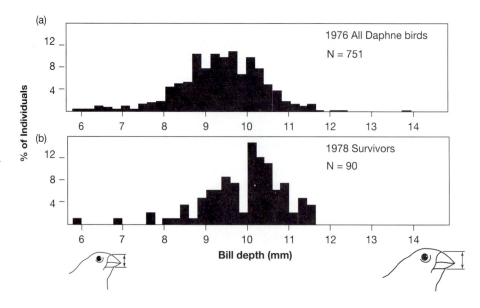

Testing Postulate 2: Is Some of the Variation Among Individuals Heritable?

Finches could vary in beak depth because of differences in their environments or their genetic heritage, or both. A birds' environment could affect beak dimensions through abrasion against hard seeds or rocks, for example. Variation in the amount of food that individuals happened to have received as chicks can also lead to variation among adults. To determine whether at least part of the variability among finch beaks is genetically based, and thus capable of being passed from parents to offspring, a colleague of Peter Grant and Rosemary Grant's named Peter Boag estimated a quantity known as heritability.

In the broad sense, heritability is the proportion of the total variation observed in a population that is due to the effects of genes. In the narrow sense, heritability is the proportion of the total variation observed in a population that is due to the *additive* effects of genes. Additive genetic effects include all of the effects a gene has except those due to gene interactions, like dominance or interactions with other loci. Because it is a proportion, heritability varies between 0 and 1. We will develop the theory behind how heritability is estimated much more fully in Chapter 6. For now we simply point out that it is usually estimated by measuring the similarity between pairs of relatives. This is because similarities between relatives are caused, at least in part, by the genes they share (Box 2.1). Data are usually collected from siblings, or from parents and offspring.

Boag compared the beak depth of *G. fortis* young after they had attained adult size to the average bill depth of their mother and father. Boag found a strong correspondence between relatives. Parents with deep beaks tended to have offspring with deep beaks, and parents with shallow beaks tended to have chicks with shallow beaks (Figure 2.10). This is evidence that a large proportion of the observed variation in beak depth is genetic, and can be transmitted to offspring (Boag and Grant 1978; Boag 1983).

Testing Postulate 3: Is There an Excess of Offspring, So That Only Some Individuals Live to Reproduce?

Because they routinely censused the ground finch population over several years, the researchers were able to observe a dramatic event. In 1977 there was a severe drought at their study area. Instead of the normal 130 mm of rainfall during the wet season, only 24 mm fell. Over the course of 20 months, 84% of the Darwin's medium ground finch population disappeared (Figure 2.11a). The team inferred that most died of starvation: there was a strong correspondence with seed availability (Figure 2.11b), 38 emaciated birds were actually found dead, and none of the missing birds reappeared the following year (Boag and Grant 1981; Grant 1991). It is clear that only a fraction of the population survived to reproduce. This sort of mortality is not that unusual: Rosemary Grant has shown that 89% of *Geospiza conirostris* individuals die before they breed (Grant 1985), and Trevor Price et al. (1984) determined that an additional 19% and 25% of the *G. fortis* on Daphne Major died during subsequent drought events in 1980 and 1982.

A drought on Daphne Major produced a dramatic selection event.

BOX 2.1   Issues that complicate how heritabilities are estimated

Heritabilities are estimated by measuring the similarity of traits among closely related individuals. But relatives share their environment as well as their genes, and any correlation that is due to their shared environment will inflate the estimate of heritability. For example, it is well known that birds tend to grow larger when they have abundant food as young. But the most food-rich breeding territories are often won and defended by the largest adults in the population, and young from these territories will tend to become the largest adults in the next generation. As a result, a researcher might measure a strong relationship between parent and offspring beak and body size and claim a high heritability for the traits when in reality there is none. In this case, the real relationship is between the environments that parents and their young experienced as chicks.

In many species, this problem can be circumvented by performing what are called cross-fostering, common garden, or reciprocal-transplant experiments. In birds these experiments involve taking eggs out of their original nest and placing them in the nests of randomly assigned foster parents. Measurements in the young, taken when they are fully grown, are then compared with their biological parents. This experimental treatment removes any bias in the analysis created by the fact that parents and

offspring share environments. Unfortunately, even cross-fostering experiments cannot remove environmental effects that are due to differences in the nutrient stores of eggs or seeds. We simply have to assume that these maternal effects are too small to seriously bias the result.

Cross-fostering experiments in a wide variety of bird species have confirmed large heritabilities in most or all of the morphological characters analyzed (Smith and Dhondt 1980; Dhondt 1982; Alatalo and Lundberg 1986; Wiggens 1989). It was not possible for Boag and Grant (1978) to perform this critical experiment on the finches, however. Because the Galápagos are a national park, experiments that manipulate individuals beyond catching and marking are forbidden. To get around this limitation Peter Grant compared individuals that were raised in food-poor versus food-rich territories to see if they bred in a similar environment as adults. Their data showed no correlation between the quality of the territory a bird was raised in and the quality of the territory where it raised its young. This negative result bolstered their claim that the large heritabilities they estimated in traits like beak dimensions were real, and not an artifact of shared environmental influences (Grant 1991).

In every natural population studied, more offspring are produced each generation than will survive to breed. If a population is not increasing in size, each parent will, in the course of its lifetime, leave an average of one offspring that survives to breed. But the reproductive capacity (or biotic potential) of organisms is astonishing (Table 2.1).

Similarly, data show that in most populations some individuals are more successful at mating and producing offspring than others. This variation in reproductive success represents an opportunity for selection, too.

Testing Postulate 4: Is Survival and Reproduction Nonrandom?

Darwin's fourth claim was that the individuals who survive and go on to reproduce, or who reproduce the most, are those with the most favorable variations. Did a nonrandom, or selected, subset of the ground finch population survive the 1977 drought? By measuring the same traits they had measured in 1976 on a large and random sample of surviving birds early in 1978, the Grant team found that a distinct subset of the population had survived: those with the deepest beaks (look back at Figure 2.9b). Because the average survivor had a deeper

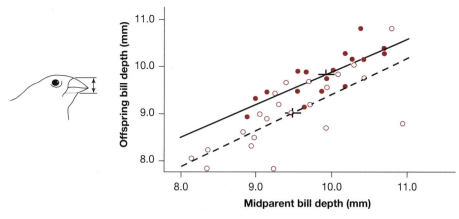

Figure 2.10   Heritability of beak depth in *Geospiza fortis*   This graph shows the relationship between the bill depth of parents and their offspring. Midparent value is the average of the maternal and paternal measurements. Using this measurement is important because male *G. fortis* are bigger than females.

The lines in the graph are the result of a statistical procedure called regression analysis. In the type of regression used here, the line is placed in a way that minimizes the sum of the squared vertical distances between each point in the plot and the line. This is called the best-fit line. If the slope of the line is flat, or 0, then there is no relationship between the two variables plotted.

The solid line and solid circles are from 1978 data, and the dashed line and open circles are from 1976 data. The results from the two different years are consistent. Both show a strong relationship between the beak depth of parents and their offspring. We can infer that the association is a product of their shared genes. From Boag (1983).

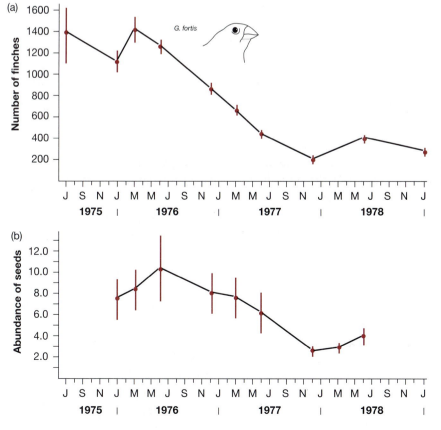

Figure 2.11   Decline of ground finch population and available seeds, during the 1977 drought (a) This graph shows the number of ground finches found on Daphne Major before, during, and after the drought. Each tick on the x-axis represents a month. The vertical lines through each data point represent a quantity called the standard error, which indicates the amount of variation in census estimates. The lines in this graph are simply drawn from point to point to make the trend easier to see.   (b) The abundance of seeds on Daphne Major before, during, and after the drought. From Boag and Grant (1981).

TABLE 2.1   Reproductive potential

This table gives the number of offspring that a single individual (or pair of individuals, for sexual species) can produce under optimal conditions, assuming that all progeny survive to breed, over various time intervals. Darwin picked the elephant for his calculations because it was the slowest breeder then known among animals.

Organism	Reproductive potential	Citation
Aphis fabae (an aphid)	524 billion in one year	Gould 1977
Elephant	19 million in 750 years	Darwin 1859
Housefly	191×10^{18} in 5 months	Keeton 1972
Mycophila speyeri (a fly that feeds on mushrooms)	20,000/square foot, in 35 days	Gould 1977
Staphylococcus aureus (a bacterium)	cells would cover the Earth 7 ft deep in 48 hours	Audesirk and Audesirk 1993
Starfish	$>10^{79}$ in 16 years*	Dodson 1960

*10^{79} is the estimated number of electrons in the visible universe.

During the drought, finches with larger, deeper beaks had an advantage.

beak than the average nonsurvivor, the average bill size in the population changed (Figure 2.12a).

But in what way were deep beaks favorable? Can we link ecological cause with evolutionary effect? The answer is yes: not only the number, but the *types* of seeds available during the 1977 drought changed dramatically (Figure 2.12b). Specifically, the large, hard fruits of an annual plant called *Tribulus cistoides* became a key food item. These seeds are ignored in normal years, but during the drought the supply of small, soft seeds quickly became exhausted (Boag and Grant 1981). Only large birds with deep, narrow beaks can crack and eat *Tribulus* fruits successfully (Price et al. 1984). In addition, large birds defend food sources more successfully during conflicts (Grant 1991). Because large size and deep beaks are positively correlated (Boag 1983), the two traits responded to selection together (Grant 1991).

There is an interesting twist to the story, however. Boag and Grant's analysis showed that individuals with relatively narrow beaks also survived better. The finches push down and to either side when cracking the *Tribulus* fruits, and narrower beaks concentrate these forces more effectively. But beak width is *positively* correlated with beak depth and large size, meaning that birds with deep beaks, which are good for applying downward force, tend also to have wider beaks, which are less effective at applying the twisting force. Presumably this correlation exists because the same genes affect beak depth and width and overall body size, making all three traits smaller or larger together. The upshot is that selection for larger size and deeper beaks resulted in wider beaks *even when narrower beaks should have been favored*. This is an important point: Because of the genetic correlations among characters, natural selection could not necessarily optimize all of the traits involved.

The 1977–1978 selection event, as dramatic as it was, was not an isolated occurrence. In 1980 and 1982 there were similar droughts, and selection again fa-

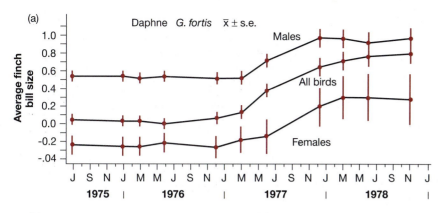

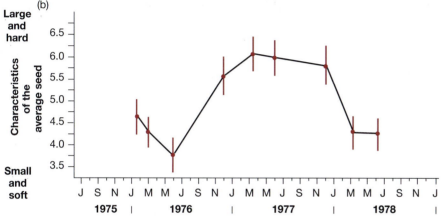

Figure 2.12   Changes in average beak depth of ground finches and seed characteristics
(a) These plots show the average beak depth of finches caught before, during, and after the drought on Daphne Major. The "All birds" line represents the average of males and females. As in Figure 2.11, the vertical lines represent standard errors and the horizontal lines are drawn between data points simply to make the trends easier to see.
(b) This graph shows changes in the hardness of seeds available on Daphne Major before, during, and after the drought. The hardness index plotted on the y-axis is a special measure created by Boag and Grant (1981).

vored individuals with large body size and deep, narrow beaks (Price et al. 1984). But then in 1983 an influx of warm surface water off the South American coast, called an *El Niño* event, created a wet season with 1359 mm of rain on Daphne Major. This dramatic environmental change (almost 57 times as much rain as in 1977) led to a superabundance of small, soft seeds and strong selection for *smaller* body size (Gibbs and Grant 1987). Larger birds were favored in drought conditions, and smaller birds in wet years. Natural selection—as we pointed out in our anlysis of HIV evolution in Chapter 1—is dynamic.

Did Evolution Occur?

The changes observed in finch beaks are examples of directional selection. This term is appropriate because natural selection pushed the distribution of traits in one direction (and then back again, in this case). But we said earlier that evolution was a *response* to selection. The data in Figure 2.9 show selection, but not evolution. Selection produces a distinct change in trait distributions within a generation, while evolution is a change in trait distributions between generations. Did evolution occur in the Galápagos finches? The answer is yes: The offspring of birds surviving the 1977 drought were significantly larger, on average, than the population that existed before the drought (Table 2.2).

TABLE 2.2   Evolutionary response to selection

These data summarize changes in the population means of body and beak traits in *Geospiza fortis,* before and after the 1976–1977 drought. "SE" is an abbreviation for standard error, which quantifies the amount of variation around the average, or mean, value. The delta, Δ, column shows the differences between generations.

	Before Selection 1976		*Next Generation* 1978		
	Mean	**SE**	**Mean**	**SE**	**Δ**
Weight (g)	16.06	0.06	17.13	0.13	+1.07
Wing length (mm)	67.88	0.10	68.87	0.20	+0.99
Tarsus length (mm)	19.08	0.03	19.29	0.07	+0.21
Bill length (mm)	10.63	0.03	10.95	0.06	+0.32
Bill depth (mm)	9.21	0.03	9.70	0.06	+0.49
Bill width (mm)	8.58	0.02	8.83	0.05	+0.25
Sample size	634		135		

Source: Grant and Grant 1995

2.4   The Nature of Natural Selection

Although the theory of evolution by natural selection can be stated concisely and demonstrated in the field, it can be difficult to understand thoroughly. One reason is that it is essentially a statistical process: a change in the trait distributions of populations. Statistical thinking does not come naturally to most people, and there are a number of widely shared naive notions about natural selection that are positively misleading. In this section our goal is to cover some key points about how selection does and does not operate.

Natural Selection Acts on Individuals, but Its Consequences Occur in Populations

When HIV strains were selected by exposure to AZT, or finch populations were selected by changes in seed availability, none of the selected individuals (virions or finches) changed in any way. They simply lived through the selection event while others died, or reproduced more than competing virions or birds. What changed after the selection process were the characteristics of the populations of virions and finches, not the affected individuals themselves. The effort of cracking *Tribulus* seeds did not make finch beaks become deeper and narrower, and the effort of transcribing RNA in the presence of AZT did not change the amino acid composition of the reverse transcriptase active site. Instead, the average beak depth and body size in the finch population increased because smaller finches died, and the average active site sequence in reverse transcriptase changed because certain mutants did a better job of making new virions. Careful inspection of the histograms in Figure 2.9 will help drive this point home.

Natural Selection Acts on Phenotypes, but Evolution Consists of Changes in Gene Frequencies

Finches with large bodies and deep, narrow beaks would have been favored during the drought even if all the variation in the population had been environmental in origin (that is, if heritabilities had been zero). But no evolution would have occurred. The phenotype distribution (as in Figure 2.9) would have changed, but not the distribution of alleles responsible for variation in body size and beak shape (that is, the genotype distribution). And unless the genotype distribution changes, the phenotype distribution will go right back to where it was before the selection event, as soon as the environment changes back.

Because evolution is the *response* to selection, it occurs only when the selected traits have a genetic basis. If the finches had gotten more muscular during the drought because of effort expended in defending food sources and heavier because of their success in doing so, there would have been no evolutionary consequences. Those phenotypic changes would not have been passed on to their offspring. (The idea that phenotypic changes *can* be passed on to offspring is called Lamarckian evolution. We will look at this hypothesis more closely in section 2.7.)

Natural Selection is Backward-, Not Forward-Looking

Each generation is a product of selection by the environmental conditions that prevailed in the generation before. The offspring of the HIV virions and finches that underwent natural selection are better adapted to environments dominated by AZT and drought conditions than their parent's generation was. If the environment changed again during their lifetime, however, these offspring would not necessarily be adapted to the new conditions.

There is a common misconception, however, that organisms can be adapted to future conditions, or that selection can look ahead in the sense of anticipating environmental changes during future generations. This is impossible. Evolution is always a generation behind any changes in the environment.

Natural Selection Acts on Existing Traits

Natural selection can select only from the variations that already exist in a population. Selection cannot, for example, instantly create a new and optimal beak for cracking *Tribulus* fruits or a reverse transcriptase molecule that has no affinity for AZT. Instead, in each generation there will be new mutations in the population that produce new variants for selection to act upon. Some of these may produce beaks that are slightly deeper and narrower, or reverse transcriptase molecules with active sites less likely to bind AZT. If drought conditions or AZT therapy continues, those mutants will be selected; in the next generation adaptation to those environments would improve. In this way natural selection continually improves fitness.

Even though it acts only on existing traits, natural selection can be a creative process that leads to completely new characteristics. This seems paradoxical, but novel features can evolve because selection is able to "repurpose" existing behaviors, structures, or genes for new functions. Stephen Jay Gould (1980), for example, has reviewed the work of D. Dwight Davis on the anatomy of the

Giant Panda's thumb. Pandas use this structure like a sixth finger when eating their favorite food, bamboo. They pass the stalks through the slot between the thumb and their other five digits to strip off the leaves and expose the shoots, which they eat. But it is not a true thumb at all. Anatomically, the thumb bone is a highly modified radial sesamoid, which in closely related species is a component of the wrist. Knowing how natural selection works in contemporary populations, we surmise that when panda populations first began exploiting bamboo, there was strong selection for elongated radial sesamoid bones. Thanks to continued selection over many generations, the average length of the radial sesamoid bone in pandas increased dramatically, until it reached its present thumb-like proportions. A trait that is used in a novel way and is eventually elaborated by selection into a completely new structure, like the radial sesamoid of the ancestral panda, is known as a preadaptation.

Natural Selection is Directed and "Purposeful," but Not Progressive

Evolution by natural selection is sometimes characterized as a random or chance process, but nothing could be further from the truth. Evolution is both directed and purposeful: It is directed by the environment and purposeful in the sense of increasing adaptation to that environment.

But as the HIV and finch examples demonstrate, the process of nonrandom selection is completely free of any entity's conscious intent. Darwin actually came to regret coining the phrase "naturally selected," because people thought the word *selection* implied a conscious act or choice by some entity. Nothing of the sort happens.

Also, although evolution has tended to increase the complexity, degree of organization, and specialization of organisms through time, it is not progressive in the everyday sense of leading toward some predetermined goal. Evolution makes organisms better only in the sense of increasing their adaptation to their environment. There is no inexorable trend toward higher forms of life. For example, contemporary tapeworms have no digestive system, and have actually evolved to be simpler than their ancestors (Figure 2.13). But a progressivist view of evolution dies hard. Even Darwin had to remind himself to "never use the words *higher* or *lower*" when discussing evolutionary relationships.

Fitness is Not Circular

The theory of evolution by natural selection is often criticized—by nonbiologists—as tautological, or circular in its reasoning (see Gould 1977, Kitcher 1982). That is, after reviewing Darwin's four postulates one could claim, "Of *course* individuals with favorable variations are the ones that survive and reproduce, because the theory defines *favorable* as the ability to survive and reproduce."

Here is the key: The word *favorable,* although a convenient shorthand, is misleading. The only requirement for natural selection to work is for *certain* variants to do better than others, as opposed to random ones. As long as nonrandom subsets of the population survive better and leave more offspring, evolution will result. Recall the HIV and finch examples, where research not only determined

(a)

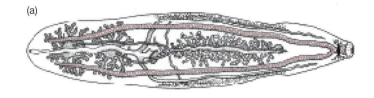

(b)

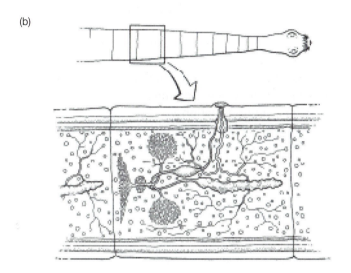

Figure 2.13   **Secondary loss of complex traits**
(a) This is the trematode worm *Opisthorchis sinensis,* with its digestive tract shown in color.   (b) Tapeworms are close relatives of trematodes, but have lost their digestive tract during the course of their evolution and absorb nutrients directly from the environment. The detailed illustration shows how each segment of a tapeworm is filled with reproductive structures.

that nonrandom groups survived a selection event but also uncovered *why* those groups were favored.

It should also make sense by now that Darwinian fitness is not an abstract or untestable quantity. Fitness can be measured in nature. This is done by counting the offspring that an individual produces during its lifetime, or by observing the ability of an individual to survive a selection event and comparing the individual's performance to that of other individuals in the population. These are independent, measurable, and objective criteria for assessing fitness.

Natural Selection Acts on Individuals, Not Groups

One of the most pervasive misconceptions about natural selection, especially selection on animal behavior, is that individual organisms will do things for the good of the species. Self-sacrificing, or altruistic, acts do occur in nature: Prairie dogs give alarm calls when predators approach, and lion mothers will nurse cubs that are not their own. But selection cannot favor traits unless they increase the bearer's fitness relative to competing individuals. If a gene existed that produced a truly altruistic behavior—that is, behavior that reduced the bearer's fitness and increased the fitness of others—it would be strongly selected against. As we will see in Chapter 16, every altruistic behavior that has been studied in detail has been found to increase the altruist's fitness either because the beneficiaries of the behavior are close genetic relatives (as in prairie dogs), or because the beneficiaries reciprocate (as in nursing lions).

The idea that animals will do things for the good of the species is so ingrained, however, that we will make the same point a second way. Consider lions again. Lions live in social groups called prides. Coalitions of males fight to take over prides. If a new group of males defeats the existing pride males in combat, the newcomers quickly kill all of the pride's nursing cubs. These cubs are unrelated to them. Killing the cubs increases the new males' fitness because pride females become fertile again sooner and will conceive offspring by the new males (Packer and Pusey 1983). Infanticide is widespread in animals. Clearly, behavior like this does not exist for the good of the species. Rather, infanticide exists because under certain conditions it enhances the fitness of individuals who perform the behavior relative to individuals who do not.

2.5 The Importance of Time

We introduced natural selection through two examples where molecular or morphological traits changed over extremely short intervals. As a result, it should be clear that selection can lead to dramatic changes in enzyme function and in body size and shape. In the grand scheme of things, however, the changes we have used as examples are relatively small: a few amino acid substitutions in reverse transcriptase, or a few millimeters of depth in a beak. To understand the full scope of evolution, it is critical to grasp another crucial factor: time.

Time is a difficult concept for most of us, simply because evolutionary and geological time scales are so far removed from our everyday experience. For a human being, trying to understand deep time is akin to comprehending the sensory world of an electric eel. But grasping it is crucial, because it speaks to a fundamental question in evolutionary biology: Can the small changes in trait distributions and gene frequencies that we have reviewed thus far explain differences in traits so great that we use them to distinguish genera, families, orders, and classes of organisms? That is, can evolution by natural selection, as we described it operating on finch beaks, also explain the difference between eagles and hummingbirds, or between redwoods and black-eyed susans?

Evolutionary biology is generally studied on two different scales, referred to as microevolution and macroevolution.

In evolutionary biology, small changes that occur within populations are referred to as microevolution, while large changes that distinguish higher level taxonomic groups are termed macroevolution. Changes in reverse transcriptase and finch beak depth are microevolutionary events; the origin of retroviruses and the evolution of flight in birds are macroevolutionary. Clearly, macroevolutionary changes can only be explained in terms of microevolutionary processes if enormous spans of time have passed.

By the mid-19th century Hutton and Lyell and their followers had established, beyond a reasonable doubt, that the Earth was old. But how old? How much time has passed since life on Earth began? Has there been enough time for natural selection to produce the incredible diversity of life?

Marie Curie's discovery of radioactivity, early in this century, gave us a way to answer these questions. Some isotopes of naturally occurring elements are unstable. They decay at a constant rate until they reach a stable state, becoming either different elements or different isotopes of the same element. We measure each isotope's decay rate in units called half-lives. One half-life is the amount of time it takes for 50% of the parent isotope present to decay into its daughter isotope (Figure 2.14). The decay rate is a function only of how many radioactive atoms are present. It is not affected by temperature, moisture, or any other envi-

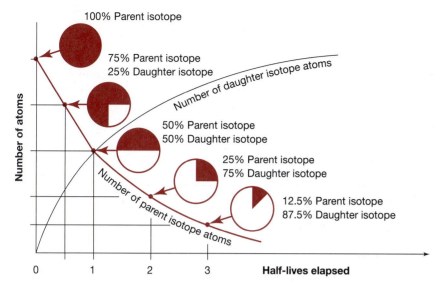

Figure 2.14 Radioactive decay
Many radioactive isotopes decay through a series of intermediates until a stable daughter isotope is produced. Researchers measure the ratio of parent isotope to daughter isotope in a rock sample, then use a graph like this to convert the measured ratio to the number of half-lives elapsed. Multiplying the number of half-lives that have passed by the number of years it takes for a half-life to elapse yields an estimate for the absolute age of the rock.

ronmental factor. As a result, radioactive isotopes function as natural clocks (Box 2.2).

Because of their long half-lives, potassium–argon and uranium–lead systems are the isotopes of choice for determining the age of the Earth. The next question becomes: What rocks should we date to determine when the Earth first formed? Our best models of earth formation predict that the planet was molten for much of its early history, which makes this a difficult issue. If we assume that all of the components of our solar system were formed at the same time, however, two classes of candidate rocks become available: moon rocks and meteorites. Because both uranium–lead and potassium–argon dating systems place the age of the moon rocks brought back by the Apollo astronauts at 4.53 billion years, and because virtually every meteorite that has been dated yields an age of 4.6 billion years, we can infer that the Earth is about 4.6 billion years old.

And how long has life on Earth been evolving? The oldest fossils yet discovered are impressions of bacteria-like cells (see Figure 11.2). These are found in 3.5-billion-year-old rocks from Western Australia (Schopf 1993). The oldest chemical evidence of life is in rocks from Greenland that are over 3.8 billion years old (Mojzsis et al. 1996). Although the Earth was apparently lifeless for most of its first 800 million years, 3.8 billion years is still an enormous amount of time for evolution to occur. This quantity is important for a simple reason: Natural selection may produce small changes each generation, such as a few amino acid substitutions or millimeters of beak depth, but multiplied by 3.8 billion, small changes become large.

The Earth is about 4.6 billion years old. The first organisms evolved about 3.8 billion years ago.

2.6 On the Origin of Species

Ironically, Darwin's masterpiece did not emphasize the subject that inspired its title. Instead, Darwin's primary focus was on changes that occur within populations over time. He wrote much less extensively about differentiation between populations and the formation of new species. It is far from obvious how

Box 2.2   A closer look at dating with radioisotopes

Radiometric dating allows us to assign absolute ages to rocks. Here is how the technique works: First, the half-life of a radioactive isotope is determined by putting a sample in an instrument that records the number of decay events over time. For long-lived isotopes, of course, we have to extrapolate from data collected over a short time interval. Then the ratio of parent to daughter isotope in a sample of rock is measured, often with an instrument called a mass spectrometer. Once we know the half-life of the parent isotope and the current ratio of parent to daughter isotope, we can calculate the number of years that have passed since the rock was formed.

A critical assumption here is that we know the ratio of parent to daughter isotope when the rock was formed. This assumption can be tested. Potassium-argon dating, for example, is an important system for dating rocks of volcanic origin. We can predict that initially there will be zero of the daughter isotope, argon-40, present, because argon-40 is a gas that bubbles out of liquid rock and only begins to accumulate after solidification. We can check recent lava flows to make sure that this is true. Expressed as a ratio of percentages, the ratio of potassium-40 to argon-40 in newly minted basalts, lavas, and ashes is, as predicted, 100:0 (see Damon 1968, Faure 1986).

Of the many radioactive atoms present in the Earth's crust, the isotopes listed in Table 2.3 are the most useful. This is because they are common enough to be present in measurable quantities, and stable in the sense of not readily migrating into or out of rocks after their initial formation. If the molecules did move, it would throw off our estimate of the surrounding rock's age.

To choose an isotope suitable for dating rocks and fossils of a particular age, geochronologists and paleontologists look for a half-life short enough to allow a measurable amount of daughter isotope to accumulate, but long enough to ensure that a measurable quantity of parent isotope is still left. In many instances more than one isotope system can be used on the same rocks or fossils, providing an independent check on the date.

TABLE 2.3   Parent and daughter isotopes used in radiometric dating

Method	Parent isotope	Daughter isotope	Half life of parent (years)	Effective dating range (years)	Materials commonly dated
Rubidium-strontium	Rb-87	Sr-87	47 billion	10 million–4.6 billion	Potassium-rich minerals such as biotite, potassium muscovite, feldspar, and hornblende; volcanic and metamorphic rocks
Uranium-lead	U-238	Pb-206	4.5 billion	10 million–4.6 billion	Zircons, uraninite, and uranium ore such as pitchblende; igneous and metamorphic rock
Uranium-lead	U-235	Pb-207	713 million	10 million–4.6 million	
Thorium-lead	Th-232	Pb-208	14.1 billion	10 million–4.6 million	Zircons, uraninite
Potassium-argon	K-40	Ar-40	1.3 billion	100,000–4.6 billion	Potassium-rich minerals such as biotite, muscovite, and potassium feldspar; volcanic rocks
Carbon-14	C-14	N-14	5730	100–100,000	Any carbon-bearing material, such as bones, wood, shells, charcoal, cloth, paper, and animal droppings

natural selection alone could make a population split into two or more separate species. What role does natural selection play in speciation?

Appropriately enough, our first answers to this question came from studies done on Darwin's finches. An ornithologist named David Lack (1945, 1947; see also Grant 1981a) pointed out that because the islands are volcanic in origin and only 2 to 3 million years old, the first finch population on the archipelago was undoubtedly made up of colonists from South or Central America. Lack also argued that there had been only one colonization event, because the 14 finch species are much more similar to one another than they are to any species on the mainland.

The first step in the diversification, or radiation, of Darwin's finches into the species we see today was thus the physical isolation of the mainland and island populations. Differentiation that begins with the geographic separation of populations is referred to as allopatric (different homeland) speciation.

The mainland and island populations of finches occupied habitats with different competitors, distinct wet and dry season regimes, and dissimilar food sources. Even within the Galápagos archipelago itself, plant communities and food sources vary markedly from island to island. As a result, when finches dispersed to occupy Cocos Island and the various Galápagos islands, they would have encountered new environments and food sources. The upshot is that dispersal and colonization events would have produced a series of finch populations occupying markedly distinct habitats.

Just as year-to-year changes in food availability led to changes in the beak shape of *Geospiza fortis* on Daphne Major during the 1970s and 1980s, different selective regimes on different islands would lead to divergence in beak size and shape. If selection were strong enough and the populations isolated for sufficient time, finches that occupied different islands would diverge strongly. Even if subsequent migration events brought finches from different islands into contact with one another again, they would be distinct enough that interbreeding would be rare. We would then consider the populations to be separate species.

This hypothesis for the radiation of Darwin's finches suggests that speciation is a product of two processes: physical isolation and divergent natural selection. We will explore the process of speciation in much more detail in Chapter 9. Our goal here is simply to recognize that natural selection is fundamental to explaining both adaptation and diversification.

New species can result when populations are geographically separated, and natural selection causes them to diverge.

2.7 The Evolution of Darwinism

Because evolution by natural selection is the general organizing feature of living systems, Darwin's theory ranks as one of the great ideas in intellectual history. Its impact on biology is analogous to that of Newton's laws on physics, Copernicus's Sun-Centered Theory of the Universe on astronomy, and the Theory of Plate Tectonics on geology. In the words of evolutionary geneticist Theodosius Dobzhansky, "Nothing in biology makes sense except in the light of evolution."

For all its power, though, Darwinism was not universally accepted by biologists until some 70 years after it was initially proposed. There were three serious problems with the theory, as originally formulated by Darwin, that had to be resolved.

1. Because Darwin knew nothing about mutation, he had no idea how variability was generated in populations. Put another way, he did not know the mechanism behind postulate 1. It was not until the early 1900s, when geneticists like Thomas Hunt Morgan began experimenting with fruit flies, that we appreciated the continuous and universal nature of mutation. Morgan and colleagues showed that mutations occur in every generation and in every trait. Until then, the prevalent view was that the amount of variability in populations was strictly limited, and that natural selection would grind to a halt when variability ran out.

2. Because Darwin knew nothing about genetics, he had no idea how variations are passed on to offspring. It was not until Mendel's experiments with peas were rediscovered and verified, 35 years after their original publication, that we understood how parental traits are passed on to offspring. The laws of segregation and independent assortment confirmed the mechanism behind postulate 2.

Until then, many biologists proposed that genes acted like pigments in paint. These advocates of blending inheritance argued that favorable mutations would simply merge into existing traits and be lost. In 1867 a Scottish engineer named Fleeming Jenkin published a mathematical treatment of blending inheritance, along with a famous thought experiment concerning the offspring of white-skinned and black-skinned people. For example, if a black-skinned sailor became stranded on an equatorial island inhabited by white-skinned people, Jenkins' model predicted that no matter how advantageous black skin might be (for example, in reducing skin cancer), the population would never become black because traits like skin color blended. If the black-skinned sailor had children by a white-skinned woman, their children would be brown-skinned. If they, in turn, had children with white-skinned people, their children would be light-brown-skinned, and so on. Conversely, if a white-skinned sailor became stranded on a northern island inhabited by black-skinned people, blending inheritance argued that no matter how advantageous white skin might be (for example, in facilitating the synthesis of Vitamin D from UV light), the population would never become white. Under blending inheritance new variants are swamped, and new mutations diluted until they cease to have a measurable effect. For natural selection to work, favorable new variations have to be passed on to offspring intact, and remain discrete.

We understand now, of course, that phenotypes blend in some traits like skin color, but genotypes never do. Jenkins' hypothetical population would, in fact, become increasingly darker- or lighter-skinned if selection were strong and mutation continually added darker- or lighter-skinned variants to the population, via changes in the genes involved in the production of melanin, (Figure 2.15).

Darwin himself struggled with the problem of inheritance, eventually adopting an entirely incorrect view based on the work of Jean-Baptiste Lamarck. Lamarck was a great French biologist of the early 19th century who proposed that species evolve through the inheritance of changes wrought in individuals. Lamarck's idea was a breakthrough for two reasons: It recognized that species have changed through time, and proposed a mechanism for producing change. His theory was wrong, however, because offspring do not inherit phenotypic changes acquired by their parents. If finches build up muscles defending *Tribulus* seeds, their offspring are not more powerful; if giraffes stretch their necks reaching for leaves in treetops, it has no consequence for the neck length of their offspring.

Darwin came to believe that the inheritance of acquired characters was widespread, however, and proposed an elaborate mechanism, called pangenesis, based on particles migrating from various body tissues to the germ cells, where they would influence inheritance. This idea was discredited by the German developmental biologist August Weismann. Weismann published observations showing that germline and somatic cells, at least in animals, are separated early in development and do not interact.

3. Lord Kelvin, the foremost physicist of the nineteenth century, published an important series of papers in the early 1860s calculating the age of the Earth at a maximum of 15–20 million years. Kelvin's analyses were based on measurements of the sun's heat and the current temperature of the Earth. Because fire was the only known source of heat at the time, Kelvin assumed that the sun was combusting like an enormous lump of coal. This had to mean that the sun was gradually burning down, releasing progressively less heat with each passing millennium. Likewise, both geologists and physicists had come to believe that the surface of the Earth was gradually cooling. This was based on the assumption that the Earth was changing from a molten state to a solid one by radiating heat to the atmosphere. This assumption seemed to be supported by measurements of progressively higher temperatures

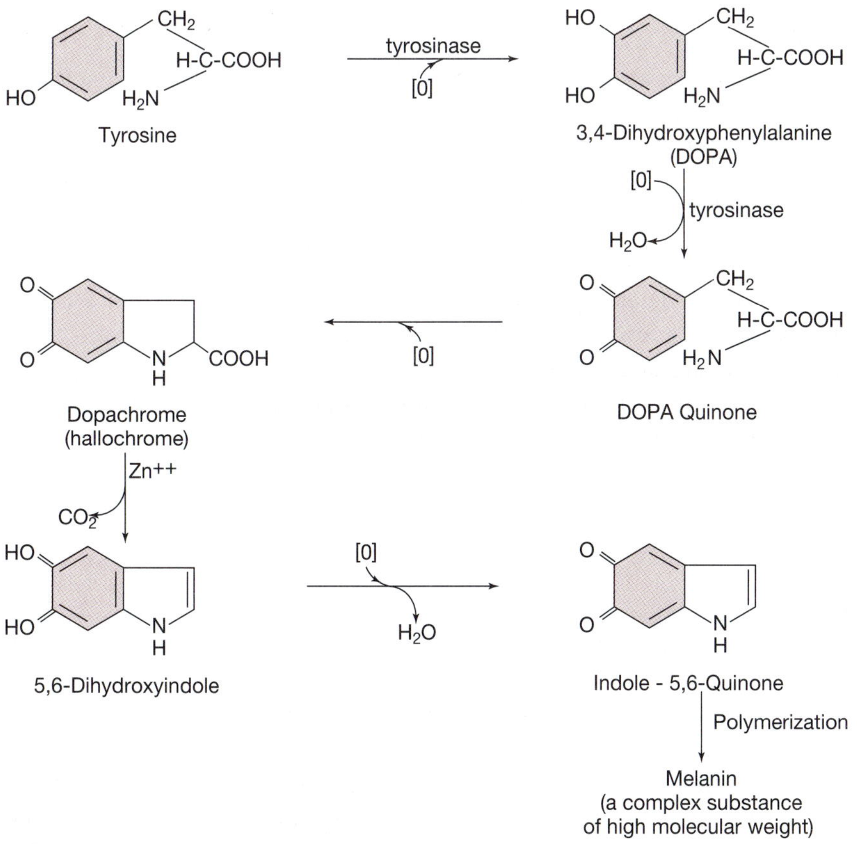

Figure 2.15 **Why blending inheritance does not occur**
This diagram shows the biochemical pathway for melanin, the dark pigment in human skin. The pathway involves a series of steps starting with phenylalanine or tyrosine, which are common dietary amino acids. The enzymes responsible for catalyzing each reaction shown are coded for by several different genes. Because different alleles of these genes have different levels of activity, and because the pathway to melanin involves numerous steps, skin color is a product of the activities of many genes and alleles. In this way the actions of genes blend to form a phenotype. Because each allele is distinct, however, each enzyme's effect on the biochemical pathway is passed on to offspring intact. This is why there is no such thing as blending inheritance.

deeper down in mineshafts. These data allowed Kelvin to calculate the rate of radiant cooling.

The bottom line from Kelvin's calculations was that the transition from a hot to cold sun and hot to cold earth created a narrow window of time when life on earth was possible. The window was clearly too narrow to allow the gradual changes of Darwinism to accumulate, and strongly supported a role for instantaneous and special creation in explaining adaptation and diversity.

The discovery of radioactive isotopes early in the 20th century, of course, changed all that. Kelvin's calculations were unassailable, but his assumptions were dead wrong. Geologists and physicists quickly confirmed that the Earth's heat is a by-product of radioactive decay and not radiant cooling, and that the sun's energy is from nuclear fusion, not combustion.

The Modern Synthesis

Understanding variability, inheritance, and time was so difficult that the first 70 years of evolutionary biology were characterized by turmoil (Eiseley 1958; Provine 1971; Mayr 1980, 1991). But between 1932 and 1953 a series of landmark books were published in systematics, botany, population genetics, and paleontology that successfully integrated genetics with Darwin's four postulates and led to a reformulation of the theory of evolution. This restatement, known as the Modern Synthesis or the Evolutionary Synthesis, was a consensus grounded in two propositions:

The Modern Synthesis resolved decades of controversy over the validity of evolution by natural selection.

- Gradual evolution results from small genetic changes that are acted upon by natural selection.
- The origin of species and higher taxa, or macroevolution, can be explained in terms of natural selection acting on individuals, or microevolution.

With the synthesis, Darwin's original four postulates—and their outcome—could be restated along the following lines:

1. As a result of mutation creating new alleles and segregation and independent assortment shuffling alleles into new combinations, individuals within populations are variable for nearly all traits.
2. Individuals pass their genes on to their offspring intact and independently of other genes.
3. In most generations, more offspring are produced than can survive.
4. The individuals that survive and go on to reproduce, or who reproduce the most, are those with the alleles and allelic combinations that best adapt them to their environment.

The outcome is that alleles associated with higher fitness increase in frequency from one generation to the next.

This View of Life

Darwin ended the introduction to the first edition of *On the Origin of Species* with a statement that still represents the consensus view of evolutionary biologists (Darwin 1859, page 6; see also Gould 1983, page 13): "Natural Selection has been the main but not exclusive means of modification." We now think of modification, or evolutionary change, primarily in terms of changes in the frequencies of the alleles responsible for traits like beak depth and AZT resistance. We are more keenly aware of other processes that cause evolutionary change in addition to natural selection. But the Darwinian view of life, as a competition between individuals with varying abilities to survive and reproduce, has proven correct in almost every detail.

As Darwin wrote in his concluding sentence (page 490): "There is grandeur in this view of life, with its several powers, having been originally breathed into a few forms or into one; and that, whilst this planet has gone cycling on according to the fixed law of gravity, from so simple a beginning endless forms most beautiful and most wonderful have been, and are being, evolved."

2.8 The Debate over Scientific Creationism

Scientific controversy over the fact of evolution ended in the early 1900s, when the evidence we reviewed early in this chapter simply overwhelmed the critics. Whether natural selection was the primary process responsible for adaptation and diversity was still being challenged until the 1930s, when the works of the Modern Synthesis provided a mechanistic basis for Darwin's four postulates and unified micro- and macroevolution. Evolution by natural selection is now considered the great unifying idea in biology. Although scientific debate about the validity of evolution by natural selection ended well over a half-century ago, a political and philosophical controversy in the United States and Europe still continues (Holden 1995; Kaiser 1995).

History of the Controversy

The Scopes Monkey Trial of 1925 (see Gould 1983, essay 20) is perhaps the most celebrated event in a religious debate that has raged since Darwin first published. John Scopes was a biology teacher who assigned reading about Darwinism in violation of the State of Tennessee's Butler Act, which prohibited the teaching of evolution in public schools. William Jennings Bryan, a famous politician and a fundamentalist orator, was the lawyer for the prosecution; Clarence Darrow, the most renowned defense attorney of his generation, represented Scopes. Although Scopes was convicted and fined one dollar, the trial was widely perceived as a great triumph for evolution. Bryan had suggested, on the witness stand, that the six days of creation described in Genesis 1−2:4 may each have lasted far longer than 24 hours. This was considered a grave inconsistency and a huge blow to the integrity of the creationist viewpoint. But far from ending the debate over teaching evolution in U.S. schools, the Scopes trial was merely a way station.

The Butler Act, in fact, stayed on the books until 1967; it was not until 1968, in *Epperson v. Arkansas,* that the U.S. Supreme Court struck down laws prohibiting the teaching of evolution on the basis of the U.S. Constitution's separation of church and state. In response, fundamentalist religious groups in the United States reformulated their arguments as creation science and demanded equal time for what they insisted was an alternative theory for the origin of species.

By the late 1970s, 26 state legislatures were debating equal time legislation (Scott 1994); Arkansas and Louisiana passed such laws only to have them struck down in state courts. The Louisiana law was then appealed all the way to the U.S. Supreme Court, which decided in 1987 *(Edwards v. Aquillard)* that because creationism is essentially a religious idea, teaching it in the public schools was a violation of the first amendment. Two justices, however, formally wrote that it would still be acceptable for teachers to present alternative theories to evolution (Scott 1994).

One response from opponents of evolution has been to drop the words *creation* and *creator* from their literature and call either for equal time for teaching that no evolution has occurred, or for teaching a proposal called Intelligent Design Theory, which infers the presence of a creator from the perfection of adaptation in contemporary organisms (Scott 1994; Schmidt 1996). The complexity and perfection of organisms is actually a time-honored objection to evolution by natural selection. Darwin was aware of it; in the *Origin* he devoted a section of his chapter titled "Difficulties on Theory" to *Organs of extreme perfection.* How can natural selection, by sorting random changes in the genome, produce elaborate and integrated traits like the vertebrate eye?

Perfection and Complexity in Nature

The English cleric William Paley, writing in 1802, used the metaphor of finding and studying a complex and accurate watch, and from it inferring the existence of a highly skilled watchmaker, to promote the Theory of Special Creation. With respect to the natural world, Paley offered the perfection of the vertebrate eye. He compared this structure to the wonderful watch, and asked his readers to infer the existence of a purposeful and perfect Creator. This logic, still used by creationists today, is called the Argument from Design (Dawkins 1986).

Because we perceive perfection and complexity in the natural world, evolution by natural selection seems to defy credulity. There are actually two concerns here. The first question is how random changes can lead to order. Mutations are chance events, so the generation of variation in a population is random. But the selection of those variants, or mutants, is nonrandom: It is directed in the sense of increasing fitness. And adaptations—structures or behavior that increase fitness—are what we perceive as highly ordered, complex, or even perfect in the natural world. But there is nothing conscious or intelligent about the process. The biologist Richard Dawkins captured this point by referring to natural selection as a "blind watchmaker."

A second, and closely related, concern is: How can complex, highly integrated structures like the eye evolve through the Darwinian process of gradual accumulation of small changes? Each step would have to incrementally increase the fitness of individuals in the population. Darwinism predicts that complex structures have evolved through a series of intermediate stages, or graded forms. Is this true? For example, when we consider a structure like the eye, do we find a diversity of forms, some of which are more complex than others?

The answer to this question is yes. In some unicellular species there are actually subcellular organelles with functions analogous to the eye. The eyespots of a group of protozoans called Euglenoids, for example, contain light-absorbing molecules that are shaded on one side by a patch of pigment. When these molecules absorb light they undergo structural changes. Because light can reach them from one side only, a change in the light-absorbing molecule contains useful information about where light is coming from. Some dinoflagellates even have a subcellular, lens-like organelle that can concentrate light on a pigment cup (Eckert and Randall 1983). It is unlikely that these single-celled protists can form an image, however, because they are not capable of neural processing. Rather, their eye probably functions in transmitting information about the cell's depth in the water column, helping the cell orient and swim toward light.

More complex eyes have a basic unit called the photoreceptor. This is a cell that contains a pigment capable of absorbing light. The simplest type of multicellular eye, consisting of a few photoreceptor cells in a cup or cup-like arrangement (Figures 2.16a and 2.16b), is found in a wide diversity of taxa, including flatworms, polychaetes (segmented worms in the phylum Annelida), some crustaceans (the shrimps, crabs, and allies), and some vertebrates. These organs are used in orientation and day-length monitoring (Willson 1984; Brusca and Brusca 1990). Slightly more complex eyes (Figure 2.16c) have optic cups with a narrow aperture acting as a lens (Gould 1994), and may be capable of forming images in at least some species. These are found in a few nemerteans (ribbon worms) and annelids (segmented worms), copepod crustaceans, and abalone and nautiloids (members of the phylum Mollusca). The most complex eyes (Figure 2.16d) fall into two functional categories, based on whether the photoreceptor cells are arrayed on a retina that is is concave like the eyes of vertebrates and octopus, or convex like the compound eyes of insects and other arthropods (Goldsmith 1990). These eyes have lenses, and in most cases are capable of forming images.

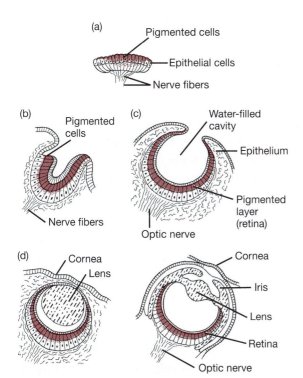

Figure 2.16   **Variation in mollusc eyes**   (a) A pigment spot; (b) a simple pigment cup;   (c) the simple optic cup found in abalone;   (d) the complex lensed eyes of a marine snail called *Littorina* and the octopus. Pigmented cells are shown in color.

It is important to recognize, however, that the simpler eyes we have just reviewed do not represent intermediate forms on the way to more advanced structures. The eyespots, pigment cups, and optic cups found in living organisms are contemporary adaptations to the problem of sensing light. They are *not* ancestral forms. It is, however, sensible to argue that the types of eyes we have reviewed form an evolutionary pathway (Gould 1983, essay 1). That is, it is conceivable that eyes like these formed intermediate stages in the evolution of the vertebrate, octopus, and insect eye. This is exactly what Darwin argued in his section on "organs of extreme perfection." (To learn more about the evolution of the eye, see Salvini-Plawen and Mayr 1977; Nilsson and Pelger 1994; Quiring et al. 1994.)

Existing variation in eyes demonstrates that intermediate forms of complex structures are possible.

Other Objections

We have found four additional arguments that creation scientists use regularly, and have added responses from an evolutionary perspective (see Gish 1979; Kitcher 1982; Futuyma 1983; Gould 1983 essays 19, 20, 21; Dawkins 1986; Swinney 1994).

1. Evolution by natural selection is unscientific because it is not falsifiable and because it makes no testable predictions. Because each of Darwin's four postulates are independently testable, the theory meets the classical criterion that ideas must be falsifiable to be considered scientific. Also, the claim that evolutionary biologists do not make predictions is not true. Paleontologists routinely (and correctly) predict which strata will bear fossils of certain types (a spectacular example was that fossil marsupial mammals would be found in Antarctica); Grant and Grant have used statistical techniques based on evolutionary theory to correctly predict the amount and direction of change in finch characteristics during selection events in the late 1980s and early 1990s (Grant and Grant 1993, 1995). Scientific creationism, on the other hand, amounts to an oxymoron; in the words of one of its leading advocates, Dr. Duane Gish (1978, p. 42): "We cannot discover by scientific investigations anything about the creative processes used by God."

2. The Earth was created as little as 6,000–8,000 years ago. There has not been enough time for Darwinian evolution to produce the adaptation and diversity observed in living things. Creation scientists present short-earth theories and argue that most geological landforms and strata resulted from the flood during the time of Noah (for example, Gish 1978; Swinney 1994). Most simply disbelieve the assumptions behind radiometric dating and deny the validity of the data. The assumption of uniformitarianism in the evolution of life and landforms is also rejected by creation scientists. Again, we quote Gish (1978, p. 42): "We do not know how God created, what processes He used, *for God used processes which are not now operating anywhere in the natural universe*" (emphasis original).

The assumptions of radiometric dating have been tested, however, and demonstrated to be correct (see Box 2.2). Radiometric dating has demonstrated that rock strata differ in age, and that the Earth is well over 4 billion years old.

3. Because organisms progress from simpler to more complex forms, evolution violates the Second Law of Thermodynamics. Although the Second Law has been stated in a variety of ways since its formulation in the late 19th century, the most general version is that "Natural processes tend to move toward a state of greater disorder" (Giancoli 1995). The Second Law is focused on the concept of entropy. This is a quantity that measures the state of disorder in a system. The second law restated in terms of entropy is that "The entropy of an isolated system never decreases. It can only stay the same or increase" (Giancoli 1995).

The key to understanding the Second Law's relevance to evolution is the word *isolated*. The Second Law is true only for closed systems. Living things, however, live in an open system: the Earth, where photosynthetic organisms capture the radiant energy of the sun and convert it to chemical energy that they and other living things can use. Because energy is constantly being added to living systems, the Second Law does not apply to their evolution.

4. No one has ever seen a new species formed, so evolution is unproven. And because evolutionists say that speciation is too slow to be directly observed, evolution is unprovable and thus based on faith. Although speciation is a slow process, it *is* ongoing and can be studied. In Chapter 9 we will explore one of the best-studied examples: the divergence of apple maggot flies into separate host races. The two forms of these flies lay their eggs in different fruits, which serve as a food source for their maggots. As a result of divergent natural selection on traits like food preferences and timing of breeding, marked genetic differences between the two populations are beginning to emerge. This research is recording key events early in the process of one species splitting into two separate species.

Because insects are far and away the most diverse, or speciose, of all living taxonomic groups, and because most insects are specialist plant eaters, what is happening in apple maggot flies is of general interest: They may now be undergoing a speciation process that has occurred many times in the course of evolution.

What Motivates the Controversy?

The evidence for evolution continues to mount. Starting with the discovery of the universal genetic code in the 1960s and continuing through recent sequencing studies, molecular biologists have found that genetic homologies are even more pervasive than structural homologies (Figures 2.17a and 2.17b). For decades, in fact, evolution by natural selection has been considered one of the best documented and most successful theories in the biological sciences. Furthermore, many Christians see no conflict between belief in evolution and religious faith. For example, in 1996 Pope John Paul II acknowledged that Darwinian evolution was a firmly established scientific result, and stated that accepting Darwinism was compatible with traditional Christian understandings of God. If the fact of evolution and the validity of natural selection are utterly uncontroversial, and if belief in evolution is compatible with belief in God, then why does the creationist debate continue?

During a discussion about whether material on evolution should be included in high school textbooks, Alabama school board member David Byers said, "It's foolish and naive to believe that what children are taught about who they are, how they got here, doesn't have anything to do with what they conclude about

Data from molecular biology have shown that structural and developmental homologies are caused by genetic homologies.

First base	Second base							Third base
	U		**C**		**A**		**G**	
U	UUU	Phenylalanine	UCU	Serine	UAU	Tyrosine	UGU	Cysteine	**U**
	UUC	Phenylalanine	UCC	Serine	UAC	Tyrosine	UGC	Cysteine	**C**
	UUA	Leucine	UCA	Serine	UAA	Stop	UGA	Stop	**A**
	UUG	Leucine	UCG	Serine	UAG	Stop	UGG	Tryptophan	**G**
C	CUU	Leucine	CCU	Proline	CAU	Histidine	CGU	Arginine	**U**
	CUC	Leucine	CCC	Proline	CAC	Histidine	CGC	Arginine	**C**
	CUA	Leucine	CCA	Proline	CAA	Glutamine	CGA	Arginine	**A**
	CUG	Leucine	CCG	Proline	CAG	Glutamine	CGG	Arginine	**G**
A	AUU	Isoleucine	ACU	Threonine	AAU	Asparagine	AGU	Serine	**U**
	AUC	Isoleucine	ACC	Threonine	AAC	Asparagine	AGC	Serine	**C**
	AUA	Isoleucine	ACA	Threonine	AAA	Lysine	AGA	Arginine	**A**
	AUG	Start (Methionine)	ACG	Threonine	AAG	Lysine	AGG	Arginine	**G**
G	GUU	Valine	GCU	Alanine	GAU	Aspartic Acid	GGU	Glycine	**U**
	GUC	Valine	GCC	Alanine	GAC	Aspartic Acid	GGC	Glycine	**C**
	GUA	Valine	GCA	Alanine	GAA	Glutamic Acid	GGA	Glycine	**A**
	GUG	Valine	GCG	Alanine	GAG	Glutamic Acid	GGG	Glycine	**G**

Figure 2.17   Gene homologies   (a) In almost every organism studied, the same nucleotide triplets, or codons, specify the same amino-acid bearing transfer RNAs. That the genetic code is nearly universal strongly implies that it originated very early in the history of life, and is additional evidence for the relatedness of all living things. Exceptions to the universal code do occur, however. We will explore this phenomenon in more detail in Chapter 7.

why they are here and what their obligations are, if, in fact, they have any obligations, and how they should live" (National Public Radio 1995). This statement suggests that, for some creationists, the controversy is not about the validity of the scientific evidence or its compatibility with religion. Instead, the concern seems to be about what evolution means for human morality and behavior.

Creationists and evolutionists share the desire that children should grow up to become morally responsible adults. Creationists fight evolution because they believe it is morally dangerous. Evolutionary biologists, on the other hand, believe that children should learn what science says about how living things came to be, and let them sort out the moral implications, if any, on their own.

Summary

Before Darwin began work on the origin of species, many scientists had become convinced that species had changed through time. The evidence for the fact of evolution included data on homologies in structures and developmental stages in related organisms, close relationships among species occupying adjacent

Paired domain

```
       1         10        20        30        40        50        60
Consensus  GQGGVNQLGGVFVNGRPLPNHIRQKIVELAHHGIRPCDISRQLRVSHGCVSKILGRYYETGSIRP
Aniridia   SHS.............DST........S.A......I.Q..N...........
mPax–6     SHS.............DST........S.A......I.Q..N...........
qPax–6     SHS.............DST........S.A......I.Q..N...........
pax[zf-a]  SHS.............DST........S.A......I.Q..N...........
ey         GHS.........G....DST........S.A......I.Q..N...........
```

```
       66   70        80        90        100       110       120       130
Consensus  GAIGGSKPRQVATPDVVKKIAEYKRENPGMFAWEIRDRLLKEGVCDNDTVPSVSSISRILRNKFGK
Aniridia   R........-....E..S...Q...C.SI.........S....T..NI......N.V...LASE
mPax–6     R........-....E..S...Q...C.SI.........S....T..NI......N.V...LASE
rPax–6                          -C.SI.........S....T..NI......N.V...LASE
qPax–6     R........-....E..S...Q...C.SI.........S....T..NI......N.V...LASE
pax[zf-a]  R........-....E..G...Q...C.SI.........S....T..NI......N.V...LASE
ey         R........-...AE..S..SQ...C.SI.........Q.N..T..NI......N.V...LAAQ
```

Homeodomain (paired–type)

```
       1         10        20        30        40        50        60
Consensus  QRRNRTTFTQEQLEALERAFERTHYPDIFTREELAQRINLTEARVQFWFSNRRAKWRRQE
Aniridia   LQ....S.....I....KE........V.A..R..AK.D.P...I.V.........E.
mPax–6     LQ....S.'...I....KE........V.A..R..AK.D.P...I.V.........E.
rPax–6     LQ....S.....I....KE........V.A..R..AK.D.P...I.V.........E.
qPax–6     LQ....S.....I....KE........V.A..R..AK.D.P...I.V.........E.
pax[zf-a]  LQ....S.....I....KE........V.A..R..AK.D.P...I.V.........E.
ey         LQ....S..ND.IDS..KE........V.A..R..GK.G.P...I.V.........E.
```

Figure 2.17 (b) This chart shows the amino acid sequences of two sections (called the paired domain and homeodomain) of a protein involved in the development of the eye (Quiring et al. 1994). The genes that code for the protein go by various names in different taxa: *Aniridia* is from humans, *mPax-6* from mouse, *rPax-6* from rat, *qPax-6* from quail, *pax(zf-a)* from zebrafish, and *ey* from fruit flies. The consensus sequence for all genes in this gene family is given in the top row. Boxes indicate shared amino acids characteristic of *Pax-6*, dots indicate the same amino acid as the consensus, and dashes show deletions. Astonishingly, the amino acid sequence identity in these two domains is well over 90% between flies and the vertebrates, even though they last shared a common ancestor hundreds of millions of years ago.

It is important to recognize that the structural and developmental homologies shown in Figure 2.1 are caused by the types of genetic homologies illustrated here.

islands, vestigial structures, the confirmation of extinction, and an ancient age for the Earth. Darwin and Wallace independently realized that the process of natural selection provided a mechanism for this pattern, which they termed descent with modification.

Evolution by natural selection is the logical outcome of four facts: (1) Individuals vary in most or all traits; (2) some of this variation is genetically based and can be passed on to offspring; (3) more offspring are born than can survive to breed, and of those that do breed, some are more successful than others; and (4) the individuals that reproduce the most are a nonrandom, or more fit, subset of the general population. This selection process causes changes in the genetic makeup of populations. Because life on earth started at least 3.8 billion years ago, there has been a great deal of time for natural selection to produce the diversity of species we see today.

Questions

1. Analogy and homology are important concepts used in comparing species. Traits are analogous if they have similar functions but are derived, evolutionarily and developmentally, from different source structures. A classic example of analogous structures is insect wings and bat wings. Traits are homologous if they are derived, evolutionarily and developmentally, from the same source structure. Which of the following pairs of structures are analogous, and which are homologous?
   a. The dorsal fins of a porpoise and a salmon
   b. The jointed leg of a ladybird beetle and a robin
   c. A siamang ape's tail and a human's coccyx
   d. The bright red bracts (modified leaves) of a poinsettia and the green leaves of a rose
   e. The bright red bracts of a poinsettia and the red petals of a rose

2. In the early 20th century, radiometric dating allowed geologists to assign absolute ages to most fossil-bearing strata. The absolute dates turned out to be entirely consistent with the relative dating done in the early 19th century. What does this fact say about the assumptions behind relative dating listed on page 33?

3. In everyday English, the word *adaptation* means "an adjustment to environmental conditions." How is the evolutionary definition of *adaptation* different from the everyday English sense?

4. Think about how the finch bill data demonstrate Darwin's postulates. For instance, what would Figure 2.9 have looked like if bill depth was not variable? What would Figures 2.10 and 2.12a have looked like if bill depth was variable, but the variation was not heritable? In Figure 2.10, why is the line drawn from 1978 data, after the drought, higher on the y-axis than the line drawn from 1976 data, before the drought?

5. Suppose that you are starting a long-term study of a population of annual, flowering plants isolated on a small island. Reading some recent papers has convinced you that global warming is real and will lead to significant, long-term changes in the amount of rain the island receives. Outline the observations and experiments you would need to do in order to document whether natural selection occurs in your study population over the course of your research. What traits would you measure, and why?

6. At the end of an article on how mutations in variable number terminal repeat (VNTR) sequences of DNA are associated with disease, Krontiris (1995, p. 1683) writes: "the VNTR mutational process may actually be positively selected; by culling those of us in middle age and beyond, evolution brings our species into fighting trim." Can this hypothesis be true, given that selection acts on individuals? Explain.

7. Many working scientists are generally uninterested in the history of their fields. Did the historical development of Darwinism, reviewed in section 2.7, help you understand the theory better? Why or why not? Do you think it is important for practicing scientists to spend time studying history?

8. Recently the Alabama School Board, after reviewing high school biology texts, voted to require that this disclaimer he posted on the inside front cover of the approved book (National Public Radio 1995):

   This textbook discussed evolution, a controversial theory some scientists present as a scientific explanation for the origin of living things, such as plants, animals, and humans. No one was present when life first appeared on Earth; therefore, any statement about life's origins should be considered as theory, not fact.

Do you accept the last sentence in this statement? Does the insert's point of view pertain to other scientific theories such as the Cell Theory, the Atomic Theory, the Theory of Plate Tectonics, and the Germ Theory of Disease?

9. A 1991 Gallup poll of U.S. adults found that 47% of the respondents believed that God created man within the last 10,000 years (Root-Bernstein, 1995). Given the evidence reviewed here for the fact of evolution, comment on why so few people in the United States accept it.

Exploring the Literature

10. During the last fifty years, many viruses, bacteria, fungi, and insects have evolved resistance to drugs, herbicides, fungicides, or pesticides. These are outstanding examples of evolution in action. In several of these cases, we know the molecular mechanisms of the evolutionary changes involved. To explore this topic further, look up the following papers. Think about how the evidence from these studies compares with the evidence for evolution in Darwin's finches and HIV.

Cohen, M.L. 1992. Epidemiology of drug resistance: implications for a post-antimicrobial era. *Science* 257:1050−1055.

Davies, J. 1994. Inactivation of antibiotics and the dissemination of resistance genes. *Science* 264:375−382.

Van Rie, J., W.H. McGaughey, D.E. Johnson, B.D. Barnett, and H. Van Melleart. 1990. Mechanism of insect resistance to the microbial inseticide *Bacillus thuringiensis. Science* 247:72−74.

11. It seems unlikely that selection of traits "for the good of the group" can occur. However, there are a few evolutionary biologists who believe that under unusual conditions, group selection of altruistic behaviors may in fact be possible. Look up the following papers to learn more about this controversial topic:

Morell, V. 1996. Genes vs. Teams: Weighing group tactics in evolution. *Science* 273:739−740. (News perspective.)

Wilson, D.F., and E. Sober. 1994. Reintroducing group selection to the human behavioral sciences. *Behavioral Brain Sciences* 17:585−609.

Wilson, D.S., and L.A. Dugatkin. 1997. Group selection and assortative interactions. *American Naturalist* 149:336−351.

Citations

Abbott, I., L.K. Abbott, and P.R. Grant. 1977. Comparative ecology of Galápagos ground finches (*Geospiza* Gould): Evaluation of the importance of floristic diversity and interspecific competition. *Ecological Monographs* 47:151−184.

Alatalo, R.V., and A. Lundberg. 1986. Heritability and selection on tarsus length in the pied flycatcher (*Fidecula hypoleuca). Evolution* 40:574−583.

Audesirk, G., and T. Audesirk. 1993. *Biology: Life on Earth* (New York: Macmillan).

Boag, P.T. 1983. The heritability of external morphology in Darwin's ground finches *(Geospiza)* on Isla Daphe Major, Galápagos. *Evolution* 37:877−894.

Boag, P.T., and P.R. Grant. 1978. Heritability of external morphology in Darwin's finches. *Nature* 274:793−794.

Boag, P.T., and P.R. Grant. 1981. Intense natural selection in a population of Darwin's finches (Geospizinae) in the Galápagos. *Science* 214:82−85.

Bowman, R.I. 1961. Morphological differentiation and

adaptation in the Galápagos finches. *University of California Publications in Zoology* 58:1–302.

Brusca, R.C., and G.J. Brusca. 1990. *Invertebrates* (Sunderland, MA: Sinauer).

Chernicoff, S. and R. Venkatakrishnan. 1995. *Geology: An Introduction to Physical Geology* (New York: Worth Publishers).

Damon, P.E. 1968. Potassium-argon dating of igneous and metamorphic rocks with applications to the Basin ranges of Arizona and Sonora. In Hamilton, E.I., and R.M. Farquhar, eds., *Radiometric Dating for Geologists* (London: Interscience Publishers).

Darwin, C. 1859. *On the Origin of Species by Means of Natural Selection* (London: John Murray).

Darwin, C. 1862. *The Various Contrivances by Which Orchids are Fertilized by Insects* (London: John Murray).

Dawkins, R. 1986. *The Blind Watchmaker* (Essex: Longman Scientific).

Dhondt, A. 1982. Heritability of blue tit tarsus length from normal and cross-fostered broods. *Evolution* 36:418–419.

Dodson, E.O. 1960. *Evolution: Process and Product* (New York: Reinhold Publishing).

Eckert, R. and D. Randall. 1983. *Animal Physiology* (New York: W.H. Freeman).

Eiseley, L. 1958. *Darwin's Century* (Garden City, NY: Anchor Books).

Faure, G. 1986. *Principles of Isotope Geology* (New York: John Wiley & Sons).

Futuyma, D.J. 1983. *Science on Trial: The Case for Evolution* (New York: Pantheon).

Giancoli, D.C. 1995. *Physics: Principles with Applications* (Englewood Cliffs, NJ: Prentice Hall).

Gibbs, H.L., and P.R. Grant. 1987. Oscillating selection on Darwin's finches. *Nature* 327:511–513.

Gish, D. T. 1978. *Evolution: The Fossils Say No!* (San Diego: Creation-Life Publishers).

Goldsmith, T.H. 1990. Optimization, constraint, and history in the evolution of eyes. *Quarterly Review of Biology* 65:281-322.

Gould, S.J. 1977. *Ever Since Darwin: Reflections in Natural History* (New York: W.W. Norton).

Gould, S.J. 1980. *The Panda's Thumb* (New York: W.W. Norton).

Gould, S.J. 1983. *Hen's Teeth and Horse's Toes* (New York: W.W. Norton).

Gould, S.J. 1994. Common pathways of illumination. *Natural History* 12:10–20.

Grant, B.R. 1985. Selection on bill characters in a population of Darwin's finches: *Geospiza conirostris* on Isla Genovesa, Galápagos. *Evolution* 39:523–532.

Grant, B.R., and P.R. Grant. 1989. *Evolutionary Dynamics of a Natural Population* (Chicago: University of Chicago Press).

Grant, B.R., and P.R. Grant. 1993. Evolution of Darwin's finches caused by a rare climatic event. *Proceedings of the Royal Society of London,* Series B 251:111–117.

Grant, P.R. 1981a. Speciation and adaptive radiation on Darwin's finches. *American Scientist* 69:653–663.

Grant, P.R. 1981b. The feeding of Darwin's finches on *Tribulus cistoides* (L.) seeds. *Animal Behavior* 29:785–793.

Grant, P.R. 1986. *Ecology and Evolution of Darwin's Finches* (Princeton: Princeton University Press).

Grant, P.R. 1991. Natural selection and Darwin's finches. *Scientific American* October:82–87.

Grant, P.R. and B.R. Grant. 1995. Predicting microevolutionary responses to directional selection on heritable variation. *Evolution* 49:241-251.

Grant, P.R., B.R. Grant, J.N.M. Smith, I.J. Abbott, and L.K. Abbott. 1976. Darwin's finches: Population variation and natural selection. *Proceedings of the National Academy of Sciences, USA* 73:257-261.

Holden, C. 1995. Alabama schools disclaim evolution. *Science* 270:1305.

Kaiser, J. 1995. Dutch debate tests on evolution. *Science* 269:911.

Keeton, W.T. 1972. *Biological Science* (New York: W.W. Norton).

Kitcher, P. 1982. *Abusing Science: The Case Against Creationism* (Cambridge, MA: MIT Press).

Krontiris, T.G. 1995. Minisatellites and human disease. *Science* 269:1682-1683.

Lack, D. 1945. The Galápagos finches (Geospizinae): A study in variation. *Occasional Papers of the California Academy of Sciences* 21:1–159.

Lack, D. 1947. *Darwin's Finches* (Cambridge University Press).

Mayr, E. 1980. Prologue. In Mayr, E., and W.B. Provine, eds., *The Evolutionary Synthesis* (Cambridge, MA: Harvard University Press).

Mayr, E. 1991. *One Long Argument: Charles Darwin and the Genesis of Modern Evolutionary Thought* (Cambridge, MA: Harvard University Press).

Mojzsis, S.J., G. Arrhenius, K.D. McKeegan, T.M. Harrison, A.P. Nutman, and C.R.L. Friend. 1996. Evidence for life on Earth before 3,800 million years ago. *Nature* 384:55-59.

National Public Radio. 1995. Evolution disclaimer to be placed in Alabama textbooks. Morning Edition, Transcript #1747, Segment #13.

Nilsson, D.-E., and S. Pelger. 1994. A pessimistic estimate of the time required for an eye to evolve. *Proceedings of the Royal Academy of London,* Series B 256:53–58.

Packer, C., and A.E. Pusey. 1983. Adaptations of female lions to infanticide by incoming males. *American Naturalist* 121:716–728.

Price, T.D., P.R. Grant, H.L. Gibbs, and P.T. Boag. 1984. Recurrent patterns of natural selection in a population of Darwin's finches. *Nature* 309:787–789.

Provine, W.B. 1971. *The Origins of Theoretical Population Genetics* (Chicago: University of Chicago Press).

Quiring, R., U. Walldorf, U. Kloter, and W.J. Gehring. 1994. Homology of the *eyeless* gene of *Drosophila* to the *small eye* gene in mice and *aniridia* in humans. *Science* 265:785–789.

Root-Bernstein, R.S. 1995. "Darwin's Rib." *Discover* September 1995.

Salvini-Plawen, L.v., and E. Mayr. 1977. On the evolution of photoreceptors and eyes. *Evolutionary Biology* 10:207–263.

Schmidt, K. 1996. Creationists evolve new strategy. *Science* 273:420–422.

Schopf, J.W. 1993. Microfossils of the early archean apex chert: New evidence of the antiquity of life. *Science* 260:640–646.

Scott, E.C. 1994. The struggle for the schools. *Natural History* 7:10–13.

Smith, J.N.M., and A. Dhondt. 1980. Experimental confirmation of heritable morphological variation in a natural population of song sparrows. *Evolution* 34: 1155–1159.

Swinney, S. 1994. "Evolution: Fact or Fiction." (Kansas City, MO: 1994 Staley Lecture Series, KLJC Audio Services).

Wiggins, D.A. 1989. Heritability of body size in cross-fostered tree swallow broods. *Evolution* 43:1808–1811.

Willson, M.F. 1984. *Vertebrate Natural History* (Philadelphia: Saunders).

Methods of Evolutionary Analysis

Our goals in Chapters 1 and 2 were to provide an overview of the questions that evolutionary biologists ask and to introduce natural selection as the primary agent responsible for producing adaptation and diversity. To complete our introduction, we turn now from defining what evolutionary biology is to exploring how the science is done.

In this chapter we introduce experimental design, the comparative method, and the fossil record. These are three of the major tools and approaches that evolutionary biologists use to answer questions about adaptation and diversity.

The case studies reviewed here will do more than just highlight important principles about scientific methods, however. In addition to introducing how evolutionary biologists go about answering questions, we need to explore how they go about asking them. This is a critical element, because asking tightly focused and stimulating questions is an essential part of the scientific enterprise. Questions drive the research. The examples presented in this chapter illustrate markedly different ways of coming up with interesting questions, and some of the more prominent strategies for answering them.

3.1 Experimental Approaches

Experiments are one of the most powerful tools in science, for a simple reason: They allow us to isolate and test the effect that single, well-defined factors have on the phenomenon in question. We have already had a chance to review two natural experiments: the changes in seed availability that led to selection on beak and body size in Galápagos finches, and the emerging contrasts in patterns of HIV transmission in different parts of the world. In this section we review two examples of planned experiments. Although our focus is on the process of planning and interpreting experiments, or the how-to's, these case studies will also illustrate the diversity of subjects that can be addressed experimentally. We begin with a suite of experiments inspired by some unusual behavior in flies.

What is the Function of the Wing Markings and Wing-Waving Display of the Tephritid Fly *Zonosemata*?

The tephritid fly *Zonosemata vittigera* has distinctive dark bands on its wings. When disturbed, the fly holds its wings perpendicular to its body and waves them up and down (Figure 3.1). Entomologists had noticed that this display seemed to mimic the leg-waving, territorial threat display of jumping spiders (species in the family Salticidae). These entomologists suggested that, because jumping spiders are fast and have a nasty bite, a fly mimicking a jumping spider might be avoided by a wide variety of other predators. Erick Greene and colleagues (1987) had a different idea. Because jumping spiders are *Zonosemata's* major predators, Greene et al. proposed that the fly uses the wing coloration and wing-waving display to intimidate the jumping spiders themselves. The fly, in other words, is a sheep in wolf's clothing. Mimicry of a predator's behavior by its own prey had never before been recorded.

Both mimicry explanations are plausible enough to be interesting, but they are really just stories at this point. Can these hypotheses be tested rigorously? Greene and his co-workers (1987) set out to do so experimentally.

The first step in any analysis is to phrase the question as precisely as possible. In this case: Do the wing markings and the wing waving of *Zonosemata vittigera* mimic the threat displays that jumping spiders use on each other, and thereby allow the flies to escape predation? Stating a question precisely makes it easier to design an experiment that will provide a clear answer.

Good experimental designs test the predictions made by several alternative hypotheses.

Figure 3.1 **A sheep in wolf's clothing?** This photograph shows the tephritid fly *Zonosemata vittigera* (right) facing one of its predators, the jumping spider *Phidippus apacheanus* (left) (Erick Greene, University of Montana).

Greene et al.'s next step was to list alternative explanations for the behavior. Good research programs are thorough in testing as many competing hypotheses as possible. Note that each of the following is a viable and biologically realistic competing explanation, not an implausible straw man proposed simply to give the impression of rigor.

H$_1$: The flies do not mimic jumping spiders. This is a distinct possibility, because other fly species have dark wing bands and wing-flicking displays that do not deter predators. In many species the flies use their markings and displays during courtship.

H$_2$: The flies mimic jumping spiders, but the flies behave like spiders to deter *other,* nonspider predators. Other fly predators that might be intimidated by a jumping spider, or a jumping spider mimic, include other kinds of spiders, assassin bugs, preying mantises, and lizards.

H$_3$: The flies mimic jumping spiders, and this mimicry functions specifically to deter predation by the jumping spiders themselves.

To create flies capable of testing these alternatives, Greene and his co-workers transplanted wings between houseflies *(Musca domestica),* which have clear and unmarked wings, and *Zonosemata* (Table 3.1). They did this by cutting the wings off house-flies and using Elmer's glue to attach them to the wing stubs of amputated *Zonosemata* (Figure 3.2). Remarkably, the surgically altered *Zonose-*

Figure 3.2 **Experimental treatment of *Zonosemata* flies** This *Zonosemata* fly has had its wings removed surgically, then replaced with wings from a housefly. *Zonosemata* treated in this way are able to display and fly normally (Erick Greene, University of Montana).

TABLE 3.1 Experimental treatments used in testing tephritid fly mimicry

This table shows predicted outcomes when flies with different treatments encounter different predators in the experiment by Greene and colleagues. An *X* in a cell indicates the prediction that flies will be stalked and attacked or killed. Note that each hypothesis makes a unique suite of predictions. (The predictions listed for hypotheses 2 and 3 assume that *Zonosemata's* wing markings and wing waving are both necessary for effective mimicry.)

		A	B	C	D	E
	Treatment	*Zonosemata* untreated	*Zonosemata* with own wings reglued	*Zonosemata* with housefly wings	Housefly with *Zonosemata* wings	Housefly untreated
	Purpose	Test effect of wing markings plus wing waving	Control for effects of operation	Test effect of wing waving but no wing markings	Test effect of wing markings but no wing waving	Test effect of no wing markings and no wing waving
Hypothesis	**Predator**	\multicolumn{5}{c}{**Predicted outcome when fly meets predator**}				
H$_1$: No mimicry	Jumping spider	X	X	X	X	X
	Other	X	X	X	X	X
H$_2$: Mimicry deters other predators	Jumping spider	X	X	X	X	X
	Other	Left alone	Left alone	X	X	X
H$_3$: Mimicry deters jumping spiders	Jumping spider	Left alone	Left alone	X	X	X
	Other	X	X	X	X	X

mata flies were able to wave their new wings normally and even fly. Greene and colleagues also performed the reciprocal transplant, placing *Zonosemata* wings on houseflies.

As Table 3.1 shows, these surgical manipulations created five experimental groups. The five treatments distinguish among the three hypotheses, because each hypothesis makes a different suite of predictions about what will happen in encounters between predators and flies. The treatments also allow the researchers to determine whether *both* the wing markings and the wing-waving display are important in mimicry. This is a powerful experimental design.

To make the test, Greene et al. needed to measure the response of jumping spiders and other predators to the five types of experimental individuals. When confronted with a test individual, did the spiders retreat, stalk and attack, or kill? To answer this question the researchers starved 20 jumping spiders from 11 different species for two days. Then they presented one of each of the five experimental prey types to each spider, in random order. The researchers made these presentations in a test arena, and recorded each jumping spider's most aggressive response during a five-minute interval. There was a clear difference: Jumping spiders tended to retreat from flies that gave the wing-waving display with marked wings, but attacked flies that lacked either wing markings, wing waving, or both (Figure 3.3). When the researchers tested treatments A, C, and E against other predators (nonsalticid spiders, assassin bugs, mantises, and whiptail lizards), all of the test flies were captured and eaten. When Greene et al. placed flies before these nonsalticid predators, there was not even an appreciable difference in time-to-capture among the three treatment groups. Comparison of Figure 3.3 to Table 3.1 shows that these results are consistent with hypothesis H_3, but inconsistent with hypotheses H_1 and H_2. Thus Greene et al.'s experiment provides strong support for the hypothesis that tephritid flies mimic their own predators to avoid being eaten by them (see also Mather and Roitberg 1987).

In terms of experimental design, the Greene et al. study illustrates several important points:

Control groups provide a contrast to treatment groups. Control individuals may have no treatment at all, or experience a treatment that tests the effects of experimental conditions predicted not to affect the outcome.

- Defining and testing effective control groups is critical. In Greene et al.'s study, groups A and B (Table 3.1 and Figure 3.3) served as controls. These individuals demonstrated that the wing surgery itself had no effect on the behavior of the flies or the spiders. Thus when the *Zonosemata* in group C were attacked and eaten by jumping spiders, Green and colleagues could be sure that this was because the flies no longer had markings on their wings, not simply because their wings had been cut and glued.

- All of the treatments (controls and experimentals) must be handled exactly alike. It was critical that Greene et al. used the same test arena, the same time interval, and the same definitions of predator response in each test. Using standard conditions allows a researcher to avoid bias and increase the precision of the data (Figure 3.4). What problems could arise if a different test arena were used for each of the five treatment groups?

- Randomization is an important technique for equalizing other, miscellaneous effects among control and experimental groups. In essence, it is another way to avoid bias. For example, Greene et al. presented the different kinds of test flies to the spiders and other predators in random order. What problems

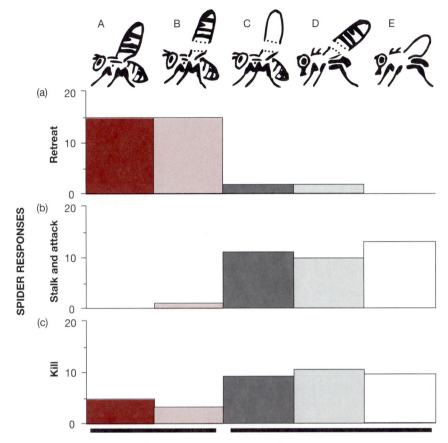

Figure 3.3 **Tephritid flies mimic jumping spiders to avoid predation**   These bar graphs show how jumping spiders responded to the five treatment and control groups listed in Table 3.1. The vertical axis represents the number of jumping spiders, from a total of 20, that showed each type of maximum response to each type of fly. Thus, for group A flies (unaltered *Zonosemata*), 15 of the jumping spiders retreated from their test flies (graph a), and five of the spiders killed their test flies (graph c). For group B flies (*Zonosemata* with their wings cut and reglued), 15 of the jumping spiders retreated from their flies, two stalked and attacked but did not kill their flies, and three killed their flies. From Greene et al. (1987)

Note that most of the flies that waved marked wings (groups A and B) survived their encounters with spiders unscathed; whereas most of the flies that lacked markings or waving or both were attacked, and many were killed. The heavy black bars at the bottom of the graphic identify treatment groups where the spider responses were statistically indistinguishable from one another. Groups A and B were indistinguishable from each other, but different from groups C, D, and E. Because each spider was presented with one fly of each type, the sample size (usually noted as N or n for "number") in each treatment group was 20.

could arise if they had presented the five types of flies in the same sequence to each spider?

- Repeating the test on many individuals is essential. It is almost universally true in experimental (and observational) work that larger sample sizes are better. This is because the goal of an experiment is to estimate a quantity. In this case, the quantity was the likelihood that a fly will be attacked by jumping spiders as a function of its ability to wave marked wings.

Larger sample sizes are critical to the success of experiments, but researchers have to trade off the costs and benefits of collecting ever-larger datasets.

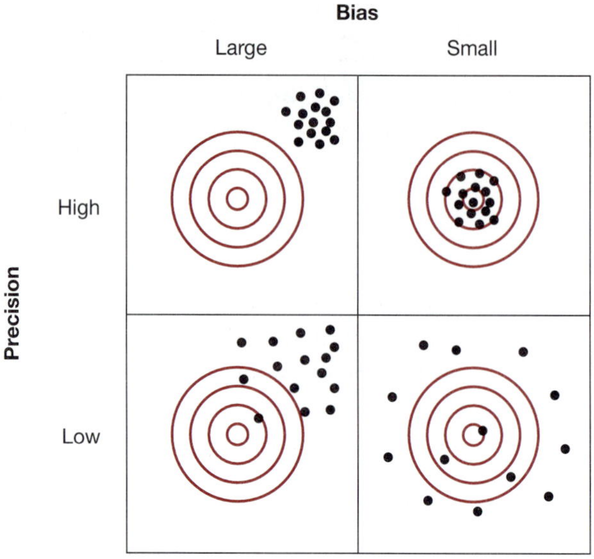

Figure 3.4 **Bias and precision**   When designing an experiment or set of observations with the goal of estimating some quantity, it is important to avoid bias and maximize precision. In this cartoon the quantity being estimated in a study is represented by the bull's-eye, and the data points collected are represented by dots. Techniques like standardizing the experimental conditions and randomizing other factors help to minimize bias and increase precision. Note that our ability to measure precision depends on having a large number of data points.

Replicated experiments or observations do two things:

- They reduce the amount of distortion in the estimate caused by unusual individuals or circumstances. For example, four of the ten *Zonosemata* with marked wings that were attacked were pounced on and killed before they even had a chance to display (groups A and B in Table 3.1 and Figure 3.3). Because Greene et al. were using standardized conditions it was not acceptable to simply throw out these data points, even though they might represent bad luck. If events like this really do represent bad luck they will be rare, and will not bias the result as long as the sample size is large.

- Replicated experiments allow researchers to understand how precise their estimate is by measuring the amount of variation in the data. Knowing how precise the data are allows the use of statistical tests. Statistical tests, in turn, allow us to quantify the probability that the result we observed was simply due to chance (Box 3.1).

In sum, Greene et al.'s experimental design was successful because it allowed independent tests of the effect that predator type, wing type, and wing display have on the ability of *Zonosemata* flies to escape predation. The next study we review demonstrates how the same principles of experimental design can be applied to a very different kind of question. The Greene et al. study focused on testing hypotheses about adaptation; the following work tests a hypothesis about diversification.

Did Stickleback Species Diverge Because of Competition?

The second application of experimental methods we examine is an innovative study of evolution in threespine sticklebacks. These small fish are common throughout the temperate zone. In northwestern North America, marine forms repeatedly invaded freshwater habitats after the glaciers retreated at the end of the last ice age (Kassen et al. 1995). The freshwater threespine sticklebacks have diversified, forming different species.

Box 3.1   A primer on statistical testing

The fundamental goal of many experimental and observational studies is to estimate the value of some quantity in two groups, such as a treatment group and a control group, and to determine if there is a difference in the quantity between the groups. In the studies we have reviewed thus far, researchers have estimated quantities like the depth of finch beaks and how frequently flies are attacked by spiders. The groups we want to compare in these examples consist of finches before and after the drought, and flies with marked or unmarked wings.

As our example, we will focus on a comparison between the flies in groups B and C of Greene et al.'s experiment (Table 3.1 and Figure 3.3). To simplify the discussion, we will lump together the outcomes "stalked and attacked" and "killed" to form a single category, "attacked." When the researchers presented a group B fly to each of 20 jumping spiders, 15 of the spiders retreated and five attacked the fly. In contrast, when the researchers presented to each of the same 20 spiders a group C fly, one of the spiders retreated and 19 attacked the fly. It certainly looks as though jumping spiders are less aggressive toward flies waving marked wings (group B) than toward flies waving unmarked wings (group C).

Once we have measured the quantity in each group and have observed a difference between groups, the statistical question becomes: Is the difference real, or could it simply be due to random variation? It is conceivable that if we tested a much larger population of spiders and flies, we would discover is that, in fact, spiders respond the same way to flies waving unmarked wings as they do to flies waving marked wings. Under this scenario, the apparent difference we observed in the experiment was just a chance result. An analogy can be found in tossing coins. Imagine you have two fair coins. It is conceivable that you could toss the first coin 20 times and get 15 heads and 5 tails, then toss the second coin 20 times and get 1 heads and 19 tails. It would appear that the coins are different, but the truth is that if you tossed both coins a very large number of times, you would discover that the coins are the same. (Note that this analogy is imperfect:

The true rate at which a fair coin gives heads is 0.5, whereas the true rate at which spiders attack flies might be anywhere from 0 to 1). What we want to know is: What is the probability that we could get a difference in spider behavior as large as the one we observed, if the truth were that spiders actually respond the same way to both kinds of fly?

Answering this question requires a statistical test. The first step in a statistical test is to specify the null hypothesis. This is the hypothesis that there is actually no difference between the groups. In our example, the null hypothesis is that the presence or absence of wing markings does not affect the way jumping spiders respond to flies. According to this hypothesis, the true frequency of attack is the same for flies with markings on their wings as for flies without markings on their wings.

The second step is to calculate a value called a test statistic. A test statistic is a number that characterizes the magnitude of the difference between the groups. More than one test statistic might be appropriate for Greene et al.'s data. Greene and colleagues chose a test statistic that compares the actual rates of retreat and attack observed in the experiment to the rates of retreat and attack that would have been expected if the null hypothesis were true.

The third step is to determine the probability that chance alone could have made the test statistic as large as it is. In other words, if the null hypothesis were true, and we did the same experiment many times, how often would we get a value for the test statistic that is larger than the one we actually got? The answer comes from a reference distribution. This is a mathematical function that specifies the probability, under the null hypothesis, of each of all the possible values of the test statistic. Often it is possible to look up the answer in a statistical table in a book, or to have a computer calculate it. For Greene et al.'s data and test statistic, the probability that chance alone would have made the test statistic this large is considerably less than 0.01. In other words, if the null hypothesis were true, and if Greene et al. repeated their experiment many times, they would have gotten a value of the test statistic larger than the one they actually got in fewer than

BOX 3.1   *(continued)*

one in one hundred experiments. This means that the null hypothesis is probably wrong, and that there is a real difference in how jumping spiders respond to flies waving marked versus unmarked wings.

The fourth and final step is to decide whether to consider the outcome of the experiment statistically significant. By convention, scientists generally consider the value of a test statistic significant if its probability under the null hypothesis is less than one in twenty, or 0.05. By this criterion Greene et al.'s result is significant with room to spare. In other words, when Greene and colleagues claimed to have demonstrated that flies must have markings on their wings to deter jumping-spider attacks, Greene and colleagues were taking a chance of less than one in a hundred of later being proved wrong by someone else who might repeat their experiment. That chance was low enough for them to claim that their

result is statistically significant. If there is more than a 5% probability that the difference observed is due to chance, the convention is to accept the null hypothesis of no real difference between groups.

In scientific papers the probability of finding the observed differences by chance is reported as a P value, where the P stands for probability. In the original paper published by Greene et al., for example, the caption to the figure that is our Figure 3.3 includes the phrase "P's < 0.01."

Statistical tests are based on explicit models of the processes that produced the data and of the design of the experiment. Many different types of random processes are modeled statistically; when analyzing data it is essential to know enough about statistics to be able to choose a model appropriate to the data collected in a particular study.

The experiment we review addresses a difficult controversy in evolutionary biology. This controversy involves a phenomenon called character displacement, or what Darwin called divergence of character. This is a pattern commonly observed in plants and animals. Populations of two closely related species are often different, morphologically or behaviorally, where the species occur together, but indistinguishable where the species occur alone (Figure 3.5).

For years ecologists and evolutionary biologists have debated why character displacement occurs (see Brown and Wilson 1956; Grant 1972; Diamond et al. 1989; Grant 1994). Competition, especially for food, has been a leading hypothesis. The idea is that natural selection favors divergent forms in areas where closely related species overlap and compete for the same resources. The logic behind this hypothesis is sound. If the ability to find food limits the survival and reproduction of individuals in the overlapping populations, then the individuals in one species that are most unlike the individuals in the other species will have to compete with the other species the least. These individuals will be able to harvest more food relative to other members of their own species. As a result, the most divergent individuals will experience higher fitness, and each population will evolve to be less like the other.

Correlational studies cannot test causation.

Proponents of the competition hypothesis of character displacement argue that competition has been a key element in promoting adaptive radiations. An adaptive radiation is the diversification of an ancestral species into many different forms adapted to different habitats or life-styles (the Galápagos finches illustrated in Figure 2.7 are a prime example). The broader implication is that competition for food is a major force in evolution, because it leads to the diversification of closely related groups.

The competition hypothesis has been difficult to test, however. It is a plausible explanation for character displacement, but plausibility is not the same as evidence. A key point is that the pattern we observe is only a correlation. The association between how much the geographic ranges of two populations overlap and how different the two populations are is clear, but correlation is not causation. More formally, the pattern of character displacement we observe is only *consistent* with the hypothesis that competition was the process that caused it. By itself, it provides that is only weak evidence for the competition hypothesis.

Critics of the competition hypothesis of character displacement point out that no one has observed evolutionary changes in characters over time in response to competition. Instead we have only snapshots of presumed results: the measurements of the characters' current states and the geographic pattern in which the states are more divergent where populations overlap. But what was the direction of change? Were the populations similar when they occupied different habitats but then became dissimilar after they occupied the same habitat, as the competition hypothesis predicts? Or did the ancestral populations occupy the same habitat originally and look quite different, then later converge in form after their ranges had separated? And was the presence of the putative competitor the real reason for evolutionary change, or was some other environmental factor involved? What is the functional significance of the characters in question? Are they genetically based, or simply different phenotypic responses to competition?

Dolph Schluter (1994) sought to address these questions by designing an experimental test of the competition hypothesis. To do the test he used two species in the threespine stickleback *(Gasterosteus aculeatus)* complex (Figure 3.6). Threespine sticklebacks are common residents of freshwater lakes in the Pacific Northwest. When they occur in the same lake, one of the two species, called the benthic form because it lives near the bottom of lakes, is large and deep-bodied (limnologists call the bottom of a body of water the benthos). The other species, called the limnetic form because it lives in open-water habitats, is small and slim (the term *limnetic* comes from the Greek word *limne,* for pool). When they occur alone, however, both the benthic and the limnetic species are ecologically and morphologically intermediate (Schluter and McPhail 1992; Day et al. 1994). These stickleback species conform to the classical pattern of character displacement.

As noted earlier, food is usually proposed as the key ecological variable driving character displacement under the competition model. In sticklebacks, an important structure used in gathering food is the gill raker. Gill rakers are protuberances along the gill arches that sieve small prey items from the water passing over the gills. In fish, the number and length of gill rakers correlates strongly with how dependent a species is on eating plankton. For example, the benthic species of threespine sticklebacks hunts invertebrates common near lake bottoms and captures them by biting. Its gill rakers are few and short. The limnetic species, in contrast, sieves microscopic plankton floating near the surface. It has numerous, long gill rakers. Might these differences, which are pronounced only when the species occur together, evolve because of fitness differences that arise during competition for food?

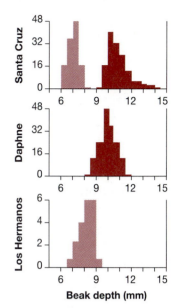

Figure 3.5 **Character displacement** These histograms plot measurements of beak depth in different populations of two of Darwin's finches, called *Geospiza fuliginosa* (light bars) and *G. fortis* (dark bars), in the Galápagos. On the island of Santa Cruz the two species occur together. Here the beak depth of the two species is markedly different. In contrast, only *G. fortis* live on the island called Daphne, and only *G. fuliginosa* live on Los Hermanos. On these islands the beak depth of the two species is much more similar. There is, in fact, considerable overlap in the distributions. This is an example of the pattern called character displacement (Schluter et al. 1985; see also Lack 1983).

Figure 3.6 **Benthic and limnetic species of threespine sticklebacks** These individuals are from the top down limnetic male, limnetic female, benthic female, and benthic male. Note that the benthic species, which lives along the bottom of lakes and hunts by biting invertebrates, is larger and deeper-bodied than the limnetics, who live high in the water column and feed by sieving plankton (Dolph Schluter, University of British Columbia).

Schluter set up his experiment to mimic an event he thinks was common in the recent evolution of threespine sticklebacks: the invasion by a limnetic form of a lake already occupied by a solitary, intermediate form. Schluter set the experiment up in two 23m × 23m ponds on the campus of the University of British Columbia. He divided each of the ponds in half, and in each of the resulting four pond halves he put 1,800 individuals from a stickleback population native to Cranby Lake, British Columbia. The Cranby population is solitary and thus intermediate in body size and shape and in gill raker size and number. Then in two of the pond halves, Schluter also put 1,200 individuals of a limnetic species native to Paxton Lake, British Columbia. He let the experiment run for three months, then harvested all the fish and measured them.

Schluter's prediction was that in the experimental sides of the ponds, the Cranby individuals most like the introduced limnetic competitors—that is, those with the most and longest gill rakers—would show the detrimental effects of competition for food by growing more slowly. This would create a relative fitness advantage for the Cranby individuals that were least like the introduced limnetic competitors. In other words, Schluter predicted that where they were forced to live with limnetic fish, Cranby individuals with the fewest and most reduced gill rakers would grow the fastest. Presumably, they would survive and reproduce better as a consequence.

The data from the two ponds are shown in Figure 3.7. Cranby individuals on the experimental sides of the ponds show a definite depression in growth rate from competition with limnetic forms. Individuals with the most gill rakers have the slowest growth rates. This is strong evidence that competition can create natural selection that favors divergent phenotypes.

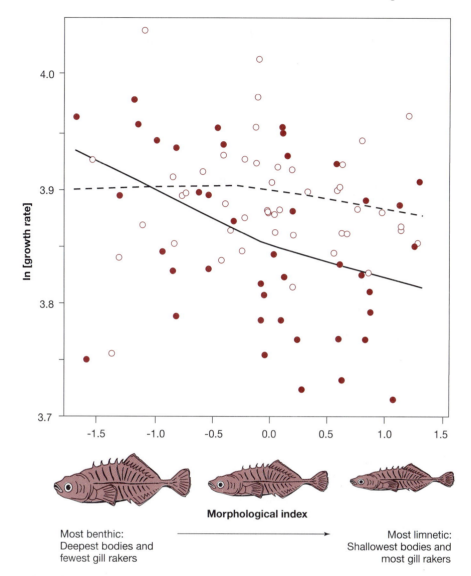

Morphological index

Most benthic:
Deepest bodies and
fewest gill rakers

Most limnetic:
Shallowest bodies and
most gill rakers

Figure 3.7 Growth of threespine sticklebacks with and without competition
Growth rate is plotted on the y-axis of this graph; the x-axis plots a morphological index
created by Schluter. This index ranges from deep-bodied forms with few gill rakers on
the left to slimmer forms with more gill rakers on the right. The open circles on the
graph represent Cranby individuals in the control sides of the ponds, living in the absence of limnetic competitors. The solid circles represent Cranby individuals from the experimental sides of the ponds, living in the presence of limnetic competitors. The best-fit
curves on the graph are the result of a statistical procedure called nonlinear regression
analysis. These regression curves represent a function that minimizes the total amount of
distance between the data points and the line. The dashed curve is the best fit through
the open circles (control individuals); the solid curve is the best fit through the closed circles (experimental individuals). The null model in this study predicts that competition
will have no effect on growth rate. This means that the two best-fit curves should be
identical. In fact, the two curves are significantly different. Furthermore, the curve for the
fish living with limnetic competitiors shows that those individuals that are most like the
competitors have the slowest growth rates (Schluter 1994).

Well-designed experiments are efficient because they produce a measurable result in a short period of time.

The Schluter study demonstrates the power of the experimental approach. Two points are important here. First, by creating an intense episode of competition, Schluter was able to measure an effect on an important component of fitness (growth rate) in a single season. Under natural conditions, the same process may have taken years to produce the same effect. Second, previous studies of character displacement had relied on making correlations between existing geographic ranges and morphological differences. As we noted earlier, correlational studies can be problematic because they do not test causation and are often subject to alternative interpretations. Because Schluter's experiment isolated competition as the only variable differing between groups, and because he measured the resulting change directly, Schluter has provided some of the most convincing evidence to date that character displacement by competition for food actually occurs in nature.

3.2   The Comparative Method

In section 3.1 we highlighted the limitations of correlational studies. In this section we highlight their strengths.

The comparative method is based on examining many different species and looking for correlations between their traits and specific environmental variables.

Biologists frequently compare species or populations and look for associations between traits and ecological or other variables as a way of testing hypotheses about adaptation. This approach is called the comparative method (see Ridley 1983; Huey 1987; Harvey and Pagel 1991; Harvey and Purvis 1991; Lauder et al. 1995; Harvey et al. 1996). For example, there is a strong correlation between the presence of long, nectar-bearing tubes in flowers and the incidence of pollination by hummingbirds. Recognizing this pattern, we infer that the tubes evolved as an adaptation that makes hummingbird pollination more effective.

This style of research is necessary because it is often impossible to experimentally recreate the conditions that prevailed in the past, as Dolph Schluter was able to do with sticklebacks. How could we experimentally duplicate the conditions that led the ancestors of larkspurs to develop a long, nectar-bearing tube, and hope to see a response in the experimental population in our lifetime? It is much more efficient to look for associations between the morphology of flowers and pollinator type, and argue that natural selection is responsible for the correlations we observe.

This research strategy is especially valuable when it is used to test the predictions made by hypotheses about adaptation. A classic example is work done in the early 1980s on a phenomenon known as sperm competition. In many plants and animals, females routinely receive pollen grains or sperm from more than one male. These pollen grains and sperm then compete among one another for the privilege of fertilizing offspring. Paul Harvey and A. H. Harcourt (1984; see also Harcourt et al. 1981) used the comparative method to test a prediction about a male adaptation for sperm competition in primates (apes and monkeys).

In Primates, Is Testes Size Correlated with the Type of Mating System?

Mating systems in primates vary from monogamous pairing to single male troops (where one male dominates matings, inseminating most of the females in the troop) to multi-male (where each female is inseminated by several males). Harvey and Harcourt reasoned that sperm competition would occur most often and most intensely in species with multi-male breeding systems. They predicted that species with multi-male breeding systems would have much larger testes, in relation to their overall body size, than species with monogamous or single-male systems. This prediction follows from a fact and an assumption. The fact is that in primates ejaculate size (number of sperm produced) correlates strongly with testes size. The assumption is that males producing large ejaculates will be more successful at fertilizing eggs when sperm from different males compete.

To test the prediction about relative testes size, Harcourt and Harvey compiled data on breeding system, average adult male body weight, and average adult testes weight in primates. Then they plotted testes size against male body weight and calculated the best-fit line through the data (Figure 3.8). This line represents the average testes weight predicted for any given male body size. The vertical distance from each data point to the line measures the relative testes size of each species. Relative testes size is higher than expected if the point is above the line (that is, males in such species have large testes for their body size), and lower than expected if the point is below the line (males in such species have small testes for their body size).

Harvey and Harcourt compared the relative testes sizes for taxa with multi-male vs. monogamous and single-male breeding systems statistically and found a clear difference: Males have larger testes for their body size in primate species with multi-male breeding systems, as predicted by the sperm competition hypothesis. The comparative approach allowed us to test a hypothesis that would be difficult to address experimentally.

But as Harvey and Harcourt were well aware, there is a serious problem with their analysis. To understand what this problem is and learn how it can be solved, we turn to a study of plant mating systems.

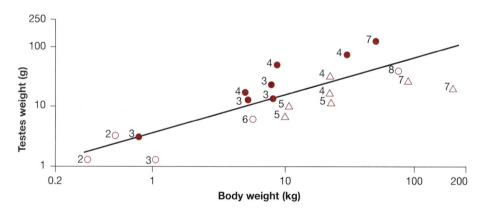

Figure 3.8 Body weight vs. testes weight in primates Each data point on this graph represents the average value for male body weight and testes weight calculated for the species within a primate genus that have the same breeding system (• = multi-male, Δ = single-male, o = monogamous). The numbers next to each point indicate the number of species used to calculate that point. Note that the scale is logarithmic on both axes. The line, which is a result of a regression analysis, represents the testes size that is predicted for primates of a particular body weight (Harvey and Harcourt 1984).

Why Are Plant Breeding Systems Correlated with Mode of Seed Dispersal?

Michael Donoghue (1989) was interested in a widely cited hypothesis and data set about the evolution of breeding systems in plants. Plants can be dioecious, meaning that separate sexes occur on separate individuals, monoecious, meaning that both sexes are present in the same individual, or perfect-flowered, meaning that both sexes are present in the same flower. A comparative study had been published that looked at the association between breeding system and seed type in gymnosperms (pines, yews, and gingkoes) and found a strong pattern: Most monoecious species have dry, wind-dispersed seeds while most dioecious species have seeds encased in fleshy, animal-dispersed fruits (Table 3.2).

Givnish (1980, 1982) and Bawa (1980) proposed that the evolution of animal dispersal selected for dioecy and created this pattern. The logic here involves two steps. The first step is that a mutant plant that invested more in female function, by developing a fleshy coat around its seeds, would achieve higher fitness than its nonmutant neighbors. The proposition is that birds would be attracted to the fruits, disperse the seeds farther than wind is capable of, and lead to more mutant offspring being successfully established.

Once animal dispersal had evolved, then a new class of mutants that produced only (fruit encased) seeds might spread in the population. This would happen if four conditions were satisfied: (1) Female-only plants were able to invest more in fruit than could plants with both sexes, (2) this investment increased the attractiveness of the female-only plants to seed dispersers, (3) the increased attractiveness led to higher fitness, and (4) the fitness benefit was disproportionately large relative to the investment they made (larger, for example, than they could have gained by investing more in male function). Once specialized seed producers began to spread in the population, then a second kind of mutation would also be favored (in different individuals): a mutation that caused individuals to invest only in pollen. The evolution of dioecy would result.

Now suppose that Givnish or Bawa wanted to test this adaptive explanation experimentally. To begin, they would have to find (or create, through genetic manipulation or a breeding program) a monoecious population where most individuals produced dry seeds, except for a few mutants that produced fruit-encased seeds. Then they would need to document the existing mode of seed

---

TABLE 3.2   Breeding system and seed type in gymnosperms

Each data point in this table represents a species. If there were no association between breeding system and seed type, the fraction of dioecious plants that are wind dispersed would be the same as the fraction of monoecious plants that are wind dispersed. The test statistic and reference distribution used for this data set (see Box 3.1) specify how probable it would be, just by chance if the null hypothesis were true, to get fractions as different as—or more different than—those reported here. This probability is less than 0.01% ($P < 0.00001$). The data are from Givnish (1980; see also Donoghue 1989).

	Monoecious	Dioecious
Dry seeds (wind dispersed)	339	18
Seeds in fleshy fruits (animal dispersed)	45	402

dispersal, perhaps by confirming that dry seeds were distributed by wind but that the mutant fleshy seeds were spread by birds. Then they would have to wait, several thousand years perhaps, to see if selection gradually increased the percentage of fleshy, animal-dispersed fruits in the population. Finally, they would have to wait again to see if the fruit-making population would evolve dioecy, as predicted. Our conclusion: Running the experiment is impractical. As we have noted, in cases like this it is much more productive to study the existing variation in traits across species and search for meaningful patterns. This is why Givnish employed the comparative approach to produce the data in Table 3.2. These data strongly support the hypothesis that dioecy evolves in animal-dispersed plants.

The comparative method is an efficient alternative when experiments are impractical.

Donoghue was one of several researchers to recognize a problem with the analysis, however. To understand this problem, study the hypothetical phylogeny drawn in Figure 3.9a. Note that all of the groups with both dioecious breeding systems (marked with an asterisk) and fleshy propagules (represented by filled bars) are found on neighboring branches in the tree. If this diagram represented the actual phylogeny of gymnosperms, it would be sensible to argue that the reason groups with animal-dispersed seeds have dioecious breeding systems is simple: They shared a common ancestor that exhibited both traits. In fact, using parsimony as a criterion for analyzing change in the hypothetical phylogeny *requires* us to infer that dioecy arose just once, and not 35 different times.

Dioecy might have evolved in the common ancestor by exactly the scenario that Givnish and Bawa proposed, but the upshot is that we really only have one data point instead of 35 when we construct a statistical test for an association between mating system and the mode of seed dispersal. This is because all statistical testing procedures assume that the data points being analyzed are independent. Statistical independence, in turn, means that the value of a data point has no influence on the value of any other point. But because related species derive many of their characteristics from their common ancestors, their characteristics are not independent.

This lack of independence is a serious issue in the use of comparative methods. This is the problem that we raised when reviewing the study of sperm competition in primates. That dataset implied that natural selection was vastly more important than common ancestry in explaining the association between testes size and mating system. Our confidence in this conclusion would be seriously shaken if it turned out that all the species with extra-large testes and multi-male breeding systems were each others' closest relatives. The simplest explanation for the pattern in Figure 3.8 would then be that the present-day species with both large testes and a multi-male breeding system were all descended from a common ancestor that had these two traits, not that a multi-male breeding system selects for males with large testes.

Because they share a common ancestor, closely related groups do not represent independent data points.

The only way to solve the problem, and determine whether a trait has evolved independently in close relatives, is to know the phylogeny of the group (Felsenstein 1985). Donoghue did. He and his colleagues had completed an analysis of seed plant phylogeny based on characteristics other than breeding system and seed morphology. Knowing the phylogeny allows a researcher to count how many times a trait has evolved independently (Figure 3.9a).

In a comparative study, knowing the phylogeny allows a researcher to control for the effects of shared ancestry.

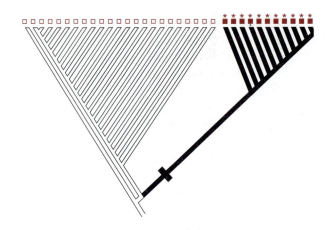

Figure 3.9 **Using the comparative method in conjunction with a phylogeny**   (a) In determining whether there is an association between mating system and seed type in plants, this tree represents a worst-case scenario. The boxes at the tips of this tree represent hypothetical taxa. Asterisks indicate dioecious groups, while unmarked taxa are monoecious or perfect. Filled branches represent lineages with fleshy fruits, and empty branches those with dry seeds. The black bar across the branch indicates the origin of dioecy. Note that this tree includes 35 taxa, representing the genera and families analyzed by Donoghue (1985), rather than the 402 species Givnish used in his analysis (see Table 3.2).

To test the hypothesis rigorously, then, Donoghue redid the analysis. He broadened the Givnish study to include angiosperms (flowering plants) and reformulated the question in a phylogenetic context. Specifically, he set out to test an important prediction made by the Givnish-Bawa hypothesis: that dioecy evolved *after* fleshy fruits.

Here is what he found. Using the estimate of the phylogeny drawn in Figure 3.9b, which has breeding system and seed type mapped onto the branches and tips, Donoghue counted the number of times that dioecy evolved after fleshy propagules were present in a plant lineage. Unfortunately, seed type is unclear in several branches that are represented only by fossil forms. If we assume that all of these questionable taxa have fleshy fruits, the count is that dioecy arose seven times in clades that already had fleshy fruits and only two times in lineages with dry fruits. If we do not make this assumption, the support for the Givnish-Bawa hypothesis is much weaker or even nonexistent.

Because of the uncertainty about seed types in fossil groups, we were left unsure about whether the hypothesis is correct. But clearly, the evidence is not nearly as overwhelming as the original papers implied. The more interesting result of Donoghue's study, in fact, has nothing to do with the original question about the proposed association between seed type and breeding system. Instead, the phylogenetic analysis makes it clear that dioecy evolved just a few times relative to the enormous number of dioecious species with animal-dispersed seeds. These lineages exploded, producing a huge diversity of species. Now the question becomes why? Is there something about being dioecious and having animal-dispersed seeds that makes plants more likely to split into independent populations and speciate, or end up less likely to go extinct? Doing a comparative study in a phylogenetic context has turned our attention to a completely different issue, and one that may be just as or even more interesting than the original question. That, more than a tidy test of a prediction, is the message of Donoghue's work.

Powerful as they are, all of the research strategies that we have examined thus far deal only with extant organisms and the current function of traits. How do we go about analyzing the origin and history of adaptations, and understanding patterns in the diversification of life through time?

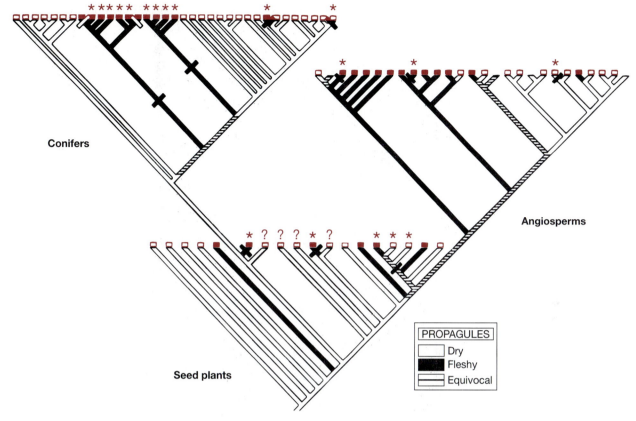

Figure 3.9 (b) An actual phylogeny of seed plants, with breeding system and seed type mapped. The unusual way in which this tree is displayed, with the three groups separated, is not important (the tree was estimated with a method that does not estimate branch lengths, so the author extended some just to make the figure easier to read and to highlight that the diagram is actually a composite of three separate studies). The taxa at the tips of the tree are marked to indicate their breeding system, which *is* important. Asterisks indicate dioecious taxa, question marks indicate fossil taxa where the breeding system has not been established, and unmarked taxa are monoecious or perfect. Donoghue (1989) used parsimony to infer where changes in propagule type had taken place on the tree, and marked the stems appropriately: Open indicates dry, black indicates fleshy, and striped is equivocal (meaning that dry and fleshy are equally parsimonious). Reading up from the base of the tree, it is clear that parsimony supports monoecy or perfect flowers as the ancestral state. The black bars across branches indicate the origin of dioecy; open bars across branches mean that there has been a reversion to monoecy or perfect flowers.

3.3 Reconstructing History

Evolutionary biology's mandate is to explain the amazing adaptation and diversity of life. This makes evolution an explicitly historical science. Why? First, we cannot explain current patterns of species diversity unless we understand the events leading up to the present. We want to know why some lineages flourished, while others declined or died out. Second, we do not really understand the contemporary function of an adaptation unless we know how the trait originated and changed through time. Understanding present form and present function is an obvious first step in analyzing an organic design, but we also want to know how the particular structures we see today, out of all the possible solutions to a problem, came to be (see Lauder and Liem, 1989).

But how do scientists do history? Experiments and the comparative method are largely irrelevant here. How do researchers analyze the origins of an adaptation? What types of evidence and analyses help us understand when and why a particular group of plants or animals diversified?

We will review two prominent research programs to illustrate how the work of paleobiology is done. The evolution of the mammalian middle ear demonstrates how changes in a particular trait are investigated. The radiation of insects shows how researchers examine the history of biodiversity.

How Did the Mammalian Ear Evolve?

One of the mammalian ear's outstanding features is a group of three bones called the ear ossicles. They are named the malleus (hammer), incus (anvil), and stapes (stirrup). Their function is to transmit wave energy from the eardrum (tympanic membrane) in the outer ear to the oval window of the cochlea in the inner ear (Figure 3.10). Why have three bones instead of just one? Their lever action amplifies the force transmitted, increasing the sensitivity of hearing. The nearly thirty fold reduction in transmission area, from the large tympanic membrane to the small oval window, also serves to amplify the signal. These ossicles are a major reason why mammals hear so well.

But where did these three little bones come from in the first place? To know, we need to:

To understand adaptation we need to understand how traits originated and changed through time.

- Establish the ancestral condition
- Understand the transformational sequence, or how and why the characters changed through time

If we can accomplish these steps, we will understand both the proximate and ultimate mechanisms of evolutionary change.

The logical place to start our analysis is with an animal called *Acanthostega gunnari* (Figure 3.11). This is among the oldest known tetrapods, or four-limbed

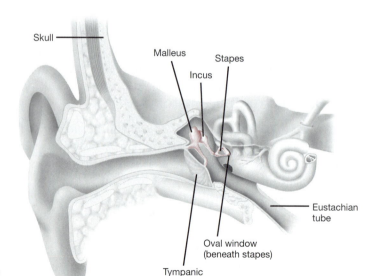

Figure 3.10 **The mammalian middle ear**   This drawing of a human ear shows the location of the three ear ossicles: the malleus, incus, and stapes.

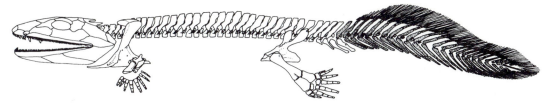

Figure 3.11   An early tetrapod   This is a reconstruction of *Acanthostega gunnari,* recovered from Upper Devonian deposits in east Greenland. *Acanthostega* is among the oldest tetrapods ever found. Tetrapods (literally, the "four-footed") are a taxonomic group that includes the amphibians, reptiles, and mammals. Early tetrapods were the first land-dwelling vertebrates. The rocks present at sites where *Acanthostega* fossils are found indicate that the animal lived in shallow-water environments. Impressions in the fossils indicate that they probably had both lungs and internal gills (see Coates and Clack 1991).

vertebrates. A partially air-breathing swamp-dweller, *Acanthostega* was one of the first animals to encounter the problem of hearing airborne sounds. Its ancestor belonged to a group of fish called the crossopterygians, which went extinct in the Lower Permian (Carroll 1988). Although the crossopterygian fish had none of the three ear ossicles, *Acanthostega* had a stapes. It was the first of the middle ear bones to appear in the fossil record (Clack 1989), and may have functioned in hearing airborne sounds.

What is the evidence for this claim? Fossilized skulls show that one end of the *Acanthostega* stapes fits into a hole in the side of the braincase that connects to the inner ear, while the other end fit into a notch in the skull near an opening called the spiracle (Clack 1994). In later fossil tetrapods and in some extant amphibians, this skull notch holds the tympanic membrane, or eardrum (Figure 3.12a). Because the form of the stapes is homologous in *Acanthostega* and later groups, we can argue that its function—transmitting airborne sound—is homologous too (Lombard and Bolt 1979; Clack 1983).

But did the stapes appear out of nowhere? How can we determine the ancestral state of this evolutionary innovation? This is a very general problem in paleontological analyses. Recall that in Chapter 2 we reviewed the claim that the "thumb" of the Giant Panda evolved from an elongated wrist bone. What is the evidence for assertions like these?

The key to understanding the origin of a derived trait is to establish homology with a trait present in the ancestor. For example, in form and in position, the stapes of *Acanthostega* is homologous to the hyomandibula of crossopterygian and extant fish. The hyomandibula is a bone that functions as a brace between the jaw and braincase (Figure 3.12b). Muscles attached to this bone pump the cheeks. This pumping action, in turn, ventilates the gills and opens and closes the spiracle. In modern lungfishes the pumping action ventilates the lungs with air. Because fossils show that *Acanthostega* had both lungs and internal gills, it is reasonable to infer that muscles attached to its stapes were involved in respiration (Clack 1989; Coates and Clack 1991; Clack 1994).

All of these facts make *Acanthostega* a classic transitional form. That is, it is an intermediate between the fish and the tetrapods (amphibians, reptiles, and mammals). Its stapes was a modification of the hyomandibula that still functioned in respiration. But it was also a bone that happened to be in the right place

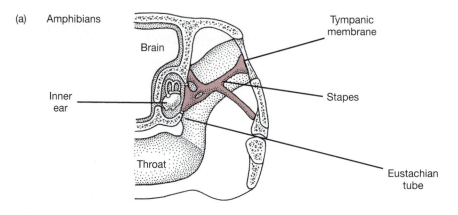

(a)   Amphibians

Brain

Tympanic membrane

Inner ear

Stapes

Throat

Eustachian tube

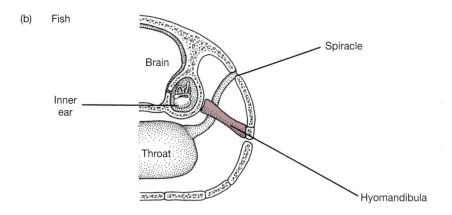

(b)   Fish

Spiracle

Brain

Inner ear

Throat

Hyomandibula

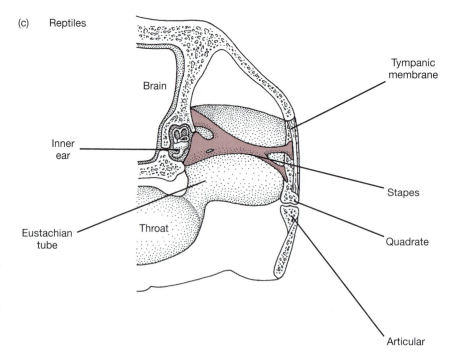

(c)   Reptiles

Tympanic membrane

Brain

Inner ear

Stapes

Eustachian tube

Throat

Quadrate

Articular

Figure 3.12 The location of the stapes in various vertebrates
Each of these illustrations is a cross-section of the head, showing the location and form of bones and other structures discussed in the text. Modified from Romer, A.S. 1995. *The Vertebrate Body* (Philadelphia: W.B. Saunders).

(anatomically) at the right time (when vertebrates first ventured out onto land) to be co-opted for use as a sound transmitter. The hyomandibula was a preadaptation for hearing.

We actually have another line of evidence that the stapes is homologous to the hyomandibula, in addition to the similarity in form and placement of the two bones in *Acanthostega* and fish. In 1837, long before Darwin established an evolutionary interpretation of homology, the German anatomist C.B. Reichert (see Gould 1993, essay 6) examined mammal embryos and determined that during development, the stapes originates from the second gill arch. In fish, this same embryonic structure becomes the hyomandibula. These developmental homologies are exactly what we would predict if the two structures represent ancestral and descendant states. Combined with the morphological data in adults, the arguments in favor of the two structures sharing a common ancestry become overwhelming.

Thus far we have been able to establish the embryological origin and the ancestral condition of the stapes. But what of the other ear ossicles, the malleus and incus? *Acanthostega* does not have them. Nor do reptiles or amphibians, including the extinct forms ancestral to the mammals (Allin 1975). All of these groups transmit sound from the tympanic membrane to the inner ear directly through the stapes (Figure 3.12a,c).

The malleus and incus first appear in fossil mammals. But from where? Where were the malleus and incus found in the ancestors of mammals? In position, they are homologous with two jawbones—called the articular and quadrate—found in reptiles, amphibians, and early mammals. In fact, the malleus, incus, and stapes of modern mammals still develop, as embryonic tissues, in exactly the same positions in which they are found in adult fossils from the group ancestral to mammals (Figure 3.13).

Homology between traits observed in different species can be established by measuring similarities in their (1) location or form, (2) developmental pathways, or (3) genetic basis.

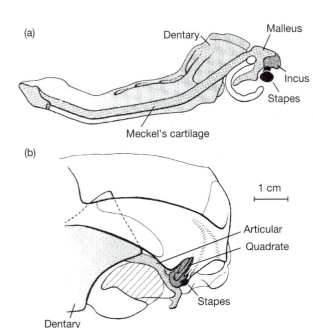

Figure 3.13 Developmental and structural homologies in bones of the mammalian middle ear (a) In mammal embryos the malleus is formed from the posterior (hindmost) part of Meckel's cartilage. This cartilage also forms the lower jawbone (dentary). (b) This graphic shows the position of the articular (malleus), quadrate (incus), and stapes in the adult non-mammalian cynodont *Thrinaxodon,* one of the ancestors of modern mammals.

Note that in relation to the jaw, the position of the three middle-ear bones of modern mammal embryos (a) is identical to their location in adults of the ancestors of mammals (b). Modified from Pough et al. 1996, *Vertebrate Life* (Upper Saddle River, NJ: Prentice Hall).

Developmentally, the malleus and incus also originate as part of the jaw structure. In early-stage mammal embryos, the cells destined to become the malleus and incus derive from the first gill arch. In fish that have jaws this gill arch produces the upper and lower mandible. In mammal embryos these cells create a structure called Meckel's cartilage, which forms the lower jaw (Figure 3.13a). The malleus forms at the posterior end of Meckel's cartilage and then detaches; the incus forms from a nearby structure (Allin 1975). Again, the evidence for homology is overwhelming.

This leads us to the second point in our analysis: examining how the malleus and incus changed through time. How were jawbones transformed into ear ossicles? It is logical to start where we left off: with the ancestors of the mammals. This is a group called the cynodonts. In cynodonts (as in modern amphibians, reptiles, and birds), the jaw joint is formed by the quadrate and articular; the stapes is the only bone directly associated with hearing. The cynodont stapes happened to lie right next to the articular, however (Allin 1975; see Figures 3.12c, and 3.13b). Examining cynodont fossils through time shows the following changes in the jaw:

- The part of the lower jaw near the hinge became larger, and one of the major muscles responsible for closing the jaw changed its area of attachment from the angular bone to the lower jaw.
- A later cynodont genus, called *Diarthrognathus,* has the ancestral jaw articulation involving the quadrate and articular bones *and* a derived articulation between the lower and upper jawbones (Figure 3.14a). *Diarthrognathus,* like *Acanthostega,* is an intermediate form. It shows both ancestral and derived states of a character (Colbert and Morales 1991).
- Early mammals have *only* the lower jaw-upper jaw articulation. The quadrate and articular bones are no longer involved. This is a key step, because the two bones are now free to assume a new function or disappear, depending on the direction of natural selection.
- In later mammals the quadrate and articular bones articulate with the stapes. Consequently they are renamed the incus and malleus. They are now located away from the jaw joint and function only in the transmission of sound (Figure 3.14b).

Transitional forms have both the ancestral and derived conditions of a trait.

What aspects of natural selection help account for all this change? Though traditional views emphasize the importance of improved efficiency in biting and chewing during the remodeling of the jaw (see Manley 1972), Allin argues that improved hearing played a role in favoring mutants with ear ossicles detached from the jaw. His hypothesis is based on the idea that the survival of small, insectivorous, and probably nocturnal early mammals may have depended on an ability to hear high-pitched insect sounds. Because it is hard to hear while chewing, ossicles detached from the jaw worked better (Allin 1975).

To summarize: Early in mammal evolution three bones changed function, reduced in size, and moved away from an articulation with the jaw. Currently, research on ear evolution is focused on understanding the developmental and genetic mechanisms of these changes (see Rowe 1996; Smith et al. 1997). As we

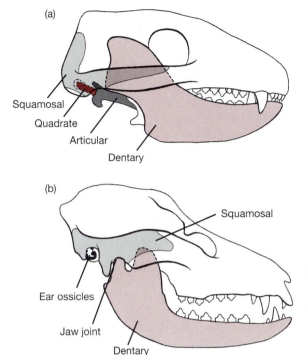

Figure 3.14 **Change in mammal jaw articulation through time** The jaw articulation of an early mammal, called *Diarthrognathus,* is shown in (a). The diagram shows that the joint involved both the quadrate and articular *and* the dentary and squamosal bones. In later mammals (b), the jaw articulation is made only by the dentary and squamosal. The ear ossicles have moved back and away from the jaw.

will see in Chapter 12, biologists are using molecular techniques to explore the genetics of macroevolutionary changes like these.

We have touched on a series of fundamental methods in paleobiology: identifying homologies at the structural and developmental levels, analyzing transformational sequences, and identifying intermediate forms. But what about the other aspect of evolutionary change: lineage diversification and the origin of species? What techniques can we bring to studying the history of speciation and extinction?

Why Are There So Many Different Kinds of Insects?

In sheer numbers of individuals and species, the insects have to be considered one of the most successful clades in the history of life. Conrad Labandeira and John Sepkoski (1993) have performed the most thorough and innovative study of the insect radiation to date. Their research focused on testing a widely accepted hypothesis: that insects diversified in response to the rise of the angiosperms (flowering plants).

The raw data in this study are the ages and identities of insect fossils. Drawing on 472 different sources in the literature, Labandeira and Sepkoski compiled the most complete database on insect diversification through time ever assembled. Many of the source publications are purely descriptive studies: catalogs produced by researchers who went into the field, collected fossils, and named and dated the taxa they found. In most cases, the dates placed on fossils are reported as a stage in a particular geological time period.

The fossil record documents how life forms have changed through time.

Natural history studies like these are the basic building blocks of paleontology. Collectively, they have created the body of knowledge we call the fossil record. And they are fundamental. Without knowing the what and when of evolutionary change, we cannot ask the how and why questions that drive the science.

Labandeira and Sepkoski had a fundamental methodological question to resolve, however, before they could synthesize and interpret the record they had compiled. What taxonomic level should they choose as the unit of analysis?

The taxonomic hierarchy is an artificial device we use to organize the diversity of life. This system groups living things in categories from the general to the specific. The traditional categories are kingdom, phylum, class, order, family, genus, species, and population. Populations, as we will see in Chapters 5 and 6, are a fundamental evolutionary unit. They are the part of the taxonomic hierarchy that changes in response to natural selection and other evolutionary processes. But because populations can fuse and lose their evolved differences, species are the normal unit of evolutionary analysis.

Labandeira and Sepkoski had a difficult issue to attend to, however: In practice, different researchers use different criteria to identify fossil insect species and genera. If Labandeira and Sepkoski chose to do their study at the level of the species or genus, then, they would have to reexamine every specimen reported in their sources firsthand and create a common set of criteria for classifying them. If they took the published reports at face value, it is virtually certain that they would be comparing markedly different groups.

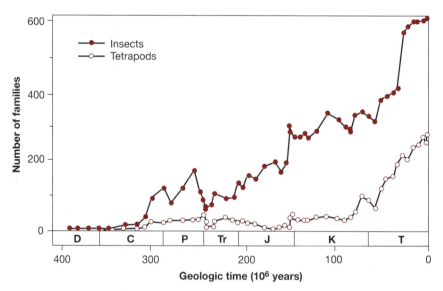

Figure 3.15 **Scattergram of family-level insect diversity through time**   After the class Insecta originated in the Devonian ("D" in the figure's time scale), the number of insect families increased more or less linearly from the Carboniferous (C) through the Permian (P). The number of families crashed at the end of the Permian, then steadily increased from the Triassic (Tr) through the Jurassic (J), Cretaceous (K), and Tertiary (T). Notice the contrast with the pattern for tetrapod families.

For this plot Labandeira and Sepkoski (1993) calculated diversity with the "range-through" method, meaning that they recorded a family as present in all intervals from its first through last appearance in the fossil record. The lines in this graph are not regression lines; they simply connect the points.

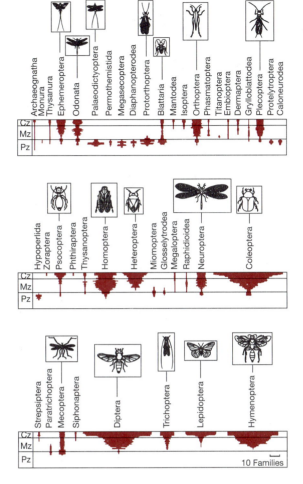

Figure 3.16 Spindle diagram of family-level insect diversity through time This graph (Labandeira and Sepkoski 1993) shows the number of insect families from the Paleozoic (Pz; Silurian through Permian periods), to the Mesozoic (Mz; Triassic through Cretaceous), to the Cenozoic eras (Cz; Tertiary through Recent). Each "spindle" represents an insect order, with the number of families plotted horizontally. A scale bar is given at the lower right. The illustrations at the top of selected spindles show an adult representative of that order (note that the illustrations are not to scale).

As a result, they decided instead to study changes in insect diversity through time using families as the unit of analysis. This is a common strategy among paleontologists. In this case it made sense, for two reasons: (1) There is a strong consensus among researchers about what constitutes an insect family, so Labandeira and Sepkoski could be confident that data points drawn from different studies are comparable, and (2) insect families represent clear ecological units. That is, because insect families represent groups of species that feed and live in similar ways, they are interesting to interpret. And because other paleontological studies had shown a strong relationship between the number of families and the number of species present, they could assume that analyzing families would reflect something real about the underlying species diversity.

Genera, families, or orders are the most common units of analysis in paleobiology.

A plot of the presence of 1,263 insect families versus geologic time (Figures 3.15 and 3.16) shows several marked patterns. Perhaps the most startling is a major extinction event at the end of the Paleozoic era (Permian period), 245 million years ago (Ma). Eight of 27 insect orders went extinct; another three were drastically reduced at the end of the Permian and disappeared soon thereafter. To appreciate the importance of this event, imagine the impact on the radiation of mammals if rodents or primates went extinct, on birds if falcons or songbirds were wiped out, or flowering plants if grasses or asters disappeared (these are all examples of orders). Extinctions of this magnitude fundamentally alter the course of evolution. The end-Permian extinction was, in fact, the most massive of all time; as we will see in Chapter 13, the majority of *all* species alive were wiped out.

Labandeira and Sepkoski organized their next analysis, of diversification subsequent to the end-Permian extinction, around a fundamental question: Was the rise in number of families through the Mesozoic, shown in Figures 3.15 and 3.16, due to low extinction rates or rapid origination (that is, speciation) rates? To answer this question they created a Lyellian survivorship curve, which plots the proportion of taxa (in this case, families) present in any time interval that are still alive today. Time is plotted on the x-axis; the y-axis shows the fraction of taxa that were alive at time x that are still alive today. In groups with high extinction rates, Lyellian curves fall steeply as the data go back through time. Lyellian curves for tetrapods, insects, and marine bivalves are shown in Figure 3.17. The insect record, although intermediate between the other two groups, generally shows low extinction rates, especially from the late Cretaceous to the present. Why? It could be that insect families are so species-rich that they seldom go extinct. But the fact is, we do not know. Like most good research, this work raises as many questions as it answers.

Finally, Labandeira and Sepkoski tested a long-standing hypothesis about insect diversification: That the rich ecological and species diversity of insects is due to the diversification of flowering plants. The logic behind this idea is that the radiation of flowering plants expanded the assortment of plant leaf and flower tissues available, and that natural selection favored the evolution of new insect species and families able to exploit them.

Labandeira and Sepkoski's first task in testing the hypothesis was to create a null model. That is, they needed to predict the pattern that would be expected if insect diversification were *not* tied to angiosperm evolution. They chose exponential expansion, based on the following logic: A "normal" course of diversification would proceed like compounding interest, with each increment in the number of taxa present increasing the number of taxa that could be produced in

The fossil record is routinely used to test predictions made by competing hypotheses about the pattern of evolutionary change.

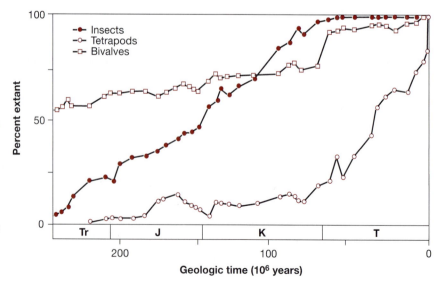

Figure 3.17 **Proportion of families alive today vs. time**   This is an example of a Lyellian survivorship curve (Labandeira and Sepkoski 1993). A steeper slope indicates a higher extinction rate. The data are plotted for stratigraphic ages in the geologic record (as in Figure 2.6), and the abbreviations are as in Figure 3.16. Bivalves include the clams and mussels; extant means "still living."

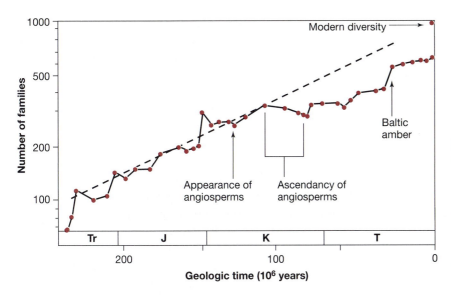

Figure 3.18 **Logarithm of number of insect families vs. time**   The dashed line indicates the null model of exponential increase (see the text for explanation). The jump in the curve labeled "Baltic amber" is due to the extraordinarily diverse fossil insects preserved in the Baltic amber deposits of Europe; abbreviations are as in Figure 3.15 (Labandeira and Sepkoski 1993).

every subsequent time interval. If a researcher plots the logarithm of number of taxa against time, then, this null model of diversification produces a straight line. (Note that the slope of the line will vary depending on the rate at which new taxa are created.) In contrast, the angiosperm evolution hypothesis predicts that there should be an abrupt increase in insect diversity when the angiosperms first appear.

The expectation from the null model and the actual data are shown in Figure 3.18. The first appearance of flowering plants and the period of their rapid diversification, as recorded in fossils, are also indicated on the graph. Contrary to the commonly accepted hypothesis, there is no correlation between insect diversification and angiosperm evolution; rather, the data indicate that the family-level insect radiation actually *slowed* as flowering plants diversified. The radiation of insect families falls off an exponential rate of increase just as flowering plants exploded. Why? Labandeira and Sepkoski consider three alternative explanations:

- Extinction events in the Mesozoic era wiped out enough insect families to remove the positive effect of angiosperm evolution. This seems unlikely, however, because insect families show little, if any, effects from a massive and widespread extinction that occurred in other taxonomic groups at the Cretaceous-Tertiary boundary (this is the "K–T extinction" that wiped out the dinosaurs; see Figures 3.15 and 3.16).

- The falloff in the curve of Figure 3.18 represents a saturation effect. Perhaps the real story is that insect families radiated in response to the evolution of seed plants, not just angiosperms, and that insects had already filled most available ecological niches by the Mesozoic. Note that the other major group of seed plants, the gymnosperms (including the pines, yews, and gingkoes), originated long before the angiosperms.

- Flowering plant diversification promoted species diversity, but not family-level diversity, in insects. If true, this hypothesis would mean that the Labandeira and Sepkoski study simply did not record the appropriate data, and missed the relevant signal.

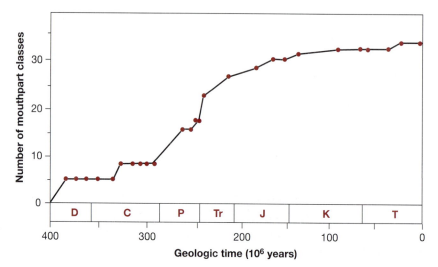

Figure 3.19   Number of insect mouthpart classes through time Angiosperms first appeared in the early Cretaceous (K); their rapid diversification occurred during the late Cretaceous (Labandeira and Sepkoski 1993).

To look at these last two ideas more closely, Labandeira and Sepkoski considered the evolution of insect mouthpart structures. These are the most important tools insects have for harvesting leaf and flower resources. The researchers measured mouthpart characteristics for 1,365 living species of insects and grouped them statistically into 34 different classes. Each of these represented a distinct morphology and method of feeding. Then they mapped these classes onto fossil forms. And the result? By the mid-Jurassic, before angiosperms are recorded, the vast majority of mouthpart classes had already appeared (Figure 3.19). Only a handful of the 34 mouthpart classes appeared after the angiosperms expanded. This is strong evidence that the ecological radiation of insects, recorded at the family level, occurred before angiosperm evolution. But determining whether insect *species* became more diverse because of flowering plants (the last idea of the three proposed here) awaits further research.

3.4   Strategies for Asking Interesting Questions

As this chapter demonstrates, evolutionary biology is an enormously diverse field. Its practitioners use experiments, comparative methods, and analyses of the fossil record to produce new insights into the mechanisms behind adaptation and diversity. In Chapter 5 we add mathematical modeling to the toolkit used in evolutionary analyses. Chapter 7 emphasizes DNA sequencing and other molecular methods, and in Chapter 10 we expand on the role that phylogenetic analyses play.

Advances in these techniques for collecting and analyzing data are a fundamentally important part of evolutionary analysis. But equally significant is the ability to ask interesting questions. Coming up with stimulating questions and *Creativity in science produces new* innovative ways to answer them form the basis for creativity in science.
techniques and questions, and alter-
native explanations for phenomena.
Each of the examples reviewed in this chapter illustrates a different approach to generating questions. We started with an unusual display made by tephritid flies, and the simple desire to explain what we see. This is science inspired by observation. Darwin, for example, was a master at finding and recognizing something unusual or striking in nature and coming up with a mechanistic and

testable explanation. We saw this in his pursuit of a rationale for the distribution of Galápagos mockingbirds or the similarity between glyptodonts and armadillos. The observations can come from a multitude of sources: sequence data from a recently discovered gene, a museum drawer filled with fossil snails, or the flight patterns of bees as they work from flower to flower.

Scientists seldom base their research programs on a single source of inspiration, however, and all researchers have distinctive styles of asking questions and going about their work. Here are some general methods, boiled down to a list of aphorisms, that successful evolutionary biologists have used to organize and stimulate their thinking.

- **Question conventional wisdom.** It is often untested. That is what made Labandeira and Sepkoski's study of insect diversification so stimulating. Because vast numbers of present-day insects are closely associated with flowering plants, the idea that insects diversified in response to angiosperm evolution was so compelling that many people accepted the hypothesis at face value.

- **Define and directly test the predictions made by hypotheses.** This is perhaps *the* classical approach to science—what most people are thinking of when they refer to the scientific method. We saw this strategy repeatedly in the examples reviewed in this chapter. It was Greene et al.'s approach to distinguishing among alternative explanations for the wing markings of tephritid flies, and Schluter's contribution to our understanding of competition's role in character displacement.

- **Question and test the assumptions underlying hypotheses.** That is why Donoghue's reformulation of an earlier study was informative. He tested the assumption of evolutionary independence, and found that it did not hold up.

- **Study natural history.** Descriptive studies can lead to the discovery of new patterns that need explanation. Some things in nature just leap out and demand explanation, like the discovery of a stapes in *Acanthostega*. Some of the most compelling science happens when a researcher simply picks an organism and sets out to learn as much as possible about it.

In addition to creativity, science also requires an attitude characterized by what might be called toughness of mind. The scientific mindset is based on critical thinking and standards of evidence based on the quality of data and the clarity, consistency, and testability of logic. Independent thinking is key. We ask our readers, as students new to evolutionary biology, to practice it. Good students are not sponges, but filters. If this chapter has been successful, it will encourage critical thinking about the ideas and evidence presented in the rest of this book. With that, our introduction to evolutionary biology is complete. We are ready to delve in.

Summary

In this chapter we introduce methods used in evolutionary biology. Strong research programs start with a well-defined question and test a series of competing explanations. At least one of these competing explanations should be a null model, which specifies the outcome that is expected by chance.

Experiments are powerful because they allow researchers to isolate and test the effects of a single variable on the phenomenon in question. Because the response to this single variable is measured directly, experiments allow tests of causation. And because a researcher can artificially heighten the intensity of test conditions, experiments can produce a measurable response efficiently. Controls, randomization procedures, standard conditions, and replicated experiments are all essential to an effective experimental design.

The comparative method is a research strategy based on looking for associations among species between the presence of particular traits and specific ecological or other variables. Although it is correlational in nature, the comparative method is efficient and is often the only effective research avenue when experimentation is impossible. Results from comparative studies are often used to infer adaptation or to test the predictions of adaptionist hypotheses. Because species are not statistically independent units of analysis, however, comparative studies must be done in a phylogenetic context. Knowing the phylogeny allows a researcher to count the number of times that the trait in question has evolved independently.

Paleontologists use a wide variety of techniques to interpret the fossil record and reconstruct the history of traits and lineages. Determining homologies in ancestors and descendants from morphological, developmental, or genetic evidence is essential for reconstructing transitional series and understanding how traits have changed through time in response to natural selection. Graphing changes in the species or family-level diversity of a particular lineage through time and looking for correlations with mass extinction events, climatic change, or changes in other taxonomic groups can be an effective way of inferring which factors have determined the history of a group.

Questions

1. Why was it important that Greene et al. tested tephritid flies whose wings had been cut off and then glued back on?

2. Schluter's experimental protocol has been criticized (see Murtaugh et al. 1995). One of the objections is that the experimental and control groups differed in overall population density—experimental ponds had 3,000 fish (1,800 intermediate-morphology fish mixed with 1,200 limnetic fish), while control ponds had just 1,800 fish (all intermediate-morphology). Schluter replies that the difference is irrelevant because density and competition have contrasting effects: Density affects all phenotypes, while competition affects only those individuals whose morphology overlaps with the limnetic forms. Which point of view do you agree with? How could you redesign the experiment to eliminate this concern?

3. When compared across animal taxa, many organs increase in size as body weight increases in an exponential relationship. That is, they follow the function $y = ax^b$ where y is organ weight, x is body weight, and a and b are constants. Taking the log of both sides gives $\log(y) = b\log(x) + \log(a)$. This is a linear relationship with slope *b*, and is the function plotted in Figure 3.8. Remember that Harcourt and Harvey used only the deviations from this allometric relationship in the analysis described in the text. Why? Why not just compare mating system to testes size without worrying about the effect of body size?

4. In reconstructing the history of the mammalian ear, we described both fossil evidence and embryological evidence. In many cases, we do not have the luxury of having both sets of evidence. Examples are the evolution of bat wings, for which there is embryological data but few fossils, and the evolution of *Triceratops* neck frills, for which there is fossil data but little embryology. In your opinion, is it necessary to have both kinds of data to determine the evolutionary history of a trait? Would fossil data alone (or embryological data alone) be enough to convince you that mammal middle-ear bones evolved from the hyomandibula, quadrate, and articular of early vertebrates? What would you think if the two data sets had contradicted each other?

5. Early tetrapods apparently did not have tympanic membranes, for there is no obvious place on their skulls where a tympanic membrane would have fit. In addition, the stapes in these animals was a relatively heavy bone and thus was unlikely to transmit high-frequency vibrations well. Yet the stapes was clearly adapted to transmit vibrations to the inner ear. Early tetrapods are thought to have had a sprawled body posture, with their large heads often resting directly on the ground. With this in mind, speculate on what sort of "hearing" abilities *Acanthostega* might have had. Are your ideas testable?

6. In your judgment, how robust is Labandeira and Sepkoski's analysis to new information? For example, if a spectacular new fossil insect fauna was reported and added 12 new families, all of which are still alive today, to those already recorded from the early Cretaceous, would it materially alter the conclusions you have drawn from examining Figure 3.15 to Figure 3.19?

7. The data point in Figure 3.18 marked "Modern diversity" shows the number of insect families alive today (or "extant"). This point is high because only 63% of extant families are represented in the fossil record. If there is no sampling bias in the fossil record, this figure indicates that each data point in Figure 3.15 through Figure 3.17 represents only 63% of the families present at the time. Does this fact change any of your conclusions from the study?

8. An exercise used in some graduate training programs is to have students list 20 questions that they would like to answer. Groups of students then discuss the questions and help each other sort out which would be the most interesting to pursue. There are many criteria for deciding that a question is interesting. Is it new? Does it address a large or otherwise important issue? Would pursuing it lead to other questions? Is it feasible, or would development of a new technique make it feasible? Try this exercise yourself.

Exploring the Literature

9. An important aspect of evaluating scientific papers is to consider other explanations for the data that the authors might have overlooked. See if you can think of alternate explanations for the data presented in the following papers:

Benkman, C.W., and A.K. Lindholm. 1991. The advantages and evolution of a morphological novelty. *Nature* 349:519–520.

Soler, M., and Møller, A.P. 1990. Duration of sympatry and coevolution between the great spotted cuckoo and its magpie host. *Nature* 343:748–750.

Finally, see the following review of Soler and Møller's work for an example of how scientific criticism can result in better science by all involved.

Lotem, A., and Rothstein, S.I. 1995. Cuckoo-host coevolution: From snapshots of an arms race to the documentation of microevolution. *Trends in Ecology and Evolutionary Biology* 10:436–437.

10. The fossil record has traditionally been the major source of data for asking questions about evolutionary history. Unfortunately the fossil record is incomplete, and cannot answer all questions. Newer techniques of molecular and genetic analysis are sometimes seen as more useful. As a result, paleontologists have recently had to defend the "usefulness" of the fossil record. Read the following papers to explore this issue more thoroughly:

Foote, M. 1996. Perspective: Evolutionary patterns in the fossil record. *Evolution* 50:1–11.

Benton, M.J., and G.W. Storrs. 1994. Testing the quality of the fossil record: Paleontological knowledge is improving. *Geology* 22:111–114.

Norell, M.A., and M.J. Novacek. 1992. The fossil record and evolution: Comparing cladistic and paleontologic evidence for vertebrate history. *Science* 255:1690–1693.

Citations

Allin, E.F. 1975. Evolution of the mammalian middle ear. *Journal of Morphology* 147:403–438.

Bawa, K.S. 1980. Evolution of dioecy in flowering plants. *Annual Review of Ecology and Systematics* 11:15–39.

Brown, W.L. Jr., and E.O. Wilson. 1956. Character displacement. *Systematic Zoology* 5:49–64.

Carroll, R.L. 1988. *Vertebrate Paleontology and Evolution* (New York: W.H. Freeman).

Clack, J.A. 1983. The stapes of the Coal Measures embolomere *Pholiderpeton scutigerum* Huxley (Amphibia: Anthracosauria) and otic evolution in early tetrapods. *Zoological Journal of the Linnean Society* 79:121–148.

Clack, J.A. 1989. Discovery of the earliest-known tetrapod stapes. *Nature* 342:425–427.

Clack, J.A. 1994. Earliest known tetrapod braincase and the evolution of the stapes and fenestra ovalis. *Nature* 369:392–394.

Coates, M.I., and J.A. Clack. 1991. Fish-like gills and breathing in the earliest known tetrapod. *Nature* 352:234–236.

Colbert, E.H., and M. Morales. 1991. *Evolution of the Vertebrates.* (New York: Wiley-Liss).

Day, T., J. Pritchard, and D. Schluter. 1994. A comparison of two sticklebacks. *Evolution* 48:1723–1734.

Diamond, J., S.L. Pimm, M.E. Gilpin, and M. LeCroy. 1989. Rapid evolution of character displacement in myzomelid honeyeaters. *American Naturalist* 134:675–708.

Donoghue, M.J. 1989. Phylogenies and the analysis of evolutionary sequences, with examples from seed plants. *Evolution* 43:1137–1156.

Felsenstein, J. 1985. Phylogenies and the comparative method. *American Naturalist* 125:1–15.

Fischman, J. 1995. Why mammal ears went on the move. *Science* 270:1436.

Givnish, T.J. 1980. Ecological constraints on the evolution of breeding systems in seed plants: Dioecy and dispersal in gymnosperms. *Evolution* 34:959–972.

Givnish, T.J. 1982. Outcrossing versus ecological constraints in the evolution of dioecy. *American Naturalist* 119:849–865.

Gould, S.J. 1993. *Eight Little Piggies* (New York: W.W. Norton).

Gould, S.J., and E.S. Vrba. 1982. Exaptation—a missing term in the science of form. *Paleobiology* 8:4–15.

Grant, P.R. 1972. Convergent and divergent character displacement. *Biological Journal of the Linnean Society* 4:39–68.

Grant, P.R. 1994. Ecological character displacement. *Science* 266:746–747.

Greene, E., L.J. Orsak, and D.W. Whitman. 1987. A tephritid fly mimics the territorial displays of its jumping spider predators. *Science* 236:310–312.

Harcourt, A.H., P.H. Harvey, S.G. Larson, and R.V. Short. 1981. Testis weight, body weight and breeding system in primates. *Nature* 293:55–57.

Harvey, P.H., and A.H. Harcourt. 1984. Sperm competition, testes size, and breeding systems in primates. In R.L. Smith, ed., *Sperm Competition and the Evolution of Animal Mating Systems* (New York: Academic Press), 589–600.

Harvey, P.H., and A. Purvis. 1991. Comparative methods for explaining adaptations. *Nature* 351:619–624.

Harvey, P.H., and M.D. Pagel. 1991. *The Comparative Method in Evolutionary Biology* (Oxford University Press).

Harvey, P.H., A.J.L. Brown, J. Maynard Smith, and S. Nee. 1996. *New Uses for New Phylogenies* (Oxford University Press).

Huey, R.B. 1987. Phylogeny, history, and the comparative method. In M.E. Feder, A.F. Bennett, W. Burggren, and R.B. Huey, eds., *New Directions in Ecological Physiology* (Cambridge University Press), 76–198.

Kassen, R., D. Schluter, and J.D. McPhail. 1995. Evolutionary history of threespine sticklebacks (*Gasterosteus* spp) in British Columbia: Insights from a physiological clock. *Canadian Journal of Zoology* 73:2154–2158.

Labandeira, C.C., and J.J. Sepkoski, Jr. 1993. Insect diversity in the fossil record. *Science* 261:310–315.

Lack, D. 1983. *Darwin's Finches* (Cambridge University Press).

Lauder, G.V., and K.F. Liem.1989. The role of historical factors in the evolution of complex organismal functions. In Wake, D.B., and G. Roth, eds., *Complex Organismal Functions: Integration and Evolution in Vertebrates* (Chichester, England: John Wiley & Sons Ltd.), 63–78.

Lauder, G.V., R.B. Huey, R.K. Monson, and R.J. Jensen. 1995. Systematics and the study of organismal form and function. *BioScience* 45:696–704.

Lombard, R.E., and J.R. Bolt. 1979. Evolution of the tetrapod ear: An analysis and reinterpretation. *Biological Journal of the Linnean Society* 11:19–76.

Manley, G.A. 1972. A review of some current concepts of the functional evolution of the ear in terrestrial vertebrates. *Evolution* 26:608–621.

Mather, M.H., and B.D. Roitberg. 1987. A sheep in wolf's clothing: Tephritid flies mimic spider predators. *Science* 236:308–310.

Murtaugh, P.A., J. Bernardo, W.J. Resetarits, Jr., A.E. Dunham, and D. Schluter. 1995. Criteria for testing character displacement. *Science* 268:1065–1067.

Pough, F.H., J.B. Heiser, and W.N. McFarland. 1996. *Vertebrate Life,* 4th ed. (Upper Saddle River, NJ: Prentice Hall).

Ridley, M. 1983. *The Explanation of Organic Diversity: The Comparative Method and Adaptations for Mating* (Oxford University Press).

Rowe, T. 1996. Coevolution of the mammalian middle ear and neocortex. *Science* 273:651–654.

Schluter, D. 1994. Experimental evidence that competition promotes divergence in adaptive radiation. *Science* 266:798–801.

Schluter, D. 1995. Adaptive radiation in sticklebacks: Trade-offs in feeding performance and growth. *Ecology* 76:82–90.

Schluter, D., T.D. Price, and P.R. Grant. 1985. Ecological character displacement in Darwin's finches. *Science* 227:1056–1058.

Schluter, D., and J.D. McPhail. 1992. Ecological character displacement and speciation in sticklebacks. *American Naturalist* 140:85–108.

Smith, K.K., A.F.H. van Nievelt, and T. Rowe. 1997. Comparative rates of development in *Monodelphis* and *Didelphis*. *Science* 275:683–684.

MECHANISMS OF EVOLUTIONARY CHANGE

In the first three chapters we focused on what evolution is. By the end of Chapter 2 we had defined evolution as changes in gene frequencies within populations, and throughout Part I we focused on natural selection as the force responsible for most evolutionary change. But selection is not the whole story; Part II offers a broader perspective on evolutionary processes. In the next six chapters we explore how evolution happens.

Chapters 4, 5, and 6 introduce the suite of forces responsible for generating genetic variation among individuals and for changing the frequencies of alleles within populations. There are four such forces: mutation, natural selection, migration, and genetic drift. These are termed microevolutionary processes because they produce changes at the level of populations and species. Chapter 7 covers the processes of molecular evolution and describes patterns of change at the level of the DNA sequence. Chapter 8 focuses on the environments that organisms adapt to and on the nature of adaptation. Chapter 9 examines how the four microevolutionary processes can cause populations to diverge and form new species.

These six chapters should provide a solid understanding of how and why evolution occurs, and lay the groundwork for an in-depth look at what evolution has produced: change through time. That is the subject of Part III.

Mutation: The Origin of New Genes and Alleles

Mutations are the raw material of evolution. They are the ultimate source of the heritable variation acted upon by natural selection and other evolutionary processes. Without mutation there are no new genes, no new alleles, and eventually, no evolution.

Our goal in this chapter is to investigate the mechanisms responsible for generating new alleles and new genes. We begin by reviewing processes that produce point mutations, or single-base changes in DNA sequences. These are the mechanisms that produce new alleles. Later in the chapter we consider larger-scale changes that can produce new genes, change the organization of individual chromosomes, or alter the number of chromosome sets in a species. We do not, however, attempt an encyclopedic review of mutations that can affect gene sequences and organization. The roster of mutation types, especially at the level of chromosomes, is simply too large. Instead, we focus on the subset of mutations that have the greatest evolutionary impact (Table 4.1).

4.1 Where New Alleles Come From

Most new alleles are derived from already existing alleles, typically as a result of point mutations. Point mutations are single-base substitutions caused by one of two processes: random errors in DNA synthesis, or random errors in the repair of sites damaged by chemical mutagens or high-energy radiation. Both types of changes are caused by DNA polymerase.

If during normal synthesis or repair synthesis, DNA polymerase mistakenly substitutes a purine (A or G) for another purine, or a pyrimidine (T or C) for another pyrimidine, the point mutation is called a transition. If a purine is substituted for a pyrimidine or a pyrimidine for a purine, the mutation is called a transversion. Of the two kinds of point mutation, transitions are far more

TABLE 4.1 Types of mutation with significant evolutionary impact
This table summarizes the types of mutation reviewed in this chapter.

Name	Description	Cause	Result
Point mutation	Base-pair substitutions in DNA sequences	Chance errors during synthesis by DNA polymerase, or during repair of DNA damaged by chemical mutagens or radiation	Both silent and replacement substitutions create new alleles
Chromosome inversion	Flipping and reannealing of a chromosome segment	Double-strand breaks caused by ionizing radiation	Tighter linkage, because inversion heterozygotes cannot recombine normally
Gene duplication	Duplication of a short stretch of DNA	Unequal crossing over during meiosis (see Fig. 4.3)	Redundant genes are free to mutate and perhaps gain new function
Polyploidization	Addition of a complete set of chromosomes	Errors in meiosis and/or (in plants) mitosis	Possibly, a new species

common (routinely outnumbering transversions by 2:1 or more) simply because transitions are easier mistakes for polymerase to make. Transitions cause much less disruption in the DNA helix during synthesis.

Depending on the nature and location of the substituted base, point mutations are classified in two ways: as transitions or transversions and as replacement or silent substitutions.

If either type of base substitution occurs in the coding region of a gene, it changes the codon read by RNA polymerase. Look back at the genetic code (Figure 2.17a). Changes in the first or second positions of a codon almost always change the amino acid specified by the resulting mRNA, while changes in the third position frequently produce no change at all. Point mutations that result in an amino acid change are called replacement (or nonsynonymous) substitutions; those that result in no change are called silent site (or synonymous) substitutions. Both create new alleles. Because many replacement substitutions produce a measurable change in protein function and thus the organism's phenotype, they are exposed to natural selection. Silent site substitutions are not exposed to selection based on protein function.

The first replacement substitution ever characterized on a molecular level was the mutation in human hemoglobin that results in the sometimes fatal disease sickle-cell anemia. Hemoglobin is the oxygen-carrying molecule found in red blood cells. In 1949 Linus Pauling's lab reported that people suffering from sickle-cell anemia had a form of hemoglobin different from that of people without the disease. Later, Vernon Ingram (1958) showed that the difference between the hemoglobins was due to a single amino acid change at position number 6 in the β chain of the hemoglobin molecule, which is 146 amino acids long. Instead of having glutamic acid at this position, the sickling allele has valine. The base substitution that causes this replacement is an adenine for a thymine (a transversion) at nucleotide 2 in the codon for amino acid 6.

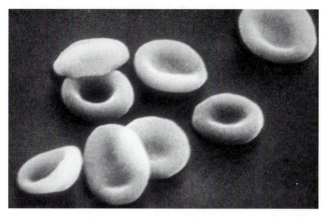 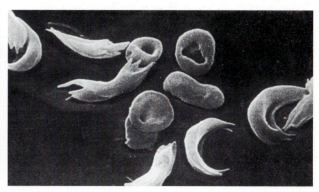

Figure 4.1 **Normal and sickling forms of human red blood cells**   (a) This photo shows normal red blood cells. (b) This photo shows sickled cells. (Photo Researchers, Inc.)

This tiny change in the hemoglobin molecule's primary structure produces a dramatic change in phenotype (Figure 4.1). The mutant hemoglobin molecules tend to crystalize, forming long fibers. Hemoglobin is found inside red blood cells; when hemoglobin crystalizes inside these cells the fibers distort the normally disc-shaped cells into sickles. The sickled cells get stuck in capillaries, blocking the flow of blood, depriving tissues of oxygen, and causing severe pain. Sickled red cells are also more fragile than normal cells and more rapidly destroyed. This continuous loss of red blood cells causes anemia.

The sickling and anemia are most severe in people homozygous for the mutant allele, because all of their hemoglobin molecules are prone to crystalization. Heterozygous people get some sickling, particularly when the concentration of dissolved oxygen in their blood gets low. But as an unexpected benefit, these people are resistant to malaria. The resistance apparently results because in heterozygotes, red blood cells containing malaria parasites are much more prone to sickling than unparasitized cells. This means that red cells containing parasites are selectively destroyed. The mutant allele is thus beneficial in environments where malaria is common, but deleterious in populations where malaria is rare. We will return to this aspect of the sickle-cell story in Chapter 5.

The sickle-cell mutation created a new allele that is beneficial in some environments. This is unusual, because the overwhelming majority of replacement substitutions create new alleles that are either deleterious in all environments or have little to no effect on fitness (Keightley 1994). It is not surprising that many of the random changes in the amino acid sequences of proteins reduce their ability to function, because most proteins have been under selection for millions of years. We would not expect a random change to improve a protein's function, any more than we would expect a random change in a computer's circuitry to improve processing performance. But it is important to recognize that replacement and silent substitutions produce a wide range of effects on fitness: from highly deleterious to inconsequential (or "neutral") to beneficial. We will return to this point in Chapters 5 and 7.

In coding sequences, the fitness effects of replacement substitutions range from highly deleterious to beneficial. Most mutations have very small effects on fitness.

Mutation Rates

How often are new alleles formed? Many of our best data on mutation rates are for loss-of-function mutations. In these types of mutations, the lack of normal protein product leads to an easily recognizable phenotype. For example, researchers may survey a large human population and count the incidence of an autosomal-dominant syndrome like Achondroplasia (dwarfism) or an X-linked recessive like Hemophilia A (in which blood clotting is impaired). The idea is to pick a trait that is easy to detect and whose transmission genetics allow researchers to identify new mutations. For example, an individual with Achondroplasia, neither of whose parents has the condition, must represent a new mutation. From data like these, a researcher can report mutation rates in units of per gene per generation.

The problem with this method is that loss-of-function mutations result from *any* process that inactivates a gene. Genes can be knocked out by base-pair substitutions that produce a chain-terminating codon or a dysfunctional amino acid sequence, frameshifts (additions or deletions of one or two base pairs that disrupt the reading sequence of codons), the insertion of a mobile genetic element, chromosome rearrangements, and so on. As a result, we have scant data on what is perhaps the most interesting type of mutation rate for evolutionary analyses: replacement substitutions per codon per generation. Also, many interesting mutations (most replacement substitutions, for example) are not detectable in these types of screens because their effects are too subtle. And surveys that score phenotypes in adult offspring miss loss-of-function mutations that lead to the death of individuals early in development.

Mutation rates vary widely, both among genes and among organisms.

Even with these limitations, we can still say some interesting things about mutation rates. For example, rates of phenotypically detectable mutations vary by 500-fold among genes within species (Table 4.2a), and by as much as five orders of magnitude, or 100,000-fold, among species (Table 4.2b). Mutation rates are very low on a per-gene basis, but there are so many loci (approximately 100,000 in humans, for example) that perhaps 10% of all gametes carry a phenotypically detectable mutation. This is a large percentage, especially considering that the rates in Table 4.2b undoubtedly underestimate the number of replacement substitutions and thus new alleles. It is conceivable, or even probable, that the majority of all offspring carry at least one new allele somewhere in their genome.

Selection on Mutation Rates

The mutation rate itself is clearly a trait with heritable variation. Frances Gillin and Nancy Nossal (1976a, b), for example, investigated single-base substitutions in the DNA polymerase of bacteriophage T4 (a virus that parasitizes bacteria). Some of the mutations they isolated decreased the rate at which polymerase made errors during DNA replication, and reduced the overall mutation rate. Other mutations in polymerase increased the error rate, and heightened the overall mutation rate.

In a key finding, Gillin and Nossal also determined that the more error-prone polymerase mutants were significantly faster than the more accurate form

TABLE 4.2 Variation in mutation rates among genes and species

(a) Rates of mutation to recessive phenotypes among genes in corn. L.J. Stadler (1942) bred a large number of corn plants and scored offspring for a series of recessive conditions.

Gene	Number of gametes tested	Number of mutations found	Average number of mutations per million gametes
$R \longrightarrow r$	554,786	273	492.0
$I \longrightarrow i$	265,391	28	106.0
$Pr \longrightarrow pr$	647,102	7	11.0
$Su \longrightarrow su$	1,678,736	4	2.4
$Y \longrightarrow y$	1,745,280	4	2.2
$Sh \longrightarrow sh$	2,469,285	3	1.2
$Wx \longrightarrow wx$	1,503,744	0	0.0

(b) These data, summarizing mutation rates for a variety of genes and species, are taken from R. Sager and F.J. Ryan, *Heredity* (New York: John Wiley, 1961).

Organism	Mutation	Value	Units
Bacteriophage T2 (bacterial virus)	Lysis inhibition $r \rightarrow r^+$ Host range $h^+ \rightarrow h$	1×10^{-8} 3×10^{-9}	*Rate:* mutant genes per gene replication
Escherichia coli (bacterium)	Lactose fermentation $lac \rightarrow lac^+$ Histidine requirement $his^- \rightarrow his^+$ $his^+ \rightarrow his^-$	2×10^{-7} 4×10^{-8} 2×10^{-6}	*Rate:* mutant cells per cell division
Chlamydomonas reinhardtii (alga)	Streptomycin sensitivity $str^s \rightarrow str^r$	1×10^{-6}	
Neurospora crassa (fungus)	Inositol requirement $inos^- \rightarrow inos^+$ adenine requirement $ad^- \rightarrow ad^+$	8×10^{-8} 4×10^{-8}	*Frequency* per asexual spore
Drosophila melanogaster (fruit fly)	Eye color $W \rightarrow w$	4×10^{-5}	
Mouse	Dilution $D \rightarrow d$	3×10^{-5}	
Human to autosomal dominants	Huntington's chorea Nail-patella syndrome Epiloia (predisposition to a type of brain tumor) Multiple polyposis of large intestine Achondroplasia (dwarfism) Neurofibromatosis (predisposition to tumors of nervous system)	0.1×10^{-5} 0.2×10^{-5} $0.4-0.8 \times 10^{-5}$ $1-3 \times 10^{-5}$ $4-12 \times 10^{-5}$ $3-25 \times 10^{-5}$	*Frequency* per gamete
to X-linked recessives	Hemophilia A Duchenne's muscular dystrophy	$2-4 \times 10^{-5}$ $4-10 \times 10^{-5}$	
in bone-marrow tissue-culture cells	Normal $\rightarrow$ azaguanine resistance	7×10^{-4}	*Rate:* mutant cells per cell division

There is a trade-off between the speed and accuracy of DNA polymerase.

of the enzyme. This means that there is a trade-off between the speed and accuracy of DNA replication.

John Drake (1991) has proposed that natural selection optimizes this trade-off. He concluded this after comparing mutation rates calculated on a per-base-pair, per-generation basis in a series of well-studied genes in four bacteriophages (viruses that parasitize bacteria), the bacterium *Escherichia coli,* the yeast *Saccharomyces cervisiae,* and the filamentous fungus *Neurospora crassa.* Drake found an inverse relationship between genome size and the average mutation rate per base pair (Figure 4.2). This implies that the number of mutations per genome per generation is relatively constant for the organisms in his study. The general implication is that all viruses, bacteria, and yeasts have the same mutation rate, when considered on a per-genome basis. Some observers are calling this pattern "Drake's rule."

Drake's interpretation is that the constant, genome-wide mutation rate he observed represents an optimal value based on the trade-off between the speed and accuracy of DNA polyermase. Drake assumes that these organisms are under strong selection to minimize their mutation rate and maximize their growth rates. He proposes that microbes with small genomes cannot replicate any faster without producing an unacceptable number of mutations. Microbes with large genomes, in contrast, cannot replicate any slower without reducing their growth rates and imposing an unacceptable fitness cost.

It is not clear that Drake's rule holds for larger, longer-lived species, however. For example, Edward Klekowski and Paul Godfrey (1989) studied mutation rates in the mangrove *Rhizophora mangle.* This is a long-lived tree with an unusual trait: The seedlings germinate on the parent. Klekowski and Godfrey counted the number of albino offspring germinating on normal parents to calculate the rate of mutation to albinism (a loss-of-function mutation). The rate was 25 times greater in the trees than the rate previously calculated for mutations to albinism in annual plants like barley and buckwheat.

Klekowski and Godfrey's explanation for their result focuses on the fact that in plants, germline cells differentiate late in development. This means that somatic cells will accumulate mutations through many cell divisions before differentiating and undergoing meiosis. As a result, long-lived plants will have higher mutation rates per generation than short-lived plants, and will not follow Drake's rule. The hypothesis here is that the key factor influencing variation in mutation rates among large organisms is life span, not genome size. Perhaps Drake's rule holds only for microbes with small genomes and rapid replication rates. Only additional research will tell.

Before we leave the topic of mutation rates, it is important to recognize that DNA polymerase is not the only protein involved in the accuracy of DNA synthesis. Point mutation rates depend on three processes: (1) discrimination by DNA polymerase when placing a nucleotide on the newly synthesized strand, (2) proofreading by DNA polymerase (the molecule has a $3'-5'$ exonuclease activity in addition to its $5'-3'$ polymerizing activity, so it can back up, delete mismatched bases, and replace the correct nucleotide), and (3) mismatch repair that takes place postsynthesis or after DNA has been damaged by chemical mutagens or radiation.

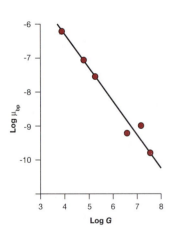

Figure 4.2   The relationship between mutation rate and genome size   Genome size ("G" on the x-axis) is plotted here as the logarithm of total base pairs. The mutation rate (the y-axis) is given in units of per base-pair per generation. Each data point represents one of the organisms in Drake's (1991) study. Note that in molecular and population genetics, the Greek letter μ ("mu") is routinely used to stand for mutation rate.

Research on postsynthesis and damage repair has been intense, because mutations in mismatch repair systems have been implicated in cancer and aging (see Chapter 17). There are several different mismatch repair systems, and in mammals at least 30 different proteins are involved (Bohr et al. 1987; Modrich 1991; Holmquist and Filipski 1994; Mellon et al. 1996). At least some of the repair systems are highly conserved; mismatch repair genes in humans were first identified through their homology with genes in yeast and *E. coli* (Friedberg et al. 1995). In the bacteria *Escherichia coli* and *Salmonella enteritidis,* mutations in these loci produce strains with mutation rates 100 to 1,000 times higher than normal (LeClerc et al. 1996). The message is that the efficiency of DNA mismatch repair, like the error rate of DNA polymerase, is a trait with heritable variation.

In addition to this general picture of complex and ancient systems, two other outstanding generalizations have emerged about DNA repair systems: Coding regions are repaired much more efficiently than noncoding regions (Bohr et al. 1985), and several of the repair systems work on transcriptionally active genes only. As a result, accuracy is greatest where mutations could be most damaging.

Repair systems replace bases damaged by chemical mutagens or radiation. In somatic cells, environmentally induced damage to DNA can lead to cancer and other diseases. In germline cells, damaged DNA can lead to mutations that cause birth defects.

4.2 Where New Genes Come From

The creation of new alleles is fairly straightforward, but where do new genes come from? As with new alleles, several kinds of mutations can create new genes, and we will review only a subset. Gene duplications are probably the most important source of new genes. Duplications result from a phenomenon known as unequal cross-over, diagrammed in Figure 4.3. Unequal cross-over is a chance mistake caused by the proteins involved in managing recombination (crossing over) during meiosis.

As Figure 4.3 shows, one of the products of unequal cross-over is a redundant stretch of DNA. The genome now has an extra copy of the sequences

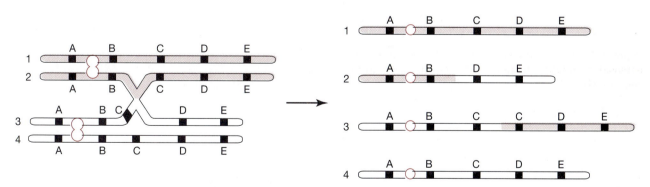

Figure 4.3   Unequal cross-over and the origin of gene duplications   The letters and bars on each chromatid in the diagram indicate the location of loci; the open circles indicate the location of the centromere. The chromatids on the left have synapsed, but cross-over has occurred at nonhomologous points. As a result, one of the cross-over products has a duplication of gene C and one a deletion of gene C. Unequal cross-over events like this are thought to be the most common mechanism for producing gene duplications.

located in the duplicated segment. Because the parent copy still produces a normal product, the redundant sequences are free to accumulate mutations without consequences to the phenotype. The new sequence might even change function over time, and become a new locus. This is an important point. Because it creates additional DNA, gene duplication is the first mechanism we have encountered that results in entirely new possibilities for gene function. Duplication events have been important in evolution, as the globin gene family illustrates.

Duplicated genes can (1) retain their original function and provide an additional copy of the parent locus, (2) gain a new function through mutation and selection, or (3) become functionless pseudogenes.

Gene Duplication Events in the Globin Gene Family

Most families of closely related genes—like the histones, globins, and collagens (Table 4.3)—and genes that exist in multiple copies, like the rRNAs, are thought to have originated in gene duplication events. Many of these gene families also contain nonfunctional sequences called pseudogenes. Pseudogenes do not produce working protein but retain a high sequence identity with the functional genes nearby.

In humans the globin gene family consists of two major clusters of loci. These genes code for the protein subunits of hemoglobin. The groups are the α-globin cluster on chromosome 16 and the β-globin cluster on chromosome 11. Hemoglobin is made up of an iron-binding heme group surrounded by four protein subunits, with two subunits coming from each of the two globin families.

Both the α- and β-like families have multiple loci (Figure 4.4), each of which is expressed at a different time in development (Figure 4.5). In first-trimester human embryos, for example, hemoglobin is made up of two ζ chains and two ϵ chains, while in adults it is made up of two α chains and two β chains (recall that the sickle-cell mutation occurs in one of the β chains). Different combinations of globin polypeptides result in hemoglobin molecules with important functional differences. For example, fetal hemoglobin has a higher affinity for oxygen molecules than adult hemoglobin, which enhances oxygen transfer from the mother to the embryo.

TABLE 4.3   Some gene families

In this table, the "Number of duplicate genes" column refers to the number of loci in various gene families. These loci are presumed to be the result of duplication events. They have high sequence homology and code for products with closely related functions.

Family	Number of duplicate genes	Family	Number of duplicate genes
Common proteins		*Vertebrates*	
Actins	5–30	Globins (many species)	
Tubulins (α and β)	5–15	α	1–3
Myosin, heavy chain	5–10	β-like	≥ 50
Histones	100–1000	Ovalbumin (chicken)	3
Keratins	>20	Vitellogenin (frog, chicken)	5
Heat-shock proteins	3	Immunoglobulins, variable regions (many species)	>500
Insects		Transplantation antigens (mouse and human)	50–100
Eggshell proteins (silk moth and fruit fly)	>50		

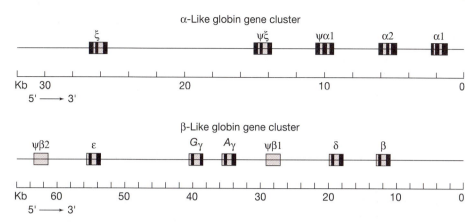

Figure 4.4   Physical maps of the globin gene families in humans   These diagrams show where the five members of the α-like globin gene family are positioned on chromosome 16 of humans, and where the seven members of the β-like globin gene family are positioned on chromosome 11 of humans. "Kb" refers to the number of kilobases separating the loci. The three coding sequences in the α-like family are designated α1, α2, and ζ; the loci marked ψζ and ψα1 are pseudogenes. The five coding sequences of the β-like family are designated ε, ᴳγ, ᴬγ, δ, and β. The loci marked ψβ2 and ψβ1 represent pseudogenes.

Strong but circumstantial evidence suggests that each locus in this family is a product of a gene duplication event. The high structural similarity of transcription units among loci, including a remarkable correspondence in the length and position of their exons and introns (Figure 4.6), supports the duplication hypothesis. It is extremely unlikely that such high structural resemblance could

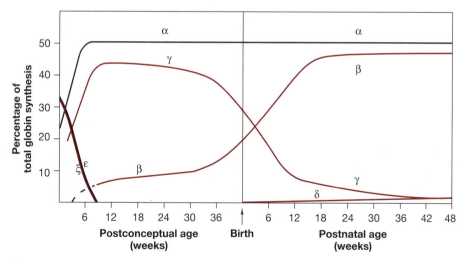

Figure 4.5   Timing of expression differs among members of the globin gene families   This graph shows changes in the expression of globin genes from the α and β families in humans, during pregnancy and after birth. In embryos, hemoglobin is made up of ζ-globin from the α-like gene cluster and ε-globin from the β-like gene cluster. In the fetus, hemoglobin is made up of α-globin from the α-like gene cluster and γ-globin from the β-like gene cluster. In adults, hemoglobin is made up of α-globin from the α-like gene cluster and β-globin from the β-like gene cluster. Each of these hemoglobins has important functional differences.

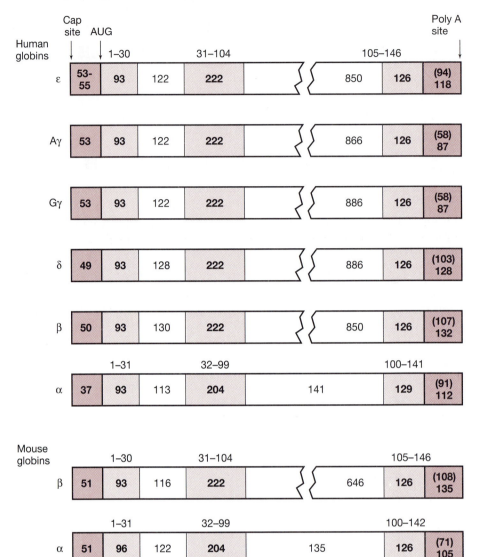

Figure 4.6 **Transcription units in the globin gene family** In these diagrams, the dark boxes represent coding sequences that are untranslated, the light boxes stand for coding sequences that are translated, and the white segments signify introns. The numbers inside the boxes denote the number of nucleotides present in the primary transcript, while the numbers above the boxes give the amino acid positions in the resulting polypeptide. AUG is the start codon. The lengths and positions of introns and exons in loci throughout the α- and β-like clusters are virtually identical.

occur in loci that do not share a recent common ancestor. The duplication hypothesis is also supported by the observation of high sequence identity among globin loci.

Structural similarity and sequence identity is common in other gene families as well, and sequence identity can be extremely high in recently duplicated loci. The amino acid sequences of the red and green visual pigment genes found in humans, for example, are 96% identical (Nathans et al. 1986). These two loci are found close together on the X chromosome, while the locus for the blue pigment is located on chromosome 7. New World monkeys, however, have only one X-linked pigment gene. Only Old World monkeys, the great apes, and humans have a third pigment gene. As a result, we can infer that the duplication event leading to the third pigment occurred less than 35–40 million years ago. This is when fossils indicate that the split between New and Old World primates occurred. We can also infer that human males with a mutated form of the red or green pigment gene experience the same color vision of our male primate ancestors.

Other Mechanisms for Creating New Genes

In addition to duplication, there are other mechanisms that can create new genes or radically new functions for duplicated genes. One example—called overprinting—results from point mutations that produce new start codons and new reading frames for translation. Paul Keese and Adrian Gibbs (1992) investigated the phenomenon in the tymoviruses, which cause mosaic disease in a variety of plants. The tiny genome of one tymovirus, turnip yellow mosaic virus, appears in Figure 4.7. Two of the virus's three genes overlap, meaning that they are translated from different reading frames in the same stretch of nucleotides. Keese and Gibbs estimated the phylogeny of five tymoviruses and nine close relatives based on amino acid sequences in their replicase protein (Figure 4.8), and found that the tymoviruses form their own branch on the phylogeny. The tymoviruses are also the only species in the phylogeny with an overlapping gene overprinted on the replicase gene.

Based on these data, Keese and Gibbs propose that the replicase gene is ancestral, and that the overlapping gene was created in the common ancestor of the tymoviruses. Their hypothesis is that a mutation created a new start codon, in a different reading frame, in the stretch of nucleotides encoding the replicase protein. They note that evolution at the new locus is likely to be tightly constrained, because any mutation that improves the function of the overlapping protein would probably be deleterious to the function of the replicase protein. Based on their survey of the literature on viral genomes, they also propose that overprinting has been a common mechanism for creating new genes during viral evolution.

Charles Langley and colleagues have investigated the origin of new genes by yet another mechanism. The ancestral gene they studied, found in fruit flies from the genus *Drosophila,* codes for the enzyme alcohol dehydrogenase (*Adh*). This locus is located on chromosome 2. Langley et al. (1982) discovered a similar locus on chromosome 3 in two (and only two) species of flies, *D. teissieri* and *D. yakuba*. Jeffs and Ashburner (1991) sequenced this chromosome 3 locus and

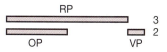

Turnip yellow mosaic virus

Figure 4.7 Genome of the turnip yellow mosaic virus This tiny virus has three genes: replicase protein (RP), which is cleaved after translation into two separate proteins that function in the replication of the virus; virion protein (VP), a coat protein; and an overlapping protein (OP) overprinted in a different reading frame on the replicase gene. Although OP functions in the spread of the virus, its exact role is unknown. The numbers 3 and 2 on the right indicate different reading frames for translation. From Keese and Gibbs (1992).

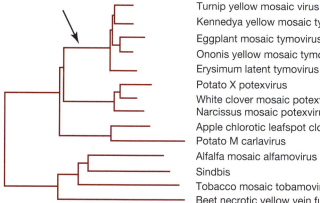

Turnip yellow mosaic virus
Kennedya yellow mosaic tymovirus
Eggplant mosaic tymovirus
Ononis yellow mosaic tymovirus
Erysimum latent tymovirus
Potato X potexvirus
White clover mosaic potexvirus
Narcissus mosaic potexvirus
Apple chlorotic leafspot closterovirus
Potato M carlavirus
Alfalfa mosaic alfamovirus
Sindbis
Tobacco mosaic tobamovirus
Beet necrotic yellow vein furovirus

Figure 4.8 Estimated phylogeny of the tymoviruses and their relatives Of the species represented in this figure, only the tymoviruses have overprinted genes. This implies that the mutation which lead to an overprinted gene occurred somewhere along the branch indicated by the arrow. From Keese and Gibbs (1992).

found that it lacks the introns found in the *Adh* gene on chromosome 2. Jeffs and Ashburner propose that the new locus on chromosome 3 was created when a messenger RNA from the *Adh* gene was reverse-transcribed, and the resulting complementary DNA (cDNA) was inserted into chromosome 3. As we will see in Chapter 7, this mechanism of gene duplication is not unusual. Reverse transcriptase is common in the nuclei of eukaryotic cells.

In many genomes, reverse transcription of mRNAs and insertion of the resulting DNA at a new location is an important source of new DNA.

The question now becomes: Does this new locus have some function, or is it merely a pseudogene? Long and Langley (1993) sequenced the DNA in the chromosome 3 locus from a number of individuals in both *D. teissieri* and *D. yakuba,* with the goal of analyzing alleles of the new locus that had arisen by point mutation. They discovered that most of the alleles differed from each other only in silent site substitutions. This implies that natural selection is acting to conserve the amino acid sequence of a protein encoded by the new locus. In contrast, the common pattern in pseudogenes is that replacement substitutions are as common as silent site substitutions. As a result, Long and Langley concluded that the new locus is a functional gene. They named it *jingwei,* after the protagonist in a Chinese reincarnation myth.

Long and Langley were able to isolate mRNA transcribed from the *jingwei* gene. Upon sequencing the mRNA, they found that the gene contains additional exons not found in its *Adh* ancestor. These additional exons were apparently annexed from a flanking region of chromosome 3 after the *Adh* reverse transcript was inserted. *Jingwei* is thus a hybrid locus, stitched together from pieces of genes on two different chromosomes. Although this mechanism of gene duplication sounds exotic, it illustrates a very general point: Genomes are dynamic. The amount, location, and make-up of the genetic material changes through time.

4.3   Chromosome Alterations

A wide variety of changes can occur in the gross morphology of chromosomes. Some of these mutations affect only gene order and organization; others produce duplications or deletions that affect the total amount of genetic material. Chromosomal alterations can involve the entire DNA molecule or just segments.

Inversions

Chromosome inversions involve much larger stretches of DNA than the mutation types we reviewed in sections 4.1 and 4.2. They also produce very different consequences. Inversions result from a multistep process that starts when ionizing radiation causes two double-strand breaks in a chromosome. After breakage a chromosome segment can detach, flip, and reanneal in its original location. As Figure 4.9 shows, gene order along the chromosome is now inverted.

Inversions change gene order and lessen the frequency of crossing over. . . .

What is the evolutionary impact? Inversions affect a phenomenon known as genetic linkage. Linkage is the tendency for alleles of different genes to assort together at meiosis. For obvious reasons, genes on the same chromosome tend to be more tightly linked (that is, more likely to be inherited to-

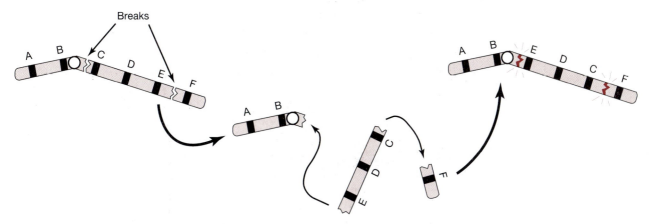

***Figure 4.9*   Chromosome inversion**   Inversions result when a chromosome segment breaks in two places, flips, and reanneals. Note that after the event, the order of the genes labeled C, D, and E is inverted.

gether) than genes on nonhomologous chromosomes. Similarly, the closer together genes are on the same chromosome, the tighter the linkage. Crossing over at meiosis, on the other hand, breaks up allele combinations and reduces linkage (see Chapter 6).

Because inverted sequences cannot align properly with their normal homolog during synapsis, a crossing-over event that takes place within an inversion results in the duplication or loss of chromosome regions and the production of dysfunctional gametes. When inversions are heterozygous, successful crossing-over events are extremely rare. The upshot is that alleles inside the inversion are locked so tightly together that they are inherited as a single "supergene."

Inversions are especially common in insects. Are they important in evolution? Consider a suite of inversions found in populations of *Drosophila subobscura*. This fruit fly is native to Western Europe, North Africa, and the Middle East, and has six chromosomes. Five of these are polymorphic for at least one inversion (Prevosti et al. 1988), meaning that chromosomes with and without the inversions exist. Biologists have known since the 1960s that the frequencies of these inversions vary regularly with latitude and climate. This type of regular change in the frequency of an allele or an inversion over a geographic area is called a cline. Several authors have argued that different inversions must contain specific combinations of alleles that function well together in cold, wet weather or hot, dry conditions. But is the cline really the result of natural selection on the supergenes? Or could it be an historical accident, caused by differences in the founding populations, long ago?

A natural experiment has settled the issue. In 1978 *D. subobscura* showed up in the New World for the first time, initially in Puerto Montt, Chile, and four years later in Port Angeles, Washington, USA. The North American population is almost certainly derived from the South American one: Of the 80 inversions present in Old World populations, precisely the same subset of 19 are found in both Chile and Washington State. Also, *Drosophila* are frugivores, Chile is a major fruit exporter, and Port Angeles is a seaport. Within a few years of their arrival on each continent, the *D. subobscura* populations had expanded extensively

. . . as a result, genes inside inversions tend to be inherited as a unit.

along each coast and developed the same clines in inversion frequencies found in the Old World (Figure 4.10). The clines are even correlated with the same general changes in climate type: from wet marine environments, to Mediterranean climates, to desert and dry steppe habitats (Prevosti et al. 1988; Ayala et al. 1989). This is strong evidence that the clines result from natural selection, and are not due to historical accident.

Which genes are locked in the inversions, and how do they affect adaptation to changes in climate? We do not know. But in the lab, *D. subobscura* lines that are bred for small body size tend to become homozygous for the inversions found in the dryer, hotter part of the range (Prevosti 1967). This is a hint that alleles inside the inversions may affect body size, with natural selection favoring large flies in cold, wet climates and small flies in hot, dry areas. Research into this natural experiment is continuing. In the meantime, the fly study illustrates a key point about inversions: They are an important class of mutations because they affect selection on groups of alleles. We will return to the topic of selection on multiple loci in Chapter 6.

Polyploidization

The final type of mutation that we will examine occurs at the largest scale possible: entire sets of chromosomes. Instead of being haploid (N) or diploid (2N), polyploid organisms can be tetraploid (4N) up to octoploid (8N) or higher. Polyploidy can result from several different events, but perhaps the most frequent is errors at meiosis that result in diploid gametes. The union of a diploid gamete with a normal haploid gamete results in a sterile offspring, because triploid cells cannot go through meiosis properly. The homologous chromosomes cannot synapse correctly, so all of the gametes produced end up with the wrong number of chromosomes. But if two diploid gametes happen to fuse they can form a viable, tetraploid (4N) offspring. Probably the most common way in which two diploid gametes come to fuse is when individuals containing both male and female reproductive structures (hermaphrodites) undergo faulty meiosis, then self-fertilize.

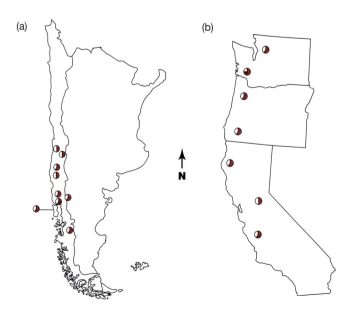

Figure 4.10 Clines in the frequency of an inversion The filled parts of the pie charts drawn along the west coast of South America (a) and North America (b) represent the frequency of a chromosomal inversion called E_{st} found in the fruit fly *Drosophila subobscura*. In each hemisphere the frequency of E_{st} increases steadily with increasing latitude. In both cases the cline in the inversion frequency also correlates with changes from hot, dry environment to cool, moist climates. A large series of inversions on different chromosomes show the same pattern, which is also found in Europe (Prevosti et al. 1988; Ayala et al. 1989).

In plants, polyploidy can also result from an error in mitosis. Specifically, nondisjunction at anaphase can produce a polyploid cell line. If these cells later differentiate into germline cells and undergo meiosis, and if the diploid gametes self-fertilize, viable polyploid offspring result.

Polyploidy is important because it can result in new species being formed. As we have mentioned, tetraploid individuals cannot successfully mate with diploid individuals, because fusion of a diploid gamete with a haploid gamete results in triploid offspring that are sterile. New polyploid mutants can mate only with themselves or their own offspring. As a result, polyploids often become isolated from their parent population. And genetic isolation, as we will see in Chapter 9, is the essence of speciation.

Populations of polyploid inviduals are often isolated, genetically, from their parental species.

Polyploidy is common in plants and rare in animals. Nearly *half* of all angiosperm (flowering plant) species are polyploid, as are the vast majority of the ferns. But in animals polyploidy is largely limited to groups that are self-compatible hermaphrodites (like earthworms and some flatworms) or capable of producing offspring without fertilization, through a process called parthenogenesis (which occurs in some species of beetles, sow bugs, moths, shrimp, goldfish, and salamanders).

Summary

Mutations range from single base-pair substitutions to the duplication of entire chromosome sets, and vary in impact from no amino acid sequence change to single amino acid changes to gene creation and genome duplication.

Point mutations result from errors made by DNA polymerase during DNA synthesis or errors made by DNA repair enzymes after sequences have been damaged by chemical mutagens or radiation. Point mutations in first and second positions of codons frequently result in replacement substitutions that lead to changes in the amino acid sequence of proteins. Point mutations in third positions of codons usually result in silent substitutions that do not lead to changes in the amino acid sequence of proteins. Point mutations create new alleles.

Mutation rates vary both among genes within genomes and among species. Both DNA polymerase and the many loci involved in mismatch repair exhibit heritable variation. As a result, mutation rate is a trait that can respond to natural selection and other evolutionary processes.

The most common sources of new genes are duplication events that result from errors during crossing over. A duplicated gene can diverge from its parent sequence and become a new locus with a different function, or a nonfunctional pseudogene. New genes can also arise when mRNAs are reverse-transcribed and inserted into the genome in a new location.

Chromosome alterations form a large class of mutations. Chromosome inversions have interesting evolutionary implications because they reduce the frequency of recombination between loci within the inversion. As a result, alleles within inversions tend to be inherited as a group instead of independently. Polyploidy is a condition characterized by duplication of an entire chromosome set. Polyploidy is common in plants, and is important because polyploid individuals are genetically isolated from the population that gave rise to them.

Now that we know something about where alleles and genes come from, we are ready to shift our focus and ask a different question: What determines the fate of these new alleles and genes in a population? This is the subject of Chapter 5.

Questions

1. The evolutionary biologist Graham Bell (1997) has said, "Most mutations are not very deleterious. The assault on adaptedness is not carried out primarily by a storm of mutations that kill or maim, but rather by a steady drizzle of mutations with slight or inappreciable effects on health and vigor." This statement is supported by classical experiments in fruit flies (see Keightley 1994) and other organisms. Yet virtually all known human mutations cause severe disease. Is it possible that mutations in humans are qualitatively different from mutations in other organisms? Or has our traditional perception of mutations, as highly deleterious, been colored by intensive studies on a small subset of human mutations? For a gene like β globin, how would you go about determining how often mutations with small effects occur?

2. Schlager and Dickie (1971) set out to determine the rate of coat-color mutations in mice. They spent seven years studying five coat-color genes in over seven million mice, examining thousands of brother-sister matings from 28 inbred strains. For each gene they studied two rates: (1) the rate at which a normal gene mutated to a form resulting in loss of function, and (2) the rate at which a mutant gene would then mutate back to the normal form. They found that in 67,395 tested crosses, the "albino" gene mutated from normal (colored) to nonfunctional (albino) just three times, which is a mutation rate of 44.5×10^{-6} mutations per gene per generation. In 839,447 tested crosses, the "dilute" gene mutated from normal to nonfunctional 10 times, a mutation rate of 11.9×10^{-6}. In all the coat color genes, the reverse mutation rates (for example, from albino to colored) were around 2.5×10^{-6}. Think about the different mutations that can all result in loss of function of a single gene, versus the mutations that can result in regain of function. Why were the mutation rates for regain of function always lower than the mutation rates for loss of function?

3. Voelker et al. (1980) studied 1,000 lines of flies for 220 generations to estimate the mutation rate of a certain protein. In 3,111,598 tested crosses, they found 16 flies with new replacement substitutions mutations in that protein (detected by a slight change in its electrical charge), for a mutation rate of 5.1×10^{-6} mutations per gene per generation. However, only 4 of these mutations resulted in altered protein function; the other 12 mutations did not detectably affect protein function. If Voelker et al. had simply been measuring loss of function, would their estimate of mutation rate have been lower or higher? If they had been able to measure silent site substitutions as well, would their estimate of mutation rate have been lower or higher? Do you think that by measuring changes in protein charge, they were definitely able to measure all the replacement substitutions that occurred? What do you think is the most informative measure of mutation rate?

4. In this chapter, we introduced the consequences of mutations in two sorts of traits: changes in phenotypic traits such as hemoglobin structure, and changes in mutation rate itself. To clarify the distinction between these two kinds of traits, look at the following list. Which of these proteins can affect mutation rate itself? Which do not affect mutation rate, but instead affect some other trait of the organism?

Protein	Example of a Mutation in the Protein
• β-globin	• Increased tendency to sickle
• Mismatch repair proteins	• Increased repair of damaged DNA
• Melanin (coat-color protein)	• Red coat color instead of black coat color
• Growth hormone	• Dwarfism or gigantism
• DNA polymerase	• Increased speed and decreased accuracy during DNA replication

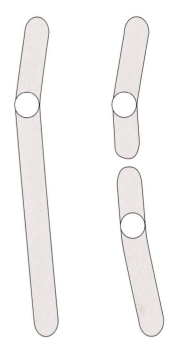

5. The discovery of overprinting shows that it is possible for one stretch of DNA to code for two entirely different proteins. To examine this phenomenon further, suppose we discovered a tiny gene, 12 base pairs long, that codes for a polypeptide just 3 amino acids long:

   • DNA: ACUGCUGUCUAA
   • Amino acids: thr-ala-val-stop

   Now suppose this organism would benefit greatly if it had another little polypeptide composed of the amino acids leu-leu-ser. Would it be possible to produce this protein if the organism started transcribing the gene from the second base pair? Look back at the genetic code in Figure 2.17a to help answer this question. Suppose further that the organism would benefit even more if instead of leu-leu-ser, it made *pro*-leu-ser. What mutations would be necessary to do this? Would they destroy the amino acid sequence of the original protein? In general, what sort of mutations can occur in the original protein that will *not* alter its amino acids, but *will* allow changes in amino acids to occur in the other, overprinted protein?

6. Chromosome number can evolve by smaller-scale changes than duplication of entire chromosome sets. For example, domestic horses have 46 chromosomes per dipoid set (2n = 46) while Przewalski's horse, an Asian subspecies, has 48. Przewalski's horse is thought to have evolved from an ancestor with 2n = 46 chromosomes. The question is: Where did its extra chromosome originate? It seems unlikely that an entirely new chromosome pair was created *de novo* in Przewalski's horse. To generate a hypothesis explaining where the new chromosome in Przewalski's horse came from, examine Figure 4.11. This drawing shows how certain chromosomes synapse in the hybrid offspring of a domestic horse–Przewalski's horse mating.

Figure 4.11   **Source of an extra chromosome in Przewalski's horse**   It is possible to create hybrids between Przewalski's horse and the domestic horse. This diagram shows how, in hybrid offspring, two chromosomes from Przewalski's horse synapse with one chromosome from the domestic horse during meiosis. The remaining chromosomes show normal 1:1 pairing.

Exploring the Literature

7. The directed-mutation hypothesis has been the most controversial topic in recent research on mutation. This hypothesis, inspired by experimental work on the bacterium *Escherichia coli,* maintains that organisms can generate specific types of mutations in response to particular environmental challenges. For example, if the environment grew hotter over time, the hypothesis maintains that organisms would respond by specifically generating mutations in genes involved in coping with hot temperatures. This implies that mutations do not occur randomly, but are directed by the environment. Look up the following papers to learn more about the controversy.

Cairns, J., J. Overbaugh, and S. Miller. 1988. The origin of mutants. *Nature* 335:142–145.

Foster, P.L., and J.M. Trimarchi. 1994. Adaptive reversion of a frameshift mutation in *Escherichia coli* by simple base deletions in homopolymeric runs. *Science* 265:407–409.

Galitski, T., and J.R. Roth. 1995. Evidence that F plasmid transfer replication underlies apparent adaptive mutation. *Science* 268:421–423.

Lenski, R.E., and J.E. Mittler. 1993. The directed mutation controversy and neo-Darwinism. *Science* 259:188–194.

Radicella, J.P., P.U. Park, and M.S. Fox. 1995. Adaptive mutation in *Escherichia coli*: A role for conjugation. *Science* 268:418–420.

Rosenberg, S.M, S. Longerich, P. Gee, and R.S. Harris. 1994. Adaptive mutations by deletions in small mononucleotide repeats. *Science* 265:405–407.

Citations

Ayala, F.J., L. Serra, and A. Prevosti. 1989. A grand experiment in evolution: The *Drosophila subobscura* colonization of the Americas. *Genome* 31:246–255.

Bell, G. 1997. *Selection: The Mechanism of Evolution* (New York: Chapman & Hall).

Bohr, V.A., D.H. Phillips, and P.C. Hanawalt. 1987. Heterogeneous DNA damage and repair in the mammalian genome. *Cancer Research* 47:6426–6436.

Bohr, V.A., C.A. Smith, D.S. Okumoto, and P.C. Hanawalt. 1985. DNA repair in an active gene: Removal of pyrimidine dimers from the DHRF gene of CHO cells is much more efficient than in the genome overall. *Cell* 40:359–369.

Drake, J.W. 1991. A constant of rate of spontaneous mutation in DNA-based microbes. *Proceedings of the National Academy of Sciences, USA* 88:7160–7164.

Friedberg, E.C., G.C. Walker, and W. Siede. 1995. *DNA Repair and Mutagenesis* (Washington D.C.: ASM Press).

Gillin, F.D., and N.G. Nossal. 1976a. Control of mutation frequency by bacteriophage T4 DNA polymerase I. The *ts* CB120 antimutator DNA polymerase is defective in strand displacement. *Journal of Biological Chemistry* 251:5219–5224.

Gillin, F.D., and N.G. Nossal. 1976b. Control of mutation frequency by bacteriophage T4 DNA polymerase II. Accuracy of nucleotide selection by L8 mutator, CB120 antimutator, and wild type phage T4 DNA polymerases. *Journal of Biological Chemistry* 251:5225–5232.

Holmquist, G.P., and J. Filipski. 1994. Organization of mutations along the genome: A prime determinant of genome evolution. *Trends in Ecology and Evolution* 9:65–69.

Ingram, V.M. 1958. How do genes act? *Scientific American* 198:68–76.

Jeffs, P., and M. Ashburner. 1991. Processed pseudogenes in *Drosophila*. *Proceedings of the Royal Society of London, Series B* 244:151–159.

Keese, P.K., and A. Gibbs. 1992. Origins of genes: "Big bang" or continuous creation? *Proceedings of the National Academy of Sciences, USA* 89:9489–9493.

Keightley, P.D. 1994. The distribution of mutation effects on viability in *Drosophila melanogaster*. *Genetics* 138:1315–1322.

Klekowski, E.J., Jr., and P.J. Godfrey. 1989. Ageing and mutation in plants. *Nature* 340:389–391.

Langley, C.H., E. Montgomery, and W.F. Quattlebaum. 1982. Restriction map variation in the *Adh* region of *Drosophila*. *Proceedings of the National Academy of Sciences, USA* 78:5631–5635.

LeClerc, J.E., B. Li, W.L. Payne, and T.A. Cebula. 1996. High mutation frequencies among *Escherichia coli* and *Salmonella* pathogens. *Science* 274:1209–1211.

Long, M., and C.H. Langley. 1993. Natural selection and the origin of *jingwei*, a chimeric processed functional gene in *Drosophila*. *Science* 260:91–95.

Mellon, I., D.K. Rajpal, M. Koi, C.R. Boland, and G.N. Champe. 1996. Transcription-coupled repair deficiency and mutations in the human mismatch repair genes. *Science* 272:557–560.

Modrich, P. 1991. Mechanisms and biological effects of mismatch repair. *Annual Review of Genetics* 25:229–253.

Nathans, J., D. Thomas, and D.S. Hogness. 1986. Molecular genetics of human color vision: The genes encoding blue, green, and red pigments. *Science* 232:193–202.

Pauling, L., H.A. Itano, S.J. Singer, and I.C. Wells. 1949. Sickle-cell anemia, a molecular disease. *Science* 110:543–548.

Prevosti, A. 1967. Inversion heterozygosity and selection for wing length in *Drosophila subobscura*. *Genetical Research Cambridge* 10:81–93.

Prevosti, A., G. Ribo, L. Serra, M. Aguade, J. Balaña, M. Monclus, and F. Mestres. 1988. Colonization of America by *Drosophila subobscura*: Experiment in natural populations that supports the adaptive role of chromosomal-inversion polymorphism. *Proceedings of the National Academy of Sciences, USA* 85:5597–5600.

Schlager, G., and M.M. Dickie. 1971. Natural mutation rates in the house mouse. Estimates for 5 specific loci and dominant mutations. *Mutation Research* 11:89–96.

Short, R.V., A.C. Chandley, R.C. Jones, and W.R. Allen. 1974. Meiosis in interspecific equine hybrids. II. The Przewalski horse/domestic horse hybrid. *Cytogenetics and Cell Genetics* 13:465–478.

Stadler, L.J. 1942. Some observations on gene variability and spontaneous mutation. *Spragg Memorial Lectures* (East Lansing: Michigan State College).

Voelker, R.A., H.E. Schaffer, and T. Mukai. 1980. Spontaneous allozyme mutations in *Drosophila melanogaster*: Rate of occurrence and nature of the mutants. *Genetics* 94:961–968.

Mendelian Genetics in Populations: The Mechanisms of Evolution

We noted in Chapter 2 that Darwin lacked an accurate model of the mechanism responsible for variation and heredity, the first two postulates of his theory of evolution by natural selection. This gap in understanding produced considerable confusion for Darwin, his contemporaries, and subsequent generations of evolutionary biologists (see Provine 1971). Gregor Mendel published his theory of genetics in 1865, but was widely ignored. When biologists finally rediscovered Mendel's work in 1900, they immediately recognized its importance for understanding evolution. It took them another 30 years, however, to achieve a synthesis of Mendelian genetics and Darwinian evolution. At the heart of this synthesis is the theory of population genetics.

A population is a group of interbreeding individuals and their offspring. Populations typically persist over many generations. Population genetics tracks the fate, across generations, of Mendelian genes in populations. Population genetics is concerned with whether a particular allele or genotype will become more common or less common in a population over time, and why. In other words, population genetics follows allele and genotype frequencies. From a population-genetic perspective, evolution is change across generations in the frequencies of alleles.

One of the reasons population genetics is a successful theory is that it is built around a null model, the Hardy-Weinberg equilibrium principle. This null model specifies algebraically, under the simplest possible assumptions, what will happen across generations to the frequencies of alleles and genotypes. What will happen under the null model is this: Allele and genotype frequencies will not change. Populations will not evolve. Violations of the simple assumptions of the null model, in the form of selection, mutation, migration, and genetic drift, can cause allele frequencies to change. Population genetics thus identifies the mechanisms of population evolution.

In Chapter 4 we explored the modern view of variation. Our goals in this chapter are to (1) develop the basic theory of population genetics and show how it identifies the mechanisms of evolution; and (2) show by example how evolutionary biologists use population genetics, in the field and the laboratory, to understand evolution and adaptation.

5.1 Mendelian Genetics in Populations: The Hardy-Weinberg Equilibrium Principle

Darwin's theory of genetics involved blending inheritance, which Fleeming Jenkin showed is incompatible with evolution by natural selection. Mendel discovered that inheritance is particulate; population genetics shows that Mendelism is compatible with evolution by natural selection.

When evolutionary biologists rediscovered Mendel's research on peas, the fundamental principles of heredity finally became widely known. The most important of these principles for evolutionary biology is Mendel's law of segregation. The law of segregation holds that each individual carries two unblending copies of each gene, and that each gamete receives only one copy of each gene chosen at random from the individual's two. Different versions of a gene are called alleles. The law of segregation allows biologists to predict the fate of alleles within families. If an *Aa* female mates with an *Aa* male, for example, they will on average have *AA*, *Aa*, and *aa* offspring in a 1:2:1 ratio. Does the law of segregation allow us to predict the fate of alleles in populations as well as families? What, for example, will be the genotype ratios in a population in which alleles *A* and *a* are present in equal numbers? Will allele *A*, and the genotypes that contain it, become more common over time?

The argument we must develop to answer such questions is somewhat lengthy, so we will give the conclusion at the outset to show where we are going. Mendel's law of segregation does, indeed, allow us to determine the fate of alleles in populations. If we assume a population in which:

Five assumptions describe life, death, and reproduction in the simplest hypothetical population possible.

1. there is no selection (that is, all individuals have equal rates of survival and equal reproductive success);
2. there are no mutations converting one allele to another or creating new alleles;
3. there is no migration of individuals into or out of the population;
4. there are no random events that cause some individuals to pass on more of their genes than do other individuals;
5. individuals choose their mates at random;

then the relative abundance of different alleles will not change across generations. The population will not evolve. If one or more of the first four assumptions is violated—if there is selection, mutation, migration, or random drift—then the relative abundance of different alleles may change. The population may evolve. Selection, mutation, migration, and random genetic drift are the four forces of evolution. This conclusion is the primary message of the chapter.

Nonrandom mating, a violation of the fifth assumption, does not alter the relative abundance of alleles. Nonrandom mating is not, by itself, a mechanism of evolution. Nonrandom mating can, however, alter the relative abundance of genotypes. As we will see, nonrandom mating can therefore indirectly influence the course of evolution.

Yule's Numerical Example

Having stated the conclusion we are headed toward, we can now go back to the beginning and trace the route from Mendel to the mechanisms of evolution (for a full history, see Provine 1971). Among the first researchers to consider Mendelian genetics in the context of populations was G. Udny Yule. In 1902 Yule published a paper in which he explored how Mendelian segregation affects the frequencies of alleles at a single locus in a population. Yule started by imagining a population established from the offspring of *AA* homozygotes mated with *aa* homozygotes. The first generation of the population consists entirely of *Aa* heterozygotes. Yule then imagined letting these heterozygotes breed at random among themselves. A useful mental trick is to picture the process of random mating happening like this: We take all the gametes (eggs and sperm) produced by all the adults in the population and mix them together in a barrel (without fertilization). This barrel is known as the gene pool. We then draw gametes from the gene pool at random, like lottery tickets, and pair them to make zygotes. In the first generation of Yule's population, half the gametes in the gene pool carry allele *A*. The other half of the gametes carry allele *a*. The fraction of the gametes in the gene pool that carry a particular allele is called the frequency of the allele. Thus the frequency of allele *A* is 0.5, and the frequency of allele *a* is 0.5. When we draw gametes from this gene pool to make the zygotes that will become the members of the second generation, what fraction of them will be *AA*, what fraction *Aa*, and what fraction *aa*?

We can approach this question by considering the prospects for a single zygote. When we draw two gametes from the gene pool and pair them to make a zygote, what is the probability of producing an *AA* homozygote? The probability that the egg will be *A* is 0.5, and the probability that the sperm will be *A* is 0.5. Thus the probability of producing an *AA* homozygote is $0.5 \times 0.5 = 0.25$ (see Box 5.1). Because the probability that any individual zygote will be *AA* is 0.25, one quarter of the entire population of zygotes will be *AA*.

When we draw two gametes from the gene pool and pair them, what is the probability of producing an *Aa* heterozygote? One way this can happen is if we draw an egg that is *A* (an event with probability 0.5) and a sperm that is *a* (an event with probability 0.5). Thus the probability of getting a heterozygote this

---

Box 5.1   Combining probabilities

The combined probability that two independent events will occur together is equal to the product of their individual probabilities. For example, the probability that a tossed penny will come up heads is $\frac{1}{2}$. The probability that a tossed nickel will come up heads is also $\frac{1}{2}$. If we toss both coins together, the outcome for the penny is independent of the outcome for the nickel. Thus the probability of getting heads on the penny and heads on the nickel is $\frac{1}{2} \times \frac{1}{2} = \frac{1}{4}$.

The combined probability that either of two mutually exclusive events will occur is the sum of their individual probabilities. When rolling a die we can get a one or a two (among other possibilities), but we cannot get both at once. Thus the probability of getting either a one or a two is $\frac{1}{6} + \frac{1}{6} = \frac{1}{3}$.

Sperm

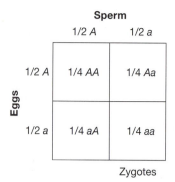

Figure 5.1 **Allele frequencies of $\frac{1}{2}$ A and $\frac{1}{2}$ a give genotype frequencies of $\frac{1}{4}$ AA, $\frac{1}{2}$ Aa, and $\frac{1}{4}$ aa** The fractions along the left and top edges of the box represent the frequencies of A and a eggs and sperm in the gene pool. The fractions inside the box represent the genotype frequencies among zygotes formed by drawing gametes at random from the gene pool.

Yule showed that if a dominant allele and a recessive allele are each at frequency 0.5, neither allele will change its frequency across generations.

first way is $0.5 \times 0.5 = 0.25$. A second way to get a heterozygote is by drawing an egg that is *a* (probability = 0.5) and a sperm that is *A* (probability = 0.5). Thus the probability of getting a heterozygote this second way is $0.5 \times 0.5 = 0.25$. The overall probability of getting a heterozygote by either the first way or the second way is $0.25 + 0.25 = 0.5$ (see Box 5.1). Thus, one half of the entire population of zygotes will be *Aa*.

When we draw two gametes from the gene pool and pair them, what is the probability of producing an *aa* homozygote? It is $0.5 \times 0.5 = 0.25$. One quarter of the entire population of zygotes will be *aa*.

We have now established the genotype frequencies in the second generation in Yule's population: One quarter of the individuals will be *AA*, half the individuals will be *Aa*, and one quarter of the individuals will be *aa* (Figure 5.1). This is not surprising; it is the same ratio of offspring genotypes we would expect in a single mating of heterozygous parents.

The genotype frequencies in Yule's population have changed from 1.0 *Aa* in the first generation to 0.25 *AA*, 0.5 *Aa*, and 0.25 *aa* in the second generation. Will the genotype frequencies have changed again in the third generation? This is the generation that results from random mating among the second-generation adults. The first step in answering this question is to calculate the frequencies of the *A* and *a* alleles in the second generation's gene pool.

We start with allele *A*. Gametes carrying allele *A* can be produced by *AA* adults or by *Aa* adults. Because *AA* individuals constitute a quarter of the adult population, they will contribute a quarter of the gametes to the gene pool. All of these gametes will carry allele *A*. So we know right away that the first quarter of the gametes in the gene pool will be *A*. *Aa* individuals constitute half the adult population, so they will contribute half the gametes to the gene pool. Half of the gametes the *Aa*'s contribute, or a second quarter of all the gametes in the gene pool, will be *A*. We now know that $\frac{1}{4} + \frac{1}{4} = \frac{1}{2}$ of the gametes in the gene pool will be *A*.

Now we do the same calculation for allele *a*. Allele *a* will appear in half of the gametes made by the *Aa*'s, or a quarter of the gene pool, plus all the gametes made by the *aa*'s, another quarter of the gene pool. Thus the total fraction of *a*-bearing gametes will be half the gene pool.

In other words, in the second generation's gene pool the frequency of allele *A* is 0.5 and the frequency of allele *a* is 0.5 (Figure 5.2). The frequency of the gametes has not changed between the first generation's gene pool and the second generation's gene pool. Because of this, the results of random draws and pairings from the second generation's gene pool to produce the third-generation individuals will be exactly the same as the results we got when we drew at random from the first generation's gene pool to make the second-generation individuals. The genotype frequencies in the third generation will be 0.25 *AA*, 0.5 *Aa*, and 0.25 *aa*.

Yule concluded that after a single generation of random mating his imaginary population had reached an unchanging, or equilibrium, state. From then on, neither the allele frequencies nor the genotype frequencies would change. The population would not evolve.

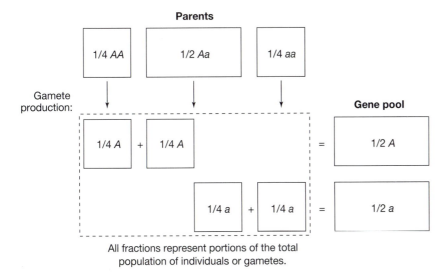

Parents

| 1/4 AA | 1/2 Aa | 1/4 aa |

Gamete production:

| 1/4 A | + | 1/4 A | | = | 1/2 A |

| | 1/4 a | + | 1/4 a | = | 1/2 a |

Gene pool

All fractions represent portions of the total
population of individuals or gametes.

Figure 5.2   **Genotype frequencies of**
$\frac{1}{4}$ *AA*, $\frac{1}{2}$ *Aa*, **and** $\frac{1}{4}$ *aa* **give allele fre-**
quencies of $\frac{1}{2}$ *A* **and** $\frac{1}{2}$ *a*   The area of
all boxes is proportional to the fraction of
the population or gene pool they represent. The top row of boxes represents the
genotype frequencies in the parents. The
boxes below the parents represent the
gametes they contribute to the gene
pool. The boxes on the right represent
the total fraction of *A* and *a* gametes in
the gene pool.

Yule's conclusion was both ground-breaking and correct, but he took it a bit
too literally. He believed that allele frequencies of 0.5 *A* and 0.5 *a*, and genotype
frequencies of 0.25 *AA*, 0.5 *Aa*, and 0.25 *aa,* represented the only equilibrium
state for a two-allele system. For example, Yule believed that if a single *A* allele
appeared as a mutation in a population whose gene pool otherwise consisted
only of *a*'s, then the *A* allele would automatically increase in frequency until it
constituted one half of the gene pool. Yule argued this claim during the discussion that followed a talk given in 1908 by R. C. Punnett (of Punnett square
fame). Punnett thought (correctly) that Yule was wrong, but Punnett did not
know how to prove it. So Punnett took the problem to his mathematician
friend G. H. Hardy, who produced the proof in short order (Hardy 1908).
Hardy simply repeated Yule's calculation in the general case, using variables in
place of the specific allele frequencies that Yule had assumed. Hardy's calculation
of the general case showed that *any* allele frequencies will be in equilibrium after only a single generation of random mating.

The General Case

For the general case we again work with an imaginary population. We are concerned with a single locus with two alleles: A_1 and A_2. We use capital letters
with subscripts because we want our calculation to cover cases in which the alleles are codominant as well as cases in which they are dominant/recessive. The
three possible diploid genotypes are $A_1A_1, A_1A_2,$ and A_2A_2. The gene pool will
contain some frequency of A_1 gametes and some frequency of A_2 gametes. We
will call the frequency of A_1 alleles in the gene pool p and the frequency of A_2
alleles in the gene pool q. There are only two alleles in the population, so
p + q = 1. The probability of drawing an A_1 gamete from the gene pool is p
and the probability of drawing an A_2 gamete is q.

Now we calculate the frequencies of the three offspring genotypes that result
when we draw and pair gametes at random from the gene pool:

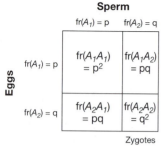

Sperm

	fr(A_1) = p	fr(A_2) = q
Eggs fr(A_1) = p	fr(A_1A_1) = p^2	fr(A_1A_2) = pq
fr(A_2) = q	fr(A_2A_1) = pq	fr(A_2A_2) = q^2

Zygotes

Figure 5.3 **Allele frequencies of p and q give genotype frequencies of p^2, 2pq, and q^2** The length of the line segments along the left and top edges of the box represent the frequences of A_1 (= p) and A_2 (= q) in the population. The length of one side of the entire box is p + q = 1. The areas of the regions inside the box represent the genotype frequencies among zygotes produced by random pairings of gametes. The total area of the box is $(p + q)^2 = 1 = p^2 + 2pq + q^2$. This last expression also gives the genotype frequencies of the zygotes.

Hardy showed that any alleles at any frequencies will remain stable in frequency across generations.

- **A_1A_1 homozygotes:** The probability of drawing an A_1 egg is p, and the probability of drawing an A_1 sperm is p. So the probability of producing an A_1A_1 zygote is p × p = p^2. Thus the frequency of A_1A_1 homozygotes in the next generation is p^2.

- **A_1A_2 heterozygotes:** The probability of drawing an A_1 egg and an A_2 sperm is p × q = pq. The probability of drawing an A_2 egg and an A_1 sperm is q × p = pq. Therefore, the frequency of A_1A_2 heterozygotes in the next generation is 2pq.

- **A_2A_2 homozygotes:** The probability of drawing an A_2 egg is q, and the probability of drawing an A_2 sperm is q. So the frequency of A_2A_2 homozygotes in the next generation is q × q = q^2.

We have gone from knowing the allele frequencies in the first generation's gene pool to knowing the genotype frequencies in the second generation. Figure 5.3 shows these calculations geometrically. The figure shows that the sum of the genotype frequencies, $p^2 + 2pq + q^2$, is equal to 1. This same result can be demonstrated algebraically by substituting (1 − p) for q. The genotype frequencies in a population must always sum to one.

We now calculate the allele frequencies in the second generation's gene pool. Yule thought that unless p = q = 0.5, then the allele frequencies in the second generation's gene pool would be different from what they were in the first generation's gene pool.

We start with the A_1 allele: A_1A_1 individuals constitute a proportion p^2 of the population. All the gametes they make are A_1. So the first p^2 of the gametes in the gene pool carry the A_1 allele. A_1A_2 individuals make up 2pq of the population, and one-half of the gametes they make are A_1. This means that an additional proportion pq of the gametes in the pool are A_1. Thus the total proportion of A_1 gametes in the second generation's gene pool is $p^2 + pq$. Substituting (1 − p) for q gives $p^2 + p(1 − p) = p$.

Now we consider the A_2 allele: Half of the gametes made by A_1A_2 individuals are A_2, which gives pq as the frequency of A_2 gametes so far. Finally, A_2A_2 individuals are q^2 of the population and all the gametes they make are A_2, adding an additional proportion of q^2 gametes that are A_2. Thus the total proportion of A_2 gametes is pq + q^2 = (1 − q)q + q^2 = q.

Figure 5.4 shows these calculations geometrically. The allele frequencies in the second generation's gene pool are exactly the same as they were in the first generation's gene pool. This means that the genotype frequencies will be exactly the same in the third generation as they were in the second generation. The allele frequencies p and q can be stable at any values at all between 0 and 1, as long as p + q = 1. In other words, any allele frequencies will be in equilibrium, not just p = q = 0.5 as Yule thought.

This is a profound result. In Chapter 2 we defined evolution as change in allele frequencies in populations. The calculations we just performed show, given simple assumptions, that in populations with Mendelian genetics allele frequencies do not change.

We have presented this result as the work of Hardy (1908). It was derived independently by Wilhelm Weinberg (1908) and has become known as the

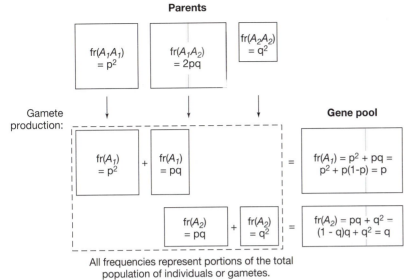

Parents

Figure 5.4 Genotype frequencies of p^2, $2pq$, and q^2 give allele frequences of p and q The area of all boxes is proportional to the fraction of the population or gene pool they represent. The top row of boxes represents the genotype frequencies in the parents. The boxes below the parents represent the gametes they contribute to the gene pool. The boxes on the right represent the total fraction A_1 and A_2 gametes in the gene pool.

Hardy–Weinberg equilibrium principle. (Some evolutionary biologists refer to it as the Hardy-Weinberg-Castle equilibrium principle, because William Castle [1903] worked a numerical example and stated the general equilibrium principle nonmathematically five years before Hardy and Weinberg explicitly proved the general case [see Provine 1971].) The Hardy-Weinberg equilibrium principle yields two fundamental conclusions:

- **Conclusion 1:** The allele frequencies in a population will not change, generation after generation.
- **Conclusion 2:** If allele frequencies in a population are given by p and q, genotype frequencies will be given by p^2, $2pq$, and q^2.

Hardy and Weinberg's result establishes a null expectation for what will happen to allele and genotype frequencies in populations. Given simple assumptions, populations will not evolve.

We get an analogous result if we generalize the analysis from the two-allele case to the usual case of a population containing many alleles of a gene. If multiple alleles exist, the frequency of any homozygous genotype A_iA_i is p_i^2, and the frequency of any heterozygous genotype A_iA_j is $2p_ip_j$, where p_i is the frequency in the gene pool of allele A_i, and p_j is the frequency in the gene pool of allele A_j. If, for example, there are three alleles with frequencies p_1, p_2, and p_3 (such that $p_1 + p_2 + p_3 = 1$), then the genotype frequencies are given by $(p_1 + p_2 + p_3)^2 = p_1^2 + p_2^2 + p_3^2 + 2p_1p_2 + 2p_1p_3 + 2p_2p_3$. Furthermore, the allele frequencies do not change generation after generation.

It may seem puzzling that in a book about evolution we have devoted so much space to a proof that apparently shows that evolution does not happen. What makes the Hardy-Weinberg equilibrium principle so useful is that it rests on a specific set of simple assumptions. When one or more of these assumptions is violated, the Hardy-Weinberg conclusions no longer hold. The crucial assumptions are:

1. There is no selection. All members of the model population survive at equal rates and contribute equal numbers of gametes to the gene pool. When this assumption is violated—when individuals with some genotypes survive and reproduce at higher rates than others—the frequencies of alleles may change from one generation to the next.

2. There is no mutation. In the model population, no copies of existing alleles are converted by mutation into other existing alleles, and no new alleles are created. When this assumption is violated, and, for example, some alleles have higher mutation rates than others, allele frequencies may change from one generation to the next.

3. There is no migration. No individuals move into or out of the model population. When this assumption is violated, and individuals with some alleles move into or out of the population at higher rates than individuals with other alleles, allele frequencies may change from one generation to the next.

4. There are no random events that cause individuals with some genotypes to pass more of their alleles to the next generation than others. We avoided this type of random event by assuming that we drew alleles from the gene pool in their actual frequencies p and q, and not some biased manner caused by chance sampling error. Another way to state this assumption is that the model population is infinitely large. When this assumption is violated, and by chance some individuals contribute more alleles to the next generation than others, allele frequencies may change from one generation to the next. This kind of allele frequency change is called genetic drift.

5. Individuals choose their mates at random. We drew and paired gametes at random from the model population's gene pool. When this assumption is violated—when, for example, individuals prefer to mate with other individuals of the same genotype—allele frequencies do not change from one generation to the next. Genotype frequencies may change, however. Such shifts in genotype frequency, in combination with a violation of one of the other four assumptions, can influence the evolution of populations.

Hardy and Weinberg's analysis identifies an explicit set of conditions under which populations will not evolve. It therefore also identifies an explicit set of forces that can cause evolution.

By furnishing a list of specific ideal conditions under which populations will not evolve, the Hardy-Weinberg equilibrium principle identifies the set of forces that can cause evolution in the real world. This is the sense in which the Hardy-Weinberg equilibrium principle serves as a null model. Biologists can measure allele and genotype frequencies in nature, and determine whether the Hardy-Weinberg conclusions hold. If the allele frequencies change from generation to generation, or if the genotype frequencies cannot, in fact, be predicted by multiplying the allele frequencies, then one or more of the Hardy-Weinberg model's assumptions is being violated. Such a discovery does not, by itself, tell us which assumptions are being violated, but it does tell us that further research may be rewarded with interesting discoveries. In the following sections we consider how violating each of the assumptions affects the two Hardy-Weinberg conclusions, and explore empirical research on the mechanisms of evolutionary change.

5.2 Selection

Selection is differential reproductive success. Selection happens when individuals with some phenotypes survive to reproductive age at higher rates than individuals with other phenotypes, or when individuals with some phenotypes produce more offspring during reproduction than individuals with other phenotypes. Selection can lead to evolution when phenotypes are heritable—that is, when certain phenotypes are associated with certain genotypes.

Population geneticists often assume that phenotypes are determined strictly by genotypes. They might, for example, think of pea plants as being either tall or short, such that individuals with the genotypes *TT* and *Tt* are tall and individuals with the genotype *tt* are short. Such a view is at least roughly accurate for some traits, including the examples we use here. When phenotypes fall into discrete classes that appear to be determined strictly by genotypes, we can think of selection as if it acts directly on the genotypes. We can then assign a particular level of lifetime reproductive success to each genotype. In reality most phenotypic traits are not, in fact, strictly determined by genotype. Pea plants with the genotype *TT*, for example, vary in height. This variation is due to genetic differences at other loci and to differences in the environments in which the pea plants grew. We consider such complications in Chapter 6. For the present, however, we adopt the simple view.

Our task in this section is to incorporate selection into the Hardy-Weinberg analysis. We begin by asking whether selection can change the frequencies of alleles in the gene pool from one generation to the next. In other words, can selection violate conclusion 1 of the Hardy-Weinberg equilibrium principle?

Adding Selection to the Hardy-Weinberg Analysis: Changes in Allele Frequencies

We start with a numerical example that shows that selection can indeed change the frequencies of alleles. Imagine a population of individuals in which there is a single locus that affects survival, with two alleles. Assume that the alleles are codominant, and designate them *B* and *B′*. Assume that the initial frequency of each allele in the gene pool is 0.5. After random mating, we get genotype frequencies for *BB*, *BB′*, and *B′B′*, of 0.25, 0.5, and 0.25. The calculation will be simpler if we give the population a finite size, so imagine that the population consists of 1,000 individuals: 250 *BB*, 500 *BB′*, and 250 *B′B′*. We will start following these individuals when they are juveniles. They will grow up simultaneously; those that survive to adulthood will breed once to produce the next generation and then die.

We incorporate selection by stipulating that the genotypes differ in their rates of survival to adulthood. All of the *BB* individuals survive, 90% of the *BB′* individuals survive, and 80% of the *B′B′* individuals survive. There are now 900 individuals in the population: 250 *BB*, 450 *BB′*, and 200 *B′B′*. The new frequency of the *BB* genotype is $\frac{250}{900} = 0.278$; the new frequency of *BB′* is $\frac{450}{900} = 0.5$; and the new frequency of *B′B′* is $\frac{200}{900} = 0.222$.

When these adults produce gametes, the frequency of the *B* allele in the new gene pool is the frequency of *BB* plus half the frequency of *BB′*: $0.278 + \frac{1}{2}(0.5) = 0.528$. This is an increase in the frequency of allele *B* compared to the previous generation, by an increment of 2.8%. The frequency of the *B′* allele in the new gene pool has dropped to half the frequency of *BB′* plus the frequency of *B′B′*: $\frac{1}{2}(0.5) + 0.222 = 0.472$. Violation of the no-selection assumption has resulted in violation of conclusion 1 of the Hardy-Weinberg analysis (Figure 5.5). The population has evolved in response to selection.

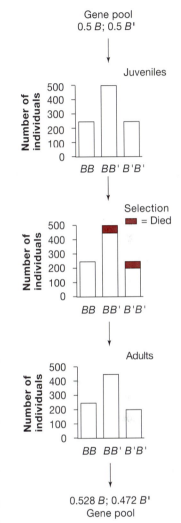

Figure 5.5 Selection can cause allele frequencies to change across generations This figure follows the imaginary population described in the text from one generation's gene pool to the next generation's gene pool. The bar graphs show the number of individuals of each genotype in the population at any given time. Selection, in the form of differences in survival among juveniles, causes the frequency of allele *B* to increase.

Natural selection is a potent force of evolution in Mendelian populations. Alleles that confer high reproductive success become more common over time.

We used strong selection to make a point in our numerical example. Rarely does selection in nature produce a change in allele frequency this large in a single generation. If selection continues for many generations, however, small changes in allele frequency in each generation can add up to substantial changes over the long run. Figure 5.6 illustrates the cumulative change in allele frequencies that can be wrought by selection. The figure is based on a model population similar to the one we used in the numerical example, except that the initial allele frequencies are 0.01 for *B* and 0.99 for *B'*. The heavy solid line shows the change in allele frequencies under our strong selection scheme. The frequency of allele *B* rises from 0.01 to 0.99 in less than 100 generations. Under weaker selection schemes the frequency of *B* rises more slowly, but still inexorably. (See Box 5.2 for a general algebraic treatment incorporating selection into the Hardy-Weinberg analysis.)

Empirical Research on Allele Frequency Change by Selection

Douglas Cavener and Michael Clegg (1981) documented a cumulative change in allele frequencies over many generations in a laboratory-based natural selection experiment on the fruit fly *(Drosophila melanogaster)*. Fruit flies, like most other animals, make an enzyme that breaks down ethanol, the poisonous active ingredient in beer, wine, and rotting fruit. This enzyme is called alcohol dehydrogenase, or ADH. Cavener and Clegg worked with populations of flies that had two alleles at the *Adh* locus: *Adh^F* and *Adh^S*. (The *F* and *S* refer to whether the protein encoded by the allele moves quickly or slowly through an electrophoresis gel [see Box 9.1 in Chapter 9]).

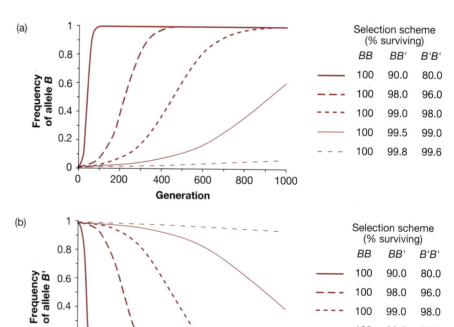

Figure 5.6   Persistent selection can produce substantial changes in allele frequencies over time Each curve shows the change in allele frequency over time under a particular selection intensity. Graph (a) shows the frequency of allele *B* over time, when *B* is selected for as shown in the key at right. Graph (b) shows the same events, except this time they are plotted as the change in the frequency of allele *B'*. The frequencies of *B* and *B'* sum to one, so the frequency of *B'* falls as the frequency of *B* rises.

Box 5.2 A general treatment of selection

We return to our population with a gene pool in which allele A_1 is at frequency p and allele A_2 is at frequency q. We draw and pair gametes at random to make offspring of genotypes A_1A_1, A_1A_2, and A_2A_2 at frequencies p^2, 2pq, and q^2.

We incorporate selection by imagining that A_1A_1 individuals survive at rate w_{11}, A_1A_2 individuals survive at rate w_{12}, and A_2A_2 individuals survive at rate w_{22}. All individuals that survive produce the same number of offspring. Therefore a genotype's survival rate is proportional to the genotype's lifetime reproductive success, or fitness. We thus refer to the survival rates as fitnesses. The average fitness for the whole population, $\overline{w}$, is $p^2 w_{11} + 2pq\, w_{12} + q^2\, w_{22}$. [To see this, note that we can calculate the average of the numbers 1, 2, 2, and 3 as $\frac{(1 + 2 + 2 + 3)}{4}$ or as $(\frac{1}{4} \times 1) + (\frac{1}{2} \times 2) + (\frac{1}{4} \times 3)$. Our expression for the average fitness is of the second form: We multiply the fitness of each genotype by its frequency in the population and then sum the results.]

We now calculate the genotype frequencies after selection (right before this generation's gametes go into the gene pool). The new frequency of A_1A_1 individuals is $p^2 w_{11}/\overline{w}$; the new frequency of A_1A_2 individuals is $2pq w_{12}/\overline{w}$; and the new frequency of A_2A_2 individuals is $q^2 w_{22}/\overline{w}$. (We have to divide by the average fitness in each case to ensure that the new frequencies still sum to 1.)

The next step is to calculate the allele frequencies in this new generation's gene pool:

- For the A_1 allele: A_1A_1 individuals contribute $p^2 w_{11}/\overline{w}$ of the gametes, all of them A_1. A_1A_2 individuals contribute $2pq w_{12}/\overline{w}$ of the gametes,

half of them A_1. So the new frequency of A_1 is $(p^2 w_{11} + pqw_{12})/\overline{w}$.

- For the A_2 allele: A_1A_2 individuals contribute $2pqw_{12}/\overline{w}$ of the gametes, half of them A_2. A_2A_2 individuals contribute $q^2 w_{22}/\overline{w}$ of the gametes, all of them A_2. So the new frequency of A_2 is $(pqw_{12} + q^2 w_{22})/\overline{w}$.

Readers should confirm that the new frequencies of A_1 and A_2 sum to 1. Also note that if we think of selection in terms of differences in reproduction rather than survival, and defer its incorporation until the production of gametes, the end result is the same for the new allele frequencies.

The change in the frequency of allele A_1 from one generation to the next is the new frequency of allele A_1 minus the old frequency of A_1:

$$\Delta p = [(p^2 w_{11} + pqw_{12})/\overline{w}] - p =$$
$$(p^2 w_{11} + pqw_{12} - p\overline{w})/\overline{w} =$$
$$(p/\overline{w})(pw_{11} + qw_{12} - \overline{w})$$

The final expression is a useful one, because it shows that the change in frequency of allele A_1 is proportional to $(pw_{11} + qw_{12} - \overline{w})$. The expression $(pw_{11} + qw_{12} - \overline{w})$ is equal to the average fitness of allele A_1 when paired at random with other alleles $(pw_{11} + qw_{12})$ minus the average fitness of the population. In other words, if the average A_1-carrying individual has higher-than-average fitness, then allele A_1 will increase in frequency.

The change in the frequency of allele A_2 from one generation to the next is

$$\Delta q = [(pqw_{12} + q^2 w_{22})/\overline{w}] - q =$$
$$(q/\overline{w})(pw_{12} + qw_{22} - \overline{w})$$

The scientists maintained two experimental populations of flies on food spiked with ethanol, and two control populations of flies on normal, nonspiked food. The researchers picked the breeders for each generation at random. This is why we are calling the project a natural selection experiment: Cavener and Clegg set up different environments for their different populations, but the researchers did not themselves directly manipulate the survival or reproductive success of individual flies.

Every several generations Cavener and Clegg took a random sample of flies from each population, determined their *Adh* genotypes, and calculated the allele frequencies. The results appear in Figure 5.7. The control populations showed

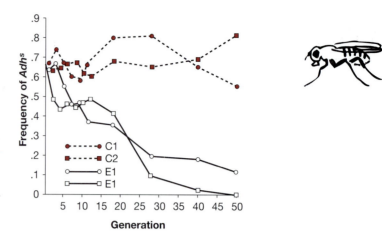

Figure 5.7 Frequencies of the *Adh^S* allele in four populations of fruit flies over 50 generations The filled symbols and dashed lines represent control populations living on normal food; the open symbols and solid lines represent experimental populations living on food spiked with ethanol. From Cavener and Clegg (1981).

Researchers studying allele frequencies in lab populations have found evidence that natural selection causes evolution.

no large or consistent long-term change in the frequency of the *Adh^S* allele. The experimental populations, in contrast, showed a rapid and largely consistent decline in the frequency of *Adh^S* (and, of course, a corresponding increase in the frequency of *Adh^F*). Hardy-Weinberg conclusion 1 appears to hold in the control populations, but is clearly not in force in the experimental populations.

Can we identify for certain which of the assumptions of the Hardy-Weinberg analysis is being violated? The only difference between the two kinds of populations is that the experimentals have ethanol in their food. This suggests that it is the assumption of no selection that is being violated in the experimental populations. Flies with the *Adh^F* allele appear to have higher lifetime reproductive success (higher fitness) than flies with the *Adh^S* allele when ethanol is present in the food. Cavener and Clegg note that this outcome is consistent with the fact that alcohol dehydrogenase extracted from *Adh^F* homozygotes breaks down ethanol at twice the rate of alcohol dehydrogenase extracted from *Adh^S* homozygotes. Whether flies with the *Adh^F* allele have higher rates of survival or produce more offspring is unclear.

Christine Chevillon and colleagues (1995) recently reported evidence for a similar effect of selection—a change in the frequencies of alleles for a protein of known function—in free-living populations of a mosquito *(Culex pipiens)*. The protein Chevillon et al. studied is acetylcholinesterase (ACE). ACE is an enzyme that breaks down the neurotransmitter acetylcholine after acetylcholine has carried a nerve impulse across a synapse. ACE is the target of the insecticide chlorpyrifos. Chlorpyrifos prevents ACE from breaking down acetylcholine, resulting in overexcitation of the nervous system and eventual death.

Researchers studying allele frequencies in wild populations have found evidence that natural selection causes evolution.

Chevillon et al. chose ACE because they knew that an allele called *Ace^R* makes the mosquitoes that carry it resistant to chlorpyrifos. Chevillon et al. also knew that chlorpyrifos has been used more extensively in some geographic areas than in others. The scientists caught—and determined the *Ace* genotypes of—a sample of mosquitoes at each of nine locations along a 54-km transect perpendicular to the Mediterranean coast of France. Chlorpyrifos had been used to control mosquitoes at the first four collection sites on the transect for 22 years, but had never been used at the last five sites. Chevillon et al. predicted that the frequency of the *Ace^R* allele would be higher where the mosquitoes had been exposed to chlorpyrifos than where they had not. In other words, the researchers predicted that the frequency of the Ace^R allele would decrease along the transect in response to violation of the no-selection assumption of the Hardy-Weinberg analysis. As controls, Chevillon et al. also measured the frequencies of alleles for five other mosquito enzymes with no known role in resistance to chlorpyrifos. The researchers predicted that allele frequencies for these other enzymes would show no systematic changes along the transect.

The results confirmed Chevillon et al.'s predictions exactly (Figure 5.8). Only *Ace* showed a signficant change in allele frequencies along the transect, with Ace^R at higher frequency where chlorpyrifos had been used. Chevillon et al. concluded that selection due to chlorpyrifos had caused an increase in Ace^R frequency in the exposed populations.

Adding Selection to the Hardy-Weinberg Analysis: The Calculation of Genotype Frequencies

The calculations and examples we just discussed show that selection can cause allele frequencies to change across generations. Selection invalidates conclusion 1 of the Hardy-Weinberg analysis. We now consider how selection affects conclusion 2 of the Hardy-Weinberg analysis. In a population under selection, can we still calculate the genotype frequencies by multiplying the allele frequencies?

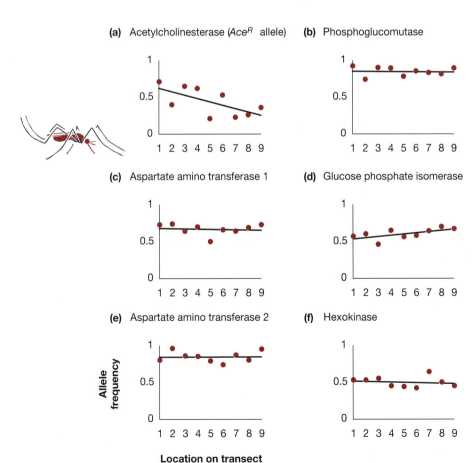

(a) Acetylcholinesterase (Ace^R allele)

(b) Phosphoglucomutase

(c) Aspartate amino transferase 1

(d) Glucose phosphate isomerase

(e) Aspartate amino transferase 2

(f) Hexokinase

Allele frequency

Location on transect

Figure 5.8 **Allele frequecies of six mosquito enzyme genes along a transect** The dots in graph (a) show the frequencies of the Ace^R allele at nine sites along a geographic transect. The graph also includes the best-fit line for allele frequency versus site number. The other graphs show the frequencies of the most common allele of each of five other mosquito enzymes, and their best-fit lines. Only the Ace^R allele shows a systematic change in frequency along the transect ($P < 0.05$); Ace^R is at higher frequency in populations exposed to chlorpyrifos (sites 1−4) than in populations not exposed to the insecticide (sites 5−9). Graphs drawn from data presented in Chevillon et al. (1995).

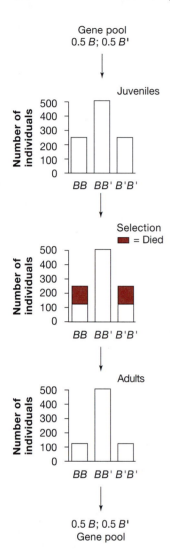

Gene pool
0.5 *B*; 0.5 *B'*

Juveniles

Selection
■ = Died

Adults

0.5 *B*; 0.5 *B'*
Gene pool

Figure 5.9 **Selection can change genotype frequencies so that they cannot be calculated by multiplying the allele frequencies**

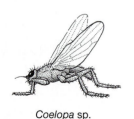

Coelopa sp.
Figure 5.10 **A seaweed fly**

We often cannot. As before, we use a population with two alleles at a locus affecting survival: *B* and *B'*. We assume that the initial frequency of each allele in the gene pool is 0.5. After random mating, we get genotype frequencies for *BB*, *BB'*, and *B'B'* of 0.25, 0.5, and 0.25. Again, we imagine that the population consists of 1,000 individuals: 250 *BB*, 500 *BB'*, and 250 *B'B'*.

Now we incorporate selection by stipulating that the genotypes differ in survival to adulthood such that all of the heterozygotes survive, but only half of each kind of homozygote survives. This kind of selection is known as heterozygote superiority, or heterosis. What are the genotype frequencies among the surviving adults? The population now contains 750 individuals: 125 *BB*, 500 *BB'*, and 125 *B'B'*. The new frequency of *BB* is $\frac{125}{750} = 0.167$; the new frequency of *BB'* is $\frac{500}{750} = 0.667$; and the new frequency of *B'B'* is $\frac{125}{750} = 0.167$. Notice that the allele frequencies for *B* and *B'* are both still 0.5. Selection has invalidated conclusion 2 of the Hardy–Weinberg analysis: We can no longer accurately predict the genotype frequencies by multiplying the allele frequencies (Figure 5.9). There are fewer homozygotes and more heterozygotes than expected. (See Box 5.2 for a general algebraic treatment of the effect of selection on the Hardy–Weinberg analysis.)

Heterozygote superiority is not the only kind of selection that disrupts Hardy–Weinberg genotype frequencies. At the end of our first numerical example for selection (Figure 5.5), we also could not calculate the exact genotype frequencies from the allele frequencies.

We used strong selection in our heterozygote superiority example to make a point. In fact, selection is rarely strong enough to produce, in a single generation, a detectable violation of Hardy–Weinberg conclusion 2. Even if it does, a single bout of random mating will immediately put the genotypes back into Hardy–Weinberg equilibrium. Nonetheless, a few studies have found violations of Hardy–Weinberg conclusion 2 that seem to be the result of selection.

Empirical Research on Selection and Genotype Frequencies

One such example is summarized in Table 5.1. R. K. Butlin and colleagues (1982) collected seaweed flies (Figure 5.10) from seven locations on the English coast in different years, and determined the flies' genotypes for alleles of the alcohol dehydrogenase gene. Nearly every population sample showed an excess of *BD* heterozygotes, and a deficit of *BB* and *DD* homozygotes, compared to Hardy–Weinberg expectations (see Box 5.3). Thus at least one of the Hardy–Weinberg assumptions is being violated. Can we tell which assumption it is?

The seven sites Butlin et al. surveyed are widely separated and almost certainly isolated from each other. Thus exchange of genes among the populations (see section 5.4) is unlikely to adequately explain the uniformity of the pattern from place to place and year to year. Because the pattern is so consistent, Butlin et al. believe that natural selection is responsible. Their hypothesis is that *BD* heterozygotes survive at a higher rate, or have more offspring, than *BB* and *DD* homozygotes. The alcohol dehydrogenase gene is embedded in a chromosomal inversion, so it is not clear whether selection is favoring alcohol dehydrogenase heterozygotes *per se,* heterozygotes at a nearby locus, or heterozygotes for a supergene in the inversion (see Chapter 4).

TABLE 5.1 Alcohol dehydrogenase genotypes in populations of seaweed flies

Butlin et al. (1982) collected seaweed flies *(Coelopa frigida)* in different years at seven different sites on the English coast. *BB, BC, BD, CC, CD,* and *DD* are genotypes for alcohol dehydrogenase (ADH). The number for each population in each genotype column indicates the number of individuals of that genotype found; the + or − sign indicates whether there were more or fewer of that genotype than expected under Hardy-Weinberg equilibrium. χ^2 values for each population are given in the second column from the right (see Box 5.3). The last column gives the statistical significance of each test (N.S. = not significant).

Comparison of **Adh** *genotypes with Hardy-Weinberg expectations*

Site	Year	BB	BC	BD	CC	CD	DD	χ^{2*}	P
Beer	1974	12−	9−	64+	9+	17−	35−	9.0	< 0.05
	1979	25−	17+	160+	3+	17−	59−	28.6	< 0.01
Portland	1974	10−	5−	50+	4+	16+	30−	5.0	N.S.
	1979	33−	15−	119+	0−	28+	58−	6.5	< 0.05
	1980 (pre-adult sample)	41−	18−	172+	8+	23−	92−	10.8	< 0.01
	1980 (adult sample)	104−	36−	334+	12+	50−	110−	30.5	< 0.01
Flamborough	1975	23+	10−	69+	3+	20+	52−	0.7	N.S.
	1979	44+	30−	96+	11+	33−	51+	1.0	N.S.
Robin Hood's Bay	1975	25−	11−	97+	4+	23+	56−	4.1	N.S.
	1979	35−	47+	117+	7+	25−	42−	15.2	< 0.01
St. Mary's Island	1975	42−	13−	188+	8+	16−	87−	16.7	< 0.01
	1979	21−	19−	118+	6+	42+	62−	12.1	< 0.01
	1980 (pre-adult sample)	48−	23−	143+	9+	35−	57−	8.7	< 0.05
	1980 (adult sample)	150−	104−	424+	18+	129+	198−	8.6	< 0.05
Rustington	1976	48−	9−	206+	8+	31−	127−	13.2	< 0.01
	1979	35−	18−	147+	1−	23+	57−	13.9	< 0.01
Barn's Ness	1980 (pre-adult sample)	92−	31−	286+	15+	44−	98−	25.9	< 0.01
	1980 (adult sample)	95−	37−	244+	11+	36−	92−	10.0	< 0.01

* The χ^2 values are calculated from five genotypes. The *CC* data are omitted because most populations have expectations of fewer than five *CC* individuals.

Table 5.2 summarizes a second example in which selection may be creating an excess of heterozygotes. Therese Markow and colleagues (1993) determined the genotypes of 122 Havasupai people (a group of Native Americans that lives in Arizona) at two loci in the major histocompatibility complex (MHC). These genes, *HLA-A* and *HLA-B,* are important in immune system function. At both loci, Markow and colleagues documented a significant excess of heterozygotes compared to Hardy-Weinberg expectations. Two explanations are possible. One is that Havasupai individuals are choosing mates with HLA genotypes different than their own (see section 5.6). Claus Wedekind and colleagues (1995) have evidence that humans can distinguish MHC genotypes based on smell, and may prefer mates whose MHC genotypes are different (see Chapter 18). The other explanation is that Havasupai fetuses, children, and/or adults have a higher probability of survival if they are heterozygous at *HLA-A* and *HLA-B.* This would be consistent with work by Ober et al. (1992), who found increased rates of fetal loss in human couples homozygous at selected MHC loci. In other words, the Hardy-Weinberg assumption being violated in the Havasupai could be either the random-mating assumption or the no-selection assumption. We challenge the reader to design a study that could determine which assumption it is.

BOX 5.3 Generating expected values using the Hardy-Weinberg Principle, and comparing observed and expected values using the χ^2 (Chi-square) test

Here we use a small subset of the data in Table 5.1 to illustrate one method for determining whether genotype frequencies deviate significantly from Hardy-Weinberg equilibrium. The first line of the table reports that in 1974 at Beer, England, Butlin and colleagues (1982) collected 12 seaweed flies with the genotype *BB*, 64 flies with the genotype *BD*, and 35 flies with the genotype *DD*. We will take these 111 flies as the entire sample, ignoring the flies with other genotypes. From the numbers in our reduced dataset, we will calculate the frequencies of alleles *B* and *D* and determine whether the genotype frequencies observed are those we would expect according to the Hardy-Weinberg equilibrium principle. There are five steps:

1. Calculate the allele frequencies. The sample of 111 flies is also a sample of 222 alleles. All 24 of the alleles carried by the 12 *BB* flies are *B*, as are 64 of the alleles carried by the *BD* flies. Thus the frequency of allele *B* is $\frac{(24 + 64)}{222} = 0.396$. The frequency of allele *D* is $\frac{(64 + 70)}{222} = 0.604$.

2. Calculate the genotype frequencies expected under the Hardy-Weinberg principle, given the allele frequencies calculated in step 1. According to the Hardy-Weinberg equilibrium principle, if the frequencies of two alleles are p and q, then the frequencies of the genotypes are p², 2pq, and q². Thus the expected frequencies of genotypes *BB*, *BD*, and *DD* are $0.396^2 = 0.157$; $2 \times 0.396 \times 0.604 = 0.478$; and $0.604^2 = 0.365$.

3. Calculate the expected number of individuals of each genotype under Hardy-Weinberg equilibrium. This is simply the expected frequency of each genotype multiplied by the number of flies, 111, in the sample. The expected values are 17.4 *BB* flies; 53.1 *BD* flies; and 40.5 *DD* flies. The numbers of flies expected are different from the number of flies actually observed (12, 64, and 35). The actual sample contains more heterozygotes and fewer homozygotes than expected. Is it plausible that this large a difference between expectation and reality could arise by chance? Or is the difference statistically significant? Our null hypothesis is that the difference is simply due to chance.

4. Calculate a test statistic. We will use a test statistic that was devised in 1900 by Karl Pearson. It is called the chi-square (χ^2). The chi-square is defined as follows:

$$\chi^2 = \sum \frac{(\text{observed} - \text{expected})^2}{\text{expected}}$$

where the symbol Σ indicates a sum taken across all the classes considered. In our data there are three classes: the three genotypes. For our dataset:

$$\chi^2 = \frac{(12 - 17.4)^2}{17.4} + \frac{(64 - 53.1)^2}{53.1} + \frac{(35 - 40.5)^2}{40.5} = 4.66$$

5. Determine whether the value of the test statistic is significant. The chi-square is defined in such a way that the chi-square gets larger as the difference between the observed and expected values gets larger. How likely is it that we could get a chi-square as large as 4.66 by chance? Most statistical textbooks have a table that provides the answer. In Zar (1984) this table is called "Critical values of the chi-square distribution."

To use this table, we need to calculate a number called the degrees of freedom for the test statistic. The number of degrees of freedom for the chi-square is equal to the number of classes minus the number of independent values we calculated from the data for use in determining the expected values. For our chi-square there are three classes: the three genotypes. We calculated two values from the data for use in determining the expected values: the total number of flies, and the frequency of allele *B*. (We also calculated the frequency of allele *D*, but it is not independent of the frequency of allele *B*, because the frequency of *D* is one minus the frequency of *B*). Thus the number of degrees of freedom is 1. (Another formula for the degrees of freedom in chi-square tests for Hardy-Weinberg equilibrium is df = k − 1 − m, where k = the number of classes, and m = the number of independent allele frequencies estimated from the data).

According to the table in the statistics book, the critical value of chi-square for 1 degree of freedom and P = 0.05 is 3.841. This means that there is a 5% chance under the null hypothesis of getting chi-square ≥ 3.841. The probability under the null hypothesis of getting chi-square ≥ 4.66 is therefore less than 5%. We reject the null hypothesis, and assert that our value of chi-square is statistically significant at P < 0.05.

The chi-square test tells us that the alleles of the *Adh* locus in the seaweed fly population are not at Hardy-Weinberg equilibrium. This indicates that one or more of the assumptions of the Hardy-Weinberg analysis has been violated. It does not, however, tell us *which* assumptions are being violated, or how.

Different Patterns of Selection Have Different Effects on Allele Frequencies

The first pattern of selection we considered had one allele favored over the other. In our first numerical calculation (Figure 5.5), for example, allele *B* conferred higher rates of survival than allele *B'*. Over time, this led to an increase in the frequency of *B* and a decrease in the frequency of *B'* (Figure 5.6). Ultimately *B'* was lost from the population. After *B'* is lost, the population is said to be fixed for allele *B*. The same pattern of selection is apparent in the experimental populations in the fruit fly study by Cavener and Clegg (Figure 5.7). This pattern of selection tends to reduce the amount of genetic variation among individuals in a population.

The second pattern of selection we considered was one in which heterozygotes were favored over either homozygote. In our second numerical calculation (Figure 5.9), for example, *BB'* individuals had higher rates of survival than either *BB* or *B'B'* individuals. This did not alter the frequencies of the alleles; instead, it tended to hold the frequencies of both alleles at intermediate levels. The same pattern of selection may be occurring in the Havasupai. Heterozygote superiority tends to maintain genetic variation among individuals in a population.

TABLE 5.2 Genotypes at two HLA loci in the Havasupai Native Americans

This table gives the number observed and the number expected with each genotype in a sample of 122 people. From Markow et al. (1993).

	Observed	Expected	Statistical significance
HLA-A			
Homozygotes	38	48.43	
Heterozygotes	84	73.57	P = 0.0317
HLA-B			
Homozygotes	21	29.56	
Heterozygotes	101	92.44	P = 0.0405

Although natural selection is a potent force of evolution, it can also be a potent force preventing evolution.

Many other patterns of selection are possible. Our final example in this section concerns a pattern called frequency-dependent selection. Like heterozygote superiority, frequency-dependent selection tends to promote the long-term stability of allele frequencies and maintain genetic variation. Frequency-dependent selection does so, however, by first favoring one allele and then favoring the other. In our example of frequency-dependent selection, the rarest genotypes always enjoy the highest fitness—precisely because they are rare.

Frequency-Dependent Selection in Scale-Eating Fish

Our example of frequency-dependent selection comes from Michio Hori's research (1993) on the scale-eating fish, *Perissodus microlepis,* in Africa's Lake Tanganyika. As its name suggests, the scale-eating fish makes its living by biting scales off other (live) fish. The scale-eater attacks from behind, grabbing scales off the victim's flank. Stranger still, within the species *P. microlepis* there are right-handed (dextral) fish, whose mouths are twisted to the right, and left-handed (sinistral) fish, whose mouths are twisted to the left (Figure 5.11). Hori showed that to a first approximation, handedness is determined by a single locus with two alleles. Right-handedness is dominant over left-handedness.

Hori observed attacks on prey fish used as lures, and in addition examined scales recovered from the stomachs of scale-eating fish. These observations show that right-handed fish always attack their victim's left flank, and left-handed fish always attack their victim's right flank. (Readers can visualize the reason for this by first twisting their lips to the right or to the left, then imagining trying to bite scales off the flank of a fish.) The prey species are wary and alert, and scale eaters are successful in only about 20% of their attacks.

Hori reasoned that if right-handed scale eaters were more abundant than left-handers, then prey species would be more vigilant for attacks from the left. This would give left-handed scale eaters, who attack from the right, an advantage in their efforts to catch their prey unawares. Left-handed scale eaters would get more food than right-handers, have more offspring, and pass on more of their left-handed genes. This would increase the frequency of left-handed scale eaters in the population.

After left-handed scale eaters had become more abundant than right-handers, then prey fish would start to be more vigilant for attacks from the right.

Figure 5.11 **Scale-eating fish (Perissodus microlepis) from Lake Tanganyika**   (top) A right-handed (dextral) fish, shown from both sides. The mouth of this fish twists to the fish's right. (bottom) A left-handed (sinistral) fish, shown from both sides. The mouth of this fish twists to the fish's left. These two individuals belong to the same species and are members of the same population (Dr. Michio Hori).

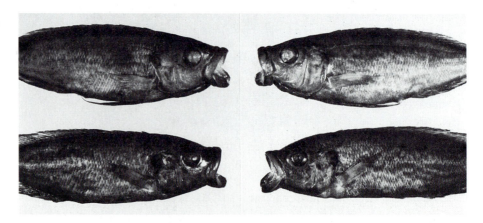

This would give right-handed scale eaters, who attack from the left, an advantage. The right-handers would get more food and have more offspring, and the frequency of right-handers in the population would increase.

The result is that left-handed fish and right-handed fish should be just about equally abundant in the population at any given time. The dominant right-handed allele should be held at a frequency of just under 0.3, and the recessive left-handed allele at a frequency of just over 0.7 (Why?). This phenomenon, in which the rare phenotype is favored by natural selection, is an example of frequency-dependent selection.

Hori's fish story is an elegant hypothesis, but is it true? Hori used a gill net to catch a sample of scale-eating fish in Lake Tanganyika every year or two for 11 years. The frequencies of the two phenotypes indeed appear to oscillate around 0.5 for each (Figure 5.12). In some years slightly more than half the fish were left-handed, but always a year or two later the pendulum had swung back, and slightly more than half the fish were right-handed.

Is it possible that the frequencies of the two forms were just drifting at random rather than being held near 0.5 by selection? In three different years Hori examined adult fish that were known to be breeding because they were caught in the act (probably of brooding their young) by scuba-diving scientists. The frequencies of handedness in these samples oscillated too, but the most abundant handedness among the breeders was always the opposite of the most abundant handedness in the population as a whole (the breeders are indicated by the squares in Figure 5.12). In other words, at any given time the rare form seems to be doing more breeding. This is consistent with Hori's frequency-dependent selection hypothesis.

Furthermore, Hori has evidence to support his hypothesis about the mechanism of selection. He examined bite wounds on a prey species in 1980, when left-handed scale eaters were more common, and in 1983, when right-handed scale eaters were more common. In both years, the prey had more bite marks inflicted by whichever of the two forms was rarer at the time

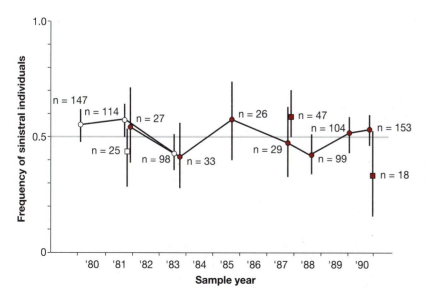

Figure 5.12 **Frequency of left-handed (sinistral) fish over time** Circles represent scale-eating fish captured with a gill net; filled versus open circles represent fish caught at different locations. Squares (in 1981, 1987, and 1990) represent actively breeding adults, selectively captured by scuba-diving scientists. The numbers (n) report the sample sizes. From Hori (1993).

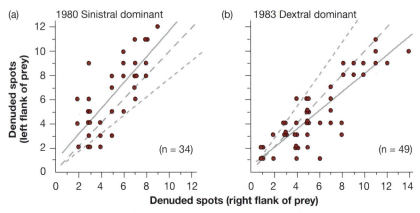

***Figure 5.13* Numbers of bite marks on the prey of scale-eating fish**   Each graph shows the number of bite marks on the left flank versus the right flank. The solid line in each graph is the best-fit line through the actual data. The long-dashed line shows where the best-fit line would be if the victims had equal numbers of bite marks on both sides. The short-dotted line shows where the best-fit line would be if the victims had numbers of bite marks on each side in proportion to the abundance in the lake of scale eaters that attack from that side.   (a) In 1980, when there were more left-handed (sinistral) scale eaters in the lake, victims had significantly more bite marks on their left flanks ($P < 0.001$). Right-handed fish, though rarer, made successful attacks more often.   (b) In 1983, when there were more right-handed (dextral) scale eaters in the lake, victims had more bite marks on their right flanks ($P < 0.001$). Left-handed fish, though rarer, made successful attacks more often. From Hori (1993).

(Figure 5.13). This is consistent with the idea that the prey are indeed more vigilant for attacks from whichever form of scale eater is more common. It appears that Hori's hypothesis is correct, and that over the long term the allele, genotype, and phenotype frequencies in scale-eating fish are being held constant by frequency-dependent selection.

In summary, selection is a difference in lifetime reproductive success among individuals with different phenotypes. Such a difference in lifetime reproductive success, or fitness, can result from a difference in survival, or reproductive output, or both. When particular phenotypes are associated with particular genotypes, selection can disrupt the Hardy-Weinberg genotype frequencies and cause changes in allele frequencies across generations. Some patterns of selection tend to decrease genetic variation; other forms of selection tend to maintain genetic variation. The ultimate source of genetic variation is mutation. We now ask whether mutation can also function as a mechanism of evolution.

5.3   Mutation

In Chapter 4 we presented mutation as the source of all new alleles and genes. In its capacity as the ultimate source of all genetic variation, mutation provides the raw material for evolution. Here we consider the importance of mutation as a force of evolution. How effective is mutation at changing allele frequencies over time? How strongly does mutation affect the conclusions of the Hardy-Weinberg analysis?

Adding Mutation to the Hardy-Weinberg Analysis: Mutation as an Evolutionary Force

Mutation by itself is generally not a potent evolutionary force. To demonstrate why, we go back to our imaginary gene pool with two alleles: A_1 and A_2. Imagine that A_1 is the wild-type allele and A_2 is a loss-of-function mutation. Let μ be the rate (per generation) at which A_1 is converted to A_2. Back mutations that restore function are much less common than loss-of-function mutations, so we will ignore mutations that convert A_2 to A_1. Furthermore, imagine that the mutations all happen while the gametes are mixed together in the gene pool. If the frequency of A_1 was p when the gametes went into the gene pool, then the frequency of A_1 when we draw the gametes out is $p* = p - \mu p$. Likewise, if the frequency of A_2 was q when the gametes went into the gene pool, then the frequency of A_2 when we draw the gametes out is $q* = q + \mu p$. Note that $p*$ and $q*$ still add up to 1. If we calculate the genotype frequences for the next generation, they come out to $(p - \mu p)^2$ for A_1A_1; $2(p - \mu p)(q + \mu p)$ for A_1A_2; and $(q + \mu p)^2$ for A_2A_2.

Among the highest mutation rates reported for any gene in any organism is $\mu = 0.0007$ per generation (See Table 4.2 in Chapter 4). Thus to a first approximation, in any real population $\mu p = 0$. Substituting 0 for μp in every expression in the previous paragraph gives allele frequencies of p and q and genotype frequencies of p^2, $2pq$, and q^2. Thus from one generation to the next, mutation has virtually no effect on either conclusion 1 or conclusion 2 of the Hardy-Weinberg analysis.

But virtually no effect is not the same as exactly no effect. Could mutation of A_1 to A_2, at the rate of μ every generation for many generations, eventually result in an appreciable change in allele frequencies? The change in the frequency of A_1 produced by mutation across a single generation is the new allele frequency minus the old allele frequency:

$$\Delta p = p* - p = (p - \mu p) - p = -\mu p.$$

This equation shows that the decrease in the frequency of A_1 alleles is proportional to both the mutation rate and the proportion of A_1's that are still around to mutate.

How fast does the change in allele frequencies proceed? After n generations, the frequency of A_1 is approximately

$$p_n = p_0 e^{-\mu n}$$

where e is the base of the natural logarithms, p_0 is the frequency of A_1 in generation zero, and p_n is the frequency of A_1 in generation n. [Readers who have taken calculus can derive this relationship by starting with $dp/dg = -\mu p$, where dg represents one generation, which we will take to be an infinitesimal length of time. Dividing both sides by p and multiplying both sides by dg gives $(1/p)dp = -\mu \, dg$. Now integrate the left side from frequency p_0 to p_n and the right side from generation 0 to n. Solve the resulting equation for p_n.]

We now assume that the mutation rate of A_1 to A_2 is 2×10^{-6} per generation. This is the highest value among the mutation rates reported for viral, bacterial, algal, or fungal genes in Table 4.2 (Chapter 4). After 5,000 generations of change from a starting point of $p = 1$ and $q = 0$, the frequency of A_1 is

$$p_{5,000} = (1)e^{(-0.000002)(5,000)} = e^{(-0.01)} = 0.99$$

In principle, mutation can be a force of evolution. In practice, mutation is usually not, by itself, a potent force of evolution.

Ninety-nine percent of the alleles in the population are still A_1. For most genes, mutation by itself is inefficient at changing allele frequencies. (Many of the corn and human genes whose mutation rates are reported in Table 4.2 of Chapter 4 appear to have been selected for study because their mutation rates are unusually high. Readers who calculate $p_{5,000}$ for these genes will discover that in some cases the mutation rate is high enough that mutation by itself, given time, can cause substantial change in gene frequencies.)

Mutation and Selection

Although mutation by itself usually cannot cause appreciable changes in allele frequencies, this does not mean that mutation is unimportant in evolution. In combination with selection (see section 5.2), mutation becomes a potent evolutionary force. This point is demonstrated by a recent experiment conducted in Richard Lenski's lab (Lenski and Travisano 1994; Elena et al. 1996). Lenski and co-workers studied the evolution of a strain of *Escherichia coli* that is incapable of recombination (here, recombination means conjugation and exchange of DNA among cells). For *E. coli* populations of this strain, mutation is the only source of genetic variation. The researchers started 12 replicate populations with single cells placed in a glucose-limited, minimal salts medium—a demanding environment for these bacteria. After allowing each culture to grow to about 5×10^8 cells, Lenski and colleagues removed an aliquot (containing approximately 5 million cells) and transferred it to fresh medium. The researchers performed these transfers daily for 1,500 days, or approximately 10,000 generations.

At intervals throughout the experiment, the researchers froze a sample of the cells being transferred for later analysis. Because *E. coli* are preserved but not killed by freezing, Lenski and colleagues could take ancestors out of the freezer and grow them up in a culture flask with an equivalent number of cells from descendant populations. These experiments allowed the team to directly measure the relative fitness of ancestral and descendant populations, as the growth rate of each under competition. In addition to monitoring changes in fitness over time in this way, the Lenski team measured cell size.

During the course of the study, both fitness and cell size increased dramatically in response to natural selection. The key point for our purposes is that these increases occurred in jumps (Figure 5.14). The step-like pattern resulted

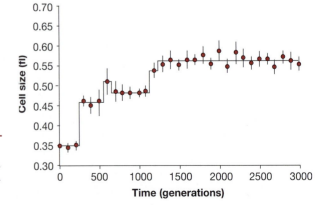

Figure 5.14 Changes over time in cell size of experimental *E. coli* populations Each point on the plot represents the average cell size in the 12 replicate populations. The vertical lines are error bars; 95% of the observations fall within the range indicated by the bars. From Elena et al. (1996).

from a simple process: the occurrence of beneficial mutations that swept rapidly through the population. Each new mutation enabled the bacteria that carried it to divide at a faster rate. The frequency of the mutants quickly increased as they out-reproduced the other members of the population. Eventually, each new mutation became fixed in the population. The time from the appearance of each mutation to its fixation in the population was so short that we cannot see it in the figure. Most of the beneficial mutations caused larger cell size. Thus the plot of cell size over time also shows abrupt jumps. During the time periods between the extremely rare occurrences of these beneficial mutations, all of the 12 replicate populations stood still. *Why* larger cells were beneficial in the nutrient-poor laboratory environment is a focus of current research.

Mutation's most important role is in supplying the variation that allows natural selection to be a potent force of evolution.

The experiment by Lenski and colleagues reinforces one of the messages of Chapter 4. Without mutation, evolution would eventually grind to a halt. Mutation is the ultimate source of genetic variation.

Mutation–Selection Balance

Unlike the mutations that led to increased cell size in Lenski et al.'s *E. coli* populations, most mutations are deleterious. Selection acts to eliminate such mutations from populations. Deleterious alleles persist, however, because they are continually created anew. When the rate at which copies of a deleterious allele are being eliminated by selection is exactly equal to the rate at which new copies are being created by mutation, the frequency of the allele is at equilibrium. This situation is called mutation-selection balance. What is the frequency of the deleterious allele at equilibrium?

Imagine that A_1 is a dominant allele and A_2 is a deleterious recessive, with frequencies of p and q. Let μ be the rate at which A_1 alleles are converted into A_2 alleles by forward (loss-of-function) mutations. We can define the relative fitness of A_1A_1 homozygotes, and A_1A_2 heterozygotes, as

$$w_{11} = w_{12} = 1$$

We can then express the reduced fitness of A_2A_2 homozygotes as

$$w_{22} = 1 - s$$

where s is called the selection coefficient. The value of s can be anything from 0 to 1. As s gets closer to 1, the relative fitness of A_2A_2 homozygotes gets closer to 0. Thus a bigger selection coefficient means stronger selection against the A_2 allele.

We assume that q, the frequency of A_2, is small; and that back mutation is negligible. Under these assumptions, the frequency of q at equilibrium is approximated by

$$\hat{q} = \sqrt{\frac{\mu}{s}}$$

(See Box 5.4 for a derivation.) This expression captures with economy what intuition tells us about the mutation-selection balance. If the selection coefficient is small (the allele is only mildly deleterious) and the mutation rate is high, then the equilibrium frequency of allele A_2 will be relatively high. If the selection coefficient is large (the allele is highly deleterious) and the mutation rate is low, then the equilibrium frequency of allele A_2 will be low.

Box 5.4 Equilibrium allele frequency under mutation-selection balance

We assume a randomly mating population, and consider a single locus with two alleles. Allele A_1 is at frequency p, and allele A_2 is at frequency q. In Box 5.2 we derived an expression for the change in q from one generation to the next as a result of selection:

$$\text{selection } \Delta q = (q/\overline{w})(pw_{12} + qw_{22} - \overline{w})$$

where $\overline{w}$ is the mean fitness, w_{12} is the fitness of genotype A_1A_2, and w_{22} is the fitness of genotype A_2A_2.

Now we add mutation. Let μ be the rate of mutation from A_1 to A_2, and ν be the rate of mutation from A_2 to A_1. The change in q from one generation to the next as a result of mutation is

$$\text{mutation } \Delta q = \mu p - \nu q$$

The total change in q from one generation to the next, due to selection and mutation, is

$$\Delta q = \text{selection } \Delta q + \text{mutation } \Delta q$$

At equilibrium, when mutation and selection exactly cancel each other, $\Delta q = 0$, meaning that

$$\text{selection } \Delta q = -(\text{mutation } \Delta q), \text{ or}$$

$$(q/\overline{w})(pw_{12} + qw_{22} - \overline{w}) = -(\mu p - \nu q) \quad \textbf{(1)}$$

This expression is completely general. It can accommodate any genetic scheme (dominant/recessive alleles, heterozygote superiority, etc.) and any mutation rates. In principle all we have to do is specify the fitness of each genotype and the mutation rates, and then we can solve for q. The resulting expression gives us the equilibrium frequency of allele A_2. In practice the algebra can get complicated, and it helps to make some simplifying assumptions.

Here we will imagine that allele A_1 is dominant, and allele A_2 is a recessive loss-of-function mutation. In accord with this scheme, we assign fitnesses as follows: $w_{11} = 1$; $w_{12} = 1$; $w_{22} = (1 - s)$. With these genotype fitnesses, the mean fitness is given by

$$\overline{w} = p^2 + 2pq + q^2(1 - s)$$

Substituting $(1 - q)$ for p and simplfying gives

$$\overline{w} = (1 - q)^2 + 2(1 - q)q + q^2(1 - s)$$
$$= 1 - 2q + q^2 + 2q - 2q^2 + q^2 - sq^2 = 1 - sq^2$$

Substituting this result and the genotype fitnesses into the left side of equation 1 and simplifying gives

$$[q/(1 - sq^2)][p + q(1 - s) - (1 - sq^2)]$$
$$= [q/(1 - sq^2)][1 - q + q - sq - 1 + sq^2]$$
$$= [q/(1 - sq^2)][-sq + sq^2]$$
$$= [q/(1 - sq^2)][sq(-1 + q)]$$
$$= [q/(1 - sq^2)][sq(-p)]$$
$$= (-spq^2)/(1 - sq^2)$$

Equation 1 is thus now

$$(-spq^2)/(1 - sq^2) = -(\mu p - \nu q) \quad \textbf{(2)}$$

At this point, we make two simplifying assumptions that will make it easy to solve for q. First, we assume that q is small. If q is small, as would be the case if A_2 is strongly selected against, then $(1 - sq^2)$ is approximately equal to 1. Second, we assume that back-mutations from A_2 to A_1 occur rarely enough that we can ignore them. This amounts to assuming that $\nu = 0$. With these assumptions, we have

$$-spq^2 = -\mu p$$

Dividing both sides by $(-p)$ and solving for q gives the equilibrium frequency of A_2 under mutation-selection balance:

$$\hat{q} = \sqrt{\frac{\mu}{s}}$$

If A_2 is lethal, then $s = 1$, and the equilibrium frequency of A_2 is the square root of the mutation rate.

As an exercise, readers may want to consider the case in which A_2 is a deleterious dominant. Under this scenario, the analog of equation 2 is

$$(-sqp^2)/(1 - s + sp^2) = -(\mu p - \nu q)$$

If we assume that (1) A_2 is rare, so that p and p^2 are approximately equal to 1; and (2) back-mutation is negligible, then we get

$$\hat{q} = \frac{\mu}{s}$$

If A_2 is lethal, then $s = 1$, and the equilibrium frequency of A_2 is equal to the mutation rate.

This analysis of the mutation-selection balance can provide insights into human genetic diseases. Joseph Felsenstein (1997) considered the population genetics of cystic fibrosis. Cystic fibrosis is caused by recessive alleles of the gene, on chromosome 7, for a protein called the cystic fibrosis transmembrane conductance regulator (Klug and Cummings 1997). This protein regulates the movement of chloride ions across the cell membrane. Individuals homozygous for recessive alleles produce an excess of thick mucus in their lungs. Until recently most such individuals died before reaching reproductive age. About 1 in 2,000 people of northern European ancestry have cystic fibrosis.

Can we explain the facts about cystic fibrosis using the Hardy-Weinberg principle and the mutation-selection balance? Let c represent the recessive cystic fibrosis allele, and let q represent its frequency. We know that the frequency of cc individuals is 1/2,000 or 0.0005. If we assume that the cystic fibrosis alleles are in Hardy-Weinberg equilibrium, then

$$q = \sqrt{0.0005} = 0.0224$$

If we assume that c is maintained at this frequency by mutation-selection balance, and if we assume that s = 1 (meaning that cc individuals have zero fitness), then we have

$$0.0224 = \sqrt{\frac{\mu}{1}}$$

Squaring both sides gives $\mu = 0.0005$. This would be an unusually high mutation rate. The fact that the data and our assumptions imply a mutation rate this high suggests that our assumptions may be incorrect. Either our characterization of the selection scheme is incorrect (perhaps there is heterozygote superiority), or the alleles are not in mutation-selection balance (the frequency of c could be high because of genetic drift in a growing population—see section 5.5; Thompson and Neel 1997). We challenge readers to devise a research program that would resolve this issue.

In summary, mutation creates the genetic diversity that is the raw material for evolution. In principle, mutation can also produce changes in allele frequencies across generations. In practice, mutation's role as the source of genetic variation is usually more important than its role as a mechanism of evolution.

5.4 Migration

Migration, in the evolutionary sense, is the movement of alleles between populations. This use of the term migration is distinct from its more familiar meaning, which refers to the seasonal movement of individuals. To evolutionary biologists, migration means gene flow: the transfer of alleles from the gene pool of one population to the gene pool of another population. Migration can be caused by anything that moves alleles far enough to go from one population to another. Mechanisms of gene flow range from the occasional long-distance dispersal of juvenile animals to the transport of pollen, seeds, or spores by wind, water, or animals. The actual amount of migration among populations in different species varies enormously, depending on how mobile individuals or propagules are at various stages of the life cycle.

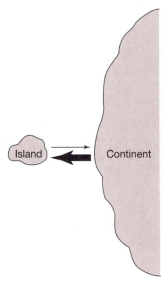

Figure 5.15 **The one-island model of migration** The arrows in the diagram show the relative amount of gene flow between the island and continental populations. Alleles arriving on the island from the continent represent a relatively high fraction of the island gene pool, whereas alleles arriving on the continent from the island represent a relatively small fraction of the continental gene pool.

Adding Migration to the Hardy-Weinberg Analysis: Migration as an Evolutionary Force

To investigate the effects of migration on the two conclusions of the Hardy-Weinberg analysis, we consider a simple model of migration called the one-island model. Imagine two populations: one on a continent, and the other on a small island offshore (Figure 5.15). Because the island population is so small relative to the continental population, any migration from the island to the continent will be inconsequential for allele and genotype frequencies on the continent. So migration, and the accompanying gene flow, is effectively one way, from the continent to the island. As usual, consider a single locus with two alleles, A_1 and A_2. Can migration from the continent to the island take the genotype frequencies on the island away from Hardy-Weinberg equilibrium?

To see that the answer is yes, imagine that before migration the frequency of A_1 on the island is 1.0 (that is, A_1 is fixed in the island population). The genotype frequencies are thus 1.0 for A_1A_1, 0 for A_1A_2, and 0 for A_2A_2. Now imagine that the continental population is fixed for allele A_2, and that a group of adults migrates from the continent to the island, so that after migration 80% of the island population is from the island, and 20% is from the continent. The new genotype frequencies are 0.8 for A_1A_1, 0 for A_1A_2, and 0.2 for A_2A_2. These frequencies are not consistent with Hardy-Weinberg conclusion 2; there is an excess of homozygotes and a deficit of heterozygotes. A single bout of random mating will, of course, put the population into Hardy-Weinberg equilibrium for genotype frequencies.

Migration has also changed the allele frequencies in the island population, violating Hardy-Weinberg conclusion 1. Before migration the island frequency of A_1 was 1; after migration the frequency of A_1 is 0.8. The island population has evolved as a result of migration (Figure 5.16).

Empirical Research on Migration and Allele Frequencies: The Balance Between Migration and Selection

The water snakes of Lake Erie provide an empirical example of migration from a mainland population to an island population. These snakes *(Nerodia sipedon)* live on the mainland surrounding Lake Erie and on the islands in the lake (Figure 5.17). Individuals vary in appearance, ranging from strongly banded to unbanded (Figure 5.18). To a rough approximation, color pattern is determined by a single locus with two alleles, with the banded allele dominant over the unbanded allele (King 1993a).

On the mainland virtually all the water snakes are banded, whereas on the islands many snakes are unbanded (Figure 5.19). The difference in the composition of mainland versus island populations appears to be the result of natural selection caused by predators. On the islands, the snakes bask on limestone rocks at the water's edge. Following up on earlier work by Camin and Ehrlich (1958), King (1993b) showed that among very young snakes unbanded individuals are more cryptic on island rocks than are banded individuals. The youngest and smallest snakes are presumably most vulnerable to predators. King (1993b) used mark-recapture studies, among other methods, to show that on the islands unbanded snakes indeed have higher rates of survival than banded snakes.

If selection favors unbanded snakes on the islands, then we would expect that the island populations would consist entirely of unbanded snakes. Why is this not the case? The answer, at least in part, is that in every generation several banded snakes move from the mainland to the islands and interbreed with the island snakes (for an algebraic treatment, see Box 5.5).

Migration as a Homogenizing Evolutionary Force Across Populations

Migration of water snakes from the mainland to the islands makes the island population more similar to the mainland population than it otherwise would be. This is the general effect of migration: It tends to homogenize allele frequencies across populations. In the water snakes, this homogenization is opposed by selection.

How far would the homogenization go if selection did not oppose it? To answer this question, we first develop an expression for the change in the allele frequency on the island from one generation to the next. We then use this expression to determine the equilibrium allele frequency. Let m stand for the fraction of the island population that arrives from the continent each generation. Let p_I denote the frequency of allele A_1 on the island and p_C its frequency on the continent. The frequency of allele A_1 in the next generation, p_I^*, is equal to

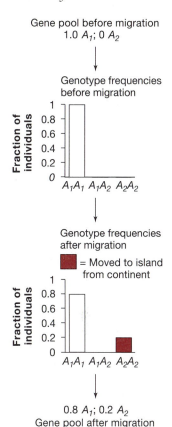

Figure 5.16 **Migration can change both genotype and allele frequencies** This figure follows the imaginary island population described in the text as migrants arrive from the continent.

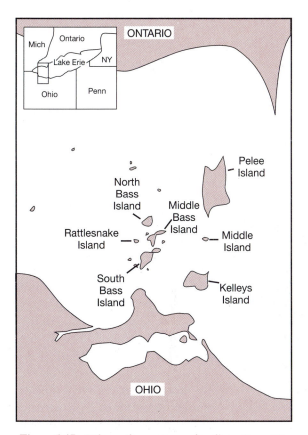

Figure 5.17 **Where the water snakes live** From King and Lawson (1995).

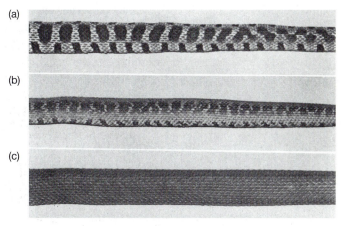

Figure 5.18 **Color-pattern variation in Lake Erie water snakes** Snake (a) is strongly banded; snake (c) is unbanded; snake (b) is intermediate (Richard B. King).

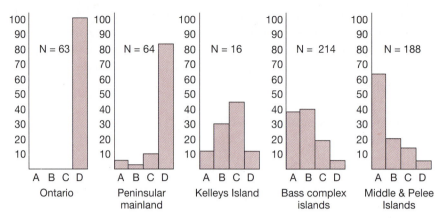

Figure 5.19   **Variation in color pattern within and between populations**
These histograms show frequency of different color patterns in various populations. Category A snakes are unbanded; category B and C snakes are intermediate; category D snakes are strongly banded. Snakes on the mainland tend to be banded; snakes on the islands tend to be unbanded or intermediate. From Camin and Ehrlich (1958).

the proportion of A_1 alleles in this generation on the island plus the proportion that just arrived from the mainland:

$$p_I* = (1 - m)p_I + m\, p_C$$

The change on the island from one generation to the next is thus

$$\Delta p_I = p_I* - p_I = (1 - m)p_I + m\, p_C - p_I = m(p_C - p_I)$$

The equilibrium condition is no change in p_I. That is, $\Delta p_I = 0$. If we set our rightmost expression for Δp_I equal to zero, and solve for p_I, we get the value of p_I at which, even if migration continues, p_I does not change:

$$p_I = p_C$$

In other words, without any opposing force, migration will eventually equalize the frequencies of the island and mainland populations. Inspection of the rightmost expression for Δp_I shows that the rate of homogenization is proportional to both the rate of migration, m, and the difference between p_C and p_I.

Migration can be a potent force of evolution. In practice, migration is most important in preventing populations from diverging.

Barbara Giles and Jérôme Goudet (1997) documented the homogenizing effect of gene flow on allele frequencies among populations of red bladder campion, *Silene dioica*. Red bladder campion is an insect-pollinated perennial wildflower (Figure 5.21a). The populations that Giles and Goudet studied occupy islands in the Skeppsvik Archipelago, Sweden. These islands are mounds of material deposited by glaciers during the last ice age and left underwater when the ice melted. The area on which the islands sit is rising at a rate of 0.9 cm per year. As a result of this geological uplift, new islands are constantly rising out of the water. The Skeppsvik Archipelago thus contains dozens of islands of different ages.

Red bladder campion seeds are transported by wind and water, and the plant is among the first to colonize new islands. Campion populations grow to sizes of several thousand individuals. There is gene flow among islands as a result of both seed dispersal and the transport of pollen by insects. After a few hundred years, campion populations are invaded by other species of plants and by a polli-

Box 5.5 Selection and migration in Lake Erie water snakes

As described in the main text, the genetics of color pattern in Lake Erie water snakes can be roughly approximated by a single locus, with a dominant allele for the banded pattern and a recessive allele for the unbanded pattern (King 1993a). Selection by predators on the islands favors unbanded snakes. If the fitness of unbanded individuals is defined as 1, then the relative fitness of banded snakes is between 0.78 and 0.90 (King and Lawson 1995). So why has selection not eliminated banded snakes from the islands? Here we calculate the effect that migration has when it introduces new banded alleles into the island population every generation.

King and Lawson (1995) lumped all the island snakes into a single population, because snakes appear to move among islands much more often than they move from the mainland to the islands. King and Lawson used genetic techniques to estimate that 12.8 snakes move from the mainland to the island every generation. The scientists estimated that the total island snake population is between 523 and 4,064 individuals, with a best estimate of 1,262. This means that migrants represent a fraction of 0.003 to 0.024 of the population each generation, with a best estimate of 0.01.

With King and Lawson's estimates of selection and migration, we can now calculate the equilibrium allele frequencies in the island population, at which the effects of selection and migration exactly balance each other. Let A_1 represent the dominant allele for the banded pattern, and A_2 the recessive allele for an unbanded pattern. Let p represent the frequency of A_1 in the gene pool, and q the frequency of A_2. Following Box 5.2, we create individuals by random mating, then let selection act. After selection (but before migration), the new frequency (q*) of allele A_2 is

$$q* = (pqw_{12} + q^2w_{22})/\overline{w}$$

where w_{12} is the fitness of A_1A_2 heterozygotes, w_{22} is the fitness of A_2A_2 homozygotes, and $\overline{w}$ is the mean fitness of all individuals in the population, given by $p^2 w_{11} + 2pq w_{12} + q^2 w_{22}$. For our first calculation, we use $w_{11} = w_{12} = 0.84$, and $w_{22} = 1$. A relative fitness of 0.84 for banded snakes is the midpoint of the range in which King

and Lawson (1995) estimated the true value to fall. This gives

$$q* = \frac{(0.84pq + q^2)}{(0.84p^2 + 2(0.84)pq + q^2)}$$

Substituting $(1 - q)$ for p gives

$$q* = \frac{(0.84(1 - q)q + q^2)}{(0.84(1 - q)^2 + 2(0.84)(1 - q)q + q^2)}$$

$$= \frac{(0.84q + 0.16q^2)}{(0.84 + 0.16q^2)}$$

Now we allow migration, with the new migrants representing—in this first calculation—0.01 of the island's population (King and Lawson's best estimate). None of the new migrants carries allele A_2, so the new frequency of A_2 is

$$q* = 0.99(0.84q + 0.16q^2)/(0.84 + 0.16q^2)$$

The change in q from one generation to the next is

$$\Delta q = q* - q$$
$$= [0.99(0.84q + 0.16q^2)/(0.84 + 0.16q^2)] - q$$

Plots of Δq as a function of q appear in Figure 5.20. Look first at the solid curve (b). This curve is for the function we just calculated. It shows that if q is greater than 0.05 and less than 0.94 in this generation, then q will be larger in the next generation (Δq is positive). If q is less than 0.05 or greater than 0.94 in this generation, then q will be smaller in the next generation (Δq is negative). The points where the curve crosses the horizontal axis, where $\Delta q = 0$, are the equilibrium points. The upper equilibrium point is stable: If q is less than 0.94, then q will rise in the next generation; if q is greater than 0.94, then it will fall in the next generation. Thus a middle-of-the-road prediction, given King and Lawson's estimates of selection and gene flow, is that the equilibrium frequency of the unbanded allele in the island population will be 0.94.

Curve (a) is a high-end estimate; it uses fitnesses of 0.78 for A_1A_1, 0.78 for A_1A_2, and 1 for A_2A_2, and a migration rate of 0.003 (0.3% of every generation's population are migrants). It predicts an equilibrium frequency of $\hat{q} = 0.99$. Curve (c) is a low-end estimate; it uses fitnesses of 0.90 for A_1A_1, 0.90 for A_1A_2, and 1 for A_2A_2, and a migration rate of

Box 5.5 *(continued)*

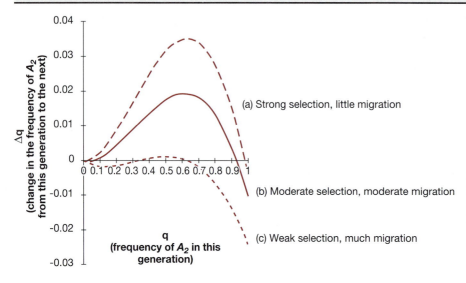

Figure 5.20 **The combined effects of selection and migration on allele frequencies in island water snakes** The curves show Δq as a function of q for different combinations of migration and selection. See text for details.

0.24 (2.4% of every generation's population are migrants). It predicts an equilibrium frequency of $\hat{q} = 0.64$.

King and Lawson's best estimate of the actual frequency of A_2 is 0.73. This value is toward the low end of our range of predictions. Our calculation is a relatively simple one, and leaves out many factors, including recent changes in the population sizes of both the water snakes and their predators, as well as recent changes in the frequencies of banded versus unbanded snakes. For a detailed treatment of this example, see King and Lawson (1995).

nator-borne disease. Establishment of new seedlings ceases, and populations dwindle as individuals die.

Giles and Goudet predicted that young populations, having been founded by the chance transport of small numbers of seeds, would vary in their allele frequencies at a variety of loci (we will consider why in section 5.5). Populations of intermediate age should be more homogeneous in their allele frequencies as a result of migration—that is, gene flow among populations via seed dispersal

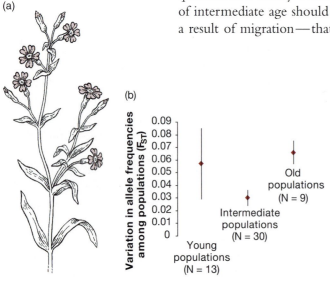

Figure 5.21 **Variation in allele frequencies among populations of red bladder campion,** *Silene dioica* (a) Red bladder campion, a perennial wildflower. (b) Giles and Goudet's (1997) measurements of variation in allele frequencies among populations. The diamonds represent values of F_{ST} (see text); the lines represent standard errors (larger standard errors represent less certain estimates of F_{ST}). There is less variation in allele frequencies among intermediate age populations than among young populations (P = 0.05). There is more variation in allele frequencies among old populations than among intermediate populations (P = 0.04). After Giles and Goudet (1997).

and pollen transport. Finally, the oldest populations, structured mainly by the fortuitous survival of a few remaining individuals, should again become more variable in their allele frequencies.

The researchers tested their predictions by collecting leaves from many individual red bladder campions on 52 islands of different ages. By analyzing proteins in the leaves, Giles and Goudet determined each individual's genotype at six enzyme loci. They divided their populations by age into three groups: young, intermediate, and old. For each of these groups, they calculated a test statistic called F_{ST}. A value for F_{ST} refers to a group of populations, and reflects the variation in allele frequencies among the populations in the group. The value of F_{ST} can be anywhere from 0 to 1. Larger values represent more variation in allele frequency among populations.

The results confirm Giles and Goudet's predictions (Figure 5.21b). There is less variation in allele frequencies among populations of intermediate age than among young and old populations. The low diversity among intermediate populations probably reflects the homogenizing influence of gene flow. The higher diversity of young and old populations probably represents genetic drift, the subject of the next section.

In summary, migration is the movement of alleles from population to population. Within a single population, migration can cause allele frequencies to change from one generation to the next. For small populations receiving immigrants from large source populations, migration can be a potent mechanism of evolution. Across groups of populations, gene flow tends to homogenize allele frequencies. Thus migration tends to resist the evolutionary divergence of populations.

5.5   Genetic Drift

In Chapter 2 we refuted the misconception that evolution by natural selection is a random process. To be sure, Darwin's mechanism of evolution depends on the generation of random variation by mutation. Mutation is random in the sense that when mutation substitutes one amino acid for another in a protein, it does so without regard to whether the change will improve the protein's ability to function. Natural selection itself, however, is anything but random. It is precisely the nonrandomness of selection in sorting among mutations that leads to adaptation. We are now in a position to revisit the role of chance in evolution. Arguably the most important insight from population genetics is that natural selection is not the only mechanism of evolution. Among the nonselective mechanisms of evolution, there is one that is absolutely random. That mechanism is genetic drift. Genetic drift does not lead to adaptation, but it does lead to changes in allele frequencies.

Drift results from the vagaries of sampling. Recall that in our derivation of the Hardy-Weinberg principle, we assumed that when making zygotes we drew alleles from the gene pool in their actual frequencies p and q. In reality, this almost never happens. Imagine a population in which alleles A_1 and A_2 are both at frequencies of 0.5. We will draw alleles at random from the gene pool to make 10 zygotes. These 10 zygotes will constitute the entire population for the next generation. We can simulate the process of drawing alleles from the gene pool by flipping a coin. Heads represents A_1; tails represents A_2. We are flipping an actual coin as we write. The ten individuals are:

$$A_1A_1 \quad A_1A_1 \quad A_1A_1 \quad A_1A_2 \quad A_2A_1$$
$$A_2A_2 \quad A_1A_1 \quad A_2A_2 \quad A_1A_2 \quad A_1A_2$$

In populations of finite size, random variation in survival and reproductive success can cause evolution.

Counting the alleles, we have A_1 at a new frequency of 0.6 and A_2 at a frequency of 0.4. The population has evolved. Counting the genotypes, we have A_1A_1 at a new frequency of 0.4, heterozygotes at a frequency of 0.4, and A_2A_2 at a frequency of 0.2. We cannot predict the new genotype frequencies by multiplying the new allele frequencies. Both conclusions of the Hardy-Weinberg analysis have been violated. This is genetic drift.

Fundamentally, genetic drift results because any real population has a finite number of individuals. If we kept tossing our coin to build up a bigger population of zygotes, the outcome would get closer and closer to that predicted by the Hardy-Weinberg principle. To illustrate this, we used a computer to simulate drawing and pairing gametes to make 250 zygotes (Figure 5.22). As before, the frequency of A_1 in the gene pool was 0.5. As the cumulative number of zygotes we made got larger, the cumulative frequency of allele A_1 in the next generation settled toward the expected value of 0.5. Genetic drift is a much more powerful evolutionary mechanism in small populations than in large populations.

Sampling Processes and the Founder Effect

Small population sizes frequently occur during the founding of new populations. New populations are sometimes established by small groups of individuals that move, or are moved, to a new location. The allele frequencies in the new population are likely, simply by chance, to be different than they are in the old population. This is called the founder effect. The founder effect is a direct result of sampling. For example, if 25 different alleles are present at a single locus in a continental population of insects, but just 10 individuals are on a log that rafts to a remote island, the probability is zero that the new island population will have all of the allelic diversity present in the continental population. If by chance any of the founding individuals are homozygotes, allele frequencies in the new population will have shifted even more dramatically. In any founder event, some degree of random genetic differentiation is almost certain between old and new

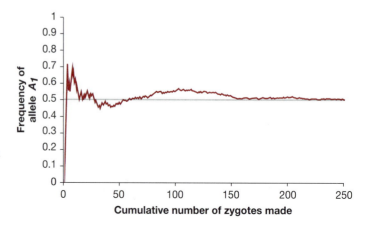

***Figure 5.22* A simulation of drawing alleles from the gene pool**   Initially the new frequency of allele A_1 fluctuates considerably. As the number of zygotes made increases, however, the new frequency of A_1 settles toward the expected value of 0.5.

populations. In other words, the founding of a new population by a small group of individuals typically represents not only the establishment of a new population but also the instantaneous evolution of differences between the new population and the old population.

Peter Grant and Rosemary Grant (1995) watched the establishment of a new population of large ground finches *(Geospiza magnirostris)* in the Galápagos. Grant and Grant worked on the island of Daphne Major for 22 years. Each year, large ground finches visited the island, with anywhere from 10 to 50 juvenile birds arriving after the breeding season and staying through the dry months. For the first several years of Grant and Grant's research, the visiting birds all left the island before the next breeding season began. The visiting finches were all members of some other island's population. Then in the fall of 1982, three males and two females—apparently induced by early rains—stayed on Daphne Major to breed. These five birds formed pairs (one female bred with two different males), built eight nests over the course of the breeding season, and fledged 17 young during early 1983.

The five 1982–1983 breeders established a new population. Since 1983, large ground finches have bred on Daphne Major every year, with the exception of three drought years. By 1993 the Daphne Major breeding population included 23 males and 23 females. Through at least 1992, the majority of the Daphne Major breeders had been hatched on the island. They were natives of the new population.

Is the newly founded finch population genetically different from the source population? Although Grant and Grant do not have direct data on allele frequencies of specific genes, they do have extensive morphological measurements of the 238 ground finches that visited Daphne Major prior to the colonization event, and of the five original Daphne Major breeders and 22 of their offspring. Grant and Grant assume that the 238 nonbreeding visitors are representative of the source population, and compare them to the members of the newly founded population. In at least two morphological traits, bill width and bill shape in relation to body size, the members of the new population were significantly larger than the source population. Research on the new population of large ground finches, and on other Darwin's finches, suggests that these traits are heritable. Thus it appears that the founding event created a new population measurably different from the source population. Evolution had occurred, not through selection but by a random sampling process. This was genetic drift in the form of the founder effect.

Founder effects are often seen in genetically isolated populations of humans. For example, the Amish population of eastern Pennsylvania is descended from a group of about 200 European settlers who came to the United States in the 18th century. One of the founders—either the husband or wife (or both) in a couple named King—was a carrier for Ellis-van Creveld syndrome. Ellis-van Creveld syndrome is a rare form of dwarfism caused by a recessive allele on chromosome 4 (Bodmer and McKie 1995). The frequency of this allele is about 0.001 in most populations, but it is about 0.07 in the present-day Amish (Postlethwait and Hopson 1992). The high frequency of the allele in the Amish population is almost certainly not due to any selective advantage conferred by the allele in either heterozygotes or homozygotes. Instead, the high frequency of the allele results simply from chance: The allele was at a high frequency in the small population of founders and continued to drift upward in subsequent generations.

Random Fixation of Alleles

The power of genetic drift as a mechanism for changing allele frequencies is manifest when drift's effects are compounded over many generations. In our numerical example, we started with alleles A_1 and A_2 at frequencies of 0.5 each and tossed a coin to simulate drawing gametes from the gene pool. The next generation, composed of 10 individuals, had allele frequencies of 0.6 for A_1 and 0.4 for A_2. We could simulate drawing alleles from this generation's gene pool by flipping a biased coin that comes up heads 60% of the time and tails 40% of the time. This is difficult to arrange with real coins, but easy with a computer. Figure 5.23 shows the results of 100 generations of drift in simulated populations of different sizes. Two patterns are evident. First, genetic drift has a more dramatic effect on allele frequencies in small populations than large populations. Second, given sufficient time, genetic drift can produce substantial changes in allele frequencies even in large populations.

If genetic drift is the only evolutionary force at work in a population—if there is no selection, mutation, or migration—then allele frequencies wander at random between 0 and 1. Sooner or later an allele's frequency will hit one or the other of these boundaries. If the allele's frequency hits 0, then the allele is lost forever (we are assuming that mutation or migration does not reintroduce it). If the allele's frequency hits 1 then the allele is said to be fixed, also forever. Sewall Wright (1931) realized that if some allele gets fixed in such a population, the probability that it will be any particular allele is equal to that particular allele's initial frequency.

Peter Buri (1956) studied the random fixation of alleles in small laboratory populations of the fruit fly, *Drosophila melanogaster*. Adopting an approach that had been used by Kerr and Wright (1954), Buri established 107 populations of flies, each with 8 females and 8 males. All the founding flies were heterozygotes for alleles of an eye-color gene called brown. All the flies had the same genotype: bw^{75}/bw. Thus in all 107 populations, the initial frequency of the bw^{75} allele was 0.5. Buri maintained these populations for 19 generations. For every population in every generation, Buri kept the population size at 16 by picking 8 females and 8 males at random to be the breeders for the next generation.

What results would we predict? If neither allele bw^{75} nor allele bw confers a selective advantange over the other, then we expect the frequency of allele bw^{75} to wander at random by genetic drift in every population. Nineteen generations should be enough, in populations of 16 individuals, for many populations to become fixed for one allele or the other. Because allele bw^{75} has an inititial frequency of 0.5, we expect this allele to be lost about as often as it becomes fixed.

Buri's results confirm these predictions (Figure 5.24). The frequency of the bw^{75} rose in some populations and fell in others. As a result, the distribution of allele frequencies, initially clustered tightly around a frequency of 0.5, rapidly spread out. Populations started becoming fixed in the fourth generation. As the allele frequency hit 0 or 1 in more and more populations, the distribution of allele frequencies became U-shaped. By the end of the experiment, bw^{75} had been lost in 30 populations, and fixed in 28. The 30:28 ratio of losses to fixations is very

Genetic drift is a more potent force of evolution in small populations.

Genetic drift leads to the random fication of some alleles, and the random loss of others.

(a)
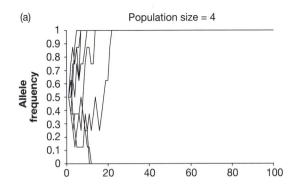

(b)
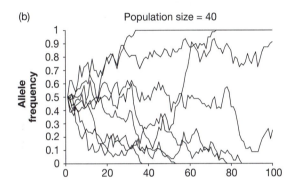

(c)
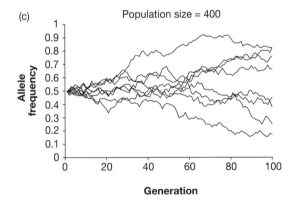

Figure 5.23 **Simulations of 100 generations of genetic drift**
Each plot shows the frequency of an allele (A_1) over time in eight simulated populations. In plot (a), each population contains 4 individuals each generation. All populations became fixed for one allele or the other in less than 25 generations. In plot (b), each population contains 40 individuals. Six of the populations had become fixed for one allele or the other after 100 generations. In plot (c), each population contains 400 individuals. No populations became fixed in 100 generations, but drift created substantial variation among populations in allele frequency.

close to the 1:1 ratio we would predict under the hypothesis of genetic drift. During Buri's experiment there was dramatic evolution in nearly all 107 of the fruit fly populations, but natural selection had nothing to do with it.

Is the random fixation of alleles important in nature? Undoubtedly, the answer is yes. Populations are subject to a wide variety of random processes. Fires or storms or floods can wipe out vast numbers of individuals, but leave small areas untouched, by happenstance. This creates a founder effect, also known in this context as a bottleneck, as the population recovers from the disaster. Furthermore some alleles, and perhaps many, simply are not under selection all the time, and instead occasionally or routinely drift in frequency. Theoretician

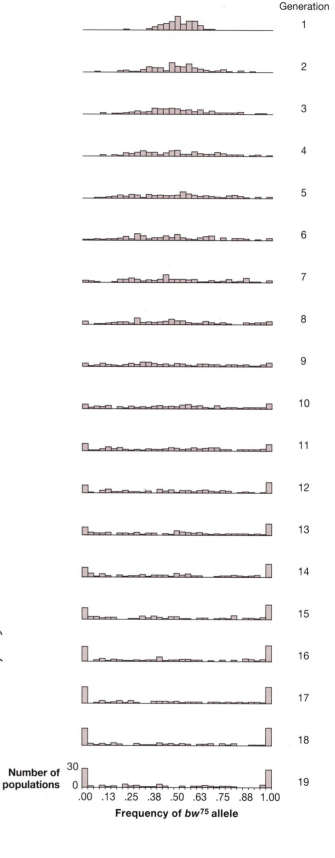

Figure 5.24 Nineteen generations of genetic drift in 107 populations of 16 fruit flies Each line is a histogram summarizing the allele frequencies in all 107 populations in a particular generation. The horizontal axis represents the frequency of the bw^{75} allele, and the vertical axis represents the number of populations showing each frequency. The frequency of bw^{75} was 0.5 in all populations in generation 0 (not shown). After one generation of genetic drift, most populations still had an allele frequency near 0.5, although one population had an allele frequency as low as 0.22 and another had an allele frequency as high as 0.69. Over the next several generations, the distribution of allele frequencies spread out as many populations drifted away from frequencies near 0.5. In generation 4, the frequency of allele bw^{75} hit 1 in a population for the first time. In generation 6, the frequency of allele bw^{75} hit 0 in a population for the first time. By the end of the experiment 30 populations had become fixed at a frequency of 0, and 28 had become fixed at a frequency of 1 (bottom line). Throughout the experiment, however, the distribution of allele fequencies remained symmetrical around 0.5. From data in Buri (1956), after Ayala and Kiger (1984).

Motoo Kimura has in fact argued that at the level of DNA sequences almost all evolutionary change happens as a result of genetic drift. We will return to Kimura's neutral theory of evolution in Chapter 7.

In summary, genetic drift is a nonadaptive mechanism of evolution. Simply as a result of chance, allele frequencies can change from one generation to the next. Genetic drift is most powerful in small populations.

5.6 Nonrandom Mating

The final assumption of the Hardy–Weinberg analysis is that individuals in the population mate at random. In this section we relax that assumption and allow individuals to mate nonrandomly. Nonrandom mating does not, by itself, cause evolution. Nonrandom mating can nonetheless have profound indirect effects on evolution.

The most common type of nonrandom mating, and the kind we will focus on here, is inbreeding. Inbreeding is mating among genetic relatives. The effect of inbreeding on the genetics of a population is to increase the frequency of homozygotes compared to what is expected under the Hardy–Weinberg assumptions. To show how this happens, we will consider the most extreme example of inbreeding: self-fertilization, or selfing. Imagine a population in Hardy–Weinberg equilibrium with alleles A_1 and A_2 at initial frequencies of 0.5 each. The frequency of each type of homozygote is 0.25, and the frequency of heterozygotes is 0.5 (Figure 5.25). If a population starts out at Hardy–Weinberg genotype frequencies but then all individuals self, homozygous parents will produce all homozygous offspring, while heterozygous parents will produce half homozygous and half heterozygous offspring. As a result, the proportion of the population that is homozygous increases each generation while the proportion that is heterozygous is halved. Conclusion 2 of the Hardy–Weinberg analysis is violated when individuals self: We cannot predict the genotype frequencies by multiplying the allele frequencies. The general case under selfing is shown algebraically in Table 5.3.

Nonrandom mating can cause genotype frequencies to change. Inbreeding increases the frequency of homozygotes, and decreases the frequency of heterozygotes.

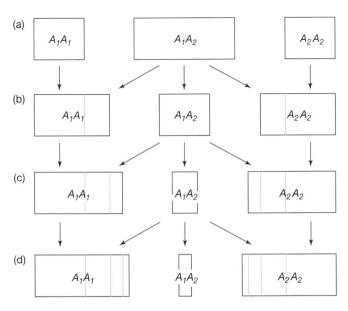

(a) A_1A_1 A_1A_2 A_2A_2

(b) A_1A_1 A_1A_2 A_2A_2

(c) A_1A_1 A_1A_2 A_2A_2

(d) A_1A_1 A_1A_2 A_2A_2

Figure 5.25 **Selfing increases the frequency of homozygotes** Row (a) represents, with the area of each box, the genotype frequencies in a population with two alleles, each at frequency 0.5, in Hardy–Weinberg equilibrium. Row (b) represents the genotype frequencies in the next generation that will result if every individual in the population mates only with itself. If selfing continues, homozygotes consititute an ever-larger fraction of the population every generation; see rows (c) and (d).

TABLE 5.3 Changes in genotype frequencies with successive generations of selfing

The frequency of allele A_1 is p and the frequency of allele A_2 is q. Note that allele frequencies do not change from generation to generation—only the genotype frequencies. After Crow (1983).

Generation	A_1A_1	Frequency of A_1A_2	A_1A_2
0	p^2	2pq	q^2
1	$p^2 + (pq/2)$	pq	$q^2 + (pq/2)$
2	$p^2 + (3\,pq/4)$	pq/2	$q^2 + (3\,pq/4)$
3	$p^2 + (7\,pq/8)$	pq/4	$q^2 + (7\,pq/8)$
4	$p^2 + (15\,pq/16)$	pq/8	$q^2 + (15\,pq/16)$

What about Hardy–Weinberg conclusion 1? Do the allele frequencies change from generation to generation under inbreeding? We can find out by calculating the frequency of allele A_1 in the gene pool produced by the population shown in the last row of Table 5.3. The frequency of allele A_1 in the gene pool is equal to the frequency of A_1A_1 adults in the population ($p^2 + \frac{15pq}{16}$) plus half the frequency of A_1A_2 adults (pq/8). That gives

$$p^2 + \frac{15pq}{16} + \frac{1}{2}\left(\frac{pq}{8}\right) = p^2 + \frac{15pq}{16} + \frac{pq}{16} = p^2 + pq$$

Nonrandom mating does not, by itself, cause allele frequencies to change.

Now substitute $(1 - p)$ for q to give $p^2 + p(1 - p) = p$. This is the same frequency for allele A_1 that we started out with at the top of Table 5.3. Although inbreeding does cause genotype frequencies to change from generation to generation, it does not cause allele frequencies to change. Inbreeding by itself, therefore, is not a mechanism of evolution. As we will see, however, inbreeding can have important evolutionary consequences.

Empirical Research on Inbreeding: The Malaria Parasite

As an example of how Hardy–Weinberg analysis is used to detect inbreeding in nature, we consider a recent study of a species of malaria parasite in New Guinea (Paul et al. 1995). The life cycle of this protozoan, *Plasmodium falciparum,* alternates between stages that live in mosquitoes and stages that live in humans (Figure 5.26). The only diploid part of the parasite's life cycle occurs in the mosquito; there, a stage called the oocyst resides in the midgut wall. The other stages, which infect the human liver and red blood cells and include the cells transmitted to mosquitoes, are haploid.

R. E. L. Paul and colleagues (1995), working in Karen Day's lab, estimated allele and genotype frequencies for three *P. falciparum* protein-encoding genes: *MSP-1, MSP-2,* and *GLURP.* The scientists first dissected female mosquitoes and isolated the malaria oocysts encased in their stomach linings. The researchers then extracted DNA from the oocysts and analyzed it directly for allelic variation. All three protein-encoding loci were polymorphic, meaning that each had more than one allele. The data provide an estimate for the frequency of different alleles in the population (Table 5.4), and counts of the number of homozygotes and heterozygotes in the sample (Table 5.5).

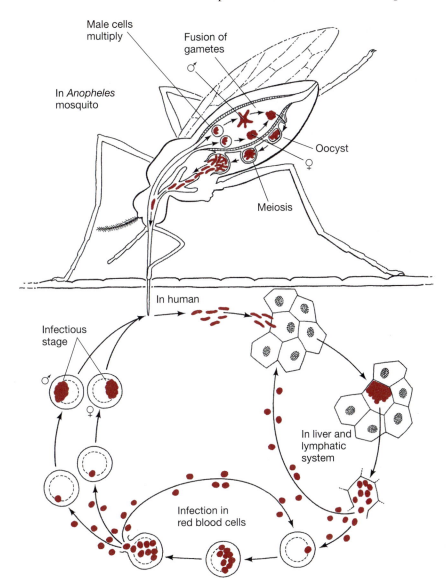

Male cells multiply

Fusion of gametes

In *Anopheles* mosquito

Oocyst

Meiosis

In human

Infectious stage

In liver and lymphatic system

Infection in red blood cells

Figure 5.26 **Life cycle of the malaria parasite *Plasmodium falciparum***

Now the question is, do the allele and genotype frequencies the researchers measured conform to those expected under Hardy-Weinberg conditions? The answer, resoundingly, is no (Table 5.5). There is an enormous excess of homozygotes in the parasite population, and a corresponding deficit of heterozygotes. The data contradict conclusion 2 of the Hardy-Weinberg analysis, which means that one or more of the Hardy-Weinberg assumptions is being violated. The most likely candidate is the assumption of random mating. One reason is that, as we have seen, violation of the other assumptions is rarely expected to give genotype frequencies as far away from expectation as Paul and colleagues found. Another reason is that the biology of the *P. falciparum* malaria parasite suggests that inbreeding may be common in this species.

Here is the logic: Years ago, W. D. Hamilton (1967) observed that unusual sex ratios are common when a single female parasite or parasitoid colonizes a new host. In the fig wasps Hamilton was studying, for example, females colonizing

TABLE 5.4 Allele frequencies for three polymorphic genes in the malaria parasite *Plasmodium falciparum*

MSP-2		MSP-3		GLURP	
Allele	**Observed Frequency**	**Allele**	**Observed Frequency**	**Allele**	**Observed Frequency**
A_1	0.02	A_1	0.02	A_1	0.07
A_2	0.06	A_2	0.26	A_2	0.42
A_3	0.18	A_3	0.19	A_3	0.28
A_4	0.27	A_4	0.32	A_4	0.08
A_5	0.12	A_5	0.12	A_5	0.15
A_6	0.08	A_6	0.09		
A_7	0.05				
A_8	0.07				
A_9	0.09				
A_{10}	0.06				

Source: Calculated from Figure 1 in Paul et al. (1995).

figs alone tended to produce many more female than male young. Hamilton went on to develop the mathematics showing why this strategy evolved, and predicted that it would occur in any parasite where single foundresses are common (Hamilton 1967, 1979; see also Read et al. 1992). The essence of the argument is that in species in which brother-sister mating is the rule, females will have more grand-offspring if they produce only enough males to ensure that all their daughters will be fertilized.

What does this have to do with the malaria parasite? Paul and colleagues observed a female sex ratio bias (more females than males) in the infectious stages of *P. falciparum* in New Guinea. The researchers have proposed that the phenomenon Hamilton discovered is at work in the parasite population they studied. Although malaria is common in New Guinea, the rate of transmission is relatively low. This means that every human infected tends to have only one to a few *P. falciparum* genotypes present. In addition, it appears that each cycle of infection in a human's red blood cells, which produces the type of *Plasmodium* cells picked up by mosquitoes, may be restricted to cells of a single genotype. The result is that each mosquito tends to become infected with one, or at most two, genotypes of

TABLE 5.5 The observed number of homozygotes and heterozygotes at three *P. falciparum* loci
In each case, the observed number of individuals with a particular kind of genotype is compared to the number expected under Hardy-Weinberg conditions of random mating and no mutation, selection, migration, or genetic drift. In all three cases the differences are statistically significant (P < 0.01; see Box 5.3).

	(a) MSP-2		(b) MSP-1		(c) GLURP	
	Observed	**Expected**	**Observed**	**Expected**	**Observed**	**Expected**
Homozygotes	55	10	38	9	40	12
Heterozygotes	9	54	1	30	1	29

P. falciparum cells. In effect, all the cells that the mosquito picks up are offspring of a single female. This infectious stage is where Paul et al. found the female-biased sex ratio. The few male cells that occur multiply inside the mosquito and then fuse with the female cells—often their own siblings—to form diploid oocysts, resulting in the homozygote excess reported in Table 5.4 and Table 5.5.

In summary, Paul et al. (1995) found a large excess of homozygotes in a population of *P. falciparum*. This finding indicates that one or more of the Hardy-Weinberg assumptions is being violated. Only nonrandom mating can easily produce homozygote excesses as large as that which Paul et al. found. Furthermore, other clues suggest that the *P. falciparum* population is inbred. Paul et al. conclude that the homozygote excess they found in *P. falciparum* is the result of nonrandom mating in the form of inbreeding.

General Analysis of Inbreeding

So far our treatment of inbreeding has been limited to self-fertilitization and sibling mating. But inbreeding can also occur as matings among more distant relatives, such as cousins. Inbreeding that is less extreme than selfing produces the same effect as selfing—it increases the proportion of homozygotes—but at a slower rate. For a general mathematical treatment of inbreeding, population geneticists use a conceptual tool called the coefficient of inbreeding. This quantity is symbolized by F, and is defined as the probability that the two alleles in an individual are identical by descent (meaning that both alleles came from the same ancestor allele in some previous generation). First we incorporate F into the Hardy-Weinberg analysis, then we discuss how F can be measured.

To add F to the Hardy-Weinberg analysis, we return to our imaginary population with two alleles at a single locus: A_1 and A_2. The frequency of A_1 in the gene pool is p, and the frequency of A_2 is q. We now calculate the genotype frequencies in the next generation by drawing gametes from the gene pool and pairing them, keeping in mind that this time the gene pool is not thoroughly mixed. Once we have drawn the first gamete from the gene pool, we can think of the rest of the gene pool as consisting of two fractions: the fraction $(1 - F)$ that is not identical by descent; and the fraction F that is identical by descent.

- **A_1A_1 homozygotes:** There are two ways to get an A_1A_1 homozygote. The first way is to draw an egg that is A_1 (an event with probability p) and a sperm that is A_1 by chance, rather than by common ancestry. The frequency of unrelated A_1 sperm in the gene pool is p(1 − F), so the probability of getting a homozygote by chance is p × p(1 − F) = $p^2(1 - F)$. The second way to get a homozygote is to draw an egg that is A_1 (an event with probability p) and a sperm that is A_1 because of common ancestry (an event with probability F). The probability of getting a homozygote this way is pF. The probability of getting an A_1A_1 homozygote by either the first way or the second way is their sum: $p^2(1 - F) + pF$.

- **A_1A_2 heterozygotes:** There are two ways to get an A_1A_2 heterozygote. The first way is to draw an egg that is A_1 (an event with probability p) and a sperm that is A_2 by chance rather than common ancestry (the probability of doing this is q(1 − F)). So the probability of getting a heterozygote this first way is pq(1 − F). The second way is to draw an egg that is A_2 (probability: q) and a sperm that is A_1 (probability: p(1 − F)). The probability of getting

F, the coefficient of inbreeding, provides a measure of the extent to which inbreeding in a population has increased the frequency of homozygotes and decreased the frequency of heterozygotes compared to Hardy-Weinberg expectations.

a heterozygote the second way is qp(1 − F). The probability of getting a heterozygote by either the first way or the second way is the sum of their individual probabilities: 2pq(1 − F).

- **A_2A_2 homozygotes:** We can get an A_2A_2 homozygote either by drawing an A_2 egg (probability: q) and a sperm that is A_2 by chance (probability: q(1 − F)), or by drawing an A_2 egg (probability: q) and a sperm that is A_2 by descent from a common ancestor (probability: F). The overall probability of getting an A_2A_2 homozygote is $q^2(1 − F) + qF$.

Readers can verify these calculations by substituting the values F = 0, which gives the original Hardy–Weinberg genotype ratios, and F = 0.5, which represents selfing and gives the ratios shown for generation 1 in Table 5.3. The same logic applies when many alleles are present in the gene pool. Then the frequency of any homozygote A_iA_i is given by $p_i^2(1 − F) + p_iF$ and any heterozygote A_iA_j by $2p_ip_j(1 − F)$, where p_i is the frequency of allele A_i and p_j is the frequency of allele A_j.

The last expression we derived states that the fraction of individuals in a population that are heterozygotes (known as the population's heterozygosity) is proportional to 1 − F. If we compare the heterozygosity of an inbred population, Het_F, with that of a random mating population, Het_0, then the relationship will be

$$Het_F = Het_0(1 − F)$$

Computing F

To measure the degree of inbreeding in actual populations, we need a way to calculate F. Doing this directly requires a pedigree—a diagram showing the geneological relationships of individuals. Figure 5.27 shows a pedigree leading to a female who is the daughter of half-siblings. There are two ways this female could receive alleles that are identical by descent. One is that she could receive two copies of her grandmother's "triangle" allele (Figure 5.27a). This will happen if the grandmother passes the triangle allele to her daughter and to her son, and the daughter passes it to the granddaughter, and the son passes it to the granddaughter. The total probability of this scenario is $\frac{1}{16}$. The second way is that she could receive two copies of her grandmother's "diamond" allele (Figure 5.27b). The total probability of this scenario is $\frac{1}{16}$. The probability that the daughter of half-siblings will have two alleles identical by descent by either the first scenario or the second scenario is $\frac{1}{16} + \frac{1}{16} = \frac{1}{8}$. Thus, F for an offspring of half-siblings is $\frac{1}{8}$.

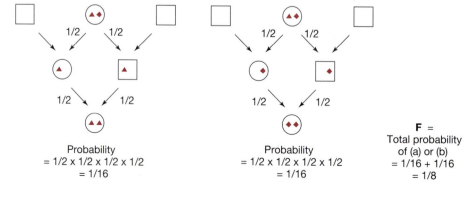

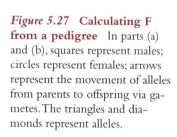

Figure 5.27 **Calculating F from a pedigree** In parts (a) and (b), squares represent males; circles represent females; arrows represent the movement of alleles from parents to offspring via gametes. The triangles and diamonds represent alleles.

(a)

1/2   1/2

1/2   1/2

Probability
= 1/2 x 1/2 x 1/2 x 1/2
= 1/16

(b)

1/2   1/2

1/2   1/2

Probability
= 1/2 x 1/2 x 1/2 x 1/2
= 1/16

F =
Total probability
of (a) or (b)
= 1/16 + 1/16
= 1/8

Inbreeding Depression

Although inbreeding does not directly change allele frequencies, it can still affect the evolution of a population. Among the most important consequences of inbreeding for evolution is inbreeding depression.

Inbreeding depression usually results from the exposure of deleterious recessive alleles to selection. To see how this works, consider the extreme case illustrated by loss-of-function mutations. These alleles are often recessive, because a single wild-type allele can still generate enough functional protein, in most instances, to produce a normal phenotype. Even though they may have no fitness consequences at all in heterozygotes, loss-of-function mutations can be lethal in homozygotes. By increasing the proportion of individuals in a population that are homozygotes, inbreeding increases the frequency with which deleterious recessives affect phenotypes. Inbreeding depression refers to the effect these alleles have on the average fitness of offspring in the population.

Studies on humans have shown that inbreeding does, in fact, expose deleterious recessive alleles, and data from several studies consistently show that children of first cousins have higher mortality rates than children of unrelated parents (Table 5.6). Strong inbreeding depression has also been frequently observed in captive populations of animals (for example, Hill 1974; Ralls et al. 1979).

By increasing the frequency of homozygotes, inbreeding can expose deleterious recessive alleles to selection.

Observational data from natural populations have been harder to come by. Inbreeding depression can be directly demonstrated only in long-term studies of marked populations, where pedigrees are complete enough to estimate F. Then can researchers compare the fitnesses of inbred versus non-inbred individuals. But interesting studies on inbreeding depression, from a wide variety of organisms, are starting to accumulate.

Long-term studies in two separate populations of a bird called the great tit *(Parus major)* have shown that inbreeding depression can have strong effects on reproductive success. When Paul Greenwood and co-workers (1978) defined inbred matings as those between first cousins or more closely related individuals, they found that the survival of inbred nestlings was much lower than that of outbred individuals. Similarly, A. J. van Noordwijk and W. Scharloo (1981)

TABLE 5.6 Inbreeding depression in humans
This table reports data from several studies on the mortality rate (with the absolute number of deaths investigated in parentheses) among children of first cousin marriages versus children of unrelated parents. In every study, children of first cousins had a higher mortality rate. The reduced health of inbred individuals is inbreeding depression. From Stern (1973; original sources therein).

Deaths	Period	Children of first cousins	Children of nonrelatives
Young children (U.S.)	Before 1858	22% (2,778)	16% (837)
Children under 20 (U.S.)	18th–19th cent.	17% (672)	12% (3,184)
Children under 10 (U.S.)	1920–1956	8.1% (209)	2.4% (167)
At/before birth (France)	1919–1950	9.3% (743)	3.9% (2,745)
Children (France)	1919–1950	14% (674)	10% (515)
Children under 1 (Japan)	1948–1954	5.8% (4,947)	3.5% (12,077)
Children 1–8 (Japan)	1948–1954	4.6% (326)	1.5% (544)

showed that in an island population of tits, there is a strong relationship between the level of inbreeding in a pair and the number of eggs in a clutch that fail to hatch (Figure 5.28). More recently, Keller et al. (1994) found that inbred individuals in a population of song sparrows in British Columbia were much more likely than outbred individuals to die during population crashes (Figure 5.29).

Perhaps the most powerful studies of inbreeding depression concern flowering plants, in which the inbreeding can be studied experimentally. In many angiosperms, selfed and outcrossed offspring can be produced from the same parent, through hand pollination. In experiments like these, inbreeding depression can be defined as $\delta = 1 - (w_s/w_o)$, where w_s and w_o are the fitnesses of selfed and outcrossed progeny, respectively. This definition makes levels of inbreeding depression comparable across species. Three patterns are starting to emerge from experimental studies.

First, inbreeding effects are often easiest to detect when plants undergo some sort of environmental stress. For example, when Michele Dudash (1990) compared the growth and reproduction of selfed and outcrossed rose pinks *(Sabatia angularis),* the plants showed some inbreeding depression when grown in the greenhouse or garden, but their performance diverged more strongly when they were planted in the field. Lorne Wolf (1993) got a similar result with a waterleaf *(Hydrophyllum appendiculatum):* Selfed and outcrossed individuals had equal fitness when grown alone, but differed significantly when grown under competition. And in the common annual called jewelweed *(Impatiens capensis),* McCall et al. (1994) observed the strongest inbreeding effects on survival when an unplanned insect outbreak occurred during the course of their experiment.

Second, inbreeding effects are much more likely to show up later in the life cycle (not, for example, during the germination or seedling stage). This pattern is striking (Figure 5.30a and b). Why does it exist? Wolfe (1993) suggests that maternal effects—specifically, the mother's influence on offspring phenotype through provisioning of seeds—can mask the influence of deleterious recessives until later in the life cycle.

Third, inbreeding depression varies among family lineages. Michele Dudash and colleagues (1997) compared the growth and reproductive performance of inbred versus outcrossed individuals from each of several families in two annual populations of the herb *Mimulus guttatus.* Some families showed inbreeding de-

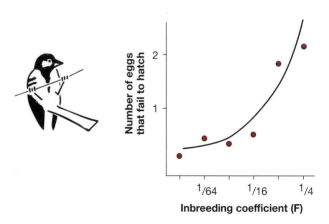

Figure 5.28 **Inbreeding increases egg failure in great tits**
From van Noordwijk and Scharloo (1981).

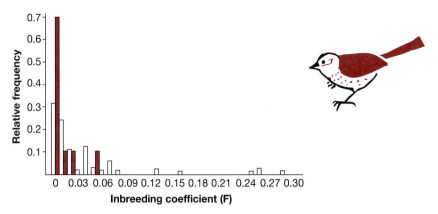

Figure 5.29 **Inbred song sparrows are less likely to survive a population crash** During their long-term study, Keller et al. (1994) observed several periods of difficult environmental conditions during which most individuals in their study populations died. The open bars show the percentage of birds that died with a given value of F, and the filled bars the percentage that survived with a given value of F. As an example of how to read this figure, note that 70% of the survivors had an F of 0, while only 32 percent of the nonsurvivors had an F of 0.

pression; others showed no discernable effect of type of mating; still others showed improved performance under inbreeding.

Given the theory and data we have reviewed on inbreeding depression, it is not surprising that animals and plants have evolved mechanisms to avoid it (see Box 5.6). But under some circumstances inbreeding may be unavoidable. In

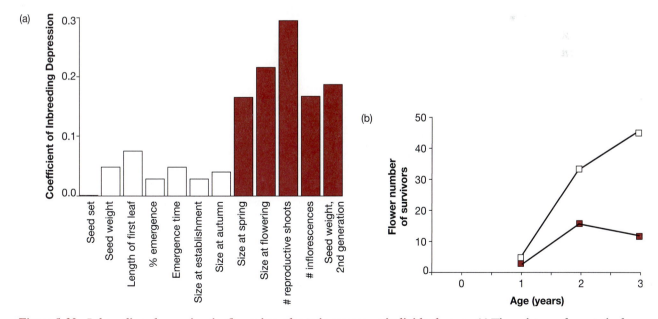

Figure 5.30 **Inbreeding depression in flowering plants increases as individuals age** (a) These data are for waterleaf, a biennial. The open bars show data from the first year of growth; the filled bars indicate traits expressed in the second year (when the plants mature, flower, and die). Inbreeding depression is much more pronounced in the second year than the first. From Wolfe (1993).   (b) This graph compares the number of flowers produced (a measure of fitness) as a function of time for outcrossed (open boxes) versus selfed (filled boxes) individuals in *Lobelia cardinalis,* a perennial in the bluebell family. The disparity in performance increases with time, indicating that inbreeding depression becomes more pronounced with age. From Johnston (1992).

BOX 5.6   Mechanisms of inbreeding avoidance

The mechanisms animals and plants use to avoid inbreeding can be grouped into three general categories: mate choice, mating systems, and dispersal. We look at each in turn.

Mate choice

Mate choices that avoid inbreeding can take place before or after the fusion of gametes. In both cases, the key is for organisms to be able to recognize relatedness. Behavioral mechanisms of recognition typically work before fertilization takes place, and molecular mechanisms after.

What are some examples of behavioral mechanisms? In social animals like the acorn woodpecker and black-tailed prairie dog, young females will not come into breeding condition while their father is still in the group and performing most of the copulations (Koenig and Pitelka, 1979; Hoogland 1982, 1992). In humans, Shepher's (1971) classic study showed that unrelated human children who grew up together on Israeli kibbutzim, spending each day interacting like siblings, were extremely unlikely to marry one another later (see also Wolf 1970).

Molecular controls over close inbreeding are also widespread. Over 40% of flowering plants, for example, have genetically controlled self-incompatibility systems. If the pollen that falls on a stigma belongs to the same individual, these systems produce immune reactions that inhibit pollen germination or limit pollen tube growth (see Ebert et al. 1989; Lee et al. 1994). And there is increasing evidence that an analogous phenomenon may take place in vertebrates. Here the self-recognition system is based on the MHC (major histocompatibility complex) loci. Recall from our discussion of immune system function in Chapter 1 that infected cells display pieces of bacterial or viral protein on their surfaces, for possible recognition by cytotoxic T lymphocytes. The membrane proteins that display these antigens are created by a cluster of genes called the major histocompatibility complex, or MHC. The supposition is that the more diversity in the proteins encoded by the MHC, the more foreign peptides can be bound and responded to, which increases resistance to disease. Data from a variety of recent human studies (for example, Gill 1992; Ober et al. 1992) show an increased probability of spontaneous abortion (and fertility problems) when mates share key MHC alleles, suggesting that inbreeding at these loci is selected against strongly.

Wayne Potts and co-workers (1991) performed a ground-breaking experiment demonstrating that MHC genotypes can, in fact, affect mate choice *prior* to fertilization, at least in mice. Mice and rats can distinguish MHC alleles on the basis of odor in urine. Potts' team typed MHC alleles in adult mice that were later released into an experimental barn. After the mice set up territories and bred, the research team typed the MHC alleles present in the offspring. When they did, they found an excess of heterozygotes compared with random mating expectations (Table 5.7). Why? First, females clearly preferred to settle on territories held by a male with an unlike MHC genotype. Second, females frequently performed extraterritorial copulations, and tended to do so with males that had MHC genotypes unlike that of the male on their territory.

The message of these studies is that mate choice—either before or after fertilization—can

TABLE 5.7 Extraterritorial matings in mice produce more MHC heterozygotes than expected

When Potts et al. (1991) compared the observed number of MHC heterozygotes to the number expected if females had mated with the territorial male, χ^2 tests showed a significant difference (P < 0.001). This suggests that females seek out copulations with males who have MHC genotypes unlike that of their territorial male.

	Heterozygotes		*Homozygotes*	
	Observed	**Expected**	**Observed**	**Expected**
Number of offspring	137	118.6	26	44.4

avoid inbreeding both at the level of the entire genome and at the level of specific loci.

Mating system

Although it is difficult or impossible to prove that inbreeding avoidance was the primary cause of their evolution, several mating systems in seed plants seem designed to increase outcrossing (Darwin 1876, 1877). These include

- **Dioecy:** The sexes are housed on different plants.
- **Monoecy:** Male and female flowers on a plant are separate.
- **Heterostyly:** In many species that have perfect flowers (both male and female structures in the same flower), differences in the lengths of the style and anther filaments make pollen transfer within the flower unlikely. Botanists sometimes refer to heterostyly as morphological self-incompatibility.

Avocado flowers accomplish the same end with a different means: They open first when their female structures are receptive to pollen, then close and reopen the next day when only the male parts are mature. Even if inbreeding avoidance was not the cause for their evolution, inbreeding avoidance is an effect of traits like these.

Dispersal

The dispersal of juvenile animals away from the site of their hatching or birth is often thought of as an inbreeding avoidance mechanism. As with plant mating systems, though, this assertion is almost impossible to

test rigorously. Certainly, definite patterns in dispersal exist. In many birds, young females disperse while males spend their entire lives close to their natal territory; mammals often show the opposite pattern.

The situation is more complex in plants, though, and research has shown that far from being an effective inbreeding avoidance mechanism, pollen dispersal is so limited in many species that inbreeding is actually routine. Experiments conducted by Barbara Schaal (1980) on the Texas bluebonnet, *Lupinus texensis,* show this limited dispersal particularly well. Schaal used allozyme electrophoresis to type plants homozygous for one of three alleles of the phosphoglucose isomerase-1 *(Pgi-1)* locus: a slow, medium, and rapidly migrating allele. Schaal then planted 91 plants 0.6m apart in a hexagonal array, with seven *Pgi-1^f* plants in the center, *Pgi-1^s* plants lined up on one margin, and *Pgi-1^m* plants making up the remainder. After recording bumblebee flights as an estimate of pollen flow, she quantified pollen dispersal rigorously by typing the *Pgi-1* locus in the resulting seeds. The results are given in Figure 5.31a. The distribution of alleles shows that most pollen moves a minimal distance, with a very few grains moving much farther. Schaal observed the same type of distribution when she recorded seed dispersal, which in this species is accomplished through the explosive rupture of the seed capsule after drying (Figure 5.31b). The message of these experiments is that plants growing near one another in nature are typically closely related. Allele movement is so low that inbreeding is considerable, even in an obligate outcrosser like lupine.

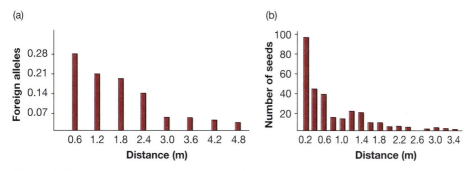

Figure 5.31 Pollen flow and seed dispersal in *Lupinus texensis*   (a) The bars indicate pollen flow, measured as the percentage of *Pgi-1^f* or *Pgi-1^s* alleles detected in the progeny of *Pgi-1^m* plants located a given distance away.   (b) Seed dispersal, measured as the number of seeds that landed a given distance away from their parent plants. From Schaal (1980).

small populations, for example, the number of potential mates for any particular individual is limited. If a population is small and remains small for many generations, and if the population receives no migrants from other populations, then eventually all the individuals in the population will be related to each other even if mating is random. Small populations thus eventually become inbred, and the individuals in them may suffer inbreeding depression. This can be a problem for rare and endangered species, and it creates a challenge for the managers of captive breeding programs (see Chapter 19).

In summary, nonrandom mating does not, by itself, alter allele frequencies. It is not, therefore, a mechanism of evolution. Nonrandom mating does, however, alter the frequencies of genotypes. Nonrandom mating can thereby change the distribution of phenotypes in a population, and alter the pattern of natural selection and the evolution of the population.

5.7 The Adaptive Landscape

We close our consideration of Mendelian genetics in populations by introducing the notion of the adaptive landscape. This concept, formulated in 1932 by Sewall Wright, integrates all the forces of evolution we have discussed so far (see question 15 at the end of this chapter). We take as our example the allelic diversity in human hemoglobin (following Templeton 1982). In our section on point mutation in Chapter 4, we mentioned two alleles of hemoglobin: the allele for regular hemoglobin, called *A*, and the allele for sickle-cell hemoglobin, *S*. As we noted earlier, when malaria is absent, selection strongly favors allele *A*. *AA* individuals get no sickling; *AS* individuals get some sickling; *SS* individuals get severe sickling. When malaria is present, however, selection maintains the *S* allele at low frequency, because selection favors the *AS* genotype (Figure 5.32). Heterozygotes get only mild sickling, and are resistant to malaria.

It happens that there is a third allele, *C*. This allele appears to confer resistance to malaria without severe sickling. It is, however, recessive to both *A* and *S*. Based on data from Livingstone (1967), Cavalli-Sforza and Bodmer (1971) estimated the fitnesses of all possible genotypes when malaria is present. A summary of genotypes, phenotypes, and fitness estimates appears in Table 5.8. Note that Cavalli-Sforza and Bodmer were able to estimate the fitnesses, particularly that of *CC* individuals, with only limited confidence (the 99% confidence in-

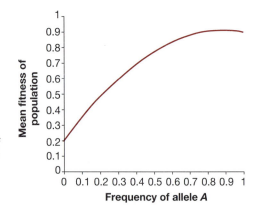

Figure 5.32 **Population mean fitness as a function of the frequency of allele *A* in a malaria zone** When malaria is present, and the only hemoglobin alleles are *A* and *S*, the population mean fitness (see Box 5.2) is highest when the frequency of the *A* allele is 0.89 and the frequency of the *S* allele is 0.11. Under natural selection, the population will evolve toward these frequencies. (Readers can confirm this by writing the equation for the change in the frequency of allele *A* from one generation to the next [see Box 5.2 and Table 5.8]. When the frequency of *A* is less than 0.89, it will be larger in the next generation. When the frequency of *A* is greater than 0.89, it will be smaller in the next generation).

TABLE 5.8 Genotypes, phenotypes, and fitnesses for human hemoglobin

Note: All fitnesses estimates assume the presence of malaria. From Cavalli-Sforza and Bodmer (1971).

Genotype	Phenotype	Fitness
AA	Susceptible; no sickling	0.9
AC	Susceptible; no sickling	0.9
CC	Resistant; little sickling	1.3
AS	Resistant; mild sickling	1.0
CS	Sickling and anemia	0.7
SS	Severe sickling and anemia	0.2

terval for the fitness of *CC* runs from 1.02 to 1.60). This example should thus be taken primarily as an illustration of genetic concepts, and only secondarily as a description of medical possibilities.

We now calculate, based on Cavalli-Sforza and Bodmer's fitness estimates, the mean fitness of a human population in a malaria zone as a function of the frequencies of the three hemoglobin alleles. Let p be the frequency of allele *A*; q be the frequency of allele *C*; and r the frequency of allele *S*. Following Box 5.2, the mean fitness is

$$\overline{w} = p^2 w_{AA} + q^2 w_{CC} + r^2 w_{SS} + 2pq\, w_{AC} + 2pr\, w_{AS} + 2qr\, w_{CS}$$

where w_{XX} = the fitness of individuals with genotype *XX*. Substituting the fitness values given in Table 5.8, we get

$$\overline{w} = p^2\, 0.9 + q^2\, 1.3 + r^2\, 0.2 + 2pq\, 0.9 + 2pr\, 1.0 + 2qr\, 0.7$$

A graph of $\overline{w}$ as a function of p and q appears in Figure 5.33a and 5.33b. The frequency p is represented along one bottom edge of the box, and q is represented along the other. Note that for any values of p and q we can determine the value of r by subtraction, because p + q + r = 1. The mean fitness of the population is represented along the vertical edge of the box. Look first at Figure 5.33a, top. This graph shows a triangular sheet. One corner is anchored near the bottom of the forward edge of the box; another corner is anchored at the top of the left edge of the box; the third corner arches away into the distance, and is anchored about three-fourths of the way up the right edge of the box. This surface is the adaptive landscape for hemoglobin allele frequencies in a malaria zone. At any given time, a population sits at a single point on this surface, with particular values of p, q, and r, and a particular mean fitness; the landscape shows the universe of possibilities for the population. Two things are immediately apparent. First, the population would have highest mean fitness if q were 1.0 (and both p and r were 0). This point, at which allele *C* is fixed, is in the upper left corner. Second, the population would have lowest fitness if r were 1.0 (and both p and q were 0). This point, at which allele *S* is fixed, is at the forward edge of the box. Figure 5.33b, top, shows the same landscape viewed from behind; it is what a person looking at Figure 5.33a, top, would see after walking around the right edge of the box and looking back. Mostly what is visible is the underside of the

An adaptive landscape shows the mean fitness in a population over all possible combinations of allele frequencies.

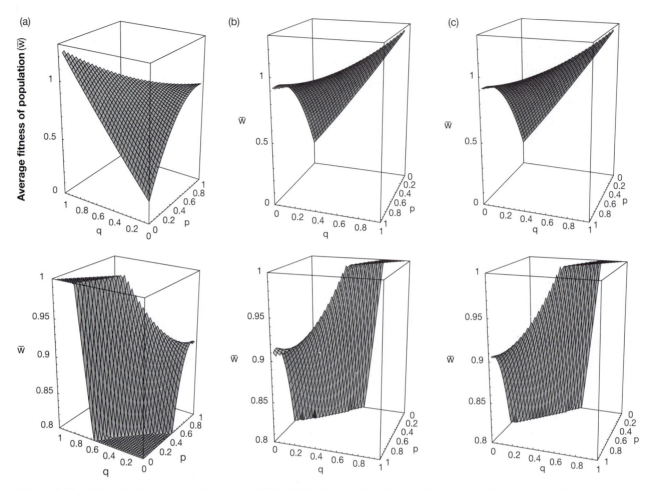

Figure 5.33 **Adaptive landscapes for hemoglobin allele frequencies in a malaria zone** See text for explanation.

surface; only at the left edge, near the 1 on the $\overline{w}$ axis, does the top of the surface appear.

In Figure 5.33a and b, bottom, the $\overline{w}$ axis runs from 0.8 to 1.0; that is why the surface now hits the floor at the forward edge of the box and the ceiling at the upper left. Expanding the middle of the $\overline{w}$ axis this way reveals another feature of the adaptive landscape: a small peak, located at p = 0.89, q = 0, and r = 0.11, near the right edge in Figure 5.33a, bottom, and near the left edge in Figure 5.33b, bottom. A population sitting on this peak would enjoy a higher mean fitness than if it were sitting at any of the immediately surrounding points (which form a saddle), but considerably lower mean fitness than if it were sitting at the high corner where q = 1.

Note that the topography of the adaptive landscape depends crucially on the environment. In the absence of malaria, the highest peak would be at the point p = 1, q = 0, r = 0; that is, without malaria the population would achieve highest mean fitness if allele *A* were fixed. Resistance confers no advantage when malaria is not present, and the genotype *CC* results in at least a small amount of sickling. This means that without malaria, allele *C* will be selected against.

Now imagine a population that has never experienced malaria and is sitting at the point p = 1, q = 0, r = 0. Imagine that malaria appears and the adaptive landscape shifts to that shown in Figure 5.33a and b. Perhaps the population moves into a malaria area, or the local climate changes to create a habitat favorable for the mosquitoes that spread the disease. What happens to the allele frequencies in the population? The short-term answer is nothing. We stipulated that the allele *A* is fixed in the population. With no other alleles present, selection cannot change the allele frequencies. So we need to allow mutation to create a few *C* alleles and a few *S* alleles. With mutation, the frequency p is not quite 1, q is not quite 0, and r is not quite 0.

Now what happens? Selection changes the allele frequencies in the population in such a way that the mean fitness of the population increases. In other words, the population climbs one of the peaks in the adaptive landscape. But which peak? Note that when the *S* and *C* alleles are rare, virtually all of them will occur in *AS* and *AC* heterozygotes. Everyone else in the population is an *AA* homozygote. As shown in Table 5.8, *AS* heterozygotes have higher fitness than *AA* homozygotes, but *AC* heterozygotes do not. Because the *C* allele is recessive to the *A* allele, selection cannot see the *C* allele, or act to increase its frequency, until it reaches a high enough frequency that it starts to appear in an appreciable number of *CC* homozygotes. When *S* and *C* are rare, selection acts to increase the frequency of the *S* allele, but not the *C* allele. The population climbs the lower peak, ending up at the spot where p = 0.89, q = 0, and r = 0.11. This is not much of an improvement in mean fitness over the spot where the population started.

Because selection pushes populations upward, never downward, populations tend to climb peaks of high mean fitness on the adaptive landscape. . . .

Once the population climbs to the top of the lower peak, it is stuck there. Even if mutation creates a few *C* alleles, all will still appear in heterozygotes, and selection will not act to increase the frequency of *C*. What forces could knock the population off the lower peak, and put it in a spot from which it could climb the higher peak, toward fixation of *C*? One force is migration. If enough individuals carrying allele *C* moved into the population from elsewhere, they might increase the frequency of *C* enough that *CC* homozygotes begin to appear. Another force is drift. If the population is small enough, then sampling variation in some lucky generation might increase the frequency of *C* enough that *CC* homozygotes begin to appear.

. . . But they may get trapped on relatively low local peaks.

A final force that could do the trick is inbreeding. Inbreeding has a curious effect: Because it systematically increases the frequency of homozygotes and decreases the frequency of heterozygotes, it changes the shape of the adaptive landscape itself. This is because population mean fitness is a function of genotype frequencies. Figure 5.33c, top and bottom, shows the adaptive landscape when the inbreeding coefficient (F) for the population is 0.1. The lower of the two peaks has been flattened, making it much easier for the population to climb the higher peak.

The sickle-cell example demonstrates that we need all the forces we have discussed in this chapter to understand evolution in the real world. Mutation provides the genetic variation that is the raw material for evolutionary change. Selection drives populations toward greater adaptation. Migration, gene flow, drift, and mating patterns can all influence which route populations travel toward greater adaptation, and on which adaptive peaks populations come to rest.

Adaptive landscapes provide a way to integrate all the forces of evolution.

Summary

Population genetics represents a synthesis of Mendelian genetics and Darwinian evolution, and is concerned with the mechanisms that cause allele frequencies to change from one generation to the next. The Hardy-Weinberg equilibrium principle is a null model that provides the conceptual framework for population genetics. It shows that under simple assumptions—no selection, no mutation, no migration, no genetic drift, and random mating—allele frequencies will not change. Furthermore, genotype frequencies can be calculated from allele frequencies.

When any one of the first four assumptions is violated, allele frequencies may change across generations. Selection, mutation, migration, and genetic drift are thus the four mechanisms of evolution. Nonrandom mating does not cause allele frequencies to change, and is thus not a mechanism of evolution. It can, however, alter genotype frequencies and thereby affect the course of evolution.

Population biologists can measure allele and genotype frequencies in real populations. Thus biologists can test whether allele frequencies are stable across generations and whether the genotype frequencies conform to Hardy-Weinberg expectations. If either of the conclusions of the Hardy-Weinberg analysis is violated, it means that one or more of the assumptions does not hold. The nature of the deviation from Hardy-Weinberg expectations does not, by itself, identify the faulty assumption. We can, however, often infer which mechanisms of evolution are at work based on other characteristics of the populations under study.

Questions

1. Black color in horses is governed primarily by a recessive allele at the *A* locus. *AA* and *Aa* horses have nonblack coats, such as bay, while *aa* horses have black coats. A few years ago, a reader of the Internet newsgroup "rec.equestrian" asked why there are relatively few black horses of the Arabian breed. One response was "Black is a rare color because it is recessive. More Arabians are bay or grey because those colors are dominant." What is wrong with this explanation? (Assume that the assumptions of the Hardy-Weinberg model are valid for the *A* locus, which was probably true at the time of this discussion.) More generally, what does the Hardy-Weinberg model show us about the impact that an allele's dominance or recessiveness has on its frequency? That is, should an allele become more common (or less common) simply because it is dominant (or recessive)?

2. We used Figure 5.7 as an example of how the frequency of an allele (*Adh^S* in fruit flies) does not change in unselected (control) populations, but does change in response to selection. However, look again at the unselected control lines in Figure 5.7. The frequency of the *Adh^S* allele in the two control populations did not change very much, but in fact it did change a little, moving up and down over time. Does the Hardy-Weinberg model hold for these control populations? Which assumption of the Hardy-Weinberg model is most probably being violated? If this experiment were done again, what simple change in experimental design would reduce this undesirable deviation from Hardy-Weinberg equilibrium?

3. Chevillon et al.'s 1995 study on acetylcholinesterase (ACE) in mosquitoes showed that *Ace^R* frequency is correlated with the history of insecticide use at certain geo-

graphical locations (Figure 5.8). The obvious explanation is that insecticide use caused an increase in Ace^R frequency. However, in your opinion, do the data show that this is definitely what happened? Does the correlation show that Ace^R actually increased from a lower to a higher frequency within populations exposed to insecticide? Does it show that insecticide use is definitely the direct cause of the higher Ace^R frequency in populations exposed to insecticide? What are some possible alternative explanations for the data? How could you test your ideas?

4. Most animal populations have a 50:50 ratio of males to females. This does not have to be so; it is theoretically possible for parents to produce predominantly male offspring, or predominantly female offspring. Imagine a monogamous population with a male-biased sex ratio of, say, 70% males and 30% females. Which sex will have an easier time finding a mate? As a result, which sex will probably have higher average fitness (lifetime reproductive success)? Which parents will have higher fitness—those that produce mostly males, or those that produce mostly females? Now imagine the same population with a female-biased sex ratio, and answer the same questions. What pattern of selection is probably maintaining the 50:50 sex ratio seen in most populations?

5. Jared Diamond, a physiologist and evolutionary biologist, has suggested that people heterozygous at the cystic fibrosis allele might have increased resistance to cholera. Cholera typically kills by altering chloride channel permeability in the gut lining, causing severe dehydration due to diarrhea; and cystic fibrosis affects chloride channels. If Diamond is correct, how might this explain the puzzlingly high "mutation rate" for the cystic fibrosis gene that we calculated in section 5.3 following Felsenstein (1997)?

6. Remote oceanic islands are famous for their endemic species—unique forms that occur nowhere else (see Quammen 1996 for a gripping and highly readable account). Consider migration as a mechanism of evolution. If the evolution of a population into a new species entails the population's genetic divergence from other populations, why should endemics be so common on remote islands?

7. Bodmer and McKie (1995) review several cases, in addition to Ellis-van Creveld syndrome in the Amish, in which genetic diseases occur at unusually high frequency in populations that are, or once were, relatively isolated. An enzyme deficiency called hereditary tyrosinemia, for example, occurs at an unusually high rate in the Chicoutimi region north of Quebec city in Canada. A condition called porphyria is unusually common in South Africans of Dutch descent. Why are genetic diseases so common in isolated populations? What else do these populations all have in common?

8. The Amish population in Pennsylvannia now numbers well over 10,000. What evolutionary forces are presently acting on the allele for Ellis-van Creveld syndrome in this population? Why has the allele remained at relatively high frequency?

9. Reconsider the question about endemic species on remote islands (question 6) in the context of genetic drift. How do plant and animal species become established on remote islands? Does the process of establishment help explain why endemic species are common on remote islands? Are island endemics more likely to evolve in some kinds of plants and animals than others?

10. As we have seen, inbreeding can reduce offspring fitness by exposing deleterious recessive alleles. However, some animal breeders practice generations of careful inbreeding within a family, or "line breeding," and surprisingly many of the line-bred animals—from champion dogs to prize cows—have normal health and fertility. How can it be possible to continue inbreeding for many generations without causing

inbreeding depression due to recessive alleles? (Hint: Responsible animal breeders do not breed from animals known to carry deleterious recessive alleles.) More generally, if a small population continues to inbreed for many generations, what will happen to the frequency of the deleterious recessive alleles in the population over time?

11. In the mid-1980s, conservation biologists reluctantly recommended that zoos should *not* try to preserve captive populations of all the endangered species of large cats. For example, some biologists recommended ceasing efforts to breed the extremely rare Asian lion, the beautiful species seen in Chinese artwork. In place of the Asian lion, the biologists recommended increasing the captive populations of other endangered cats, such as the Siberian tiger and Amur leopard. By reducing the number of species kept in captivity, the biologists hoped to increase the captive population size of each species to several hundred, preferably at least 500. Why did the conservation biologists think that this was so important as to be worth the risk of losing the Asian lion forever?

12. In this chapter we saw that in many cases, allele frequencies in small populations change at different rates than in large populations. As a review, state whether the following processes will typically have greater, smaller, or similar effects on the rate of evolution in small versus large populations:

    Selection
    New mutations per individual
    New mutations per generation in the whole population
    Migration
    Genetic drift
    Inbreeding

    How would our analysis of the adaptive landscape for sickle-cell anemia change in a small versus large population?

Exploring the Literature

13. In the example of the scale-eating fish, we saw that frequency-dependent selection tends to maintain an even ratio of left-handed fish to right-handed fish. See the following references for some interesting cases of possible frequency-dependent selection in other species. How plausible do you find each scenario?

    Raymond, M., D. Pontier, A.B. Dufour, and A.P. Møller. 1996. Frequency-dependent maintenance of left-handedness in humans. *Proceedings of the Royal Society of London,* Series B 263:1627–1633.

    Sinervo, B., and C.M. Lively. 1996. The rock-paper-scissors game and the evolution of alternative male strategies. *Nature* 380:240–243.

    Smithson, A., and M.R. MacNair. 1996. Frequency-dependent selection by pollinators: Mechanisms and consequences with regard to behaviour of bumblebees *Bombus terrestris* (L.) (Hymenoptera: Apidae). *Journal of Evolutionary Biology* 9:571–588.

14. We mentioned in section 5.6 that inbreeding depression is a concern for biologists trying to conserve endangered organisms with small population sizes. Geneticists have recently discovered that inbreeding depression varies among environments and among families. For papers that explore the implications of this discovery for conservation efforts, see:

    Pray, L.A., J.M. Schwartz, C. J. Goodnight, and L. Stevens. 1994. Environmental dependency of inbreeding depression: Implications for conservation biology. *Conservation Biology* 8:562–568.

    Pray, L.A., and C.J. Goodnight. 1995. Genetic variation in inbreeding depression in the red flour beetle *Tribolium castaneum*. *Evolution* 49:176–188.

15. The version of the adaptive landscape that we presented in section 5.7 is actually somewhat different from the original version of the concept that Sewall Wright presented in 1932. Furthermore, there is even a third common interpretation of the adaptive landscape idea. For a discussion of the differences among the three versions, see Chapter 9 in:

Provine, W.B. 1986. *Sewall Wright and Evolutionary Biology* (Chicago: University of Chicago Press).

For Sewall Wright's response to Provine's history, see:

Wright, S. 1988. Surfaces of selective value revisited. *The American Naturalist* 131:115–123.

Wright's original 1932 paper is reprinted in Chapter 11 of:

Wright, S. 1986. *Evolution: Selected papers,* William B. Provine, ed. (Chicago: University of Chicago Press).

Citations

Much of the population genetics material in this chapter is modeled after presentations in the following:

Crow, J.F. 1983. *Genetics Notes* (Minneapolis, MN: Burgess Publishing).

Felsenstein, J. 1997. *Theoretical Evolutionary Genetics* (Seattle, WA: ASUW Publishing, University of Washington).

Griffiths, A.J.F., J.H. Miller, D.T. Suzuki, R.C. Lewontin, and W.M. Gelbert. 1993. *An Introduction to Genetic Analysis* (New York: W.H. Freeman).

Templeton, A.R. 1982. Adaptation and the integration of evolutionary forces. In R. Milkman, ed., *Perspectives on Evolution* (Sunderland, MA: Sinauer), 15–31.

Here is the list of all other citations in this chapter:

Ayala, F.J., and J.A. Kiger, Jr. 1984. *Modern Genetics* (Menlo Park, CA: Benjamin/Cummings).

Bodmer, W., and R. McKie. 1995. *The Book of Man* (New York: Scribner).

Buri, P. 1956. Gene frequency in small populations of mutant *Drosophila. Evolution* 10:367–402.

Butlin, R.K., P.M. Collins, S.J. Skevington, and T.H. Day. 1982. Genetic variation at the alcohol dehydrogenase locus in natural populations of the seaweed fly, *Coelopa frigida. Heredity* 48:45–55.

Camin, J.H., and P.R. Ehrlich. 1958. Natural selection in water snakes (*Natrix sipedon* L.) on islands in Lake Erie. *Evolution* 12:504–511.

Castle, W.E. 1903. The laws of heredity of Galton and Mendel, and some laws governing race improvement by selection. *Proceedings of the American Academy of Arts and Sciences* 39:223–242.

Cavalli-Sforza, L.L., and W.F. Bodmer. 1971. *The Genetics of Human Populations* (San Francisco: W.H. Freeman and Company).

Cavener, D.R., and M.T. Clegg. 1981. Multigenic response to ethanol in *Drosophila melanogaster. Evolution* 35:1–10.

Chevillon, C., N. Pasteur, M. Marquine, D. Heyse, and M. Raymond. 1995. Population structure and dynamics of selected genes in the mosquito *Culex pipiens. Evolution* 49:997–1007.

Darwin, C. 1876. *The Effects of Cross and Self Fertilization in the Vegetable Kingdom* (London: John Murray).

Darwin, C. 1877. *The Different Forms of Flowers on Plants of the Same Species* (London: John Murray).

Dudash, M.R. 1990. Relative fitness of selfed and outcrossed progeny in a self-compatible, protandrous species, *Sabatia angularis* L. (Gentianaceae): A comparison in three environments. *Evolution* 44:1129–1139.

Dudash, M.R., D.E. Carr, and C.B. Fenster. 1997. Five generations of enforced selfing and outcrossing in *Mimulus guttatus:* Inbreeding depression variation at the population and family level. *Evolution* 51:54–65.

Ebert, P.R., M.A. Anderson, R. Bernatzky, M. Altschuler, and A.E. Clarke. 1989. Genetic polymorphism of self-incompatibility in flowering plants. *Cell* 56:255–262.

Elena, S.F., V.S. Cooper, and R.E. Lenski. 1996. Punctuated evolution caused by selection of rare beneficial mutations. *Science* 272:1802–1804.

Giles, B.E., and J. Goudet. 1997. Genetic differentiation in *Silene dioica* metapopulations: Estimation of spatiotemporal effects in a successional plant species. *The American Naturalist* 149:507–526.

Gill, T.J. III. 1992. Influence of MHC and MHC-linked genes on reproduction. *American Journal of Human Genetics* 50:1–5.

Grant, P.R., and B.R. Grant. 1995. The founding of a new population of Darwin's finches. *Evolution* 49:229–240.

Greenwood, P.J., P.H. Harvey, and C.M. Perrins. 1978. Inbreeding and dispersal in the great tit. *Nature* 271:52–54.

Hamilton, W.D. 1967. Extraordinary sex ratios. *Science* 156:477–488.

Hamilton, W.D. 1979. Wingless and fighting males in fig wasps and other insects. In M.S. Blum and N.A. Blum, eds., *Sexual Selection and Reproductive Competition in Insects* (New York: Academic Press), 167–220.

Hardy, G.H. 1908. Mendelian proportions in a mixed population. *Science* 28:49–50.

Hill, J.L. 1974. *Peromyscus:* Effect of early pairing on reproduction. *Science* 186:1042–1044.

Hoogland, J.L. 1982. Prairie dogs avoid close inbreeding. *Science* 215:1639–1641.

Hoogland, J.L. 1992. Levels of inbreeding among prairie dogs. *The American Naturalist* 139:591–602.

Hori, M. 1993. Frequency-dependent natural selection in the handedness of scale-eating cichlid fish. *Science* 260:216–219.

Johnston, M. 1992. Effects of cross and self-fertilization on progeny fitness in *Lobelia cardinalis* and *L. siphilitica. Evolution* 46:688–702.

Keller, L., P. Arcese, J.N.M. Smith, W.M. Hochachka, and S.C. Stearns. 1994. Selection against inbred song sparrows during a natural population bottleneck. *Nature* 372: 356–357.

Kerr, W.E., and S. Wright. 1954. Experimental studies of the distribution of gene frequencies in very small populations of *Drosophila melanogaster*. I. *Forked. Evolution* 8:172–177.

King, R.B. 1987. Color pattern polymorphism in the Lake Erie water snake, *Nerodia sipedon insularum. Evolution* 41:241–255.

King, R.B. 1993a. Color pattern variation in Lake Erie water snakes: Inheritance. *Canadian Journal of Zoology* 71:1985–1990.

King, R.B. 1993b. Color-pattern variation in Lake Erie water snakes: Prediction and measurement of natural selection. *Evolution* 47:1819–1833.

King, R.B., and R. Lawson. 1995. Color-pattern variation in Lake Erie water snakes: The role of gene flow. *Evolution* 49:885–896.

Klug, W.S., and M.R. Cummings. 1997. *Concepts of Genetics,* 5th ed. (Upper Saddle River, NJ: Prentice Hall).

Koenig, W.D., and F.A. Pitelka. 1979. Relatedness and inbreeding avoidance: Counterplays in the communally nesting acorn woodpecker. *Science* 206:1103–1105.

Lee, H.-S., S. Huang, and T.-h. Kao. 1994. S proteins control rejection of incompatible pollen in *Petunia inflata. Nature* 367:560–563.

Lenski, R.E., and M. Travisano. 1994. Dynamics of adaptation and diversification: A 10,000-generation experiment with bacterial populations. *Proceedings of the National Academy of Sciences, USA* 91:6808–6814.

Livingstone, F.B. 1967. *Abnormal Hemoglobins in Human Populations* (Chicago: Aldine).

Markow, T., P.W. Hedrick, K. Zuerlein, J. Danilovs, J. Martin, T. Vyvial, and C. Armstrong. 1993. HLA polymorphism in the Havasupai: Evidence for balancing selection. *American Journal of Human Genetics* 53:943–952.

McCall, C., D.M. Waller, and T. Mitchell-Olds. 1994. Effects of serial inbreeding on fitness components in *Impatiens capensis. Evolution* 48:818–827.

Mendel, G.J. 1865. Versuche uber Pflanzenhybriden (Experiments in plant hybridisation). *Verh. Naturf. Ver. in Brunn, Abhandlungen,* IV. English translation in Bennett, J.H., editor. 1965. Experiments in plant hybridisation; Mendel's original paper in English translation, with commentary and assessment by Sir Ronald A. Fisher, together with a reprint of W. Bateson's biographical notice of Mendel (Edinburgh: Oliver & Boyd).

Ober, D., S. Elias, D.D. Kostyu, and W.W. Hauck. 1992. Decreased fecundability in Hutterite couples sharing HLA-DR. *American Journal of Human Genetics* 50:6–14.

Paul, R.E.L., M.J. Packer, M. Walmsley, M. Lagog, L.C. Ranford-Cartwright, R. Paru, and K.P. Day. 1995. Mating patterns in malaria parasite populations of Papua New Guinea. *Science* 269:1709–1711.

Postlethwait, J.H., and J.L. Hopson. 1992. *The Nature of Life,* 2nd ed. (New York: McGraw-Hill).

Potts, W.K., C.J. Manning, and E.K. Wakeland. 1991. Mating patterns in seminatural populations of mice influenced by MHC genotype. *Nature* 352:619–621.

Provine, W.B. 1971. *The Origins of Theoretical Population Genetics* (Chicago: The University of Chicago Press).

Purves, W.K., and G.H. Orians. 1987. *Life: The Science of Biology* (Sunderland, MA: Sinauer).

Quammen, D. 1996. *The Song of the Dodo* (New York: Touchstone).

Ralls, K., K. Brugger, and J. Ballou. 1979. Inbreeding and juvenile mortality in small populations of ungulates. *Science* 206:1101–1103.

Read, A.F., A. Narara, S. Nee, A.E. Keymer, and K.P. Day. 1992. Gametocyte sex ratios as indirect measures of outcrossing rates in malaria. *Parasitology* 104:387–395.

Schaal, B.A. 1980. Measurement of gene flow in *Lupinus texensis. Nature* 284:450–451.

Shepher, J. 1971. Self-imposed incest avoidance and exogamy in second generation kibbutz adults. *Archives of Sexual Behavior* 1:293–307.

Stern, C. 1973. *Principles of Human Genetics* (San Francisco: W. H. Freeman & Co.).

Thompson, E.A., and J.V. Neel. 1997. Allelic disequilibrium and allele frequency distribution as a function of social and demographic history. *American Journal of Human Genetics* 60:197−204.

van Noordwijk, A.J., and W. Scharloo. 1981. Inbreeding in an island population of the great tit. *Evolution* 35:674−688.

Wedekind, C., T. Seebeck, F. Bettens, and A. Paepke. 1995. MHC dependent mate preferences in humans. *Proceedings of the Royal Society of London,* Series B 260:245−249.

Weinberg, W. 1908. Ueber den nachweis der vererbung beim menschen. *Jahreshefte des Vereins für Vaterländische Naturkunde in Württemburg* 64:368−382. English translation in Boyer, S. H. 1963. Papers on Human Genetics (Englewood Cliffs, NJ: Prentice Hall).

Wolf, A.P. 1970. Childhood association and sexual attraction: A further test of the Westermarck hypothesis. *American Anthropologist* 72:503−515.

Wolfe, L.M. 1993. Inbreeding depression in *Hydrophyllum appendiculatum:* Role of maternal effects, crowding, and parental mating history. *Evolution* 47:374−386.

Wright, S. 1931. Evolution in Mendelian populations. *Genetics* 16:97−159.

Wright, S. 1932. The roles of mutation, inbreeding, crossbreeding, and selection in evolution. *Proceedings of the 6th International Congress of Genetics* 1:356−366.

Yule, G.U. 1902. Mendel's laws and their probable relations to intra-racial heredity. *New Phytologist* 1:193−207; 222−238.

Zar, J.H. 1984. *Biostatistical Analysis* (Englewood Cliffs, NJ: Prentice Hall).

Evolution at Multiple Loci: Linkage, Sex, and Quantitative Genetics

In Chapter 5 we identified the mechanisms of evolution by exploring population genetics. In our treatment of population genetics, our models and examples concerned traits that are (or appear to be) controlled by alleles at a single locus. We examined the evolution of only one such trait at a time. In real organisms, of course, the number of loci is in the hundreds or thousands. Furthermore, most traits of organisms are determined by the combined influence of alleles at many loci. In Chapter 6 we take our models of the mechanics of evolution closer to real organisms by considering two or more loci simultaneously.

Our first step in that direction involves considering two loci at a time. The conceptual model we use is a direct extension of the Hardy-Weinberg analysis introduced in Chapter 5. The two-locus model uncovers no new mechanisms of evolution, but it does reveal a second kind of equilibrium in addition to Hardy-Weinberg equilibrium. This second kind of equilibrium is called linkage equilibrium. When a population is in linkage equilibrium, evolution at one locus is independent of evolution at other loci, and can thus be analyzed with the tools we used in Chapter 5. When a population is not in linkage equilibrium— when it is in linkage disequilibrium—selection acting at one locus can cause evolution at other loci. Linkage disequilibrium is a somewhat abstract concept, but the effort it takes to understand it will be rewarded by insight into the adaptive significance of one of the most striking and puzzling characteristics of organisms: sexual reproduction.

The two-locus model is an improvement over the one-locus model, but it is still inadequate for analyzing the evolution of traits determined by the combined effects of alleles at large numbers of loci. When studying such traits, we usually do not know the identities of the loci involved. Quantitative genetics is the branch of evolutionary biology that provides tools for analyzing such traits.

Quantitative genetics allows us to estimate the degree to which a trait is heritable and the strength of selection on the trait. Estimates of these two quantities allow us to predict the evolutionary response to selection.

6.1 Evolution at Two Loci: Linkage Equilibrium and Linkage Disequilibrium

In this section, we expand the one-locus version of the Hardy-Weinberg analysis to consider two loci simultaneously. We start by defining linkage equilibrium and outlining where our discussion is headed. Imagine a locus A with alleles *A* and *a*, and a locus B with alleles *B* and *b*. These loci are part of the genome of a population of random-mating individuals. If, in this population, an individual's genotype at locus A is independent of the individual's genotype at locus B, then the population is in linkage equilibrium. This is the case if the frequencies of *BB*, *Bb*, and *bb* are the same among *AA* individuals as they are among *Aa* individuals and *aa* individuals (Figure 6.1a). If an individual's genotype at locus A is not independent of the individual's genotype at locus B, then the population is in linkage disequilibrium. This is the case if the frequencies of *BB*, *Bb*, and *bb* are different among *AA* individuals than they are among *Aa* individuals or *aa* individuals (Figure 6.1b).

Linkage disequilibrium is the nonrandom association of genotypes at different loci.

The major concepts in our discussion of linkage equilibrium and linkage disequilibrium are as follows:

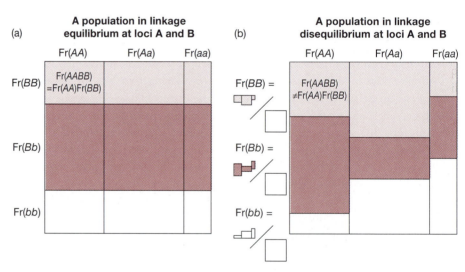

Figure 6.1 **Populations in linkage equilibrium and linkage disequilibrium**   Fr(X) is the frequency in the population of genotype X.   (a) The frequencies of *BB*, *Bb*, and *bb* are the same among *AA*, *Aa*, and *aa* individuals. We can calculate the frequency of any two-locus genotype, such as *AABB*, by multiplying the frequencies of the constituent single-locus genotypes.   (b) The frequencies of *BB*, *Bb*, and *bb* are different among *AA* versus *Aa* versus *aa* individuals. We cannot calculate the frequency of two locus genotypes by multiplying the frequencies of the constituent single-locus genotypes.

- Linkage disequilibrium can be created by physical linkage, by selection on multilocus genotypes, and by genetic drift.

- Linkage (dis)equilibrium is an important factor in the evolution of populations, because it determines whether selection at one locus affects allele and genotype frequencies at the other locus. If the population is in linkage equilibrium, then allele and genotype frequencies at locus B are unaffected by selection on genotypes at locus A. If the population is in linkage disequilibrium, then allele and genotype frequencies at locus B can be altered by selection on genotypes at locus A.

- Linkage disequilibrium is reduced by sexual reproduction with random mating.

Linkage Disequilibrium and Physical Linkage

Linkage disequilibrium takes its unwieldy name from one of the mechanisms that can produce it: the physical linkage of genes on chromosomes. One of the first examples of physical linkage was reported in 1905 by William Bateson, Edith R. Saunders, and Reginald C. Punnett (Bateson 1913; Punnett 1928). Bateson and colleagues had been doing breeding experiments on sweet peas, looking at two loci, each with two alleles. One locus controls flower color, with purple (*P*) dominant over red (*p*); the other locus controls pollen shape, with long (*L*) dominant over round (*l*). In one of their experiments, Bateson and colleagues crossed pure-breeding purple-long parents *(PPLL)* with red-round parents *(ppll)* to get double heterozygotes *(PpLl)*, then crossed these dihybrids with more red-round individuals:

$$PpLl \times ppll$$

The offspring of this cross constitute a population in which the genotype frequencies at the color locus should be 0.5 for *Pp* and 0.5 for *pp*. (To see this, complete a Punnett square for just the color locus.) Likewise, the genotype frequencies at the pollen locus should be 0.5 *Ll* and 0.5 *ll*. If the population is in linkage equilibrium, then the genotypes at the color locus are randomly distributed with respect to the genotypes at the pollen locus. In other words, an individual's genotype at one locus is independent of its genotype at the other. We can calculate the predicted frequency of any two-locus genotype, such as *Ppll*, simply by multiplying the frequencies of the constituent one-locus genotypes, in this case *Pp* and *ll*. Thus the frequency of *Ppll* should be 0.5 × 0.5 = 0.25. Four 2-locus genotypes are possible in this population *(PpLl, Ppll, ppLl, ppll)*, each with a predicted linkage-equilibrium frequency of 0.25. This is the same prediction we would make with a 4 × 4 Punnett square.

Here are the genotype frequencies that Bateson and colleagues actually found in the population of offspring from the cross:

PpLl	*Ppll*	*ppLl*	*ppll*
0.44	0.06	0.06	0.44

The frequency of the single-locus genotype *Pp* is as predicted: 0.44 + 0.06 = 0.5. So are the frequencies of the other single-locus genotypes, *pp*, *Ll*, and *ll*. But the frequencies of the two-locus genotypes are not

Linkage disequilibrium can occur when two loci are nearby on the same chromosome.

as predicted. The frequency of *PpLl* is not equal to the frequency of *Pp* times the frequency of *Ll* (that is, $0.44 \neq 0.5 \times 0.5$). In other words, the population is in linkage disequilibrium. The reason is that the color locus and the pollen locus are on the same chromosome. The original *PpLl* dihybrids got from their *PPLL* parents a chromosome carrying *P* and *L*, and from their *ppll* parents a chromosome carrying *p* and *l*. A more informative way to write their genotype is *PL/pl*. The *PL/pl* dihybrids made gametes that were mostly *PL* and *pl*.

The reason the *PL/pl* dihybrids did not make gametes that were exclusively *PL* and *pl*, yielding an offspring population that was half *PpLl* and half *ppll,* was that some crossing over occurred during meiosis. This crossing over produced gametes with the genotypes *Pl* and *pL*. These gamete genotypes are new combinations of the alleles at the two loci; they did not exist in this population before they were created by a mating between unrelated individuals followed by meiosis with crossing over. Thus the genetic effect of sexual reproduction with meiosis and crossing over is called genetic recombination. Genetic recombination reduced the level of linkage disequilibrium in the population compared to what it otherwise would have been.

Note that linkage equilibrium and Hardy-Weinberg equilibrium are not the same thing. If we let the individuals in the offspring population mate at random among themselves, the population will be in Hardy-Weinberg equilibrium at each single locus in the next generation. But the population still will not be in linkage equilibrium across the two loci. The rate at which the population will achieve linkage equilibrium depends on the rate of crossing over between the pollen locus and the color locus. We explain why later in this section.

Linkage Disequilibrium and Selection

Linkage disequilibrium can arise in a population even when the loci in question are on different chromosomes. Linkage disequilibrium can be created by selection on multilocus genotypes. Consider an example in garden peas based on a pair of unlinked loci that were studied by Gregor Mendel: tall (*T*) versus short (*t*) plants, and round (*R*) versus wrinkled (*r*) seeds. Imagine a random-mating pea population in which all four alleles are at a frequency of 0.5, with each locus in Hardy-Weinberg equilibrium, and the two loci in linkage equilibrium. Here are the frequencies of all possible single-locus genotypes, organized by phenotype:

Population A:

Tall		Short	Round		Wrinkled
	Tt			*Rr*	
TT	*Tt*	*tt*	*RR*	*Rr*	*rr*
$\frac{1}{4}$	$\frac{2}{4}$	$\frac{1}{4}$	$\frac{1}{4}$	$\frac{2}{4}$	$\frac{1}{4}$

Here are the frequencies of all possible two-locus genotypes, organized by phenotype:

Population A:

Tall, Round				Tall, Wrinkled		Short, Round		Short, Wrinkled
			TtRr					
			TtRr					
	TTRr	TtRR	TtRr		Ttrr		ttRr	
TTRR	TTRr	TtRR	TtRr	TTrr	Ttrr	ttRR	ttRr	ttrr
$\frac{1}{16}$	$\frac{2}{16}$	$\frac{2}{16}$	$\frac{4}{16}$	$\frac{1}{16}$	$\frac{2}{16}$	$\frac{1}{16}$	$\frac{2}{16}$	$\frac{1}{16}$

Notice that the frequency of any two-locus genotype is the product of the frequencies of the constituent one-locus genotypes. For example, the frequency of *TtRr* is equal to the frequency of *Tt* times the frequency of *Rr*: $\frac{4}{16} = \frac{2}{4} \times \frac{2}{4}$.

To see how selection on multilocus genotypes can create linkage disequilibrium, remove all the tall-wrinkled and short-round pea plants from the population. The new two-locus genotype frequencies are:

Population B:

Tall, Round				Tall, Wrinkled	Short, Round	Short, Wrinkled
			TtRr			
			TtRr			
	TTRr	TtRR	TtRr			
TTRR	TTRr	TtRR	TtRr			ttrr
$\frac{1}{10}$	$\frac{2}{10}$	$\frac{2}{10}$	$\frac{4}{10}$			$\frac{1}{10}$

Counting single-locus genotypes in population B gives the single-locus frequencies:

TT	**Tt**	**tt**	**RR**	**Rr**	**rr**
$\frac{3}{10}$	$\frac{6}{10}$	$\frac{1}{10}$	$\frac{3}{10}$	$\frac{6}{10}$	$\frac{1}{10}$

In population B we can no longer calculate the frequencies of the two-locus genotypes by multiplying the frequencies of the constituent one-locus genotypes. For example, the frequency of *TtRr* is not equal to the frequency of *Tt* times the frequency of *Rr* (that is, $\frac{4}{10} \neq \frac{6}{10} \times \frac{6}{10}$). The genotype *TtRr* is at a higher frequency than expected under the null hypothesis of linkage equilibrium. Also, the frequency of *Ttrr* is not equal to the frequency of *Tt* times the frequency of *rr*. *Ttrr* is at a lower frequency than expected. Selection favoring peas that are either tall and round or short and wrinkled has created linkage disequilibrium.

Linkage disequilibrium can result from selection on multilocus genotypes.

Linkage Disequilibrium and Drift

Linkage disequilibrium can also be created by genetic drift. Like selection, drift can create linkage disequilibrium even when the loci in question are on different chromosomes. To see how drift can create linkage disequilibrium, go back to population A. Imagine that the population is small and that chance events kill all the peas with the rarest phenotype, short and wrinkled. Here are the new two-locus genotype frequencies:

Population C:

Tall, Round				Tall, Wrinkled		Short, Round		Short, Wrinkled
			TtRr					
			TtRr					
	TTRr	*TtRR*	*TtRr*		*Ttrr*		*ttRr*	
TTRR	*TTRr*	*TtRR*	*TtRr*	*TTrr*	*Ttrr*	*ttRR*	*ttRr*	
$\frac{1}{15}$	$\frac{2}{15}$	$\frac{2}{15}$	$\frac{4}{15}$	$\frac{1}{15}$	$\frac{2}{15}$	$\frac{1}{15}$	$\frac{2}{15}$	

Counting single-locus genotypes in population C gives the single-locus frequencies:

TT	**Tt**	**tt**	**RR**	**Rr**	**rr**
$\frac{4}{15}$	$\frac{8}{15}$	$\frac{3}{15}$	$\frac{4}{15}$	$\frac{8}{15}$	$\frac{3}{15}$

Linkage disequilibrium can result from genetic drift.

We can no longer calculate the frequencies of the two-locus genotypes by multiplying the frequencies of the constituent one-locus genotypes. For example, the frequency of *TtRr* is not equal to the frequency of *Tt* times the frequency of *Rr* (that is, $\frac{4}{15} \neq \frac{8}{15} \times \frac{8}{15}$), and the frequency of *ttrr* is not equal to the frequency of *tt* times the frequency of *rr*. Drift has created linkage disequilibrium.

Linkage Disequilibrium and Evolution

Whether a population is in linkage equilibrium or linkage disequilibrium is of crucial importance to evolution. When a population is in linkage equilibrium, then selection at any given locus has no effect on genotype frequencies at other loci. In other words, selection at any given locus is independent of selection at all other loci. When a population is in linkage disequilibrium, however, selection at any one of the loci in disequilibrium also changes the genotype frequencies at the other loci.

To see that under linkage equilibrium selection is independent at each locus, go back again to population A, which is in linkage equilibrium. This time, select only on the seed locus by removing all the wrinkled peas:

Population D:

Tall, Round				Tall, Wrinkled	Short, Round		Short Wrinkled
			TtRr				
			TtRr				
	TTRr	*TtRR*	*TtRr*			*ttRr*	
TTRR	*TTRr*	*TtRR*	*TtRr*		*ttRR*	*ttRr*	
$\frac{1}{12}$	$\frac{2}{12}$	$\frac{2}{12}$	$\frac{4}{12}$		$\frac{1}{12}$	$\frac{2}{12}$	

The single-locus genotypes in population D are as follows:

TT	**Tt**	**tt**	**RR**	**Rr**
$\frac{3}{12} = \frac{1}{4}$	$\frac{6}{12} = \frac{1}{2}$	$\frac{3}{12} = \frac{1}{4}$	$\frac{4}{12} = \frac{1}{3}$	$\frac{8}{12} = \frac{2}{3}$

Comparing population D to population A, we can see that selection at the seed locus has altered the genotype frequencies at the seed locus, but not at the height locus.

To see that under linkage *disequilibrium* selection at one locus *can* alter genotype frequencies at the other locus, go back to population C, which is in linkage disequilibrium. Select on the seed locus by removing all the wrinkled peas:

When two loci are in linkage disequilibrium, selection at one locus can cause evolution at the other locus.

Population E:

Tall, Round				Tall, Wrinkled	Short, Round	Short, Wrinkled
			TtRr			
			TtRr			
	TTRr	*TtRR*	*TtRr*			*ttRr*
TTRR	*TTRr*	*TtRR*	*TtRr*		*ttRR*	*ttRr*
$\frac{1}{12}$	$\frac{2}{12}$	$\frac{2}{12}$	$\frac{4}{12}$		$\frac{1}{12}$	$\frac{2}{12}$

In population E the single-locus genotypes are as follows:

TT	*Tt*	*tt*	*RR*	*Rr*
$\frac{3}{12}$	$\frac{6}{12}$	$\frac{3}{12}$	$\frac{4}{12}$	$\frac{8}{12}$

Comparing population E to population C, we can see that selection on the seed locus has altered the genotype frequencies at both the seed locus and the height locus.

Our numerical examples have illustrated that (1) linkage disequilibrium can arise from physical linkage, selection, or genetic drift; (2) if a population is in linkage disequilibrium, then selection at one locus can cause evolution at another locus; and (3) linkage disequilibrium is reduced by sexual reproduction. We now develop an algebraic model of evolution at two loci. This model allows us to identify more precisely how sexual reproduction reduces linkage disequilibrium.

The Two-Locus Version of Hardy-Weinberg Analysis

When we performed the one-locus version of the Hardy-Weinberg analysis in Chapter 5, we were concerned with allele frequencies in the gene pool. Starting with these allele frequencies, we imagined random mating to make zygotes. Then we imagined that the zygotes grew to adulthood and then made gametes. We recalculated the allele frequencies in the new gene pool, and took note of whether the allele frequencies had changed. We will follow the same steps for the two-locus version.

For the two-locus version of the Hardy-Weinberg analysis, we are primarily concerned with gamete frequencies, not allele frequencies. Imagine we have two loci: locus A with alleles *A* and *a*; and locus B with alleles *B* and *b*. (We do not intend these symbols to necessarily imply a dominant/recessive relationship between alleles. Instead, we use them because they make the equations easier to read than alternative notations.) Given these loci and alleles, there are four possible gamete genotypes: *AB, Ab, aB,* and *ab*. In the one-locus version of Hardy-Weinberg, we assigned the variables p and q to represent the frequencies of *A* and *a*. We do the same here, and let s and t represent the frequencies of *B* and *b*. Our primary concern, however, is with gamete frequencies. We will let g_{AB}, g_{Ab}, g_{aB}, and g_{ab} represent the frequencies of the four gametes. Each of these frequencies can take any value between zero and one, subject to the constraint that $g_{AB} + g_{Ab} + g_{aB} + g_{ab} = 1$.

We are now in a position to define linkage equilibrium and linkage disequilibrium in terms of gamete and allele frequencies. This definition is exactly analogous to the definition we gave above in terms of two-locus and single-locus genotype frequencies. The population is in linkage equilibrium if the genotype of a gamete at locus A is independent of the genotype of the gamete at locus B.

Figure 6.2 illustrates this definition of linkage (dis)equilibrium geometrically. Figure 6.2 also illustrates the definition of a quantity called the coefficient of linkage disequilibrium (D):

D is a measure of the amount of linkage disequilibrium between two loci.

$$D = g_{AB}g_{ab} - g_{Ab}g_{aB} \tag{1}$$

The coefficient of linkage disequilibrium provides a measure of how close a population is to linkage equilibrium. When a population is in linkage equilibrium, $g_{AB} = ps$, $g_{ab} = qt$, $g_{Ab} = pt$, and $g_{aB} = qs$, so

$$D = (ps \times qt) - (pt \times qs) = 0$$

When a population is in linkage disequilibrium, $D \neq 0$. The further the population is from linkage equilibrium, the further D is from zero. The maximum value D can assume is 0.25, when *AB* and *ab* are the only gametes present and each is at a frequency of 0.5. The minimum value D can assume is -0.25, when *Ab* and *aB* are the only gametes present and each is at a frequency of 0.5. The reason we have taken the time to define D will become clear shortly.

Using the definition in Figure 6.2, we can describe intuitively how sexual reproduction with random mating reduces linkage disequilibrium. Imagine the population is at the extreme of disequilibrium; the only gametes present are *AB* and *ab*, each at a frequency of 0.5. Under random mating, half the zygotes in the population will be *AB/ab*. When these zygotes grow up and produce gametes by meiosis, some of the gametes they make will be *Ab* and *aB*. Thus *AB* and *ab* are no longer the only gamete genotypes in the population, and linkage disequilibrium has been reduced from its maximum value.

By working through the two-locus version of Hardy-Weinberg, we can show algebraically that sexual reproduction with random mating reduces linkage disequilibrium. We can also show that if a population is in linkage equilibrium, the gamete frequencies do not change across generations. We make the same assumptions about our model population that we made in Chapter 5: There is no selection, no mutation, no migration, no genetic drift, and the individuals in the population mate at random.

(a) **A population in linkage equilibrium at loci A and B**

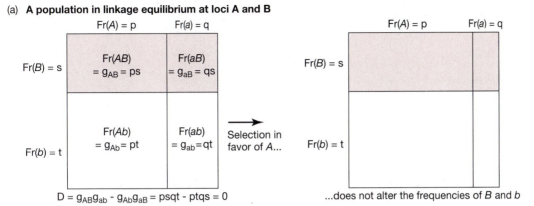

$$D = g_{AB}g_{ab} - g_{Ab}g_{aB} = psqt - ptqs = 0$$

...does not alter the frequencies of B and b

(b) **A population in linkage disequilibrium at loci A and B**

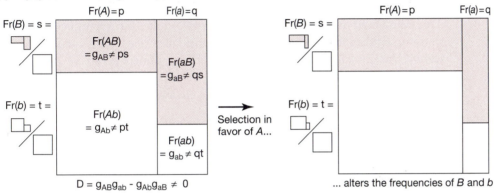

$$D = g_{AB}g_{ab} - g_{Ab}g_{aB} \neq 0$$

... alters the frequencies of B and b

Figure 6.2 **Linkage equilibrium defined at the level of gamete and allele frequencies**   (a) The frequency of allele B is the same in A individuals as it is in a individuals. We can calculate the frequency of any gamete, such as AB, by multiplying the frequencies of the constituent alleles. For example, $g_{AB} = ps$. (b) The frequency of allele B is different in A individuals than in a individuals. We cannot calculate the frequencies of gametes by multiplying the frequencies of the constituent alleles. For example, $g_{AB} \neq ps$.

From Gamete Pool to Zygotes via Random Mating

We have created a population of individuals producing four kinds of gametes, AB, Ab, aB, and ab, at population frequencies g_{AB}, g_{Ab}, g_{aB}, and g_{ab}. We allow random mating by the same process we used in Chapter 5. We take all the gametes produced by all the adults in the population and mix them together (without fertilization) in a barrel. We will call this barrel the gamete pool. Then we draw gametes from the gamete pool at random, and pair them to make zygotes. The probability of making a zygote with the genotype AB/AB is the probability of drawing an AB egg ($= g_{AB}$) multiplied by the probability of drawing an AB sperm ($= g_{AB}$), which gives $g_{AB}g_{AB}$. The probability of making a zygote with the genotype AB/Ab is the probability of drawing an AB egg and an Ab sperm plus the probability of drawing an Ab egg and an AB sperm: $2g_{AB}g_{Ab}$. In all, there are 10 possible zygote genotypes. Together with their frequencies, they are

- AB/AB
- $g_{AB}g_{AB}$

- AB/Ab
- $2g_{AB}g_{Ab}$

- AB/aB
- $2g_{AB}g_{aB}$

- AB/ab
- $2g_{AB}g_{ab}$

- Ab/Ab
- $g_{Ab}g_{Ab}$

- Ab/aB
- $2g_{Ab}g_{aB}$

- Ab/ab
- $2g_{Ab}g_{ab}$

- aB/aB
- $g_{aB}g_{aB}$

- aB/ab
- $2g_{aB}g_{ab}$

- ab/ab
- $g_{ab}g_{ab}$

Readers may want to confirm, algebraically or with a numerical example, that the zygote frequencies sum to 1.

From Zygotes to the Next Generation's Gamete Pool via Meiosis

The zygotes we just made grow up, become adults, and reproduce. Now we want to calculate the gamete frequencies in the next generation's gene pool, and see whether they have changed. We start with the gamete genotype AB, and use the symbol g_{AB}' to represent its frequency in the new gamete pool. The adult genotypes that can produce AB gametes are enclosed in boxes:

• AB/AB • $g_{AB}g_{AB}$	• AB/Ab • $2g_{AB}g_{Ab}$	• AB/aB • $2g_{AB}g_{aB}$	• AB/ab • $2g_{AB}g_{ab}$
	• Ab/Ab • $g_{Ab}g_{Ab}$	• Ab/aB • $2g_{Ab}g_{aB}$	• Ab/ab • $2g_{Ab}g_{ab}$
		• aB/aB • $g_{aB}g_{aB}$	• aB/ab • $2g_{aB}g_{ab}$
			• ab/ab • $g_{ab}g_{ab}$

We will explain shortly why some boxes are shaded. For now, note a new twist that did not factor into the one-locus version of Hardy-Weinberg. The two loci we are following might be on different chromosomes, or they might be on the same chromosome (that is, they might be linked).

The Effect of Physical Linkage on Meiosis

If the loci are on different chromosomes, then during meiosis the alleles are distributed to gametes according to Mendel's law of independent assortment. An Ab/aB individual will produce as many AB gametes as Ab gametes. If the two loci are on the same chromosome, then during meiosis the alleles at the two loci will segregate together, except for cases in which there is crossing over between the loci. An Ab/aB individual may produce more Ab gametes than AB gametes, the exact ratio depending on the rate of crossing over.

   For the genotypes in the open boxes, it does not matter whether the loci are linked, and it does not matter whether there is crossing over. All the gametes made by AB/AB individuals will be AB. Half of the gametes made by AB/Ab individuals will be AB, as will half of the gametes made by AB/aB individuals.

When Physical Linkage Matters, the Crossing-Over Rate is Equal to the Recombination Rate, r

For the genotypes in the shaded boxes, it does matter whether the genes are linked. Assume for the moment that the loci are linked. We will use the variable r to represent the rate of crossing over. The r comes from *recombination,* which describes the genetic effect of crossing over in creating new combinations of alleles on chromosomes. The value of r can be anywhere from 0 to $\frac{1}{2}$. A value of r = 0 represents loci so close together that crossing over never occurs between them. A value of $\frac{1}{2}$ represents loci so far apart that they behave as if they were on different chromosomes. (Even when the loci are on different chromosomes, *Ab*/*aB* individuals produce "recombinant" *AB* gametes only half the time.) Thus our use of the variable r lets us describe both linked and unlinked pairs of loci with the same set of equations.

The variable r lets us calculate the number of *AB* gametes made by the genotypes in the shaded boxes. Consider gamete production by the *Ab*/*aB* individuals. They will produce *AB* gametes only when crossing over occurs. When crossing over occurs, half of the gametes they produce are *AB*. Therefore, the fraction of the gametes in the gamete pool that are *AB* and come from *Ab*/*aB* individuals is equal to (r) $(\frac{1}{2})$ $2g_{Ab}g_{aB}$. Now consider the *AB*/*ab* individuals. They produce *AB* gametes only when crossing over does not occur. When crossing over does not occur, half of the gametes they produce are *AB*. The fraction of the gametes in the gamete pool that are *AB* and come from *AB*/*ab* individuals is equal to $(1 - r)(\frac{1}{2})2g_{AB}g_{ab}$.

The Frequency of AB Gametes in the New Gamete Pool

We can now write the total frequency of *AB* in the new gamete pool:

$$g_{AB}' = g_{AB}g_{AB} + (\tfrac{1}{2})2g_{AB}g_{Ab} + (\tfrac{1}{2})2g_{AB}g_{aB}$$
$$+ (r)(\tfrac{1}{2})2g_{Ab}g_{aB} + (1 - r)(\tfrac{1}{2})2g_{AB}g_{ab}$$

Every $(\frac{1}{2})$ is canceled by a 2, so we have

$$g_{AB}' = g_{AB}g_{AB} + g_{AB}g_{Ab} + g_{AB}g_{aB} + (r)g_{Ab}g_{aB} + (1 - r)g_{AB}g_{ab}$$
$$= g_{AB}g_{AB} + g_{AB}g_{Ab} + g_{AB}g_{aB} + (r)g_{Ab}g_{aB} + g_{AB}g_{ab} - (r)g_{AB}g_{ab}$$

Rearranging terms gives

$$g_{AB}' = g_{AB}g_{AB} + g_{AB}g_{Ab} + g_{AB}g_{aB} + g_{AB}g_{ab} - (r)g_{AB}g_{ab} + (r)g_{Ab}g_{aB}$$

Factoring g_{AB} out of the first four terms and r out of the last two terms gives

$$g_{AB}' = g_{AB}(g_{AB} + g_{Ab} + g_{aB} + g_{ab}) - r(g_{AB}g_{ab} - g_{Ab}g_{aB})$$

The sum $(g_{AB} + g_{Ab} + g_{aB} + g_{ab})$ equals 1, so we have

$$g_{AB}' = g_{AB} - r(g_{AB}g_{ab} - g_{Ab}g_{aB})$$

The expression $(g_{AB}g_{ab} - g_{Ab}g_{aB})$ is familiar. It is D, the coefficient of linkage disequilibrium we defined in equation 1. Thus we have

$$g_{AB}' = g_{AB} - rD \tag{2}$$

We will leave it to the reader to derive the new frequencies of the other three gametes, which are as follows:

$$g_{Ab}' = g_{Ab} + rD \qquad \text{(3)}$$

$$g_{aB}' = g_{aB} + rD \qquad \text{(4)}$$

$$g_{ab}' = g_{ab} - rD \qquad \text{(5)}$$

Conclusions

The two-locus version of the Hardy-Weinberg analysis has led us to the following conclusion: If a population is in linkage equilibrium (that is, if D = 0), and if the assumptions of the analysis hold, then equations 2–5 show that the gamete frequencies will not change from one generation to the next. If, however, the population is in linkage disequilibrium, then equations 2–5 show that the gamete frequencies will change from one generation to the next. The amount of change is proportional to the rate of recombination between the two loci. The first population geneticist to report this result was H. S. Jennings (1917).

We address one final question with the two-locus model: What happens to the level of linkage disequilibrium over many generations? We can answer this question by writing an expression for D', the linkage disequilibrium in the next generation, and simplifying it. By definition (equation 1),

$$D' = g_{AB}'g_{ab}' - g_{Ab}'g_{aB}'$$

Substitution of the expressions we just derived (equations 2–5) into this equation gives

$$D' = [(g_{AB} - rD)(g_{ab} - rD)] - [(g_{Ab} + rD)(g_{aB} + rD)]$$

$$= [g_{AB}g_{ab} - g_{AB}rD - g_{ab}rD + (rD)^2] - [g_{Ab}g_{aB} + g_{Ab}rD + g_{aB}rD + (rD)^2]$$

$$= g_{AB}g_{ab} - g_{AB}rD - g_{ab}rD + (rD)^2 - g_{Ab}g_{aB} - g_{Ab}rD - g_{aB}rD - (rD)^2$$

Canceling and rearranging terms gives

$$D' = g_{AB}g_{ab} - g_{Ab}g_{aB} - g_{AB}rD - g_{ab}rD - g_{Ab}rD - g_{aB}rD$$

$$= (g_{AB}g_{ab} - g_{Ab}g_{aB}) - rD(g_{AB} + g_{ab} + g_{Ab} + g_{aB})$$

Finally, the expression $(g_{AB}g_{ab} - g_{Ab}g_{aB})$ is equal to D (equation 1), and the expression $(g_{AB} + g_{ab} + g_{Ab} + g_{aB})$ is equal to 1, so we have

$$D' = D - rD = D(1 - r)$$

Because $(1 - r)$ is always a number between $\frac{1}{2}$ and 1, the value of D gets closer to zero every generation. The conclusion is that a random-mating population in linkage disequilibrium moves closer to linkage equilibrium every generation (Figure 6.3). The rate at which the population reaches linkage disequilibrium is proportional to the rate of recombination between the two loci in question.

Sexual reproduction with random mating reduces linkage disequilibrium.

We have established that the effect of sexual reproduction with random mating is to reduce the level of linkage disequilibrium in populations. If a particular multilocus genotype is less frequent than expected under linkage equilibrium, then random mating tends to increase its frequency. If a particular multilocus genotype is more frequent then expected, then random mating tends to reduce its frequency. We can now consider the question of whether sexual reproduction itself is favored by natural selection, and why.

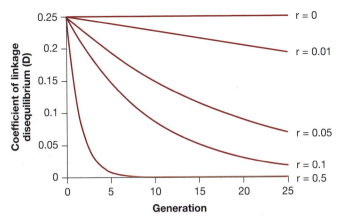

Figure 6.3 With sexual reproduction and random mating, linkage disequilibrium falls over time This graph shows the level of linkage disequilibrium between two loci over 25 generations in a random-mating population. The population starts with linkage disequilibrium at its maximum possible value, 0.25. Each curve shows the decline in linkage disequilibrium, according to the equation $D' = D(1 - r)$, for a different value of r. With r = 0.5, which corresponds to the free recombination of loci on different chromosomes, the population reaches linkage equilibrium in less than 10 generations. With r = 0.01, which corresponds to closely linked loci, linkage disequilibrium persists for many generations. After Hedrick (1983).

6.2   The Adaptive Significance of Sex

Sexual reproduction is complicated, costly, and dangerous. An individual attempting to reproduce sexually must find a member of the opposite sex and induce him or her to cooperate. This takes time and energy, and often increases the individual's risk of being killed by a predator. Engaging in sex may also expose an individual to sexually transmitted diseases. Why do organisms not avoid all the trouble and risk, and simply reproduce asexually instead? This question sounds odd to human ears, because we do not have a choice: We inherited from our ancestors the inability to reproduce by any other means than sex. But many organisms do have a choice, at least in a physiological sense: They are capable of both sexual and asexual reproduction, and regularly switch between the two.

Most aphid species, for example, have spring and summer populations composed entirely of asexual females. These females feed on plant juices and, without the participation of males, produce live-born young genetically identical to their mothers (Figure 6.4). This mode of reproduction, in which offspring develop from unfertilized eggs, is called parthenogenesis. In the fall aphids change modes, producing males and sexual females. These mate, and the females lay overwintering eggs from which a new generation of parthenogenetic females hatch the following spring.

Many species are capable of both sexual and asexual reproduction.

Many other organisms are capable of both sexual and asexual reproduction. Examples include *Volvox* and hydra (Figure 6.5), and the many species of plants that can reproduce both by sending out runners and by developing flowers that exchange pollen with other individuals.

Which Reproductive Mode Is Better: Sexual or Asexual?

The existence of two different modes of reproduction in the same population raises the question of whether one reproductive mode will replace the other over time. John Maynard Smith approached this question by imagining a sexual population in which a mutation occurs that allows females to reproduce parthenogenetically (see Maynard Smith 1978). Maynard Smith made two assumptions:

Figure 6.4   **Asexual reproduction in an aphid**   This aphid is giving birth to a daughter, produced by parthenogenesis, that is genetically identical to its mother. (P.J. Bryant/University of California at Irvine/Biological Photo Service/Terraphotographics)

1. The number of offspring a female can make depends only on the amount of food she can gather, not on her mode of reproduction.

2. The probability that an offspring will itself survive to reproduce does not depend on whether that offspring was produced sexually or parthenogenetically.

Maynard Smith also noted that all the offspring of parthenogenetic females are themselves female, whereas the offspring of sexual females are a mixture of both sexes (typically equal numbers of males and females). This simple fact, together with the two assumptions, means that parthenogenetic females will constitute a

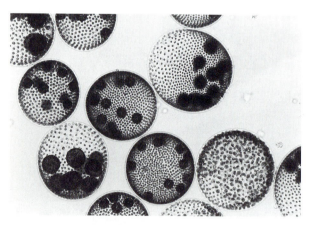

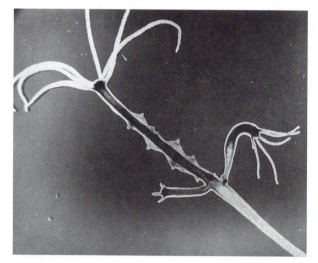

Figure 6.5   **Organisms with two modes of reproduction**   (a) *Volvox aureus,* a freshwater alga. Each large sphere is a single adult individual. Before maturity, any individual has the potential to develop as a sexual male, a sexual female, or an asexual. The (entirely visible) individual at lower right is a male. The randomly oriented disks are packets of sperm. The large individual above and to his left is a female. Each of the dark fuzzy spheres inside her is an encysted zygote. The individual directly to the left of the male is an asexual. Each of the dark spheres inside is an offspring, developing by mitosis into a clone of the parent. (Jon Herron) (b) A hydra. This individual is reproducing both sexually and asexually. The crown of tentacles at the upper left surround the hydra's mouth. Along the body below the mouth are rows of testes. Below the testes are two asexual buds. (Dr. Ralph Buchsbaum)

larger and larger fraction of the population every generation (Figure 6.6). Maynard Smith's argument is presented algebraically in Box 6.1. His conclusion was that asexual reproduction is twice as good as sexual reproduction. As a result, Maynard Smith's argument is often characterized as having identified a twofold cost of sex. According to Maynard Smith's model, parthenogenetic females should quickly take over any population in which they appear.

And yet such asexual takeovers do not seem to have happened very often. The vast majority of multicellular species are sexual, and there are many species, like aphids, *Volvox,* and hydra, in which sexual and asexual reproduction stably coexist. It was Maynard Smith's point to demonstrate that these facts represent a paradox for evolutionary theory. In this sense Maynard Smith's argument is a null hypothesis, devised to help identify the mechanism responsible for an interesting pattern.

Obviously, sex must confer benefits that allow it to persist in spite of the strong reproductive advantage offered by parthenogenesis. But what are these benefits? The mathematical logic of Maynard Smith's model is correct, so the benefits of sex must lie in the violation of one or both of Maynard Smith's assumptions. The first assumption, that the number of offspring a female can make does not depend on whether she is sexual or asexual, is violated in species in which males provide resources or parental care. Species in which female reproductive success is limited by male parental care certainly exist. Examples include humans, many birds, and pipefish (see Chapter 15). However, species with male parental care are the minority. In most species (most mammals, most insects, etc.) males contribute only genes. A general advantage to sex is thus more likely to be found in the violation of the second assumption, that the probability that a female's offspring will survive does not depend on whether she produces them asexually or sexually.

The persistence of sex is a paradox because a simple model shows that asexual females should rapidly take over any population.

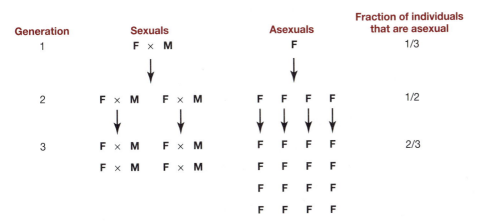

Figure 6.6 The reproductive advantage of asexual females Imagine a population founded by three individuals: a sexual female, a sexual male, and an asexual female. Every generation each female produces four offspring, after which the parents die. All offspring survive to reproduce. Half the offspring of sexual females are female; the other half are male. All the offspring of asexual females are, of course, female. Under these simple assumptions, the fraction of individuals in the population that are asexual females increases every generation. After John Maynard Smith (1978).

BOX 6.1   John Maynard Smith's model of sexual versus asexual reproduction, and the twofold cost of sex

Imagine a population of sexual organisms, with the same number of males and females. A mutation appears that causes females to reproduce parthenogenetically, so that their offspring are clones of themselves. Assume:

1. The number of eggs each female lays, k, is the same for both kinds of females.
2. The probability that an egg survives to breed, S, is the same for both kinds of offspring.

The following tables calculate changes in the number of individuals from one generation to the next.

	Number of adults in this generation	Number of eggs	Number of adults in next generation
Partheno-genetic females	n	kn	Skn
Sexual females	N	(1/2) kN	(1/2) SkN
Sexual males	N	(1/2) kN	(1/2) SkN

(Sexual males fertilize the kN eggs made by sexual females; half these eggs produce sexual females and half produce sexual males.)

	This generation	Next generation
Total population	n + 2N	$Skn + SkN$ $= Sk(n + N)$
Fraction parthenogenetic	$\dfrac{n}{(n + 2N)}$	$\dfrac{Skn}{Sk(n + N)}$ $= \dfrac{n}{(n + N)}$

The fraction of the population that is parthenogenetic always increases between this generation and the next generation. If n is small (say, 1) compared to N (say, 50), then the fraction of asexuals nearly doubles from this generation to the next. After Maynard Smith (1978).

R.L. Dunbrack and colleagues (1995) tested this second assumption experimentally. They showed that the assumption is wrong, at least under the conditions of their experiments. Dunbrack and colleagues studied laboratory populations in the flour beetle *Tribolium castaneum*. In each of a series of trials, the researchers established a mixed population with equal numbers of red beetles and black beetles. The beetles of one color were designated as the sexual strain, and the beetles of the other color were designated to be the "asexual" strain. For example, in half of the trials the red beetles were the sexual strain, and the black beetles the asexual strain. We will stay with these designations for the rest of our description of the protocol.

Flour beetles are not actually capable of asexual reproduction, so the researchers had to manage the black population so that it was numerically and evolutionarily equivalent to a population with actual asexual reproduction. Every generation, the researchers counted the adults of the black strain and threw them out. They then replaced each one of these black adults with three new black adults from a reservoir population of pure-bred black beetles unexposed to competition with red beetles. This procedure effectively gave the black strain a threefold reproductive advantage over the red strain. Thus the black strain was analogous to an asexual subpopulation in which every generation is genetically identical to the generation before, but in which individuals enjoy an even greater reproductive advantage than an actual asexual strain would have in nature. Because every generation of the black population was (except for drift) genetically identical to the generation before, the black strain could not evolve in response to selection imposed by competition with red.

The red adults were allowed to breed among themselves and remain in the experimental culture. They thus constituted a sexual population that could evolve in response to competitive interactions with the black (asexual) strain.

Dunbrack and colleagues added an environmental challenge for the mixed populations by spiking the flour in which the beetles lived with the insecticide malathion. This imposed selection on the flour beetle populations that favored the evolution of insecticide resistance. Finally, the researchers used a clever procedure, the details of which need not concern us here, to prevent the red and black beetles from mating with each other. The researchers maintained the experiment for 30 generations, which took 2 years.

If John Maynard Smith's model were correct, then in each trial the asexual strain should have occupied an ever-increasing fraction of the population, until eventually the sexual strain was eliminated altogether. Maynard Smith's first assumption is built into the experiment; in fact, asexual individuals in the experiment produced *more* offspring each generation than sexual individuals. The only way null prediction, that asexuals will take over, could turn out to be wrong is if Maynard Smith's second assumption is incorrect.

Dunbrack and colleagues performed eight replicates of their experiment. Four were as we described, with red as the sexual (evolving) strain and black as the asexual (nonevolving) strain. Each of these replicates used a different concentration of malathion. The other four replicates used black as the sexual (evolving) strain and red as the asexual (nonevolving) strain. Again, each replicate used a different concentration of malathion. The researchers also performed a control for each of the eight replicates. In the controls neither the red nor the black beetles were allowed to evolve, but one or the other color had a threefold reproductive advantage.

The results appear in Figure 6.7. In the control cultures (Figure 6.7b and d), the outcome was always consistent with Maynard Smith's model: The strain with the threefold reproductive advantage quickly eliminated the other strain. In the experimental cultures, however, the outcome was always contrary to the prediction of Maynard Smith's model. Initially, the asexual strain appeared to be on its way to taking over, but within about 20 generations, depending on the concentration of malathion, the evolving sexual strain recovered. Eventually, the evolving sexual strain completely eliminated the nonevolving asexual strain, in spite of the asexual strain's threefold reproductive advantage.

We can conclude that assumption 2 of Maynard Smith's model is incorrect. Over time spans of just a few generations, descendants produced by sexual reproduction achieve higher fitness than descendants produced by asexual reproduction. The next question is, why?

The only inherent difference between offspring that a female makes sexually versus asexually is that asexual offspring are genetically identical to their mother and to each other, whereas sexual offspring are genetically different from their mother and from each other. Most theories about the benefits of

At least under some conditions, descendants produced by sexual reproduction achieve higher fitness than descendants produced by asexual reproduction.

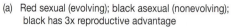

(a) Red sexual (evolving); black asexual (nonevolving);
    black has 3x reproductive advantage

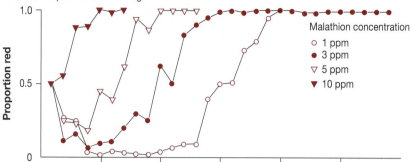

(b) Red control: Both colors asexual (nonevolving);
    black has 3x reproductive advantage

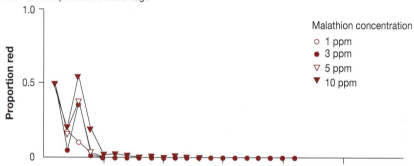

(c) Black sexual (evolving); red asexual (nonevolving);
    red has 3x reproductive advantage

Figure 6.7   An experimental test of assumption 2 of Maynard Smith's model   Each panel shows the relative frequency, in a mixed population, of the flour beetle strain that has been placed at a reproductive disadvantage but (in a and c) reproduces sexually and evolves. The four time series in each panel represent cultures treated with different concentrations of malathion (ppm = parts per million). Transfer is analogous to generation.   (a) Red is at a reproductive disadvantage, but has sexual reproduction and evolves. Red eventually eliminates black. (b) Red is at a reproductive disadvantage, and does not evolve. Red is quickly eliminated by black. (c) Black is at a reproductive disadvantage, but has sexual reproduction and evolves. Black quickly eliminates red.   (d) Black is at a reproductive disadvantage, and does not evolve. Black is eventually eliminated by red. From Dunbrack et al. (1995).

(d) Black control: Both colors asexual (nonevolving);
    red has 3x reproductive advantage

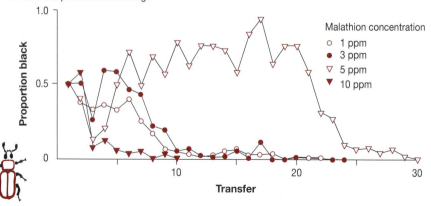

sex concern reasons why females that produce genetically diverse offspring will see more of them survive and reproduce than will females that produce genetic copies of themselves.

We should note at this point that there is a tremendous diversity of theories about the advantages of sex. We have space here to focus only on population genetic models and tests, and a small number of them at that. For a more comprehensive overview of the field, see Michod and Levin (1988).

Sex in Populations Means Genetic Recombination

When population geneticists talk about sex, what they usually mean, and what we mean here, is reproduction involving (1) meiosis with crossing over; and (2) matings between unrelated individuals, such as occur during random mating. The consequence of these processes acting in concert is genetic recombination. If we follow a particular allele through several generations of a pedigree, every generation the allele will be part of a different multilocus genotype. For example, a particular allele for blue eyes may be part of a genotype that includes genes for blond hair in one generation, whereas the blue-eye allele's directly descended copy may be part of a genotype that includes genes for brown hair in the next generation. At the population-genetic level of analysis, recombination has exactly one consequence: It shuffles multilocus genotypes. Another way to state this sole consequence of sexual reproduction is this: Sex reduces linkage disequilibrium. This was a central conclusion of section 6.1.

In a population genetic analysis, sex has exactly one effect: It reduces linkage disequilibrium.

The reduction of linkage disequilibrium is the only effect of sex at the level of population genetics (Felsenstein 1988). In populations that are already at linkage equilibrium, sex has no effect. If sex has no effect, it can confer no benefits. Therefore any population-genetic model for the evolutionary benefits of sex must at minimum include two things: (1) a mechanism that eliminates particular multilocus genotypes, thereby creating linkage disequilibrium; and (2) a reason why genes that tend to recreate those multilocus genotypes and thereby reduce linkage disequilibrium (by promoting sex) are favored. Based on this analysis, Joseph Felsenstein neatly divides nearly all population-genetic models for the benefit of sex into two general theories. These general theories are distinguished by the evolutionary force they posit for the creation of linkage disequilibrium. Some models posit genetic drift as the factor that creates linkage disequilibrium; other models posit natural selection.

Genetic Drift, in Combination with Mutation, Can Make Sex Beneficial

According to the drift theory of sex, mutation and drift create problems that sex can solve. Imagine, for example, that an asexual female sustains a deleterious genetic mutation in her germ cells. She will pass the mutation to all of her offspring, which in turn will pass it to all of their offspring. The female's lineage is thus forever hobbled by the deleterious mutation. The only escape is if one of her descendants is lucky enough to experience the back-mutation, an unlikely event. If the female were sexual, however, she could produce mutation-free offspring immediately, simply by mating with a mutation-free male.

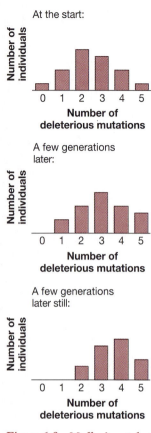

Figure 6.8   Muller's ratchet: asexual populations accumulate deleterious mutations   Each histogram shows a snapshot of a finite asexual population. In any given generation, the class with the fewest deleterious mutations may be lost by drift. Because forward-mutation to deleterious alleles is more likely than back-mutation to wild-type alleles, the distribution slides inexorably to the right. After Maynard Smith (1988a).

Sex can be advantageous because it reduces linkage disequilibrium created by drift.

The role of drift in this scenario becomes clear when we scale the model up to the level of populations. Muller's ratchet is the most famous drift model; it argues that asexual populations are doomed to accumulate deleterious mutations. H. J. Muller (1964) pictured a finite asexual population in which individuals occasionally sustain deleterious mutations. Because the mutations envisioned by Muller are deleterious, they will be selected against. The frequency of each mutant allele in the population will reflect the mutation rate, the strength of selection, and genetic drift (see Chapter 5).

At any given time, Muller's population may include individuals that carry no mutations, individuals that carry one mutation, individuals that carry two mutations, and so on. Because the population is asexual, we can think of these groups as distinct subpopulations, and plot the relative number of individuals in each one in a histogram (Figure 6.8). The number of individuals in each group may be quite small, depending on the size of the entire population, and on the balance between mutation and selection (see Chapter 5). The group with zero mutations is the one whose members on average enjoy the highest fitness; but if this group is small, then in any given generation chance events may conspire to prevent the reproduction of all individuals in the group. If this happens just once, then the zero-mutation subpopulation is lost, and the members of the one-mutation group are now the highest-fitness individuals. The only way the zero-mutation group will reappear is if a member of the one-mutation group sustains the back-mutation that converts into a no-mutation individual.

With the demise of the zero-mutation group, the members of the one-mutation subpopulation enjoy the highest mean fitness. But this group too may be quite small, and may be lost by chance in any given generation. Again the loss of the group by drift is much easier than its recreation by back-mutation. As the ratchet clicks away, and highest-fitness group after highest-fitness group is lost from the population, the average fitness of the population declines over time. The burden imposed by the accumulating mutations is known as the genetic load. Eventually the genetic load carried by the asexual population becomes so high that the population goes extinct.

Sex breaks the ratchet. If the no-mutation group is lost by chance in any given generation, it can be quickly reconstituted by recombination and outcrossing. If two individuals mate, each carrying a single copy of a deleterious mutation, $\frac{1}{4}$ of their offspring will be mutation free. In Muller's view, the genes responsible for sex are maintained in populations because they help to create zero-mutation genotypes. As these zero-mutation genotypes increase in frequency, the genes for sex increase in frequency with them—in effect going along for the ride.

In Muller's scenario, linkage disequilibrium is created by drift. Particular multilocus genotypes are at lower-than-linkage-equilibrium frequencies because chance events have eliminated them. These missing multilocus genotypes are the no-mutation genotype, then the one-mutation genotype, and so on. Sex reduces linkage disequilibrium by recreating the missing genotypes.

Haigh (1978, reviewed in Maynard Smith 1988a) developed and explored an explicit mathematical model of Muller's ratchet. Not surprisingly, the most critical parameter in the model is the population size. In populations of 10 or fewer individuals drift is a potent mechanism of evolution and the ratchet turns rapidly. In populations of more than 1,000 drift is a weak mechanism of evolution and the ratchet does not turn at all. Also important are the mutation rate and the impact of deleterious mutations. The ratchet operates fastest with mildly deleterious mutations. This is because severely deleterious mutations are eliminated by selection before drift can carry them to fixation.

Lin Chao (1990) tested Muller's ratchet experimentally. Because the ratchet operates over many generations, Chao chose a class of organisms with a very short generation time: bacteriophage viruses. Because the ratchet operates fastest when the mutation rate is high, Chao chose an RNA virus. RNA viruses carry their genetic information in RNA instead of DNA. The error rate in RNA replication is several orders of magnitude higher than the error rate in DNA replication. Because the ratchet operates fastest in small populations, Chao put his viruses through repeated population bottlenecks of a single individual.

Chao started the experiment with 20 clonal strains of the RNA bacteriophage $\phi6$. Chao grew the viral strains in cultures of the bacterium *Pseudomonas phaseolicola,* and put the viral strains through 40 growth cycles. Each cycle started with a single virus, which then reproduced to create a clonal population of about 8 billion viruses. Chao then picked a single virus at random from this population and used it to start the next cycle. After 40 such cycles, Chao measured the change in fitness in each strain by measuring its population growth rate relative to its clonal ancestor, which Chao had stored in the freezer (another advantage of viruses as an experimental system). The 20 viral strains sustained a substantial loss in fitness over the course of the experiment. Their average fitness at the end was only 78% that of their ancestors. The 95% confidence interval for the mean fitness was 0.671 to 0.884, indicating that the loss of fitness was statistically significant. Chao's interpretation was that the viral strains had accumulated deleterious mutations via Muller's ratchet.

Muller's ratchet thus works both in principle and in practice. A problem with models that posit drift as the source of linkage disequilibrium is that the benefits ascribed to sex accumulate only over the long term. If an asexual female appeared in a sexual population, it would take many generations for Muller's ratchet to catch up with her descendants, and lower their fitness enough to drive them to extinction. In the meantime, the asexual female's descendants would enjoy the twofold reproductive benefit identified by Maynard Smith. Yet the rarity of asexual species suggests, and the experiment by Dunbrack and colleagues demonstrates, that the advantage of sex accrues over just a few generations. This reasoning has prompted a search for the short-term benefits of sex.

Selection Imposed by a Changing Environment Can Make Sex Beneficial

To see the logic of the changing-environment theories for sex, first imagine an asexual female and a sexual female living in a constant environment. If these females themselves survived to reproduce, and their offspring live in the same environment, then the offspring of the asexual female will probably survive to reproduce too. After all, they will receive exactly their mother's already-proven genotype. The genetically diverse offspring of the sexual female, however, may or may not survive to reproduce, depending on the nature of the genetic differences between their mother and themselves. By this reasoning, in a constant environment asexual reproduction is a safer bet.

In a variable environment, however, all bets are off. If the environment is changing in such a way that the asexual female herself might not survive in the new conditions, then her offspring will have poor prospects too. If the environment changes for a sexual female, however, there is always a chance that some of her diverse offspring will have genotypes that allow them to thrive in the new conditions. Some changing-environment theories focus on changes in the physical environment, whereas others focus on changes in the biological environment. Note that all changing-environment theories assume trade-offs, such that genotypes that do relatively well in some environments necessarily do relatively poorly in others.

A. H. Sturtevant and K. Mather (1938, reviewed in Felsenstein 1988) were the first to consider an explicit population-genetic model of varying selection. They imagined a population in which selection favors some multilocus genotypes in some generations (say, genotypes $A-B-$ and $aabb$ in a two-locus model) and other genotypes ($A-bb$ and $aaB-$) a few generations later. (The dashes indicate that the second allele could be either of the possibilities.) Thus the population would alternate between a selection regime that generates linkage disequilibrium with positive values of D and a selection regime that generates linkage disequilibrium with negative values of D (section 6.1, equation 1). Under these conditions, sex could be favored for its ability to recreate the genotypes that were recently selected against but are now favored. As in Muller's ratchet, the genes for sex get a ride to high frequency in the high-fitness multilocus genotypes they help to create.

Sex can be advantageous because it reduces linkage disequilibrium created by selection.

The variable pattern of selection required by the changing-environment theories can be imposed either by physical factors in the environment or by biological interactions (see Chapter 8). For many years, the physical environment was considered a poor candidate for selection in favor of sex, because simple versions of the varying-selection model require the environment to behave extremely capriciously. Imagine, for example, that $A-$ is favored in hot environments and aa in cold environments, whereas $B-$ is favored in dry environments while bb is favored in wet environments. For a varying environment to favor sex, then in some generations all available habitats to which individuals might move must be either hot-dry or cold-wet, while a few generations later all habitats must be either hot-wet or cold-dry (Maynard Smith 1978). When all the environments are either hot-wet or cold-dry, selection will create linkage disequilibrium with positive values of D. There will be an excess of AB and ab gametes, and a deficit of Ab and aB gametes, compared to equilibrium expectations. When all the environments are either hot-wet or cold-dry, selection will create linkage disequilibrium with negative values of D. There will be a deficit of AB and ab gametes and an excess of Ab and aB gametes. Environmental change this rapid and dramatic is implausible. However, Maynard Smith (1980; 1988b) showed that when several loci of small effect combine to determine the phenotype, then a plausibly slow pattern of environmental change can favor sex.

Currently, the most popular version of the varying-selection theory of sex involves not the physical environment, but biological interactions: specifically, the evolutionary arms races between parasites and their hosts (reviewed in Seger and

Hamilton 1988). Parasites and their hosts are locked in a perpetual struggle, with the host evolving to defend itself and the parasite evolving to evade the host's defenses. It is relatively easy to imagine that a parasite would select in favor of some multilocus host genotypes in some generations, and in favor of other multilocus host genotypes in other generations. Figure 6.9 presents a scenario for such an arms race.

Curtis Lively (1992) investigated whether parasites, in fact, select in favor of sex in their hosts. Lively studied the freshwater snail *Potamopyrgus antipodarum.* This snail, which lives in lakes and streams throughout New Zealand, is the host of over a dozen species of parasitic trematode worms. The trematodes typically castrate their host by eating its gonads. In an evolutionary sense, castration is equivalent to death: It prevents reproduction. Trematodes thus exert on the snail populations strong selection pressure for resistance to infection. Most populations of the snail contain two kinds of females: obligately sexual females that produce a mixture of male and female offspring, and obligately parthenogenetic females

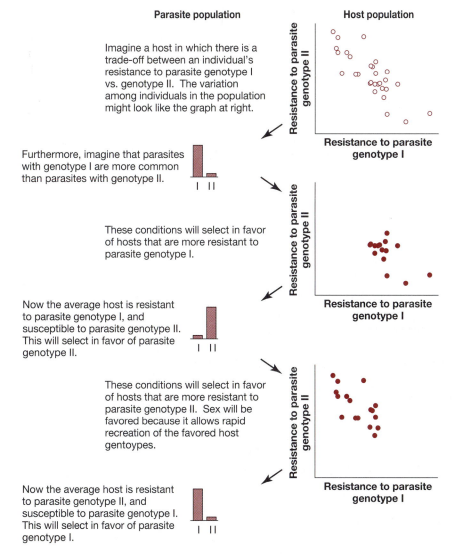

Parasite population

Host population

Imagine a host in which there is a trade-off between an individual's resistance to parasite genotype I vs. genotype II. The variation among individuals in the population might look like the graph at right.

Furthermore, imagine that parasites with genotype I are more common than parasites with genotype II.

These conditions will select in favor of hosts that are more resistant to parasite genotype I.

Now the average host is resistant to parasite genotype I, and susceptible to parasite genotype II. This will select in favor of parasite genotype II.

These conditions will select in favor of hosts that are more resistant to parasite genotype II. Sex will be favored because it allows rapid recreation of the favored host gentoypes.

Now the average host is resistant to parasite genotype II, and susceptible to parasite genotype I. This will select in favor of parasite genotype I.

Figure 6.9 **A host parasite arms race can make sex beneficial**
Hosts resistant to parasite genotype I are necessarily susceptible to parasite genotype II, and vice versa. As the parasite population evolves in response to the hosts, it first selects for hosts resistant to parasite genotype I, then for hosts resistant to parasite genotype II. Genes for sex ride to high frequency in the currently-more-fit genotypes they help to create.

whose daughters are clones of their mother. (Note that both kinds of female must have an ovary to reproduce; the difference is that the eggs of the parthenogenetic females do not have to be fertilized.) The proportion of sexuals versus asexuals varies from population to population. So does the frequency of trematode infection. If an evolutionary arms race between the snails and the trematodes selects in favor of sex in the snails (see Figure 6.9), then sexual snails should be more common in the populations with higher trematode infection rates.

Lively took samples of snails from 66 lakes, and determined the sex of each snail and whether it was infected with parasites. Lively used the frequency of males in each population as an index of the frequency of sexual females. (His logic was that sexual males are produced only by sexual females.) Lively found that a higher proportion of the females are indeed sexual in more heavily parasitized populations (Figure 6.10). This result is consistent with the varying-selection theory of sex.

Lively notes that because his study was observational, alternative explanations for the association he found should be considered. For example, if:

1. Trematode infection rates are higher in more dense populations of snails, because high host density facilitates parasite transmission; and

2. The frequency of parthenogenetic females is higher in less dense populations of snails, because the real benefit of parthenogenesis is that it allows females to reproduce even when mates are hard to find;

then these two effects in combination would produce a positive association between the frequency of sexuals and the frequency of trematode infection. Lively rejected this alternative explanation by showing that although there is a positive correlation between infection rate and snail density (effect 1 is true), there is also a positive correlation between the frequency of parthenogenetic

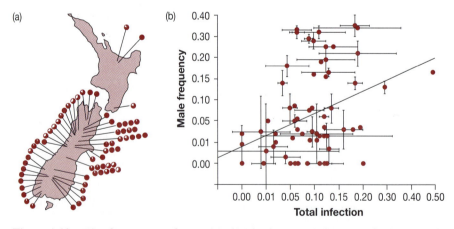

Figure 6.10   **The frequency of sexual individuals in populations of a host snail is positively correlated with the frequency of its trematode parasites**   (a) The map shows the locations of the 66 lakes Lively sampled. In each population's pie diagram, the white slice represents the frequency of males.   (b) The scatterplot shows the frequency of males in each population versus the proportion of snails infected with trematodes. The graph includes a best-fit line. Males are more frequent in populations in which more snails are infected. From Lively (1992).

females and snail density (effect 2 is false). After considering this and other alternative explanations, Lively concludes that the simplest explanation for the pattern he found is that the trematodes indeed select in favor of sexual reproduction by the snails.

In summary, in the context of population genetics the effect of sex is to reduce linkage disequilibrium. A population-genetic model for the adaptive value of sex therefore must have two components: a mechanism for the creation of linkage disequilibrium, and a reason why selection favors traits that tend to reduce linkage disequilibrium. There are two classes of models for sex. In the first class, genetic drift creates linkage disequilibrium. Sex is then favored because it helps to recreate high-fitness genotypes lost by drift. In the second class, natural selection creates linkage disequilibrium. Then the pattern of selection changes, and sex is favored because it helps to recreate the now-favorable genotypes recently selected against. The various scenarios favoring sex are mutually compatible with each other. It is likely that varying selection imposed by changes in both the biological and physical environments combines with Muller's ratchet to create an advantage for sex that is greater than the advantage any one factor can produce alone.

6.3   Selection on Quantitative Traits

The conceptual tools we developed in Chapter 5, and the tools we have developed so far in Chapter 6, allow us to analyze the evolution of traits controlled by one or two loci. Most traits in most organisms, however, are determined by the combined effects of many different loci, and are also influenced by the environment. Examples include human height, lizard sprint speed, and plant flower size. Such traits show continuous variation among individuals, and are called quantitative traits. (This is in contrast to qualitative traits like scale-eater handedness, in which a fish is either right-handed or left-handed.) We need tools that allow us to analyze and understand the genetics and evolution of quantitative traits, even when we do not know the identities of the many specific genes involved. These tools are the province of quantitative genetics, and are the subject of this last section of the chapter.

Recall the basic tenets of Darwin's theory of evolution by natural selection: If there is heritable variation among the individuals in a population, and if there is differential survival and/or reproductive success among the variants, then the population will evolve. Quantitative genetics includes tools for measuring heritable variation, tools for measuring differential survival and/or reproductive success, and tools for predicting the evolutionary response to selection.

Quantitative genetics allows us to analyze evolution by natural selection in traits controlled by many loci.

Measuring Heritable Variation

Imagine a population of organisms in which there is continuous variation among individuals in some trait. For example, imagine a population of humans in which there is continuous variation among individuals in height. Continuously variable traits are typically normally distributed, so that a histogram of a trait has the familiar bell-curve shape. For example, in a typical human population a few people are very short, many people are more or less average in height, and a few people are very tall (Figure 6.11). We want to know: Is height heritable?

It is worth thinking carefully about exactly what this question means. Questions about heritability are often expressed in terms of nature versus nurture. But such questions are meaningful only if they concern comparisons among individuals. It makes no sense to focus only on the student on the far left of Figure 6.11a and ask, without reference to the other individuals, whether this student is 4 feet 10 inches tall because of his genes (nature) or because of his environment (nurture). He had to have both genes and an environment to be alive and thus to be any height at all. He did not get 3 feet of his height from his genes, and 1 foot 10 inches of his height from his environment, so that $3' + 1'10'' = 4'10''$. He is 4 feet 10 inches tall because of his genes and his environment. Within this single individual we cannot, even in principle, disentangle the influence of nature and nurture. To do so, we would have to make two clones of the student, one with genes but no environment and the other with an environment but no genes, and see how tall each becomes. This is obviously impossible.

The only kind of question it makes sense to ask is a comparative one: Is the shortest student shorter than the tallest student because they have different genes, or because they grew up in different environments, or both? This is a question we can answer. In principle, for example, we could take a clone of the short student and raise it in the environment experienced by the tall student. If this clone still grew up to be 4 feet 10 inches tall, then we would know that the

(a)

(b)

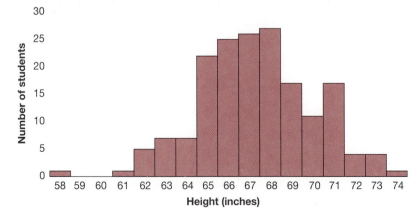

Figure 6.11 **Normally distributed variation in a trait** (a) A photograph, published in the Journal of Heredity in 1914 by Albert Blakeslee, of a group of students at Connecticut Agricultural College sorted by height. The arrangement of the students forms a living histogram.   (b) A graphical histogram representing the distribution of heights among the students shown in (a).

difference between the shortest and tallest students is due entirely to differences in their genes. If the clone grew up to be 6 feet 2 inches tall, then we would know that the difference between the shortest and tallest students is due entirely to differences in their environments. In fact, the clone would probably grow up to be somewhere between 4′10″ and 6′2″. The difference between the two students is probably partly due to differences in their genes, and partly due to differences in their environments. Considering the whole population rather than just two individuals, we can ask: What fraction of the variation in height among the students is due to variation in their genes, and what fraction of the variation in height is due to variation in their environments?

The fraction of the total variation in a trait that is due to variation in genes is the heritability of the trait. The total variation in a trait is referred to as phenotypic variation, and is symbolized by V_P. Variation among individuals that is due to variation in their genes is called genetic variation, and is symbolized by V_G. Variation among individuals due to variation in their environments is called environmental variation, and is symbolized by V_E. Thus we have: heritability $= V_G/V_P = V_G/(V_G + V_E)$. More precisely, this fraction is known as the broad-sense heritability, or degree of genetic determination. We will define the narrow-sense heritability shortly. Heritability is always a number between 0 and 1.

The first step in a quantitative genetic analysis is to partition the total phenotypic variation (V_P) into a component due to genetic variation (V_G) and a component due to environmental variation (V_E).

Before wading any deeper into symbolic abstractions, we note the simple truth that if the variation among individuals is due to variation in their genes, then offspring ought to resemble their parents. It is easy, in principle, to check whether they do. We first make a scatterplot with offspring trait values represented on the y-axis, and their parents' trait values on the x-axis (Figure 6.12). We have two parents for every offspring, so we use the midparent value, which is the average of the two parents. If we have more than one offspring for each family, we use a midoffspring value as well. We then draw the best-fit line through the data points. If offspring do not resemble their parents, then the slope of the best-fit line through the data will be near zero (Figure 6.12a); this is evidence that the variation among individuals in the population is due to variation in their environments, not variation in their genes. If offspring strongly resemble their parents, the slope of the best-fit line will be near one (Figure 6.12c and d); this is evidence that variation among individuals in the population is due to variation in their genes, not variation in their environments. Most traits in most populations fall somewhere in the middle, with offspring showing a moderate resemblance to their parents (Figure 6.12b); this is evidence that the variation among individuals is partly due to variation in their environments and partly due to variation in their genes. (For another example, look back at the scatterplot in Figure 2.10, which analyzes the heritability of beak depth in Darwin's finches.)

It we determine the best-fit line through the data using the method of least-squares linear regression, then the slope represents a version of the heritability symbolized by h^2 and known as the narrow-sense heritability. Least-squares linear regression is the standard method taught in introductory statistics texts and used by statistical software packages to determine best-fit lines. (For readers familiar with statistics, it may prevent some confusion if we note here that h^2 is *not* the fraction of the variation among the offspring explained by variation in

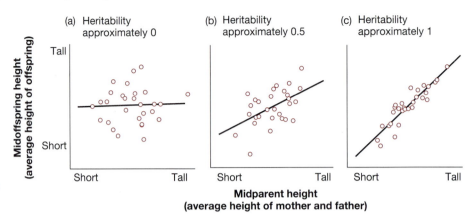

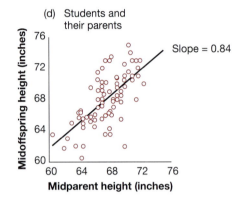

Figure 6.12   Scatterplots showing offspring height as a function of parent height Each of the top three scatterplots shows data for a hypothetical population, and each includes a best-fit line through the data. (a) In this population, offspring do not resemble their parents. (b) In this population, offspring bear a moderate resemblance to their parents.   (c) In this population, offspring strongly resemble their parents.   (d) This graph shows data for an actual population of students in a recent evolution course at a university in the Pacific Northwest, USA.

The heritability, h^2, is a measure of the (additive) genetic variation in a trait.

the parents. That quantity would be r^2. Instead, h^2 is an estimate of the fraction of the variation among the *parents* that is due to variation in their genes.)

To explain the difference between narrow-sense heritability and broad-sense heritability, we need to distinguish between two components of genetic variation: additive genetic variation versus dominance genetic variation. Additive genetic variation (V_A) is variation among individuals due to the additive effects of genes, whereas dominance genetic variation (V_D) is variation among individuals due to gene interactions such as dominance (Box 6.2). The total genetic variation is the sum of the additive and dominance genetic variation: $V_G = V_A + V_D$. The broad-sense heritability, defined earlier, is V_G/V_P. The narrow-sense heritability, h^2, is $V_A/V_P = V_A/(V_A + V_D + V_E)$. When evolutionary biologists mention heritablity without noting whether they are using the term in the broad or narrow sense, they almost always mean the narrow-sense heritability. We will use the narrow-sense heritability in the rest of this discussion. It is the narrow-sense heritability, h^2, that allows us to predict how a population will respond to selection.

When estimating the heritability of a trait in a population, it is important to keep in mind that offspring can resemble their parents for reasons other than the genes offspring inherit (see Box 2.1). Environments run in families too. Among humans, for example, some families exercise more than others, and different families eat different diets. Our estimate of the heritability will be accurate only if we can make sure that there is no correlation between the environ-

Box 6.2   Additive genetic variation versus dominance genetic variation

To simplify this discussion, we will analyze genetic variation at a single locus with two alleles as though we were analyzing a quantitative trait. We further assume that there is no environmental variation in the trait in question: An individual's phenotype is determined solely and exactly by its genotype. The alleles at the locus are A_1 and A_2; each is at frequency 0.5, and the population is in Hardy-Weinberg equilibrium. We imagine two situations: (1) the alleles are codominant; and (2) allele A_2 is dominant over allele A_1.

(1) Alleles A_1 and A_2 are codominant. A_1A_1 individuals have a phenotype of 1. In A_1A_2 and A_2A_2 individuals, each copy of allele A_2 adds 0.5 to the phenotype. At left in Figure 6.13a is a histogram showing the distribution of phenotypes in the population. At center and right are scatterplots that allow us to analyze the genetic variation in the population. The x-axis represents the genotype, calculated as the number of copies of allele A_2. The y-axis represents the phenotype. The horizontal gray line shows the mean phenotype for the population (= 1.5). The plot at center shows that the total genetic variation (V_G) is a function of the deviations of the data points from the population mean (black arrows). We can quantify V_G by calculating the sum of the squared deviations. The plot at right shows the best-fit line through the data points. The addi-

tive genetic variation (V_A) is that fraction of the total genetic variation that is explained by the best-fit line (dark gray arrows). In this case, the best-fit line explains all of the genetic variation, so $V_G = V_A$. There is no dominance genetic variation.

(2) Allele A_2 is dominant. A_1A_1 individuals have a phenotype of 1. The effects of substituting copies of A_2 for copies of A_1 are not strictly additive: The first copy of A_2 (which makes the genotype A_1A_2) changes the phenotype from 1 to 2. The second copy of A_2 (which makes the genotype A_2A_2) does not alter the phenotype any further. At left in Figure 6.13b is a histogram showing the distribution of phenotypes in the population. At center and right are scatterplots that allow us to analyze the genetic variation in the population. The plot at center shows that the total genetic variation (V_G) is a function of the deviations of the data points (black arrows) from the population mean (gray line; = 1.75). The plot at right shows the best-fit line through the data points. The additive genetic variation (V_A) is that fraction of the total genetic variation that is explained by the best-fit line (dark gray arrows). The dominance genetic variation (V_D) is that fraction of the total genetic variation left unexplained by the best-fit line (light gray arrows). In this case, the best-fit line explains only part of the genetic variation, so $V_G = V_A + V_D$.

(a) No dominance. Phenotypes: $A_1A_1 = 1$; $A_1A_2 = 1.5$; $A_2A_2 = 2$

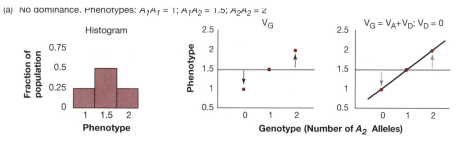

(b) Complete dominance. Phenotypes: $A_1A_1 = 1$; $A_1A_2 = 2$; $A_2A_2 = 2$

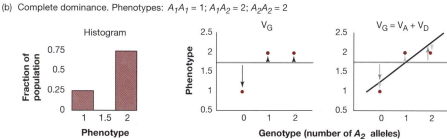

Figure 6.13 **Additive genetic variation versus dominance genetic variation in a trait controlled by two alleles at a single locus.**

ments experienced by parents and the environments experienced by offspring. We obviously cannot do so in a study of humans. In an animal study, however, we could collect all the offspring at birth, then foster them at random among the parents. In a plant study, we could place seeds at random locations in a field.

There are a variety of other methods for estimating the heritability besides calculating the slope of the best-fit line for offspring versus parents. For example, studies of twins can be used. The logic of twin studies works as follows. Monozygotic (identical) twins share their environment and all of their genes, whereas dizygotic twins share their environment and half of their genes. If the heritability is high, and variation among individuals is due mostly to variation in genes, then monozygotic twins will be more similar to each other than are dizygotic twins. If the heritability is low, and variation among individuals is due mostly to variation in environments, than monozygotic twins will be as different from each other as dizygotic twins. For a detailed introduction to methods of measuring heritability, see Falconer (1989). Data on the heritability of traits are frequently misinterpreted, particularly when the species under study is humans (see Box 6.3).

Measuring the Strength of Selection

In the preceding paragraphs we developed techniques for measuring the heritable variation in quantitative traits. The next tenet of Darwin's theory of evolution by natural selection is differential survival and/or reproductive success. We now discuss techniques for measuring differential success—that is, for measuring the strength of selection. Once we can measure both heritable variation and the strength of selection, we will be able to predict the evolutionary changes in response to selection.

Selection on a particular trait is systematic variation in the survival and/or reproductive success of individuals with some values of the trait compared to individuals with other values of the trait. Measuring the strength of selection means determining how much the winners differ from the losers in the trait of interest.

In selective breeding experiments, the strength of selection is easy to calculate. Imagine, for example, that we were trying to breed lab mice with longer tails. (Perhaps we are interested not just in whether tail length can evolve, but also in whether longer tails will result from the addition of new vertebrae, or from the elongation of extant vertebrae.) Every generation, we would measure the tail lengths of all the mice in our population. We would then pick, say, the one-third of the mice whose tails are the longest, and let them breed among themselves to make the next generation of offspring. The strength of selection is the difference between the mean tail length of the breeders and the mean tail length of the entire population (Figure 6.17a on page 213). This measure of selection is called the selection differential, and is symbolized by S.

The selection differential, S, is a measure of the strength of selection on a trait.

There is a second measure of the strength of selection that is useful because of its broad applicability. This method yields a measure called the selection gradient (Lande and Arnold 1983). We will develop it using the hypothetical selective breeding experiment from the preceding paragraph. The first step is to assign fitnesses to the mice in our population. We will think of fitness as survival to reproductive age. In our population, $\frac{1}{3}$ of the mice survived long enough to

BOX 6.3 The bell-curve fallacy, and other misinterpretations of heritability

Any measurement of heritability is specific to a particular population in a particular environment. As a result, heritability tells us nothing about the causes of differences between populations that live in different environments. We can illustrate this point with a thought experiment. Pea aphids reproduce asexually during spring and early summer. Imagine that we collect a single pea aphid from each of a dozen different pea fields and bring them into the lab. Together, these dozen aphids form our laboratory population. Because the aphids came from different fields, they are genetically different from each other. Our population contains genetic variation. But because each aphid within the population reproduces asexually (by mitosis), each aphid's offspring are genetically identical to their mother and to each other. In other words, the descendants of any one of our original aphids constitute a clone. We can create duplicates of our original population by taking one member from each of our dozen clones.

Imagine that we set up two such duplicate populations. We rear one population in a carefully controlled environment on pea plants, and the aphids grow to large sizes. Because all the aphids in this population shared the same environment, any differences among them in size are entirely due to genetic variation. When grown on peas, the heritability of size in our lab population is 1.

Now we rear the other duplicate aphid population on clover, and the aphids in this population do not get so big (they are pea aphids, after all). Nonetheless, all the aphids in this population shared the same environment, so any differences among them in size will be due genetic variation. When grown on clover, the heritability of size in our lab population is 1.

We have high heritability within both populations, and a difference in mean size between populations. Does this mean that the population reared on peas is genetically superior to the clover population with respect to size? Of course not; we intentionally set up the populations in a way that made them identical in genetic composition. The fact that heritability is high in each population tells us nothing

about the cause of differences between the populations, because the populations were reared in different environments. (For an actual experiment similar to the one we have described, see Via 1991.)

The mistaken notion that heritability can tell us something about the causes of differences between populations has been particularly persistent in studies of human intelligence. In 1994, Charles Murray and Richard J. Herrnstein sold thousands of copies of their book, *The Bell Curve.* Murray and Herrnstein claimed that the difference in average IQ score between African Americans and European Americans is due to genetic differences between these groups (our analysis of their argument is based on an extract of their book published in the New Republic [Murray and Herrnstein 1994]).

Murray and Herrnstein take note of the point we just made with pea aphids. They state: "Most scholars accept that I.Q. in the human species as a whole is substantially heritable, somewhere between 40 percent and 80 percent, meaning that much of the observed variation in I.Q. is genetic. And yet this information tells us nothing for sure about the origin of the differences between groups." Having said that, however, Murray and Herrnstein proceed to develop the following erroneous argument.

Based on various sources, Murray and Herrnstein assume that the average IQ of African Americans is 85, that the average IQ of European Americans is 100, and that the variance (a statistical measure of the variation) in each group is 225. Bell curves representing these assumptions appear in Figure 6.14a. Murray and Herrnstein further assume that the heritability of IQ in each group is 0.6. There are reasons to dispute each of Murray and Herrnstein's assumptions, but we will grant them here—for the sake of argument only. (There are also reasons to dispute whether IQ tests measure anything meaningful, but we leave that argument to others.)

Next Murray and Herrnstein imagine what the bell curves for IQ would look like if all the genetic variation among individuals within each population were removed. In other words, they imagine that all

Box 6.3 *(continued)*

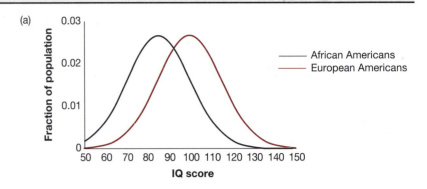

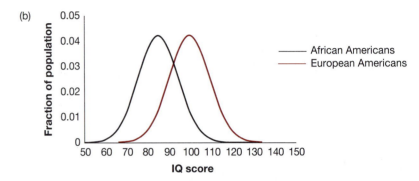

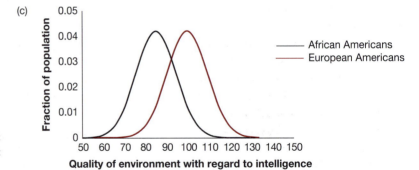

Figure 6.14 **Illustrations of Murray and Herrnstein's erroneous argument concerning IQ and ethnicity**   See text for explanation.

African Americans have been made genetically identical to the average African American, and all European Americans have been made genetically identical to the average European American. On the assumption that 60% of the variation within each group is due to genetic variation, this leaves 40% of the original variation within each group. Bell curves representing this thought experiment appear in Figure 6.14b.

Now Murray and Herrnstein consider the proposition that the difference between the average IQ of the African Americans and the average IQ of the European Americans in Figure 6.14b is solely due to differences in environment. Under this proposition, Murray and Herrnstein say, we could replace the label "IQ score" with the label "Quality of environment with regard to intelligence," as shown in Figure 6.14c. Murray and Herrnstein find it implausible that the difference in the quality of the environment experienced by

African Americans versus European Americans is as great as that shown Figure 6.14c. They conclude that at least part of the difference between the mean IQ score of African Americans versus European Americans must be due to genetic differences between the groups.

There are at least two serious flaws in Murray and Herrnstein's argument. First, by replacing the label on the horizontal axis in Figure 6.14b with the one in Figure 6.14c, Murray and Herrnstein are implicitly assuming that there is a linear relationship between environment and IQ. This assumption is almost certainly wrong.

Second, Murray and Herrnstein's argument from incredulity amounts to rhetorical technique, not science. A scientific approach to Murray and Herrnstein's hypothesis would be to conduct a common garden experiment: rear European Americans and African Americans together in an environment typically experienced by European Americans, and then compare their IQ scores. This design, and the reciprocal experiment, in which everyone is reared in an environment typically experienced by African Americans, are shown in Figure 6.15.

We obviously cannot do this experiment with humans. But experiments like it have been done with plants and animals. For example, Clausen, Keck, and Hiesey (1948) conducted a series of common garden experiments with the plant *Achillea* (Figure 6.16). Plants in this genus collected from low-altitude populations make more stems than plants collected from high-altitude populations (Figure 6.16a). Is the difference between the low- versus high-altitude plants due to differences in their genes or differences in their environments? When plants from low altitude and plants from high altitude are grown together at low altitude, the low-altitude plants make more stems (Figure 6.16b). This result is consistent with the hypothesis that plants from low altitude are genetically programmed to make more stems. When plants from low altitude and plants from high altitude are grown together at high altitude, however, the high-altitude plants make more stems (Figure 6.16c). This result was wholly unanticipated in the experimental design. It reveals genetic differences between low- and high-altitude plants in the way each responds to the environment. It also reveals that each population of plants is superior in its own environment of origin.

What would happen if we did this kind of experiment with African Americans and European Americans? No one has the slightest idea. It is misleading to say that high heritabilities for IQ within groups tell us "nothing for sure about the origin of the differences between groups" (Murray and Herrnstein 1994). In fact, high heritabilities within groups tell us nothing *at all* about the origin of differences between groups.

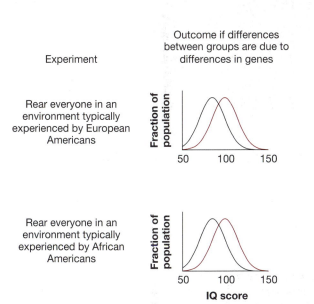

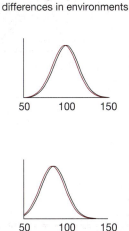

Figure 6.15 **An experiment that would test Murray and Herrnstein's claim** The left column describes two experimental treatments. The middle and right columns show the predicted outcomes under the hypothesis that differences between groups are due to differences in genes versus the predicted outcomes under the hypothesis that differences between groups are due to differences in environments.

Box 6.3   *(continued)*

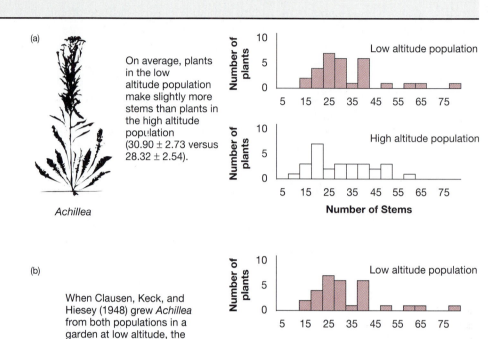

(a)

Achillea

On average, plants in the low altitude population make slightly more stems than plants in the high altitude population (30.90 ± 2.73 versus 28.32 ± 2.54).

(b)

When Clausen, Keck, and Hiesey (1948) grew *Achillea* from both populations in a garden at low altitude, the plants from the low altitude population made more stems (30.90 ± 2.73 versus 7.21±1.08).

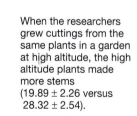

Figure 6.16 **Data from experiments by Clausen, Keck, and Hiesey (1948)** (a) A comparison between *Achillea* populations from low altitude (San Gregorio, California) versus high altitude (Mather, California). (b) Plants from low and high altitude grown in a common garden at low altitude (Stanford, California).   (c) Plants from low and high altitude grown in a common garden at high altitude (Mather, California).

(c)

When the researchers grew cuttings from the same plants in a garden at high altitude, the high altitude plants made more stems (19.89 ± 2.26 versus 28.32 ± 2.54).

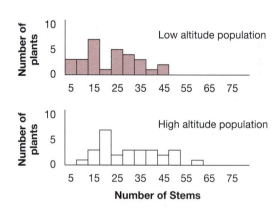

Finally, it is worth noting that heritability also tells us nothing about the role of genes in determining traits that are shared by all members of a population. There is no variation among humans in number of noses. The heritability of nose number is undefined, because $V_A/V_P = 0/0$. This obviously does not mean that our genes are not important in determining how many noses we have.

So what good *does* it do us to measure the heritability of a trait? Precisely and only this: It allows us to predict whether a population will respond to selection on a trait. We discuss how in the section beginning on page 215.

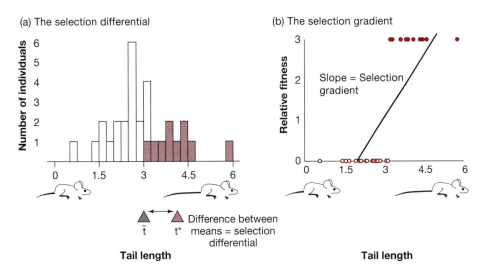

Figure 6.17   Measuring the strength of selection   (a) The histogram shows the variation in tail length in a fictional population of lab mice. The filled bars represent the mice chosen as breeders for the next generation. The gray triangle ($\bar{t}$) indicates the average tail length for the entire population; the filled triangle (t^*) indicates the average tail length for the breeders. The difference between these two means is the selection differential.   (b) A scatterplot for the same fictional population of mice showing relative fitness (see text) as a function of tail length. Filled dots represent mice chosen as breeders for the next generation. The scatterplot includes the best-fit line.

reproduce. (This does not necessarily mean that we actually killed the short-tailed mice, just that we removed them from the breeding population; as far as our breeding population is concerned, we did not let them breed so they did not survive long enough to reproduce.) The short-tailed $\frac{2}{3}$ of the mice have a fitness of 0, and the long-tailed $\frac{1}{3}$ have a fitness of 1. The mean fitness of the population is 0.33 (if, for example, there are 30 mice in the population, the mean is $[(20 \times 0) + (10 \times 1)]/30 = 0.33$). We next calculate each mouse's relative fitness by dividing its absolute fitness (0 or 1) by the mean fitness (0.33). The short-tailed mice have a relative fitness of 0; the long-tailed mice have a relative fitness of 3. Finally, we make a scatterplot of relative fitness as a function of tail length, and calculate the slope of the best-fit line (Figure 6.17b). The slope of the best-fit line is the selection gradient.

It may not appear at first glance that the selection gradient and the selection differential have much to do with each other. In fact they are closely related, and each can be converted into the other. If we are analyzing selection on a single trait, then the selection gradient is equal to the selection differential divided by the variance in tail length (see Box 6.4). One advantage of the selection gradient is that we can calculate it for any measure of fitness, not just survival. We might, for example, measure fitness in a natural population of mice as the number of offspring weaned. If we first calculate each mouse's relative fitness (by dividing its number of offspring by the mean number of offspring), then plot relative fitness as a function of tail length and calculate the slope of the best-fit line, then that slope is the selection gradient.

[Another advantage of the selection gradient is that it can be used to analyze selection on two or more characters at once (see Lande and Arnold 1983). If

Box 6.4   The selection gradient and the selection differential

The selection differential is an intuitively straight-forward measure of the strength of selection: It is the difference between the mean of a trait among the survivors and the mean of the trait among the entire population. The selection gradient, among other advantages, can be calculated for a wider variety of fitness measures. The fact that two measures of selection are closely related lends the selection gradient some of the intuitive appeal of the selection differential.

Here we show that in our example on tail length in mice (Figure 6.17), the selection gradient for tail length (t) is equal to the selection differential for tail length divided by the variance of tail length. Imagine that we have 30 mice in our population. First, note that the selection differential is

$$S = t^* - \bar{t}$$

where t^* is the mean tail length of the 10 mice we kept as breeders, and $\bar{t}$ is the mean tail length of the entire population of 30 mice.

The selection gradient is the slope of the best-fit line for relative fitness (w) as a function of tail length. The slope of the best-fit line in linear regression is given by the covariance of y and x divided by the variance of x:

$$\text{slope} = \text{cov}(y, x)/\text{var}(x)$$

The covariance of y and x is defined as

$$\text{cov}(y, x) = \frac{1}{n} \sum_{i=1}^{n} (y_i - \bar{y})(x_i - \bar{x})$$

and the variance of x is defined as

$$\text{var}(x) = \frac{1}{n} \sum_{i=1}^{n} (x_i - \bar{x})^2$$

where n is the number observations, $\bar{y}$ is the mean value of y, and $\bar{x}$ is the mean value of x. The selection gradient for t is therefore:

$$\text{selection gradient} = \text{cov}(w, t)/\text{var}(t)$$

Thus, what we need to show is that $\text{cov}(w, t) = t^* - \bar{t}$.

Because (by definition) the mean relative fitness is 1, we have:

$$\text{cov}(w, t) = \frac{1}{30} \sum_{i=1}^{30} (w_i - 1)(t_i - \bar{t})$$

$$= \frac{1}{30} \sum_{i=1}^{30} (w_i t_i) - \frac{1}{30} \sum_{i=1}^{30} (w_i \bar{t})$$

$$- \frac{1}{30} \sum_{i=1}^{30} (t_i) + \frac{1}{30} \sum_{i=1}^{30} (\bar{t})$$

$$= \frac{1}{30} \sum_{i=1}^{30} (w_i t_i) - \bar{t} - \bar{t} + \bar{t}$$

$$= \frac{1}{30} \sum_{i=1}^{30} (w_i t_i) - \bar{t}$$

$$= t^* - \bar{t}$$

[To see that

$$\frac{1}{30} \sum_{i=1}^{30} (w_i t_i) = t^*$$

note that for the first 20 mice $w_i = 0$, and for the last 10 mice $w_i = 3$. This means that

$$\frac{1}{30} \sum_{i=1}^{30} (w_i t_i) = \frac{1}{30} \sum_{i=21}^{30} (3t_i)$$

$$= \frac{3}{30} \sum_{i=21}^{30} (t_i)$$

$$= \frac{1}{10} \sum_{i=21}^{30} (t_i) = t^*.]$$

two characters, x and y, are correlated with each other, then the selection differentials (the mean of each character before and after selection) can be misleading. Selection favoring larger values of x can lead to an increase in y, even though selection directly on y might by itself favor a decrease in y. Calculating the two-dimensional selection gradient allows us to disentangle direct versus indirect selection on each trait.]

Predicting the Evolutionary Response to Selection

Once we know the heritability and the selection differential, we can predict the evolutionary response to selection. Here is the equation for doing so:

$$R = h^2 S$$

where R is the predicted response to selection, h^2 is the heritability, and S is the selection differential.

The response to selection, R, can be predicted from h^2 and S.

The logic of this equation is shown graphically in Figure 6.18. This figure shows a scatterplot of midoffspring values as a function of midparent values, just like the scatterplots in Figure 6.12. The scatterplot in Figure 6.18 represents a population of 30 families. The plot includes a best-fit line, whose slope estimates the heritability (h^2). The difference between P^* and $\overline{P}$ is the selection differential (S) that we would have applied to this population had we picked as our breeders only the 10 pairs of parents with the largest midparent values. The difference between O^* and $\overline{O}$ is the evolutionary response (R) we would have gotten as a result of selection.

The slope of a line can be calculated as the rise over the run. We have a rise of $(O^* - \overline{O})$ over a run of $(P^* - \overline{P})$, so

$$h^2 = (O^* - \overline{O})/(P^* - \overline{P})$$

In other words, $h^2 = R/S$. Or, $R = h^2 S$.

We now have a set of tools for studying the evolution by natural selection of multilocus traits. We can estimate how much of the variation in a trait is due to variation in genes, quantify the strength of selection that results from differences in survival or reproduction, and put these two together to predict how much the population will change from one generation to the next.

Alpine Skypilots and Bumblebees

As an example of the kinds of questions evolutionary biologists ask and answer with quantitative genetics, we review Candace Galen's (1996) research on flower size in the alpine skypilot *(Polemonium viscosum)*. The alpine skypilot is a

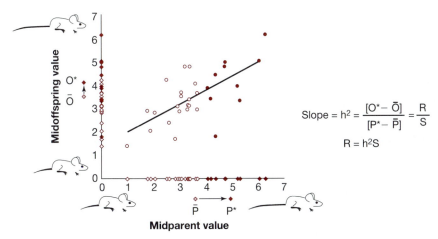

Slope $= h^2 = \dfrac{[O^* - \overline{O}]}{[P^* - \overline{P}]} = \dfrac{R}{S}$

$R = h^2 S$

Figure 6.18 **The response to selection is equal to the heritability times the selection differential** The midoffspring and midparent values are both indicated as dots on the scatterplot and as diamonds on the y and x axes. The filled symbols represent the 10 families with the largest midparent values. $\overline{P}$ is the average midparent value for the entire population; P^* is the average midparent value of the families with the largest midparent values. $\overline{O}$ is the average midoffspring value for the entire population; O^* is the average midoffspring value for the families with the largest midparent values. After Falconer (1989).

perennial Rocky Mountain wildflower (Figure 6.19). Galen studied skypilots on Pennsylvania Mountain in Colorado, including populations growing at the timberline and populations growing in the higher-elevation tundra. At the timberline, skypilots are pollinated by a diversity of insects, including flies, small solitary bees, and some bumblebees. In the tundra, skypilots are pollinated almost exclusively by bumblebees. Tundra skypilots have flowers that are on average 12% larger in diameter than those of timberline skypilots. Previously, Galen (1989) had documented that larger flowers attract more visits from bumblebees, and that skypilots that attract more bumblebees produce more seeds.

Galen wanted to know whether the selection on flower size imposed by bumblebees is responsible for the larger flowers of tundra skypilots. If it is, she also wanted to know how long would it take for selection by bumblebees to increase the average skypilot flower size in a population by 12%—the difference between timberline and tundra flowers.

Galen worked with a small-flowered timberline population. Galen's first step was to estimate the heritability of flower size. She measured the flower diameters of 144 skypilots and collected their seeds. She germinated the seeds in her laboratory, and planted the 617 resulting seedlings at random locations in the same habitat their parents lived in. Seven years later, 58 of the seedlings had matured and flowered, so that Galen could measure their flowers. This allowed Galen to plot offspring flower diameter (corolla flare) as a function of maternal parent flower diameter (Figure 6.20). The slope of the best-fit line is approximately 0.5. For reasons beyond the scope of this discussion, the slope of the best-fit line for offspring versus a single parent (as opposed to the midparent) is an estimate of $\frac{1}{2}h^2$ (see Falconer 1989). Thus the heritability of flower size in the timberline skypilot population is approximately $2 \times 0.5 = 1$. Note, however, that the graph in Figure 6.20 shows considerable scatter. Galen's statistical analysis indicated that she could safely conclude that the heritability of flower size is between 0.2 and 1. In other words, at least 20% of the variation in skypilot flower size is due to additive variation in genes.

Galen's next step was to estimate the strength of the selection differential imposed by bumblebee pollinators (recall that bumblebees prefer to visit larger flowers, and that more bumblebee pollinators means more seeds). Galen built a large screen-enclosed cage at her study site, transplanted 98 soon-to-flower

(a)

(b)

Figure 6.19 **An alpine skypilot** *(Polemonium viscosum)* **(a) and a bumblebee** *(Bombus* **sp.) (b).** [(a) Richard Parker/Photo Researchers, Inc.; (b) Stephen Dalton/Photo Researchers, Inc.]

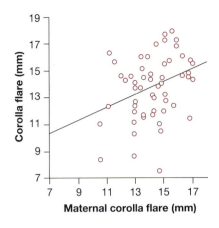

***Figure 6.20* Estimating the heritability of flower size (corolla flare) in alpine skypilots**   The scatterplot shows offspring corolla flare as a function of maternal plant corolla flare for 58 skypilots. The slope of the best-fit line is 0.5. From Galen (1996).

skypilots into it, and added bumblebees. The cage kept the bumblebees in and all other pollinators out. When the caged skypilots flowered, Galen measured their flowers. Later she collected their seeds, germinated them in the lab, and planted the resulting seedlings at random locations back out in the parental habitat. Six years later, Galen counted the number of surviving offspring that had been produced by each of the original caged plants. Using the number of surviving 6-year-old offspring as her measure of fitness, Galen plotted relative fitness as a function of flower size, and calculated the slope of the best-fit line (Figure 6.21). The slope of this line, 0.13, is the selection gradient. Multiplying the selection gradient by the variance in flower size (5.66) gives the selection differential: S = 0.74 mm. The average flower size was 14.2 mm. Thus the selection differential can also be expressed as $\frac{0.74}{14.2} = 0.05$, or 5%. Roughly speaking, this means that when skypilots attempt to reproduce by enticing bumblebees to visit them, the plants that win have flowers that are 5% larger than those of the average plant in the population.

Galen performed two control experiments to confirm that bumblebees select in favor of larger flowers. In one control, she pollinated skypilots by hand (without regard to flower size); in the other, she allowed skypilots to be pollinated by all other natural pollinators except bumblebees. In neither control was there any

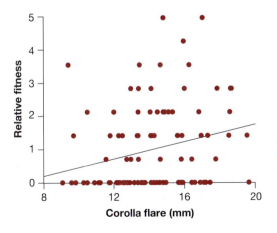

***Figure 6.21* Estimating the selection differential in alpine skypilots pollinated by bumblebees**   The scatterplot shows relative fitness (number of surviving 6-year-old offspring divided by average number of surviving 6-year-old offspring) as a function of maternal flower size (corolla flare). The slope of the best-fit line is 0.13. Prepared with data provided by Candace Galen.

(a) Offspring of hand-pollinated plants

(b) Offspring of bumblebee-pollinated plants

Figure 6.22 **Measuring the evolutionary response to selection in alpine skypilots** The histograms show the distribution of flower size (corolla flare) in the offspring of hand-pollinated skypilots (a; average = 13.1 mm) and bumblebee-pollinated skypilots (b; average = 14.4 mm). Redrawn from Galen (1996).

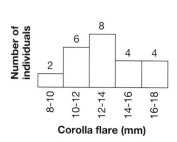

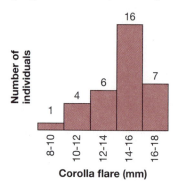

relationship between flower size and fitness. Only bumblebees select for larger flowers.

Galen's data allow her to predict how the population of timberline skypilots should respond to selection by bumblebees. The scenario she imagines is that a population of timberline skypilots that had been pollinated by a variety of insects moves (by seed dispersal) to the tundra, where the plants are now pollinated only by bumblebees. Using the low-end estimate that $h^2 = 0.2$, and the estimate that $S = 0.05$, the predicted response to selection is $R = h^2S = 0.2 \times 0.05 = 0.01$. Using the high-end estimate that $h^2 = 1$, and the estimate that $S = 0.05$, the predicted response to selection is $R = 1 \times 0.05 = 0.05$. In other words, a single generation of selection by bumblebees should produce an increase of 1 to 5% in the average flower size of a population of timberline skypilots.

Galen's prediction is therefore that flower size will evolve rapidly under selection by bumblebees. Is this prediction correct? Recall the experiment described earlier, in which Galen reared offspring of timberline skypilots that had been pollinated by hand and offspring of timberline skypilots that been pollinated by bees. Galen calculated the mean flower size of each group and found that the offspring of bumblebee-pollinated skypilots had flowers that were on average 9% larger than those of hand-pollinated skypilots (Figure 6.22). The prediction is correct: Skypilots show a strong and rapid response to selection. In fact, the response is even larger than predicted.

Galen's conclusion is that the 12% difference in flower size between timberline and tundra skypilots can be plausibly explained by the fact that timberline skypilots are pollinated by a diversity of insects, whereas tundra skypilots are pollinated almost exclusively by bees. Timberline skypilots can set seed even if bumblebees avoid them, but tundra skypilots cannot. Furthermore, it would take only a few generations of bumblebee-only pollination for a population of timberline skypilots to evolve flowers that that are as large as those of tundra skypilots.

Modes of Selection

Galen found that selection by bumblebees favors larger flower size in alpine skypilots. This pattern of selection, in which fitness consistently increases (or decreases) with the value of a trait, is known as directional selection (Figure 6.23).

Directional selection on a continuous trait changes the average value of the trait in the population. In the hypothetical population shown in Figure 6.23, the mean phenotype before selection was 6.9, whereas the mean phenotype after selection was 7.4.

Before leaving the topic of selection on quantitative characters, we note that other modes of selection are possible (Figure 6.23). In stabilizing selection, individuals with intermediate values of a trait have highest fitness. Stabilizing selection on a continuous trait trims off the tails of the trait's distribution. In the hypothetical population shown in Figure 6.23, there is less variation around the mean after stabilizing selection than before. Stabilizing selection is common; an empirical example appears in Figure 6.24.

In disruptive selection, individuals with extreme values of a trait have the highest fitness. Disruptive selection on a continuous trait trims off the top of the trait's distribution. In the hypothetical population shown in Figure 6.23, a higher proportion of the population has phenotypes distant from the mean after disruptive selection than before. Disruptive selection is relatively rare; an empirical example appears in Figure 6.25.

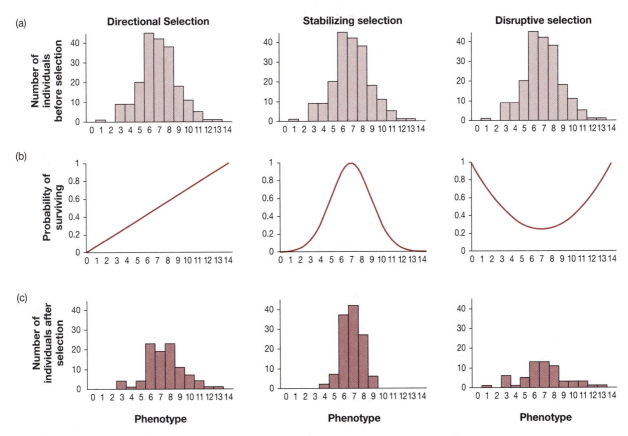

Figure 6.23 Three modes of selection Each column represents a mode of selection. The graphs in row (a) are histograms showing the distribution of a phenotypic trait in a hypothetical population before selection. The graphs in row (b) show different patterns of selection; they plot the probability of survival (a measure of fitness) as a function of phenotype. The graphs in row (c) are histograms showing the distribution of the phenotypic trait in the survivors. After Cavalli-Sforza and Bodmer (1971).

Figure 6.24 **Stabilizing selection on birthweight in humans** The horizontal axis represents birthweight. The left vertical axis and the histogram show the distribution of birthweights in a sample of 13,730 babies. The average birthweight is about 7 pounds. The right vertical axis and the data points with best-fit curve show mortality rate as a function of birthweight (on a logarithmic scale). The mortality rate is much higher among very small and very large babies than among babies of average size. (Note that in this figure, fitness is plotted as probability of mortality, whereas in Figure 6.23, fitness is plotted as probability of surviving.) The optimum birthweight is that with the lowest mortality rate. It is very close to the population average. Natural selection on birthweight in this population tends to hold the population average at a constant value. From Cavalli-Sforza and Bodmer (1971) and reference therein.

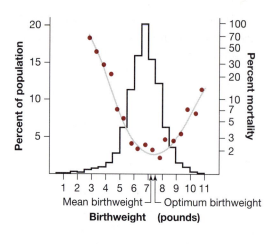

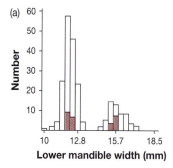

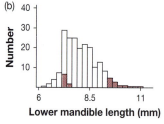

Figure 6.25 **Disruptive selection on bill size in the black-bellied seedcracker (Pyrenestes o. ostrinus)** Each graph shows the distribution of lower bill widths (a) or lengths (b) in a population of black-bellied seedcrackers, an African finch. The open portion of each bar represents juveniles that did not survive to adulthood; the filled portion represents juveniles that did survive. The survivors were those individuals with bills that were either relatively large or relatively small. (Note that there appears to be an element of stabilizing selection at work here too: Except in the case of birds with extremely long bills, the birds with the most extreme phenotypes did not survive.) From Bates Smith (1993).

Summary

Extension of the Hardy–Weinberg analysis to a genetic system involving two loci reveals a second kind of genetic equilibrium in populations. This is linkage equilibrium. When a population is in linkage equilibrium, genotypes at one locus are independent of genotypes at the other locus. Under Hardy–Weinberg assumptions, gamete frequencies do not change in a population in linkage equilibrium. Furthermore, selection on one locus has no effect on allele or genotype frequencies at the other locus. In contrast, when a population is in linkage disequilibrium, the genotypes at one locus are nonrandomly associated with genotypes at the other locus. Even under Hardy–Weinberg assumptions, in a population in linkage disequilibrium gamete frequencies change from generation to generation. Furthermore, selection on one locus can alter allele and genotype frequencies at the other locus.

Linkage disequilibrium can result from physical linkage, selection on multilocus genotypes, and genetic drift. Under Hardy–Weinberg assumptions, gamete frequencies change in a way that moves populations toward linkage equilibrium. This reduction in linkage disequilibrium is the result of the genetic recombination that happens during sexual reproduction.

The fact that sexual reproduction with random mating reduces linkage disequilibrium provides a key to understanding why sexual reproduction persists in populations. Simple theoretical arguments suggest that asexual reproduction should sweep to fixation in any population in which it appears. Empirical observations and experiments indicate, however, that sex confers substantial benefits. These benefits can be found in the population-genetic consequences of sex. When drift or selection has reduced the frequency of particular multilocus genotypes below their expected levels under linkage equilibrium, sexual reproduction can be favored because it recreates the missing genotypes.

Most traits involve many more than two loci, and we often do not know the identity of these loci. Quantitative genetics gives us tools for analyzing the evolution of such traits. Heritability can be estimated by examining similarities among relatives. The strength of selection can be measured by analyzing the relationship between phenotypes and fitness. When we know both the heritability of a trait and the strength of selection on the trait, we can predict the evolutionary response of the population.

Questions

1. In horses, the basic color of the coat is governed by the E locus. *EE* and *Ee* horses are black, while *ee* horses are a reddish chestnut. A different locus, the R locus, can cause "roan," a scattering of white hairs throughout the basic coat color. However, the roan allele has a serious drawback: *RR* embryos die during fetal development. *Rr* embryos survive and are roan, while *rr* horses survive and are not roan. The E locus and the R locus are on the same chromosome and are tightly linked. Suppose that several centuries ago, a Spanish galleon with a load of conquistadors' horses was shipwrecked near a large island. Just by chance, the only horses that survived the shipwreck and swam to shore were 10 roan chestnuts and 10 non-roan blacks. On the island, the horses interbred with each other and established a wild population. The island environment exerts no direct selection on the E locus. If you were to travel to this island today, would you expect to find more chestnut horses, or more black horses? Why?

2. Imagine a population of pea plants that is in linkage equilibrium for the height locus (T = tall, t = short) and the seed coat locus (R = round, r = wrinkled). What sort of selection event would create linkage disequilibrium? For example, will selection at just one locus (for example, all short plants die) create linkage disequilibrium? What about selection at two loci (for example, all short plants die, and all plants with wrinkled peas die)? What about selection on a certain combination of genotypes at two loci (for example, only plants that are both short and have wrinkled peas die)?

3. Now imagine a population of peas that is already in linkage disequilibrium at the height and seed coat loci. Will selection at just one locus affect evolution at both loci? How is your answer different from your answer to question 2, and why?

4. Figure 6.2 shows that when $g_{AB} = ps$, $g_{Ab} = pt$, $g_{aB} = qs$, and $g_{ab} = qt$, then $D = 0$. Show that when $D = 0$, $ps = g_{AB}$. (Hint: p, the frequency of allele A, is equal to $g_{AB} + g_{Ab}$. Likewise, s, the frequency of allele B, is equal to $g_{AB} + g_{aB}$. Multiply these quantities and simplify the expression. Knowing that $D = g_{AB}g_{ab} - g_{Ab}g_{aB} = 0$ will allow you to make a key substitution.)

5. In the beetle evolution experiment, Dunbrack et al. did not actually use sexual beetles and asexual beetles. Instead, they used two colors of sexual beetles, but forced one color of beetles to grow in population size as if it were asexual. They also did the experiment again, with the other color as "asexual." Why was it important that the researchers ran the experiment both ways? Actual asexual beetles would reproduce twice as fast as sexual beetles, but in the experiment, the authors made the asexual beetles reproduce three times as fast. Why do you think they did this? The researchers' simulated asexual population was not allowed to evolve at all in response to selection imposed by competition. How is this different from what would have happened in an actual population of asexual beetles? Do you think Dunbrack et al.'s experiment is a valid test of asexual reproduction versus sexual reproduction? If not, briefly describe the next experiment that you think Dunbrack et al. should do to follow up on this topic.

6. In a recent paper, Spolsky and colleagues (1992) reported genetic evidence that asexually reproducing lineages of a salamander have persisted for about 5 million years. Is this puzzling? Why or why not? Speculate about what sort of environment these asexual salamanders live in, and whether their population sizes are typically small (say, under 100) or large (say, over 1,000).

7. *Volvox* are abundant and active in lakes during the spring and summer. During winter they are inactive, existing in a resting state in encysted zygotes called zygospores. During most of the spring and summer, *Volvox* reproduce asexually; but at times they switch and reproduce sexually instead. When would you predict that *Volvox* would be sexual: spring, early summer, or late summer? Explain your reasoning.

8. Suppose you are telling your roommate that in biology class, you learned that within any given human population, height is highly heritable. Your roommate, who is studying nutrition, says "That does not make sense, because just a few centuries ago, most people were shorter than they are now, and it is clearly because of diet. If most variation in human height is due to genes, how could diet make such a big difference?" Your roommate is obviously correct that poor diet can dramatically affect height. How do you explain this apparent paradox to your roommate?

9. Now consider heritability in more general terms. Suppose heritability is extremely high for a certain trait in a certain population. There are two important questions: First, can the trait be strongly affected by the environment despite the high heritability value? To answer this question, suppose that all the individuals within a certain population have been exposed all their lives to the *same level* of a critical environmental factor. Will the heritability reflect that the environment is very important?

10. Second, can the heritability value itself change if the environment changes? To answer this question, imagine that the critical environmental factor changes, such that different individuals are now exposed to *different levels* of this environmental factor. What happens to variation in the trait in the whole population? What happens to the heritability?

11. Galen's skypilot study suggests that bumblebees could cause a timberline skypilot population to evolve flowers as large as those of tundra skypilots in just a few generations. Galen's research involved a timberline skypilot population. What experiments or observations would determine whether bumblebees are still exerting selection for large flower size today in tundra skypilots? What experiments would determine whether flower size is still evolving to become even larger?

Exploring the Literature

12. Many human pathogens, including bacteria and eukaryotes, are capable of both asexual reproduction and genetic recombination (that is, sex in the population-genetic sense). The frequency of recombination in a pathogen population can have medical implications. (Think, for example, about how fast resistance to multiple antibiotics will evolve in a population of bacteria that does versus a population that does not have recombination.) How can we tell whether a given pathogen population is engaging in genetic recombination or is predominantly clonal? Genetic recombination is such a powerful force in reducing linkage disequilibrium that the amount of disequilibrium in a pathogen population provides a powerful clue. See:

    Maynard Smith, J., N.H. Smith, M.O'Rourke, and B.G. Spratt. 1993. How clonal are bacteria? *Proceedings of the National Academy of Sciences, USA* 90:4384–4388.

    Burt, A., D.A. Carter, G.L. Koenig, T.J. White, and J.W. Taylor. 1996. Molecular markers reveal cryptic sex in the human pathogen *Coccidioides immitis*. *Proceedings of the National Academy of Sciences, USA* 93:770–773.

    Go, M.F., V. Kapur, D.Y. Graham, and J.M. Musser. 1996. Population genetic analysis of *Helicobacter pylori* by multilocus enzyme electrophoresis: Extensive allelic diversity and recombinational population structure. *Journal of Bacteriology* 178:3934–3938.

    Gräser, Y. et al. 1996. Molecular markers reveal that population structure of the human pathogen *Candida albicans* exhibits both clonality and recombination. *Proceedings of the National Academy of Sciences, USA* 93:12473–12477.

    Jiménez, M., J. Alvar, and M. Tibayrenc. 1997. *Leishmania infantum* is clonal in AIDS patients too: Epidemiological implications. *AIDS* 11:569–573.

13. How far and how fast can directional selection on a quantitative trait shift the distribution of the trait in a population? For one answer, see:

    Weber, K.E. 1996. Large genetic change at small fitness cost in large populations of *Drosophila melanogaster* selected for wind tunnel flight: Rethinking fitness surfaces. *Genetics* 144:205–213.

14. For another example of the use of quantitative genetics to estimate heritability, measure the strength of selection, and predict the evolutionary response, see:

    Grant, P.R., and B.R. Grant. 1995. Predicting microevolutionary responses to directional selection on heritable variation. *Evolution* 49:241–251.

Citations

Please note that much of the population and quantitative genetics in this chapter is modeled after presentations in:

Cavalli-Sforza, L.L., and W.F. Bodmer. 1971. *The Genetics of Human Populations* (San Francisco: W. H. Freeman and Company).

Falconer, D.S. 1989. *Introduction to Quantitative Genetics* (New York: John Wiley & Sons).

Felsenstein, J. 1997. *Theoretical Evolutionary Genetics* (Seattle, WA: ASUW Publishing, University of Washington).

Felsenstein, J. 1988. Sex and the evolution of recombination. In R.E. Michod and B.R. Levin, eds. *The Evolution of Sex* (Sunderland, MA: Sinauer), 74–86.

Hartl, D.L. 1981. *A Primer of Population Genetics* (Sunderland, MA: Sinauer).

Here is the listing of all citations in this chapter:

Bateson, W. 1913. *Mendel's Principles of Heredity* (Cambridge: Cambridge University Press).

Bates Smith, T. 1993. Disruptive selection and the genetic basis of bill size polymorphism in the African finch *Pyrenestes. Nature* 363:618–620.

Blakeslee, A.F. 1914. Corn and men. *Journal of Heredity* 5:511–518.

Cavalli-Sforza, L.L., and W.F. Bodmer. 1971. *The Genetics of Human Populations* (San Francisco: W. H. Freeman and Company).

Chao, L. 1990. Fitness of RNA viruses decreased by Muller's ratchet. *Nature* 348:454–455.

Clausen, J., D.D. Keck, and W.M. Hiesey. 1948. Experimental studies on the nature of species. III. Environmental responses of climatic races of *Achillea*. (Washington, DC: Carnegie Institution of Washington Publication No. 581), 45–86.

Dunbrack, R.L., C. Coffin, and R. Howe. 1995. The cost of males and the paradox of sex: An experimental investigation of the short-term competitive advantages of evolution in sexual populations. *Proceedings of the Royal Society of London,* Series B 262:45–49.

Falconer, D. S. 1989. *Introduction to Quantitative Genetics* (New York: John Wiley & Sons).

Felsenstein, J. 1988. Sex and the evolution of recombination. In R.E. Michod and B.R. Levin, eds. *The Evolution of Sex* (Sunderland, MA: Sinauer), 74–86.

Galen, C. 1989. Measuring pollinator-mediated selection on morphometric floral traits: Bumblebees and the alpine sky pilot, *Polemonium viscosum. Evolution* 43:882–890.

Galen, C. 1996. Rates of floral evolution: Adaptation to bumblebee pollination in an alpine wildflower, *Polemonium viscosum. Evolution* 50:120–125.

Haigh, J. 1978. The accumulation of deleterious mutations in a population: Muller's ratchet. *Theoretical Population Biology* 14:251–267.

Hedrick, P.W. 1983. *Genetics of Populations* (Boston: Science Books International).

Jennings, H.S. 1917. The numerical results of diverse systems of breeding, with respect to two pairs of characters, linked or independent, with special relation to the effects of linkage. *Genetics* 2:97–154.

Lande, R., and S.J. Arnold. 1983. The measurement of selection on correlated characters. *Evolution* 37:1210–1226.

Lively, C.M. 1992. Parthenogenesis in a freshwater snail: Reproductive assurance versus parasitic release. *Evolution* 46:907–913.

Maynard Smith, J. 1978. *The Evolution of Sex* (Cambridge: Cambridge University Press).

Maynard Smith, J. 1980. Selection for recombination in a polygenic model. *Genetical Research, Cambridge* 35:269–277.

Maynard Smith, J. 1988a. The evolution of recombination. In R.E. Michod and B.R. Levin, eds. *The Evolution of Sex* (Sunderland, MA: Sinauer), 106–125.

Maynard Smith, J. 1988b. Selection for recombination in a polygenic model—the mechanism. *Genetical Research, Cambridge* 51:59–63.

Michod, R.E., and B.R. Levin, eds. 1988. *The Evolution of Sex* (Sunderland, MA: Sinauer).

Muller, H.J. 1964. The relation of recombination to mutational advance. *Mutation Research* 1:2–9.

Murray, C., and R.J. Herrnstein. 1994. Race, genes and I.Q.—An apologia. *The New Republic* (October):27–37.

Punnett, R.C., ed. 1928. *Scientific Papers of William Bateson* (Cambridge: Cambridge University Press).

Seger, J., and W.D. Hamilton. 1988. Parasites and sex. In R.E. Michod and B.R. Levin, eds. *The Evolution of Sex* (Sunderland, MA: Sinauer), 176–193.

Spolsky, C.M., C.A. Phillips, and T. Uzzell. 1992. Antiquity of clonal salamander lineages revealed by mitochondrial DNA. *Nature* 356:706–710.

Sturtevant, A.H., and K. Mather. 1938. The interrelations of inversions, heterosis, and recombination. *American Naturalist* 72:447–452.

Via, S. 1991. The genetic structure of host plant adaptation in a spatial patchwork: Demographic variability among reciprocally transplanted pea aphid clones. *Evolution* 45: 827–852.

Molecular Evolution

In Chapters 5 and 6 we looked at Mendelian genetics in populations. Now we shift our attention slightly, from studying changes in allele frequencies to examining changes in the genes and gene products themselves—that is, in DNA and proteins. The advent of techniques for studying proteins and DNA opened up an entirely new arena for the study of evolutionary processes. Since its founding in the early 1960s, the field of molecular evolution has embraced three topics: the evolution of macromolecules, reconstructing evolutionary relationships, and the origin of life (Li and Graur 1991). In this chapter we focus on the first of these issues. Our goal is to explore how mutation, selection, and drift act on the diversity of genes and alleles. In Chapter 10, we explore molecular techniques for inferring phylogenies; Chapter 11 is devoted to the origin and early evolution of life.

Early studies in molecular genetics and evolution produced a fundamental surprise: Far from being a relatively homogenous collection of sequences that code for proteins, the genome is a bestiary of sequence types. Only a fraction of the eukaryotic genome consists of single-copy genes that code for proteins and conform to our traditional idea of a locus. Instead, in numerical terms most eukaryotic genomes are dominated by moderately to highly repetitive sequences, many of which do not code for proteins (Box 7.1). In this chapter we investigate the origin of several major classes of coding and noncoding sequences, and ask how selection and drift affect their evolution.

The chapter closes by introducing another insight from traditional and molecular genetic studies: Genomes are not completely cohesive units in which all loci cooperate to maximize the fitness of the individual. Many genomes are riddled with parasitic sequences, which add nothing to the fitness of the organism. Selfish genes, which maximize their own fitness at the expense of other genes in the same genome, have also been found in numerous organisms.

BOX 7.1   Segments of the genome

Some of our first insights into the overall structure of the eukaryotic genome came through DNA renaturation experiments. In these protocols, genomic DNA is purified from cells and sheared into small pieces. The DNA is heated to 90°C, which breaks the hydrogen bonds between complementary strands, and then cooled to 60°. As the temperature falls, the single-stranded DNA begins to re-form hydrogen bonds with complementary sequences. The percent of single-stranded DNA remaining at various times during the course of the experiment can then be measured. This is done by running a series of aliquots over a hydroxyapatite column. Hydroxyapatite is a form of calcium phosphate that adsorbs double-stranded DNA but lets single strands pass through. The speed and extent of DNA reassociation, and thus the amount of single-stranded DNA remaining, depends on the concentration of complementary strands and the amount of time that has passed. The key parameter in these experiments is thus the initial concentration of DNA (C_0) times time, or C_0t (pronounced "cot"). C_0t is usually reported on a logarithmic scale.

Plots of the percentage of single-stranded DNA remaining versus C_0t are simple in bacteria. The data reported in Figure 7.1a, for an experiment run with DNA from *E. coli,* are typical. Note that at low values of C_0t virtually no reassociation occurs. Then, between a C_0t of about 1 and about 20 in this experiment, reassociation occurs rapidly. At higher values of C_0t almost all of the DNA is again double-stranded. Because the concentration of single-stranded DNA is now very low, the rate of reassociation tails off. The linear relationship in the middle of the curve is what we expect to observe if each gene is present in one copy.

C_0t curves from eukaryotes are more complex (Figure 7.1b). They typically show three different segments, each similar in shape to the single curve observed in bacteria. The three segments observed in eukaryotes correspond to the reassociation of sequences that are present in high, moderate, and single-copy number. One of our challenges in this chapter is to understand the origins and evolutionary dynamics of these highly repetitive, moderately repetitive, and single-copy sequences.

A second point to emerge from early studies of genome structure was that the amount of DNA in each of the three sequence fractions shows extreme variation among organisms. Examine the data in Table 7.1, and note that the total amount of DNA found in the haploid genomes of different organisms shows no correlation with their morphological complexity. This is surprising. If the genome consisted only of single-copy coding sequences, we would expect a strong correlation between an organism's DNA content and its degree of developmental and morphological complexity. As a consequence, the lack of correlation was dubbed the C-value paradox. It results from variation in the amount of noncoding, or parasitic DNA in the genomes of different species.

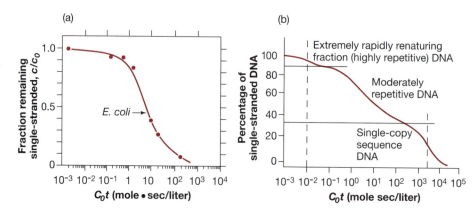

*Figure 7.1 C_0t **curves**
(a) This graph reports the fraction of single-stranded DNA remaining as a function of C_0t in *Escherichia coli*. This curve is typical for bacteria, and indicates that most of these genomes consist of single-copy sequences.   (b) This graph reports the fraction of single-stranded DNA remaining as a function of C_0t in a typical eukaryote.

TABLE 7.1 C values of eukaryotes

The total amount of DNA in a haploid genome is called the C value. Genome sizes show extreme variation within and between taxonomic groups. In flowering plants, for example, haploid genomes range by a factor of 2,500, from 50,000 to 125 billion kilobases (kb).

Species	C value (kb)
Navicola pelliculosa (diatom)	35,000
Drosophila melanogaster (fruitfly)	180,000
Paramecium aurelia (ciliate)	190,000
Gallus domesticus (chicken)	1,200,000
Erysiphe cichoracearum (fungus)	1,500,000
Cyprinus carpio (carp)	1,700,000
Lampreta planeri (lamprey)	1,900,000
Boa constrictor (snake)	2,100,000
Parascaris equorum (roundworm)	2,500,000
Carcarias obscurus (shark)	2,700,000
Rattus norvegicus (rat)	2,900,000
Xenopus laevis (toad)	3,100,000
Homo sapiens (human)	3,400,000
Nicotiana tabaccum (tobacco)	3,800,000
Paramecium caudatum (ciliate)	8,600,000
Schistocerca gregaria (locust)	9,300,000
Allium cepa (onion)	18,000,000
Coscinodiscus asteromphalus (diatom)	25,000,000
Lilium formosanum (lily)	36,000,000
Pinus resinosa (pine)	68,000,000
Amphiuma means (newt)	84,000,000
Protopterus aethiopicus (lungfish)	140,000,000
Ophioglossum petiolatum (fern)	160,000,000
Amoeba proteus (amoeba)	290,000,000
Amoeba dubia (amoeba)	670,000,000

Source: Li and Graur (1991)

Even alleles at the same locus can be in conflict with one another. This look at parasitic genes, selfish genes, and genomic conflict illustrates several new ideas in evolutionary analysis.

We now launch our inquiry into evolution at the level of the DNA sequence. We begin with the question of how natural selection and genetic drift interact to produce evolutionary changes in the genes that code for proteins.

7.1 Single-Copy Sequences

Single-copy loci code for proteins that catalyze biochemical reactions, participate in signaling or information transfer, form structures, or regulate the transcription of other genes. In Chapter 4 we reviewed data indicating that mutation rates vary widely among these protein-coding loci. Reading down the rows of data in Table 7.2 confirms that substitution rates vary as well. When homologous coding sequences from humans and mice are compared, some loci are found to be nearly identical while others have undergone rapid divergence. Equally striking is that substitution rates vary *within* genes. The most important pattern is illustrated by comparing the columns of data in Table 7.2: Silent

In almost every type of coding sequence compared, substitution rates are higher at silent sites than at replacement sites.

sites have diverged much more rapidly than replacement sites in all 36 genes examined.

Clearly, the rate of molecular evolution varies widely among sites within genes and among genes within genomes. Is this variation primarily due to natural selection or genetic drift?

A key to answering this question lies in the following observation: Genes that are responsible for the most vital cellular functions appear to have the lowest rates of replacement substitutions. Histone proteins, for example, interact with DNA to form structures called nucleosomes. These protein-DNA complexes are a major feature of the chromatin fibers typical of interphase cell nuclei. Changes in the amino acid sequences of histones would presumably disrupt the structural integrity of the nucleosome, with potentially far-reaching consequences for the efficiency of DNA transcription and synthesis. In contrast, genes that are less vital to the cell, and thus under less stringent functional constraints, show more rapid rates of replacement substitutions (Table 7.2).

Rapid rates of evolution in silent sites, compared to replacement sites in the same genes, reinforce the idea that molecular evolution is dominated by functional constraints. Because silent sites do not lead to changes in amino acid sequences, they have relatively little effect on the phenotype. Silent site changes are consequently much less likely than replacement changes to affect fitness negatively and be eliminated by natural selection. Therefore, substitution rates are higher at silent sites. The duplicated sequences called pseudogenes, which we introduced in Chapter 4, are presumably under no constraints at all, because they do not code for proteins. When compared to silent sites in the functioning sequences from which they are derived, pseudogenes exhibit still higher rates of evolution (Li et al. 1981; Li and Graur 1991).

The general messages are that variation in the rate of molecular evolution among genes and sequences appears to correlate strongly with the degree of functional constraint, and that rates of molecular evolution are strongly influenced by selection against deleterious mutations, or what is called negative selection. Recall from Chapter 4 that when mutations occur, alleles are produced that have a wide distribution of phenotypic effects. These effects range from lethal or highly deleterious to neutral to highly advantageous. The fate of an allele depends on where it falls along this continuum. In the histones, virtually all replacement mutations are highly deleterious and are rapidly eliminated by negative selection. The strength of negative selection varies from locus to locus and site to site, however. This variation is undoubtedly responsible for many of the differences in rates of molecular evolution among loci and sites recorded in Table 7.2.

But what about the mutations that do increase to fixation? Clearly, not all mutations are highly deleterious and eliminated by negative selection. For a population of a given size, is the fate of mutations that increase in frequency dictated primarily by the action of drift or by positive selection? Thinking again about the distribution of allelic effects, we expect that selection will be the dominant evolutionary process when mutations fall at either end of the spectrum. Near the middle, where mutations have little to no effect on fitness, selection will be much less powerful and genetic drift will explain most changes in allele frequencies. In essence, then, the question is whether the substitutions we observe at the level of the DNA sequence occurred because they produced a

Natural selection against deleterious mutations is called negative selection. Natural selection favoring beneficial mutations is called positive selection.

TABLE 7.2 Rates of nucleotide substitution vary among genes and among sites within genes

These data report rates of replacement and silent substitutions in a series of protein-coding genes compared between humans and either mice or rats. The data are expressed as the average number of substitutions per site per billion years, plus or minus a statistical measure of uncertainty called the standard error.

We introduced the distinction between replacement and silent sites in Chapter 4. To review, recall that bases in DNA are read in groups of three, called codons. There are 64 different codons in the genetic code. But only 20 different amino acids need to be specified, which means that the code is redundant. As a result, base pair substitutions may or may not lead to amino acid sequence changes that are exposed to natural selection. DNA sequence changes that do not result in amino acid changes are called silent site (or synonymous) substitutions; sequence changes that do result in an amino acid change are called replacement (or nonsynonymous) substitutions.

Gene	L	Replacement rate ($\times 10^9$)	Silent rate ($\times 10^9$)
Histones			
Histone 3	135	0.00 ± 0.00	6.38 ± 1.19
Histone 4	101	0.00 ± 0.00	6.12 ± 1.32
Contractile system proteins			
Actin α	376	0.01 ± 0.01	3.68 ± 0.43
Actin β	349	0.03 ± 0.02	3.13 ± 0.39
Hormones, neuropeptides, and other active peptides			
Somatostatin-28	28	0.00 ± 0.00	3.97 ± 2.66
Insulin	51	0.13 ± 0.13	4.02 ± 2.29
Thyrotropin	118	0.33 ± 0.08	4.66 ± 1.12
Insulin-like growth factor II	179	0.52 ± 0.09	2.32 ± 0.40
Erythropoietin	191	0.72 ± 0.11	4.34 ± 0.65
Insulin C-peptide	35	0.91 ± 0.30	6.77 ± 3.49
Parathyroid hormone	90	0.94 ± 0.18	4.18 ± 0.98
Luteinizing hormone	141	1.02 ± 0.16	3.29 ± 0.60
Growth hormone	189	1.23 ± 0.15	4.95 ± 0.77
Urokinase-plasminogen activator	435	1.28 ± 0.10	3.92 ± 0.44
Interleukin I	265	1.42 ± 0.14	4.60 ± 0.65
Relaxin	54	2.51 ± 0.37	7.49 ± 6.10
Hemoglobins and myoglobin			
α-globin	141	0.55 ± 0.11	5.14 ± 0.90
Myoglobin	153	0.56 ± 0.10	4.44 ± 0.82
β-globin	144	0.80 ± 0.13	3.05 ± 0.56
Apolipoproteins			
E	283	0.98 ± 0.10	4.04 ± 0.53
A-I	243	1.57 ± 0.16	4.47 ± 0.66
A-IV	371	1.58 ± 0.12	4.15 ± 0.47
Immunoglobulins			
Ig V_H	100	1.07 ± 0.19	5.66 ± 1.36
Ig γ1	321	1.46 ± 0.13	5.11 ± 0.64
Ig k	106	1.87 ± 0.26	5.90 ± 1.27
Interferons			
α1	166	1.41 ± 0.13	3.53 ± 0.61
β1	159	2.21 ± 0.24	5.88 ± 1.08
γ	136	2.79 ± 0.31	8.59 ± 2.56

Source: Li and Graur (1991)

fitness advantage, or whether they are neutral with respect to fitness and just happened to be fixed by drift.

The Neutral Theory of Molecular Evolution

Population geneticist Motoo Kimura (see Kimura 1968, 1977, 1983, 1986; Kimura and Ohta 1974) claimed that the vast majority of base substitutions that become fixed in populations, at all loci and in all types of sequences, are neutral with respect to fitness. In Kimura's view, genetic drift dominates evolution at the level of the DNA sequence. If correct, the hypothesis means that positive natural selection is largely inconsequential as an explanation for molecular evolution.

Kimura modeled the evolution of neutral mutations as follows:

- If there are N individuals in a diploid population, then there are 2N copies of each gene present in that population.
- All of the 2N copies present in the current population are descendants of a single allele that existed at some time in the distant past. Conversely, of the 2N copies currently in existence, only one will become the ancestor of all copies present in the population at some point in the distant future.
- If all 2N copies of the gene are selectively equivalent, or neutral in their effect on the bearer's fitness, then each has an equal chance of becoming the allele that becomes fixed in the population. This chance is equal to $\frac{1}{2N}$.
- In every generation, mutation will introduce new neutral alleles to the population. If ν is the mutation rate per gene per successful gamete, then $2N\nu$ new mutants will be introduced into the population each generation.
- Based on the preceding statements, the rate at which new neutral mutants become fixed in the population is equal to $(2N\nu)(\frac{1}{2N})$, or simply ν.

The neutral theory models the fate of alleles whose selection coefficient is 0.

This derivation means that the rate of sequence evolution, if all alleles are neutral, is simply equal to the mutation rate. This is an elegant result. It is also astonishing, for two reasons:

- Positive natural selection is excluded. The theory's central claim is that the vast majority of base substitutions in replacement and silent sites are neutral. The overall rate of evolution will be equal to the neutral mutation rate only when this proposition is true. Proponents of the neutral theory go further, however, by pointing out that even if a small proportion of nonneutral mutations occur in a population, they are likely to be deleterious and rapidly eliminated by natural selection. Thus, ν will represent the maximum rate of evolutionary change measured. This prediction is supported by the observation that pseudogenes have the highest substitution rates recorded among loci and sites in the nuclear genome. Because they do not code for proteins, mutations in pseudogenes should be completely neutral with respect to fitness and increase to fixation solely in response to genetic drift. For this reason, pseudogenes are considered a paradigm of neutral evolution (Li et al. 1981).
- The size of the population has no role. In Chapter 5 we showed that genetic drift is far more effective at changing the frequency of already existing alleles in small populations than in large ones. But Kimura's model

shows that for strictly neutral mutations, the rate of fixation of novel alleles due to drift does not depend on population size.

Although Kimura (1983) emphasized that his model was by no means antagonistic to the theory of evolution by natural selection, he did insist that the dominant evolutionary forces at the level of the DNA sequence are mutation, chance fixation, and negative selection. Proponents of the neutral theory acknowledge that advantageous substitutions do occur, but argue that they are extremely rare—so rare, in fact, that they are insignificant when compared to the vast number of neutral changes observed at the molecular level.

The neutral theory was a direct challenge to orthodox Darwinism, which held that virtually all evolutionary change was due to positive selection on advantageous mutations. This challenge was dramatized in the title of a follow-up article to Kimura's original publication. A paper called "Non-Darwinian evolution" (King and Jukes 1969) stated the neutralist position in vivid terms: Most evolution at the sequence level has little or nothing to do with positive natural selection. Is this true?

Testing the Neutral Theory I: Rates of Evolution in MHC Genes

Recall from our discussion of immune system function in Chapter 1 that infected cells display pieces of bacterial or viral protein on their surfaces, for possible recognition by cytotoxic T lymphocytes. The membrane proteins that display these antigens are created by a cluster of genes called the major histocompatibility complex, or MHC. The part of an MHC protein that binds to the foreign peptide is called the antigen recognition site (ARS). Austin Hughes and Masatoshi Nei (1988) set out to test the neutral theory by studying sequence changes in the ARS of several different MHC loci in humans and mice.

Specifically, Hughes and Nei wanted to compare the number of base substitutions that have occurred in silent versus replacement sites within the ARS. According to the neutral theory, silent site substitutions will always be more common than replacement substitutions because most replacement contributions will be damaging and thus eliminated by negative selection. The rate of silent site substitutions should approximate v and represent the highest rate of molecular evolution possible in a coding sequence. The data in Table 7.2 are consistent with this pattern. But is it true for the ARS of MHC genes in mice and humans?

The answer is no. When Hughes and Nei compared alleles from the MHC complexes of 12 different humans and counted the number of differences observed in silent versus replacement sites, they found that there have been significantly more replacement site than silent site changes. The same pattern occurred in the ARS of mouse MHC genes, although the differences were not as great. This pattern could only result if the replacement changes were selectively advantageous. This is because positive selection would cause the replacement changes to spread through the population much more quickly than neutral alleles can spread by chance. Hughes and Nei found this pattern only in the ARS, however; other exons showed no difference or more silent site changes. But this work provides a clear example of a gene segment where neutral substitutions do not predominate (see also Hughes and Nei 1989; Tanaka and Nei 1989).

Testing the Neutral Theory II: Alcohol Dehydrogenase Alleles in Fruit Flies

John McDonald and Marty Kreitman (1991) created a different test of the neutral theory. Remember, Kimura's model assumes that all replacement substitutions observed when we compare gene sequences, either within species or between species, are neutral. Replacement substitutions are expected to occur more slowly than silent site substitutions, because most replacement mutations are eliminated by negative natural selection. In contrast, all neutral mutations have an equal chance of spreading in the population. What McDonald and Kreitman realized is that although these rates of replacement and silent site change will be different, the neutral theory predicts that in any particular gene the *ratio* of replacement to silent site substitutions should be constant. Specifically, McDonald and Kreitman expected that the ratio of replacement to silent site substitutions in between-species comparisons of a particular gene should be the same as the ratio observed within species.

They tested this prediction by comparing sequence data from the alcohol dehydrogenase *(Adh)* gene of 12 *Drosophila melanogaster,* 6 *D. simulans,* and 12 *D. yakuba* individuals. In an attempt to sample as much within-species variation as possible, the individuals chosen for the study were from geographically widespread locations. McDonald and Kreitman lined up the *Adh* sequences from each individual in the study and identified sites where a base differed from the most commonly observed nucleotide, or what is called the consensus sequence. The researchers counted differences as fixed if they were present in all individuals from a particular species, and as polymorphisms if they were present in only some individuals from a particular species. Differences that were fixed in one species and polymorphic in another were counted as polymorphic.

McDonald and Kreitman found that 29% of the differences that were fixed between species were replacement substitutions. Within species, however, only 5% of the polymorphisms in the study represented replacements. Rather than being the same, these ratios show an almost sixfold, and statistically significant, difference (P = 0.006). This is strong evidence against the neutral model's prediction. McDonald and Kreitman's interpretation is that the replacement substitutions fixed between species are selectively advantageous. They suggest that these mutations occurred after *D. melanogaster, D. simulans,* and *D. yakuba* had diverged, and spread rapidly to fixation due to positive selection.

Assessing the Neutral Theory

The neutralist controversy is a debate over the relative importance of selection and drift in explaining molecular evolution.

Although we have just reviewed two tests that provide strong evidence against the neutral theory, the jury is still out on how general the model is. No one is arguing that substitutions in pseudogenes and at many silent sites are selectively advantageous—they clearly are neutral and behave according to the Kimura model. But claims about the relative importance of selection versus drift in explaining replacement substitutions continue to generate controversy. Because these claims are based on the relative frequency of chance events versus positive selection, it will take numerous tests, with many different genes in many different organisms, to decide whether neutral evolution is as important as Kimura and his co-workers originally claimed.

Two things *are* clear: First, the neutral theory furnishes a null model that makes precise and testable predictions about how molecular evolution occurs in the absence of positive selection; and second, based on these predictions we now have rigorous tests capable of detecting positive selection at the level of the DNA sequence (see Berry 1996; Ohta and Kreitman 1996). Even if the neutral theory proves to be less important than originally claimed, it is furnishing a framework for extensive empirical testing and further theoretical work.

Before closing our review of how selection and drift act on single-copy genes, we need to examine two situations in which positive selection can affect the fate of mutations at silent sites.

Codon Bias

Most of the 20 amino acids are encoded by more than one codon. We have emphasized that changes among redundant codons do not cause changes in the amino acid sequences of proteins, and have implied that these silent site changes are neutral with respect to fitness. If this were strictly true we would expect that codon usage would be random, and that each codon in a suite of synonymous codons would be present in equal numbers throughout the genome of a particular organism. But early sequencing studies confirmed that codon usage is highly nonrandom (Table 7.3). This phenomenon is known as codon bias.

Several important patterns have emerged from studies of codon bias. In every organism studied to date, codon bias is strongest in highly expressed genes—like those for the proteins found in ribosomes—and weak to nonexistent in rarely expressed genes. In addition, the suite of codons that are used most frequently correlates strongly with the most abundant species of tRNA in the cell (Figure 7.2).

The leading hypothesis to explain codon bias is natural selection for translational efficiency (Sharp and Li 1986; Sharp et al. 1988; Li and Graur 1991; Akashi 1994, 1996). The logic here is that if a "synonymous" substitution in a highly expressed gene creates a codon that is rare in the pool of tRNAs, the mutation will be selected against. The selective agent is the speed and accuracy of translation. Speed and accuracy are especially important when the proteins encoded by particular genes are turning over rapidly, and the corresponding genes have to be transcribed continuously. It is reasonable, then, to observe the strongest codon bias in highly expressed genes.

Selection against certain synonymous substitutions represents a form of negative selection; it slows the rate of molecular evolution. The upshot is that in genomes where codon bias occurs, not all redundant substitutions are "silent" with respect to natural selection. Indeed, strictly neutral mutations in highly expressed coding sequences may be rare or nonexistent. Neutral evolution may be limited to moderately or rarely expressed single-copy genes, pseudogenes, and other noncoding sequences.

Hitchhiking

Another phenomenon affecting the rate and pattern of change in coding sequences is referred to as hitchhiking, or a selective sweep. Hitchhiking can

TABLE 7.3 Codon bias in prokaryotes and eukaryotes

This table reports the relative frequencies of different RNA codons found in four different species: the bacterium *E. coli*, baker's yeast *(Saccharomyces cerevisiae),* the fruit fly *Drosophila melanogaster,* and humans. The rows in the table give the relative frequencies of codons for four amino acids specified by redundant codons and one amino acid specified by a single codon. If each codon were used equally in each genome, the relative frequencies would all be 1. Deviations from 1.00 indicate codon bias.

The amino acids listed are leucine, valine, isoleucine, phenylalanine, and methionine. The columns marked "High" and "Low" report relative codon usage in genes that are abundantly transcribed versus rarely transcribed. In every case reported here, codon bias is more extreme in highly expressed genes.

The columns marked "G + C" and "A + T" under "Humans" report relative codon usage in genes with third-positions rich in G or C versus genes with third-positions rich in A or T. In every case reported here, codon bias is more pronounced in GC-rich genes.

Amino acid	Codon	*Escherichia coli*		*Saccharomyces cerevisiae*		*Drosophila melanogaster*		Human	
		High	**Low**	**High**	**Low**	**High**	**Low**	**G + C**	**A + T**
Leu	UUA	0.06	1.24	0.49	1.49	0.03	0.62	0.05	0.99
	UUG	0.07	0.87	5.34	1.48	0.69	1.05	0.31	1.01
	CUU	0.13	0.72	0.02	0.73	0.25	0.80	0.20	1.26
	CUC	0.17	0.65	0.00	0.51	0.72	0.90	1.42	0.80
	CUA	0.04	0.31	0.15	0.95	0.06	0.60	0.15	0.57
	CUG	5.54	2.20	0.02	0.84	4.25	2.04	3.88	1.38
Val	GUU	2.41	1.09	2.07	1.13	0.56	0.74	0.09	1.32
	GUC	0.08	0.99	1.91	0.76	1.59	0.93	1.03	0.69
	GUA	1.12	0.63	0.00	1.18	0.06	0.53	0.11	0.80
	GUG	0.40	1.29	0.02	0.93	1.79	1.80	2.78	1.19
Ile	AUU	0.48	1.38	1.26	1.29	0.74	1.27	0.45	1.60
	AUC	2.51	1.12	1.74	0.66	2.26	0.95	2.43	0.76
	AUA	0.01	0.50	0.00	1.05	0.00	0.78	0.12	0.64
Phe	UUU	0.34	1.33	0.19	1.38	0.12	0.86	0.27	1.20
	UUC	1.66	0.67	1.81	0.62	1.88	1.14	1.73	0.80
Met	AUG	1.00	1.00	1.00	1.00	1.00	1.00	1.00	1.00

Source: Sharp et al. (1988)

occur when strong positive selection acts on a particular amino acid substitution. As a favorable mutation increases in frequency, neutral or even mildly deleterious mutations closely linked to the favored site will increase in frequency along with the beneficial locus. These linked mutations are swept along by selection and can actually "hitchhike" to fixation. Note that this process can occur if and only if recombination does *not* break up linkage between the hitchhiking sites and the site that is under selection.

Perhaps the best example of hitchhiking discovered to date was found on the fourth chromosome of fruit flies. The fourth chromosome in *Drosophila* is remarkable because no recombination occurs along its entire length. As a consequence, the entire chromosome represents a linkage group and is inherited like a single gene.

Andrew Berry and colleagues (1991) sequenced a 1.1-kb region of the fourth chromosome in ten *Drosophila melanogaster* and nine *D. simulans.* This chromosome region includes the introns and exons of a gene expressed in fly embryos called *cubitus interruptus Dominant* (*ci*^{D}). Berry et al. found a remarkable

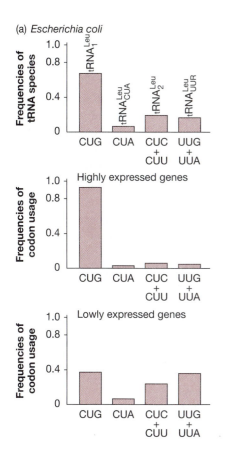

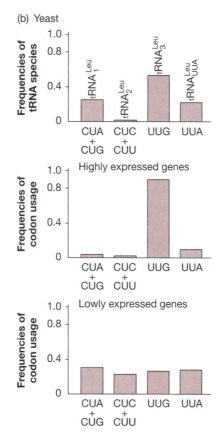

Figure 7.2 **Codon bias correlates with the relative frequencies of tRNA species** The bar chart in the top row of both (a) and (b) shows the frequencies of four different tRNA species that carry leucine in *E. coli* (a) and the yeast *Saccharomyces cerevisiae* (b). The bar charts in the middle and bottom rows report the frequency of the mRNA codons corresponding to each of these tRNA species in the same organisms. The mRNA codons were measured in two different classes of genes: those that are highly transcribed (middle) and those that are rarely transcribed (bottom). The data show that codon usage correlates strongly with tRNA availability in highly expressed genes, but not at all in rarely expressed genes. From Li and Graur (1991).

pattern in the sequence data: No differences were observed among the *D. melanogaster* individuals surveyed. The entire 1.1 kb of sequence was identical in the ten individuals. Further, only one base difference was observed among the *D. simulans* in the study. This means that there was almost no variation within species, or polymorphism, in this region. In contrast, a total of 54 substitutions were found when the same sequences were compared between the two species. A key observation is that genes on other chromosomes, surveyed in the same individuals, show normal amounts of polymorphism. These latter data serve as a control, and confirm that the lack of variation observed in and around the ci^D locus is not caused by an unusual sample of individuals. Rather, there is something unusual about the fourth chromosome in these flies.

Berry et al. suggest that selective sweeps recently eliminated all or most of the variation on the fourth chromosome in each of these two species. Their argument is that an advantageous mutation anywhere on the fourth chromosome will eliminate all within-species polymorphism as it increases to fixation. New variants, like the one substitution observed in the *D. simulans* sampled, will arise only through mutation.

In this way, selective sweeps create a "footprint" in the genome: a startling lack of polymorphism within linkage groups. Similar footprints have been found in other chromosomal regions where the frequency of recombination is low, including the *ZFY* locus of the human Y chromosome (Dorit et al. 1995) and a variety of loci in other fruit flies (Begun and Aquadro 1992;

Martín-Campos et al. 1992; Langley et al. 1993; Kliman and Hey 1993; Hanfstingl et al. 1994; Hilton et al. 1994; Hudson 1994; Hamblin and Aquadro 1996).

Has hitchhiking produced all of these regions of reduced polymorphism? The answer is probably not. Another process, called background selection, can produce a similar pattern (Charlesworth et al. 1993). Background selection results from negative selection against deleterious mutations, rather than positive selection for advantageous mutations. Like hitchhiking, it occurs in regions of reduced recombination. The idea here is that selection against deleterious mutations will remove closely linked neutral substitutions and produce a reduced level of polymorphism.

Background selection and hitchhiking are contrasting processes that lead to the same pattern.

Although the processes called hitchhiking and background selection are not mutually exclusive, their effects can be distinguished in at least some cases. Hitchhiking results in dramatic reductions in polymorphism as an occasional advantageous mutation quickly sweeps through a population. Background selection causes a slow, steady decrease in polymorphism as frequent deleterious mutations remove individuals from the population. The current consensus is that hitchhiking is probably responsible for the most dramatic instances of reduced polymorphism in linked regions—for example, where sequence variation is entirely eliminated—while background selection causes the less extreme cases.

The general message from this discussion of neutral evolution, codon bias, and drift is that evolutionary change in single-copy genes is a complex interaction of negative selection, positive selection, and drift. We turn now to examining how these processes affect the evolution of duplicated gene segments.

7.2 Duplicated Coding Sequences

In Chapter 4 we introduced the process called unequal cross-over and focused on how it results in duplicated gene segments (see Figure 4.3). Unequal cross-over is an important mutational process because it is a major mechanism for producing new genes. New loci are important because they can serve as the raw material for evolutionary innovation. The question we address here is: Once an additional stretch of sequence has been created by unequal cross-over, what happens to it?

Once a sequence is duplicated, mutation, selection, and drift will operate on it independently of their action on the parent sequence. Several fates are possible for duplicated sequences.

Several fates are possible for duplicated loci. We have already mentioned that if mutations in the duplicated segment make transcription impossible, the sequence becomes a nonfunctional pseudogene. But if the sequence continues to encode a functioning RNA or protein, it can (1) diverge and acquire a completely new function, (2) diverge and acquire a related function, or (3) evolve in concert with the original locus. How these events transpire is the subject of this section.

New Functions for Duplicated Segments: Lysozyme *c* and α-Lactalbumin

Lysozyme *c* is an enzyme found in a diverse array of animals including mammals, birds, reptiles, and insects. (Other lysozymes are found in even more diverse groups, including bacteriophages.) The protein is produced in a variety of tissues and is an important component of secretions like tears, saliva, and nasal mucus. The enzyme's activity was first observed by the bacteriologist Alexander

Fleming, who later discovered penicillin. While fighting a head cold, Fleming let several drops of his nasal mucus drip onto a culture plate teeming with bacteria. When he checked the plate later, he discovered that bacteria in and around the drops had been killed. He later confirmed that the active agent in the mucus was lysozyme *c*. The enzyme's function is antibiotic: It destroys the cell walls of bacteria.

α-Lactalbumin, in contrast, is an enzyme found only in mammals, where it is localized to a single organ: the mammary gland. The protein is not an enzyme, but is a key regulatory component in the synthetic pathway leading to lactose, the principal sugar found in milk.

Although their functions are unrelated, lysozyme *c* and α-lactalbumin have long been proposed as homologous loci that arose by gene duplication (see Prager and Wilson 1988). This inference was made on the basis of high sequence identity, close correspondence in how the two DNA sequences are organized, and similarities in the three-dimensional configuration of the resulting proteins (Figure 7.3). The hypothesis of homology was supported in an innovative experiment performed by Izumi Kumagai and colleagues (1992). These researchers cut a copy of exon 2 from the lysozyme gene of a chicken and used it to replace the exon 2 region in an α-lactalbumin gene from a goat. The proteins produced by this synthetic, hybrid gene showed clear enzymatic activity: They were able to shear the cell walls of bacteria. Remarkably, the exon transplant had changed the nonenzymatic α-lactalbumin gene into a lysozyme-like antibiotic gene.

These two loci may be the best example found to date in which a duplicated sequence has diverged from the ancestral sequence, due to selection and drift, and acquired a completely new function. The following paragraphs are a reminder that it is much more common for duplicated sequences to retain functions related to the ancestral sequence.

Goat α-lactalbumin gene

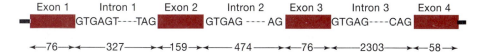

Hen lysozyme gene

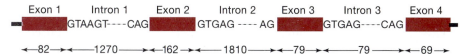

Figure 7.3 **Homologies in α-lactalbumin and lysozyme *c***   This diagram shows the organization and size of introns and exons in the α-lactalbumin gene of goats and the lysozyme *c* gene of chickens. The black boxes indicate exons, and the thin lines represent introns; the nucleotides given are found at the start and end of each intron. The numbers refer to the base pairs present. DNA sequence identity in exons 1−4 of the two genes is 30%, 51%, 46% and 29%, respectively. From Kumagai et al. (1992).

Gene Families

In Chapter 4 we reviewed the history, organization, and functions of two well-characterized families of loci: the globin family of genes in mammals and the eye-pigment genes of primates. Many examples of gene families like these exist. All of them share the following characteristics: They are thought to originate in duplication events, they show high sequence and/or structural homology, and they have related, but distinct, functions. Over the course of evolution, numerous loci have been duplicated and have diverged from their ancestral sequence under the influence of mutation, selection, and drift. These duplicated loci acquired new functions closely related to that of the ancestral sequence.

The immunoglobulin superfamily of genes may be the most spectacular example of this phenomenon. These loci represent a family of gene families—a branching lineage of duplicated genes that encompasses most of the important mammalian loci involved in fighting disease. The superfamily includes gene families like the Class I and Class II MHC loci, the immunoglobulins, and the T-cell receptor genes. All of these loci are thought to have originated in a single ancestral gene that existed hundreds of millions of years ago (Hood et al. 1985). To illustrate the importance of gene families in evolution, we will look at one family within the superfamily—the immunoglobulin loci—in more detail.

When we reviewed the immune system's reaction to invasion by a virus in Chapter 1, we mentioned that a type of white blood cell called a B cell produces antibodies. These are proteins that cling to epitopes derived from virions (or bacteria, or any other foreign body). This reaction is extremely specific—each antibody reacts with only 5–10 different epitopes—and marks virions and bacteria for destruction by macrophages. If an antibody does not recognize the epitope presented by a particular virus or bacterium, the invader can multiply without response. The diversity of epitopes produced by viruses and bacteria is immense, however. How does the immune system produce antibody proteins capable of binding to them all?

Gene duplication events are an important part of the answer. Antibodies are proteins made up of four polypeptide subunits (Figure 7.4). These subunits, in turn, are encoded by loci that have been duplicated numerous times in the course of evolution. The polypeptide subunit called the heavy chain, for example, consists of amino acid sequences that are encoded by three different subfamilies of genes. These loci are the V (variable), D (diversity), and J (joining) genes. Humans have 300 different variable loci, perhaps 25 diversity loci, and 4 joining loci (Figure 7.4). Each of these copies has a different sequence of nucleotides and produces a different polypeptide product. A heavy-chain polypeptide, then, consists of 1 of the 300 different V sequences, 1 of the 25 different diversity sequences, and 1 of the four different joining sequences. This means that $300 \times 25 \times 4 = 30,000$ different types of heavy chains can be produced by these subfamilies of duplicated genes. Because analogous variability exists at the gene families responsible for producing the other polypeptides in an antibody, *millions* of different antibody proteins can be produced by a single individual. Again, this diversity exists because gene duplication events, followed by sequence divergence, have produced large numbers of copies of each gene in the subfamilies of the immunoglobulin family.

Duplicated genes that diverge slightly in function are an important source of genetic variability. The presence of duplicated gene families makes the generation of antibody diversity possible.

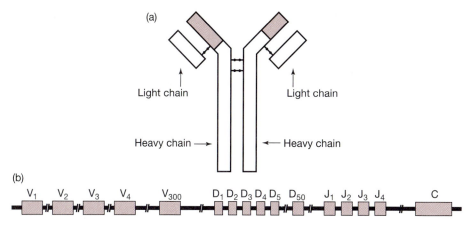

Figure 7.4   Organization of the heavy-chain immunoglobulin gene family
(a) This cartoon of an antibody molecule shows that the protein is made up of four polypeptide subunits: two heavy chains and two light chains. The colored portion of the heavy chain represents the variable region, which is encoded by the genes diagrammed in (b).
(b) In humans, the genes responsible for producing the heavy chain segment of an antibody molecule reside on the long arm of chromosome 14. Three gene families are involved in making the variable region of a single heavy-chain polypeptide. These families are called the variable (V), diversity (D), and joining (J) loci. This diagram shows how the 300 V loci, 25 D loci, and 4 J loci are arranged. The subscripts next to each locus indicate that the nucleotide sequence of each V, D, and J gene is different. Because of their high sequence and structural homology, we infer that all of the V, D, and J loci probably originated from duplication events. The C region diagrammed to the right represents the exon for the constant region of the heavy chain. The constant region is the white part of the heavy chains diagrammed in (a).

Before leaving the topic of gene families, we should note that one of the most surprising observations about the genome of baker's yeast *(Saccharomyces cerevisiae),* once it had been completely sequenced, was the high degree of redundancy present among coding sequences. Numerous genes, often scattered around the yeast genome, have homologous sequences and appear to code for the same or very similar proteins. Do these loci represent duplicated genes that have been transposed to new locations in the genome? Are the functions of the homologous sequences really identical?

Although these questions are still unanswered, the yeast genetics community is now transferring its efforts from genome sequencing to research programs focused on understanding the function of coding sequences. The expectation of some leading researchers (for example, Goffeau et al. 1996) is that the "redundant" members of these prospective gene families will have subtle differences important to the function of yeast cells in their natural environment: the skins of grapes bound for the wine press. The divergence of duplicated sequences may be an important aspect of this organism's evolutionary history.

Multiple-Copy Coding Sequences

Ribosomal rRNAs, some tRNAs, and the several histone genes are also examples of gene families. Two characteristics distinguish them from the gene families we just reviewed:

- Loci in these families produce the same product (Figure 7.5). Instead of diverging, these duplicated segments represent multiple copies of functionally identical genes. High copy numbers are thought to be advantageous in these loci because rRNAs, tRNAs, and histone proteins are needed in enormous quantities, especially early in development when cells are growing rapidly. The hypothesis is that a single locus would be unable to meet the demand for these transcripts.

- DNA sequences at these loci are completely or nearly identical (see Elder and Turner 1995). In the African clawed frog *(Xenopus laevis),* there are over 450 copies of the 18S and 28S rRNA genes in each haploid genome. The coding sequences of each of these alleles is *exactly* the same. This remarkable observation is called concerted evolution because the loci involved appear to be evolving in concert.

In the remainder of this section, we first consider two processes that could be responsible for sequence homogenization in multiple-copy loci. Then we examine an innovative study of concerted evolution in progress.

Unequal Cross-Over

We have pointed out that unequal cross-over events are probably the primary source of duplicated genes. But an important model demonstrates that repeated unequal cross-over events can also lead to concerted evolution when multiple-copy loci are arranged in tandem repeats—meaning that they are lined up tip-to-tail along the chromosome.

Inequal cross-over can lead to gene duplications. When it occurs frequently among genes that have already been duplicated, it can lead to concerted evolution.

Figure 7.6 demonstrates how frequent unequal cross-over events in tandemly repeated sequences can lead to sequence homogenization. An important point here is that unequal cross-over is a drift-like process (Smith 1976). Once a particular copy of a tandemly repeated array has been duplicated by unequal cross-

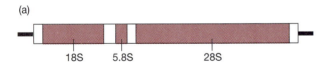

(a) Genome source	Number of rRNA repeats
Human mitochondrion	1
Mycoplasma capricolum	2
Escherichia coli	7
Saccharomyces cerevisiae	~140
Drosophila melanogaster	130–250
Human	~300
Xenopus laevis	400–600

(b) Species	Number of repeats
Sea urchin	300–1,000
Drosophila melanogaster	110
Xenopus laevis	20–50
Mouse	10–20
Chicken	10
Human	30–40

Figure 7.5   **Multiple-copy coding sequences**   (a) The diagram shows the order and relative size of rRNA genes in yeast. The colored sections represent exons, the white sections introns, and the black lines noncoding sequences called non-transcribed spacers. This set of genes is often tandemly repeated. The table specifies the number of rRNA repeats found in different organisms.   (b) The diagram shows the sequence of histone genes found in fruit flies. Each of the histone genes numbered 1–4 codes for a different protein. The colored sections represent exons and the black lines non-transcribed spacer regions. This set of genes is usually tandemly repeated. The table specifies the number of histone repeats found in the genomes of different organisms.

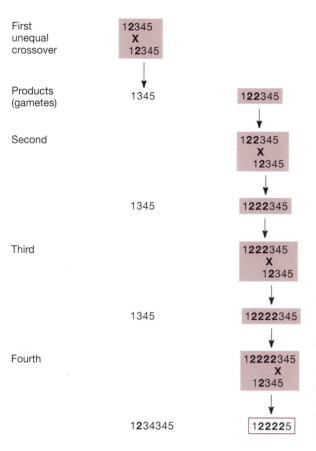

First unequal crossover

12345
X
12345

Products (gametes)

1345

122345

Second

122345
X
12345

1345

1222345

Third

1222345
X
12345

1345

12222345

Fourth

12222345
X
12345

1234345

122225

Figure 7.6 **Unequal crossing over can lead to concerted evolution** In this diagram the numbers 1–5 represent multiple-copy loci arranged in tandem repeats, such as rRNA genes. The X's indicate crossing-over events between homologous or sister chromatids during meiosis. The cross-overs diagrammed here are unequal because the chromatids have not lined up correctly. As a result, the cross-over event results in daughter chromatids with a duplicated or a deleted locus. When this process is frequent, the tandem repeats are homogenized and we observe concerted evolution.

over, like the copy numbered 2 in Figure 7.6, it is much more likely to be duplicated again in subsequent unequal cross-over events. If the duplicated copy has a selective advantage, it will serve to enhance the efficiency of this sorting process.

Gene Conversion

Gene conversion is a second process that leads to sequence homogenization. Like unequal cross-over, it depends on crossing-over events between homologous chromatids. The difference is that unequal cross-over occurs when recombination occurs at the wrong location, while gene conversion occurs when cross-over products need to be repaired.

The leading model for how crossing over proceeds at the molecular level predicts that the resulting daughter chromatids will have regions of "heteroduplex DNA" (see Box 7.2). In heteroduplex segments, each of the paired strands originates on a different chromosome. If mutation has created a different allele at the locus where crossing over occurred, the heteroduplex sequences will not be identical (this situation is diagrammed in Figure 7.8a). The mismatched pairs that result (see Figure 7.8b) will be recognized and repaired by enzymatic machinery in the cell. As Figures 7.8c and d show, this repair event can result in one allele being converted to the other. This event homogenizes the two loci. Although the repair process is probably random, gene conversion will lead to concerted evolution most efficiently when a bias exists in one of two processes: the direction of repair, or the selective advantage of the two repair products.

Box 7.2   A model for the molecular mechanism of recombination

The leading model for how recombination occurs at the level of the DNA sequence begins with an enzyme that nicks synapsed chromatids at homologous sites (Figures 7.7a and b). Single strands of DNA then migrate from one chromatid to the other and pair with their complementary bases on the homologous chromosome. This event creates hybrid sequences called heteroduplexes (Figure

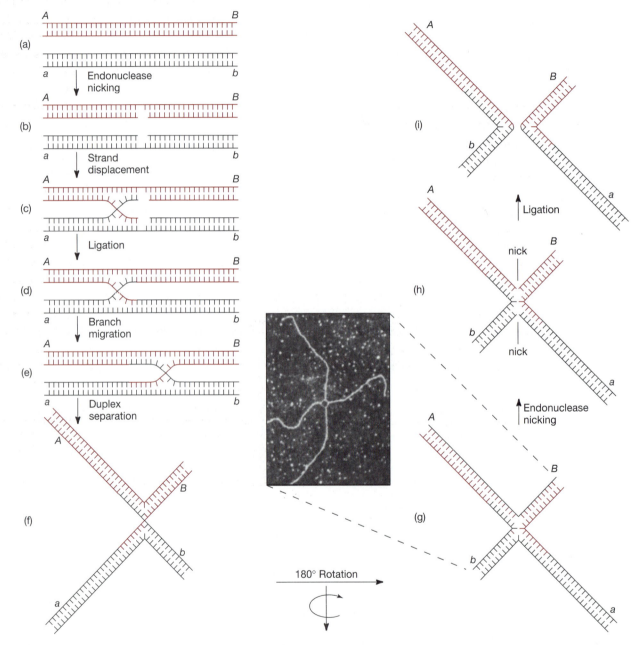

Figure 7.7 **Proposed interactions among homologous sequences during crossing-over** In this diagram, A and a and B and b represent alleles at homologous loci. The text for Box 7.2 explains each of the steps illustrated here. The photograph at step (g) is an electron micrograph of a "chi-structure" observed in *E. coli* plasmid DNA (a plasmid is an extrachromosomal loop of DNA found in many bacteria).

7.7c). After the nicked ends are ligated, the cross-bridged "Holliday structure" tends to migrate down the chromosome as stresses on the double helices are resolved by breakage and reformation of hydrogen bonds between nucleotides (Figures 7.7d and e). This creates a larger region of heteroduplex DNA. As the chromosomes separate, a cross or chi form can result if one of the strands rotates 180°

(Figure 7.7f). These structures have actually been observed under electron microscopy (Figure 7.7g). This rotation results in allelic exchange when each chromatid is again nicked and ligated to complete the recombination event (Figures 7.7h and i).

A number of the enzymes involved in the recombination process have now been characterized in *E. coli*. For an entree to this literature, see Voloshin et al. 1996.

Evidence for Concerted Evolution

In an informative study of concerted evolution in natural populations, David Hillis and colleagues (1991) examined rRNA genes in a group of asexual lizard species in the genus *Heteronotia*. Based on the gross morphology of their chromosomes, each of these asexual species is thought to be the product of a hybridization event between two sexual species. Each asexual daughter species is triploid. Two of its chromosome sets come from one of its sexual ancestors and one chromosome set from the other sexual ancestor. The sexual species were discovered so recently that they have not yet been given Latin names, and have temporarily been designated CA6 and SM6.

Hillis et al. reasoned that if concerted evolution has occurred in the rDNA cluster of the asexual daughter species, they should have either CA6 rDNA or SM6 rDNA, but not both. To test this prediction they examined rDNA types in 109 asexual individuals. The sample represented 40 independent lineages created by different hybridization events. When the researchers counted the distribution of sequences diagnostic for SM6 and CA6 rDNA in these individuals, they found that 53 of the triploids had only SM6 rDNA while 56 had a mix of SM6 and CA6 rDNA. Of the 53 lizards with SM6 sequences, 32 had karyotypes (chromosome morphologies) indicating that they had a *diploid* CA6 complement and a *haploid* SM6 genome. In these individuals, then, the SM6 rDNA had converted CA6 rDNAs even though they were originally outnumbered 2:1. Further, none of the lizards had only CA6 rDNA, and of the 56 individuals with both rDNA types, SM6 sequences were present in much higher copy number in 51. The data strongly support the hypothesis that concerted evolution has occurred (and may still be occurring) in these rDNAs. And because concerted evolution seems to have favored SM6 sequences in many independent lineages, it is highly probable that they confer a selective advantage in hybrid individuals.

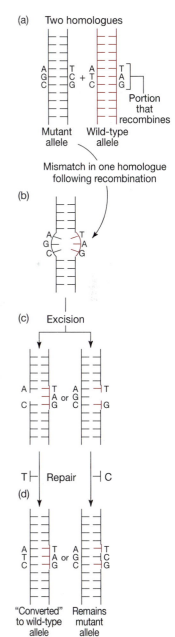

Figure 7.8   Resolution of heteroduplexes can lead to gene conversion   Note that the mutant and wild-type alleles diagrammed here differ by a single base-pair substitution. If this site were part of the heteroduplex created by a crossing-over event, it would create a base-pair mismatch that would be repaired after recombination is complete. Repair is a two-step process: One of the mismatched bases is excised, and then the correct matching base is inserted. As the diagram shows, repair can "convert" the mutant allele to the wild-type sequence.

7.3 Highly Repetitive Sequences

We have examined how mutation, selection, and drift affect the evolution of several different classes of coding sequences. But most eukaryotic genomes consist primarily of noncoding sequences. These come in various types. Sequences that flank coding regions do not produce a product but are critical to transcriptional control and form part of the locus. Sequences located between exons, called introns, are also part of the locus but are spliced out of transcripts before they are translated. Whether these intervening sequences have a function that benefits the gene is unclear; in Chapter 11 we will discuss the origin of introns in detail.

In sheer numbers, however, the most important class of noncoding DNA may be the highly repetitive sequences. These are short, tandemly repeated sequences that do not code for gene products. Instead, they make up the vast majority of the chromosome regions that remain highly condensed throughout the cell cycle. These heterochromatic regions tend to dominate the center and ends of chromosomes, and are distinguished from euchromatic regions by their morphology. The euchromatic regions of the chromosome uncoil during interphase of the cell cycle, and contain the vast majority of coding sequences.

Heterochromatic regions of eukaryotic chromosomes are dominated by highly repetitive sequences.

The highly repetitive sequences found in heterochromatin are important parts of the genome. In *Drosophila melanogaster,* 30–35% of the genome is heterochromatic in typical cells, including all of the Y chromosome, half of chromosome 4, 40% of the X chromosome, and 25% of chromosomes 2 and 3. Not all of this heterochromatin is made up of highly repetitive sequences, however: Some middle repetitive sequences and some single-copy coding loci also occur (Weiler and Wakimoto 1995). But typically, highly repetitive sequences are so abundant and so homogenous in their composition that when a sample of genomic DNA is centrifuged and separated in a density gradient, the highly repetitive fraction forms a distinctive band. This observation inspired an alternative name for these sequences: satellite DNA.

Highly Repetitive Sequences in Euchromatin: Micro- and Minisatellites

Highly repetitive sequences can also be found scattered among euchromatic regions of the genome. These variable-number tandem repeats, or VNTRs, come in two general classes. "Microsatellites" are di-, tri-, tetra-, or pentanucleotide tandem repeats (CACACA . . . would be an example of a dinucleotide repeat) that can be located at the centromere or distributed in euchromatin (see Jarne and Lagoda 1996). Tandemly repeated "minisatellite" sequences 30- to 60-bp (base pairs) in length are widely distributed in eukaryotic genomes; they were first found in the introns of mammalian myoglobin genes (Jeffries et al. 1985). Each particular series of micro- or minisatellite repeats is inherited in Mendelian fashion and is considered to be a genetic locus.

The number of repeats at micro- and minisatellite loci often varies between individuals of the same species or even individuals in the same family. This extreme variability is due to frequent mutation, caused by unequal cross-over events or because DNA polymerase slips during replication. Although point mutations do occur, changes in copy number are the dominant type of mutation in highly repetitive sequences. The total mutation rate at these loci is among the highest ever recorded (see Jarne and Lagoda 1996).

These length variants are now attracting intense interest from researchers in a wide variety of disciplines. Although alleles at most micro- and minisatellite loci appear to be neutral with respect to fitness, changes in the copy number of certain microsatellites have recently been implicated as causative agents in several serious human diseases and conditions, including the fragile-X syndrome of mental retardation, Huntington disease, and the most common form of adult muscular dystrophy (see Sutherland and Richards 1995). Variation in minisatellite copy number is so pervasive in humans that virtually every individual has a unique repeat-number allele at one or more minisatellite locus. This variability is the basis of DNA fingerprinting, now widely used in paternity analysis and forensics (see Lewontin and Hartl 1991; Amos and Pemberton 1993; Avise 1994). We will explore the use of these sequences in more detail in Chapter 15.

Centromeres and Telomeres

Centromeres and telomeres are specialized structures on the chromosome that are made up primarily of highly repetitive sequences. Both regions are heterochromatic. Centromeres are the location of the kinetochore: a plate-like structure that binds the spindle microtubules and regulates chromosome movements during mitosis and meiosis (Pluta et al. 1995). Telomeres form the ends of chromosomes and protect the DNA from degradation (see Greider 1993; Zakian 1995). Because they make linear chromosomes, mitosis, and meiosis possible, centromeres and telomeres represent a key innovation in the evolution of eukaryotes (Shaw 1994).

The centromeres of humans are made up of tandem arrays of a highly repeated 171-bp sequence. This sequence shows concerted evolution. Because repeat number is highly variable, indicating frequent deletions and duplications of the sequence, the most likely mechanism for the concerted evolution of these highly repeated sequences is unequal crossing over.

The highly repetitive sequences that make up telomeres are also well characterized in a wide variety of taxa (Table 7.4), although copy number and telomere length can vary from chromosome to chromosome. Perhaps the most

TABLE 7.4 Highly repetitive sequences found in telomeres

Organism	Sequence
Protozoa: *Tetrahymena*	CCCCAA
Slime molds: *Physarum*	CCCTAA
Fungi: *Neurospora*	CCCTAA
Invertebrates: *Bombyx* and other insects	CCTAA
Vertebrates	CCCTAA
Plants: *Arabidopsis*	CCCTAAA

Source: Zakian (1995)

unusual feature of telomeres is that they are synthesized by a telomere-specific reverse transcriptase called telomerase (Blackburn 1990; Zakian 1995).

In sum, highly repetitive DNAs are now well characterized in their distribution and abundance. Functionally, the sequences appear to be essential elements of both centromeres and telomeres, and thus serve as an important structural component of the chromosome. Their origins remain obscure, however, and we still have much to learn about their evolutionary dynamics. We have no evidence that the highly repetitive sequences scattered elsewhere in the genome serve a positive function in the fitness of the cell, and there is mounting evidence that mutations affecting copy number at these loci can occasionally have serious deleterious consequences. Finally, because highly repetitive sequences within a genome normally show sequence identity, it is probable that these loci undergo concerted evolution.

7.4 Organelle Genomes

In this section we move outside the nucleus and consider the evolution of genomes present in the chloroplasts and mitochondria of eukaryotes. Both of these organelles contain DNA that codes for functioning proteins, rRNAs, and tRNAs. The structure and organization of these genomes differs markedly from nuclear DNA (nDNA), however. This is largely because both organelles originated as bacteria that took up residence in eukaryotic cells early in the history of life. Chloroplasts are derived from the cyanobacteria and mitochondria from the purple bacteria. We will review the evidence for this claim, called the endosymbiotic hypothesis, when we consider the early evolution of life in Chapter 11. Here we describe the organization and gene content of these genomes and consider three questions about how they evolve.

Chloroplasts and mitochondria have their own DNA, and are derived from bacteria that lived symbiotically in the cytoplasm of early eukaryotic cells.

Chloroplast DNA

In chloroplasts the energy captured from photons is used to power the synthesis of sugars. Many of the proteins used in photosynthesis are encoded by chloroplast DNA (cpDNA). This is a circular double helix that lacks the histones and other chromosomal proteins found in nuclear DNA. Algal and plant cells typically contain many chloroplasts; each organelle normally has 20–80 copies of cpDNA.

Several cpDNAs have now been completely sequenced (see Wolfe et al. 1991). The 156-kb cpDNA genome of tobacco, for example, contains a total of 113 genes. The functions of these loci fall into two broad classes: sequences required for gene expression, and sequences required for photosynthesis. The loci involved in transcription and translation include four RNA polymerase subunits, 21 ribosomal proteins, a translation initiation factor, 30 tRNAs, and four rRNAs. The 29 genes involved in photosynthesis code for many of the proteins found in the thylakoid of the organelle, where photon capture takes place. Compared to nDNA, relatively few noncoding sequences are present.

The genome size and gene content of tobacco cpDNA appears to be typical of plants from other taxonomic groups (see Clegg et al. 1994), indicating that the information content of the plastid genome has been conserved during evolution. Even the order of genes along the chromosome appears to be highly conserved, if not identical (Palmer 1987). Further, all cpDNA genomes examined to date use the same genetic code as nDNA.

The inheritance of chloroplast genomes is normally uniparental. In most flowering plants, all of the cpDNA present in embryos is derived from the mother; but in at least some gymnosperms, inheritance is paternal. Biparental inheritance has also been recorded in some flowering plants (Palmer 1987). Recombination between cpDNA copies is extremely rare, having been recorded only in the green alga *Chlamydomonas*. This means that in most organisms the entire cpDNA genome is inherited like a single allele.

Relative to plant nuclear genes, sequence evolution in cpDNA is slow. Compared to silent sites surveyed in a sample of nuclear genes, silent sites in a sample of cpDNA loci evolve about one-quarter as fast (Table 7.5). Because of its high copy number in leaf cells and uniparental inheritance, cpDNA has been an important tool in reconstructing the evolutionary relationships of plants (for example, Chase et al. 1993).

The gene content and size of the chloroplast genome is highly conserved, and the rate of sequence change is low relative to the nuclear genome.

Structure and Evolution of mtDNA

In mitochondria the energy stored in sugars is used to power the synthesis of adenosine triphosphate (ATP). Many of the proteins used in cellular respiration are coded by mitochondrial DNA (mtDNA). Like cpDNA, this organelle genome is usually a circular double helix that lacks chromosomal proteins. The number of mitochondria per cell varies from four in some unicellular yeasts to thousands in the muscle cells of vertebrates; in vertebrates each organelle has $5-10$ copies of mtDNA.

The mitochondrial DNAs of plants, fungi, and animals are very different from one another. Among plant groups the sequence of genes on the molecule

TABLE 7.5 Rates of nucleotide substitution in plant organelle and nuclear genes

The column marked K_S reports the average number of substitutions observed per silent site, and K_A is the average number of substitutions per replacement site. The column marked L_S gives the number of silent sites examined, and L_A the number of replacement sites examined. Based on times of divergence between monocots and dicots provided by the fossil record, the average divergence in plant mtDNAs reported here is about 1.7×10^9 substitutions per silent site per year (for additional data see Wolfe et al. 1987, 1989).

Genomes	K_S	L_S	K_A	L_A
Comparison between monocot and dicot species				
Chloroplast genes	0.58 ± 0.02	4,177	0.05 ± 0.00	14,421
Mitochondrial genes	0.21 ± 0.01	1,219	0.04 ± 0.00	4,380
Comparison between maize and wheat or barley				
Nuclear genes	0.71 ± 0.04	1,475	0.06 ± 0.00	5,098
Chloroplast genes	0.17 ± 0.01	2,068	0.01 ± 0.00	7,001
Mitochondrial genes	0.03 ± 0.01	413	0.01 ± 0.00	1,526

Source: Li and Graur (1991)

and the total genome size show extreme variation (see Hanson and Folkerts 1992; Gray 1992). Within the melon family alone, mtDNA genome size varies between 300 kb and 2,400 kb. Aspects of genome structure like the location and length of intervening sequences, or even whether the molecule is circular or linear, are also highly variable among plant species. Recombination is frequent in plant mtDNAs (Hanson and Folkerts 1992), and sequence divergence between species is low (Table 7.5). The rate of silent substitutions in plant mtDNA is less than one-third the rate in cpDNA (Wolfe et al. 1987; 1989). Mitochondrial DNA evolves slowly in plants.

In the mtDNA of fungi, genome size is also highly variable. Among species in the filamentous ascomycetes, for example, genome size varies between 26.7 kb and 115 kb. In fungal mtDNA, coding sequences have introns and the genetic code is modified from the standard system we reported in Chapter 2. In all species investigated to date, the codon UGA specifies tryptophan in fungal mtDNA instead of "stop." In some species AUA specifies methionine instead of isoleucine, and CUN codes for threonine instead of leucine (see Clark–Walker 1992; Grivell et al. 1993).

A remarkable feature of animal mtDNA is that the genome is highly compact: There are no introns and non-coding sequences are limited in extent. In the mouse an astonishing 94% of the mtDNA genome codes for functional RNA. In general, gene order and genome size are moderately to highly conserved in the mtDNA of animals (Table 7.6). The mutation rate is high, however. The mitochondrial genome is synthesized by the γ form of DNA polymerase, which lacks any proofreading capability (reviewed in Brown 1983). Further, the enzymes that repair mismatches or damaged sites in nDNA do not exist in mitochondria.

The rate of sequence divergence is also high in animal mtDNA. Wes Brown and colleagues (1982) have estimated that silent sites in mammalian mtDNA average 5.7×10^{-8} substitutions per site per year, versus the 1.7×10^{-9} substitutions per site per year estimated to occur in silent sites of plant mtDNAs. This highlights a general contrast between plant and animal mtDNAs: Plant mtDNA is characterized by high variation in genome structure but little sequence divergence, while animal mtDNA is characterized by conservation of genome structure but rapid sequence divergence. The rate of mtDNA evolution in animals is 5–10 times higher than the average rate of nDNA evolution. In Chapter 10 we will explore how this rapid rate of evolution has made mtDNA a valuable tool in reconstructing the evolutionary history of recently diverged animal species.

The mitochondrial genome of animals also shows some discrepancies from the standard genetic code: AUA codes for methionine instead of isoleucine and, as in fungal mtDNA, UGA codes for tryptophan instead of a stop codon.

Animal mtDNA genomes have three additional characteristics that are shared across lineages (see Brown 1983; Wilson et al. 1985; Moritz et al. 1987):

- **Conservation of information content.** The best-characterized mitochondrial genomes specify 22 tRNAs, 2 rRNAs, and 13 mRNAs. The tRNAs and rRNAs are required to translate the mitochondrial mRNAs, which code for proteins involved in cellular respiration.

The size, structure, and rate of sequence change in plant, fungal, and animal mtDNA varies enormously, but the information content is highly conserved.

TABLE 7.6 Size of the mitochondrial genome in selected organisms

Organism	Size (kb)
Mouse	16.2
Human	16.6
Frog	18.4
Fruit fly	18.4
Baker's yeast	84.0
Pea	110.0

Because alterations in the genetic code are observed only in the mitochondria of selected taxa, these changes undoubtedly occurred long after the evolution of endosymbiosis.

- **Maternal inheritance.** In studies where crosses have been made between parents with distinctive mtDNAs, no copies of the paternal allele have been found in offspring.
- **No recombination.** Like many cpDNAs, the mtDNA genome in animals is inherited as a single allele.

Evolution of Organelle Genomes

Having completed a brief introduction to the organelle genomes of eukaryotes, we are ready to address three prominent questions about their evolution.

How Does Sequence Homogenization Occur?

Within individuals there is generally little or no sequence or structural variation in cpDNA or mtDNA. Even though they are present in enormously high copy number, organelle DNAs are nearly or completely identical within individuals (see Gillham et al. 1985; Wilson et al. 1985; Palmer 1987). How can this be?

The simplest explanation is that the organelle population is derived from an extremely restricted number of copies present in the embryo. The hypothesis is that organelle genomes go through a bottleneck every generation that reduces their variability within individuals. A mature mammalian egg typically contains 10^5 mitochondria, however, and we do not know of developmental mechanisms that would lead to an extreme reduction in the organelle genomes actually transmitted to the offspring.

Another possibility is that sequence homogenization results from the rapid turnover of mtDNA genomes in somatic cells. Mitochondria have a short life span in the cell and are constantly turning over. This results in an intense, drift-like sampling process among mtDNA lineages, with many variants being eliminated and one or a few variants increasing in number by chance.

The two explanations we have offered here are not mutually exclusive, however. The more important point is that both implicate genetic drift as the primary agent responsible for sequence homogenization.

Why Is the Inheritance of Organelle Genomes Usually Uniparental?

The answer to this question is brief: We do not know. The mechanism of uniparental inheritance in cpDNA and mtDNA is a mystery, the variation in inheritance of cpDNA (that is, whether inheritance is paternal or maternal) is unexplained, and we do not know why, in terms of selection and drift, the pattern exists.

Recall that except for some plant mtDNAs, recombination is rare or nonexistent in organelle genomes; as a consequence, mtDNA and cpDNA are inherited like a single allele. If we think of the entire organelle genome in this way, uniparental inheritance means that one copy of the allele—either the mother's or the father's—achieves zero fitness. This asymmetry in transmission suggests that a conflict should exist between maternal and paternal copies of the genome. Conflict between the maternal and paternal copies of the same gene is a topic we will treat in more detail in section 7.5, when we examine the phenomenon known as genomic imprinting. For the present, no one knows whether genomic imprinting exists in organelle DNA.

How Do the Nuclear and Organelle Genomes Interact?

Because both chloroplasts and mitochondria originated as free-living bacteria, the evolution of endosymbiosis meant that eukaryote lineages acquired several intact bacterial genomes. This is the first time we have encountered the movement of genetic information *between* species, or what is termed horizontal transfer (or horizontal evolution). Although endosymbiosis represents the largest horizontal transfer event in the history of life, numerous other instances have been documented (Box 7.3).

Once bacteria were inside eukaryotes, gene movement continued. Present-day organelle genomes have only a tiny fraction of the genetic information present in a bacterium (Gray 1992), implying that numerous loci have been lost completely or transferred to the nucleus. The genes that encode what may be the most abundant protein in nature, ribulose bisphosphate carboxylase (RuBPCase), are a vivid example. This enzyme catalyzes the fixation of CO_2 during the Calvin-Benson cycle, which is the central pathway in the light-independent reactions of photosynthesis. RuBPCase is made up of two subunits. The gene for the protein's small subunit is found in the nuclear genome, while the gene for the large subunit is part of the cpDNA genome (Gillham et al. 1985). The same type of evidence for gene transfers exists in mitochondria: The organelle's ribosomes, for example, are composed of rRNAs encoded by mtDNA and proteins encoded by nDNA.

Because the gene content of chloroplasts and mitochondria is highly conserved, most of these gene transfers probably took place very early in the evolution of endosymbiosis (Gillham et al. 1985; Clegg et al. 1994). Recent gene transfers have also been documented, however. For example, the gene *tufA,* which codes for a translation factor active only in the chloroplast, is found in the cpDNA of green algae but in the nuclear genome of the flowering plant *Arabidopsis.* This observation suggests that *tufA* was transferred to the nucleus after algae and higher plants diverged, except that copies of the gene exist in *both* genomes of some green algae. Thus, the more likely explanation is that the gene was duplicated and transferred to the nucleus early in plant evolution, then subsequently lost from the cpDNA of some derived lineages like flowering plants (Baldauf and Palmer 1990; Baldauf et al. 1990).

Evidence for an even more recent gene transfer involves the *cox2* locus found in the mtDNA of plants. This gene codes for one of the large subunits of cytochrome oxidase, a key component of the respiratory chain. In most plants this locus is part of the mitochondrial genome. Most legumes (members of the pea family) have a copy in both nuclear DNA and mtDNA, however, and in mung bean and cowpea the only copy is located in the nucleus. In these species the structure of the gene closely resembles the structure of an edited mRNA transcript. Because RNAs can be reverse transcribed to DNA, these facts suggest that gene transfer from the mitochondrion to the nucleus took place recently via the reverse transcription of an edited mRNA intermediate (Nugent and Palmer 1991; Covello and Gray 1992).

Research on interactions between the nuclear and organelle genomes continues. In the yeast *Saccharomyces cerevisiae,* the nuclear loci involved in the synthesis of mitochondria are scattered over many different chromosomes, yet are regulated in concert; how this regulation is accomplished is an active

Horizontal transfer of genes refers to the movement of sequences between species and across taxonomic boundaries.

Box 7.3   Evidence for horizontal transfer of genes

Horizontal transfer refers to the movement of a gene from one lineage to another, as from a bacterium to a eukaryote. The transfer is horizontal in the sense of occurring within a generation and crossing lineages, as opposed to the normal pattern in which genes are transmitted vertically—across generations and within a lineage.

How do we determine that horizontal transmission has taken place? When a phylogenetic tree is constructed based on similarities and differences in the DNA sequences of a particular gene and a species shows up next to a taxon that is obviously distantly related, it is a clue that horizontal transfer may have occurred between them. The logic here is that the gene tree (the phylogenetic relationships es-

timated from a single gene) clashes with the species tree (the true evolutionary relationships) because the gene was transferred horizontally. This situation is illustrated in Figure 7.9.

Hybridization events between eukaryote species are a mechanism of horizontal gene transfer between closely related species (for example, Sang et al. 1995). When gene transfer occurs between prokaryotes or from eukaryotes to prokaryotes or vice versa (Smith et al. 1992; Cummings 1994), the most probable mechanism is movement via transposons or plasmids (see section 7.5). The importance of horizontal transfer as a mechanism of gene acquisition and evolutionary change is still being debated.

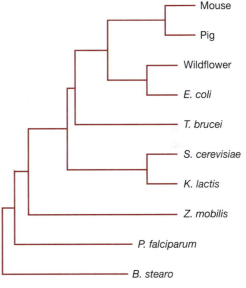

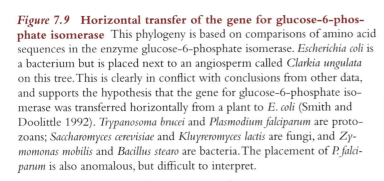

Figure 7.9 **Horizontal transfer of the gene for glucose-6-phosphate isomerase** This phylogeny is based on comparisons of amino acid sequences in the enzyme glucose-6-phosphate isomerase. *Escherichia coli* is a bacterium but is placed next to an angiosperm called *Clarkia ungulata* on this tree. This is clearly in conflict with conclusions from other data, and supports the hypothesis that the gene for glucose-6-phosphate isomerase was transferred horizontally from a plant to *E. coli* (Smith and Doolittle 1992). *Trypanosoma brucei* and *Plasmodium falciparum* are protozoans; *Saccharomyces cerevisiae* and *Kluyveromyces lactis* are fungi, and *Zymomonas mobilis* and *Bacillus stearo* are bacteria. The placement of *P. falciparum* is also anomalous, but difficult to interpret.

area of research (see Grivell et al. 1993). Many other questions remain about the evolution of these cohabiting genomes. Why have some genes, but not all, been transferred from organelles to the nucleus? Is there a selective advantage to having certain genes located in each genome, or were the movements merely chance events? Is reverse transcription of mRNAs the usual mechanism of transfer, or are other processes involved? Finally, a number of bacteria live parasitically in the cytoplasm of present-day eukaryotes (see Werren 1994; Werren et al. 1995). Why did the ancestors of chloroplasts and mitochondria evolve a symbiotic relationship, when parasitism can also be a successful strategy?

7.5   Genomic Parasites, Selfish Genes, and Genomic Conflict

We have now examined the evolution of a remarkably diverse collection of genes. Although they differ widely in their response to mutation, selection, and drift, all of these sequences have a critical feature in common: They have a positive function in the cell and contribute to the fitness of the organism. If the "goal" of evolution is to maximize the number of gene copies transmitted to the next generation, the loci we have been discussing achieve that goal by increasing the reproductive success of the individual where they reside—that is, by contributing to the survival and mating success of the organism. These loci also follow Mendel's rules: They transmit copies of themselves to 50% of the gametes produced by the individual, and transmit just one copy of themselves per gamete.

Not all genes conform to this picture, however. Some sequences have no fitness benefit for the organism but are transmitted anyway. These loci parasitize the fitness benefits provided by other genes. Other loci transmit more than one copy of themselves to the next generation by transposing to new sites within the genome. Still others transmit themselves to more than 50% of the gametes produced by the individual. These types of genes spread at the expense of competing alleles at the same locus.

Instead of cooperating to increase the fitness of the individual, some genes spread by pursuing selfish or parasitic strategies.

The general message of these loci is that the genome is not a cohesive collection of genes, all present because they contribute to the fitness of the individual. Many genomes are littered with selfish and parasitic sequences. Our goal in this section is to explore how and why these sequences evolve.

Transposable Elements

Transposable genetic elements are capable of moving from one location to another in the genome. Many types of transposable elements leave a copy behind in the original location when they move, creating a gene duplication event. In this way transposable elements increase in frequency in a population by transmitting themselves both within the genome and between generations.

The vast majority of transposable elements contain only the sequences required for transposition. That is, they do not code for proteins or RNAs that function in the cell. Because the insertion of transposable elements will occasionally disrupt coding sequences and cause deleterious mutations, transposable elements are most accurately characterized as genomic parasites.

Transposable elements are also remarkably successful. They are classified as middle repetitive sequences and are responsible for much of the variation in genome sizes recorded in Table 7.1. The key to understanding this success is to appreciate that transposable elements can increase in a population even when they harm their host slightly. This can occur because transposition events will result in more than one copy of the parasitic locus being present in many gametes. The upshot is that transposable elements can be transmitted to the next generation in greater than Mendelian proportions. The transposable elements that replicate themselves most efficiently and with the least fitness cost to the host genome are favored by natural selection and tend to spread (Charlesworth and Langley 1989).

Another key to understanding transposable elements is to appreciate that their success depends on sex (Hickey 1992). In eukaryotes, outcrossing results in haploid genomes being mixed. This presents transposable elements with new targets for transposition and allows them to spread throughout a population. In prokaryotes and other groups where outcrossing is rare and most reproduction is by fission, transposable elements in the main chromosome tend to be eliminated by selection or drift.

Transposable genetic elements can be grouped into two broad classes, based on whether they move via an RNA or DNA intermediate sequence. These Class I and Class II elements are distributed throughout eukaryotes and come in an enormous array of sizes, copy numbers, and structurally related families. The literature on mobile genetic elements is vast and growing rapidly (for reviews, see Hickey 1992; Wichman et al. 1992; Cummings 1994; Voytas 1996). Although we can only touch on this body of research here, understanding some basic types of transposable elements is important. In sheer numbers, they represent some of the most successful genes in the history of life.

Class I Elements

Class I elements, also called retrotransposons, are the product of reverse transcription events. Although the molecular mechanism of transposition is not yet fully known, we do know that movement of Class I elements occurs through a ribonucleic acid (RNA) intermediate. Transposition is also replicative, meaning that the original copy of the sequence is intact after the event. The long interspersed elements (LINEs) are retrotransposons that contain the coding sequence for reverse transcriptase and are thought to catalyze their own transposition. In mammals, LINEs are typically 6–7 kb in length and can number over 10,000 copies scattered throughout the haploid genome (Hutchison et al. 1989; Wichman et al. 1992).

Transposable elements that spread via an RNA intermediate have to be reverse transcribed, by reverse transcriptase, before being inserted into a new location.

Another important category of retrotransposons is distinguished by the presence of long terminal repeats (LTRs). LTRs are one of the hallmarks of retroviral genomes. When retroviruses insert themselves into a host DNA to initiate an infection, LTRs mark the insertion point. In corn, 10 different families of retrotransposons with LTRs have been identified, each of which is found in 10 to 30,000 copies per haploid genome (SanMiguel et al. 1996). The complete genome of baker's yeast, *Saccharomyces cerevisiae,* has been sequenced and found to contain 52 complete LTR-containing sequences called Ty elements, and 264 naked LTRs that lack the coding regions of normal retrotransposons. These "empty" LTRs are interpreted as transposition footprints, meaning that they are sequences left behind when Ty elements were somehow excised from the genome (see Boeke 1989; Goffeau et al. 1996).

Where did retrotransposons come from? One hypothesis is that LINEs and the LTR-containing retrotransposons evolved from retroviruses. In their sequence homology, retrotransposons resemble retroviruses that have lost the coding sequences required to make capsule proteins. This hypothesis proposes that retrotransposons have adopted a novel evolutionary strategy. In contrast to retroviruses, which replicate in their host cell, move on to infect new cells, and eventually infect new host individuals of the same generation, retrotransposons replicate by infecting the germline. Instead of being transmitted horizontally, they

Retrotransposons and retroviruses may be closely related.

replicate by being transmitted vertically, to the next generation of hosts. Their transmission is much slower than that of conventional retroviruses, but retrotransposons also escape detection by the immune system.

Howard Temin (1980), however, has pointed out that it is equally likely that retroviruses evolved from retrotransposons. Despite their prominence in a wide variety of genomes, the origin of retrotransposons is still obscure.

A second type of Class I element is called a retrosequence. Retrosequences do not contain the coding sequence for reverse transcriptase, but amplify via RNA intermediates that are reverse transcribed and inserted into the genome. The short interspersed elements (SINEs) of mammals are among the best-studied examples. SINEs are middle-repetitive elements that are grouped into several different families, each of which is distinguished by its sequence homology with a different functioning gene. The *Alu* family of sequences in primates, for example, is about 90% identical to the 7SL RNA gene, which is involved in transmembrane protein transport; other families are homologous with various tRNA genes (Table 7.7) SINEs are typically under 500 bp in length and lack the sequences necessary for translation of a transcribed RNA message.

Retrosequences do not code for an RNA or protein that functions in the cell, but are closely related to functioning genes.

Where did SINEs and other retrosequences come from, and how are they replicated? In most SINE families it appears that only one or a very few master copies of the locus are actively transposing, and that the remainder represent inactive copies analogous to pseudogenes (see Shen et al. 1997). As with LINEs and other Class I elements, the mechanism of this transposition is not known. We do not know how transcription of the master gene locus is regulated, where

TABLE 7.7 Some short interspersed element (SINE) families

For more information on the distribution and molecular structure of SINEs, see Deininger and Daniels (1986), Deininger (1989), and Takasaki et al. (1994).

Family	Species distribution	Copy number	% of genome
Alu	Primates	500,000	5
Alu subA	Primates	75,000	0.7
Alu type IIa	Galago	200,000	2
Alu type IIb	Galago	total	
Monomer	Galago	200,000	0.8
B1 (rodent type II)	Rodents	80,000	0.3
B2a (rodent type IIa)	Rodents	80,000	0.7
B2b (rodent type IIb)	Mouse		
Identifier (ID)	Rat	130,000	0.5
	Mouse	12,000	0.05
	Hamster	2,500	0.01
Rabbit C	Rabbit	170,000	1.7
Artiodactyl C-BCS	Goat and cow	300,000	3
art2	Goat	100,000	1.8

Source: Deininger (1989)

the reverse transcriptase comes from, or how insertion of the resulting DNA copy proceeds. We can, however, date the origin of some families. The *Alu* sequences, for example, have been found in every primate surveyed to date but do not occur in rodents. This observation suggests that the *Alu* family entered the genome of a primate ancestor after the divergence of these mammalian orders.

Class II Elements

Class II transposable elements replicate via a DNA intermediate, and are the dominant type of transposable genetic element found in bacteria. Their transposition can be replicative, as in Class I elements, or conservative. In conservative transposition the element is excised during the move so that copy number does not increase.

The first Class II elements ever described were the insertion sequences, or IS elements, discovered in the bacterium *Escherichia coli* (Table 7.8). When insertion sequences contain one or more coding sequences, they are called transposons. In addition to being inserted into the main bacterial chromosome, however, transposons are commonly inserted into plasmids. Plasmids are circular, extrachromosomal DNA sequences. They replicate independently of the main chromosome and are readily transferred from one bacterial strain to another during conjugation.

The chromosomes of bacteria are largely free of transposable elements. But transposable elements are frequently found in the extrachromosomal elements called plasmids.

Transposons encode a protein, called a transposase, that catalyzes transposition. In bacterial transposons the coding region frequently also codes for a protein that confers resistance to an antibiotic (Table 7.9). As a result, plasmid-borne transposons have been responsible for the rapid evolution of drug resistance in many groups of disease-causing bacteria. Transposons that confer antibiotic resistance are the first example we have encountered of a transposable element that creates a fitness advantage for its host.

Class II transposable elements are also found in eukaryotes. In fact, the first transposable elements ever discovered, by Barbara McClintock in the 1950s,

TABLE 7.8 Insertion sequences found in the genome of *Escherichia coli*

Some of the insertion sequences listed here are found on extra chromosomal genetic elements in *E. coli*. IS2 and IS3, for example, are located on F (or "fertility") factors, and pSC101 on a plasmid. F factors are circular DNA molecules found in bacteria that can either replicate independently or integrate into the main chromosome and replicate with it. The presence of an F factor in a bacterial cell confers the ability to donate genetic material during conjugation (bacterial sex).

Insertion sequence	Normal occurrence in *E. coli*	Length (bp)
IS1	5–8 copies on chromosome	768
IS2	5 on chromosome; 1 on F	1327
IS3	5 on chromosome; 2 on F	1400
IS4	1 or 2 copies on chromosome	1400
IS5	Unknown	1250
γ-δ (TN1000)	1 or more on chromosome; 1 on F	5700
pSC101 segment	On plasmid pSC101	200

Source: Griffiths et al. (1993)

TABLE 7.9 Transposons can carry genes conferring antibiotic resistance

This listing illustrates of the diversity and number of plasmid-borne transposons that carry antibiotic-resistance genes.

Transposon	Confers resistance to	Length (bp)
Tn1	Ampicillin	4,957
Tn2	Ampicillin	
Tn3	Ampicillin	
Tn4	Ampicillin, streptomycin, sulfanilamide	20,500
Tn5	Kanamycin	5,400
Tn6	Kanamycin	4,200
Tn7	Trimethoprim, streptomycin	14,000
Tn9	Chloramphenicol	2,638
Tn10	Tetracycline	9,300
Tn204	Chloramphenicol, fusidic acid	2,457
Tn402	Trimethoprim	7,500
Tn551	Erythromycin	5,200
Tn554	Erythromycin, spectinomycin	6,200
Tn732	Gentamicin, tobramycin	11,000
Tn903	Kanamycin	3,100
Tn917	Erythromycin	5,100
Tn1721	Tetracycline	10,900

Source: Griffiths et al. (1993)

were the *Ac* and *Ds* elements of corn. These Class II sequences code for a transposase as well as other proteins. Another example are the P elements found in *Drosophila melanogaster*. A typical fly genome contains 30–50 copies of these elements, which have insertion-sequence-like repeats at their ends and as many as three coding regions for protein products (Ajioka and Hartl 1989).

Evolutionary Impact of Transposable Elements

Geneticists and evolutionary biologists have been struggling to understand the evolutionary impact of transposable elements since they were first described. Plasmid-borne transposons are responsible for the rapid evolution of antibiotic resistance in bacteria. But we also have clear evidence that actively transposing elements frequently cause deleterious mutations. Insertion of LINE elements in humans, for example, have been responsible for tumor formation and cases of hemophilia (see Hutchison et al. 1989). It also seems that high copy numbers might grossly inflate genome sizes, with deleterious consequences for the cell. During mitosis, the time and molecular resources required to replicate a genome burdened by parasitic DNA could place a limit on growth rates, particularly in small, rapidly dividing organisms. This idea, though sensible, has yet to be confirmed. If transposable elements are highly deleterious, we would expect that hosts might evolve defense systems. But we have no evidence of cellular machinery capable of excising and destroying transposable elements once they are inserted into the host DNA.

In sum, transposable elements are prominent parts of many prokaryotic and eukaryotic genomes. They are highly variable in their copy number, molecular structure, and mode of transposition. Thanks to 50 years of research, the natural history of transposable elements is fairly well known. Efforts are now focused on understanding the molecular mechanisms of transposition, the origin of various classes of elements, and their response to selection and drift (see Condit 1990; Hickey 1992).

Transmission Distorters

Transmission distorters are loci that flout Mendel's rules. Instead of being transmitted in equal proportions to each daughter gamete, these selfish genes transmit themselves preferentially and at the expense of competing alleles. They do this by killing gametes or zygotes that do not carry a copy of themselves.

The best-studied selfish genes create "meiotic drive," or non-Mendelian ratios in gametes. The segregation distorter complex in *Drosophila melanogaster* provides an example. The two most important loci in this complex are tightly linked on the second chromosome. These are genes called *Segregation distorter (Sd)* and *Responder (Rsp)*. We will discuss two alleles at the *Sd* locus: the wild-type Sd^+, which does not cause transmission distortion, and the mutant *Sd*, which can cause transmission distortion. The two most important alleles at *Rsp* are the "sensitive" wild-type version, Rsp^s, and the mutant "insensitive" version Rsp^i. Sperm that carry the Rsp^s allele can be killed by other sperm carrying *Sd*. In contrast, sperm that carry Rsp^i are immune to the product produced by *Sd*. Table 7.10 summarizes the four combinations of alleles that are possible, and their effects.

We mentioned that the mutant *Sd* allele produces a factor (presumably a protein) that kills sperm carrying a sensitive allele at *Rsp*. Although the molecular mechanism of *Sd* action is not known, the chromosomes of affected sperm fail to condense properly and the cells never mature. Interestingly, the *Responder* locus is located in the heterochromatic region near the centromere of chromosome 2. The wild-type allele (Rsp^s) contains a high copy number of a 120-bp repeat. This allele is sensitive to the gene product of the *Sd* locus. When sperm

---

TABLE 7.10   Alleles in the segregation distorter complex and their effects

The top row in this table indicates the two alleles at the *Responder* locus of the segregation distorter complex in fruit flies. The leftmost column gives the alleles at the *Segregation distorter* locus. The entries in the table describe the fate of chromosomes containing each of the four allelic combinations possible.

	Rsp^s (wild-type responder, sensitive)	Rsp^i (mutant responder, insensitive)
Sd (segregation distorter)	*Sd* product acts on the sensitive responder on the same chromosome. Result: a suicidal chromosome.	Responder locus protects this chromosome against the *Sd* product. Result: a killer chromosome.
Sd^+ (wild-type, normal)	Responder locus susceptible. Result: this chromosome is killed when *Sd*-bearing sperm are present.	Responder locus protects this chromosome. Result: no transmission distortion, even when *Sd* sperm are present.

bearing this allele are killed, *Sd*-bearing sperm are found in greater than Mendelian proportions and transmission distortion results. In contrast, the mutant allele that is insensitive and protects sperm from *Sd* appears to be a complete deletion of the repeats. (A wide variety of alleles exist with intermediate copy numbers of the repeat and intermediate sensitivity to *Sd*. Sensitivity appears to correlate strongly with repeat number; see Temin et al. 1991; Palopoli and Wu 1996.) The normal function of the *Responder* locus is not yet known.

When one copy of the second chromosome in male fruit flies contains *Sd* and *Rsp^i* alleles and the second copy of the second chromosome contains *Sd$^+$* and *Rsp^s* alleles, more than 95% of the sperm produced are *SdRsp^i*. Instead of being found in 50% of the offspring of these males, then, the killer chromosomes are found in almost all (Crow 1979; Temin et al. 1991; Palopoli and Wu 1996). These selfish genes transmit themselves at the expense of other alleles at the same locus.

An obvious question is: If meiotic drive is so effective, why is it that *SdRsp^i* chromosomes have not swept to fixation? These types of chromosomes have been found in fly populations from all over the globe, but always in frequencies of just 1–5%. Clearly, some sort of selection is counteracting the advantage caused by transmission bias. What is keeping these selfish genes in check? To answer this question, we have to examine the fitness effects of different allelic combinations. Consider that:

- Males bearing *SdRsp^i* alleles on one of their chromosomes suffer a fitness deficit because they are less fertile than wild-type males. Males are nearly sterile when this chromosome is paired with another *SdRsp^i* chromosome (the reason is unknown). When this chromosome is paired with an *Sd$^+$Rsp^s* chromosome, half of the sperm are destroyed.
- When chromosomes bearing *SdRsp^i* are common, *Rsp^i* alleles paired with wild-type *Sd$^+$* loci should spread rapidly because they have normal fertility and are insensitive to killer chromosomes. This increase is slowed, however, because the wild-type allele *Rsp^s* has a fitness advantage over *Rsp^i*.

Brian Charlesworth and Dan Hartl (1978) created a genetic model to investigate these evolutionary dynamics. Their goal was to calculate the relative frequencies of the different gamete genotypes that are possible (see Table 7.10) through time. These gamete genotypes are *Sd$^+$Rsp^s, SdRsp^i,* and *Sd$^+$Rsp^i*. (They left out *SdRsp^s* gametes because sperm bearing this combination are suicidal and thus eliminated from the population immediately.) Charlesworth and Hartl assigned a relative fitness to each of the nine offspring genotypes that result from matings of the three gamete types. These fitness assignments were based on estimates of the fertility of each combination measured in the lab.

Charlesworth and Hartl then calculated the frequency of each gamete genotype through many generations of random mating. When they started the simulation under realistic conditions—with *Sd$^+$Rsp^s* at high frequency and the mutant alleles at very low frequency—they found that the population went through a series of oscillations that converged on an equilibrium frequency of about 4% *SdRsp^i* chromosomes (Figure 7.10). This conforms almost exactly to what we measure in nature. The model confirms that the fitness advantage conferred by transmission distortion is balanced by the selective disadvantage of reduced fertility and the spread of insensitive alleles on non-killer chromosomes.

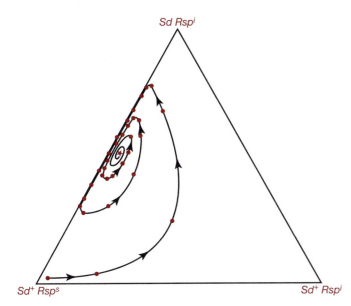

Figure 7.10 Evolutionary dynamics of the *Sd* and *Rsp* loci in *Drosophila* This triangle plot illustrates the evolutionary dynamics of three combinations of *Sd* and *Rsp* alleles. The analysis starts in the lower left corner, where the wild-type alleles are near fixation and the *Sd* and *Rsp*[i] alleles are at low frequency. The simulation ends with an equilibrium established, where *SdRsp*[i] (transmission distorting) chromosomes are at low frequency, and *Sd*[+]*Rsp*[s] and *Sd*[+]*Rsp*[i] chromomes are both at moderate frequency. Note that each corner of the triangle denotes a different allelic combination present at 100% frequency. Moving away from a particular corner indicates that the frequency of that allelic combination decreases. Thus, each point within the triangle represents a different combination of frequencies for *Sd*[+]*Rsp*[i], *Sd*[+]*Rsp*[s], and *SdRsp*[i] chromosomes. The three frequencies always add up to 1. From Charlesworth and Hartl (1978).

Genes that cause meiotic drive are widespread in nature (see Werren et al. 1988). They have been found in species as diverse as mosquitoes (e.g. Wood and Newton 1991), corn (e.g. Dawe and Cande 1996), mice (see Lenington and Heisler 1991; Hammer 1991; Lyon 1991), lily, tobacco, wasps (see Werren and Beukeboom 1993; Werren 1994), trillium, rye, and grasshoppers. In every case the meiotic drive system involves a complex of loci, including a drive gene and a target locus analogous to *Sd* and *Rsp* in flies. The gamete-killing alleles are always found in relatively low frequencies, however, due to fertility losses and the spread of alleles at a suppresser or insensitive locus. Systems of balancing selection, like that illustrated in Figure 7.10, appear to be a general feature of gene complexes that cause transmission distortion.

Selfish genes are kept in check by balancing selection.

Genomic Imprinting

Transmission distortion is not the only example in which the fitness interests of different alleles at the same locus can conflict. Consider the copies of a mammalian gene inherited from the father versus the mother. Why might these alleles be in conflict? In most mammals, females carry offspring from many different males in the course of a lifetime. Indeed, offspring with different fathers are frequently found in the same litter. Because the mother is related to each of these offspring equally, natural selection should act to equalize her investment in each. Natural selection should, on the other hand, favor fathers that can coerce the mother into investing more heavily in their offspring at the expense of offspring from other males. Male mammals have this opportunity because mothers feed their young continuously during development, as opposed to provisioning an egg, seed, or spore prior to fertilization.

Consistent with this prediction, in mammals at least some loci are biochemically marked to distinguish paternal and maternal alleles (Barlow 1995). This phenomenon is called gametic imprinting. Further, a total of 16 genes in humans and mice have now been described in which paternal or maternal alleles are differentially expressed in the embryo. The paternal allele of a hormone called insulin-like growth factor II (IGF-II), for example, is widely expressed in mice, while the maternal copy is hardly transcribed. This hormone is a general stimulant to cell division, and acts through a cell surface protein called the type-1 IGF-II receptor. However, it happens that another abundant cell surface protein in mice, called the cation-independent mannose-6-phosphate receptor (CI-MPR), also has a binding site for IGF-II (this alternative binding site is called the type-2 receptor). CI-MPR's function is completely unrelated to growth, and in mouse embryos it is transcribed only from the maternal allele.

David Haig and colleagues have proposed that this bizarre arrangement of hormones, receptors, and transcription patterns results from a tug-of-war between the interests of maternal and paternal alleles. According to this interpretation, the paternally transcribed IGF-II is selected to maximize rates of cell division. This increases growth rates and monopolizes the flow of maternal resources to the embryo. In contrast, the maternally transcribed type-2 receptor is selected to bind excess paternal hormone, mitigate the effects of IGF-II over-transcription, and equalize the flow of resources to different embryos (Haig and Graham 1991; Moore and Haig 1991).

Consistent with this interpretation, CI-MPR does *not* bind IGF-II in chickens and frogs; their embryos are provisioned before fertilization. This is a hint that the type-2 receptor evolved after the advent of pregnancy, in response to selection that favored equalization of maternal resources, consistent with the mother's fitness interests. Genomic imprinting has also been confirmed in flowering plants, and may have been important in the evolution of the nutritive tissue called endosperm (see Haig and Westoby 1989, 1991).

To conclude this section, the general message of transposable elements, transmission distorters, and genomic imprinting is that the genome is not a single, cooperative unit in which every gene acts to maximize the fitness of the group, or what we call the individual. Rather, the genome is an amalgam of sequences that pursue different strategies to maximize their own fitness.

Summary

Mutations in coding sequences create alleles with a wide variety of phenotypic effects, from highly deleterious to highly advantageous. The neutral theory specifies the rate of fixation for purely neutral alleles, and provides a null model for testing whether positive selection or drift is responsible for observed rates and patterns of molecular evolution.

Variation in the availability of tRNAs can cause selection on silent site substitutions and lead to codon bias. Selection on advantageous mutations can also sweep linked sequences to fixation, even if those sequences are neutral or slightly deleterious.

Gene duplication events can produce functionless pseudogenes, loci that gain completely new functions, families of genes that diverge and gain slightly different functions from the parent sequence, or loci that retain the same function and evolve in concert with the parent sequence. Concerted evolution occurs when sequences are homogenized by frequent unequal cross-over events or gene conversion.

Chloroplasts and mitochondria originated as endosymbiotic bacteria and represent the largest horizontal gene transfer events in evolutionary history. Although the gene order, gene organization, and rate of sequence change in chloroplast and mitochondrial DNA vary among species, the information content of these organelles is remarkably well conserved.

Highly repetitive sequences are often found in the centromere and telomeres, and are thought to be essential to the structural integrity of the eukaryotic chromosome. Some highly repetitive sequences, called mini- and microsatellites, are scattered throughout euchromatic regions. Determining the origins of highly repetitive sequences is still a focus of research.

Transposable genetic elements are often present in moderate copy number in the genomes of eukaryotes and prokaryotes. Transposable elements can move via either an RNA or DNA intermediate, and transposition can be replicative or conservative. They do not code for products that increase the fitness of the host organism, but increase in frequency by transmitting themselves in non-Mendelian proportions. As a result, they qualify as selfish or parasitic DNA. Other examples of selfish DNAs include transmission distorters like the segregation distorter complex in fruit flies. These loci perpetuate themselves at the expense of competing loci by killing gametes or zygotes that do not carry a copy of themselves.

Questions

1. Why is the rate of silent site substitutions reported in Table 7.2 generally higher than the rate of replacement substitutions? If you were attempting to reconstruct evolutionary relationships within mammals by comparing their DNA sequences, which data would be more useful: replacement substitutions in histone proteins, or in hemoglobin proteins? Why?

2. Recall that the fourth chromosome of *Drosophila melanogaster* does not recombine during meiosis. The lack of genetic polymorphism on this chromosome has been interpreted as the product of a selective sweep. If the fourth chromosome had normal rates of recombination, would you expect the level of polymorphism to be different? Why?

3. We have reviewed evidence that horizontal gene transfer, or gene exchange between different evolutionary lineages, can occur. How might this phenomenon affect studies that use divergence in DNA as a yardstick to measure the time since two lineages last shared a common ancestor?

4. Consider the possible costs and benefits of transposable elements to their hosts. Do there appear to be any costs to a eukaryote for carrying retrotransposons? Do there appear to be any benefits? Similarly, do there appear to be any costs or benefits to a prokaryote for carrying plasmids? Finally, a speculative question: Do you think it likely that either group (prokaryotes or eukaryotes) might evolve some sort of genomic defense to block the spread of transposable elements?

5. Review the evolutionary dynamics of the *Sd* allele, shown in Figure 7.10. Would this allele be more or less common if $SdRsp^i$ homozygotes were not sterile? Why?

6. This chapter described several types of genes that can spread in populations, despite decreasing or not affecting the reproductive fitness of their hosts. We have also discussed some genes that spread by "normal" natural selection—that is, by increasing the reproductive fitness of their hosts. It is important to understand the distinctions

between these very different types of evolution. As a final review, specify whether the host's fitness is increased, decreased, or unaffected for each of the genes listed below. If the allele decreases or does not affect the host, explain how it is possible for the allele to spread in the population.

Sd in a male fly
Sd in a female fly
Any allele that has spread by hitchhiking (in either sex)
A retrotransposon
A strictly neutral allele

Exploring the Literature

7. An important extension of the neutral theory of molecular evolution is called the nearly neutral model. This work predicts the evolutionary dynamics of alleles with slightly deleterious effects, which are thought to be extremely common. To learn how drift can overwhelm selection against nearly neutral alleles when population sizes are small, see:

Ohta, T., and M. Kimura. 1971. On the constancy of the evolutionary rate of cistrons. *Journal of Molecular Evolution* 1:18−25.

Ohta, T. 1973. Slightly deleterious mutant substitutions in evolution. *Nature* 246:96−98.

8. Embryo or spore killers create transmission distortion by killing offspring that do not contain a copy of themselves. To compare these loci to *Segregation distorter* and other selfish genes that kill gametes, see:

Beeman, R.W., K.S. Friesen, and R.E. Denell. 1992. Maternal-effect selfish genes in flour beetles. *Science* 256:89−92.

Hurst, L.D. 1993. *scat*[+] is a selfish gene analogous to *Medea* of *Tribolium castaneum*. *Cell* 75:407−408.

Nauta, M.J., and R.F. Hoekstra. 1993. Evolutionary dynamics of spore killers. *Genetics* 135:923−930.

Citations

Ajioka, J.W., and D.L. Hartl. 1989. Population dynamics of transposable elements. In D.E. Berg and M.M. Howe, eds. *Mobile DNA* (Washington, DC: American Society of Microbiology), 939−958.

Akashi, H. 1994. Synonymous codon usage in *Drosophila melanogaster:* natural selection and translational accuracy. *Genetics* 144:927−935.

Akashi, H. 1996. Molecular evolution between *Drosophila melanogaster* and *D. simulans:* Reduced codon bias, faster rates of amino acid substitution, and larger proteins in *D. melanogaster. Genetics* 144:1297−1307.

Amos, B., and J. Pemberton. 1993. DNA fingerprinting in non-human populations. *Current Opinions in Genetics and Development* 2:857−860.

Avise, J.C. 1994. *Molecular Markers, Natural History, and Evolution* (New York: Chapman & Hall).

Baldauf, S.L., and J.D. Palmer. 1990. Evolutionary transfer of the chloroplast *tufA* gene to the nucleus. *Nature* 344:262−265.

Baldauf, S.L., J.R. Manhart, and J.D. Palmer. 1990. Different fates of the chloroplast *tufA* gene following its transfer to the nucleus in green algae. *Proceedings of the National Academy of Sciences, USA* 87:5317−5321.

Barlow, D.P. 1995. Gametic imprinting in mammals. *Science* 270:1610−1613.

Beeman, R.W., K.S. Friesen, and R.E. Denell. 1992. Maternal-effect selfish genes in flour beetles. *Science* 256:89−92.

Begun, D.J., and C.F. Aquadro. 1992. Levels of naturally occurring DNA polymorphism correlate with recombination rates in *D. melanogaster. Nature* 356:519−520.

Berry, A. 1996. Non-non-Darwinian evolution. *Evolution* 50:462−466.

Berry, A., J.W. Ajioka, and M. Kreitman. 1991. Lack of

polymorphism on the *Drosophila* fourth chromosome resulting from selection. *Genetics* 129:1111–1117.

Blackburn, E.H. 1990. Telomeres and their synthesis. *Science* 249:489–490.

Boeke, J.D. 1989. Transposable elements in *Saccharomyces cerevisiae*. In D.E. Berg and M.M. Howe, eds. *Mobile DNA* (Washington, DC: American Society of Microbiology), 335–374.

Brown, W.M. 1983. Evolution of animal mitochondrial DNA. In M. Nei and R.K. Koehn, eds. *Evolution of Genes and Proteins* (Sunderland, MA: Sinauer), 62–88.

Brown, W.M., E.M. Prager, A. Wang, and A.C. Wilson. 1982. Mitochondrial DNA sequences of primates: Tempo and mode of evolution. *Journal of Molecular Evolution* 18:225–229.

Charlesworth, B., and D.L. Hartl. 1978. Population dynamics of the segregation distorter polymorphism of *Drosophila melanogaster*. *Genetics* 89:171–192.

Charlesworth, B., and C.H. Langley. 1989. The population genetics of *Drosophila* transposable elements. *Annual Review of Genetics* 23:251–287.

Charlesworth, B., M.T. Morgan, and D. Charlesworth. 1993. The effects of deleterious mutations on neutral molecular variation. *Genetics* 134:1289–1303.

Chase, M.W., et al., 1993. Phylogenetics of seed plants: An analysis of nucleotide sequences from the plastid gene *rbc*L. *Annals of the Missouri Botanical Garden* 80:528–580.

Clark-Walker, G.D. 1992. Evolution of mitochondrial genomes in fungi. In D.R. Wolstenholme and K.W. Jeon, eds. *Mitochondrial Genomes* (San Diego, CA: Academic Press), 89–127.

Clegg, M.T., B.S. Gaut, G.H. Learn, Jr., and B.R. Morton. 1994. Rates and patterns of chloroplast DNA evolution. *Proceedings of the National Academy of Sciences, USA* 91:6795–6801.

Condit, R. 1990. The evolution of transposable elements: Conditions for establishment in bacterial populations. *Evolution* 44:347–359.

Covello, P.S., and M.W. Gray. 1992. Silent mitochondrial and active nuclear genes for subunit 2 of cytochrome *c* oxidase *(cox2)* in soybean: Evidence for RNA-mediated gene transfer. *EMBO Journal* 11:3815–3820.

Crow, J.F. 1979. Genes that violate Mendel's rules. *Scientific American* 240:134–146.

Cummings, M.P. 1994. Transmission patterns of eukaryotic transposable elements: Arguments for and against horizontal transfer. *Trends in Ecology and Evolution* 9:141–145.

Dawe, R.K., and W.Z. Cande. 1996. Induction of centromeric activity in maize by *suppressor of meiotic drive 1*. *Proceedings of the National Academy of Sciences, USA* 93:8512–8517.

Deininger, P.L. 1989. SINEs: Short interspersed repeated DNA elements in high eucaryotes. In D.E. Berg and M.M. Howe, eds. *Mobile DNA* (Washington, DC: American Society of Microbiology), 619–636.

Deininger, P.L., and G.R. Daniels. 1986. The recent evolution of mammalian repetitive DNA elements. *Trends in Genetics* (March) 76–80.

Dorit, R.L., H. Akashi, and W. Gilbert. 1995. Absence of polymorphism at the *ZFY* locus on the human Y chromosome. *Science* 268:1183–1185.

Elder, J.F., Jr., and B.J. Turner. 1995. Concerted evolution of repetitive DNA sequences in eukaryotes. *Quarterly Review of Biology* 70:297–320.

Gillham, N.W., J.E. Boynton, and E.H. Harris. 1985. Evolution of plastid DNA. In T. Cavalier-Smith *The Evolution of Genome Size* (New York: John Wiley & Sons), 299–351.

Goffeau, A., B.G. Barrell, H. Bussey, R.W. Davis, B. Dujon, H. Feldmann, F. Galibert, J.D. Hoheisel, C. Jacq, M. Johnston, E.J. Louis, H.W. Mewes, Y. Murakami, P. Philippsen, H. Tettelin, and S.G. Oliver. 1996. Life with 6000 genes. *Science* 274:546–567.

Gray, M.W. 1992. The endosymbiont hypothesis revisited. In D.R. Wolstenholme and K.W. Jeon, eds. *Mitochondrial Genomes* (San Diego, CA: Academic Press), 233–357.

Greider, C.W. 1993. Telomeres and telomerase in small eukaryotes. In P. Broda, S.G. Oliver, and P.F.G. Sims, eds. *The Eukaryotic Genome* (Cambridge University Press), 31–42.

Griffiths, A.J.F., J.H. Miller, D.T. Suzuki, R.C. Lewnstin, and W.M. Gelbart. 1993. *An Introduction to Genetic Analysis.* (New York: W.H. Freeman).

Grivell, L.A., J.H. de Winde, and W. Mulder. 1993. Global regulation of mitochondrial biogenesis in yeast. In P. Broda, S.G. Oliver, and P.F.G. Sims, eds. *The Eukaryotic Genome* (Cambridge University Press), 321–332.

Haig, D., and C. Graham. 1991. Genomic imprinting and the strange case of the insulin-like growth factor II receptor. *Cell* 64:1045–1046.

Haig, D. and M. Westoby. 1989. Parent-specific gene expression and the triploid endosperm. *American Naturalist* 134:147–155.

Haig, D., and M. Westoby. 1991. Genomic imprinting in endosperm: Its effect on seed development in crosses between species, and between different ploidies of the same species, and its implications for the evolution of apomixis. *Philosophical Transactions of the Royal Society of London,* Series B 333:1–13.

Hamblin, M.T., and C.F. Aquadro. 1996. High nucleotide sequence variation in a region of low recombination in *Drosophila simulans* is consistent with the background selection model. *Molecular Biology and Evolution* 13:1133–1140.

Hammer, M.F. 1991. Molecular and chromosomal studies on the origin of *t* haplotypes in mice. *American Naturalist* 137:359–365.

Hanfstingl, U., A. Berry, E.A. Kellogg, J.T. Costa III, W. Rüdiger, and F.M. Ausubel. 1994. Haplotypic divergence coupled with lack of diversity at the *Arabidopsis thaliana* alcohol dehydrogenase locus: Roles for both balancing and directional selection? *Genetics* 138:811–828.

Hanson, M.R., and O. Folkerts. 1992. Structure and function of the higher plant mitochondrial genome. In D.R. Wolstenholme and K.W. Jeon, eds. *Mitochondrial Genomes* (San Diego, CA: Academic Press), 129–172.

Hickey, D.A. 1992. Evolutionary dynamics of tranposable elements in prokaryotes and eukaryotes. *Genetica* 86:269–274.

Hillis, D.J., C. Moritz, D.A. Porter, and R.J. Baker. 1991. Evidence for biased gene conversion in concerted evolution of ribosomal DNA. *Science* 251:308–310.

Hilton, H., R.M. Kliman, and J. Hey. 1994. Using hitchhiking genes to study adaptation and divergence during speciation within the *Drosophila melanogaster* species complex. *Evolution* 48:1900–1913.

Hood, L., M. Kronenberg, and T. Hunkapiller. 1985. T cell antigen receptors and the immunoglobulin supergene family. *Cell* 40:225–229.

Hudson, R.R. 1994. How can the low levels of DNA sequence variation in regions of the *Drosophila* genome with low recombination rates be explained? *Proceedings of the National Academy of Sciences, USA* 91:6815–6818.

Hughes, A.L., and M. Nei. 1988. Pattern of nucleotide substitution at major histocompatibility complex class I loci reveals overdominant selection. *Nature* 335:167–170.

Hughes, A.L., and M. Nei. 1989. Nucleotide substitution at major histocompatibility complex class II loci: Evidence for overdominant selection. *Proceedings of the National Academy of Sciences, USA* 86:958–962.

Hutchison, C.A. III, S.C. Hardies, D.D. Loeb, W.R. Shehee, and M.H. Edgell. 1989. LINEs and related retroposons: Long interspersed repeated sequences in the eucaryotic genome. In D.E. Berg and M.M. Howe, eds. *Mobile DNA* (Washington, DC: American Society of Microbiology), 593–617.

Jarne, P., and P.J.L. Lagoda. 1996. Microsatellites, from molecules to populations and back. *Trends in Ecology and Evolution* 11:424–429.

Jeffreys, A.J., V. Wilson, and S.L. Thein. 1985. Hypervariable "minisatellite" regions in human DNA. *Nature* 314:67–73.

Kimura, M. 1968. Evolutionary rate at the molecular level. *Nature* 217:624–626.

Kimura, M. 1977. Preponderance of synonymous changes as evidence for the neutral theory of molecular evolution. *Nature* 267:275–276.

Kimura, M. 1983. The neutral theory of molecular evolution. In M. Nei and R.K. Koehn, eds. *Evolution of Genes and Proteins* (Sunderland, MA: Sinauer), 208–233.

Kimura, M. 1986. DNA and the neutral theory. *Philosophical Transactions of the Royal Society of London,* Series B 312:343–354.

Kimura, M., and T. Ohta. 1974. On some principles governing molecular evolution. *Proceedings of the National Academy of Sciences, USA* 71:2848–2852.

King, J.L., and T.H. Jukes. 1969. Non-Darwinian evolution. *Science* 164:788–798.

Kliman, R.M., and J. Hey. 1993. DNA sequence variation at the period locus within and among species of the *Drosophila melanogaster* complex. *Genetics* 133:375–387.

Kumagai, I., S. Takeda, and K.-I. Miura. 1992. Functional conversion of the homologous proteins α-lactalbumin and lysozyme by exon exchange. *Proceedings of the National Academy of Sciences, USA* 89:5887–5891.

Langley, C., J. MacDonald, N. Miyashita, and M. Aguadé. 1993. Lack of correlation between interspecific divergence and intraspecific polymorphism at the suppressor of forked region in *Drosophila melanogaster* and *Drosophila simulans. Proceedings of the National Academy of Sciences, USA* 90:1800–1803.

Lenington, S., and I.L. Heisler. 1991. Behavioral reduction in the transmission of deleterious *t* haplotypes by wild house mice. *American Naturalist* 137:366–378.

Lewontin, R., and D. Hartl. 1991. Population genetics in forensic DNA typing. *Science* 254:1745–1750.

Li, W.-H., T. Gojobori, and M. Nei. 1981. Pseudogenes as a paradigm of neutral evolution. *Nature* 292:237–239.

Li, W.-H., and D. Graur. 1991. *Fundamentals of Molecular Evolution* (Sunderland, MA: Sinauer).

Lyon, M.F. 1991. The genetic basis of transmission-ratio distortion and male sterility due to the *t* complex. *American Naturalist* 137:349–358.

Lyttle, T.W. 1991. Segregation distorters. *Annual Review of Genetics* 25:511–557.

Martín-Campos, J.M., J.M. Comerón, N. Miyashita, and M. Aguadé. 1992. Intraspecific and interspecific variation at the *y-ac-sc* region of *Drosophila simulans* and *Drosophila melanogaster. Genetics* 130:805–816.

McDonald, J.H., and M. Kreitman. 1991. Adaptive protein evolution at the *Adh* locus in *Drosophila. Nature* 351:652–654.

Moore, T., and D. Haig. 1991. Genomic imprinting in mammalian development: A parental tug-of-war. *Trends in Genetics* 7:45–49.

Moritz, C., T.E. Dowling, and W.M. Brown. 1987. Evolution of animal mitochondrial DNA: Relevance for population biology and systematics. *Annual Review of Ecology and Systematics* 18:269–292.

Nugent, J.M., and J.D. Palmer. 1991. RNA-mediated transfer of the gene *coxII* from the mitochondrion to the nucleus during flowering plant evolution. *Cell* 66:473–481.

Ohta, T., and M. Kreitman. 1996. The neutralist-selectionist debate. *BioEssays* 18:673–683.

Palmer, J.D. 1987. Chloroplast DNA evolution and biosystematic uses of chloroplast DNA variation. *American Naturalist* 130:S6–S29.

Palopoli, M.F., and C.-I. Wu. 1996. Rapid evolution of a coadapted gene complex: Evidence from the *Segregation Distorter (SD)* system of meiotic drive in *Drosophila melanogaster*. *Genetics* 143:1675–1688.

Petes, T.D. 1980. Unequal meiotic recombination within tandem arrays of yeast ribosomal DNA genes. *Cell* 19:765–774.

Pluta, A.F., A.M. Mackay, A.M. Ainsztein, I.G. Goldberg, and W.C. Earnshaw. 1995. The centromere: Hub of chromosomal activities. *Science* 270:1591–1595.

Prager, E.M., and A.C. Wilson. 1988. Ancient origin of lactalbumin from lysozyme: Analysis of DNA and amino acid sequences. *Journal of Molecular Evolution* 27:326–335.

Sang, T., D.J. Crawford, and T.F. Stuessy. 1995. Documentation of reticulate evolution in peonies (Paeonia) using internal transcribed spacer sequences of nuclear ribosomal DNA: Implications for biogeography and concerted evolution. *Proceedings of the National Academy of Sciences, USA* 92:6813–6817.

SanMiguel, P., A. Tikhonov, Y.-K. Jin, N. Motchoulskaia, D. Zakharov, A. Melake-Berhan, P.S. Springer, K.J. Edwards, M. Lee, Z. Avramova, and J.L. Bennetzen. 1996. Nested retrotransposons in the intergenic regions of the maize genome. *Science* 274:765–768.

Sharp, P.M., and W.-H. Li. 1986. An evolutionary perspective on synonymous codon usage in unicellular organisms. *Journal of Molecular Evolution* 24:28–38.

Sharp, P.M., E. Cowe, D.G. Higgins, D. Shields, K.H Wolfe, and F. Wright. 1988. Codon usage patterns in *Escherichia coli, Bacillus subtilis, Saccharomyces cerevisiae, Schizosaccharomyces pombe, Drosophila melanogaster,* and *Homo sapiens:* A review of the considerable within-species diversity. *Nucleic Acids Research* 16:8207–8211.

Shaw, D.D. 1994. Centromeres: Moving chromosomes through space and time. *Trends in Ecology and Evolution* 9:170–175.

Shen, M.R., J. Brosius, and P.L. Deininger. 1997. BC1 RNA, the transcript from a master gene for ID element amplification, is able to prime its own reverse transcription. *Nucleic Acids Research* 25:1641–1648.

Smith, G.P. 1976. Evolution of repeated DNA sequences by unequal crossover. *Science* 191:528–535.

Smith, M.W., and R.F. Doolittle. 1992. Anomalous phylogeny involving the enzyme glucose-6-phosphate isomerase. *Journal of Molecular Evolution* 34:544–545.

Smith, M.W., D.-F. Feng, and R.F. Doolittle. 1992. Evolution by acquisition: The case for horizontal gene transfers. *Trends in Biochemical Sciences* 17:489–493.

Sutherland, G.R., and R.I. Richards. 1995. Simple tandem DNA repeats and human genetic disease. *Proceedings of the National Academy of Sciences, USA* 92:3636–3641.

Takasaki, N., S. Murata, M. Saitoh, T. Kobayashi, L. Park, and N. Okada. 1994. Species-specific amplication of tRNA-derived short interspersed repetitive elements (SINEs) by retroposition: A process of parasitization of entire genomes during the evolution of salmonids. *Proceedings of the National Academy of Sciences, USA* 91:10153–10157.

Tanaka, T., and M. Nei. 1989. Positive Darwinian selection observed at the variable-region genes of immunoglobulins. *Molecular Biology and Evolution* 6:447–459.

Temin, H.M. 1980. Origin of retroviruses from cellular moveable genetic elements. *Cell* 21:599–600.

Temin, R.G., B. Ganetzky, P.A. Powers, T.W. Lyttle, S. Pimpinelli, P. Dimitri, C.-I. Wu, and Y. Hiraizumi. 1991. Segregation distortion in *Drosophila melanogaster:* Genetic and molecular analyses. *American Naturalist* 137:287–331.

Voloshin, O.N., L. Wang, and R.D. Camerini-Otero. 1996. Homologous DNA pairing promoted by a 20-amino acid peptide derived from RecA. *Science* 272:868–872.

Voytas, D.F. 1996. Retroelements in genome organization. *Science* 274:737–738.

Weiler, K.S., and B.T. Wakimoto. 1995. Heterochromatin and gene expression in *Drosophila*. *Annual Review of Genetics* 29:577–605.

Werren, J. 1994. Genetic invasion of the body snatchers. *Natural History* 6:36–38.

Werren, J.H., U. Nur, and C.-I. Wu. 1988. Selfish genetic elements. *Trends in Ecology and Evolution* 3:297–302.

Werren, J.H., and L.W. Beukeboom. 1993. Population genetics of a parasitic chromosome: Theoretical analysis of PSR in subdivided populations. *American Naturalist* 142:224–241.

Werren, J.H., D. Windsor, and L. Guo. 1995. Distribution of *Wolbachia* among neotropical arthropods. *Proceedings of the Royal Society of London,* Series B 262:197–204.

Wichman, H.A., R.A. Van Den Bussche, M.J. Hamilton, and R.J. Baker. 1992. Transposable elements and the evolution of genome organization in mammals. *Genetica* 86:287–293.

Wilson, A.C., R.L. Cann, S.M. Carr, M. George, U.B. Gyllensten, K.M. Helm-Bychowski, R.G. Higuchi, S.R. Palumbi, E.M. Prager, R.D. Sage, and M. Stoneking. 1985. Mitochondrial DNA and two perspectives in evolutionary genetics. *Biological Journal of the Linnean Society* 26:375–400.

Wolfe, K.H., W.-H. Li, and P.M. Sharp. 1987. Rates of nucleotide substitution vary greatly among plant mitochondrial, chloroplast, and nuclear DNAs. *Proceedings of the National Academy of Sciences, USA* 84:9054–9058.

Wolfe, K.H., P.M. Sharp, and W.-H. Li. 1989. Rates of synonymous substitution in plant nuclear genes. *Journal of Molecular Evolution* 29:208–211.

Wolfe, K.H., C.W. Morden, and J.D. Palmer. 1991. Ins and outs of plastid genome evolution. *Current Opinions in Genetics and Development* 1:523–529.

Wood, R.J., and M.E. Newton. 1991. Sex-ratio distortion caused by meiotic drive in mosquitoes. *American Naturalist* 137:379–391.

Zakian, V. 1995. Telomeres: Beginning to understand the end. *Science* 270:1601–1607.

Agents of Selection and the Nature of Adaptations

An individual that has reproduced has won the game of evolution; it has safely escorted its genes into the next generation. To succeed, the individual had to overcome an enormous diversity of obstacles. To survive to maturity, it had to protect itself from the elements, find or capture food, evade predators, fend off parasites, repair its wounds, and both compete and cooperate with members of its own and other species. To reproduce, the individual had to gather a surplus of energy and materials with which to construct and provision its offspring; if it reproduced sexually, it also had to find a member of the opposite sex and secure his or her assistance. A difference among individuals in success at meeting any of these challenges can mean that some individuals leave more offspring than others. In other words, all these obstacles are potential agents of selection.

A trait, or integrated suite of traits, that allows its possessor to succeed in the face of one or more agents of selection is an adaptation. The purpose of this chapter is to explore the nature of adaptations to a variety of agents of selection. In Chapter 7 our focus was on DNA sequences and proteins; here it is on whole organisms. Adaptations can involve aspects of an organism's behavior, physiology, or morphology. Furthermore, adaptations can involve phenotypic plasticity—the ability of an individual to alter its behavior, physiology, or morphology depending on the environment.

We begin this chapter by considering the adaptationist program, the research effort directed toward identifying and understanding the adaptive significance of the traits of organisms. We then explore the nature of various adaptations to agents of selection in both the physical and biological environments. Next we consider phenotypic plasticity, how it can be adaptive, and how it can complicate research on adaptations. Finally, we discuss some of the evolutionary constraints that prevent populations from becoming perfectly adapted to all agents of selection.

Our goal throughout the chapter is to illustrate the thinking of evolutionary biologists doing careful research. We review projects that employ all of the techniques that we introduced in Chapter 3: experiments, the comparative method, and the reconstruction of evolutonary history.

8.1 The Adaptationist Program

Adaptations require explanation. One of the triumphs of Darwin's *On the Origin of Species* was that it provided an explanation for the existence of adaptations that was testable and that has survived testing. This explanation is, of course, evolution by natural selection. As we discussed in Chapter 5, the development of population genetics identified additional mechanisms of evolution. But none of these other mechanisms—mutation, migration, and genetic drift—can produce adaptations.

One of the major activities of evolutionary biology since Darwin's time has been to demonstrate that the traits of organisms are indeed adaptations (Mayr 1983). Roughly speaking, in order to demonstrate that a trait is an adaptation, we need first to determine what a trait is for and then show that individuals possessing the trait have higher fitness than individuals lacking it. The adaptive signficance of some traits is obvious: Eyes are manifestly devices for detecting objects at a distance by gathering and analyzing light; in many (but not all) animal species, individuals with good eyesight will be better able to find food and avoid predators than individuals with poor eyesight. Understanding the adaptive significance of other traits requires effort. This research effort is called the adaptationist program.

A plausible hypothesis about the adaptive value of a trait is the beginning of a careful study, not the end.

Pursuit of the adaptationist program requires caution (Gould and Lewontin 1979; Mayr 1983). Is is not enough simply to devise a plausible hypothesis about what a trait is for and why it is adaptive. We must challenge the hypothesis by using it to make predictions, and subject the predictions to test by collecting data.

Here is an example: Polar bears are white. Because this trait is unique among bears, we can conclude that it is derived. That is, polar bears evolved from a brown ancestor. Polar bears live in the arctic, spending much of their time on ice and snow, another trait unique among bears. Is the white pelt of the polar bear an adaptation for life in the arctic, and if so, how? The answer seems obvious: Polar bears eat seals (Stirling and Oritsland 1995), and a polar bear should be able to catch more seals if it is camouflaged on snow and ice. The polar bear's white pelt is an adaptation for hunting seals in a white environment.

Hypotheses must be tested.

The camouflage hypothesis is so plausible that it has been used by a champion of the adaptationist program as an example of the explanatory power of evolution by natural selection (Dawkins 1987, pages 38–39). The trouble is that the camouflage hypothesis is probably wrong. This we discover when we use the hypothesis to make predictions and test them. One simple prediction of the camouflage hypothesis is that polar bears ought to hunt seals in ways that take advantage of their camouflage. Sometimes they do. Ian Stirling (1974) described a bear that successfully hunted sunbathing seals by slowly swimming up to the seals with its hindquarters out of the water. Most often, however, bears hunt seals in a way that seems to depend rather little on camouflage. In 233 of the 288 hunts that Stirling watched, the bear simply waited motionlessly beside a breathing hole and waited for a seal to surface (see also Anderson and Guravich 1992). In another common hunting strategy, camouflage is completely irrelevant. The bear uses its nose to detect a seal in an under-snow lair, then rushes from 50 to 100 meters downwind to jump on the lair and kill the occupant by crushing it (Smith 1980; Furgal et al. 1996). The definitive test of the camouflage hypothesis would be to dye some polar bears brown, and then compare their hunting success to that of control bears. As far as we know, this experiment remains to be done. However, we suspect that loss of camouflage would not reduce the bears' hunting success as much as the camouflage hypothesis predicts.

Other observations suggest an alternative adaptive hypothesis. When photographed under ultraviolet (UV) light, polar bears are black (Legvine and Oritsland 1974). That is, polar bear fur absorbs UV light. Examined closely, the individual hairs on a polar bear pelt are not white at all; they are clear. Polar bear hairs simply lack the pigmented cells that occupy the core of the hair in most mammals. R. E. Grojean and colleagues (1980) examined the optical characteristics of polar bear fur. They liken the structure of a polar bear hair to a drawn quartz tube with an irregular inner surface. The hair traps incident light and reflects it along this inner surface to the bear's black skin. Grojean and colleagues calculate that a polar bear pelt transmits about 16% of the incident light to the skin. The rest of the light is scattered and reflected, which makes the bear look white. Grojean and colleagues suggest that the white pelt of the polar bear evolved as a solar heat collector.

The solar-heat-collector hypothesis has not, as far as we know, been definitively tested. However, we have made our point: We cannot uncritically accept a hypothesis about the adaptive significance of a trait simply because it is plausible. We must subject all hypotheses to rigorous tests.

Here are some other caveats to keep in mind when studying adaptations:

- Differences among populations or species are not always adaptive. It is possible that the white pelt of the polar bear is not an adaptation at all. A mutation causing white fur could have become fixed in the polar bear population by genetic drift, perhaps via the founder effect (see Chapter 5). At the molecular level, much of the variation among individuals, populations, and species may be selectively neutral (see Chapter 7).

Alternative explanations must be considered.

- Not every trait of an organism, or every use of a trait by an organism, is an adaptation. As we noted earlier, polar bears sometimes hunt seals in a way that appears to rely on the camouflage afforded by their white fur. This does not necessarily mean that the white pelt of the polar bear evolved because it improves the bear's hunting success. The white pelt may have evolved because of its properties as a solar heat collector, and is only incidentally exploited as camouflage by some of the bears when they hunt.

- Not every adaptation is perfect. According to the calculations by Grojean et al., the polar bear's pelt absorbs only about 16% of the incident light. Perhaps this is the highest efficiency possible in a solar heat collector built of fur and skin.

In the next two sections, we review various examples in which researchers have formulated and tested hypotheses about the adaptive significance of traits.

The first of these sections concerns traits that evolved in response to physical agents of selection; the second concerns traits that evolved in response to biological agents of selection.

8.2 Adaptations to the Physical Environment

The physical environment includes temperature, moisture, light, chemistry, and the various mechanical forces that buffet organisms. We will focus first on temperature. Temperature is a feature of all environments, and because life is possible within only a narrow range of temperatures, it is important to all organisms.

Temperature

The vast majority of organisms are ectothermic, which means that their body temperatures are determined by the temperatures of their environments. As Figure 8.1 demonstrates for desert iguanas *(Dipsosaurus dorsalis),* body temperature has a profound effect on the physiological performance of ectotherms. Desert iguanas can survive short exposures to body temperatures as low as 0°C and as high as about 47°C, but the lizards can function only between about 15°C and 45°C. Within this narrower range, cold lizards run and digest slowly, tire quickly, and hear poorly. As they get warmer, the lizards run and digest more quickly, tire more slowly, and hear more keenly. The lizards' physiological capacities reach a plateau in the mid-to-high 30s. Above about 45°C the lizards are too hot, and collapse.

The relationship between physiological performance and temperature is called a thermal performance curve. The shape of the desert iguana's thermal performance curves are typical of those of a wide variety of physiological processes in a diversity of organisms. (Box 8.1 explores why temperature affects physiological performance so strongly.) Given the sensitivity of physiological function to temperature, we can predict that ectotherms will exhibit at least two kinds of adaptations to temperature: behavioral thermoregulation, and a match between the thermal dependence of physiological processes and the temperature of the environment.

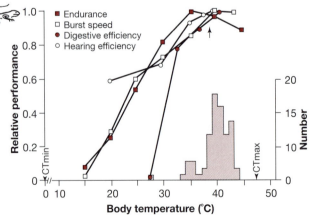

Figure 8.1 **Physiological abilities of the desert iguana *(Dipsosaurus dorsalis)* as a function of body temperature** The solid lines show locomotor endurance, burst (sprint) speed, digestive efficiency, and hearing efficiency of lizards as a function of body temperature. The shaded region is a histogram showing the distribution of body temperatures for active lizards captured in nature. The arrow indicates the body temperature chosen by lizards in the lab. CTmax is the critical thermal maximum; that is, the upper lethal temperature. CTmin is the critical thermal minimum. From Huey and Kingsolver (1989).

BOX 8.1 Why temperature affects physiology

If we look closely enough, life is a series of carefully orchestrated chemical reactions. Chemical reactions are inherently sensitive to temperature (Mahan 1975). First, consider a chemical reaction occurring in a mixture of two gases. Readers familiar with chemistry will recall two crucial facts:

1. Temperature is the kinetic energy of atoms and molecules. The higher the temperature, the faster the atoms and molecules move.
2. In order for two molecules to react with each other to form a new molecule or molecules, they must collide in the proper orientation, and with sufficient energy. The threshold energy for a chemical reaction is called the activation energy.

At any given temperature on the absolute (Kelvin) scale, the distribution of kinetic energies of a group of molecules is given by the Maxwell-Boltzmann Distribution (Figure 8.2). The figure shows two distributions; one for molecules at a low temperature, the other for molecules at a high temperature. The dashed vertical line shows the activation energy for a hypothetical chemical reaction. The area under each curve to the right of this line represents the fraction of molecules with sufficient energy to react. This area is thus proportional to the rate of reaction. In the example shown, a 45°C change in temperature roughly doubles the reaction rate. Note that as the activation energy gets higher (that is, the dashed line moves to the right), the effect on the reaction rate of a given change in temperature increases. To see this, note that if the activation energy is very low (that is, the dashed line is near the origin), then virtually all of the reactants will have enough energy to react, regardless of whether their temperature is 0°C or 45°C.

When chemical reactions happen in solution, additional factors come into play. In a gas, colliding molecules either react or immediately bounce apart. In a solution, however, reactant molecules can be held in contact with each other by the surrounding molecules of solvent. While in contact, the reactant molecules may acquire the energy and orientation they need to react. Thus the factors that influence rates of reaction in solution include not only orientation and kinetic energy, but also the duration of encounter.

Biological chemical reactions occur in solution, and with the aid of enzymes. Enzymes facilitate chemical reactions by holding the reactants at the right orientation, which lowers the activation energy, and by increasing the duration of the encounter. Because they lower the activation energy, enzymes should tend to lower the temperature sensitivity of reactions (Figure 8.2).

Enzymes, however, are themselves highly sensitive to temperature (Hochachka and Somero 1984). The ability of an enzyme to function depends on both the enzyme's three-dimensional shape and its flexibility. The initial binding of the enzyme to the reactants involves an active site on the enzyme that complements the shape of the reactants. This match

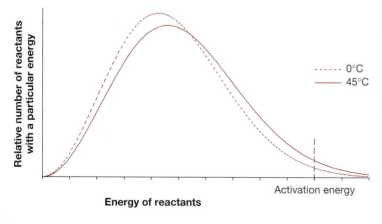

Figure 8.2 **The Maxwell-Boltzmann Distribution for a group of molecules at two different temperatures** The horizontal axis represents the kinetic energy of the reactants for a chemical reaction; the vertical axis represents the relative number of reactants at any particular kinetic energy. A few reactants have little kinetic energy, most reactants have an intermediate amount of kinetic energy, and a few reactants have high kinetic energy. When the reactants are warmer, the distribution of their kinetic energies shifts to the right.

BOX 8.1 *(continued)*

between the shape of the reactants and the enzyme may be modified during binding itself, in a manner analogous to the way a knit glove modifies its shape to conform to the hand inserted into it. After binding the reactants, enzymes often undergo further conformational change to bring the reactants into proper orientation. The three-dimensional shape and flexibility of an enzyme are determined by the weak bonds, such as hydrogen bonds and van der Waals interactions, that form between the amino acids that constitute the enzyme's primary structure. By their nature, these weak bonds are highly sensitive to temperature. If an enzyme gets too cold, it becomes too rigid to function, and if it gets too hot, it becomes too flexible.

Behavioral Thermoregulation

As the temperature of their environment changes, desert iguanas do not just passively accept the consequences. Instead, they regulate their body temperature by moving into the sun to warm up, and into the shade to cool off. Not surprisingly, desert iguanas prefer to maintain themselves at body temperatures in the high 30s (Figure 8.1, arrow). This is the center of the range of temperatures at which the iguanas perform best. After all, a lizard never knows when it will need to chase some food, or run away from a predator. In nature, of course, lizards may not always have a sufficient range of environmental temperatures to move among to maintain themselves at exactly their preferred temperature. As the shaded histogram of field body temperatures in Figure 8.1 shows, however, desert iguanas do reasonably well.

Note that although we have asserted that desert iguanas thermoregulate, the mere fact that iguanas captured in nature are usually at or near their optimal body temperature does not, by itself, prove that the lizards are active in maintaining those temperatures. It could simply be that the environments where the lizards live are always in the mid-to-high 30s. To prove behavioral thermoregulation we must show (1) that the animal in question is choosing particular temperatures more often than it would encounter those temperatures if it simply moved at random through its environment, and (2) that its choice of temperatures is adaptive.

Ray Huey and colleagues (1989) made a detailed study of the thermoregulatory behavior of the garter snake *(Thamnophis elegans)* at Eagle Lake, California. Garter snakes are affected by temperature in the same way as desert iguanas, although for garter snakes the optimal temperature, preferred temperature, and maximum survivable temperature are all a few degrees lower. Huey et al. surgically implanted several snakes with miniature radio transmitters. Each implanted transmitter emits a beeping signal that allows a biologist with a handheld receiver and directional antenna to find the implanted snake, even when the snake is hiding under a rock or in a burrow. In addition, the transmitter reports the snake's temperature by changing the rate at which it beeps.

Garter snakes in the lab prefer to stay at temperatures between 28°C and 32°C. Huey et al. found that snakes in nature do a remarkable job of thermoregulating in the same range. Figure 8.3 shows the body temperatures of two of the implanted snakes, each over the course of a 24-hour day. Both snakes

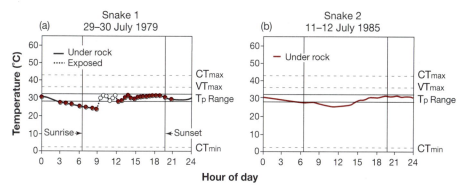

Figure 8.3 Body temperatures of garter snakes in nature (a) Snake 1 spent part of the day under a rock (filled dots and solid line), and part of the day in the sun (open dots and dashed line). (b) Snake 2 spent the entire day under a rock. VTmax is the voluntary thermal maximum, the highest temperature snakes voluntarily tolerate. Tp is the preferred temperature range, measured in the lab. CTmax and CTmin are defined in Figure 8.1. Both snakes kept their temperature near 30°C for the entire day. From Huey et al. (1989).

kept their temperature within or near the preferred range. How do the garter snakes manage to thermoregulate so well? The two snakes shown in Figure 8.3 spent the day under or near rocks. Other options include moving up and down a burrow, and staying on the surface of the ground while moving back and forth from sunshine to shade.

Huey et al. compared the relative merits of each of these thermoregulatory strategies by monitoring the environmental temperature under rocks of various sizes, and at various depths in a burrow, and by monitoring the temperature of a model snake left on the surface in the sun or shade (Figure 8.4). For a snake under a rock, the thickness of the rock proves critical. A snake under a thin rock (Figure 8.4a) would not only get dangerously cold at night, but would overheat in the daytime. (As Huey says, "The snake would die by 11 a.m., and remain dead until at least 6 p.m.") A snake under a thick rock (Figure 8.4a) would remain safe all day, but would never reach its preferred temperature. Rocks of medium thickness are just right (Figure 8.4b). By moving around under the rock, a snake under a rock of medium thickness can stay close to or within its preferred temperature range for the entire day. A snake moving up and down a burrow could do reasonably well (Figure 8.4c), but would get colder at night than a snake under a medium-sized rock. Finally, a snake on the surface could thermoregulate effectively in the daytime by moving between sun and shade, but would get dangerously cold at night. Putting these observations together, it appears that snakes have many options for thermoregulation during the daytime, as long as they avoid thin rocks or direct sun in the afternoon. At night, however, it appears that the best place for a snake to be is under a rock of medium thickness.

Most garter snakes do, in fact, retreat under rocks at night. Under the hypothesis of behavioral thermoregulation, Huey et al. predicted that snakes would choose their nighttime retreats adaptively. That is, the researchers predicted that snakes would preferentially select rocks of medium thickness. Huey et al. tested their prediction by comparing the availability of rocks of different

By monitoring temperatures of potential retreats, researchers showed that the best place for a snake to spend the night is under a rock of medium thickness.

Figure 8.4 **Environmental temperatures available to garter snakes at Eagle Lake** The graphs show the daily cycle of temperatures in various places a snake might go. (a) Under a thin (4 cm) rock it is cold at night and hot during the day. Under a thick rock (43 cm) it is cool all the time. (b) Under a rock of medium thickness (30 cm) there is virtually always a spot within the range of temperatures prefered by snakes. (c) In a burrow it is cool at night and cool-to-warm in the daytime, with the exact temperature depending on depth within the burrow. (d) On the surface it is cold at night and just-right-to-hot in the daytime, depending on whether a snake is in shade or direct sunlight. From Huey et al. (1989).

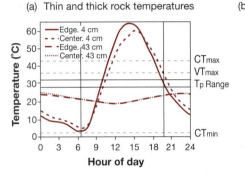

(a) Thin and thick rock temperatures

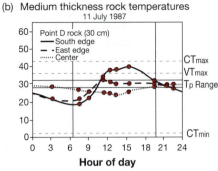

(b) Medium thickness rock temperatures
11 July 1987

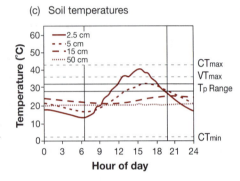

(c) Soil temperatures

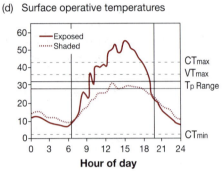

(d) Surface operative temperatures

sizes at Eagle Lake to the sizes of the rocks actually chosen as nighttime retreats by radio-implanted snakes (Table 8.1). Thin, medium, and thick rocks are equally available, so if the snakes chose their nocturnal retreats at random, they should be found equally often under rocks of each size. In fact, however, the garter snakes are almost always found under medium rocks or thick rocks. The fact that snakes avoid thin rocks is good evidence that the snakes are active behavioral thermoregulators.

The Thermal Dependence of Physiological Processes

We predicted two kinds adaptations to temperature. Behaviorial thermoregulation was the first. The second was a match between the thermal dependence of

Snakes retreat under medium rocks significantly more often than expected by chance. This result shows that snakes choose their nighttime retreats adaptively.

TABLE 8.1 Distributions of rocks available to snakes versus rocks chosen by snakes

Thin, medium, and thick rocks are equally abundant at Eagle Lake, but garter snakes retreating under rocks at night show a strong preference for rocks of medium thickness ($P < 0.05$; chi-square test with thin and thick rocks combined because of small expected values).

	Thin (<20 cm)	Medium (20–40 cm)	Thick (>40 cm)
Rocks available to snakes (N = 182):	32.4%	34.6%	33%
Rocks chosen by snakes (N = 13):	7.7%	61.5%	30.8%

Source: From Table 1 in Huey et al. (1989).

physiological processes and the temperature of the environment. Imagine that the environment in which a population lives gets hotter. The range of available temperatures in the new environment is such that individuals have difficulty thermoregulating at their optimal temperature. We can predict that the physiology of this population will evolve. Mutations will be favored that allow enzymes, and thus individuals, to function better at higher temperatures. As these mutations increase in frequency, the population will become better adapted to its new thermal environment.

Albert Bennett and colleagues (1992) tested this prediction by making direct observations of the evolution of physiological performance in response to changes in environmental temperature. Their study organism was the bacterium *Escherichia coli,* and their measure of physiological performance was growth rate. Bennett and colleagues took a strain of bacteria that had been living in the laboratory at 37°C for some 2,000 generations. The researchers took a single bacterium from this strain and let it divide to produce a colony. They then used bacteria from this colony to establish a series of experimental cultures (all of them thus genetically identical to each other). The researchers grew six of these cultures at 32°C, six at 37°C, and six at 42°C. The researchers predicted that all of the bacterial cultures would evolve, by mutation and natural selection, such that the bacterial growth rates would increase. Furthermore, the scientists predicted more dramatic evolutionary responses in the experimental cultures growing at 32°C and 42°C. These were novel temperatures, because the ancestral strain had already had 2,000 generations to evolve at 37°C.

How could the researchers tell at the end of the experiment whether a culture of bacteria was growing faster than its ancestors? Bennett et al. took advantage of the fact that *E. coli* can survive being frozen at −80°C. This allowed the researchers to store the ancestors of the evolving bacterial strains, then thaw them out and grow them in direct competition with their descendants. One member of each ancestor/descendant pair carried a genetic mutation that served as a marker. The mutation was selectively neutral under the culture conditions Bennett et al. used, but it caused the bacterial colonies to turn red (wildtype colonies are white). Thus the researchers could determine the relative growth rates of ancestor and descendant by counting the number of red versus white colonies in the culture dish. Bennett and colleagues let their experimental cultures evolve for 2,000 generations, and measured the relative growth rates of ancestors and descendants at intervals of 100 to 200 generations.

The results confirmed Bennett et al.'s predictions exactly (Figure 8.5). At all three temperatures the experimental strains evolved faster growth rates. The evolutionary divergence between ancestors and descendants was more dramatic at the novel temperature of 32°C, and especially at the novel temperature of 42°C, than at the ancestral temperature of 37°C. In later work with the same bacterial strains, Bennett and Richard Lenski (1993) showed that the improved performance of the descendant strains indeed reflects adaptation to specific temperatures. For example, the 42°C lines grew faster than their ancestors at temperatures between 40 and 42°C, but not at 37°C or any lower temperature.

To summarize, temperature strongly affects the physiological performance of organisms, and is thus a powerful agent of selection. Documented adaptations to temperature include behavioral thermal regulation, and a match between

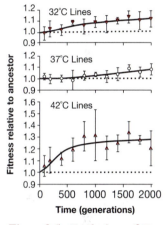

Figure 8.5 Evolution of E. coli in response to a change in environmental temperature: A natural-selection-in-the-lab experiment Each graph plots changes over 2,000 generations in the relative fitness of experimental strains (= Lines) compared to their ancestors. Each of the three plots represents evolution in strains kept at a particular temperature. Relative fitness is equal to the growth rate of the descendants divided by the growth rate of their ancestors. A value of 1 means that the ancestors and descendants grow at the same rate. Each data point represents the mean ± 95% confidence interval of six experimental cultures. The best-fit curves estimate the evolutionary trend. From Bennett et al. (1992).

the thermal dependence of physiological function and the temperature of the environment.

Chemistry

Just as all environments have a temperature, all environments also have a chemical composition. Some environments are acidic, others are alkaline; some contain abundant oxygen, others are anaerobic; some contain severe toxins, others are more benign. Our example of an adaptation to the chemical composition of the environment concerns photosynthesis and atmospheric carbon dioxide.

During photosynthesis, plants combine carbon dioxide and water to make glucose and oxygen. The net chemical reaction is:

$$6\,CO_2 + 6\,H_2O \rightarrow C_6H_{12}O_6 + 6\,O_2$$

At the heart of this process is a series of chemical reactions called the Calvin-Benson cycle. In the first step of the Calvin-Benson cycle, an enzyme called Rubisco (ribulose-1,5-bisphosphate carboxylase/oxygenase) catalyzes a reaction between CO_2 and the 5-carbon sugar RuBP (ribulose bisphosphate). This reaction produces two molecules of the 3-carbon sugar PGA (phosphoglycerate). PGA is one of the reactants for the second step of the Calvin-Benson cycle.

Rubisco also sometimes catalyzes an alternative reaction. This alternative combines RuBP with O_2 instead of CO_2, producing one molecule of PGA and one molecule of the 2-carbon sugar phosphoglycolate. Of these two products, only PGA can participate in the remainder of the Calvin-Benson cycle. Phosophoglycolate is a waste product; its breakdown results in a release of CO_2 called photorespiration. In a metabolic pathway whose function is to harvest CO_2 from the atmosphere, the generation of CO_2 is manifestly inefficient. Photorespiration also entails a second inefficiency: The plant has to expend energy to regenerate RuBP.

How serious an inefficiency is photorespiration? Using a mathematical model of photosynthesis and photorespiration, James Ehleringer and colleagues

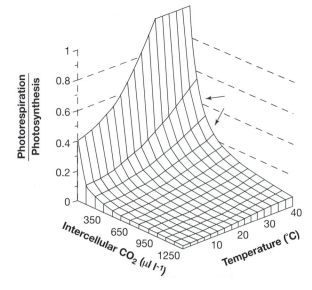

Figure 8.6   The ratio of photorespiration to photosynthesis as a function of temperature and leaf intercellular CO_2 concentration   The two horizontal axes represent CO_2 concentration and temperature. The vertical axis represents the ratio of photorespiration to photosynthesis. Small numbers for the fraction (photorespiration/photosynthesis) mean that photorespiration is a minor inefficiency; large numbers mean that it is a major inefficiency. The arrows mark the range of intercellular CO_2 concentrations that normally occur in plants at the present concentration of atmospheric CO_2. The model upon which this graph is based assumes a plant using C_3 photosynthesis (see text), and an atmospheric O_2 concentration of 21% (the current level). From Ehleringer et al. (1991).

(1991) found that the answer depends both on the temperature and on the concentration of CO_2 in the intercellular spaces within the photosynthesizing leaf (Figure 8.6). When the CO_2 concentration is high and the temperature is low, photorespiration is not a problem, because it hardly happens at all. At low concentrations of intercellular CO_2, however, especially if it is hot, photorespiration causes the loss of a substantial amount of CO_2. As the arrows in Figure 8.6 indicate, at current levels of atmospheric CO_2, plants are poised at the boundary between conditions under which photorespiration is a minor inefficiency and conditions under which photorespiration is a major inefficiency.

Ehleringer et al. note that there are two possible mechanisms by which plants might reduce the loss of CO_2 via photorespiration. The first is to modify Rubisco to make it more CO_2 specific, so that the enzyme is less likely to catalyze the reaction between O_2 and RuBP. There is some evidence that the Rubisco of flowering plants does, indeed, have a more specific affinity for CO_2 than the Rubisco of more ancestral groups of plants (Ehleringer et al. 1991). In general, however, Rubisco's functional characteristics are highly conserved. Perhaps it is simply not possible to build a Rubisco that can effectively catalyze the reaction between CO_2 and RuBP without ever combining O_2 with RuBP.

The other mechanism for reducing the loss of CO_2 via photorespiration, used by a variety of plants, is to increase the *cytoplasmic* concentration of CO_2 in the cells in which the Calvin-Benson cycle operates. One way to do this is with a CO_2 pump (Figure 8.7). The CO_2 pump can increase the concentration of CO_2 by a factor of 10 or more, thereby greatly reducing the occurrence of photorespiration. Photosynthesis using the CO_2 pump is called C_4 photosynthesis, in contrast to regular, or C_3, photosynthesis. The names refer to the number of carbon atoms in the product of the initial reaction in which CO_2 is captured.

C_4 photosynthesis appears to be derived from C_3 photosynthesis. C_4 photosynthesis occurs only in angiosperms (flowering plants), which are among the

A biochemical analysis indicates that C_4 photosynthesis is an adaptation to low concentrations of CO_2.

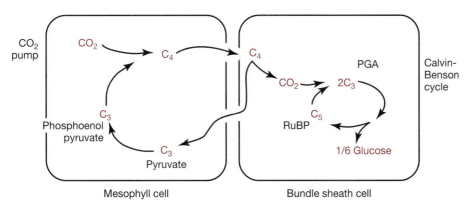

Figure 8.7 C_4 photosynthesis uses a carbon dioxide pump The bundle sheath cell (right) runs the Calvin-Benson cycle. The mesophyll cell (left) supplies CO_2 to the bundle sheath cell via a CO_2 pump. In the CO_2 pump, atmospheric CO_2 reacts with phosphoenol pyruvate to make a 4-carbon molecule (either malate or aspartic acid). The 4-carbon molecule then moves into the bundle sheath cell, where it breaks down into CO_2 and pyruvate. The CO_2 joins the Calvin-Benson cycle, and the pyruvate moves back into the mesophyll cell to be converted into phosphoenol pyruvate. Redrawn from Ehleringer et al. (1991).

youngest groups of plants. Furthermore, although angiosperms date back at least to the beginning of the Cretaceous, some 140 million years ago (Stewart 1983), the earliest plant fossil that shows the leaf anatomy typical of C_4 plants is from the Miocene, only 7 to 5 million years ago (Thomasson et al. 1986). Among living angiosperms, C_4 photosynthesis occurs in at least some members of at least 18 families (Ehleringer et al. 1991). Some pairs of C_4 taxa are distantly related, and their common ancestors do not appear to have been C_4 plants. This suggests that C_4 photosynthesis has evolved independently many times.

Is C_4 photosynthesis better than C_3 photosynthesis? That depends on the environment (Ehleringer et al. 1991). When the CO_2 concentration is low, and photorespiration becomes a problem for C_3 plants, C_4 plants are at an advantage. But the CO_2 pump uses energy, making it inefficient to run the pump when it is not needed. Furthermore, C_4 photosynthesis reaches its maximum rate at lower concentrations of CO_2 than C_3 photosynthesis. Thus when the CO_2 concentration is high, and rates of photorespiration are low, C_3 plants may be at an advantage. Another important factor is environmental moisture. Dry conditions force plants to limit water loss by closing their stomata, the pores that allow gases to move into and out of leaves. As CO_2 is used up and its concentration drops inside the closed leaves, photorespiration becomes a problem in C_3 plants, but not C_4 plants.

An experiment by F. A. Bazzaz and R. W. Carlson (1984) provides an illustration of these patterns. Bazzaz and Carlson planted seedlings of four species of annual herb together in tubs of soil, and grew them for 35 days in growth chambers equipped to control the concentration of CO_2 in the air. One of the herbs was a C_4 species, *Amaranthus retroflexus;* the other three were C_3 species, *Polygonum pensylvancium, Ambrosia artemisiifolia,* and *Abutilon theophrasti.* Bazzaz and Carlson grew the plants under six sets of conditions: dry soil with CO_2 concentrations of 300, 600, and 1,200 parts per million (ppm); and moist soil with CO_2 concentrations of 300, 600, and 1,200 ppm. After 35 days the scien-

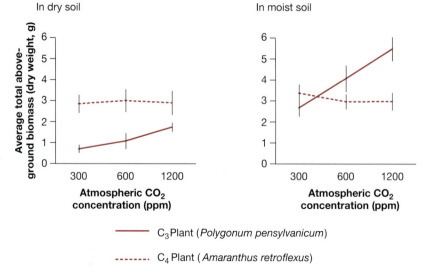

Figure 8.8 **Biomass reached after 35 days of growth at different soil moistures and CO_2 concentrations in a C_3 plant and a C_4 plant** The vertical lines show the 95% confidence intervals around the mean biomass of six plants of each species at each combination of CO_2 concentration and soil moisture. Drawn from data presented in Bazzaz and Carlson (1984).

tists harvested the above-ground parts of all the plants, and measured their dry weights.

Because *Amaranthus,* the C_4 species, and *Polygonum,* one of the C_3 species, grew to much larger sizes than the other two plants, we will focus only on *Amaranthus* and *Polygonum* in presenting the results. *Amaranthus,* the C_4 plant, was little affected by either the CO_2 concentration or the amount of moisture in the soil (Figure 8.8). In contrast, *Polygonum,* the C_3 plant, was strongly affected by both factors. Most important, the environment determined which plant was competitively superior: At low CO_2 concentrations and in dry soil, the C_4 plant grew to a larger size; at high CO_2 concentrations in moist soil, the C_3 plant grew to a larger size. C_4 photosynthesis is adaptive in some, but not all, environments.

An experiment confirmed that when the CO_2 concentration is low, or when the soil is dry, a C_4 plant grows faster than a C_3 plant.

The concentration of CO_2 in the atmosphere has changed dramatically since the angiosperms first appeared (Figure 8.9). Ehleringer et al. suggest that the decline in CO_2 concentration at the end of the Cretaceous, and again at the start of the Miocene, favored the evolution of C_4 photosynthesis by creating conditions under which photorespiration is a problem for C_3 plants. As we mentioned earlier, the earliest fossil showing C_4 leaf anatomy dates from the Miocene. Jay Quade, Thure Cerling, and colleagues have gathered additional geological evidence that C_4 plants began to flourish in some habitats at the end of the Miocene (Quade et al. 1989; Cerling et al. 1993). C_3 plants and C_4 plants incorporate into their tissues different ratios of the carbon isotopes C^{12} and C^{13}. Some of the carbon in the plants is later preserved in soil carbonates, and in the tooth enamel of mammals that ate the plants. Quade, Cerling, and colleagues found that soil carbonates and mammal tooth enamel from Pakistan, and horse tooth enamel from North America, all indicate a shift about 7 million years ago from plant communities dominated by C_3 plants to plant communities dominated by C_4 plants (Figure 8.10).

Humans are causing the concentration of CO_2 in the atmosphere to rise again, by burning fossil fuels and by razing and burning tropical forests. This increase in atmospheric CO_2 could eventually eliminate the advantage C_4 plants currently have over C_3 plants in some habitats. Increasing concentrations of CO_2 are expected to cause changes in climate as well, however, making it

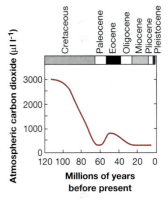

Figure 8.9 The concentration of CO_2 in the atmosphere, from the mid-Cretaceous to the present Flowering plants have existed over the entire time shown. From Ehleringer et al. (1991).

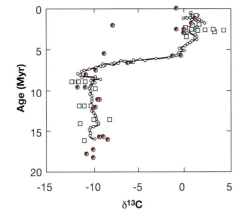

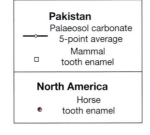

Figure 8.10 Carbon isotope ratios in soil carbonates and mammal tooth enamel over the past 20 million years $\delta^{13}C$ is a measure of the ratio of C^{12} to C^{13}. The values around -10 prior to 7 million years ago are a signature of C_3 plants; the more recent values above 0 are a signature of C_4 plants. From Cerling et al. (1993).

difficult to predict the future of competitive interactions between the two kinds of plants (Bazzaz 1990).

Mechanical Forces

All organisms are subject to mechanical forces. The causes of these mechanical forces include gravity, moving wind or water, surface tension, friction, and the movement of the organisms' own bodies. Which mechanical forces are important to an organism depends, of course, on the organism's size and on where and how it lives. Among the organisms exposed to the harshest mechanical forces are those that live on wave-swept rocks in the marine intertidal zone.

The sea urchins *Echinometra mathaei* and *Colobocentrotus atratus* (Figure 8.11) live in the intertidal zone on islands in the Indian and Pacific Oceans, including the Hawaiian Islands (Denny and Gaylord 1996). *E. mathaei* lives both in the intertidal zone and in more protected habitats, whereas *C. atratus* is an intertidal zone specialist. Within the intertidal zone, *E. mathaei* lives only in cracks and crevices that are somewhat protected, while *C. atratus* ranges everywhere, and is often fully exposed to the surf. Two other characteristics distinguish these closely related urchins. First, whereas *E. mathaei* (Figure 8.11a) is built like a typical sea urchin, with long sharp spines radiating from its body, *C. atratus* (Figure 8.11b) has an unusual shape. Its body is somewhat flattened, and most of its spines are short, blunt, and tile-like. The tile-like spines give *C. atratus* its common name, the shingle urchin. Second, *C. atratus* is much more tenaciously adhesive than its cousin. Urchins cling to the rock with numerous tube feet, similar to those of starfish. It requires more than ten times as much force to dislodge *C. atratus* from the rock than to dislodge *E. mathaei* (Gallien 1986, as cited by Denny and Gaylord 1996). By this point, readers have undoubtedly begun to suspect that the differences in shape and tenacity between *C. atratus* and *E. mathaei* are related to the fact that *C. atratus* lives in microhabitats where it is subject to hydrodynamic forces considerably greater than those experienced by *E. mathaei*.

Are the unusual features of *C. atratus* in fact adaptations for life in the wave-swept intertidal zone? An urchin clinging to a rock while a wave crashes over it feels three different forces that could potentially pull it free: drag and lift, which are caused by the movement of the water, and a force called the acceler-

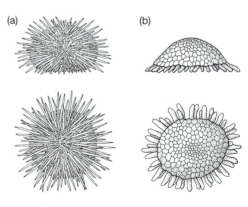

Figure 8.11 **Two species of intertidal sea urchin** Redrawn from Denny and Gaylord (1996).

Echinometra mathaei   *Colobocentrotus atratus*

ation reaction, which is caused not by the movement of the water per se, but by the fact that the water is accelerating (Figure 8.12). The generation of all three of these forces is somewhat complicated in the details (see Vogel 1981; Daniel 1984). Here we attempt to provide an intuitive understanding of where the forces come from.

To visualize drag, imagine a stream of water moving at constant velocity over a rock. The pressure in the water is the same everywhere. Onto the rock we now place an urchin (Figure 8.12a), which the water must flow around. Water is viscous, meaning that the molecules of water tend to stick to each other. Roughly speaking, this viscosity causes the water to pile up at the leading (upstream) side of the urchin, and to be slow in coming back together at the trailing (downstream) side of the urchin. This creates a high pressure area on the upstream side of the urchin, and a low pressure area on the downstream side. The difference in pressure across the urchin creates a net force tending to push the urchin downstream. This force is called drag.

To visualize lift, use the same stream of water, rock, and urchin. Note that the water has to speed up as it flows over the top of the urchin (Figure 8.12b). The faster the water moves, the lower its pressure. Thus the movement of water over the urchin creates an area of low pressure above the urchin, which tends to pull the urchin upward. This upward force is called lift.

Finally, to visualize the acceleration reaction, imagine that instead of moving at constant velocity as it passes over the urchin, the stream of water is accelerating—that is, speeding up. Whatever force is causing the water to accelerate creates a pressure gradient in the water, with higher pressure upstream and lower pressure downstream (Figure 8.12c). This pressure gradient in the water means that the upstream side of the urchin experiences a higher pressure than the downstream side, creating a net force that tends to push the urchin downstream. This force is called the acceleration reaction.

Keeping these three forces in mind, we return to the differences between the two species of sea urchin. It seems clear that the greater adhesive tenacity of the

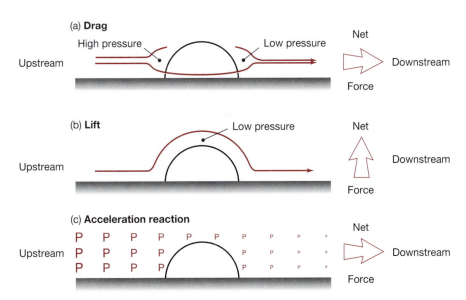

Figure 8.12 **Three forces experienced by an urchin clinging to a rock in the intertidal zone**
(a) Drag. The fine arrows represent the flow of water around the urchin. The large arrow indicates the direction of the net force due to drag.
(b) Lift. The fine arrow represents the flow of water over the top of the urchin. The large arrow indicates the direction of the net force due to lift.
(c) The acceleration reaction. The P's represent the pressure in water that is accelerating over the urchin from upstream to downstream. Bigger P's indicate higher pressure. The large arrow indicates the direction of the net force due to the acceleration reaction. See text for details.

Mechanical considerations suggested that tile-like spines are an adaptation that reduces drag on sea urchins in the waveswept intertidal zone.

shingle urchin makes it better able to cling to rocks in the face of all three wave-induced forces. The more interesting problem is the shingle urchin's flattened shape and tile-like spines. A reasonable hypothesis is that the apparently more-streamlined shape of the shingle urchin is primarily an adaptation that functions to reduce drag (Denny and Gaylord 1996). Drag, lift, and the acceleration reaction are, however, impossible to predict theoretically for all but the simplest shapes. The study of these forces remains stubbornly empirical.

Mark Denny and Brian Gaylord (1996) set out to evaluate the drag-reduction hypothesis by measuring the drag, lift, and acceleration reaction experienced by the two urchins. Denny and Gaylord collected specimens of both urchin species from Hawaii, and killed, preserved, and dried them with their spines in a natural position. To measure drag and lift on the urchins, Denny and Gaylord placed their specimens in a wind tunnel. The air moving through a wind tunnel is both less dense and less viscous than water, but by increasing the wind speed to compensate for these differences, Denny and Gaylord were able to estimate the drag and lift experienced by the urchins in moving water. Their estimates, for water moving at a speed typical of the intertidal zone, appear in Figure 8.13. Contrary to the drag-reduction hypothesis, the two urchins experience drag forces that are so similar as to be statistically indistinguishable. Furthermore, *C. atratus* (the shingle urchin) experiences lift forces that are more than twice as large as those experienced by *E. mathaei* (the typically shaped urchin). Is the fact that the *C. atratus* can cling to exposed rocks while *E. mathaei* is confined to crevices unrelated to shape, and simply the result of the shingle urchin's dramatically greater adhesive tenacity?

Denny and Gaylord's measurements of the acceleration reaction experienced by the urchins clarify the situation considerably. To measure the force imposed on the urchins by accelerating water, Denny and Gaylord measured the force that stationary water imposes on the urchins when the urchins are accelerating. Accelerating the urchins was easier than accelerating the water, and allowed the scientists to calculate the force the urchins would feel if they were stationary

Figure 8.13 Estimates of the forces experienced by sea urchins of different shapes The drag and lift coefficients reported here characterize the drag and lift experienced by urchins at a water velocity typical of the intertidal (Reynolds number = 10^5; see Denny and Gaylord 1996). The added mass coefficients characterize the acceleration reaction experienced by urchins. Based on the data and equations in Denny and Gaylord (1996).

		Drag Coefficient	Lift Coefficient	Added Mass Coefficient
Flow from anterior to posterior				
	C. atratus	0.64	0.10	0.74
	E. mathaei	0.63	0.04	2.23
Flow from left to right				
	C. atratus	0.71	0.18	0.93
	E. mathaei	0.66	0.07	2.90

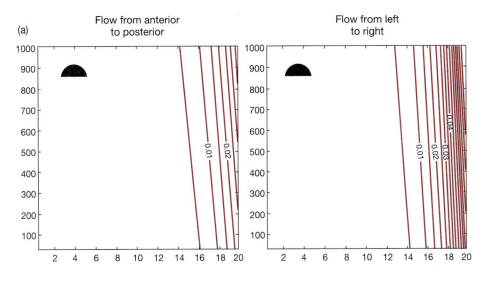

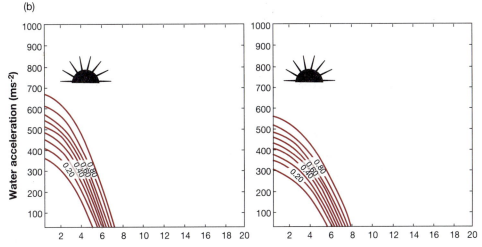

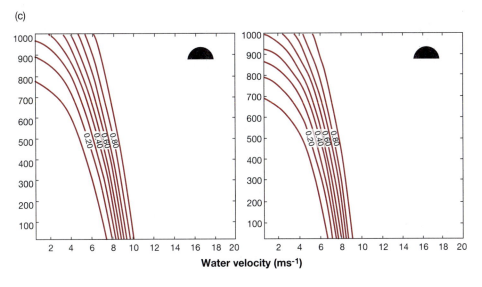

Figure 8.14 **Calculated probability of urchin dislodgment as a function of water speed and acceleration** These plots are read like topographical maps, with water velocity analogous to longitude, water acceleration analogous to latitude, and probability of dislodgment analogous to elevation. The lines in each plot run along trajectories of equal probability of dislodgment. At the lower left corner of each plot, representing low speeds and accelerations, urchins have low probabilities of dislodgment; moving to the right, or upward, representing higher speeds and greater accelerations, respectively, the urchins have higher probabilities of dislodgment. (a) *Colobocentrotus atratus* has a low probability of dislodgment even at high velocities and accelerations. (b) *Echinometra mathaei* reaches a high probability of dislodgment at moderate speeds and moderaate accelerations. (c) If *C. atratus* had an adhesive tenacity as low as that of *E. mathaei,* it would reach a high probability of dislodgment at moderate speeds and high accelerations. From Denny and Gaylord (1996).

Careful measurements of the forces on sea urchins refuted the drag-reduction hypothesis, suggesting instead that tile-like spines reduce the acceleration reaction urchins experience.

and the water were accelerating. Denny and Gaylord attached the urchins upside-down to the bottom of a sled running on rails along the top of a water tank. They used a dropped weight, attached to the sled via a cable and pulley, to accelerate the sled. After measuring the force exerted by the water on the urchins, and factoring out the drag and lift, Denny and Gaylord calculated a quantity, called the added mass coefficient, that characterizes the acceleration reaction. They found that *C. atratus* experiences an acceleration reaction that is less than 1/3 as strong as that felt by *E. mathaei* (Figure 8.13).

Having rejected the drag-reduction hypothesis, we now have the acceleration-reaction-reduction hypothesis. Denny and Gaylord calculated the probability that an urchin of each species would be knocked off a rock by water moving at different combinations of velocity and acceleration (Figure 8.14 on page 283). They did so by combining their own measurements of drag, lift, and the acceleration reaction with Gallien's (1986) measurements of adhesive tenacity. Denny and Gaylord found that *C. atratus,* the shingle urchin, is unlikely to be dislodged from a rock even at high water velocities and accelerations (Figure 8.14a). *Echinometra mathaei,* in contrast, is likely to be dislodged at even moderate velocities and accelerations (Figure 8.14b). This appears to explain why only *C. atratus* can venture into the exposed areas of the intertidal.

To determine the relative importance of adhesive tenacity versus unusual shape in giving *C. atratus* this freedom, Denny and Gaylord calculated the probability of dislodgment for a hypothetical *C. atratus* with an adhesive tenacity as low as that of *E. mathaei* (Figure 8.14c). Visual comparison of Figure 8.14c with a and b indicates that high adhesive tenacity is the most important of the adaptations that allow the shingle urchin to roam the exposed rocks of the intertidal zone. The shingle urchin's unusual shape is adaptive too, however, especially in allowing the shingle urchin to withstand high accelerations.

8.3 Adaptations to the Biological Environment

We now turn from investigating adaptations to the physical environment to testing hypotheses about adaptations to the biological environment. The interactions that may occur between an organism and its biological environment include consumption by one organism of all or part of another, competition, and mutualism. These interactions may occur between an organism and individuals of its own or other species.

Consumption

When one organism eats another, the consumer acts as a powerful agent of selection. The consumer may be a predator, an herbivore, or a parasite. The organism at risk of being eaten is under selection for improved defensive adaptations. For our example, we use a snail preyed upon by a crab.

Littorina obtusata is a small (~ 1 cm long) marine snail that lives in the intertidal zone in New England. Among this snail's predators is the crab *Carcinus maenas.* Before 1900, the crab did not occur north of Cape Cod, Massachusetts. After the turn of the century, however, the crab expanded its range northward, and is now found as far north as Nova Scotia (Seeley 1986). The crab's range expansion exposed snail populations north of Cape Cod to a new agent of selection. Robin Seeley (1986) investigated whether the snail populations evolved in response to this change.

Seeley found, in two museums, samples of pre-1900 *L. obtusata* shells collected north of Cape Cod. She compared these old shells to shells collected more recently at the same three locations, including shells from a 1915 museum collection, and shells she collected herself between 1982 and 1984. Seeley took the overall width of each shell as a measure of the shell's size; she also measured the thickness of each shell, and the height of its spire. In all the samples, Seeley found that larger shells are thicker and have higher spires (Figure 8.15). However, shells collected after 1900 have lower spires for their size (Figure 8.15 top) than shells collected before 1900. Shells collected after 1900 are also thicker for their size (Figure 8.15 bottom) than shells collected before 1900. These results suggest that after the crabs arrived, snail populations north of Cape Cod evolved lower-spired, thicker shells as a defense against crab predation.

Is a lower-spired, thicker shell actually an adaptation against predatory crabs? That is, are snails with lower-spired, thicker shells really better defended? To answer this question, Seeley took advantage of the fact that there are still some populations of *L. obtusata* in New England that have high-spired, thin shells. She collected living snails from one such population (Sipp Bay, Maine, where crabs are rare), and from a nearby population with low-spired, thick shells (Gleason Point, Maine, where crabs are common). Seeley performed two experiments: one in the lab and the other in the field.

In the lab, Seeley offered each of eight crabs a high-spired, thin-shelled snail. All eight crabs quickly crushed and ate their snails. It took them an average of 42 seconds. Seeley offered each of another eight crabs a low-spired, thick-shelled snail. Only one of these crabs was able to crush and eat its snail within 8 minutes. During that time many of the other seven crabs gave up trying.

In the field, Seeley drilled small holes in the shells of a number of snails, and used fishing line to tether the snails to seaweeds in the intertidal zone. She then

Changes in shell shape in populations of snails suggested that thick shells with low spires are more resistant to crab attack. Experiments in the lab and the field supported this hypothesis.

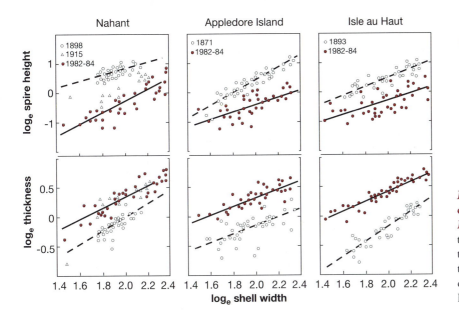

Figure 8.15 Changes in shell shape over time in the intertidal snail, *Littorina obtusata* Each panel shows the difference between samples of shells taken before 1900 (open circles) and after 1900 (triangles and filled circles) at one of three locations in New England. From Seeley (1986).

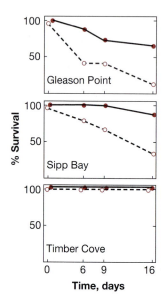

Figure 8.16 Survival of tethered snails with different shell types   Each panel shows the fraction of tethered snails not yet killed by crabs as a function of time. Open circles represent snails with high spires and thin shells; filled circles represent snails with low spires and thick shells. From Seeley (1986).

returned every few days to see which snails survived. This method allowed Seeley to distinguish between snails that were killed by crabs, and the few snails that broke free of their tethers or died in their shells. If a snail was killed by a crab, a fragment of its crushed shell remained tied to the tether. Seeley tethered the snails in pairs, with each pair including one high-spired, thin-shelled snail and one low-spired, thick-shelled snail. Seeley tethered 15 pairs at Timber Cove, where crabs appear to be absent; 15 pairs at Sipp Bay, where crabs are present but rare; and 15 pairs of snails at Gleason Point, where crabs are common. She checked on the snails after 6, 9, and 16 days. At Timber Cove, no snails died in the claws of crabs (Figure 8.16). At Sipp Bay and Gleason Point, however, many snails died (Figure 8.16). At both locations where crabs occur, crabs killed and ate the high-spired, thin-shelled snails at a much higher rate than low-spired, thick-shelled snails. These results, and those from Seeley's lab experiment, are consistent with the hypothesis that low-spired, thick shells are a defensive adaptation against predatory crabs.

Competition

Another type of biological interaction, nearly as ubiquitous as the consumption of one organism by another, is competition. Competition can involve a variety of resources including territories, food, and mates. Furthermore, competition can occur between members of the same or different species. Here we consider adaptations to a variety of forms of competition by reviewing research on a single species, the threespine stickleback, *Gasterosteus aculeatus* (Figure 8.17).

The threespine stickleback is a small fish (about 4 inches long) that lives in lakes, streams, and estuaries throughout the temperate zone. In spring, the males establish territories, which they defend against intrusion by other males. Territorial males build nests and court females. If a male succeeds in getting a female to lay her eggs in his nest, he fertilizes the eggs and tends them by himself. He continues to care for the fry for up to two weeks after they hatch, herding them together in a school and chasing off potential predators. (For a general review of the reproductive behavior of the threespine stickleback, see FitzGerald 1993.)

Competing successfully to establish a territory is vital to a male stickleback's reproductive success, because only if he holds a territory can he build a nest and court females. Continuing to exclude other males from the territory is also vital. Territorial intruders sometimes attempt to dart into a male's nest to fertilize

Figure 8.17   Threespine sticklebacks, *Gasterosteus aculeatus* (a) A female, pursued by a male. (Oxford Scientific Films/Animals Animals/ Earth Scenes)   (b) A male. (R.J. Goldstein/Visuals Unlimited)

(a)

(b)

eggs just deposited there, to eat eggs, and to steal eggs to take back to their own nests (Mori 1995). (This last fact, that male sticklebacks steal eggs from each other's nests to rear as their own, is an evolutionary puzzle. See Exploring the Literature at the end of this chapter.)

During the period of territorial defense and courtship, male threespine stick-lebacks in most populations develop a vivid red color on their bellies. Males display this red coloration to each other, and they recognize it as an aggressive territorial threat. In other words, the ability to develop a red belly appears to be an adaptation that enhances a male's success in defending his territory. In a laboratory experiment, D. W. Hagen and colleagues (1980) placed ready-to-breed male sticklebacks in a series of aquaria. Each aquarium contained one red male and one black male (we will explain shortly where the black males came from). In 30 of 46 trials (65%), the red male was able to claim a larger share of the aquarium as his territory ($P < 0.05$). In addition, red males made significantly more intrusions into the territories of black males than black males made into territories of red males ($P < 0.001$). Red-bellied males are, indeed, more successful competitors for territorial space.

Male sticklebacks compete among themselves not only to obtain territories, but also to offer the most attractive advertisements to choosy potential mates. Among other criteria, females prefer males with brighter red bellies. Field research by Gerard FitzGerald and colleagues documented that brighter males receive more eggs in their nests (Figure 8.18a). In paired-choice tests in the lab, Manfred Milinski and Theo Bakker (1990) documented directly that females prefer brighter males, and that the strength of their preference was positively correlated with the difference in brightness between the two available males (Figure 8.18b and c). These results indicate that the ability to develop a bright red belly is a dual-purpose adaptation. It not only enhances a male's ability to compete for space but also enhances his ability to attract mates. (The differences among male sticklebacks in success at getting mates, as a result of both territorial competition with other males and female choice, is an example of sexual selection. Sexual selection is the topic of Chapter 15.)

The story of stickleback bellies gets even more complex, however. Male threespine sticklebacks sometimes have to compete for territories not only with males of their own species, but with males of another species as well. Recall that we mentioned Hagen et al.'s experiments involving ready-to-breed male

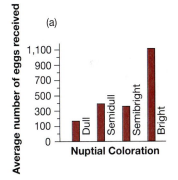

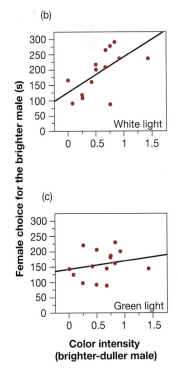

Figure 8.18 **Female threespine sticklebacks prefer more brightly colored males** (a) "Nuptial coloration" refers to the males' red bellies. In nature, brighter males receive more eggs in their nests. From FitzGerald (1993).   (b) Each of fifteen female sticklebacks had a choice of spending time near one or the other of a pair of males. The vertical axis represents the number of seconds the female spent nearer to the brighter male during a 5-minute observation period. The horizontal axis represents the difference in color intensity between the two males (as scored by students on an arbitrary scale). Under white light, the larger the difference between the two males, the more time the females spent nearer to the brighter male ($P < 0.01$). From Milinski and Bakker (1990).   (c) Under green light, which leaves the behavior of the males unaffected but renders their red mating colors nearly invisible, the females' preference for the more brightly colored of the available males largely disappears ($P = 0.75$). From Milinski and Bakker (1990).

sticklebacks that were black instead of red. These black males came from one of a few populations on Washington State's Olympic Peninsula in which all the male sticklebacks turn black instead of red during the breeding season (Figure 8.19). Laboratory breeding experiments show that whether a male turns black or red can be explained by a single Mendelian locus with two alleles—one for red, the other for black (Hagen and Moodie 1979). Neither of the alleles is dominant, so that heterozygotes are intermediate in color. How can we explain that the black allele is fixed in a few populations, given that the red allele is fixed in most others? It happens that the black allele is fixed only where, and everywhere, threespine sticklebacks co-occur with the mudminnow, *Novumbra hubbsi*.

Mudminnows have a breeding system similar to that of threespine sticklebacks. Each spring males stake out territories, build nests, and court females. The two crucial differences are that (1) mudminnows lay claim to their territories earlier in the spring than sticklebacks, so that sticklebacks living in the same streams have to elbow their way into habitats already occupied by mudminnows; and (2) male mudminnows are black.

D. W. Hagen and colleagues (1980) suspected that male sticklebacks in these streams have evolved black breeding colors because mudminnows recognize black, not red, as a territorial display.

Hagen et al. caught mudminnows, and red and black sticklebacks, and brought them into the lab. The researchers set up a series of aquaria, each containing two nest sites. They first placed a male mudminnow in each tank and allowed him to set up a territory. They next added a female mudminnow, let her mate with the male, and removed her. They then introduced a male stickleback and waited to see if he succeeded in claiming a territory in spite of aggressive resistance by the mudminnow. If the stickleback succeeded, the scientists added a female stickleback, allowed her to mate with the male, and removed her. Finally, they watched to see if the mated male stickleback managed to rear his

Mudminnows are present everywhere sticklebacks have black, not red, breeding colors. This observation suggested that black breeding colors help male sticklebacks defend territories against mudminnows.

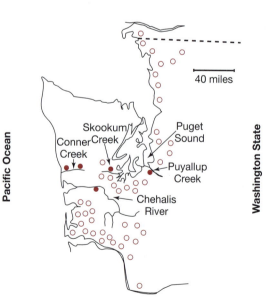

Figure 8.19 **Map of threespine stickleback populations on the Olympic Peninsula** Populations in which males turn red during the breeding season are marked by open circles; populations in which males turn black during the breeding season are marked by filled circles. From Hagen and Moodie (1979).

eggs to hatching. Hagen et al. predicted that black male sticklebacks would be more successful than red sticklebacks at claiming territories and rearing eggs.

The results confirmed Hagen et al.'s predictions (Table 8.2). Black male sticklebacks were more successful than red males on all counts: A higher fraction of black males were able to claim territories in tanks occupied by mudminnows; among those that did establish territories, the black males suffered fewer invasions and fights; and more of the black males reared eggs to hatching. These results suggest that competition with mudminnows creates strong selection in favor of the black allele where sticklebacks and mudminnows live together. In other words, the ability to turn black during the breeding season is an adaptation that enhances a male's success in interspecific competition over territories. Note that the male sticklebacks face a selective trade-off. Red breeding colors lead to higher success in competition with males of their own species, whereas black breeding colors lead to higher success in competition with mudminnows. Evidently the selection imposed on sticklebacks by interspecific competition is stronger than the selection imposed by intraspecific competition. We will return to the issue of trade-offs in section 8.5.

Lab experiments confirmed that in competition with mudminnows black sticklebacks are more successful than red sticklebacks. .

Mutualism

Life is about more than just consumption and competition. Many organisms live in cooperation with members of their own and different species. For example, many animals, humans included, harbor microorganisms in their digestive tracts that aid in digestion to the mutual benefit of both parties. Sometimes the benefits of cooperation can be substantial enough to select in favor of cooperation-specific adaptations.

Naomi Pierce and colleagues have made a detailed study of the mutualistic association between the Australian lycaenid butterfly, *Jalmenus evagoras,* and the ant, *Iridomyrmex anceps.* The larvae of the butterfly are studded with glands that exude a fluid containing about a 10% concentration of sugar, and high concentrations of amino acids (Pierce 1984; Pierce et al. 1987). These secretions are attractive to the ants. The ants tend the caterpillars in such numbers that the caterpillars are hard to see underneath the ants (Pierce et al. 1987), and the ants defend the

---

TABLE 8.2 Reproductive success of red and black male threespine sticklebacks in aquaria with territorial male mudminnows

	Black male sticklebacks	Red male sticklebacks	P
Number tested	34	34	
Number that were able to claim territories and build nests	32	17	<0.05
Average number of territorial intrusions by, and fights with, mudminnows	5.7	10	<0.001
Number that hatched fry	19	8	<0.05

Source: Based on Table 2 in Hagen et al. (1980).

caterpillars readily and aggressively against predators and parasites. The ants build satellite colonies under the Acacia trees in which the caterpillars live; virtually every tree containing caterpillars has a satellite ant colony under it.

The caterpillars' secretions appear to represent a specialized adaptation for associating with ants. To show that the secretions are adaptive, however, Pierce and colleagues had to demonstrate that attracting ants increases the fitness of the caterpillars. Indeed, the butterfly larvae reap a substantial benefit from the ants. Pierce et al. (1987) forced caterpillars to do without ants by ringing with Tanglefoot the bases of trees in which the caterpillars were living. (Tanglefoot is a sticky substance that ants and other insects cannot walk across.) Control caterpillars had their trees only partially ringed, so that ants still had access. The scientists performed the experiment at two sites. The results were dramatic. At Mt. Nebo, and especially at Canberra, the mortality rates due to predation were much higher among the ant-deprived caterpillars than among the ant-tended controls (Figure 8.20). At Canberra, it is difficult to imagine how the butterfly population could sustain itself without ants to protect its larvae.

This benefit comes at a measurable cost to the butterfly. Ants tend the butterfly's pupae as well. Because pupae do not feed, Pierce et al. (1987) were able to measure the weights and weight losses of tended and untended pupae over a five-day period. Sixteen ant-tended pupae started at a mean weight of 159 mg, and lost an average of 25 mg (16%). In contrast, 12 untended pupae started at a mean weight of 161 mg, and lost an average of 19.2 mg (12%). The difference in weight lost was statistically significant ($P < 0.01$). The extra weight lost by the tended pupae apparently represents material transferred from the pupae to the ants. The extra weight lost also means that the adult butterflies emerging from tended pupae are smaller than those coming from untended pupae. Small size puts butterflies of both sexes at a disadvantage in mating and reproduction (Elgar and Pierce 1988). The cost of associating with ants appears to be minor, however, compared to the benefit.

Is it adaptive for the ants to tend the caterpillars and pupae, or are the ants simply being manipulated? In fact, the ants reap a substantial benefit from the association too. Pierce et al. (1987) quantified this benefit by measuring the weights of ants captured on their way up into a tree containing 62 caterpillars, and ants captured on their way back down to the ant colony (Figure 8.21).

The secretions of a caterpillar appear to be designed to attract ants. Field experiments confirmed that caterpillars tended by ants suffer less predation than caterpillars not tended by ants.

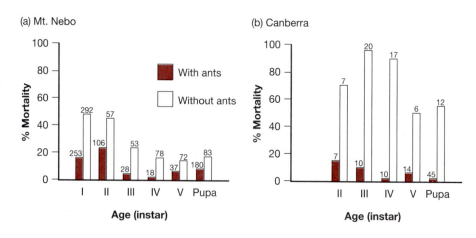

Figure 8.20 Mortality rates of caterpillars tended by ants or untended by ants at two locations   The numbers above the bars give the sample sizes. Combining the results from four experiments at the two sites, mortality due to predation was significantly lower in caterpillars tended by ants ($P < 0.000002$). From Pierce et al. (1987).

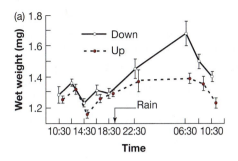

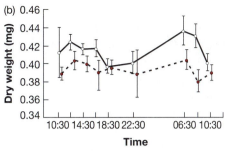

Figure 8.21 **Weights of ants captured while going up or down a tree inhabited by caterpillars** Each data point represents the mean ± standard error of many ants. Ants coming down the tree consistently weigh more than ants going up ($P < 0.01$). From Pierce et al. (1987).

These measurements indicate that each ant gains a substantial amount of weight during each trip up to tend the caterpillars. Pierce et al. estimate that over the course of a day, the ant colony derived from the caterpillars in this tree enough energy to make 100 new workers. As with the butterfly larvae, the ants reap this benefit at a cost. Individual workers invest a substantial amount of time tending the caterpillars, and the ants may sacrifice themselves to protect the caterpillars from predators. Nonetheless, the benefit of associating with caterpillars appears to amply outweigh the cost.

8.4 Phenotypic Plasticity

In sections 8.2 and 8.3 we discussed adaptations that involve behavior, physiology, and morphology. In some cases we found that a trait that is adaptive in one environment is maladaptive in another. For example, black breeding colors are adaptive for male sticklebacks living in streams with mudminnows, but maladaptive for male sticklebacks living in streams that lack mudminnows. In streams without mudminnows, red breeding colors are adaptive instead. For a male stickleback, it is his genotype that determines whether his breeding phenotype will be red or black. Often, however, individual organisms have phenotypes that are plastic. Individuals with identical genotypes may have different phenotypes if they live in different environments. Phenotypic plasticity is itself a trait that can evolve. Phenotypic plasticity may or may not be adaptive. As with the other traits we have discussed, to demonstrate that an example of phenotypic plasticity is adaptive we must first determine what it is for, then show that individuals that have it achieve higher fitness than individuals that lack it.

Phenotypic Plasticity in the Anti-Predator Defenses of a Rotifer

To introduce phenotypic plasticity, we present the rotifer *Brachionus calyciflorus*. *Brachionus calyciflorus* is a tiny multicellular animal common in ponds. Among its predators are rotifers in the genus *Asplanchna,* and the copepod *Mesocyclops edax.*

Rotifers, including *B. calyciflorus,* are ideal subjects for studying phenotypic plasticity because they reproduce asexually most of the time. This allows a scientist to rear genetically identical individuals in different environments. Following up on a discovery by P. de Beauchamp, John Gilbert (1966) showed that *B. calyciflorus* can develop either of two body types. When unexposed to *Asplanchna* (one of its predators), *B. calyciflorus* develops two pairs of anterior spines and a single pair of posterior spines (Figure 8.22a). Gilbert refers to this form as short-spined. When reared from hatching in water previously inhabited by

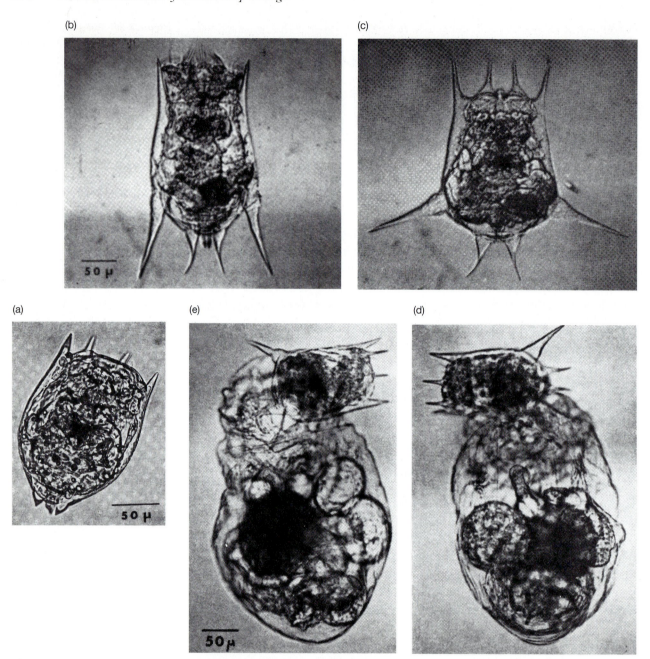

Figure 8.22 **Different body forms in genetically identical** *Brachionus calyciflorus*   (a) The short-spined form. The ante-rior end of the rotifer is facing the upper right. The animal is shaped like a wineglass (minus the stem). The four anterior spines are arrayed around the rim of the glass. Inside the rim is the mouth. Although not visible in this photograph, the mouth is sur-rounded by cilia. The beating cilia create a current that powers the rotifer as it swims about. Inside the mouth is a pair of jaws the rotifer uses to capture prey. The rotifer eats other tiny aquatic organisms.   (b) The long-spined form. The anterior end is facing upward. At the very top, in the center, some of the cilia are visible. The individual in this photograph is genetically identical to the individual in (a).   (c) This is the same individual shown in (b), with its long posterior spines flexed.   (d) Here, a long-spined individual has been captured by a larger predatory rotifer, *Asplanchna sieboli*.   (e) The individual shown in (d) escapes. Repro-duced by permission from J.J. Gilbert, 1966. Rotifer ecology and embryological induction. *Science* 151:1234–1237, March 11, 1966, p. 1235, Fig. 1. © American Association for the Advancement of Science.

Asplanchna, a short-spined *B. calyciflorus* is not visibly affected. Its asexual offspring, however, develop much longer spines than their mother had, plus a second pair of long posterior spines that were completely lacking in their mother (Figure 8.22b). When attacked by a predator, a long-spined *B. calyciflorus* holds this extra pair of spines perpendicular to its body (Figure 8.22c). The spines make their owner harder for *Asplanchna* to swallow (Figure 8.22d and e). In short, *Brachionus calyciflorus* uses a chemical released by *Asplanchna* as a cue for the development of a defense against *Asplanchna.* This is an example of phenotypic plasticity.

Gilbert (1980) showed that the long-spined morphology is an effective antipredator defense against *Asplanchna,* but not against the copepod *M. edax.* Gilbert added predators to cultures of *B. calyciflorus* with different body types, then watched through his microscope as the predators attacked their prey. When attacked by *Asplanchna sieboldi,* short-spined *B. calyciflorus* were captured about a third of the time. Once caught, the short-spined individuals never got away (Table 8.3). Long-spined individuals were captured at about twice the rate of short-spined individuals, but the long-spined individuals always got away (Table 8.3). When attacked by *M. edax, B. calyciflorus* were always captured, and once captured they were always eaten (Table 8.3).

To the extent that the long-spined form is effective against *Asplanchna,* but not against *M. edax,* it appears to be adaptive that *B. calyciflorus* grow long spines when *Asplanchna* are present. But why not simply grow the spines all the time, and be ready in case an *Asplanchna* shows up unexpectedly? The obvious hypothesis is that the spines carry some cost that makes it inefficient to make them when they are not needed. Gilbert (1980) grew both forms in the lab with different foods in various amounts, but found no evidence of differences between the two forms in survival, fecundity, or reproductive potential. Perhaps in nature *B. calyciflorus* face some third predator that, like *Asplanchna,* finds it easier to capture long-spined *B. calyciflorus,* but, unlike *Asplanchna,* can also swallow them. Gilbert's (1980) work shows that the long spines themselves are sometimes adaptive, but leaves unanswered the question of whether the phenotypic plasticity in the production of the spines is adaptive.

In many species, individuals exhibit different phenotypes when reared in different environments, a phenomenon called phenotypic plasticity.

TABLE 8.3 Predation attempts on *Brachionus calyciflorus* by two predators

All data are for adult *B. calyciflorus.*

Body form	Number of attacks	% Caught	% Escaped	% Eaten
Attacks by Asplanchna sieboldi				
Short-Spined	63	35	0	35
Long-Spined	141	62	62	0
Attacks by Mesocyclops edax				
Short-Spined	26	100	0	100
Long-Spined	19	100	0	100

Source: Based on Tables 1 and 2 from Gilbert (1980).

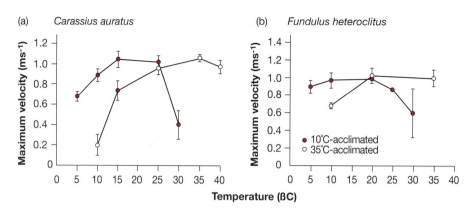

Phenotypic Plasticity of Thermal Physiology: Acclimation

Brachionus calyciflorus displays phenotypic plasticity in a morphological trait. Phenotypic plasticity can occur in physiological traits as well. Phenotypic plasticity in physiological traits is called acclimation.

Timothy Johnson and Albert Bennett (1995) documented acclimation to temperature in goldfish *(Carassius auratus)* and killifish *(Fundulus heteroclitus).* Johnson and Bennett took 20 goldfish from a single population, and divided them at random into two groups of 10. They kept one group at 10°C and the other group at 35°C. Likewise, they took 10 killifish from a single population, divided them into two groups of 5, and kept one group at 10°C and one group at 35°C. After giving the fish a month to adjust to their acclimation temperatures, Johnson and Bennett used a high-speed video camera to measure the maximum speed each fish attained when startled. They performed this test at each of several different temperatures. In both species, fish that had lived for a month at 10°C swam faster at low temperatures than fish that had lived at 35°C, whereas fish that had lived for a month at 35°C swam faster at high temperatures (Figure 8.23). In other words, the individual fish had adjusted their physiology to the temperature at which they had been living. Based on the graphs in Figure 8.23, thermal acclimation in goldfish and killifish appears to be adaptive. No individual, regardless of the temperature at which it has been raised, can swim fast at all temperatures. During acclimation, however, the thermal dependence of performance changes. Acclimated fish perform better at their acclimation temperature than do non-acclimated fish.

Phenotypic plasticity can be adaptive.

Note that acclimation involves a change in the phenotype only. Like other forms of phenotypic plasticity, acclimation takes place within individuals; it does not involve genetic change.

Phenotypic Plasticity in Behavior: Water Fleas and Fish

Phenotypic plasticity is also possible in behavior. Here we present an example of behavioral phenotypic plasticity that also demonstrates some important points about the evolution of plasticity itself.

The water flea, *Daphnia magna,* is a tiny crustacean that lives in freshwater lakes (Figure 8.24). Like the rotifers we discussed earlier, *Daphnia* reproduce

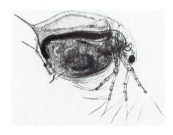

Figure 8.24 **A water flea, *Daphnia* sp.** The large branched appendages are antennae; the water flea uses them like oars for swimming. The dark object near the antennae is an eyespot. Also visible through this individual's transparent carapace are the intestine and other internal organs. This photograph is enlarged about 10 times. (Omikron/Photo Researchers, Inc.)

asexually most of the time. In other words, *Daphnia* clone themselves. This makes them ideal for studies of phenotypic plasticity, because researchers can grow genetically identical individuals in different environments and compare their phenotypes.

Luc De Meester (1996) studied phenotypic plasticity in *D. magna*'s phototactic behavior. An individual is positively phototactic if it swims toward light, and negatively phototactic if it swims away from light. De Meester measured the phototactic behavior typical of different genotypes of *D. magna*. In each single test, De Meester placed 10 genetically identical individuals in a graduated cylinder, illuminated them from above, gave the water fleas time to adjust to the change in environment, then watched to see where in the column they swam. De Meester summarized the results by calculating an index of phototactic behavior. The index can range in value from −1 to +1. A value of −1 means that all the *Daphnia* in the test swam to the bottom of the column, away from the light. A value of +1 means that all the *Daphnia* in the test swam to the top of the column, toward the light. An intermediate value indicates a mixed result.

De Meester measured the phototactic behavior of 10 *Daphnia* genotypes (also called clones) from each of three lakes. The results are indicated by the black dots in Figure 8.25. The population in each lake harbors considerable genetic variation in phototactic behavior.

De Meester also measured the phototactic behavior of the same 30 *Daphnia* genotypes in water that had been previously occupied by fish. The results are indicated by the shaded squares in Figure 8.25. *Daphnia magna*'s phototactic behavior is phenotypically plastic. In Lake Blankaart, in particular, most *Daphnia*

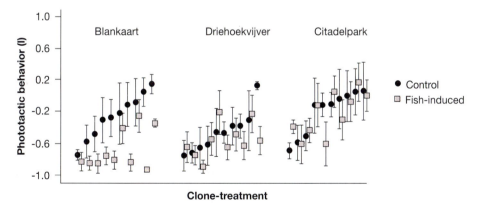

Figure 8.25 **Variation in phototactic behavior in *Daphnia magna*** Blankaart, Driehoekvijver, and Citadelpark are three lakes in Belgium. Each black dot represents the average result from three to five tests of the phototactic behavior of a single genotype (described in main text). The error bars indicate ± 2 standard errors. Genotypes (clones) whose error bars do not overlap are significantly different. Above or below each black dot is a shaded square. This shaded square represents the average result from three or four tests of the phototactic behavior of the same genotype. The difference is that this time the *Daphnia* were tested in water that had been previously occupied by fish. Lake Blankaart is home to many fish; Lake Driehoekvijver has few fish; Lake Citadelpark has no fish. From De Meester (1996).

genotypes score considerably lower on the phototactic index when tested in the presence of chemicals released by fish.

Finally, and most importantly, De Meester's results demostrate that phenotypic plasticity is a trait that can evolve. Recall that a trait can evolve in a population only if the population contains genetic variation for the trait. Each of the *Daphnia* populations De Meester studied contains genetic variation for phenotypic plasticity. That is, some genotypes in each population alter their behavior more than others in the presence versus absence of fish (Figure 8.25). Genetic variation for phenotypic plasticity is called genotype-by-environment interaction.

Phenotypic plasticity can evolve.

Has phenotypic plasticity in fact evolved in the *Daphnia* populations De Meester studied? It has. The average genotype in Lake Blankaart shows considerably more phenotypic plasticity than the average genotype in either of the other lakes. Blankaart is the only one of the lakes with a sizeable population of fish. Fish are visual predators, and they eat *Daphnia*. A reasonable hypothesis is that fish select in favor of *Daphnia* that avoid well-lit areas when fish are present.

Phenotypic Plasticity in the Antipredator Defenses of Snails

Phenotypic plasticity is a potential complication in any study of adaptation. To illustrate this point, we return to Robin Seeley's (1986) work on the snail *Littorina obtusata* (section 8.3). After studying museum specimens and conducting experiments (Figures 8.15 and 8.16), Seeley concluded that snail populations in New England have evolved thicker shells in response to predation by crabs.

Phenotypic plasticity can complicate studies of adaptation.

Geoffrey Trussell (1996) pointed out a potential problem with Seeley's conclusion. Some prey species (like *Brachionus calyciflorus*) show phenotypic plasticity in their antipredator defenses, making defensive structures when the predator is present, but not when the predator is absent. It is possible that the change in shell form that Seeley documented is simply a phenotypic response to the presence of crabs. If so, then the snail populations have may not have evolved at all; instead, they may simply have revealed their phenotypic plasticity.

To investigate this possibility, Trussell gathered live snails from populations living in locations exposed to waves, where crabs are rare, and from populations living in locations protected from waves, where crabs are common. He grew these snails in laboratory aquaria for 78 days. Some aquaria contained crabs enclosed in perforated containers; others did not. Trussell found that snails from both source populations grew thicker shells for their size when crabs were present than when crabs were absent (Figure 8.26). In other words, snail shells are indeed phenotypically plastic. Nonetheless, snails from areas where crabs are naturally common grew thicker shells in both treatments than did snails from areas where crabs are naturally rare. This indicates that there are also genetic differences between snails from the different source populations.

Trussell's results suggest that some of the change in shell form documented by Seeley may be due to phenotypic plasticity, and that some of the change is probably due to evolution as well. Elizabeth Boulding and Toby Hay (1993) showed that shell form is heritable in another species of *Littorina,* further supporting the hypothesis that natural selection via crab predation can lead to the evolution of altered shell form. Trussell's work is a valuable reminder, however, that we must keep the possibility of phenotypic plasticity in mind whenever we compare populations that live in different environments.

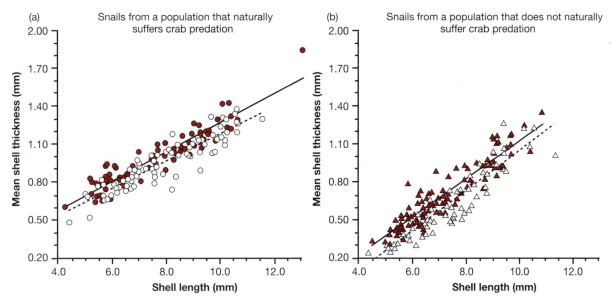

Figure 8.26 **Phenotypic plasticity and population differences in shell thickness in** *Littorina obtusata*   Each plot shows the relationship between shell thickness and overall size (shell length) for snails reared in the lab in the presence (filled symbols and solid lines) or absence (open symbols and dashed lines) of crabs.   Plot (a) is for snails collected from a population that naturally suffers crab predation;   plot (b) is for snails collected from a population that does not naturally suffer crab predation. In both populations, larger shells are thicker. In both populations, shell thickness is phenotypically plastic: Snails reared with crabs have thicker shells for their size. Nonetheless, the snails from the population normally exposed to crabs have thicker shells, regardless of treatment, than snails from the population not normally exposed to crabs. To see this, note that the best-fit lines have different *y*-intercepts. From Trussell (1996).

8.5 Trade-Offs and Constraints

It is impossible for any population of organisms to evolve optimal solutions to all agents of selection at once. We have mentioned examples of trade-offs in passing. Male threespine sticklebacks, for instance, cannot simultaneously be red, which is best for intraspecific interactions; and black, which is best for interspecific interactions. In this section, we explore additional examples of research into factors that limit adaptive evolution. These factors include trade-offs, functional constraints, and lack of genetic variation.

Organisms cannot evolve to be optimally adapted to all agents of selection.

Female Flower Size in a *Begonia:* A Trade-Off

The tropical plant *Begonia involucrata* is monoecious: There are separate male and female flowers on the same plant. The flowers are pollinated by bees. As the bees travel among male flowers gathering pollen, they sometimes also transfer pollen from male flowers to female flowers. The male flowers offer the bees a reward, in the form of the pollen itself. The female flowers offer no reward; instead they get pollinated by deceit (Ågren and Schemske 1991). Not surprisingly, bees make more and longer visits to male flowers than to female flowers.

The female flowers resemble the male flowers in color, shape, and size (Figure 8.27a). This resemblance is presumably adaptive. Given that bees avoid female flowers in favor of male flowers, the rate at which female flowers are visited should depend on the degree to which they resemble male flowers. The ability to attract pollinators should, in turn, influence fitness through female

Figure 8.27 *Begonia involucrata* (a) Male (left) and female (right) flowers. The flowers lack true petals. Instead, each has a pair of petaloid sepals (see Figure 12.26 for the parts of a generic flower). The sepals are white or pinkish. In the center of each flower is a cluster of yellow anthers or stigmas. The stigmas of female flowers resemble the anthers of males. (b) An inflorescence, or stalk bearing many flowers. Each inflorescence makes both male and female flowers. Typically, the male flowers open first, and the female flowers open later. The infloresence shown is unusual in having flowers of both sexes open at once. (Doug W. Schemske, University of Washington)

function, because seed set is limited by pollen availability. Doug Schemske and Jon Ågren (1995) sought to distinguish between two hypotheses about the mode of selection imposed by bees on female flower size:

Hypothesis 1: The more closely female flowers resemble male flowers, the more often they are visited by bees. Selection on female flowers is stabilizing, with best phenotype for females identical to the mean phenotype of males (Figure 8.28a, left).

Hypothesis 2: The more closely female flowers resemble the most rewarding male flowers, the more often they will be visited by bees. If larger male flowers offer bigger rewards, then selection on female flowers is directional, with bigger flowers always favored over smaller flowers (Figure 8.28a, right).

Schemske and Ågren made artificial flowers of three different sizes (Figure 8.28b), arrayed equal numbers of each in the forest, and watched to see how often bees approached and visited them. The results were clear: The larger the flower, the more bee approaches and visits it attracted (Figure 8.28c). Selection by bees on female flowers is strongly directional.

Taken at face value, Schemske and Ågren's results suggest that female flower size in *Begonia involucrata* is maladaptive. Selection by bees favors larger flowers, yet the female flowers are no bigger than the male flowers. One solution to this paradox is that *B. involucrata* simply lacks genetic variation for female flowers that are substantially larger than male flowers. Schemske and Ågren have no direct evidence on this suggestion; *B. involucrata* is a perennial that takes a long time to reach sexual maturity, so quantitative genetic experiments are difficult to do.

Another solution to the paradox is that focusing on individual female flowers gives biologists too narrow a view of selection. Schemske and Ågren expanded their focus from individual flowers to inflorescences (Figure 8.27b). The researchers measured the size and number of the female flowers on 74 inflorescences. They discovered a trade-off: The larger the female flowers on an

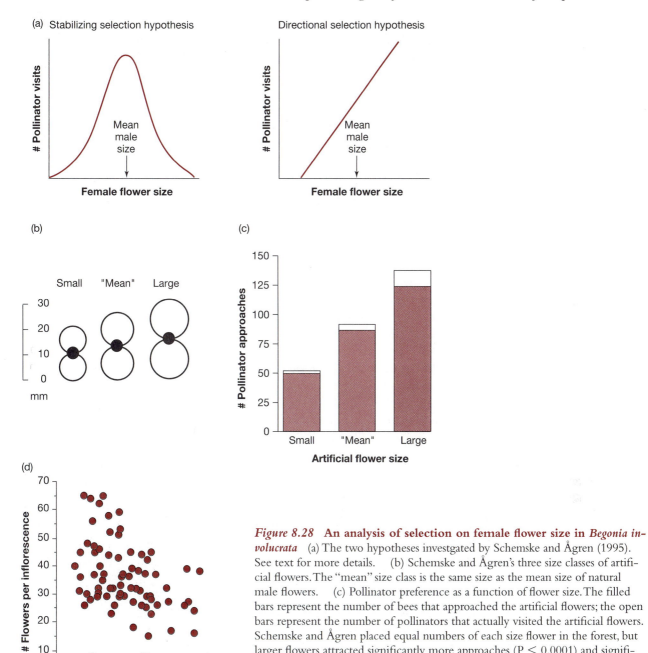

Figure 8.28 An analysis of selection on female flower size in *Begonia involucrata* (a) The two hypotheses investgated by Schemske and Ågren (1995). See text for more details.   (b) Schemske and Ågren's three size classes of artificial flowers. The "mean" size class is the same size as the mean size of natural male flowers.   (c) Pollinator preference as a function of flower size. The filled bars represent the number of bees that approached the artificial flowers; the open bars represent the number of pollinators that actually visited the artificial flowers. Schemske and Ågren placed equal numbers of each size flower in the forest, but larger flowers attracted significantly more approaches (P < 0.0001) and significantly more visits (P < 0.001) from bees.   (d) Number of female flowers per inflorescence as a function of flower size. There is a statistically significant trade-off between flower size and flower number (P = 0.003). From Schemske and Ågren (1995).

inflorescence, the fewer flowers there are (Figure 8.28d). Such a trade-off makes intuitive sense. If an individual plant has a finite supply of energy and nutrients to invest in flowers, it can slice this pie into a few large pieces or many small pieces, but not into many large pieces. Inflorescences with more flowers may be favored by selection for two reasons. First, bees may be more attracted to inflorescences with more flowers. Second, more female flowers

Adaptative evolution can be constrained by a trade-off between one advantageous trait and another.

means greater potential seed production. Schemske and Ågren hypothesize that female flower size in *B. involucrata* has been determined, at least in part, by two opposing forces: directional selection for larger flowers, and the trade-off between flower size and number.

Flower Color Change in a *Fuchsia:* A Constraint

Fuchsia excorticata is a bird-pollinated tree endemic to New Zealand (Delph and Lively 1989). Its flowers hang downward like bells (Figure 8.29a). The ovary is at the top of the bell. The body of the bell consists of the hypanthium, or floral tube, and the sepals. The style resembles an elongated clapper. It is surrounded by shorter stamens and a set of reduced petals.

The hypanthium and sepals are the most conspicuously showy parts of the flower. They remain green for about 5.5 days after the flower opens, then begin to turn red (Figure 8.29b). The transition from green to red lasts about 1.5 days, at the end of which the hypanthium and sepals are fully red. The red flowers remain on the tree for about 5 days. The red flowers then separate from the ovary at the abscission zone and drop from the tree.

Pollination occurs during the green phase and into the intermediate phase, but it is complete by the time the flowers are fully red. The flowers produce nectar on days 1 through 7 (Figure 8.29b). Most flowers have exported more than 90% of their pollen by the end of that time. The stigmas are receptive to pollen at least until the second day of the fully red phase, but rarely does pollen arriving after the first day of the red phase fertilize eggs. Not surprisingly, bellbirds and other avian pollinators strongly prefer green flowers, and virtually ignore nectarless red flowers (Delph and Lively 1985).

Why do the flowers of this tree change color? A general answer, supported by research in a variety of plants, is that color change serves as a cue to pollinators, alerting them that the flowers are no longer offering a reward (see Delph and Lively [1989] for a review). By paying attention to this cue, pollinators can increase their foraging efficiency; they do not waste time looking for nonexistent rewards. The plant benefits in return, because when pollinators forage efficiently they also transfer pollen efficiently. They do not deposit viable pollen on unreceptive stigmas, and they do not deposit nonviable pollen on receptive stigmas.

This answer is only partially satisfying, however. Why does *F. excorticata* not just drop its flowers immediately after pollination is complete? Dropping the flowers would give an unambiguous signal to pollinators that a reward is no longer being offered, and it would be metabolically much cheaper than maintaining the red flowers for several days. Retention of the flowers beyond the time of pollination seems maladaptive.

Lynda Delph and Curtis Lively (1989) consider two hypotheses for why *F. excorticata* keeps its flowers (and changes them to red) instead of just dropping them. The first is that red flowers may still attract pollinators to the tree displaying them, if not to the red flowers themselves. Once drawn to the tree, pollinators could then forage on the green flowers still present. Thus retention of the red flowers could increase the overall pollination efficiency of the individual tree retaining them. If this hypothesis is correct, then green flowers surrounded by red flowers should receive more pollen than green flowers not surrounded. Delph and Lively tested this prediction by removing red flowers from some

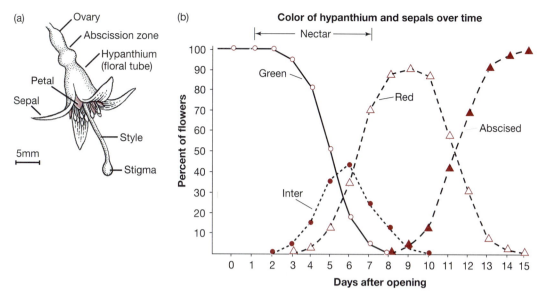

Figure 8.29 **Flower color change in *Fuchsia excorticata*** (a) A *Fuchsia excorticata* flower.   (b) The horizontal axis shows flower age, in days after opening. The vertical axis and graph lines show the percentage of flowers that are in each color phase at each age. From Delph and Lively (1989).

trees but not from others, and from some branches within trees but not from others. The researchers then compared the amount of pollen deposited on green flowers in red-free trees and branches versus red-retaining trees and branches. They found no significant differences. The pollinator-attraction hypothesis does not explain the retention of the red flowers in *F. excorticata*.

The second hypothesis Delph and Lively considered is that a physiological constraint prevents *F. excorticata* from dropping its flowers any sooner than it does. This physiological constraint is the growth of pollen tubes. After a pollen grain lands on a stigma, the pollen germinates. The germinated pollen grain grows a tube down through the style to the ovary. The pollen grain's two sperm travel through this tube to the ovary, where one of the sperm fertilizes an egg. The growth of pollen tubes takes time, especially in a plant like *F. excorticata*, which has long styles. If the plant were to drop its flowers before the pollen tubes had time to reach the ovaries, the result would be the same as if the flowers had never been pollinated at all. Delph and Lively pollinated 40 flowers by hand. After 24 hours, they plucked 10 of the flowers, dissected them, and examined them under a microscope to see whether the pollen tubes had reached the ovary. After 48 hours, they plucked and dissected 10 more flowers, and so on. The results appear in Table 8.4. It takes about three days for the pollen tubes to reach the ovary.

This result is consistent with the physiological constraint hypothesis. *F. excorticata* cannot start the process of dropping a flower until about three days after the flower is finished receiving pollen. Dropping a flower involves forming a structure called an abscission zone between the ovary and the flower (Figure 8.29a). The abscission zone consists of several layers of cells that form a division between the ovary and the flower. In *F. excorticata* the growth of the abscission

Adaptative evolution can be constrained by physiological limits on organisms.

TABLE 8.4 Pollen tube growth in *Fuchsia excorticata*

Days since pollination	1	2	3	4
Percentage of 10 flowers with pollen tubes in ovary:	0	20%	100%	100%

Source: After Delph and Lively (1989).

layer takes at least 1.5 days. Thus the plant is constrained to retain its flowers for at least 4.5 days after pollination ends. In fact, the plant retains its flowers for about 5 days. Delph and Lively suggest that flower color change in *F. excorticata* is an adaptation that evolved to compensate for the physiological constraints that necessitate flower retention. Given that the plant had to retain its flowers, selection favored individuals offering cues to pollinators that distinguish receptive versus unreceptive flowers.

Host Shifts in an Herbivorous Beetle: Constrained by Lack of Genetic Variation?

In nearly every previous chapter, we have made the point that genetic variation is the raw material for evolution by natural selection. Because natural selection is the process that produces adaptations, genetic variation is also the raw material from which adaptations are molded. Conversely, populations of organisms may be prevented from evolving particular adaptations simply because they lack the necessary genetic variation to do so.

Here is an extreme example: Pigs have not evolved the ability to fly. We can imagine that flying might well be adaptive for pigs. It would enable them to escape from predators, and to travel farther in search of their favorite foods. Pigs do not fly, however, because the vertebrate developmental program lacks genetic variation for the growth of both a trotter and a wing from the same shoulder. Other vertebrates have evolved the ability to fly, of course. But in bats and birds, the developmental program has been modified to convert the entire forelimb from a leg to a wing; in neither group does an entirely new limb sprout from the body. Too bad for pigs.

Pig flight makes a vivid example, but in the end it is a rather trivial one. The wished-for adaptation is too unrealistic. Douglas Futuyma and colleagues have sought to determine whether lack of genetic variation has constrained adaptation in a more realistic and meaningful example (Funk et al. 1995; Futuyma et al. 1995; references therein). Futuyma and colleagues studied host plant use by herbivorous leaf beetles in the genus *Ophraella*. Among these small beetles, each species feeds as larvae and adults on the leaves of one or a few closely related species of composites (plants in the sunflower family, the Asteraceae). Each species of host plant makes a unique mixture of toxic chemicals that serve as defenses against herbivores. For the beetles, the ability to live on a particular species of host plant is a complex adaptation that includes the ability to recognize the plant as an appropriate place to feed and lay eggs, as well as the ability to detoxify the plant's chemical defenses.

An estimate of the phylogeny for 12 species of leaf beetle appears in Figure 8.30. The figure also lists the host plant for each beetle species. The evolutionary history of the beetle genus has included several shifts from one host plant to another. Four of the host shifts were among relatively distantly related plants species: They involved switches from a plant in one tribe of the Asteraceae to a plant in another tribe. These shifts are indicated in the figure by changes in the shading of the phylogeny. Other shifts involved movement to a new host in the same genus as the ancestral host, or in a genus closely related to that of the ancestral host.

Each combination of a beetle species and the host plant used by one of its relatives represents a plausible evolutionary scenario for a host shift that might have happened but did not. For example, the beetle *Ophraella arctica* might have switched to the host *Iva axillaris*. Futuyma and colleagues have attempted to elucidate why some host shifts have actually happened while others have remained hypothetical. Here are two hypotheses:

Hypothesis 1: All host shifts are genetically possible. That is, every beetle species harbors sufficient genetic varation in its feeding and detoxifying mechanisms to allow at least some individuals to feed and survive on every potential host species. If a few individuals can feed and survive, they can be the founders for a new population of beetles that will evolve to become well-adapted to the new host. Because all host shifts are genetically possible, the pattern of actual host shifts has been determined by ecological factors and by chance. Ecological factors might include the abundance of the various host species within the geographic ranges of the beetle species, and the predators and competitors associated with each host species.

Hypothesis 2: Most host shifts are genetically impossible. That is, most beetle species lack sufficient genetic variation in their feeding and detoxifying mechanisms to allow any individuals to feed and survive on any but a few of the potential host species. The pattern of actual host shifts has been largely determined by what was genetically possible. Genetically possible host shifts have happened; genetically impossible host shifts have not.

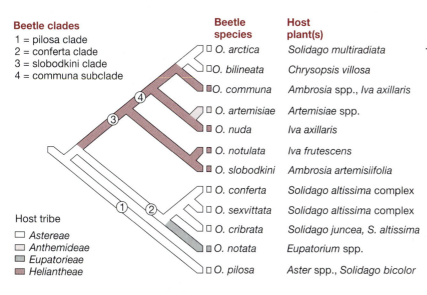

Beetle clades
1 = pilosa clade
2 = conferta clade
3 = slobodkini clade
4 = communa subclade

Host tribe
☐ Astereae
☐ Anthemideae
▨ Eupatorieae
▧ Helantheae

Beetle species	Host plant(s)
O. arctica	*Solidago multiradiata*
O. bilineata	*Chrysopsis villosa*
O. communa	*Ambrosia* spp., *Iva axillaris*
O. artemisiae	*Artemisiae* spp.
O. nuda	*Iva axillaris*
O. notulata	*Iva frutescens*
O. slobodkini	*Ambrosia artemisiifolia*
O. conferta	*Solidago altissima* complex
O. sexvittata	*Solidago altissima* complex
O. cribrata	*Solidago juncea, S. altissima*
O. notata	*Eupatorium* spp.
O. pilosa	*Aster* spp., *Solidago bicolor*

Figure 8.30 **Phylogeny of the leaf beetles, genus *Ophraella*** The numbers on branches define the major branches (clades) of the beetle evolutionary tree. The shading of branches indicates the tribes of host species. The evolutionary history of the beetle genus has included four host shifts across tribes. From Futuyma et al. (1995).

TABLE 8.5 Summary of tests for genetic variation in larval or adult feeding on potential host plants

(*a*) *Tests for genetic variation in larval or adult feeding, by relationship among host plants:*

	Genetic variation?	
Beetle tested for feeding on a plant that is . . .	**Yes**	**No**
. . . in the same tribe as the beetle's actual host:	7	1
. . . in a different tribe than the beetle's actual host:	14	17

Conclusion: Genetic variation for feeding is more likely to be found when a beetle is tested on a potential host that is closely related to its actual host ($P < 0.04$).

(*b*) *Tests for genetic variation in larval or adult feeding, by relationship among beetles:*

	Genetic variation?	
Beetle tested for feeding on a plant that is . . .	**Yes**	**No**
. . . the host of a beetle in the same major clade:	12	4
. . . the host of a beetle in a different major clade:	9	14

Conclusion: Genetic variation for feeding is more likely to be found when a beetle is tested on a potential host that is the actual host of a closely related beetle ($P < 0.03$).

Source: From Table 7 in Futuyma et al. (1995).

We have presented these hypotheses as mutually exclusive. In fact, the truth is almost certainly that the actual pattern of host shifts has resulted from a mixture of genetic constraints, ecological factors, and chance. What Futuyma and colleagues were looking for was concrete evidence that genetic constraints have been at least part of the picture.

Futuyma and colleagues used a quantitative genetic approach (see Chapter 6) to determine how much genetic variation the beetles harbor for feeding and surviving on other potential hosts. The researchers examined various combinations of four of the beetle species listed in Figure 8.30 with six of the host plants. Their tests revealed that there is little genetic variation in most beetle species for feeding and surviving on most potential host species. In 18 of 39 tests of whether larvae or adults of a beetle species would recognize and feed on a potential host plant, the researchers found no evidence of genetic variation for feeding. In 14 of 16 tests of whether larvae could survive on a potential host plant, the researchers found no evidence of genetic variation for survival.

These results suggest that hypothesis 2 is at least partially correct. Many otherwise-plausible host shifts appear to be genetically impossible. Futuyma and colleagues performed an additional test of hypothesis 2 by looking for patterns in their data on genetic variation for larval and adult feeding. If hypothesis 2 is correct, then a beetle species is more likely to show genetic variation for feeding on a potential new host if the new host is a close relative of the beetle's present host. Futuyma et al.'s data confirm this prediction (Table 8.5a). Likewise, if hypothesis 2 is correct, then a beetle species is more likely to show genetic variation for feeding on a potential new host if the new host is the actual host of one of the beetles' close relatives. Futuyma et al.'s data also confirm this prediction (Table 8.5b). Futuyma and colleagues conclude that hypothesis 2 is at least partially correct. The history of host shifts in the beetle genus *Ophraella* has been constrained by the availability of genetic variation for evolutionary change.

Adaptive evolution can be constrained by a lack of genetic variation.

8.6 The Maintenance of Genetic Variation

Evolution by natural selection requires genetic variation. However, natural selection reduces the amount of genetic variation in populations by driving some alleles to fixation and eliminating others. Thus the process of adaptive evolution bites the hand that feeds it.

Although selection often erodes the genetic variation necessary for adaptive evolution, genetic variation is maintained by mutation and by serveral different patterns of selection.

The study of *Ophraella* by Futuyma and colleagues indicates that a lack of genetic variation can constrain adaptive evolution. On the other hand, most populations of organisms contain considerable genetic variation for a great variety of traits. Furthermore, the diversity and adaptedness of organisms testifies to an enormous wealth of genetic diversity that must have existed in past populations.

Here is a brief list of some of the mechanisms that can maintain genetic variation in populations:

- Mutation. In section 8.2 we discussed a natural-selection-in-the-lab experiment by Albert Bennett and colleagues involving *E. coli*. The researchers intentionally started all of their experimental populations from genetically identical bacteria, so there was no genetic variation at all, either within or between populations. And yet over the course of 2,000 generations, all of the experimental populations showed considerable adaptive evolution. All of the genetic raw material for this evolution came from mutations that happened during the course of the experiment.
- Heterozygote advantage. In Chapter 5 we discussed the pattern of selection on the alleles of the gene for sickle-cell anemia. When malaria is present, selection favors heterozygotes. This heterozygote advantage keeps both the sickle-cell and wild-type alleles in the population. In other words, under heterozygote advantage, selection itself maintains genetic variation.
- Frequency-dependent selection. Also in Chapter 5, we discussed selection on handedness in scale-eating fish. Left-handed fish have higher fitness when right-handers dominate the population, and right-handed fish have higher fitness when left-handers dominate the population. Over the long run, the result is the same as with heterozygote advantage: Selection itself maintains genetic variation.
- Environmental heterogeneity plus gene flow. In section 8.3 we considered selection on sticklebacks. In stickleback populations that coexist with mudminnows, the allele for black male breeding colors is fixed. In stickleback populations that do not coexist with mudminnows, the allele for red breeding colors is fixed. Thus in any given population, selection acts to eliminate genetic variation. However, there are several stickleback populations in Washington State that occur on borders between areas occupied by mudminnows and areas that are mudminnow-free. These border populations receive immigrant sticklebacks from both red and black populations. Because of this gene flow, the border populations remain genetically variable for male breeding colors (Figure 8.31).

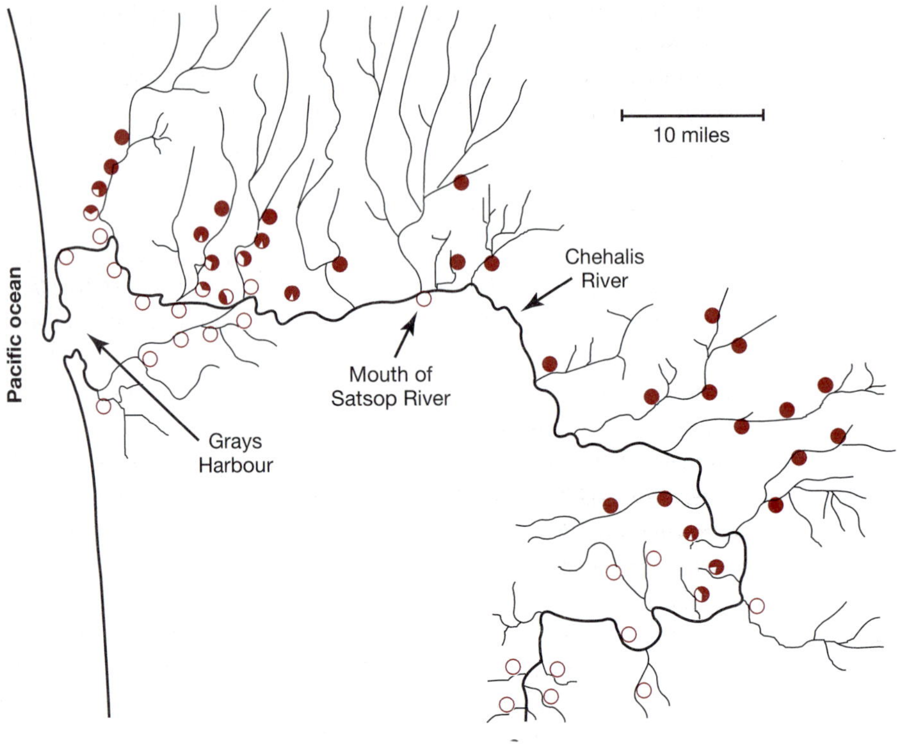

Figure 8.31 **Genetic variation for male breeding colors in stickleback populations of the Chehalis River drainage**
The white area of each pie diagram represents the frequency of the red allele in the local population; the filled area represents the frequency of the black allele. Most populations are fixed for one allele or the other. However, several populations are genetically variable. These populations are near the edges of areas occupied by mudminnows. From Hagen and Moodie (1979).

Summary

A central objective of evolutionary research is to understand the adaptive significance of the traits of organisms. Such research, called the adaptationist program, must be conducted with care. Hypotheses must be tested, and alternative explanations must considered. The result is a deep understanding of the agents of selection that buffet organisms, and of the diversity and complexity of the traits that evolve in response.

We reviewed examples of selection imposed by the physical environment. Fluctuations in environmental temperature can select on ectotherms in favor of behavioral thermoregulation and physiological adaptations. Changes in the chemistry of the atmosphere appear to have been the selective force behind modifications of photosynthesis in some flowering plants. Exposure to surf in the intertidal zone, in particular to the forces imposed by accelerating water, appears to have been responsible for the evolution of an unusual shape in a sea urchin.

We also reviewed examples of selection imposed by the biological environment. Predation by crabs has selected in favor of thicker shells in populations of a marine snail. Competition with conspecifics over territories and in advertising for mates has selected for bright breeding colors in a fish; competition with heterospecifics has selected for a change in those breeding colors in some populations. The benefits of mutualism have led to special adaptations in caterpillars that encourage tending by ants.

In many species, genetically identical individuals reared in different environments have different phenotypes. These species provide striking examples of a phenomenon called phenotypic plasticity. Phenotypic plasticity is adaptive when the phenotype elicited by a particular environment has higher fitness in that environment than other possible phenotypes. Phenotypic plasticity is itself a trait that can evolve. Phenotypic plasticity is a potential complication in any study that compares populations from different environments.

No species is perfectly adapted to all of the agents of selection it experiences. Among the factors that prevent optimal adaptation are trade-offs between one adaptation and another, constraints imposed by physiological processes, and a lack of genetic variation for adaptive evolution. Although natural selection often erodes genetic variation, certain patterns of selection tend to preserve genetic variation.

Questions

1. Geckos are unusual lizards in that they are active at night instead of during the day. Describe the difficulties a gecko would face in trying to use its behavior to regulate its temperature at night. Would you predict that geckos have an optimal temperature for sprinting that is the same, higher, or lower than that of a typical diurnal lizard? Huey et al. (1989) found that the geckos they studied had optimal temperatures that are the same as those of typical diurnal lizards (a finding in conflict with the researchers' own hypothesis). Can you think of an explanation?

2. Think about the costs and benefits of being a certain body size. For example, a mouse can easily survive a 30-foot fall. A human falling 30 feet would probably be injured, and an elephant falling 30 feet would probably be killed. Finally, a recent study of the bone strength of *Tyrannosaurus rex* revealed that if a fast-running *T. rex* ever tripped, it would probably die (Farlow, Smith, and Robinson 1995). Given these costs, why has large body size ever evolved? Can you think of some costs of small body size?

3. Milkweed plants have poisonous chemicals in their leaves that serve to defend the plants against many herbivores. Imagine a population of monarch caterpillars that has evolved a counter-defense, allowing its members to eat milkweed leaves and survive. How might natural selection act on this population, toward the solution of some other problem that caterpillars face?

4. Imagine you are an explorer who has just discovered two previously unknown large islands. Each island has a population of a species of shrub unknown elsewhere. On Island A, the shrubs have high concentrations of certain poisonous chemicals in their leaves. On Island B, the shrubs have nonpoisonous, edible leaves. The islands differ in many ways—for instance, Island A has less rainfall and a colder winter than Island B, and it has some plant-eating insects that are not found on Island B. Island A also has a large population of muntjacs, a small tropical deer that loves to eat shrubs. You suspect the muntjacs have been the selective force that has caused evolution of leaf toxins. How could you test this hypothesis? What alternative hypotheses can you think of? What data would disprove your hypothesis, and what data would disprove the other hypotheses?

5. P1 is a virus that infects bacteria. It often resides for long periods of time as a plasmid inside a bacterial cell, replicating and being transmitted to descendants of the original cell. Among the genes on P1 are two loci that form what is called an addiction module. One of the genes (called "doc," which stands for "death on cure") encodes a small protein that is both highly poisonous to bacterial cells, and stable. The other gene (called "phd," which stands for "prevent host death") encodes a protein

that serves as an antidote to the poison. This antidote is very unstable, because it is degraded by a bacterial protease. If a bacterial cell containing the P1 plasmid divides and produces a daughter cell that does not contain the plasmid, that daughter cell's protoplasm still contains the poison and the antidote. The antidote breaks down quickly, while the poison persists. The daughter cell then dies. Explain carefully why it is selectively advantageous for the P1 virus to carry these genes. (For more information on this example, see Lehnherr et al. 1993; Lehnherr and Yarmolinsky 1995).

6. A common challenge in paleontology is to identify the agent of selection that caused a new trait to spread through a lineage, many millions of years ago. One classic example is the evolution of high-crowned teeth in mammals. These teeth provide extra protection against abrasive particles in the diet, such as the silica particles in grassblades. High-crowned teeth have evolved independently in dozens of terrestrial mammals since the Miocene, including antelope, kangaroos, horses, and rabbits. Here are two hypotheses for the evolution of high-crowned teeth:

    1. It might be linked to the evolution of grasses.
    2. It might be linked to an increased amount of grit in the diet.

    Are these two hypotheses testable? What evidence would help distinguish them? (For more information on this example, see MacFadden 1997.)

7. Recently, there have been concerns about the possible adverse effects of cattle grazing on the grasslands of North America. Several studies on the arid and semi-arid grasslands of western North America have found that cattle grazing usually has negative effects. However, cattle grazing in the Midwest and Great Plains regions seems to have milder or even beneficial effects on the grasslands. Before the arrival of European settlers bison were apparently restricted to the Midwest (Balmford 1996). Can you suggest a hypothesis for why midwestern grasslands can tolerate domestic cattle, while grasslands further west apparently cannot? Is your hypothesis testable?

8. Consider skin color in humans. Does this trait show genetic variation? Phenotypic plasticity? Genotype-by-environment interaction? Give examples documenting each phenomenon. Could phenotypic plasticity for skin color evolve in human populations? How?

9. The examples of the sticklebacks (section 8.3) and *Begonias* (section 8.5) illustrated that organisms are frequently caught between opposing forces of selection. Each of the following examples also illustrates a tug-of-war between several forces of natural selection. For each example, hypothesize about what selective forces may maintain the trait described, and what selective forces may oppose it.

    A male moose grows new antlers, made of bone, each year.
    Douglas-fir trees often grow over 60 feet tall.
    A termite's gut is full of cellulose-digesting microorganisms.
    Maple trees lose all their leaves in the autumn.
    A male moth has huge antennae, which can detect female pheromones.
    A barnacle attaches itself permanently to a rock when it matures.

10. Schemske and Ågren (1995) used artificial flowers instead of real flowers in their experiment (section 8.5). What were the advantages of using artificial flowers? (There are at least two important ones.) What were the disadvantages?

11. The popular media often presents evolution as being a predictable process with a definite goal. For instance, in one "Star Trek: Voyager" episode, the captain instructs the ship's computer to extrapolate the "probable course" of evolution of hadrosaurs (a bipedal dinosaur), if hadrosaurs had been removed from Earth before the K-T extinction and allowed to evolve on another planet. What information about the hadrosaurs' new environment would have been useful for developing the best possi-

ble prediction? Given the information in this chapter and in Chapter 5 on genetic drift and agents of selection, is it theoretically possible to accurately predict the long-term course of evolution?

Exploring the Literature

12. We noted in the chapter that male sticklebacks sometimes steal eggs from other males' nests to rear as their own. Sievert Rohwer suggested that egg-stealing is a courtship strategy. Males often eat eggs out of their own nests, in effect robbing the reproductive investment made by their mates and using the proceeds to fund their own reproductive activities. Females, in consequence, should prefer to lay eggs in nests already containing the eggs of other females, thereby reducing the risk to their own. A female preference for nests already containing eggs would mean that males without eggs in their nests could increase their attractiveness by stealing eggs from other males. See:

> Rohwer, S. 1978. Parent cannibalism of offspring and egg raiding as a courtship strategy. *American Naturalist* 112:429–440.

> For a related phenomenon in birds, see:

> Gori, D.F., S. Rohwer, and J. Caselle. 1996. Accepting unrelated broods helps replacement male yellow-headed blackbirds attract females. *Behavioral Ecology* 7:49–54.

13. For a dramatic example of phenotypic plasticity in which an herbivorous insect uses the chemical defenses of its host as a cue for the development of defenses against its predators, see:

> Greene, Erick. 1989. A diet-induced developmental polymorphism in a caterpillar. *Science* 243:643–646.

14. Among the challenges faced by parasites is moving from one host to another. This challenge is a particularly potent agent of selection for parasites in which every individual must spend different parts of its life cycle in different hosts. What adaptations might you expect to find in parasites to facilitate dispersal from host to host? For dramatic examples in which parasites manipulate their hosts' behavior and/or appearance, see:

> Tierney, J.F., F.A. Huntingford, and D.W.T. Crompton. 1993. The relationship between infectivity of *Schistocephalus solidus* (Cestoda) and anti-predator behavior of its intermediate host, the three-spined stickleback, *Gasterosteus aculeatus*. *Animal Behavior* 46:603–605.

> Lafferty, K.D., and A.K. Morris. 1996. Altered behavior of parasitized killifish increases susceptibility to predation by bird final hosts. *Ecology* 77:1390–1397.

> Bakker, T.C.M., D. Mazzi, and S. Zala. 1997. Parasite-induced changes in behavior and color make *Gammarus pulex* more prone to fish predation. *Ecology* 78:1098–1104.

15. Brown-headed cowbirds *(Molothrus ater)* lay their eggs in other birds' nests, a behavior called nest parasitism. When this strategy succeeds, the host birds accept the cowbird egg as one of their own and rear the cowbird chick. When the strategy fails, the host birds recognize the cowbird egg as an imposter and eject it from the nest. Why do any host species accept cowbird eggs in their nests? Given the obvious costs involved in rearing a chick of another species, acceptance seems maladaptive. Evolutionary biologists have proposed two competing hypotheses to explain why some host species accept cowbird eggs. The evolutionary lag hypothesis posits that species that accept cowbird eggs do so simply because they have not yet evolved ejection behavior. Either the host species lack genetic variation that would allow them to evolve ejection behavior, or the host species have been exposed to cowbird nest parasitism only recently and there has not been sufficient time for such behavior to evolve. The evolutionary equilibrium hypothesis posits that host species that accept cowbird eggs do so because they face a fundamental mechanical constraint: Their bills are too small to allow them to grasp a cowbird egg, and if they tried to puncture the cowbird egg they would destroy too many of their own eggs in the process. Given this constraint, host species have evolved a strategy that makes the best of a bad situation. Think about how you would test each of the competing hypotheses. Then see:

Rohwer, S., and C.D. Spaw. 1988. Evolutionary lag versus bill-size constraints: A comparative study of the acceptance of cowbird eggs by old hosts. *Evolutionary Ecology* 1988:27−36.

Rohwer, S., C.D. Spaw, and E. Røskaft. 1989. Costs to northern orioles of puncture-ejecting parasitic cowbird eggs from their nests. *The Auk* 106:734−738.

Røskaft, E., S. Rohwer, and C.D. Spaw. 1993. Cost of puncture ejection compared with costs of rearing cowbird chicks for northern orioles. *Ornis Scandinavica* 24:28−32.

Sealy, S.G. 1996. Evolution of host defenses against brood parasitism: Implications of puncture-ejection by a small passerine. *The Auk* 113:346−355.

Given that some host species eject cowbird eggs by first puncturing the egg, then lifting it out of the nest, what adapatations would you expect to find in cowbird eggs? Would these adaptations carry any costs? See:

Spaw, C.D., and S. Rohwer. 1987. A comparative study of eggshell thickness in cowbirds and other passerines. *The Condor* 89:307−318.

Picman, J. 1997. Are cowbird eggs unusually strong from the inside? *The Auk* 114:66−73.

Citations

Ågren, J., and D. W. Schemske. 1991. Pollination by deceit in a Neotropical monoecious herb, *Begonia involucrata*. *Biotropica* 23:235−241.

Anderson, R.B., and D. Guravich. 1992. Bear baiting. *Natural History* 101:84−85.

Balmford, A. 1996. Extinction filters and current resilience: The significance of past selection pressures for conservation biology. *Trends in Ecology and Evolution* 11:193−196.

Bazzaz, F.A. 1990. The response of natural ecosystems to the rising global CO_2 levels. *Annual Review of Ecology and Systematics* 21:167−196.

Bazzaz, F.A., and R.W. Carlson. 1984. The response of plants to elevated CO_2. I. Competition among an assemblage of annuals at two levels of soil moisture. *Oecologia* 62:196−198.

Bennett, A.F., and R.E. Lenski. 1993. Evolutionary adaptation to temperature. II. Thermal niches of experimental lines of *Escherichia coli*. *Evolution* 47:1−12.

Bennett, A.F., R.E. Lenski, and J.E. Mittler. 1992. Evolutionary adaptation to temperature. I. Fitness responses of *Escherichia coli* to changes in its thermal environment. *Evolution* 46:16−30.

Boulding, E.G., and T.K. Hay. 1993. Quantitative genetics of shell form of an intertidal snail: constraints on short-term response to selection. *Evolution* 47:576−592.

Cerling, T.E., Y. Wang, and J. Quade. 1993. Expansion of C4 ecosystems as an indicator of global ecological change in the late Miocene. *Nature* 361:344−345.

Daniel, T.L. 1984. Unsteady aspects of aquatic locomotion. *American Zoologist* 24:121−134.

Dawkins, R. 1987. *The Blind Watchmaker* (New York: W. W. Norton & Company).

Delph, L.F., and C.M. Lively. 1985. Pollinator visits to floral colour phases of *Fuchsia excorticata*. *New Zealand Journal of Zoology* 12:599−603.

Delph, L.F., and C. M. Lively. 1989. The evolution of floral color change: Pollinator attraction versus physiological constraints in *Fuchsia excorticata. Evolution* 43:1252–1262.

De Meester, L. 1996. Evolutionary potential and local genetic differentiation in a phenotypically plastic trait of a cyclical parthenogen, *Daphnia magna. Evolution* 50:1293–1298.

Denny, M., and B. Gaylord. 1996. Why the urchin lost its spines: Hydrodynamic forces and survivorship in three echinoids. *Journal of Experimental Biology* 199:717–729.

Ehleringer, J.R., R.F. Sage, L.B. Flanagan, and R.W. Pearcy. 1991. Climate change and the evolution of C4 photosynthesis. *Trends in Ecology and Evolution* 6:95–99.

Elgar, M.A., and N.E. Pierce. 1988. Mating success and fecundity in an ant-tended lycaenid butterfly. In T.H. Clutton-Brock, ed. *Reproductive Success* (Chicago: University of Chicago Press), 59–75.

Farlow, J.D., M.B. Smith, and J.M. Robinson. 1995. Body mass, bone "strength indicator," and cursorial potential of *Tyrannosaurus rex. Journal of Vertebrate Paleontology* 15:713–725.

FitzGerald, G.J. 1993. The reproductive behavior of the stickleback. *Scientific American* (April):80–85.

Funk, D.J., D.J. Futuyma, G. Ortí, and A. Meyer. 1995. A history of host associations and evolutionary diversification for *Ophraella* (Coleoptera: Chrysomelidae): New evidence from mitochondrial DNA. *Evolution* 49:1008–1017.

Furgal, C.M., S. Innes, and K.M. Kovacs. 1996. Characteristics of ringed seal, *Phoca hispida,* subnivean structures and breeding habitat and their effects on predation. *Canadian Journal of Zoology* 74:858–874.

Futuyma, D.J., M.C. Keese, and D.J. Funk. 1995. Genetic constraints on macroevolution: The evolution of host affiliation in the leaf beetle genus *Ophraella. Evolution* 49:797–809.

Gallien, W.B. 1986. A comparison of hydrodynamic forces on two sympatric sea urchins: Implications of morphology and habitat. M.Sc. thesis, University of Hawaii.

Gilbert, J.J. 1966. Rotifer ecology and embryological induction. *Science* 151:1234–1237.

Gilbert, J.J. 1980. Further observations on developmental polymorphism and its evolution in the rotifer *Brachinonus calyciflorus. Freshwater Biology* 10:281–294.

Gould, S.J., and R.C. Lewontin. 1979. The spandrels of San Marco and the Panglossian paradigm: A critique of the adaptationist programme. *Proceedings of the Royal Society of London,* Series B 205:581–598.

Grojean, R.E., J.A. Sousa, and M.C. Henry. 1980. Utilization of solar radiation by polar animals: An optical model for pelts. *Applied Optics* 19:339–346.

Hagen, D.W., and G.E.E. Moodie. 1979. Polymorphism for

breeding colors in *Gasterosteus aculeatus.* I. Their genetics and geographic distribution. *Evolution* 33:641–648.

Hagen, D.W., G.E.E. Moodie, and P.F. Moodie. 1980. Polymorphism for breeding colors in *Gasterosteus aculeatus.* II. Reproductive success as a result of convergence for threat display. *Evolution* 34:1050–1059.

Hochachka, P.W., and G.N. Somero. 1984. *Biochemical adaptation* (Princeton, NJ: Princeton University Press).

Huey, R.B., P.H. Niewiarowski, J. Kaufmann, and J.C. Herron. 1989. Thermal biology of nocturnal ectotherms: Is sprint performance of geckos maximal at low body temperatures? *Physiological Zoology* 62:488–504.

Huey, R.B., and J.G. Kingsolver. 1989. Evolution of thermal sensitivity of ectotherm performance. *Trends in Ecology and Evolution* 4:131–135.

Huey, R.B., C.R. Peterson, S.J. Arnold, and W.P. Porter. 1989. Hot rocks and not-so-hot rocks: Retreat-site selection by garter snakes and its thermal consequences. *Ecology* 70:931–944.

Huey, R.B., and J.G. Kingsolver. 1993. Evolution of resistance to high temperature in ectotherms. *American Naturalist* 142:S21–S46.

Johnson, T.P., and A.F. Bennett. 1995. The thermal acclimation of burst escape performance in fish: An integrated study of molecular and cellular physiology and organismal performance. *Journal of Experimental Biology* 198:2165–2175.

Legvine, D.M., and N.A. Oritsland. 1974. Black polar bears. *Nature* 251:218–219.

Lehnherr, H., E. Maguin, S. Jafri, and M.B. Yarmolinsky. 1993. Plasmid addiction genes of bacteriophage P1: doc, which causes cell death on curing of prophage, and phd, which prevents host death when prophage is retained. *Journal of Molecular Biology* 233:414–428.

Lehnherr, H., and M.B. Yarmolinsky. 1995. Addiction protein Phd of plasmid prophage P1 is a substrate of the ClpXP serine protease of *Escherichia coli. Proceedings of the National Academy of Sciences, USA* 92:3274–3277.

MacFadden, B.J. 1997. Origin and evolution of the grazing guild in New World terrestrial mammals. *Trends in Ecology and Evolution* 12:182–187.

Mahan, B. H. 1975. *University Chemistry,* 3rd Ed. (Reading, MA: Addison-Wesley Publishing Company).

Mayr, E. 1983. How to carry out the adaptationist program? *American Naturalist* 121:324–334.

Milinski, M., and T.C.M. Bakker. 1990. Female sticklebacks use male coloration in mate choice and hence avoid parasitized males. *Nature* 344:330–333.

Mori, S. 1995. Factors associated with and fitness effects of nest-raiding in the three-spined stickleback, *Gasterosteus aculeatus,* in a natural situation. *Behavior* 132:1011–1023.

Pierce, N.E. 1984. Amplified species diversity: A case study of an Australian lycaenid butterfly and its attendant ants. In R.I. Vane-Wright and P.R. Ackery, eds. *Biology of Butterflies. XI Symposium of the Royal Entomological Society of London* (London: Academic Press), 197–200.

Pierce, N.E., R.L. Kitching, R.C. Buckley, M.F.J. Taylor, and K.F. Benbow. 1987. The costs and benefits of cooperation between the Australian lycaenid butterfly, *Jalmenus evagoras,* and its attendant ants. *Behavioral Ecology and Sociobiology* 21:237–248.

Quade, J., T.E. Cerling, and J.R. Bowman. 1989. Development of Asian monsoon revealed by marked ecological shift during the latest Miocene in northern Pakistan. *Nature* 342:163–166.

Schemske, D.W., and J. Ågren. 1995. Deceit pollination and selection on female flower size in *Begonia involucrata:* An experimental approach. *Evolution* 49:207–214.

Seeley, R.H. 1986. Intense natural selection caused a rapid morphological transition in a living marine snail. *Proceedings of the National Academy of Sciences, USA* 83:6897–6901.

Smith, T.G. 1980. Polar bear predation of ringed and bearded seals in the land-fast sea ice habitat. *Canadian Journal of Zoology* 58:2201–2209.

Stewart, W.N. 1983. *Paleobotany and the Evolution of Plants* (Cambridge University Press).

Stirling, I. 1974. Midsummer observations on the behavior of wild polar bears *(Ursus maritimus). Canadian Journal of Zoology* 52:1191–1198.

Stirling, I., and N.A. Oritsland. 1995. Relationships between estimates of ringed seal *(Phoca hispida)* and polar bear *(Ursus maritimus)* populations in the Canadian Arctic. *Canadian Journal of Fisheries and Aquatic Sciences* 52:2594–2612.

Thomasson, J.R., M.E. Nelson, and R.J. Zakrzewski. 1986. A fossil grass (Gramineae: Chloridoideae) from the Miocene with Kranz anatomy. *Science* 233:876–878.

Trussell, G.C. 1996. Phenotypic plasticity in an intertidal snail: The role of a common crab predator. *Evolution* 50:448–454.

Vogel, S. 1981. *Life in Moving Fluids* (Boston: Willard Grant Press).

Mechanisms of Speciation

All organisms alive today trace their ancestry back through time to the origin of life some 3.8 billion years ago. Between then and now, millions—if not billions—of branching events have occurred as species split and diverged. In this chapter, we examine how these branching events happened.

The material we review offers our first look at the mechanisms responsible for generating biodiversity. In Chapters 5 and 6 we investigated how mutation, natural selection, migration, and drift act to change allele frequencies within populations; now we ask how these four processes can lead to genetic differences between populations.

In addition to providing a foundation for studying the history of life, studying speciation has important practical applications. The genetic isolation of populations is the essence of the speciation process, and much of the material we explore focuses on the extent and causes of gene flow (or lack of gene flow) among differentiated groups of organisms. Understanding these topics is fundamental to creating effective strategies for preserving biodiversity, and to managing genetically engineered organisms that are released into the environment.

Along with considering these applied issues and the general problem of how evolutionary processes can isolate populations, we need to examine two additional questions: What happens when recently diverged populations come into contact and interbreed? and What genetic changes take place during differentiation? To begin our analysis we start with the field's most fundamental question: What is a species?

9.1 Species Concepts

All human cultures recognize different types of organisms in nature and name them. These taxonomic systems are based on judgments about the degrees of similarity among organisms. People intuitively group like with like. The challenge to biologists has been to move beyond these informal judgments to create a definition of species that is mechanistic and testable, and a classification system for life forms that accurately reflects evolutionary history. This has been difficult to do. In the past 30 years alone there have been at least half a dozen species concepts proposed, recurrent controversies about which definition is best, and even philosophical debates about whether the unit we call the species actually exists in nature or whether it is merely a linguistic and cultural construct (Mayr 1963; Donoghue 1985; Paterson 1985; McKitrick and Zink 1988; Cracraft 1989; Templeton 1989).

In this section we describe the pros and cons of three major species concepts and review how they are applied. Although they differ in detail, all species definitions agree on three essential points:

1. Species consist of groups of interbreeding populations.

2. Species are a fundamental unit of evolution.

Species are groups of interbreeding populations that evolve independently.

3. Species share a distinguishing characteristic, which is evolutionary independence. Evolutionary independence occurs when mutation, selection, migration, and drift operate on each species separately. This means that species form a boundary for the spread of alleles. Consequently, different species follow independent evolutionary trajectories.

The differences among species concepts center on the problem of establishing practical criteria for identifying evolutionary independence. This is a challenge, because the data available to define species varies between sexual and asexual organisms and between fossil and extant groups. The following three examples illustrate this point.

The Biological Species Concept

The biological species concept (BSC) has been the textbook definition of a species since Ernst Mayr proposed it in 1942. It is used in practice by many zoologists and is the legal definition employed in the Endangered Species Act, which is the flagship biodiversity legislation in the United States. Under the BSC, the criterion for identifying evolutionary independence is reproductive isolation. Specifically, if populations of organisms do not hybridize, or fail to produce fertile offspring when they do, then they are reproductively isolated and considered good species.

Reproductive isolation is clearly an appropriate criterion for identifying species because it confirms lack of gene flow. This is the litmus test of evolutionary independence. But although the BSC is compelling in concept and useful in some situations, it can be difficult to apply. For example, if nearby populations do not actually overlap, we have no way of knowing whether they are reproductively isolated. Instead, biologists have to make subjective judgments to the effect that, "If these populations were to meet in the future, we believe that they are divergent enough already that they would not interbreed, so we will name them different species." In these cases species designations cannot be tested with data. Further, the biological species concept can never be tested in fossil forms, is irrelevant to asexual populations, and is difficult to apply in the many plant groups where hybridization between strongly divergent populations is routine.

The Phylogenetic Species Concept

Systematists are the people responsible for classifying the diversity of life, and a growing number are promoting an alternative to the BSC called the phylogenetic or evolutionary species concept. This approach focuses on a criterion called monophyly. Monophyletic groups are defined as populations or suites of populations that contain all the known descendants of a single common ancestor (Figure 9.1). Under the phylogenetic species concept (PSC), species are defined as the smallest diagnosable monophyletic group. They are identified by estimating the phylogeny of closely related populations. Any population that forms an independent branch on the phylogeny is considered a species.

The rationale behind the PSC is that traits can become diagnostic, and thus produce independent branches on the phylogeny, only if populations have diverged in isolation from one another. Put another way, to be called separate species under the PSC, populations must have been evolutionarily independent long enough for diagnostic traits to emerge. The appeal of this approach is that it is testable: Species are named on the basis of statistically significant differences in the traits used to estimate the phylogeny.

The problem comes with putting this criterion into practice. Carefully constructed phylogenies are available for only a handful of groups thus far. Also, many biologists object to the idea that a "species-specific trait" can be anything that distinguishes populations in a phylogenetic context. Such traits can be as trivial as a silent site substitution in DNA that is fixed in one population but not in another, or a slight but measurable and statistically significant increase in hairiness on the underside of leaves in different populations.

Estimates vary, but the general opinion is that instituting the phylogenetic species concept could easily double the number of named species. Proponents

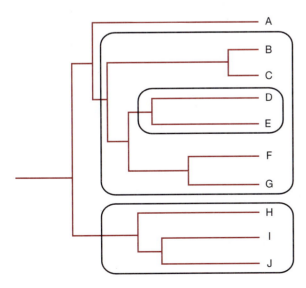

Figure 9.1 **Monophyletic groups** The subgroups circled on this phylogenetic tree are all monophyletic. There are many other monophyletic subgroups in this clade besides the ones that are circled.

of the PSC are not bothered by that prospect. They respond by saying that if this increase did occur, it would merely reflect biological reality.

The Morphospecies Concept

Paleontologists define species on the basis of morphological differences among fossils. When rigorous tests of reproductive isolation or well-estimated phylogenies are lacking, as they usually are, botanists and zoologists working on extant species do the same. The great advantage of the morphospecies concept is that it is so widely applicable. But when it is not applied carefully, species definitions can become arbitrary and idiosyncratic. In the worst-case scenario, species designations made by different investigators are not comparable.

Paleontologists have to work around other restrictions when identifying species. Fossil species that differed in color or the anatomy of soft tissues cannot be distinguished. Neither can populations that are not distinguishable in morphology but were strongly divergent in traits like songs, temperature or drought tolerance, habitat use, or courtship displays. These are called cryptic or sister species.

Given these limitations, is the morphospecies concept still useful? Specifically, are the species we identify in the fossil record analogous to those we recognize today, using much larger suites of characters?

Jeremy Jackson and Alan Cheetham (1990, 1994) have performed the most careful analysis to date of the morphological species concept. Jackson and Cheetham study speciation in fossil forms of the colonial, marine-dwelling invertebrates called cheilostome Bryozoa. Accordingly, they set out to check a critical assumption in their work: that the fossil morphospecies they have named conform to genetically differentiated species of bryozoans living today (Figure 9.2).

Jackson and Cheetham's first task was to establish that the skeletal measurements used to distinguish fossil morphospecies have a genetic basis. It is possible that these skeletal characteristics vary among bryozoan species simply because of environmental differences. To test this hypothesis, the researchers collected embryos from several extant cheilostomes and raised them in the same environment: a shallow-water habitat in the Caribbean. When they measured skeletal characters in the full-grown individuals, all but nine of the 507 offspring they raised were assigned to the correct morphospecies (that of their parents). This confirmed that the species-specific characters have a genetic basis. Then, in the critical experiment, Jackson and Cheetham surveyed variation in proteins isolated from eight species in three different genera. They found unique types of proteins in each of the named morphospecies. This result indicates that strong genetic divergence has occurred, and that no gene flow is occurring between the populations. Jackson and Cheetham interpreted these experiments as proof that morphospecies, at least in Bryozoa, correspond to independent evolutionary units.

Applying Species Concepts: The Case of the Red Wolf

Although it is probably unrealistic to insist on a single, all-purpose criterion for identifying species (Endler 1989), the major species concepts that have been

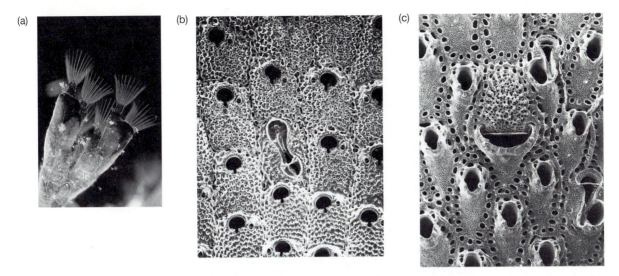

Figure 9.2 **Morphological characters used to distinguish morphospecies in cheilostome Bryozoa** Living colony members are shown in photo (a). (Kjell B. Sandved/Butterfly Alphabet, Inc.) The enlarged photos show the skeletons of *Stylopoma spongites* (b) and *Metrarabdotos tenue* (c). Each colony member occupies one of the long chambers that has an orifice near the top. To distinguish morphospecies in these genera, Jackson and Cheetham (1994) measured traits like orifice length and width and the number of pores present per 0.2mm square. (b, c) Reproduced by permission from J.B.C. Jackson and A.H. Cheetham, *Paleobiology* 20(4):407–423 (1994.)

proposed are productive when applied in appropriate situations. Consider, for example, how different species definitions have informed the controversy over the red wolf (Wayne and Gittleman 1995).

The red wolf, *Canis rufus,* is a member of the dog family native to the southeastern United States (Figure 9.3). Due to widespread hunting and clearing of forest habitats, its population had dwindled to a mere handful of individuals by the early 1970s. Many of the animals that were left showed characteristics typical of coyotes *(Canis latrans),* which had become abundant in the wolf's range. This indicated that wolves and coyotes were hybridizing extensively. Fortunately, biologists from the U.S. Fish and Wildlife Service were able to capture 14 red wolves that apparently had no coyote traits before the population went extinct in the wild. These animals bred readily in captivity, and now several hundred red wolves await reintroduction to protected habitats in the wild.

Under the BSC, however, the extensive hybridization with coyotes made the species status of the red wolf questionable. Are red wolves reproductively isolated, evolutionarily independent units? Or are they actually a population of hybrids between the gray wolf *(Canis lupus)* and coyote? Ronald Nowak (1979, 1992) studied a large series of skull and dental characters and showed that red wolves collected before 1930 were a clearly identifiable morphospecies. These individuals had characteristics intermediate between gray wolves and coyotes. Because his data showed that animals collected after 1930 were much more similar to coyotes, Nowak suggested that hybridization was a recent phenomenon caused by the inability of wolves to find mates when their population dwindled. His conclusion was that red wolves clearly qualify as a species.

Genetic studies using DNA extracted from the captive population and from wolf pelts collected before 1930 told a different story, however. Surveys of

HISTORICAL

MODERN

☐Gray wolf ☐Coyote ■Red wolf

***Figure 9.3* The red wolf**
The maps show the distribution of gray and red wolves and coyotes inferred from historical records versus current data. Note that the range of coyotes has expanded while the ranges of gray and red wolves have shrunk. (Photo by Tom McHugh/National Audubon Society (Photo Researchers, Inc.)

mitochondrial DNA variation and alleles at 10 microsatellite loci showed no diagnostic (that is, no species-specific) differences between red wolves and coyotes (Wayne and Jenks 1991; Roy et al. 1994). Instead, the genetic data strongly supported the hypothesis that red wolves are a hybrid between gray wolves and coyotes, and have no unique genetic characteristics of their own. This analysis implies that morphological data are simply not informative in this case, and that the red wolf's intermediate characteristics are the result of hybridization and not independent evolution. Under the PSC as well as the BSC, red wolves are not a distinct species.

The fate of the red wolf remains to be decided. The genetic research suggests that other species may be a higher priority for public monies committed to preserving biodiversity, and that the reintroduction program's greatest value may lie in bringing a top predator back into the ecosystem of North America's southern woodlands. Employing several criteria for identifying species can be a productive approach in clarifying conservation and evolutionary issues.

9.2 Mechanisms of Isolation

Having explored different criteria for identifying species, we now consider how species form. Speciation is often studied as a three-stage process: an initial step that creates a physical separation between populations, a second step that results in divergence in mating tactics or habitat use, and a final step that produces reproductive isolation. In this section we consider how physical separation can reduce gene flow and initiate a speciation event. In section 9.3 we ask how the remaining three evolutionary processes (genetic drift, natural selection, and mutation) can cause populations to diverge, and in section 9.4 we consider how natural selection can complete the speciation process by causing reproductive isolation.

Migration

In Chapter 5 we introduced migration as gene flow between populations, and developed models showing that migration tends to homogenize gene frequencies and reduce the differentiation of populations. Using the example of water snakes from mainland and island habitats in Lake Erie, we also introduced the idea of a balance between migration and natural selection. Recall that experiments had shown a selective advantage for unbanded snakes on island habitats. But because migration of banded forms from the mainland occurs regularly, and because banded and unbanded forms subsequently interbreed, the island populations did not completely diverge from mainland forms. Migration continually introduced alleles for bandedness, even though selection tended to eliminate them from the island populations.

Now consider a thought experiment: What would happen if changes in shoreline habitats or lake currents effectively stopped the migration of banded forms from the mainland to the islands? The island populations would then be isolated from the mainland population. Gene flow would stop and the migration-selection balance would tip. The island population would be free to differentiate as a consequence of mutation, natural selection, and drift. These forces would act on them independently of the forces acting on mainland forms.

The speciation process begins when gene flow is disrupted and populations become genetically isolated.

This scenario illustrates a classical theory for how speciation begins, called the allopatric model. This theory was developed by Ernst Mayr (1942, 1963). One of Mayr's most important hypotheses is that speciation is especially likely to occur in small populations that become isolated on the periphery of a species' range, such as the island forms of water snakes. Population genetic models have shown that speciation in peripheral populations can occur rapidly when selection for divergence is strong and gene flow is low (García-Ramos and Kirkpatrick 1997).

Physical isolation is obviously an effective barrier to gene flow, and undoubtedly the most important trigger for the second stage in the speciation process: genetic and ecological divergence. Geographic isolation can come about through dispersal and colonization of new habitats or through vicariance events, where an existing range is split by a new physical barrier (Figure 9.4). Dispersal and vicariance are thus responsible for the first step in speciation.

Geographic isolation produces genetic isolation.

Geographic Isolation through Dispersal and Colonization

One of the most spectacular radiations in the class Insecta is also a superb example of geographic isolation through dispersal. The Hawaiian drosophilids, close relatives of the fruit flies we have encountered before, number over 480 named species in two genera and an estimated 350 species yet to be formally described and named (Kaneshiro and Boake 1987). The ecological diversification in the group is almost unprecedented: Hawaiian flies can be found from sea level to montane habitats and from dry scrub to rainforests. Food sources, especially the

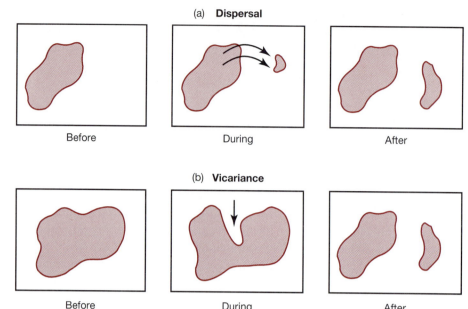

Figure 9.4 **Isolation by dispersal and vicariance**
In the diagram of dispersal (a), the arrows indicate movement of individuals. In the diagram of vicariance (b), the arrows indicate an encroaching physical feature like a river, glacier, lava flow, or new habitat.

plant material used as the medium for egg laying and larval development, also vary widely from species to species. One of the Hawaiian flies even lays its eggs in spiders, while another has aquatic larvae. In addition, many species have elaborate traits, like patterning on their wings or modified head shapes, that are used in courtship (Figure 9.5).

How did this enormous diversity come to be? The leading explanation is called the founder hypothesis. Many of the Hawaiian flies are island endemics, meaning that their range is restricted to a single island in the archipelago. The founder hypothesis maintains that this endemism resulted when small populations of flies, or perhaps even a single gravid female, would occasionally island-hop. The new population would be cut off from the ancestral species. Divergence would begin, resulting from drift and selection on the genes involved in courtship displays and habitat use.

Populations can become geographically isolated and through dispersal and colonization events . . .

The logic of the founder hypothesis is compelling, but do we have evidence, other than endemism, that these events actually occurred? Because the geology of the Hawaiian islands is well known, the hypothesis makes a strong prediction about speciation patterns in flies. The Hawaiian islands are produced by a volcanic hotspot under the Pacific Ocean. The hotspot is stationary but periodically spews magma up and out onto the Pacific plate. After islands form, continental drift carries them to the north and west. As time passes, the volcanic cones gradually erode down to atolls and submarine mountains.

The founder hypothesis makes two predictions based on these facts: (1) closely related species should almost always be found on adjacent islands, and (2) at least some sequences of branching events should correspond to the sequence in which islands were formed. Rob DeSalle and Val Giddings (1986) used sequence differences in mitochondrial DNA to estimate the phylogeny of four closely related species, and found exactly these patterns (Figure 9.6). The most recent species are found on the youngest islands, and several of the branching

Figure 9.5 **Hawaiian Drosophila** As these photos of *Drosophila nigribasis* (left) and *D. macrothrix* (right) show, Hawaiian *Drosophila* are remarkably diverse in body size, wing coloration, and other traits. (Kenneth Y. Kaneshiro, University of Hawaii)

events correspond to the order of island formation. This is strong evidence that dispersal to new habitats can trigger speciation.

The founder hypothesis is actually identical to the model we developed in Chapter 2 when we discussed speciation events in Darwin's finches. Each speciation event in these birds is thought to have begun when a flock of individuals colonized an island and became physically separated from the source population (Lack 1942; Grant and Grant 1996). This mechanism of physical isolation is not just important on oceanic islands, however. Hot springs, deep sea vents, fens, bogs, caves, mountaintops, and lakes or ponds with restricted drainage also represent habitat islands. Dispersal to novel environments is a general mechanism for initiating speciation.

Geographic Isolation through Vicariance

Vicariance events split a species' distribution into two or more isolated ranges and discourage or prevent gene flow between them. There are many possible mechanisms of vicariance, ranging from slow processes like the rise of a mountain range or a long-term drying trend that fragments forests, to rapid events like a mile-wide lava flow that bisects a snail population.

Nancy Knowlton and colleagues at the Smithsonian Tropical Research Institute in Panama have been studying a classical vicariance event: the recent separation of marine organisms on either side of Central America. We know from geological evidence that the Isthmus of Panama closed (and the land bridge between South and North America opened) about 3 million years ago. Is this enough time for speciation to occur?

. . . or vicariant events, where a physical barrier splits a species' range.

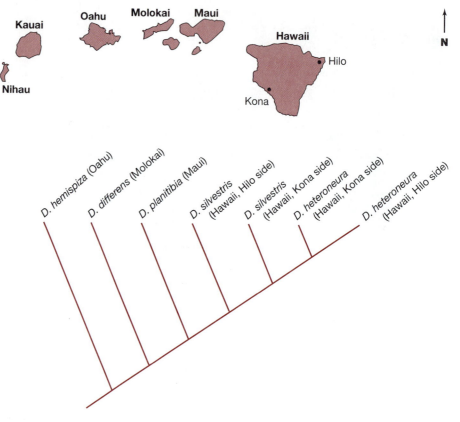

Figure 9.6 Phylogeny of some Hawaiian *Drosophila*   We know that *Drosophila silvestris, D. heteroneura, D. planitibia,* and *D. differens* are a closely related group because the size, number, and markings of their chromosomes are indistinguishable. In contrast, the chromosomes of most other Hawaiian flies are clearly distinguishable, either because they contain chromosomal rearrangements or because they have unique banding patterns visible in the enlarged chromosomes present in their salivary glands.

The phylogeny shown here, estimated from data on sequence divergence in mitochondrial DNA (DeSalle and Giddings 1986), is consistent with the hypothesis that at least some of the speciation events in this group were the result of island hopping. For phylogenetic analyses of this and other Hawaiian fly groups, see Beverly and Wilson (1985), DeSalle et al. (1986), and DeSalle and Hunt (1987).

Knowlton et al. (1993) recently looked at a series of snapping shrimp *(Altheus)* populations from either side of the isthmus (Figure 9.7a). The populations they sampled appeared to represent seven pairs of closely related morphospecies, with one member of each pair found on each side of the land bridge. The phylogeny of these shrimp, estimated from differences in their mitochondrial DNA sequences, confirms this hypothesis (Figure 9.7b). The species pairs from either side of the isthmus, reputed to be sisters on the basis of morphology, are indeed each others' closest relatives. This is consistent with the prediction from the vicariance hypothesis.

Further, when Knowlton et al. put males and females of various species pairs together in aquaria and watched for aggressive or pairing interactions, the researchers found a strong correlation between the degree of genetic distance between species pairs and how interested the shrimp were in courting. Males

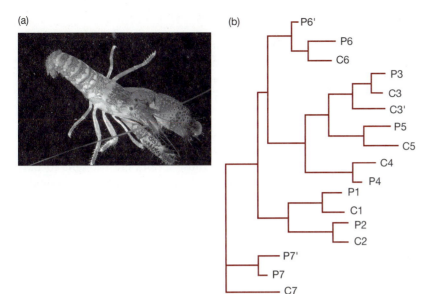

Figure 9.7 Snapping shrimp (a) This is *Alpheus malleator*, found on the Pacific side of the Panamanian isthmus. (Carl C. Hansen/Nancy Knowlton/Smithsonian Tropical Research Institute) (b) This tree was estimated from sequence divergence in mitochondrial DNA (Knowlton et al. 1993). The P or C designations with each number refer to whether that species is found in the Pacific or Caribbean. The prime (′) marks after some letters indicate cryptic species distinguished by sequence differences. In every case, the putative sister species are indeed each others' closest relative.

and females from species with greater genetic divergence, indicative of longer isolation times, were less interested in one another. Finally, almost none of the pairs that formed during the courtship experiments produced fertile clutches. This last observation confirms that the Pacific and Caribbean populations are indeed separate species under all three of the definitions we have reviewed.

One of the most interesting aspects of the study, though, was that the data contradicted a prediction made by the vicariance hypothesis. If the land bridge had formed rapidly, we would expect that genetic distances and degrees of reproductive isolation would be identical in all seven species pairs. This is not the case. For example, DNA sequence divergence between species pairs varied from about 6.5% to over 19%. What is going on? Because Knowlton and colleagues do not believe that the land bridge popped up all at once, a prediction of identical divergence would be naive. Instead, as the land rose and the ocean gradually split and retreated on either side, different shrimp populations would become isolated in a staggered fashion, depending on the depth of water each species occupied and how efficiently their larvae dispersed. The ranges of deeper-water species, or those with less-motile larvae, would be cut in two first. Consistent with this idea, the degree of genetic divergence in four of the seven species pairs is very similar. These are the lowest values observed, perhaps indicating "the final break" between the two oceans.

9.3 Mechanisms of Divergence

Dispersal and vicariance events that lead to the physical isolation of populations only create the conditions for speciation. For the process to continue, the remaining three evolutionary forces have to create divergence in the separated populations. In this section we review how genetic drift, natural selection, and mutation act on closely related populations, once gene flow between them has been reduced or eliminated.

Genetic Drift

In Chapter 5 we introduced population genetic models that quantified the major effects of genetic drift within populations: random fixation of alleles and random loss of alleles. We also reviewed data from a founder event observed by Peter and Rosemary Grant during a study of Darwin's finches in the Galápagos. Their measurements of body size in a flock of colonizing birds confirmed that the founding population was a nonrandom sample of the source population.

Drift can produce rapid genetic divergence in small, isolated populations.

Because genetic drift is a sampling process, its effects are most pronounced in small populations. This is important because most species are thought to have originated with low population sizes. Normally only tiny numbers of individuals are involved in colonization events, and peripheral populations tend to be small. As a consequence, genetic drift has long been hypothesized as the key to speciation's second stage.

A variety of genetic models have examined how drift might lead to rapid genetic differentiation in small populations (for reviews, see Carson and Templeton 1984; Barton 1989; Templeton 1996). An important result from this work is that drift's impact on founding populations is not simply through a loss of overall genetic diversity. In fact, when a population is reduced to a small size for a short period of time—the phenomenon known as bottlenecking—only very rare alleles tend to be lost due to drift (Lande 1980, 1981). For genetic diversity to be reduced dramatically, the founding population has to be extremely small.

In contrast, quantitative genetic models show that heritable variation in a population can actually *increase* as a result of bottlenecking (Goodnight 1987, 1988). This result, which is counterintuitive, has been confirmed in laboratory experiments with houseflies (Bryant et al. 1986; Bryant and Meffert 1993).

To understand how this increase in genetic variance can occur, recall from Chapters 2 and 6 that genes have both additive and epistatic, or interaction, effects. Additive effects represent gene action that is independent of the action of other alleles or loci, while epistatic effects result from a gene's interaction with alleles at other loci. In Chapter 6 we largely limited our discussion of V_G—the variation among individuals due to variation in their genes—to just the additive component. This is the component of genetic variation that is used to estimate the heritability of a trait and predict the response to selection.

During a bottleneck, however, part of the epistatic component of genetic variance is converted to additive variance. This happens because of the reduction in the number of alleles present. As an example, consider a locus A and locus B whose protein products interact to affect the phenotype. Locus A could code for a hormone and B for its receptor. In a large population, an allele at locus A is normally expressed in combination with many different alleles at locus B. At least part of the effect that locus A has in creating genetic variation depends on epistatic interactions with the different alleles at B. But in a small population, the number of alleles at B are reduced. In this population, more of A's action is due to additive effects and less to epistatic effects. This conversion of epistatic to additive variance increases the heritability of traits in small populations and consequently their response to selection.

To summarize, small populations that become isolated are almost certain to start off as a nonrandom sample of the ancestral population. Other effects—including random loss of alleles, random fixation of existing and new alleles, and the conversion of epistatic to additive variance—could then produce rapid divergence in the isolated population. (Note that these effects occur with equal efficiency in populations that have been reduced to small size by overhunting or habitat loss; see Chapter 19.)

The role of drift in speciation events has been controversial, however. Peter and Rosemary Grant (1996) point out that hundreds of small populations have been introduced to new habitats around the world in the last 150 years, due to the action of humans; but that few, if any, dramatic changes in genotypes have resulted because of genetic drift. Although genetic drift once dominated discussions of speciation mechanisms, most evolutionary biologists now take a more balanced view. Natural selection has also been shown to be an important force promoting the divergence of isolated populations.

Natural Selection

Marked genetic differences have to emerge between closely related populations for speciation to proceed beyond the first stage. Drift almost always plays a role when at least one of the populations is small. But natural selection can also lead to divergence if one of the populations occupies a novel environment.

Selection's role in the second stage of speciation is clearly illustrated by recent research on apple and hawthorn flies. These closely related insect populations are diverging because of natural selection on preferences for a crucial resource: food.

The apple maggot fly, *Rhagoletis pomonella,* is found throughout the northeastern and north-central United States (Figure 9.8). The species is a major agricultural pest, causing millions of dollars of damage to apple crops each year. The flies also parasitize the fruits of trees in the hawthorn group (species of *Crataegus*), which are closely related to apples.

Male and female *Rhagoletis* identify their host trees by sight, touch, and smell. Courtship and mating occur on or near the fruits. Females lay eggs in the fruit while it is still on the tree; the eggs hatch within two days and then develop through three larval stages in the same fruit. This takes about a month. After the fruit falls to the ground, the larvae leave and burrow a few inches into the soil. There they pupate after 3–4 days and spend the winter. Most leave this resting stage and emerge as adults the next summer, starting the cycle anew.

Apple trees clearly represent a novel food source for *Rhagoletis.* Hawthorn trees and *Rhagoletis* are native to North America, but apple trees were introduced from Europe less than 300 years ago. Our question is this: Are the flies that parasitize apples and hawthorns distinct populations? This hypothesis implies that natural selection, based on a preference for different food sources, has created two distinct races of flies. The contrasting hypothesis is that flies parasitizing hawthorns and apples are members of the same population. This hypothesis predicts that flies on hawthorns and apples interbreed freely, and that selection for exploiting different hosts has not occurred.

The hypothesis of no differentiation actually appears much more likely, because the first stage in speciation has not occurred. The two host trees and fly populations occur together throughout their ranges. Far from being isolated, at some sites hawthorn and apple trees are almost in physical contact. Marked flies

(a)

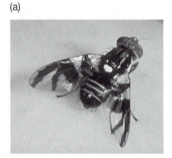

(b)

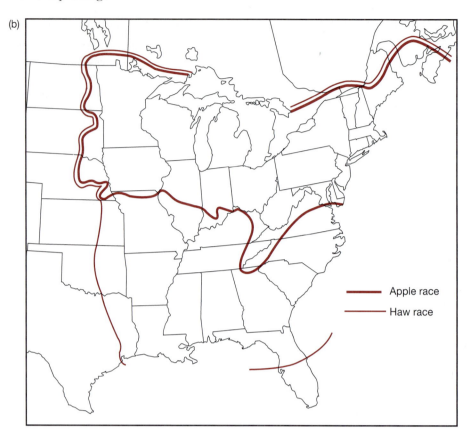

Apple race

Haw race

Figure 9.8 **Apple and hawthorn maggot flies**
(a) Apple and hawthorn races of *Rhagoletis pomonella* are indistinguishable to the eye. This is an apple fly. (Jeff Feder, University of Notre Dame)   (b) The two races are sympatric, or co-occurring, over much of their range. Modified from Feder et al. (1990).

Natural selection can cause populations to diverge, even in the absence of geographic isolation.

have been captured over a mile from the site where they were originally captured, proving that individuals do search widely for appropriate fruit to parasitize. Thus, flies from the same population might simply switch from apple to hawthorn trees and back, based on fruit availability.

To test these two hypotheses, Jeff Feder and colleagues looked at the genetic makeup of flies collected from hawthorns versus apples using a technique called protein electrophoresis (Box 9.1). Remarkably, they found a clear distinction in the two samples: Flies collected from hawthorns versus apples have statistically significant differences in the frequencies of alleles for six different enzymes (Feder et al. 1988, 1990). This is strong support for the hypothesis that hawthorn and apple flies have diverged and now form distinct populations. Even though the two races look indistinguishable, they are easily differentiated on the basis of their genotypes.

How could this have occurred? Have these flies skipped the first stage in speciation (Box 9.2)? The key is that instead of being isolated by dispersal or vicariance, apple and hawthorn flies are isolated on different host species. In experiments where individuals are given a choice of host plants, apple and hawthorn flies show a strong preference for their own fruit type (Prokopy et al. 1988). Because mating takes place on the fruit, this habitat preference should result in strong nonrandom mating. Feder and colleagues (1994) confirmed this prediction by following marked individuals in the field. They found that matings between hawthorn and apple flies accounted for only 6% of the total observed.

Box 9.1 Protein electrophoresis

Protein electrophoresis is a widely used technique for assessing the degree of genetic differentiation among populations. In this process, enzymatic proteins isolated from a series of individuals are placed in separate wells in the middle of a slab of starch gel (Figure 9.9). The gel is immersed in a buffered solution with a voltage across it. Because proteins carry electric charges, they migrate toward one of the two poles of the gel box. The distance that they migrate is a function of their charge and molecular size: Larger and less highly charged enzymes move more slowly.

After the proteins have been separated, different staining systems can be used to visualize specific enzymes. If more than one protein shows up in the same enzyme staining system in the same individual, we infer that the individual is heterozygous at the locus responsible for coding for that protein (Figure 9.9b). These allelic enzymes are referred to as allozymes.

Although protein electrophoresis is less popular now that DNA sequencing technology is widely available, it is still useful for genetic surveys of large numbers of individuals because it is relatively simple and inexpensive.

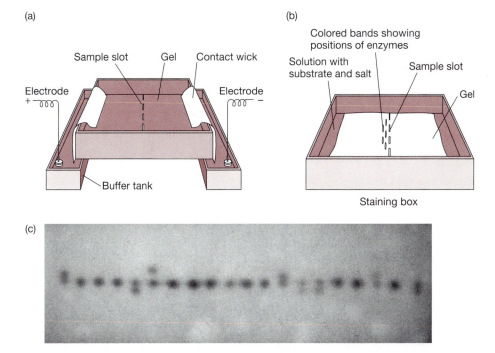

Figure 9.9 **Starch gel electrophoresis** (a) The top diagram shows a gel box ready for a run. Proteins extracted from different individuals are loaded in each well, or sample slot. Then a voltage is applied to the electrodes for several hours. (b) This diagram shows a staining box, where the gel is placed after proteins have migrated far enough that differences between them can be distinguished. (c) This photograph shows a gel stained for the PGM enzyme. If an individual has two forms of an enzyme, two bands appear in the same lane. Because we assume that these forms are the products of different alleles (Feder et al. 1989a), we can infer that these individuals are heterozygous. This gel indicates that three different alleles exist, and that eight of the 20 flies in this sample are heterozygous at this locus. (Jeff Feder, University of Notre Dame)

BOX 9.2 Sympatric and parapatric speciation

An enduring debate in speciation research has been whether physical isolation is an absolute requirement for populations to diverge. Can natural selection for divergence be so strong that populations skip the first stage in the speciation process? Careful genetic modeling suggests that the answer is yes. Joseph Felsenstein (1981) and William Rice (1987) have shown that populations can diverge even with low to moderate degrees of gene flow if two important conditions are met: Selection for divergence must be strong, and mate choice must be correlated with the factor that is promoting divergence. Exactly these conditions are met by apple and hawthorn flies. The conditions are restrictive, however, and sympatric speciation is almost certainly much less common than allopatric speciation.

Parapatric speciation is the other major mode of divergence that does not require physical isolation of populations. In parapatric speciation, strong selection for divergence causes gene frequencies in a continuous population to diverge along a gradient (Figure 9.10). Although theory demonstrates that parapatric divergence is possible (Endler 1977), we still lack well-documented examples in natural populations.

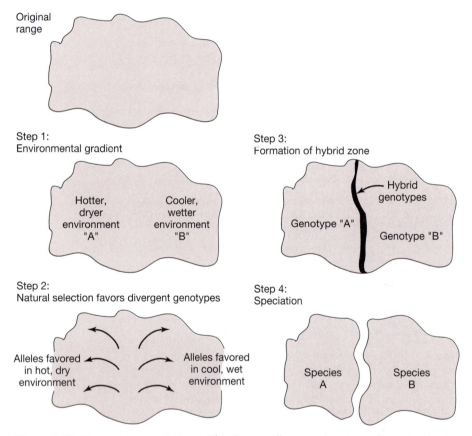

Figure 9.10 **Parapatric speciation** This diagram illustrates the series of steps involved in a mode of divergence called parapatric ("beside place") speciation.

Host plant fidelity serves as an effective barrier to mating, and few genes are exchanged between the two populations.

These experiments imply that natural selection has favored distinct habitat preferences in apple and hawthorn flies. But why? What fitness advantages accrue to *R. pomonella* individuals parasitizing apples? For natural selection to favor a host shift, flies parasitizing apples would have to survive and reproduce at least as well as flies that breed on hawthorns. But Prokopy et al. (1988) found that individual fly larvae of both races actually do much better when reared in hawthorn fruit. When these researchers switched eggs from apples to hawthorns and hawthorns to apples, they found that 52% of eggs from both races survived in hawthorn fruits versus only 27% in apples. In feeding efficiency, there is a huge disadvantage in switching to apples. This makes sense. *Rhagoletis* have probably been parasitizing *Crateagus* for thousands of generations, so we would expect that maggots are better adapted to the chemistry of hawthorn fruit.

Three factors make the switch to apples work:

- Escape from parasitoids. *Rhagoletis* are themselves parasitized by several types of wasp. Female wasps patrol ripe apples and hawthorns, find fly larvae that are feeding near the surface, thrust their ovipositor into the maggot, and lay one or more eggs. After the eggs hatch, the wasp larvae develop by feeding on the maggot (the maggot dies). But Feder (1995) did a series of experiments showing that the average level of wasp parasitism in apples was 70% less than in hawthorns. Why? Wasp ovipositors were often too short to reach fly larvae burrowing through apples, simply because apple fruits are over 30 times the size of hawthorn fruits. Also, most wasps oviposit too late in the season to find infested apples, which ripen almost a month earlier than hawthorn fruits.

- Escape from intraspecific competition. Feder et al. (1995) found an average of 20.06 oviposition marks (and thus, eggs) from *Rhagoletis* on each apple at their field site, versus just 3.71 marks on each hawthorn fruit. But because apples are so much larger, a fly larva still has 5.7 times as much fruit available, on average, in an apple versus hawthorn. This implies a quality-quantity trade-off: The food might not be as good, but there is a lot more of it.

- Escape from interspecific competition. Apple and hawthorn fruits are also parasitized by other species, such as caterpillars and weevils. The two types of fruits are victimized by these other species at equal rates. But when these competitors are present, fly survivorship drops dramatically only in hawthorns. The larger size of apples means that fly larvae have significantly fewer caterpillars and weevils to contend with. In apples, there is enough fruit to accommodate caterpillars, weevils, *and* maggot flies.

Apparently, these three fitness advantages outweigh the disadvantages of reduced larval survival and create a selective advantage for the switch to apples.

To summarize, the experiments we have reviewed demonstrate that natural selection has been responsible for strong divergence between *R. pomonella* populations. Many biologists now consider hawthorn and apple flies to be incipient species. This designation means that the populations have clearly diverged and are largely, but not completely, isolated in terms of gene flow.

To appreciate how widespread this type of natural selection may be, recall the experiment on sticklebacks that we reviewed in Chapter 3, demonstrating that competition for food occurs in closely related benthic and limnetic forms, and that natural selection favors divergent phenotypes. For other examples of divergence due to selection on food or habitat choice, primarily in insects and fish, see Pashley 1988; Feder et al. 1989b; Rice and Salt 1990; Carroll and Boyd 1992; Brown et al. 1996; Schluter 1996; Wood and Foote 1996; Pigeon et al. 1997.

Sexual Selection

Sexual selection results from differences among individuals in their ability to obtain mates. This is considered a form of selection distinct from natural selection. Male red deer, for example, fight among each other for the privilege of copulating with females. Success in fighting correlates strongly with antler size and overall body size, and evolution by sexual selection occurs because there is heritable variation in these traits. Evolutionary changes in the antler or body size of male red deer are primarily due to sexual selection.

We will explore the causes and consequences of sexual selection in much greater detail in Chapter 15, but need to introduce the topic here because of its impact on speciation's second stage. Changes in the way that a population of sexual organisms chooses or acquires mates can lead to rapid differentiation from ancestral populations (Fisher 1958; Lande 1981, 1982). Here is the key point: Because it directly affects traits involved in gene flow, sexual selection is a particularly efficient mechanism promoting divergence (West-Eberhard 1983).

Because it acts on traits that determine gene flow, sexual selection can cause rapid divergence between populations.

In the Hawaiian *Drosophila,* for example, sexual selection is thought to have been a key factor in promoting divergence among isolated populations. Many of these flies court and copulate in aggregations, called leks. In this mating system males fight for small display territories and dance or sing for females, who visit the lek to select mates. Lek breeding systems are often associated with elaborate male characters. These vary widely among Hawaiian flies. Does this imply that sexual selection has been important in speciation?

The evidence in favor of the hypothesis is tantalizing, though not conclusive. For example, males of *Drosophila heteroneura* have hammer-shaped heads with eyes mounted on stalks (Kaneshiro and Boake, 1987). Because males butt heads when fighting to stake out a courting arena on the lek, the unusual head shape appears to be a product of sexual selection (Figure 9.11). In contrast, males and females of *D. heteroneura*'s closest relative, *Drosophila silvestris,* have normally shaped heads similar in size and shape to female *D. heteroneura* (Figure 9.12). *D. silvestris* males do not head-butt. They fight on the lek by rearing up and grappling with one another. Both species are endemic to the island of Hawaii.

These facts are consistent with the following scenario:

1. In the ancestor to *silvestris* and *heteroneura* males had normal heads, courted on leks, and fought for display territories by rearing up and grappling.
2. A mutation occurred in an isolated subpopulation, which led to males with a new fighting behavior: head-butting.
3. The mutant males were more efficient in combat on leks and experienced increased reproductive success.

(a)

(b)

Figure 9.11 **Contrasting fighting strategies in Hawaiian flies** (a) Male *Drosophila heteroneura* butt heads to establish display territories on a lek. (b) Male *Drosophila silvestris* fight over display territories by rearing up and grappling with one another. (Kenneth Y. Kaneshiro, University of Hawaii)

4. The mutation increased to fixation, and additional mutations led to elaboration of the trait over time. For example, it is possible that mutations leading to stalk-mounted eyes were favored because they made eyes less prone to damage during head–butting fights.

As a result of this sequence of events, strong divergence among the populations would occur due entirely to sexual selection. The differentiation between populations would be based on the strategies and weaponry employed in male–male combat.

Formulating this type of plausible sequence is a productive way to generate testable hypotheses, but does not substitute for genetic models, experiments, or other types of evidence. Fortunately, another dataset supports a strong role for sexual selection in this spectacular insect radiation. Hampton Carson and colleagues (1982) have found populations of *D. silvestris* that may be in the early stages of speciation through geographic isolation and sexual selection. They have

(a)

(b)

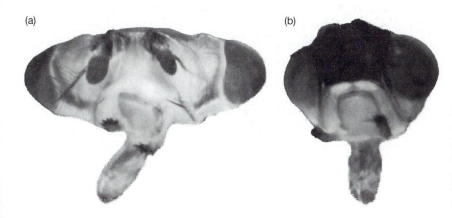

Figure 9.12 **Contrasting head shapes in closely related Hawaiian *Drosophila*** In contrast to *Drosophila heteroneura* (a), both males and females of *D. silvestris* (b) have normal-shaped heads. (Kenneth Y. Kaneshiro, University of Hawaii)

found *silvestris* populations on the north and east side of the big island (Hawaii) in which males have a third, and extra, row of long hairs, called cilia, on the tibia (lower segment) of their foreleg. The males use these hairs to stroke the female's abdomen during the final stage of courtship. Carson and Lande (1984) have shown that cilia number is heritable, which confirms that the trait can respond to sexual selection. In a critical experiment, Ken Kaneshiro and Joyce Kurihara (1981) performed mate-choice tests in the lab and found that females from two-row populations of *D. silvestris* are very unlikely to copulate with males from three-row populations. This confirms that sexual selection on cilia number could occur, and supports the hypothesis that sexual selection was a prominent cause of divergence in the isolated populations.

Mutation

In Chapter 5 we reviewed population genetic models showing that mutation, by itself, is ineffective in changing gene frequencies. For speciation, the primary role of point mutations, gene duplications, and chromosome inversions is to provide raw material for drift and selection once gene flow is reduced.

Of the many types of mutation examined in Chapter 4, large-scale chromosomal changes are the most important in causing populations to differentiate. Recall that polyploidization can result in instant reproductive isolation because of incompatibities between gametes with different chromosome numbers. An event like this bypasses the requirement for physical separation between source and founder populations. Purely on the basis of its distribution in plants, polyploidization has to be considered an important mechanism of divergence.

Populations can become genetically isolated because of differences in chromosome number.

We need to introduce a distinction, however, between two types of polyploidy. Polyploids whose chromosomes originate from the same ancestral species are called autopolyploids. Polyploids that result from hybridization events between different species are called allopolyploids. Although no exact figures are available, allopolyploids are thought to be significantly more common than autopolyploids (Soltis and Soltis 1993). We examine allopolyploids in more detail in section 9.4, when we consider hybridization events between closely related populations and the third stage in the speciation process.

How important is polyploidization as a mechanism of speciation? Unfortunately, estimates for the frequency of auto- and allopolyploid species are rarely done in a phylogenetic context, where we can infer which ploidy is ancestral and get precise counts for the number of times that polyploidization has occurred. But under the assumption that any plant with $N > 14$ is polyploid, Verne Grant (1981) has estimated that 43% of species in the flowering plant class Dicotyledonae and 58% of species in the flowering plant class Monocotyledonae are the descendants of ancestors that underwent polyploidy. Jane Masterson (1994) came up with an even higher estimate. She showed that cell size correlates strongly with chromosome number in extant plants, and then measured the size of cells in plant fossils that lived early in the radiation of flowering plants. Her data support the hypothesis that $N = 7$ or 9 or lower is the ancestral haploid number in angiosperms. This suggests that 70% of flowering plants have polyploidy in their history. Polyploidy is also common in mosses, not infrequent in green algae, and almost the rule in ferns, where 95% of species occur in clades with an ancestral polyploidization event. Grant goes so far as to say that polyploidy "is a characteristic of the plant kingdom" (1981: 289).

Changes in chromosome number less drastic than polyploidization may also be important in speciation. For example, Oliver Ryder and colleagues (1989) studied chromosome complements in a series of small African antelopes—called dik-diks—that were being displayed in North American zoos. Although zookeepers traditionally recognized just two species, Ryder's team distinguished three on the basis of chromosome number and form. Further, they were able to show that hybrid offspring between unlike karyotypes are infertile. Their research revealed a cryptic species. Two of the dik-dik species have apparently differentiated on the basis of karyotype.

It is extremely common to find small-scale chromosomal changes like these when the karyotypes of closely related species are compared. Although these mutations could be important in causing genetic divergence between populations (White 1978), much of the extensive work on chromosomal differentiation done to date is merely correlative. That is, many studies have measured chromosome differences in related species and claimed that chromosomal incompatibilities are responsible for isolation and divergence. But in many cases, it is likely that the chromosome differences arose *after* speciation had occurred due to other causes (Patton and Sherwood 1983). Until more causative links are established, we have to be cautious in interpreting the importance of small-scale karyotype differences in speciation.

9.4 Secondary Contact

We have portrayed speciation as a three-step process that begins with the isolation of populations and continues when selection, mutation, and drift create divergence. A third step may occur if recently diverged populations come back into contact and have the opportunity to interbreed. Hybridization events between recently diverged species are especially common in plants. For example, over 700 of the plant species that have been introduced to the British Isles in the recent past have hybridized with native species at least occasionally, and about half of these native/nonnative matings produce fertile offspring (Abbott 1992). Ten percent of all bird species also hybridize regularly and produce fertile offspring (Grant and Grant 1992).

In at least some cases, the fate of these hybrid offspring determines the direction of the speciation process. Will the hybrids thrive, interbreed with each of the parental populations, and eventually erase the divergence between them? Or will hybrids have new characteristics and create a distinct population of their own? And what happens when hybrid offspring have reduced fitness?

Reinforcement

The geneticist Theodosius Dobzhansky (1937) formulated one of the leading hypotheses about speciation's third stage. Dobzhanzky reasoned that if populations have diverged sufficiently in allopatry, their hybrid offspring will have markedly reduced fitness relative to individuals in the parental population. Because producing hybrid offspring reduces a parent's fitness in this case, there should be strong natural selection favoring assortative mating. That is, selection

Reinforcement is a type of selection that leads to assortative mating and the prezygotic isolation of populations.

should favor individuals that choose mates only from the same population. Selection that reduces the frequency of hybrids in this way is called reinforcement. If reinforcement occurs, it would finalize the speciation process by producing complete reproductive isolation.

The reinforcement hypothesis predicts that some sort of mechanism of premating isolation will evolve in closely related species that come into contact and hybridize. Selection might favor mutations that alter aspects of mate choice, genetic compatibility, or life history (such as the timing of breeding). Divergence in these traits prevents fertilization from occurring and results in prezygotic isolation of the two species. But populations can also remain genetically isolated in the absence of reinforcement if hybrid offspring are sterile or infertile. This possibility is known as postzygotic isolation.

Some of the best data on prezygotic isolation and the reinforcement hypothesis have been assembled and analyzed by Jerry Coyne and Allen Orr (1997). Coyne and Orr examined data from a large series of sister-species pairs in the genus *Drosophila*. Some of these species pairs live in allopatry and others in sympatry. Coyne and Orr's dataset included estimates of genetic distance, calculated from differences in allozyme frequencies, along with measurements of the degree of pre- and postzygotic isolation. When they plotted the degree of prezygotic isolation against genetic distance, which they assumed correlates at least roughly with time of divergence, they found a striking result: Prezygotic isolation evolves much faster in sympatric species pairs than it does in allopatric species pairs (Figure 9.13). This is exactly the prediction made by the reinforcement hypothesis. Postzygotic isolation, in contrast, increased with genetic distance at the same rate in both sympatric and allopatric populations, and much more slowly than prezygotic isolation. (We examine postzygotic isolation in more detail in section 9.5.)

These laboratory experiments with *Drosophila* are some of the best evidence we have in support of the reinforcement hypothesis. In contrast, field studies that have looked for evidence of reinforcement in hybridizing populations have produced mixed results. Genetic models exploring reinforcement's effectiveness have generally been unconvincing as well (see Butlin 1987). Although Dobzhansky considered reinforcement as a universal stage in speciation, this view is probably overstated. A new consensus is emerging, based on a series of recent studies like Coyne and Orr's: Reinforcement can and does occur, but it is not essential to completing speciation (Wells and Henry 1992; Liou and Price 1994; Butlin 1995; Noor 1995).

Hybridization

Reinforcement can occur when hybrid offspring have reduced fitness. But what happens to hybrid offspring that survive and reproduce well? Their fate has important consequences for speciation, and practical applications as well. For example, because several crop plants have close relatives that are serious weeds, evolutionary biologists have expressed concern about the release of genetically engineered crop plants to the wild. What benefit comes from introducing a gene for herbicide resistance into a crop species if hybridization quickly carries the allele into a weed population?

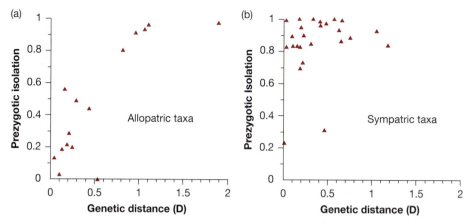

Figure 9.13 **Prezygotic isolation in allopatric versus sympatric species pairs of**
Drosophila These graphs (Coyne and Orr 1997) plot degree of prezygotic isolation versus
genetic distance in a variety of sister-species pairs from the genus *Drosophila*. Prezygotic isolation is estimated from mate-choice tests performed in the laboratory. A value of 0 indicates
that different populations freely interbreed (0% prezygotic isolation) and 1 indicates no interbreeding (100% prezygotic isolation). Genetic distance is estimated from differences in allele frequencies found in allozyme surveys. Sibling species with the same degree of overall
genetic divergence show much more prezygotic isolation if they live in sympatry.

Using an experimental system featuring crop sorghum *(Sorghum bicolor)* and
johnsongrass *(S. halepense),* Paul Arriola and Norman Ellstrand (1996) recently
confirmed that this scenario could occur. Sorghum is one of the world's most
important crops, and johnsongrass a serious weed. Arriola and Ellstrand sowed a
field with crop sorghum seeds that carried a distinctive allozyme marker and
planted seedlings of johnsongrass at various distances nearby (Figure 9.14a). After the johnsongrass plants had set seed, Arriola and Ellstrand harvested them
and raised the progeny. Protein electrophoresis of these first-generation (F_1)
plants confirmed that the allozyme marker had been carried by crop sorghum
pollen to johnsongrass plants. Significant gene flow had taken place in the crop-
to-weed direction (Figure 9.14b).

Creation of New Species through Hybridization

The concern about sorghum-johnsongrass hybrids illustrates a general point:
Hybrid offspring do not necessarily have lower fitness than their parents. If crop
sorghum plants were transformed with a gene for herbicide resistance and these
individuals hybridized with johnsongrass, it is conceivable that a few of the offspring would end up with a favorable combination of traits from each parent.
The hybrids might receive the allele for herbicide resistance from the crop parent, but a suite of genes that cause rapid growth from the weedy parent. This is
the basis for concern about "superweeds." And this concern is not idle. Both
classical and recent studies have confirmed that interspecific hybridization is a
major source of evolutionary novelty in plants (Stebbins 1950; Grant 1981; Abbott 1992; Arnold and Hodges 1995; Rieseberg 1995). In newly colonized environments or certain restricted habitats, hybrid offspring may have higher fitness
than either of the parental species. Will these hybrid populations become new
species?

Figure 9.14 Gene flow from crops to weeds through hybridization The photographs show the flowering heads of crop sorghum (top) and johnsongrass (bottom).   (a) This diagram shows the planting scheme employed by Arriola and Ellstrand (1996). Potted seedlings of johnsongrass were placed along each of four rectangles, as shown in the diagram. (The diagram is not to scale.)   (b) In this graph, the percentage of johnsongrass plants that produced at least one hybrid offspring is plotted versus the distance that johnsongrass seedlings were growing from the edge of the crop. [Photos: (top) Joe Munroe/Photo Researchers, Inc.; (bottom) A.W. Ambler/National Audubon Society/Photo Researchers, Inc.]

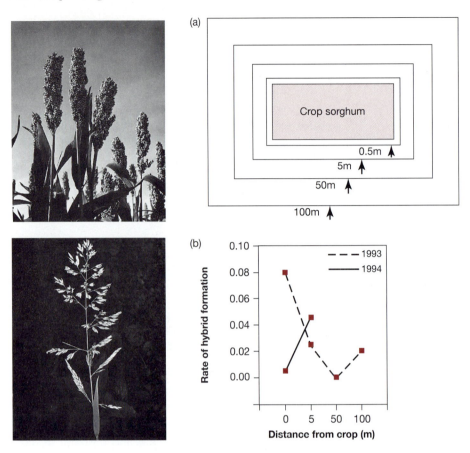

A recent experimental study of plant hybridization, conducted in Loren Rieseberg's lab (1996), has actually duplicated a natural hybridization event that led to speciation. Rieseberg and co-workers crossed the annual sunflowers *Helianthus annuus* and *H. petiolaris* to produce three lines of F$_1$ hybrids. They then either mated each line back to *H. annuus* (this is called a backcross) or sib-mated the individuals for four additional generations. Each line underwent a different sequence and combination of backcrossing and sib-mating. This protocol simulated different types of matings that may have occurred when populations of *H. annuus* and *H. petiolaris* hybridized naturally.

At the end of the experiment, Rieseberg and colleagues surveyed the three hybrid populations genetically. Their goal was to determine how similar the hybrids were to each other and to individuals in a naturally occurring species with hybrid characteristics called *H. anomalous*. To make this comparison possible, the researchers mapped a large series of species-specific DNA sequences, called randomly amplified polymorphic DNA (RAPD) markers, in the two parental forms. These markers allowed the researchers to determine which alleles from the two parental species were present in the three experimental hybrid populations and compare them to alleles present in *H. anomalous*. The results were striking: The three independently derived experimental hybrids and the natural hybrid shared an overwhelming majority of markers. The experimental and natural hybrids were almost identical.

The experiment supports the hypothesis that only certain alleles from *H. petiolis* and *H. anuus* work in combination, and that other hybrid types are inviable or have reduced fitness. This result is analogous to the superweed scenario we outlined before. The sunflower experiment showed that the genetic composition of the hybrids quickly sorted out into a very similar favorable combination. Even more remarkable, this combination of alleles was nearly the same as that produced by a natural hybridization event that occurred thousands of years ago. This puts an interesting twist on speciation's third stage: Secondary contact and gene flow between recently diverged species can result in the creation of a new, third, species.

Hybrid Zones

A hybrid zone is a region where interbreeding between diverged populations occurs and hybrid offspring are frequent. Hybrid zones are produced by two distinct situations (Hewitt 1988). (1) After secondary contact between species that have diverged in allopatry, a hybrid zone can form where they meet and interbreed. (2) During parapatric speciation, a hybrid zone can develop between populations that are diverging (Box 9.2).

We have already seen that hybrid offspring can have lower or higher fitness than purebred offspring, with very different consequences (reinforcement of parental forms or the formation of a new species). Research on hybrid zones has confirmed that a third outcome is also possible. Frequently no measurable differences can be found between the fitnesses of hybrid and pure offspring. These three possibilities can also dictate the size, shape, and longevity of hybrid zones (Endler 1977; Barton and Hewitt 1985; Hewitt 1988):

- When hybrid and parental forms are equally fit, the hybrid zone is wide. Individuals with hybrid traits are found at high frequency at the center of the zone and progressively lower frequencies with increasing distance. In this type of hybrid zone, the dynamics of gene-frequency change are dominated by drift. The width of the zone is a function of two factors: how far individuals from each population disperse each generation, and how long the zone has existed. The farther the individuals move each generation and the longer the populations are in contact, the wider the zone.

- When hybrids are less fit than purebred individuals, the fate of the hybrid zone depends on the strength of selection against them. If selection is very strong and reinforcement occurs, then the hybrid zone is narrow and short-lived. If selection is weak, then the region of hybridization is wider and longer-lived. These types of hybrid zones are an example of a selection-migration balance, analogous to the situation with water snakes in Lake Erie.

- When hybrids are more fit than purebreds, the fate of the hybrid zone depends on the extent of environments in which hybrids have an advantage. If hybrids achieve higher fitness in environments outside the ranges of the parental species, then a new species may form (as reviewed earlier). If hybrids have an advantage at the boundary of each parental population's range, then a stable hybrid zone may form. For example, many hybrid zones are found in regions called ecotones, where markedly different plant and animal communities meet. Two closely related species or populations (often called subspecies

Figure 9.15 **Hybrid sagebrush are intermediate in form between parental subspecies** On this graph, a quantity called the principal component score is plotted against the elevation where sagebrush plants were sampled. Principal component analysis (PCA) is a statistical procedure for distilling information from many correlated variables into one or two quantities that summarize the variation measured among individuals in the study. In this case, Carl Freeman and colleagues (1991) measured a large series of morphological traits in sagebrush such as height, circumference, crown diameter, and branch length. The PCA was performed to combine these many variables into a single quantity, the PCA score, that summarizes overall size and shape. Each data point represents an individual.

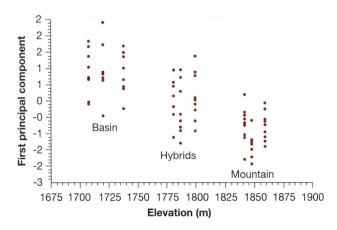

or varieties) will be found on either side of the ecotone, with a hybrid zone between them. One hypothesis is that hybrid individuals with intermediate characteristics have a fitness advantage in these transitional habitats.

To illustrate how researchers go about distinguishing between these possibilities, we review recent work on what may be the most widespread and economically important plant in the American west: big sagebrush *(Artemesia tridentata)*. A total of four subspecies of big sagebrush have been described, including two that hybridize in the Wasatch Mountains of Utah (Freeman et al. 1991, 1995; Graham et al. 1995). Basin big sagebrush *(A.t. tridentata)* is found at low elevations in river flats, while mountain sagebrush *(A.t. vaseyana)* grows at higher elevations in upland habitats. The two subspecies hybridize where they make contact at intermediate elevations.

The first task in analyzing a hybrid zone is to describe the distribution and morphology of hybrids relative to parental populations. Previous work had shown that hybrid zones between sagebrush populations are narrow—often less than the length of a football field—and Carl Freeman and colleagues (1991) found that hybrids are intermediate in form between basin and mountain subspecies (Figure 9.15). Historical records indicate that the size and distribution of

TABLE 9.1 Fitness of sagebrush hybrids

The rows in this table, listing basin through mountain populations, represent big sagebrush plants sampled along a lower-to-higher elevational gradient. N is the sample size and the numbers in parentheses are standard deviations—a measure of variation around the average value. Differences among these populations in the number of inflorescences (flowering stalks) are not statistically significant. But hybrids have significantly more flowering heads per unit inflorescence length—a measure of flower density ($P < 0.05$).

Population	*N*	Number of inflorescences		Number of flowering heads	
Basin	25	19.92	(6.16)	175.1	(124.9)
Near basin	25	17.72	(6.59)	174.4	(92.5)
Hybrid	27	20.11	(6.75)	372.7	(375.9)
Near mountain	25	17.04	(6.50)	153.7	(75.2)
Mountain	25	16.80	(6.34)	102.0	(59.4)

Source: Adapted from Table 2 in Graham et al. (1995).

TABLE 9.2 Outcomes of secondary contact and hybridization

When populations hybridize after diverging in allopatry, several different outcomes are possible. The type of hybrid zone formed and the eventual outcome depend on the relative fitness of hybrid individuals.

Fitness of hybrids	Hybrid zone	Eventual outcome
Lower than parental forms	Relatively narrow and short-lived	Reinforcement (differentiation between parental populations increases)
Equal to parental forms	Relatively wide and long-lived	Parental populations coalesce (differentiation between parental populations decreases)
Higher than parental forms	Depends on whether fitness advantage occurs in ecotone or new habitat	Stable hybrid zone or formation of new species

hybrid zones have been stable in extent for at least 2–3 sagebrush generations. This suggests that strong selection for or against hybrids is not occurring.

To assess the relative fitness of hybrid offspring versus pure forms, John Graham et al. (1995) compared a variety of fitness components in individuals sampled along an elevation gradient: seed and flower production, seed germination, and extent of browsing by mule deer and grasshoppers. Table 9.1 shows some data representative of their results. In general, hybrids show equal or even superior production of flowers and seeds, and resistance to herbivores that is equal to the mountain forms. Thus, hybrid offspring do not appear to be less fit than offspring of the parental populations.

This leaves two possibilities: The hybrid zone could be maintained by selection-mutation balance, or by positive selection on hybrids in the ecotone. The critical experiments to make this distinction are now underway. The research group is studying growth rates and other components of fitness in basin, hybrid, and mountain seedlings that have been transplanted to each others' habitats. These reciprocal transplant experiments should confirm how selection and drift interact in sagebrush zones, and add to our understanding of speciation's third stage (Table 9.2).

Having reviewed isolation, divergence, and secondary contact, we can move on to consider the genetic mechanisms responsible for these processes. Understanding the genetic basis of speciation is our focus in section 9.5.

9.5 The Genetics of Differentiation and Isolation

What degree of genetic differentiation is required to isolate populations and produce new species? The traditional view was that some sort of radical reorganization of the genome, called a genetic revolution, was necessary (Mayr 1963). This hypothesis was inspired by a strict interpretation of the biological species concept. The logic went as follows: Under the biological species concept (BSC), species are reproductively isolated if and only if hybrids are inviable or experience dramatic reductions in fitness. For this to happen, sister species would have to be genetically incompatible. Combining their alleles would produce dysfunctional development, morphology, or behavior. The general idea was that speciation involved genetic mechanisms distinct from the normal action of mutation, selection, drift, and gene flow introduced in Chapters 4–6.

Genetic models have shown that these types of large-scale changes in the genome are not only unlikely, but unnecessary, for divergence and speciation to occur (Lande 1980; Barton and Charlesworth 1984). These theoretical results have been verified by the work we reviewed in section 9.3, which demonstrates that marked differentiation between populations can result from the normal evolutionary processes of mutation, selection, and drift. And the sunflower experiment we just reviewed shows that alleles from well-differentiated species can recombine to produce a new and stable lineage. As a result, the questions that motivate current research in the genetics of speciation are focused on the number, location, and nature of genes that distinguish closely related species.

Insights from Classical Genetics

When experimenters make crosses between recently diverged taxa, it is not unusual to find that one sex in the F_1 generation is absent or sterile. These crosses provide an example of postzygotic isolation. Premating isolation does not occur in these crosses. At least some offspring are formed, but they are incapable of hybridizing further among themselves. This postzygotic isolation is important because it confirms that the populations are reproductively isolated and that the speciation process is complete.

An interesting pattern in the evolution of postzygotic isolation helps to put this phenomenon in context. When Jerry Coyne and Allen Orr (1997) plotted the relationship between degree of postzygotic isolation and genetic divergence among the sister-species pairs of *Drosophila* in their study, they found that postzygotic isolation evolves more slowly than prezgotic isolation (Figure 9.16). Specifically, when the degree of prezygotic versus postzygotic isolation is com-

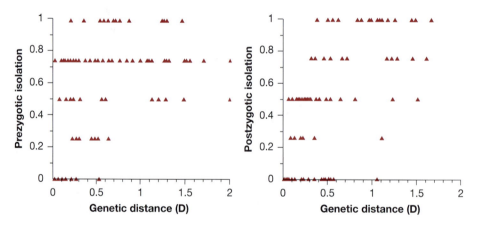

Figure 9.16 **Pre- and postzygotic isolation in *Drosophila*** These graphs (Coyne and Orr 1997) plot degree of pre- and postzygotic isolation versus genetic distance in sister-species pairs from the genus *Drosophila*. Prezygotic isolation is estimated as described in Figure 9.13. Postzygotic isolation is estimated from experimental crosses performed in the laboratory, with a value of 0 indicating that no offspring are sterile or inviable (0% postzygotic isolation) and 1 indicating that all offspring are sterile or inviable (100% postzygotic isolation). Genetic distance is estimated from differences in allele frequencies found in allozyme surveys. When only data from the most recently diverged species are analyzed (D < 0.5), sibling species with the same degree of overall genetic divergence show a statistical difference in their degree of pre- versus postzygotic isolation (P = 0.0018).

pared in the most recently diverged sister-species pairs, there is a statistically significant difference. The degree of prezygotic isolation is greater. But here is a key point: The difference is due *entirely* to sister-species pairs that exist in sympatry. When species pairs that live in allopatry are compared, there is no difference between the degree of pre- and postzygotic isolation.

To interpret this pattern, recall that we have been analyzing speciation as a three-step process. We have just examined a variety of consequences of the third stage. But what happens if recently diverged species do not come back into contact? Then there is no chance for reinforcement to evolve or for hybridization to occur. If the essence of speciation is reproductive isolation, does the process continue if populations remain allopatric?

The evolution of postzygotic isolation implies that the answer is yes. The idea here is that populations can diverge so thoroughly in their genetic makeup that hybrid offspring simply do not develop normally and are sterile or inviable as a result. When this happens reproductive isolation has evolved, even though the populations have not interbred since the original split.

Postzygotic isolation occurs when populations have diverged so thoroughly that hybrid offspring are inviable or sterile.

Identifying the genes responsible for postzygotic isolation might tell us something interesting about this aspect of the speciation process. One of our best clues in this gene hunt comes from a striking observation made in experimental crosses between recently diverged taxa: If one sex in the hybrid offspring is sterile or inviable, it is almost always the one heterozygous for the sex chromosomes (Coyne and Orr 1989). This is referred to as the heterogametic sex. In species with X and Y sex chromosomes, such as insects and mammals, males are the heterogametic sex. But in species with Z and W sex chromosomes (birds and butterflies), females are the heterogametic sex. The pattern of sterility or inviability in the heterogametic sex holds, irrespective of how sex is specified genetically (Table 9.3). The generalization is so pervasive that it has been called Haldane's rule in honor of population geneticist J.B.S. Haldane, who first published on the phenomenon in 1922.

The question is, why? What is it about having one of each sex chromosome that contributes to sterility or inviability when hybrids are formed between divergent populations? Although we still do not have a definitive answer (see

---

TABLE 9.3 Haldane's rule

In this table, the "Hybridizations with asymmetry" column indicates the number of closely related populations that have been crossed and that exhibit asymmetries in the fertility or viability of males and females in the F_1 hybrids. The "Number obeying Haldane's rule" column reports how many of these crosses showed greater fertility or viability loss in the heterogametic sex.

Group	Trait	Hybridizations with asymmetry	Number obeying Haldane's rule
Mammals	Fertility	20	19
Birds	Fertility	43	40
	Viability	18	18
Drosophila	Fertility and viability	145	141

Source: Coyne and Orr 1989.

Coyne 1994; Turelli and Orr 1995), much of the classical and recent research has focused on the X chromosome simply because the Y contains so few loci. And experimental results have confirmed that at least some of the major loci responsible for postzygotic isolation reside on the X.

Some of the best work on postzygotic isolation has utilized classical approaches to identify the location of these genes and estimate the magnitude of their effects. Researchers working with fruit flies, for example, can establish an experimental population in one species that is homozygous for a series of recessive traits whose location in the fly genome is already known. When homozygous, these recessive alleles might produce a distinctive eye color or bristle morphology. This line is then crossed with a population from a recently diverged species that does not carry the markers. The F_1 offspring of this mating are heterozygous for these traits and show a normal phenotype. Hybrid females are then backcrossed to males of the experimental line that is homozygous for the recessive alleles. The presence or absence of the recessive traits in the F_2 generation is recorded, and F_2 individuals are tested for sterility. If there is a strong correlation between the presence of a specific marker and the degree of sterility in F_2 males, it indicates that a locus near the marker causes sterility. Using exactly this protocol in crosses between *Drosophila simulans* and *D. seychellia,* Jerry Coyne and Marty Kreitman (1986) examined a series of markers scattered throughout the fly genome. They determined that at least one—and possibly two—loci on the X chromosome had the greatest effect on degree of postzygotic isolation.

In Drosophila, the genes responsible for postzygotic isolation are localized on the X chromosome.

What are the protein products of these loci, and why do they produce sterility or inviability in hybrid offspring? These are important questions about the genetics of postzygotic isolation, but they remain unanswered. The hunt for these "speciation genes," located somewhere on the X chromosome, continues.

Analyzing Quantitative Trait Loci

An innovative experimental approach called quantitative trait loci (QTL) mapping is offering an additional way to locate genes involved in divergence and measure their effects. Many or most of the morphological and behavioral differences we observe between closely related species are quantitative traits. Characteristics like body size, body shape, songs, and flower shape result from the combined effects of many loci. QTL mapping is a technology for locating genes with small, but significant, effects on these types of traits.

H.D. Bradshaw and co-workers (1995) recently used QTL mapping to investigate divergence between sister species of monkeyflowers called *Mimulus cardinalis* and *M. lewisii* (Figure 9.17). These two species hybridize readily in the lab and produce fertile offspring. They also have overlapping ranges in the Sierra Nevada of California. But no hybrids have ever been found in the field. The reason is that they attract different pollinators. *M. cardinalis* is hummingbird-pollinated and *M. lewisii* is bee-pollinated.

Flower morphology in these species correlates with the differences in the pollinators. Bees do not see well in the red part of the visible spectrum and need a platform to land on before walking into a flower and foraging. Hummingbirds, in contrast, see red well, have long, narrow beaks, and hover while harvesting nectar. *M. lewisii* has a prominent landing spot, while *M. cardinalis* has an elongated tube with a nectar reward at the end. The flower characteristics of *M. lewisii* and *M. cardinalis* conform to classical bee- and bird-pollinated colors and shapes.

Although a formal phylogenetic analysis has not been done, most species in the genus *Mimulus* are bee-pollinated. This implies that the flatter, purple, bee-pollinated flower is ancestral and the more tubular, reddish, hummingbird-pollinated form derived. The question is, what genes are responsible for the radical make-over in flower shape? Can we pinpoint the loci that have responded to selection for hummingbird pollination? QTL mapping gives researchers a way to answer these questions.

In a QTL study, the hunt for the targeted genes takes place in hybrids between the two populations or species that have diverged. For example, Bradshaw et al. first created a large population of hybrids between *M.*

QTL studies can map the loci responsible for divergence between closely related species.

(a)   (b)

Figure 9.17   **Contrasting floral traits in monkeyflowers**   *Mimulus cardinalis* (a) is hummingbird-pollinated while *M. lewisii* (b) is bee-pollinated. The following table summarizes contrasts in eight floral traits. (Courtesy of Toby Bradshaw and Doug Schemske, University of Washington. Photo by Jordan Rehm.)

Characteristic	M. cardinalis	M. lewisii
Magenta/pink pigment (anthocyanins) in petals	high	low
Yellow pigment (carotenoids) in petals	low	high
Corolla width	low	high
Petal width	low	high
Nectar volume	high	low
Nectar concentration	low	high
Stamen (male structure) length	high	low
Pistil (female structure) length	high	low

Notes:

• The yellow pigment in *M. lewisii* petals is arranged in stripes called nectar guides, which are interpreted as a "runway" for bees as they land on the wide petals.

• Nectar volume and concentration are thought to contrast in bird- and bee-flowers simply because of the enormous difference in body size (hummingbirds can drink a lot more).

• The difference in stamen and pistil length is important: In *M. cardinalis* these structures extend beyond the flower and make contact with the hummingbird's forehead as it feeds.

Figure 9.18   **F₂ hybrids between *Mimulus cardinalis* and *M. lewisii***   These individuals are the products of self-fertilization of F₁ interspecific hybrids, and show a wide variety of floral characters. (Doug W. Schemske, University of Washington)

cardinalis and *lewisii*. These F₁ individuals tend to be intermediate in flower type and not especially variable. This is because they are all heterozygous for flower size, shape, and color alleles from *M. cardinalis* and *M. lewisii*. But when these F₁ individuals are self-pollinated, their alleles for flower morphology segregate into many different combinations. Accordingly, the F₂ offspring have an astonishing variety of flower colors, shapes, and nectar rewards (Figure 9.18).

In QTL mapping, researchers try to find correlations between the size or color or shape of these F₂ individuals and genetic markers that they carry (Box 9.3). If a statistically significant association is found between the trait and the marker that has been mapped, it implies that a QTL near that marker contributes to that trait. In the study done by Bradshaw et al., the researchers found that each of the eight floral traits listed in Figure 9.17 was associated with at least one QTL. In fact, in every case there was one QTL that accounted for 25% or more of the total variation in the trait. Because this is high for a polygenic trait, Bradshaw et al. refer to these loci as genes with major effects. But it is clear that many loci contributed to divergence in flower shape and speciation.

What is the next step in this research program? The team is putting potted F₂ hybrids in the field in an effort to recreate how selection acted to produce divergence. Their goal is to measure the reproductive success of these variable floral types in the presence of both hummingbirds and bees. If these experiments are successful in quantifying selection on particular floral traits, they would join Dolph Schluter's work with sticklebacks, reviewed in section 3.3, as experimental evidence for selection gradients that contribute to speciation.

9.6 Asexual Species

Much of the research reviewed in this chapter has focused on evolutionary processes that lead to reproductive isolation. Indeed, the biological species concept treats this property as *the* criterion of speciation. But members of asexual species reproduce without exchanging genetic material. In a majority of bacteria, genetic recombination appears to be rare or absent. When gene exchange among bacteria does occur, it is limited to small segments of the genome. Further, it is often unidirectional and can occur between members of widely

Box 9.3 QTL mapping

The goal of QTL mapping is to find statistical associations between genetic markers unique to each parental species and the value of the trait in the hybrid offspring. Whenever an association is confirmed, we know that a locus at the marker, or tightly linked to it, contributes to the trait in question (Figure 9.19). By quantifying the degree of association between the QTL and the phenotype, we can also estimate how much of the observed variation is due to the alleles at each locus. That is, we can distinguish between genes that have major or minor effects on the traits we are interested in.

There are several tricks that increase the efficiency of a QTL analysis. First, it helps to have a very large series of species-specific genetic markers scattered among all of the chromosomes in the species being studied. In the monkeyflower study, Bradshaw et al. (1995) mapped 153 RAPD markers, five allozymes, and one visible trait (changes in the distribution of yellow color due to the *yup* locus). Second, it is easiest to identify statistically significant

associations between those markers and the effects of nearby genes if the traits being investigated vary dramatically. As Figure 9.18 shows, the phenotypic variation that Bradshaw et al. could measure in F_2 hybrids was remarkable. Third, it is crucial to measure these effects in a very large number of hybrid offspring in order to find statistically significant associations. Bradshaw et al. measured flower phenotypes and mapped genetic markers in 93 hybrid offspring.

QTL analysis is becoming increasingly important in agricultural research, as well as evolutionary studies. Researchers are mapping commercially important quantitative traits, like the water content of tomatoes, and using them to design breeding programs. For an entree to this literature, see Michelmore and Shaw (1988), Paterson et al. (1988), Lander and Botstein (1989), Haley and Knott (1992), and the more recent references in Bradshaw et al. (1995). For a superb overview of QTL mapping, see Tanksley (1993).

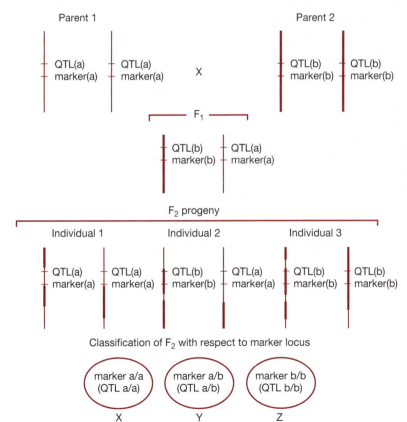

Figure 9.19 **Experimental protocol for QTL mapping** This diagram from Tanksley (1993) illustrates a simplified protocol for QTL mapping (in reality, hundreds of markers are usually used). The vertical lines represent chromosomes from each parental population. Marker (a) and marker (b) can be any type of species-specific genetic marker (RAPD loci are the most common type employed). The locus marked QTL on each chromosome represents a locus that affects the trait we are interested in. As the figure shows, F_1 hybrids are heterozygous for the QTLs and nearby markers. When these individuals are self-fertilized (or mated with sibs), the F_2 progeny will have a wide variety of genotypes and phenotypes. These F_2 progeny are then grouped according to phenotypes. For example, in the Bradshaw et al. study, hybrid offspring with reddish petals would be placed in one group (analogous to group X in the figure) and offspring with purple petals in another. The genotypes of each individual in the groups is assessed. Finally, the investigators look for statistical associations between the genotypes and phenotypes found in the different groups.

diverged taxa. What most of us think of as normal outcrossing, with reciprocal exchange of homologous halves of genomes, is unheard of in enormous numbers of organisms. And asexuality is not limited to prokaryotes. Asexual populations are not uncommon in plants and have been described in a wide variety of animal phyla.

Understanding the nature of gene flow and speciation in asexual organisms is another example of an evolutionary issue with practical applications. Genes for antibiotic resistance, which are normally found in extra-chromosomal loops of DNA called plasmids, have frequently been exchanged by widely diverged genera of bacteria. Furthermore, prokaryotes like the bacterium *Escherichia coli* are frequent targets of genetic engineering studies with industrial or agricultural applications. Research on the mechanisms and dynamics of gene exchange in bacteria may help us better manage both an enormous public health problem—the evolution of antibiotic resistance—and the release of genetically engineered organisms.

Species Concepts and Gene Flow in Bacteria

In eukaryotes, genetic exchange is generally limited to organisms whose genomes have diverged a total of 2% or less, and eukaryotic species that exchange genes are almost always classified in the same genus. In contrast, gene flow can occur between bacteria taxa whose genomes have diverged up to 16% (Cohan 1995). Bacteria that are classified as members of different phyla can and do exchange genes, via plasmids (Cohan 1994).

Recombination in bacteria has several other prominent characteristics: It is rare, unidirectional, and limited to small segments of the genome (Figure 9.20). Clearly, sexual isolation is not a prerequisite for divergence in bacteria or an appropriate criterion for identifying species. In contrast to the situation in eukaryotes that regularly outcross, gene flow in prokaryotes plays a relatively minor role in homogenizing allele frequencies in populations (Cohan 1994, 1995). Instead, the primary result of gene flow among bacteria is the spread of specific sequences with high fitness advantages. This implies that the population genetics of bacterial populations may be dominated by selective sweeps that are triggered by the acquisition, through sex, of an adaptive allele (Guttman 1997).

Frederick Cohan (1995) has proposed that the key to recognizing bacterial species is finding strains that share a large suite of selectively neutral sequences. The logic behind this proposal is based on a phenomenon we reviewed in Chapter 7: Selective sweeps homogenize neutral variation within populations. If individuals share many neutral characters, the populations are the products of the same selective sweeps and share a recent common ancestry.

Work continues on the task of creating a workable species concept for prokaryotes and on measuring the rate of recombination in natural populations of bacteria. Exploring the mechanisms of speciation in bacteria is an exciting frontier in speciation research.

Asexual Plants and Animals

Asexual species occur in lizards, salamanders, fish, earthworms, freshwater planarians, grasshoppers, weevils, cockroaches, flies, a wide variety of flowering

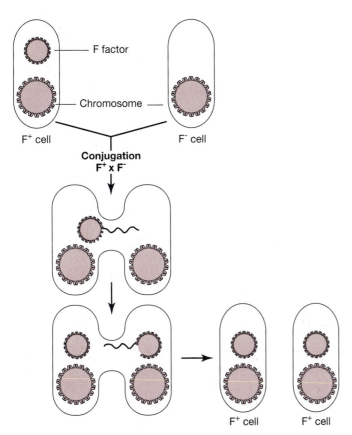

F factor

Chromosome

F⁺ cell   F⁻ cell

Conjugation
F⁺ x F⁻

F⁺ cell   F⁺ cell

Figure 9.20 **Genetic recombination in the bacterium *Escherichia coli***   Bacterial cells with an extrachromosomal loop of DNA called an F (for fertility) factor can engage in recombination. The process starts when cells with F factors form conjugation tubes with F⁻ cells. A copy of the F factor migrates through the conjugation tube, converting the recipient cell from F⁻ to F⁺. Occasionally F factors will integrate into the chromosome. These integrated sequences can later leave the chromosome, frequently taking chromosomal sequences (that is, new genes) with them. In this way, F factors can transfer genes between bacterial chromosomes.

plant families, and some ferns (White 1978; Grant 1981). In animals the most spectacular concentration of asexual species is in freshwater invertebrates called the bdelloid rotifers: No males have been found in any of the 200 or so species described to date. And some of the most widespread and numerically abundant flowering plant species belong to groups called agamic complexes. These complexes originate with a single pair of diploid, sexual ancestors whose descendants undergo repeated hybridization and polyploidization events. The result is a group of closely related asexual populations (the agamic complex). But considering the overall diversity of plant and animal life, the general picture is of the odd asexual lineage scattered among enormously more diverse sexual clades.

   This is not to say that asexual species are not interesting. The mechanisms for zygote production, for example, vary wildly. In the freshwater fish genus *Poeciliopsis,* asexual "hemiclones" consist of female *P. monacha* that mate with males of *P. occidentalis.* Although the offspring are diploid, the male's chromosomes are destroyed before meiosis so that inheritance is strictly maternal (and thus asexual). In asexual species of salamanders in the genus *Ambystoma,* eggs have to be stimulated by sperm from males of a closely related species for development to take place, even though fertilization does not occur. The eggs are already diploid because a chromosome doubling event precedes meiosis. The analogous process in plants, in which pollination is required to stimulate the development of a diploid but unfertilized egg, is called pseudogamy ("false marriage"). In other asexual plant species, seeds develop directly from somatic tissue in the parent plant without benefit of meiosis.

Chromosome studies have shown that almost all of the unisexual vertebrates, and the vast majority of asexual angiosperms and ferns, were produced through hybridization of closely related sexual species. When researchers compare the karyotypes of asexual species to closely related sexual forms, it is common to find that half the chromosome complement in the asexual individuals matches each putative parent species.

9.7 Rates of Speciation

Some central issues in speciation research are purely descriptive matters. We want to know something about the natural history of speciation rates. How fast can new species form? How quickly do new species form on average? Are there differences among taxa in how rapidly speciation occurs?

We tackle these questions in order, starting with a look at what may be the fastest radiation of species in the history of life.

Explosive Speciation in African Cichlids

The cichlids are a family of freshwater fish distributed throughout the tropics and subtropics. Of the 1,300 species found worldwide, almost 1,000 are endemic to the Rift Lakes of East Africa (Figure 9.21). Lakes Victoria, Malawi, and Tanganyika are well-dated and their cichlid faunas well-studied (Table 9.4). The Rift Lake cichlids employ feeding strategies that are astonishingly varied, even when compared to the radiation of Galápagos finches reviewed in Chapter 2. Each lake contains cichlid species that eat fish, mollusks, insect larvae, algae, zooplankton, or phytoplankton. Within each of these broad categories, species vary by method of feeding or microhabitat (for example, the depth of water where they feed). Most are brightly colored, and recent studies of breeding systems imply that sexual selection may be intense. Many of the species are similar in other respects or even cryptic morphologically, often differing only in the shape of their teeth. The majority brood their eggs, and later their young, in their mouths. (Some species even feed by clamping the snouts of female cich-

(a)

(b)

(c)

Figure 9.21 **Rift Lake cichlids** Many of the cichlid species found in the Rift Lakes of Africa are similar in size and shape. They differ primarily in color and in the morphology of their teeth.   (a) The snail dweller cichlid *(Lamprologus brevis)* of Lake Tanganyika nests in empty snail shells. (Ken Lucas/Visuals Unlimited)   (b) A Lake Malawi endemic. (Mark Boulton/Photo Researchers, Inc.)   (c) The possum cichlid *(Haplochromis livingstonii),* also from Lake Malawi. (Tom McHugh/Photo Researchers, Inc.)

TABLE 9.4   Cichlid species in the Rift Lakes of Africa

	Age of lake (years)	Total number of cichlid species	Number of endemic cichlids
Lake Victoria	<1,000,000	300	Almost all
Lake Malawi	1,000,000 to 2,000,000	500	All but 4
Lake Tanganyika	Around 2,000,000	171	Almost all

lids, forcing them to eject the young, and then eating those young—a practice known as paedophagy.)

The African cichlids have long been a textbook example of rapid speciation. But two new sources of data have refined our ideas about just how much speciation has occurred, and how fast.

Early work on the evolution of cichlids used morphological traits to delineate species and genera. These taxonomic treatments assigned the Rift Lake cichlids to a large number of different genera, and concluded that species within the same genus are often found in different lakes. This result implied that the radiation within each lake started with many different founders.

Phylogenies estimated from DNA sequence data, however, support the idea that the cichlid fauna in each lake is monophyletic (that is, descended from a single founder species). Axel Meyer and co-workers (1990) have sequenced a total of at least 803 base pairs from several genes in the mitochondrial DNA of 24 Lake Malawi species, 14 Lake Victoria species, and eight Lake Tanganyika species. The trees they have generated thus far clearly separate the species found in each lake (Figure 9.22). Thus far, no lineage seems to have descendants in more than one lake.

If the pattern of monophyly is confirmed when more species are added to the analysis, it will contrast sharply with the early morphological work. If enough

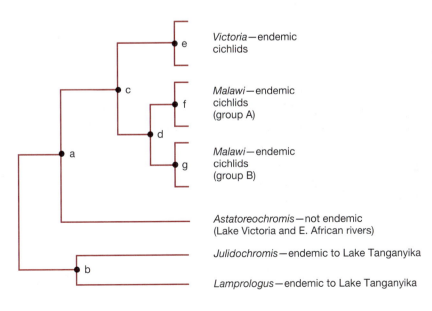

Figure 9.22 **A partial phylogenetic tree of Rift Lake cichlids**   This phylogenetic tree was estimated from sequence differences in mitochondrial DNA (Meyer et al. 1990). The 14 species of cichlids endemic to Lake Victoria, which radiated from node e, are so similar genetically that they cannot be distinguished on the tree and have to be lumped. The 24 species in the study from Lake Malawi, which radiated from node d, fall into two very closely related groups on the tree. Group A species, which radiated from node f, all live in sandy habitats, while Group B species, which radiated from node g, all live in rocky habitats. The Lake Tanganyika endemics in the study radiated from node b.

data accumulate to reverse the earlier conclusion, it means that all the living cichlids in each lake are descended from a single population. The implication would be that speciation rates have been even faster than previously thought.

The second type of data comes from sediment corings and other techniques of studying lake history. The most recent work demonstrates that Lakes Malawi and Victoria experienced dramatic reductions in depth during the Pleistocene. For example, Owen et al. (1990) found soil layers derived from sand dunes when they analyzed shallow sediment cores taken under Lake Malawi. The depth of these sandy strata indicate that the lake dropped at least 120m as recently as 500 years ago. Deeper cores and soundings, done in earlier studies, have shown that even larger regressions occurred in Lake Malawi earlier in the Pleistocene. Many of the cichlids endemic to the lake are found only on single rock outcrops, many of which would have dried up during these drought periods. Thus, species ranges might have contracted as lake levels fell and then expanded or shifted as lake levels rose again. Alternatively, drops in lake level might have been dramatic enough to produce extinction events, and the subsequent rises in lake level great enough to open up new habitats and encourage speciation. We cannot be sure.

Johnson et al. (1996) have strong evidence that Lake Victoria recently dried up *completely*. Victoria is the shallowest of the three lakes, and in sediments recovered from the deepest waters they found a soil stratum, dated at 12,400 years before present, with vertical plant rootlets instead of the horizontal rootlets found in wetland plants. The pollen recovered from this soil layer comes from grasses instead of typical lake edge or marshland plants like rushes, sedges, or cattails. What does this mean for cichlid radiation? The obvious suggestion is that the 300 species of cichlids in Lake Victoria have descended from a common ancestor in less than 10,000 years. This implies that a new fish species was created every 33 years on average.

Consistent with this hypothesis, Sturmbauer and Meyer (1992) found almost no variation in DNA sequences among cichlid species in Lake Victoria. This suggests that very little time has passed since they were part of the same population. In contrast, the cichlids they surveyed from Lake Malawi thus far show higher overall divergence; and six congeneric species from Lake Tanganyika much more still (Figure 9.23). The combination of geologic and genetic data supports the idea that speciation in Lake Victoria has occurred on time scales that both paleontologists and ecologists would call instantaneous.

Rift Lake cichlids have speciated rapidly because of frequent isolation, followed by intense natural selection and genetic drift.

Speciation in Rift Lake cichlids may be extraordinarily rapid simply because the processes we have reviewed in this chapter occur with exceptional frequency and strength. The original colonization event in each lake was apparently from a single founder species. After the original colonization, drought periods produced repeated vicariant events that isolated populations within lakes. Finally, the ecology and mating systems of cichlids provided many opportunities for divergent selection on mate-choice characteristics, habitat choice, and feeding strategies.

Before moving on, we should mention two additional points. Other fish families have undergone similar radiations in several other large freshwater lakes. There are, for example, species flocks analogous to the Rift Lake cichlids in the Philippines' Lake Lanao, Russia's Lake Baikal, Mexico's Laguna Chichancanab,

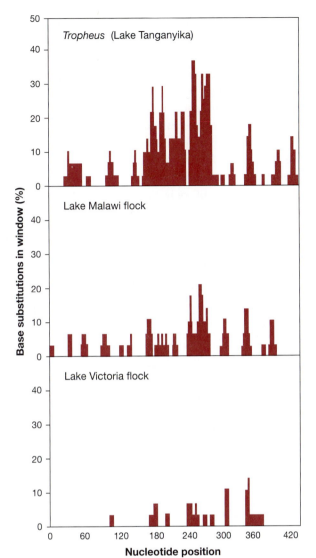

Figure 9.23 **A "sliding window analysis" of genetic divergence among cichlid species** To generate this diagram, Sturmbauer and Meyer (1992) lined up the sequences they had obtained for 54 different individuals from 21 populations of *Tropheus,* a genus endemic to Lake Tanganyika (the horizontal lines in the diagram represent the sequences). Then they divided the 442 base pairs into segments, or windows, of 9 base pairs, with 3 base pairs overlapping on either side of each window. Finally, they counted the number of base substitutions that had occurred in every window of 9 base pairs among the organisms sampled, and calculated the percentage of 9 base pairs that had experienced a substitution. They did the same thing for sequences from 45 individuals in 24 species endemic to Lake Malawi, and from 32 individuals in 14 species endemic to Lake Victoria. Much more genetic diversification has occurred in cichlids from Lake Tanganyika than in Lake Malawi or Lake Victoria populations.

and South America's Lake Titicaca. Second, the African cichlids are threatened. Introduced predators are devastating many populations, and extinctions may already have occurred.

Patterns Across Taxa

Do rates of speciation vary among taxonomic groups in a meaningful way? This is a central question in speciation research, and one that we can address with data from both living and fossil forms.

Several papers out of John Avise's lab have looked for patterns in genetic distances between closely related species pairs in different groups of living vertebrates (Avise and Aquadro 1982; Kessler and Avise 1985). These distances were either measured as the average divergence between proteins resolved in allozyme surveys or as estimates of sequence divergence based on studies of mitochondrial DNA.

If we assume that these genetic distances are roughly proportional to time, it is clear that some vertebrate groups speciate much more rapidly than others (Figure 9.24). In particular, bird species appear to be much younger, on average, than reptiles and amphibians. To a lesser extent, bird species appear younger than fish and mammals. As sequence data continue to accumulate, we should get a better idea of how strong and how general this pattern actually is.

And what does the fossil record have to say about rates of speciation? Steve Stanley (1979) pulled together a massive amount of data on this question and distilled it by constructing Lyellian curves. This is an analytical tool we introduced in Chapter 2. The first step in this analysis is to study a fauna over a well-defined time interval and record the duration of each species. The common way to estimate species durations is to use the difference between date of first and last appearance in the fossil record. To construct a Lyellian curve, then, the x-axis is broken up into regular time intervals, and the percentage of species alive at that time which are still alive today is plotted on the y-axis.

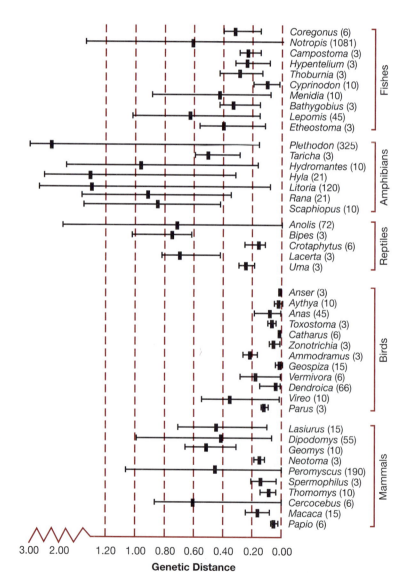

Figure 9.24 Genetic distances between closely related pairs of vertebrate species
John Avise and Charles Aquadro (1982) surveyed the literature to find studies that estimated genetic distances between very closely related species pairs using differences in allozyme frequencies. The black boxes show the mean divergence in each genus shown. The horizontal bars indicate the range (highest and lowest values) of genetic distances in the species assayed. Genetic distances in birds are smaller than in other vertebrates, suggesting that bird species are younger.

Figure 9.25 shows Lyellian curves for European mammals from the Pliocene through late Pleistocene and for bivalve mollusks (clams and oysters) from the Lower Miocene through recent times in the temperate and subtropical Pacific. The curves are markedly different: Virtually none of the European mammals present late in the Pleistocene were alive three million years ago, while almost 70% of bivalves were. To look at this a different way, Stanley points out that doubling the 50% point on the curve (that is, where half of the species are extant) gives a rough estimate for the amount of time required for a complete faunal turnover. This number also represents a rough estimate for the average duration of a species in that fauna. The curves in Figure 9.25 yield the following estimates: 1.4 My for the average length a species of European mammal survives versus 15 My for Pacific bivalves. This is an order of magnitude difference, and reinforces the conclusion we reached previously from reviewing data on genetic distances in extant taxa: Rates of speciation vary widely among taxa. To get a more graphic picture of this, Stanley constructed Lyellian curves for a wide variety of taxa and time periods and produced the dramatic diagram shown in Figure 9.26.

What causes this tremendous variation in speciation rates? To make sense of it, we need to think back to the modes of isolation and divergence presented in sections 9.2 and 9.3. For example, Stanley (1979) has proposed that terrestrial organisms speciate much more rapidly than marine forms because environmental heterogeneity is so much greater on land. His premise is that habitat heterogeneity makes physical isolation and divergence through habitat choice much more likely. The data from African cichlids we just reviewed and the rapid speciation rates shown by marine organisms like trilobites and ammonites (Figure 9.26) present a counterargument, however, and suggest that the hypothesis needs further development.

Elizabeth Vrba (1980, 1987) offers another viewpoint. She agrees that habitat choice and ecological divergence are the key to speciation rates, but proposes

Speciation rates vary dramatically among lineages.

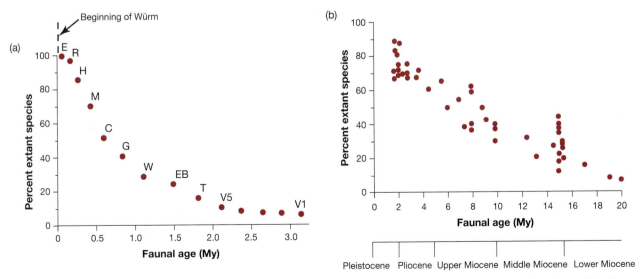

Figure 9.25 **Lyellian curves** Graph (a) shows data for European mammals. Graph (b) plots data from Pacific bivalves (Stanley 1979). Note that the two graphs have very different time scales, as indicated along each horizontal axis. The text interprets the differences in these curves.

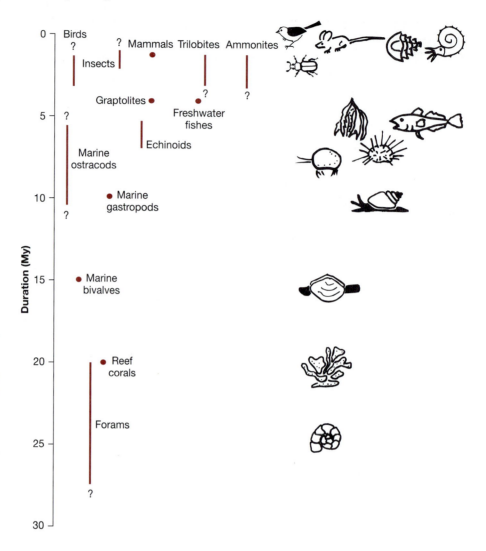

Figure 9.26 **Average species durations for a variety of animal taxa**   This diagram shows the average length of time that species from different animal taxa are found in the fossil record (there is no horizontal axis; groups are arranged only for convenience). Steve Stanley (1979) noted that species in morphologically more complex taxa tend to have shorter life spans than species in morphologically simpler taxa. The animals drawn are not to scale.

that the important contrast is between habitat generalists and specialists, instead of between terrestrial and marine forms. She collected data on African antelopes showing that from the Miocene to the present, clades that only browsed or only grazed speciated much more rapidly than clades that both browsed and grazed (Vrba 1987). Her premise is that the habitat specialists were more likely to be isolated in small areas or to diverge in sympatry in a manner analogous to apple maggot flies. But how general is this hypothesis? Does it work for cichlids? Hominids? Hawaiian *Drosophila*? We simply do not have the data yet to answer these questions.

In conclusion, we have no tidy explanation for the marked variation observed in speciation rates. Stanley considered his analyses of speciation rates across taxa "a progress report in a nascent field of endeavor" (1979: 230). With respect to estimates for species ages in living forms, John Avise and co-workers might agree. Clearly, this field is ripe for additional comparative work, and for fruitful interactions among scientists studying fossil and extant organisms.

Summary

A wide variety of species concepts have been proposed, all agreeing on three essential points: (1) Species consist of groups of interbreeding populations; (2) species are a fundamental unit of evolution; and (3) the distinguishing characteristic of species is evolutionary independence. Different species concepts are distinguished by the criteria employed for recognizing evolutionary independence.

Speciation is usually analyzed as a three-step process: physical isolation of populations caused by dispersal or vicariance, divergence based on drift or selection, and completion or elimination of divergence upon secondary contact. There are numerous exceptions to this sequence, however. In some cases selection for divergence is strong enough that populations can differentiate without physical isolation, as research into apple and hawthorn maggot flies illustrates. Mutations with large effects, like polyploidization, can also produce divergence without physical isolation. Further, a variety of outcomes are possible after secondary contact. These include formation of stable hybrid zones and creation of a new species containing genes from each of the parental forms.

The primary strategy employed in genetic analyses of speciation is to look for correlations between mapped phenotypic or molecular markers and the distribution of traits in F_2 offspring of recently diverged species. These strategies have confirmed that loci on the X chromosome are particularly important in postzygotic isolation in *Drosophila*.

Work on species concepts and speciation processes in asexual taxa is still in its infancy. Several important conclusions have already emerged from this work, however: Genetic recombination in bacteria is rare, one-way (from a donor to recipient cell), involves small segments of the genome, and can involve very widely diverged taxa.

Although work on speciation rates is also preliminary, the analyses done to date indicate that speciation rates vary widely among both fossil and extant taxa. Various hypotheses are now being explored to explain the differences among groups.

Evolution has produced a tree of life. Understanding the size, shape, and growth of this tree is a fundamental goal of evolutionary analysis. Our exploration of this topic continues in Part III: The History of Life.

Questions

1. Different species definitions tend to be advocated by people working in different disciplines. Consider the needs of a botanist working on the ecology of oak trees, a zoologist working on bird distributions, a conservation biologist working on endangered marine turtle populations, and a paleontologist working on extinction events in fossils from planktonic organisms. Which species definition would be most useful to each scientist?

2. Hybridization is proving to be very common in plants and fairly common in animals. How does this affect your ideas about the biological species concept? Suppose a species with a very small population size and many unique genetic characteristics is found to be hybridizing with a non-endangered species. Does this mean the

endangered population is not really a species? If resources committed to preservation efforts are scarce, should the endangered population be allowed to go extinct?

3. When the Panama land bridge between North and South America was completed, some North American mammal lineages crossed to South America and underwent dramatic radiations. For terrestrial species, did the completion of the land bridge represent a vicariance or dispersal event?

4. Would the recent glaciation events in northern Europe and North America have created vicariance events? If so, how? Which organisms would have been affected?

5. Apple maggot flies are not the only insects that seem to be evolving to exploit new host plants. Within the past 50 years, soapberry bug populations in the United States have diversified into host race populations distinguished by markedly different beak lengths (Carroll and Boyd 1992). These bugs eat the seeds inside soapberry fruits (Family Sapindaceae). Native and introduced varieties of soapberries differ greatly in size. Describe the experiments or observations you would make to launch an in-depth study of host race formation in these bugs. What data would tell you whether they are separate populations evolving independently, or a single interbreeding population? Many museums contain insect specimens from decades ago. What would you examine in these old specimens? What information about the host plants would be useful?

6. Habitat patchiness and dispersal ability are important factors affecting a population's tendency to diverge from other populations. An unusual example is the red crossbill, a small finch specialized for eating seeds from conifer trees. Red crossbills are nomadic and often fly thousands of kilometers each year. Despite this high mobility, there are several different "types" of crossbills with distinctive bill shapes, body sizes, and vocalizations. Each type prefers to feed on a different species of conifer, and each species of conifer is found only in certain patches of forest. Conifer cones are difficult to break open, and only birds with a certain bill size and shape can efficiently open cones of a certain conifer species. Explain how a highly mobile animal such as the red crossbill could possibly be diverging into separate species in the absence of any geographic barrier. How could you test your ideas? If crossbills could not fly, do you think speciation would occur more quickly or more slowly? If conifer species were not patchily distributed, do you think crossbill speciation would occur more quickly or more slowly? In general, how do habitat patchiness and dispersal ability interact to affect divergence?

7. The monkeyflowers studied by Bradshaw et al. tend to be found at different altitudes. The hummingbird-pollinated *Mimulus cardinalis* occurs at higher elevations and the bee-pollinated *M. lewisii* at lower elevations. In addition to the QTLs responsible for the differences in flower color and shape, it is likely that loci affecting physiological traits, like the ability to photosynthesize and grow at colder temperatures, have also diverged between the two species. Write a protocol describing how you would go about mapping these traits.

8. Only about 10% of bacterial species can be cultured in the lab. How does this fact affect our ability to measure rates of recombination in bacteria?

9. Would you expect that organisms with small population sizes would be more or less likely to speciate rapidly? Why?

10. It is estimated that 10% of living species are marine. Does this fact correlate or conflict with Steve Stanley's (1979) contention that marine habitats are much more homogenous than terrestrial ones?

11. On the graphs of genetic distances given in Figure 9.24, where would you expect the Rift Lake cichlids to fall?

Exploring the Literature

12. Many insect species carry a peculiar bacterium called *Wolbachia,* which is passed from mother to offspring in the cytoplasm of the egg. *Wolbachia* does not appear to affect adult health, but it does affect reproduction by causing "cytoplasmic incompatibility" between the sperm of infected males and the eggs of uninfected females. Offspring of these crosses die, but offspring of other crosses develop normally. As a result, *Wolbachia* can affect the population structure of its host species and may even play a role in host speciation. To learn more about this bizarre parasite and its possible role in speciation of arthropods, see:

Frank, S.A. 1997. Cytoplasmic incompatibility and population structure. *Journal of Theoretical Biology* 184:327–330.

Werren, J.H. 1994. Genetic invasion of the body snatchers. *Natural History* 6:36–38.

Werren, J.H., D. Windsor, and L. Guo. 1995. Distribution of *Wolbachia* among neotropical arthropods. *Proceedings of the Royal Society of London, Series B* 262:197–204.

Citations

Abbott, R.J. 1992. Plant invasions, interspecific hybridization and the evolution of new plant taxa. *Trends in Ecology and Evolution* 7:401–405.

Arnold, M.L., and S.A. Hodges. 1995. Are natural hybrids fit or unfit relative to their parents? *Trends in Ecology and Evolution* 10:67–71.

Arriola, P.E., and N.C. Ellstrand. 1996. Crop-to-weed gene flow in the genus Sorghum (Poaceae): Spontaneous interspecific hybridization between johnsongrass, *Sorghum halepense,* and crop sorguhm, *S. bicolor. American Journal of Botany* 83:1153–1160.

Avise, J.C., and C.F. Aquadro. 1982. A comparative summary of genetic distances in the vertebrates. *Evolutionary Biology* 15:151–185.

Barton, N.H. 1989. Founder effect speciation. In D. Otte and J.A. Endler, *Speciation and Its Consequences* (Sunderland, MA: Sinauer), 229–256.

Barton, N.H., and B. Charlesworth. 1984. Genetic revolutions, founder effects, and speciation. *Annual Review of Ecology and Systematics* 15:133–164.

Barton, N.H., and G.M. Hewitt. 1985. Analysis of hybrid zones. *Annual Review of Ecology and Systematics* 15:133–164.

Beverly, S.M., and A.C. Wilson. 1985. Ancient origin for Hawaiian Drosophilinae inferred from protein comparisons. *Proceedings of the National Academy of Sciences, USA* 82:4753–4757.

Bradshaw, H.D. Jr., S.M. Wilbert, K.G. Otto, and D.W. Schemske. 1995. Genetic mapping of floral traits associated with reproductive isolation in monkeyflowers *(Mimulus). Nature* 376:762–765.

Brown, J.M., W.G. Abrahamson, and P.A. Way. 1996. Mitochondrial DNA phylogeography of host races of the goldenrod ball gallmaker, *Eurosta solidaginis* (Diptera: Tephritidae). *Evolution* 50:777–786.

Bryant, E.H., S.A. McCommas, and L.M. Combs. 1986. The effect of an experimental bottleneck upon quantitative genetic variation in the housefly. *Genetics* 114:1191–1211.

Bryant, E.H., and L.M. Meffert. 1993. The effect of serial founder-flush cycles on quantitative genetic variation in the housefly. *Heredity* 70:122–129.

Butlin, R. 1987. Speciation by reinforcement. *Trends in Ecology and Evolution* 2:8–13.

Butlin, R. 1995. Reinforcement: An idea evolving. *Trends in Ecology and Evolution* 10:432–434.

Carroll, S.P., and C. Boyd. 1992. Host race radiation in the soapberry bug: Natural history with the history. *Evolution* 46:1052–1069.

Carson, H.L., F.C. Val, C.M. Simon, and J.W. Archie. 1982. Morphometric evidence for incipient speciation in *Drosophila silvestris* from the island of Hawaii. *Evolution* 36:132–140.

Carson, H.L., and R. Lande. 1984. Inheritance of a secondary sexual character in *Drosophila silvestris. Proceedings of the National Academy of Sciences, USA* 81:6904–6907.

Carson, H.L., and A.R. Templeton. 1984. Genetic revolutions in relation to speciation phenomena: The founding of new populations. *Annual Review of Ecology and Systematics* 15:97–131.

Charlesworth, B., R. Lande, and M. Slatkin. 1982. A neo-Darwinian commentary on macroevolution. *Evolution* 36:474–498.

Cohan, F.M. 1994. Genetic exhange and evolutionary divergence in prokaryotes. *Trends in Ecology and Evolution* 9:175–180.

Cohan, F.M. 1995. Does recombination constrain neutral divergence among bacterial taxa? *Evolution* 49:164–175.

Coyne, J.A. 1994. Rules for Haldane's rule. *Nature* 369:189–190.

Coyne, J.A., and M. Kreitman. 1986. Evolutionary genetics of two sibling species, *Drosophila simulans* and *D. sechellia. Evolution* 40:673–691.

Coyne, J.A., and H.A. Orr. 1989. Two rules of speciation. In D. Otte and J.A. Endler, eds. *Speciation and Its Consequences* (Sunderland, MA: Sinauer), 180–207.

Coyne, J.A., and H.A. Orr. 1997. "Patterns of speciation in *Drosophila*" revisited. *Evolution* 51:295–303.

Cracraft, J. 1989. Speciation and its ontology: The empirical consequences of alternative species concepts for understanding patterns and processes of differentiation. In D. Otte and J.A. Endler. *Speciation and its Consequences* (Sunderland, MA: Sinauer), 28–59.

DeSalle, R., and L.V. Giddings. 1986. Discordance of nuclear and mitochondrial DNA phylogenies in Hawaiian *Drosophila. Proceedings of the National Academy of Sciences, USA* 83:6902–6906.

DeSalle, R., L.V. Giddings, and A.R. Templeton. 1986. Mitochondrial DNA variability in natural populations of Hawaiian *Drosophila.* I. Methods and levels of variability in *D. silvestris* and *D. heteroneura* populations. *Heredity* 56:75–85.

DeSalle, R., and J.A. Hunt. 1987. Molecular evolution in Hawaiian drosophilids. *Trends in Ecology and Evolutionary Biology* 2:213–216.

Dobzhansky, T. 1937. *Genetics and the Origin of Species* (New York: Columbia University Press).

Donoghue, M.J. 1985. A critique of the biological species concept and recommendations for a phylogenetic alternative. *Bryologist* 88:172–181.

Endler, J.A. 1977. *Geographic Variation, Speciation, and Clines* (Princeton, NJ: Princeton University Press).

Endler, J.A. 1989. Conceptual and other problems in speciation. In D. Otte and J.A. Endler, eds. *Speciation and Its Consequences* (Sunderland, MA: Sinauer), 625–648.

Feder, J.L. 1995. The effects of parasitoids on sympatric host races of *Rhagoletis pomonella* (Diptera: Tephritidae). *Ecology* 76:801–813.

Feder, J.L., C.A. Chilcote, and G.L. Bush. 1988. Genetic differentiation between sympatric host races of the apple maggot fly *Rhagoletis pomonella.* Nature 336:61–64.

Feder, J.L., C.A. Chilcote, and G.L. Bush. 1989a. Inheritance and linkage relationships of allozymes in the apple maggot fly. *Journal of Heredity* 80:277–283.

Feder, J.L., C.A. Chilcote, and G.L. Bush. 1989b. Are the apple maggot, *Rhagoletis pomonella,* and the blueberry maggot, *R. mendax,* distinct species? Implications for sympatric speciation. *Entomological Experiments and Applications* 51:113–123.

Feder, J.L., C.A. Chilcote, and G.L. Bush. 1990. The geographic pattern of genetic differentiation beween host associated populations of *Rhagoletis pomonella* (Diptera: Tephritidae) in the eastern United States and Canada. *Evolution* 44:570–594.

Feder, J.L., S.B. Opp, B. Wlazlo, K. Reynolds, W. Go, and S. Spisak. 1994. Host fidelity is an effective pre-mating barrier between sympatric races of the Apple Maggot Fly, *Rhagoletis pomonella. Proceedings of the National Academy of Sciences, USA* 91:7990–7994.

Feder, J.L., K. Reynolds, W. Go, and E.C. Wang. 1995. Intra- and interspecific competition and host race formation in the apple maggot fly, *Rhagoletis pomonella* (Diptera: Tephritidae). *Oecologia* 101:416–425.

Felsenstein, J. 1981. Skepticism towards Santa Rosalia, or why are there so few kinds of animals? *Evolution* 35:124–138.

Fisher, R.A. 1958. *The Genetical Theory of Natural Selection* (New York: Dover).

Freeman, D.C., W.A. Turner, E.D. McArthur, and J.H. Graham. 1991. Characterization of a narrow hybrid zone between two subspecies of big sagebrush (*Artemisia tridentata:* Asteraceae). *American Journal of Botany* 78:805–815.

Freeman, D.C., J.H. Graham, D.W. Byrd, E.D. McArthur, and W.A. Turner. 1995. Narrow hybrid zone between two subspecies of big sagebrush *Artemisia tridentata* (Asteraceae). III. Developmental instability. *American Journal of Botany* 82:1144–1152.

García-Ramos, G., and M. Kirkpatrick. 1997. Genetic models of adaptation and gene flow in peripheral populations. *Evolution* 51:21–28.

Goodnight, C.J. 1987. On the effect of founder events on epistatic genetic variance. *Evolution* 41:80–91.

Goodnight, C.J. 1988. Epistasis and the effect of founder events on the additive genetic variance. *Evolution* 42:441–454.

Graham, J.H., D.C. Freeman, and E.D. McArthur. 1995. Narrow hybrid zone between two subspecies of big sagebrush. II. Selection gradients and hybrid fitness. *American Journal of Botany* 82:709–716.

Grant, P.R., and B.R. Grant. 1992. Hybridization of bird species. *Science* 256:193–197.

Grant, P.R., and B.R. Grant. 1996. Speciation and hybridization in island birds. *Philosophical Transactions of the Royal Society of London,* Series B 351:765–772.

Grant, V. 1981. *Plant Speciation* (New York: Columbia University Press).

Guttman, D.S. 1997. Recombination and clonality in natural populations of *Escherichia coli. Trends in Ecology and Evolution* 12:16–22.

Haley, C.S., and S.A. Knott. 1992. A simple regression method for mapping quantitative-trait loci in line crosses using flanking markers. *Heredity* 69:315–324.

Hewitt, G.M. 1988. Hybrid zones—natural laboratories for evolutionary studies. *Trends in Ecology and Evolution* 3:158–166.

Jackson, J.B.C., and A.H. Cheetham. 1990. Evolutionary significance of morphospecies: A test with cheilostome Bryozoa. *Science* 248:579–583.

Jackson, J.B.C., and A.H. Cheetham. 1994. Phylogeny reconstruction and the tempo of speciation in the cheilostome Bryozoa. *Paleobiology* 20:407–423.

Johnson, T.C., C.A. Scholz, M.R. Talbot, K. Kelts, R.D. Ricketts, G. Ngobi, K. Beuning, I. Ssemmanda, and J.W. McGill. 1996. Late Pleistocene desiccation of Lake Victoria and rapid evolution of cichlid fishes. *Science* 273:1091–1093.

Kaneshiro, K.Y., and C.R.B. Boake. 1987. Sexual selection and speciation: Issues raised by Hawaiian *Drosophila*. *Trends in Ecology and Evolution* 2:207–213.

Kaneshiro, K.Y., and J.S. Kurihara. 1981. Sequential differentiation of sexual behavior in populations of *Drosophila silvestris*. *Pacific Science* 35:177–183.

Kessler, L.G., and J.C. Avise. 1985. A comparative description of mitochondrial DNA differentiation in selected avian and other vertebrate genera. *Molecular Biology and Evolution* 2:109–125.

Knowlton, N., L.A. Weigt, L.A. Solórzano, D.K. Mills, and E. Bermingham. 1993. Divergence in proteins, mitochondrial DNA, and reproductive incompatibility across the isthmus of Panama. *Science* 260:1629–1632.

Lack, D. 1947. *Darwin's Finches.* (Cambridge: Cambridge University Press).

Lande, R. 1980. Genetic variation and phenotypic evolution during allopatric speciation. *American Naturalist* 116:463–479.

Lande, R. 1981. Models of speciation by sexual selection on polygenic traits. *Proceedings of the National Academy of Sciences, USA* 78:3721–3725.

Lande, R. 1982. Rapid origin of sexual isolation and character divergence in a cline. *Evolution* 36:213–223.

Lander, E.S., and D. Botstein. 1989. Mapping Mendelian factors underlying quantitative traits using RFLP linkage maps. *Genetics* 121:185–199.

Liou, L.W., and T.D. Price. 1994. Speciation by reinforcement of premating isolation. *Evolution* 48:1451–1459.

Masterson, J. 1994. Stomatal size in fossil plants: Evidence for polyploidy in majority of angiosperms. *Science* 264:421–424.

Mayr, E. 1942. *Systematics and the Origin of Species* (New York: Columbia University Press).

Mayr, E. 1963. *Animal Species and Evolution* (Cambridge, MA: Harvard University Press).

McKitrick, M.C., and R.M. Zink. 1988. Species concepts in ornithology. *Condor* 90:1–14.

Meyer, A., T.D. Kocher, P. Basasibwaki, and A.C. Wilson. 1990. Monophyletic origin of Lake Victoria cichlid fishes suggested by mitochondrial DNA sequences. *Nature* 347:550–553.

Michelmore, R.W., and D.V. Shaw. 1988. Character dissection. *Nature* 335:672–673.

Noor, A.M. 1995. Speciation driven by natural selection in *Drosophila*. *Nature* 375:674–675.

Nowak, R.M. 1979. North American Quaternary *Canis*. Monograph 6, Museum of Natural History, University of Kansas, Lawrence.

Nowak, R.M. 1992. The red wolf is not a hybrid. *Conservation Biology* 6:593–595.

Owen, R.B., R. Crossley, T.C. Johnson, D. Tweddle, I. Kornfield, S. Davison, D.H. Eccles, and D.E. Engstrom. 1990. Major low levels of Lake Malawi and their implications for speciation rates in cichlid fishes. *Proceedings of the Royal Society of London,* Series B 240:519–553.

Pashley, D.P. 1988. Quantitative genetics, development, and physiological adaptation in host strains of the fall army worm. *Evolution* 42:93–102.

Paterson, A.H., E.S. Lander, J.D. Hewitt, S. Peterson, S.E. Lincoln, and S.D. Tanksley. 1988. Resolution of quantitative traits into Mendelian factors by using a complete linkage map of restriction fragment length polymorphisms. *Nature* 335:721–726.

Paterson, H.E.H. 1985. The recognition concept of species. In E.S. Vrba, ed. *Species and Speciation* (Pretoria: Transvaal Museum Monograph no. 4), 21–29.

Patton, J.L., and S.W. Sherwood. 1983. Chromosome evolution and speciation in rodents. *Annual Review of Ecology and Systematics* 14:139–158.

Pigeon, D., A. Chouinard, and L. Bernatchez. 1997. Multiple modes of speciation involved in the parallel evolution of sympatric morphotypes of lake whitefish (*Coregonus clupeaformis,* Salmonidae). *Evolution* 51:196–205.

Prokopy, R.J., S.R. Diehl, and S.S. Cooley. 1988. Behavioral evidence for host races in *Rhagoletis pomonella* flies. *Oecologia* 76:138–147.

Rice, W.R. 1987. Speciation via habitat specialization: The evolution of reproductive isolation as a correlated character. *Evolutionary Ecology* 1:301–314.

Rice, W.R., and G.W. Salt. 1990. The evolution of reproductive isolation as a correlated character under sympatric conditions: Experimental evidence. *Evolution* 44:1140–1152.

Rieseberg, L.H. 1995. The role of hybridization in evolution: Old wine in new skins. *American Journal of Botany* 82:944–953.

Rieseberg, L.H., B. Sinervo, C.R. Linder, M.C. Ungerer, and D.M. Arias. 1996. Role of gene interactions in hybrid speciation: Evidence from ancient and experimental hybrids. *Science* 272:741–745.

Roy, M.S., E. Geffen, D. Smith, E.A. Ostrander, and R.K. Wayne. 1994. Patterns of differentiation and hybridization in North American wolflike canids, revealed by analysis of microsatellite DNA. *Molecular Biology and Evolution* 11:553–570.

Ryder, O.A., A.T. Kumamoto, B.S. Durrant, and K. Benirschke. 1989. Chromosomal divergence and reproductive isolation in dik-diks. In D. Otte and J.A. Endler. *Speciation and its Consequences* (Sunderland, MA: Sinauer), 180–207.

Schluter, D. 1996. Ecological speciation in postglacial fishes. *Philosophical Transactions of the Royal Society of London,* Series B 351:807–814.

Soltis, D.E., and P.S. Soltis. 1993. Molecular data and the dynamic nature of polyploidy. *Critical Reviews in the Plant Sciences* 12:243–273.

Stanley, S.M. 1979. *Macroevolution: Pattern and Process* (San Francisco: W.H. Freeman and Company).

Stebbins, G.L. 1950. *Variation and Evolution in Plants* (New York: Columbia University Press).

Sturmbauer, C., and A. Meyer. 1992. Genetic divergence, speciation and morphological stasis in a lineage of African cichlid fishes. *Nature* 358:578–581.

Tanksley, S.D. 1993. Mapping polygenes. *Annual Review of Genetics* 27:205–233.

Templeton, A.R. 1989. The meaning of species and speciation: A genetic perspective. In D. Otte and J.A. Endler. *Speciation and its Consequences* (Sunderland, MA: Sinauer), 3–27.

Templeton, A.R. 1996. Experimental evidence for the genetic-transilience model of speciation. *Evolution* 50:909–915.

Turelli, M., and H.A. Orr. 1995. The dominance theory of Haldane's rule. *Genetics* 140:389–402.

Vrba, E.S. 1980. Evolution, species, and fossils: How does life evolve? *South African Journal of Science* 76:61–84.

Vrba, E.S. 1987. Ecology in relation to speciation rates: Some case histories of Miocene-Recent mammal clades. *Evolutionary Ecology* 1:283–300.

Wayne, R.K., and J.L. Gittleman. 1995. The problematic red wolf. *Scientific American* 273:36–39.

Wayne, R.K., and S.M. Jenks. 1991. Mitochondrial DNA analysis implying extensive hybridizaton of the endangered red wolf *Canis rufus. Nature* 351:565–568.

Wells, M.M., and C.S. Henry. 1992. The role of courtship songs in reproductive isolation among populations of green lacewings of the genus *Chrysoperla* (Neuroptera: Chrysopidae). *Evolution* 46:31–42.

West-Eberhard, M.J. 1983. Sexual selection, social competition, and speciation. *Quarterly Review of Biology* 58:155–183.

White, M.J.D. 1978. *Modes of Speciation* (San Francisco: W.H. Freeman).

Wood, C.C., and C.J. Foote. 1996. Evidence for sympatric genetic divergence of anadromous and nonanadromous morphs of sockeye salmon *(Oncorhynchusnerka). Evolution* 50:1265–1279.

PART III

THE HISTORY OF LIFE

In Part II we introduced the processes responsible for evolutionary change. Part III is devoted to questions about what these processes have produced. When did life on Earth begin? What did the first living organism look like? When did multicellularity evolve? What genetic changes were involved in the diversification of multicellular life?

The two most powerful tools we have for understanding the history of life are phylogenies and the fossil record. In Chapter 10, we review how phylogenies are estimated and used; Chapter 12 introduces the nature and use of the fossil record. The early sections of these two chapters are among the most technique-intensive in the book. The reason for this emphasis is simple: Understanding the methods used to infer phylogenies and the strengths and weaknesses of the fossil record are fundamental to thinking like an evolutionary biologist, and a prerequisite for reading the primary literature on life through time.

The remainder of Part III focuses on patterns in the history of life. Chapter 11 introduces the earliest unicellular life forms and the branching events that led to the groups we call bacteria, fungi, protozoans, animals, and plants. Fossils are rare in Earth's most ancient rocks, so much of what we know about these early life forms depends on what we can infer from the phylogeny of extant groups. In Chapter 12 we investigate the evolution of multicellular life, with a special focus on how paleontologists and molecular geneticists are collaborating to explore the genetics of major evolutionary innovations. Chapter 13 covers mass extinctions. These were events that pruned the tree of life, and cleared the stage for dramatic changes in composition of the biota.

Estimating and Using Phylogenies

The evolutionary history of a group is called its phylogeny. A phylogenetic tree is a graphical summary of this history. The tree describes the pattern, and in some cases the timing, of branching events (Box 10.1). In earlier chapters we presented phylogenies for a wide variety of groups and then used them to answer questions about the pattern of evolutionary change. But we have been using these trees without asking how they were put together. The history of a group can never be known with absolute certainty — it must be estimated from data. How do we know if our estimates are correct? If there is an error in the way branches are placed, it could radically affect the types of conclusions we have reached.

These issues are not just academic, because phylogeny inference methods are being used to understand the origins of emerging viruses and the spread of disease. In the early 1990s, for example, an HIV-positive dentist in Florida was suspected of transmitting the virus to one of his patients. In response, many of his patients were tested for the presence of the virus. Several were positive. Did they get it from the dentist? This was unclear, because some had other risk factors for transmission. To answer the question, investigators from the Centers for Disease Control sequenced the *gp120* gene from viruses in infected patients and infected nonpatients, and estimated the evolutionary relationships of the viruses (Ou et al. 1992). The resulting tree clearly demarcated a "dental clade" (Figure 10.3, page 366). This defined a group of individuals that had acquired the AIDS virus from the dentist. The fact that transmission was confirmed helped promulgate extensive changes in the practice of dentistry and medicine, designed to decrease the risk of transmission from health care providers to patients and vice versa. But was the phylogeny correct? Ronald DeBry and colleagues (1993) challenged the original paper and its conclusions, arguing that the data actually

Box 10.1   A field guide to evolutionary trees

Because phylogenetic trees are an unusual (though extremely effective) way to represent data, we offer this brief guide to reading them.

First we will cover some basic vocabulary. A phylogeny consists of branches, nodes, and tips (Figure 10.1). Branches represent single taxonomic groups through time. Nodes (or forks) occur where an ancestral taxon splits into two or more daughter groups; if more than two daughter groups emerge from a node, the node is called a polytomy. Tips (or terminal nodes) are endpoints representing either extant groups or a dead end in extinction. Taxa that occupy adjacent branches on the tree are called sisters or neighbors.

Phylogenies may or may not be rooted. It is often more interesting to interpret rooted than unrooted trees because rooting establishes the order in which events occurred. In most cases rooting can be accomplished by the outgroup method (Maddison et al. 1984). The first step in this method is to choose an outgroup taxon that is not part of the monophyletic ingroup (this relationship has to be established on the basis of independent evidence). Then, after the best *unrooted* topology has been found by a computer program, the root of the tree is placed between the outgroup(s) specified by the researcher and the ingroup. The program then "grabs that branch and pulls down" to create the basal node in the tree. This node, in turn, represents the common ancestor of the outgroup and ingroup.

Trees can be oriented vertically, with the tips at the top and the root at the base, or horizontally, with the tips at the right and the root at the left. Branches can be represented by either diagonal or squared-off lines (Figure 10.2). If branch lengths are proportional to time or to the amount of genetic change, a scale or labeled axis is provided. Otherwise, branch lengths are arbitrary, and scaled only for readability.

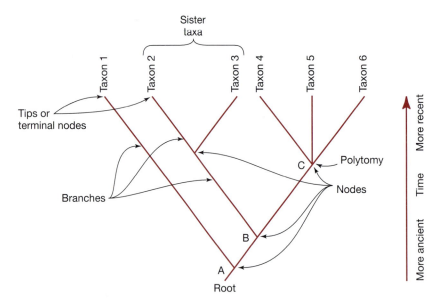

Figure 10.1 Parts of a phylogenetic tree   In this hypothetical phylogeny, node A defines a monophyletic group, or clade, comprising taxa 1−6. Node B defines a monophyletic group comprising taxa 2−6. Node C defines a monophyletic group comprising taxa 4−6.

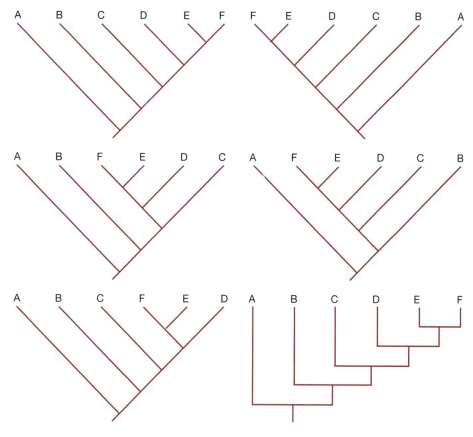

Figure 10.2 **Alternative ways of drawing the same tree** These graphs show six ways to illustrate the same evolutionary relationships. These trees all happen to be oriented vertically, so that the basal taxon is at the bottom and derived groups are toward the top, but they could also be tipped 90° and presented horizontally. From Mayden and Wiley (1992).

contained little support for the clade implicating transmission from dentist to patient. But when David Hillis and co-workers (1994) reanalyzed the original data, they confirmed both the phylogeny and its conclusions.

The task before us now is to understand how phylogenies are estimated, so we can interpret them critically and make independent judgments about their quality. The goal of this chapter is to review the vocabulary and concepts used in the discipline called phylogenetic systematics. If the presentation is successful, readers will be able to evaluate the strengths and weaknesses of the different approaches to phylogeny inference.

We begin by reviewing the types of characters used in building trees. Then we discuss criteria for identifying the best tree that can be estimated from these data, review the computer algorithms used to search for this best tree among the huge number of possible trees, and explore ways to evaluate how good that tree really is. This presentation parallels the sequence of decisions that researchers make when estimating phylogenies (Swofford et al. 1996).

But we also need to explore, in much more depth, the various ways that phylogenies can be used to answer questions about evolutionary patterns and processes. The chapter closes with examples illustrating how phylogenetic thinking is applied in contemporary research. Topics in this last section range widely, from how we should classify the diversity of life to the evolution of ants that farm fungi.

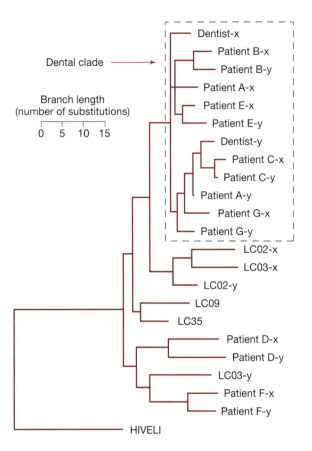

Figure 10.3 **Using a phylogeny to determine the history of HIV transmission** This tree shows the evolutionary relationships among human immunodeficiency viruses in a Florida dentist, a series of his patients, and HIV-positive individuals from the same geographic area (denoted as LC). The x and y designations refer to different HIV clones cultured from the same individual. The outgroup, called HIVELI, is from an African HIV-1 sequence.

Reading up the tree, the first branch off identifies a clade that includes a local individual (LC03-y) who was not a patient of the dentist, along with patients D and F. The existence of this branch shows that these patients did not get the virus from the dentist; it happens that these individuals had other risk factors for transmission. The clade enclosed by the dotted line, however, suggests that a virus from the dentist is ancestral to those found in patients A, B, C, E, and G. These patients did not have other risk factors for the disease, suggesting that they contracted the virus while being treated by the dentist. From Hillis et al. (1994).

10.1 Which Characters Are Best for Inferring Phylogenies?

At its most basic level, the logic of estimating evolutionary relationships is simple: The most closely related taxa should share the most traits. Naively we would say that any traits that are independent of one another, heritable, and variable among the taxa in the study can help us reconstruct who evolved from whom. These characters might be DNA sequences, the presence or absence of certain skeletal elements or flower parts, or the mode of embryonic or larval development.

Things are not nearly this simple in practice, however. Evolutionary relationships are only revealed by traits that are similar because they were derived from a common ancestor (Box 10.2). But species can also share traits because they evolved independently in the lineages leading to those species. Flippers are a derived trait found in both penguins and seals, but they did not evolve through modification in a common ancestor. Instead, the similarity we observe in the limbs of penguins and seals results from natural selection that favored the same type of structure in ancestors from very different lineages (birds and mammals). This is called convergent evolution. Putting penguins and seals next to one another on the tree of life, because we have grouped them according to a shared derived trait, would be an error. Convergent evolution can occur in any type of trait, from DNA sequences to the way embryos develop. When convergence occurs in recently diverged species instead of distant relatives, it is called parallelism.

Shared traits that are derived from a common ancestor are called synapomorphies.

These are not the only confounding issues in estimating phylogenies. Derived traits can also revert to the ancestral form due to mutation or selection. Events like these remove similarity caused by descent from a common ancestor. This wipes out the phylogenetic signal and restricts our ability to estimate evolutionary relationships.

Convergence, parallelism, and reversal are lumped under the term *homoplasy*. Homoplasy represents noise in the datasets used to reconstruct phylogenies. Convergence and parallelism create similarity that is *not* due to modifications in a common ancestor, while reversal removes similarity in species that *are* descended from a common ancestor. Some noise is inevitable in estimating any quantity. Paying careful attention to technique, to reduce measurement error, and amassing large datasets, to reduce the influence of random errors, are important in minimizing the effects of noise.

We can now return to our original question: What types of characters will be most informative for inferring evolutionary relationships? Reliable characters have several characteristics in common:

Phylogenetically informative traits share several important characteristics.

- They are independent. This is an important point, because many of the criteria for choosing a best tree require that each data point used in the study be independent of the others.
- They are homologous. We have emphasized that characters are informative only if they are derived from a common ancestor. This means that data points from different taxa must represent homologous traits.
- They minimize homoplasy. The key to this feature is understanding how rapidly the characters change relative to the age of the taxa being examined. Traits that change extremely rapidly are likely to be unreliable for inferring phylogenies. Unless the taxa being compared diverged very recently, and many taxa are included in the analysis, rapidly changing traits frequently converge and revert. A preference for conservative traits is balanced by an obvious need: Enough change has to have occurred among the taxa in the study to produce clear differences. The goal is to find characters in the group being studied that maximize variability while minimizing homoplasy.
- They are available in large numbers. If homoplasy is relatively rare in a set of characters, then the signal-to-noise ratio will be higher in larger data sets.

After reviewing how selected molecular and morphological characters meet these criteria, we will look at how they are used to estimate evolutionary history.

Molecular Characters

A wide variety of molecular characters have been used to infer phylogenies: allozymes, amino acid sequences, sites where restriction enzymes cut DNA, and DNA sequences, for example. In many cases large quantities of molecular data can be obtained relatively cheaply, and characters can be measured with low error.

The overall homology of molecular characters is established by comparing the same gene or protein across the taxa in the study. More difficult decisions about homology occur when sequence data have to be aligned. The beginning and end of genes can be determined by the start and stop codons, but small insertions or deletions occur frequently in some sequences. These cause gaps that

Box 10.2   Cladistic methods

Techniques that identify monophyletic groups on the basis of shared derived characters are called cladistic methods. Traits that are shared because they were modified in a common ancestor are called synapomorphies. Mammals, for example, share fur and lactation as traits that were derived from their common ancestor (a population from a long-extinct group called the synapsids). Synapomorphies can be identified at whatever taxonomic level a researcher might be interested in: populations, species, genera, families, orders, or classes.

Two ideas are key to understanding why this approach to phylogeny inference is valid. The first is that synapomorphies identify evolutionary branch points. Why? After a species splits into two lineages

that begin evolving independently, some of their homologous traits undergo changes due to mutation, selection, and drift. These changed traits are synapomorphies that identify populations in the two independent lineages. The second key idea is that synapomorphies are nested. That is, as you move back in time and trace an evolutionary tree backward from its tips to its root, each branching event adds one or more synapomorphies. As a result, the hierarchy described by synapomorphies replicates the hierarchy of branching events. The source of these insights is the German entomologist Willi Hennig, who began writing on cladistic approaches in the 1950s.

The cornerstone of tree building with cladistic

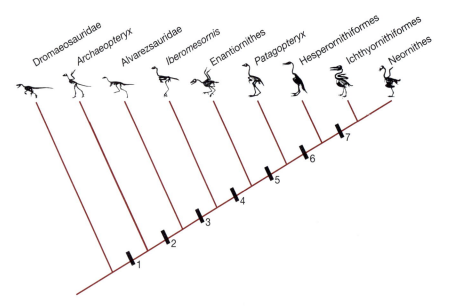

Figure 10.4 **Phylogeny of bird groups from the Mesozoic era**   Luis Chiappe (1995) used skeletal characters to infer the phylogeny of early bird lineages. *Archaeopteryx, Iberomesornis,* and *Patagopteryx* are genera; all the rest of the taxa indicated are families or higher-level groupings.

Examples of the synapomorphies defining each group include (1) fewer than 26 tail vertebrae; (2) a breastbone with a prominent ridge called a keel, (3) a ratio for the lengths of the radius and ulna (lower wing) bones that is less than 0.7, (4) a fused pelvis structure with more than eight vertebrae, (5) tail vertebrae with flanges that are reduced or absent, (6) a pelvic structure that is modified so that the pubis is parallel to the ischium and illium, and (7) a globe-shaped head.

These synapomorphies identify trends in changes to the forelimbs, hindlimbs, breastbone, tail, and pelvis of birds. The fusions, reductions in bone number, and other modifications were undoubtedly favored as adaptations for flight.

methods is to identify these shared, derived traits accurately. Identifying and evaluating synapomorphies requires judgments about the homology of traits and the direction, or polarity, of change through time. There are several ways to determine which states of a trait are ancestral and which are derived. One of the most basic and reliable is called outgroup analysis. In outgroup analysis, the character state in the group of interest (the ingroup) is compared to the state in a close relative that clearly branched off earlier (the outgroup). The outgroup is assumed to represent the ancestral state. Finding an appropriate outgroup, in turn, involves borrowing conclusions from other phylogenetic analyses or confirming an earlier appearance in the fossil record. Most investigators prefer to examine several outgroups so they can corroborate that the characters used in the analysis were not present in the "derived" state earlier in evolution.

When no convergence, parallelism, or reversal occurs, all the similarities observed in a data set are due to modifications that took place in common ancestors. A researcher would say that all synapomorphies identified are congruent. In this case, inferring the phylogeny is straightforward: The investigator simply groups the taxa according to their synapomorphies. Each branch on the tree corresponds to one or more synapomorphies that distinguish the derived groups. A phylogenetic tree inferred in this manner is called a cladogram. By convention, the location of these synapomorphies is indicated on published trees with horizontal bars across the branches. These bars are numbered, and the synapomorphies are described in a key or labels that accompany the diagram. Figure 10.4 presents an example cladogram.

When homoplasy is present, synapomorphies are not completely congruent and the strategy for inferring phylogenies changes slightly. The goal now is to find the branching pattern that minimizes homoplasy. Many or all of the possible branching patterns have to be examined and evaluated, and a best tree chosen. How this evaluation and selection are done is the subject of sections 10.2 through 10.5.

For further background on cladistic approaches, see Hennig (1979), Eldredge and Cracraft (1980), Nelson and Platnick (1981), Wiley (1981), and Mayden and Wiley (1992).

---

can make alignment difficult. Fortunately, computer programs are available that can examine many possible alignments and choose the one that requires the minimal number of base substitutions, the fewest gaps, and the smallest gaps. These routines optimize sequence alignment with the goal of maximizing the number of shared bases at conserved sites.

Defending the idea that base substitutions are independent of one another, and that they represent distinct data points for phylogenetic inference, is also reasonably straightforward. In most genes the probability that a substitution occurs at any particular site is independent of what is happening at other sites. Exceptions do occur, however. In some sequences a base substitution can lead to correlated changes elsewhere in the same gene, through hitchhiking effects (see, for example, Kreitman 1983) or compensatory mutations that restore the secondary structure of proteins or RNAs (Dixon and Hillis 1993).

Minimizing homoplasy in molecular data is a much greater challenge. "Multiple hits," or sites that have mutated more than once, are a prominent source of homoplasy. One key to minimizing the effects of multiple hits is to study a region of the genome that changes slowly relative to the group being studied. As we established in Chapter 7, some genes change much more rapidly than others. If a rapidly evolving gene is studied in taxa that diverged long ago, many of the sites will have changed more than once. These multiple hits will lead to convergence, parallelism, reversal, and, very possibly, incorrect conclusions about the phylogeny of the group. A slowly evolving region of DNA, on the other hand, is identical in recently diverged taxa and thus uninformative about phylogenetic relationships. The challenge is to match the speed of molecular evolution to the age of the group being studied.

As DNA sequencing has become cheaper and more accessible technically, it is beginning to dominate the molecular data sets used in phylogeny inference. Sequence data from mitochondrial DNA (mtDNA), for example, is a popular choice for phylogenetic studies in animals. As we noted in Chapter 7, mitochondria contain a small, circular DNA molecule containing a series of genes. With rare exceptions, this molecule is maternally inherited. There is also no recombination in animal mtDNA, meaning that the entire molecule is inherited as a single gene. Finally, animal mtDNA tends to change much faster than nuclear genes, making it a useful tool for studying relationships among genera and species.

In plants, molecular phylogenies are often estimated from sequence or restriction site data from chloroplast-DNA (cpDNA). cpDNA is maternally inherited in many flowering plants (see, for example, Doyle et al. 1990; Kohn et al. 1996), but is much more variable in size, gene order, and composition than the mtDNA of animals. Data from cpDNA have been most useful for estimating the phylogenies of genera and species within plant families.

For more ancient divergences, genes that code for ribosomal RNA (rDNA) and transfer RNA (tRNA) have proven informative. rRNA and tRNA sequences are highly conserved even among phyla of animals, plants, and bacteria. This conservatism results from tight functional constraints and negative selection on these molecules. Their function is so critical that substitutions altering that function are rare. As a result, these sequences show limited change through time.

Morphological Characters

Measurements taken from any body part and life stage can be used to reconstruct phylogenies. Examples include the size or shape of insect genitalia, the numbers of whorls in mollusc shells, and the way leaves are arranged on the stem.

How are the criteria of homology, independence, and minimal homoplasy evaluated for morphological characters? The homology of characters is determined by the criteria we introduced in Chapters 2 and 3: similarity in location, composition, and development. Establishing that characters evolve independently can be done statistically, by computing correlation coefficients or other measures that assess the tendency for different quantities to change together.

As in molecular studies, homoplasy is a much more difficult issue to assess and avoid. Some morphological characters change through time more quickly than others, and are more subject to homoplasy. An investigator selecting morphological characters to study faces the same challenge as a molecular systematist choosing an appropriate gene: finding traits that change slowly enough to be resistant to homoplasy, but rapidly enough to produce informative variation in the group being studied. With morphological data, the most common pitfall is including characters that have converged due to natural selection for similar function.

Discrete and Distance Data

The molecular and morphological data used to infer phylogenies can be measured or handled in two ways: as discrete or distance data. With discrete data only certain states are possible. The presence or absence of a restriction enzyme site is discrete. So is the nucleotide present at a particular site in a gene sequence. Distance measures, in contrast, can take a continuous range of values. For example, in the technique called DNAxDNA hybridization a researcher melts DNA helices from one species and lets these molecules react with single-stranded DNA from a different species. The percentage of hybrid molecules that have formed at a given time and temperature is recorded, and interpreted as a measure of genetic distance between the two species. Later in the chapter, we introduce another technique for measuring genetic distance, called micro-complement fixation.

Although the homology of the sequences tested in DNA hybridization and microcomplement fixation is clear, it is more difficult to assess how resistant these data are to homoplasy. Distance measures are also not independent among pairs of species. This is less of a concern because techniques for translating distance data into phylogenetic estimates do not require that the data be independent.

Discrete data can also be translated into distance values. Examples include measures of overall morphological divergence, computed as the distance between data points plotted in two or more dimensions (Figure 10.5), or the percentage of nucleotide sites that differ between two taxa (a 10% difference means that an average of 10 nucleotides have changed per 100 bases).

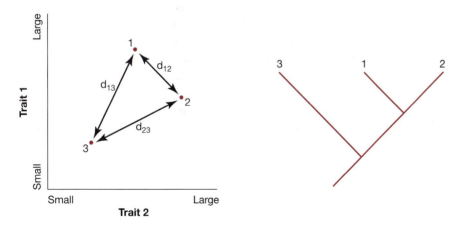

Figure 10.5   **Computing a morphological distance**   There are various algorithms for estimating how similar plants or animals are in their overall morphology. The basic approach is to measure a variety of characters, plot them on a series of axes, and then compute the average distance between each taxon in the study. In the example shown here, just two characters have been measured and graphed. The distances, d, between species 1 and 2, 2 and 3, and 1 and 3 could be used to infer the phylogenetic relationships between the taxa. For more information on distance methods, see Swofford et al. 1996.

10.2 Optimality Criteria: How Do We Choose the Best Tree from Among the Many That Are Possible?

The datasets used in phylogeny inference consist of large matrices, with the taxa in the study making up the rows and the characters that were measured making up the columns. Each cell in the matrix records the state of the character in each species. Now the question becomes: How do we decide which tree is the best one implied by these data? Clearly, we need criteria for distinguishing which trees are most likely, given the data.

Three optimality criteria are routinely employed in phylogenetic systematics. We examine each in turn.

Parsimony

Parsimony provides a clear criterion for choosing a best tree: The preferred topology is the one that minimizes the total amount of evolutionary change that has occurred. To implement a parsimony criterion, a computer program considers a particular tree topology and counts the number of changes, or steps, that would be needed to produce the character data measured. For example, Figure 10.6 shows a codon's worth of sequence data mapped onto the three topologies that are possible for four species. In this case, one of the topologies

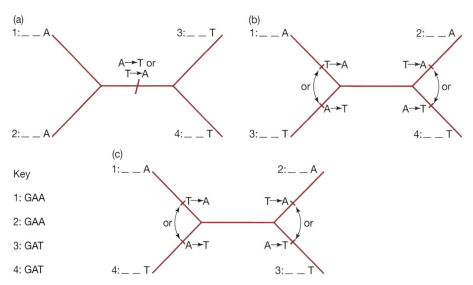

Figure 10.6 **Implementing a parsimony criterion**   This figure shows how a codon's worth of data can be mapped onto the three different branching patterns possible among four taxa (labeled 1–4 here). Phylogeny (a) implies that a minimum of one base substitution has occurred at the third position of the codon. In contrast, trees (b) and (c) require at least two changes. Under a parsimony criterion, we conclude that phylogeny (a) represents the best estimate of the evolutionary history of these four species.

Note that these trees look very different from those illustrated in Figures 10.1 and 10.2. The reason is that they are unrooted. Unrooted trees specify only the relative order of branching events. Rooted trees, like those in Figures 10.1 and 10.2, specify the oldest-to-youngest, or absolute, order of branching events. This is possible because they show where the base of the tree is relative to an outgroup.

clearly implies that less change has occurred relative to the amount of change required by the two competing topologies. Under the parsimony criterion, we conclude that the topology requiring the least change is the best estimate implied by the data.

There is a compelling rationale for invoking parsimony when inferring phylogenies. In many instances we can assume that convergence, parallelism, and reversal will be rare relative to similarity that is due to modification in a common ancestor (for important exceptions, see Felsenstein 1978, 1983). To use the terminology we introduced in Box 10.2, synapomorphies are usually more common than homoplasy. It makes sense, then, to prefer the tree that minimizes the number of steps required to produce the data observed. The most parsimonious tree should minimize the amount of homoplasy in the data.

Over the past 10 years, parsimony has been the most popular criterion for choosing among possible trees.

Maximum Likelihood

Maximum likelihood is a general methodology in the field of mathematical statistics that has recently been applied to the task of estimating phylogenies. The essence of the likelihood approach is for a researcher to ask: Given a particular model for how the process I am studying occurs, what is the probability of getting the data I measured? In the case of phylogenies estimated from DNA sequence data, the question becomes: Given a formula describing the probability that different types of base substitution will occur, and given a particular tree, how likely am I to obtain this particular set of DNA sequences?

Just as computer programs can go over a tree topology and count the number of steps required by a particular parsimony model, a computer can go over a tree topology and compute the probability of producing the observed data given the specified model of character change. The sum of the probabilities of getting each branch represents the probability of getting the observed data if the tree is correct. This probability is reported as the tree's log-likelihood. The criterion for accepting or rejecting competing tree topologies, then, is to choose the one with the highest log-likelihood.

Simulation and other theoretical studies have shown that when rates of change vary among lineages, maximum likelihood can outperform parsimony methods in recovering the correct tree (for a review, see Swofford et al. 1996). As a result, likelihood methods are becoming increasingly popular.

Distance Data

To estimate a phylogeny from distance data, computer programs are employed to cluster taxa so that the most similar forms are found close to one another on the tree. This general strategy, based on clustering taxa according to their similarities, is called a phenetic approach (Sneath and Sokal 1963). The preferred tree is the one that minimizes the total distance among taxa.

There has been a long-standing controversy over whether distance data contain reliable information that can be used to infer phylogenies. The reason for the diversity of opinion is that distance data represent an aggregate, or composite, measure that summarizes an underlying quantity, like the number of sequence differences.

A controversial aspect of phenetic methods is that it is difficult or impossible to diagnose homoplasy in distance data. The concern is that the similarity used to infer phylogenetic relationships may result from convergence, not from recent descent from a common ancestor. With discrete data it is possible, at least in principle, to check the data character by character, and assess whether natural selection could have produced similarity due to convergence.

The current consensus is that distance methods can and do yield reliable estimates of phylogeny, if the characters used to compute the distance are unlikely to demonstrate convergence (Felsenstein 1988, 1993; Huelsenbeck and Hillis 1993). With sequence data, for example, investigators often compute a genetic distance among the taxa in the study and estimate evolutionary relationships using phenetic methods, along with likelihood or parsimony approaches.

One final note before moving on: When computing measures of genetic distance between species, it is usually advisable to correct for multiple hits. With sequence data this can be done statistically, based on a standard model that specifies the probabilities of transitions and transversions. The derivation of the formula that is commonly used, though not particularly complex, is beyond the scope of this book. Interested readers should see Kimura (1980) for the original work and Wakely (1996) for an expanded treatment.

Researchers frequently use more than one optimality criterion to evaluate the trees implied by a dataset.

10.3 Models of Character Change

Once an optimality criterion is chosen, the next step is to specify how the characters in the study are expected to change through time. In this section we review some of the parsimony methods that have been developed and introduce models of character change that can be evaluated in a likelihood context. Throughout the discussion, readers should keep a general point in mind: To generate a reliable estimate of evolutionary relationships, an investigator must match the data that have been measured to the most appropriate parsimony method or model of character change.

Parsimony Methods

Parsimony methods differ in the way that changes are counted or because they specify that only certain types of changes are allowed. Different methods are applied to different types of data.

- In Fitch parsimony, all types of changes are assumed to be equally likely. We used Fitch parsimony in working the example given in Figure 10.6.

  With sequence data, invoking Fitch parsimony would require that transitions and transversions have the same probability of occurring. This is seldom true. But if an investigator trimmed the dataset to include only transversions, Fitch parsimony might be an appropriate choice.

- Under Wagner parsimony, changes between different character states are ordered. For example, if a particular character can be found in states coded 0, 1, or 2, and a particular branch shows that a 0 to 2 change occurred, Wagner parsimony counts this as two steps. Fitch parsimony counts this as one change.

When an outgroup is used to establish the direction of change from an-cestor to descendent, Wagner parsimony is equivalent to the cladistic meth-ods reviewed in Box 10.2 (Swofford et al. 1996).

- Dollo parsimony can only be used with characters coded as present or absent. Under this parsimony method, each of the characters is allowed to change from absent to present once—and only once—over the entire tree topology.

  Dollo parsimony is commonly used to model character change in restric-tion site data. Restriction enzymes are bacterial proteins that cut DNA at specific 4- or 6-base pair sequences. The enzyme EcoR I, for example, cuts DNA whenever it encounters the sequence GAATTC. Dollo parsimony is a reasonable model to employ with restriction site data because mutations are much more likely to cause the loss of a site than a gain.

- Generalized parsimony methods allow an investigator to specify *any* relation-ship between character states. For example, if a researcher observes that tran-sitions outnumber transversions by a ratio of 5:1 in a data set, the generalized parsimony scheme can be set to give transversions a weight of 5 and transi-tions a weight of 1. Then, when the computer encounters a transversion while assessing a particular topology, it contributes five steps instead of one to the total amount of change required to produce the tree.

  In effect, generalized parsimony methods offer a limitless variety of mod-els, each of which assigns a specific probability to different types of change in the characters being studied. Because of this flexibility, generalized parsimony methods are becoming increasingly popular (Swofford and Olsen 1990; Swofford and Maddison 1992).

Models of Sequence Evolution

A suite of mathematical tools has been developed to model the evolution of DNA sequences. When implementing a likelihood criterion, these models are used to specify the probability of the dataset given a tree topology. The key dis-tinction in the models is their treatment of two key parameters: whether all types of changes among bases are assumed to be equally likely, and whether each type of base is assumed to exist in equal frequencies. The more complex models allow researchers to specify that transitions and transversions occur with different probabilities, and that the four bases may be present in unequal fre-quencies. The latter parameter is useful when codon bias exists.

When choosing a model of sequence evolution to use in implementing a likelihood criterion, an investigator has to weigh an important trade-off. The more complex and realistic a model becomes, the more computationally inten-sive it is. Until recently, in fact, even simple models for sequence evolution were so time-consuming to evaluate that it was impractical to implement a likelihood approach for even moderately large datasets (say, those containing more than 8–10 species). This stumbling block is becoming less important, thanks to the increased processing speeds of personal computers. As a result, likelihood methods are becoming increasingly practical and popular. For an ex-cellent and readable introduction to the models of sequence evolution cur-rently being employed, see Swofford et al. (1996).

10.4   Algorithms for Searching Among the Many Possible Trees

After a parsimony or likelihood criterion has been selected and an appropriate model of character change specified, the task is to evaluate the many possible tree topologies and find the best one. This is difficult because the number of tree topologies that are possible becomes astonishingly large even in a moderately sized study. When four species are included, only three different branching patterns are possible (Figure 10.6). Adding a fifth species to the dataset makes the number of topologies possible jump from three to 15. A sixth taxon leads to 105 and a seventh to 945. With 10 taxa, 2×10^6 different branching patterns exist. It is fairly routine now for studies to include 50 taxa. In this case, 3×10^{74} different unrooted trees are possible.

When a study includes even a moderate number of species, the number of possible trees is so large that it is impractical to evaluate all of them.

When the number of taxa in a study is relatively low—typically fewer than 11—it is feasible to employ computer programs that evaluate all of the possible trees. This strategy is called an exhaustive search. Because it guarantees that the optimal tree implied by a particular dataset will be found, the approach is called an exact method.

An algorithm called the branch-and-bound method can also yield an exact solution to the problem of finding the best tree implied by a dataset. To implement branch-and-bound, the computer creates a search tree. This is done as follows: Starting with three taxa, the program adds a fourth and evaluates the three trees possible using the optimality criterion specified. To each of these three trees, the computer then adds a fifth taxon and evaluates the 15 resulting trees. It proceeds by adding a sixth taxon to each of these 15 trees, evaluating the 105 possible trees, adding a seventh taxon, and so on. But instead of searching all possible arrangements and duplicating an exhaustive search, the branch-and-bound algorithm ignores sets of trees that return an optimality criterion worse than what we would expect a randomized dataset to produce. For example, if one of the 105 trees generated by the addition of a sixth taxon did not meet this threshold value, the computer would not add a seventh taxon to it. In this way the branch-and-bound algorithm can dispense with large portions of the search tree. Thanks to their increased efficiency, branch-and-bound methods can produce an exact solution for datasets including 20 or more taxa.

With larger datasets even branch-and-bound algorithms become impractical, however, and heuristic searches have to be employed. Although these approaches do not guarantee that the optimal tree will be found, in practice they have been found extremely effective. One common routine is called stepwise addition. This algorithm proceeds in a similar way to branch-and-bound, except that as each taxon is added, only the best or few best trees (according to the optimality criterion being employed) are saved before adding an additional taxon. The limitation of stepwise addition and other heuristic methods is that the final output is sensitive to which three taxa are chosen to start the search, as well as the order that taxa are added subsequently. If a particular search happens to start far from the optimal arrangement, it is unlikely ever to encounter and evaluate the best tree.

To understand why heuristic searches behave this way, computer scientists offer the analogy of a nearsighted pilot who is forced to parachute from her plane in mountainous terrain, and loses her glasses in the process. Although she believes an outpost exists at the highest point in the area, she cannot see far enough to evaluate all the hilltops in the region. Instead, her only option is simply to climb the nearest hill. Once at the top, however, she has no way of knowing if it is the highest in the area. Heuristic searches are analogous because they climb the "nearest hill" identified by the first few taxa evaluated. They cannot see that a higher hill (that is, a more optimal solution) may be nearby.

Researchers cope with this problem by doing a large number of searches, starting with different taxa and adding subsequent groups randomly or in various prespecified orders. Researchers may also use a subroutine called branch swapping. When branch swapping is added to a heuristic search, the computer temporarily stops adding taxa and randomly swaps branches in the trees that have been saved, in an attempt to determine if a more optimal tree exists before continuing.

The important point here is that performing a thorough search is critical. Two recent studies of human evolution failed to conduct careful searches and,

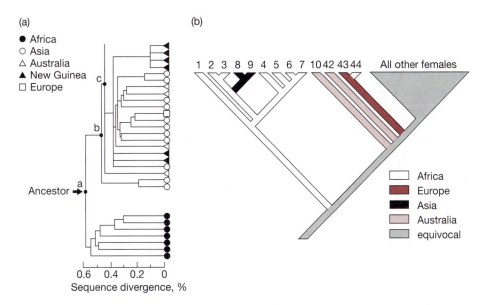

Figure 10.7   The importance of finding and interpreting the best tree   (a) Cann et al. (1987) mapped mtDNA restriction sites from a wide variety of human populations, and then used a parsimony analysis to evaluate competing tree topologies. This figure shows the region near the root of the tree they published. The ancestor was placed at point a. This location was justified by the midpoint-rooting method, which calls for placing the root at the midpoint of the longest branch of an unrooted topology (this is an alternative to outgroup rooting). Thus, the tree shows that the basal (or most ancestral) split in human populations is between Africans and non-Africans. Cann et al. used this observation to infer that *Homo sapiens* originated in Africa.   (b) When David Maddison (1991) reanalyzed the data of Cann et al., he searched the universe of possible topologies more thoroughly and found more-parsimonious trees that did not support the "Out of Africa" hypothesis for the origin of humans. The tree shown here, for example, does not support an African vs. non-African split at the base. By counting the number of migration events required by this tree, Maddison showed that it is equally parsimonious to argue that humans originated in Africa, Europe, Asia, New Guinea, or Australia.

using a parsimony criterion, reported phylogenies that turned out to be far from the best ones implied by the data. In each case, the tree published originally implied that all modern populations of humans originated in Africa (Figure 10.7a). But when Alan Templeton and colleagues (1991) and David Maddison (1991) reanalyzed the data and searched the universe of possible trees more thoroughly, they found more-parsimonious trees that did not support the claim for African origins (Figure 10.7b). The original analyses had "climbed a hill" far from the optimum. As we will see in Chapter 14, the African-origins hypothesis for the origin of *Homo sapiens* is still being tested.

10.5 Evaluating and Interpreting Trees

The final step in phylogeny inference is to evaluate the optimal topology found by the exhaustive or heuristic search. Recall that this tree *estimates* the evolutionary history of a group. The estimate was produced by a particular dataset under a specific optimality criterion and model of character change, and found with a distinct implementation of search algorithms.

A phylogenetic tree must always be interpreted as an estimate of a group's actual history.

We consider three issues in this section. First, how good is the optimal tree? That is, how much confidence do we have that this estimate is significantly better than other trees implied by the data? Second, can we assess how much statistical support particular branches within the tree have? Finally, can we evaluate the length of different branches on the tree, and understand when branching events occurred in time?

Comparing Trees

Tree-searching algorithms frequently find many different topologies at or near the optimum. Several trees may be equally parsimonious or require merely one or two additional steps, and the log-likelihoods or minimal distances implied by competing topologies can be extremely close. The question is whether these nearly optimal trees are significantly different, or whether they simply represent the type of random variation we expect to observe when estimating any quantity. If the latter is true, then these nearly optimal trees are as valid as the optimal topology. As estimates of a quantity they are indistinguishable. Clearly, these equally likely trees must be evaluated before drawing conclusions about the group's evolutionary history.

Competing trees inferred from the same dataset can be compared and evaluated with statistical tests (see Swofford et al. 1996; Huelsenbeck and Rannala 1997). Many investigators also examine the topologies of trees close to the optimal tree visually, and make an informed judgment about how different they are. If the nearly optimal trees differ in only minor ways from the optimal tree, it increases our confidence in using the most optimal tree to draw conclusions about the group's evolution. In addition, several computer programs can evaluate multiple trees and create a consensus representing the topology supported by all of the nearly optimal trees.

It is also routine for researchers to employ more than one optimality criterion or model of character change to analyze the same dataset. When the best trees produced by contrasting approaches are identical or nearly so, the analysis provides additional support for the optimal tree. When they conflict, we are left to make a judgment about which optimality criterion and model of character change is most valid in this case. We face a similar situation when comparing trees inferred from entirely different datasets. Trees produced by different studies are treated as competing estimates. When they conflict, researchers have more confidence in trees estimated with larger datasets and from characters thought to be less subject to homoplasy.

Evaluating Particular Branches

How much support exists for particular branches in a phylogeny? That is, would the tree be just as parsimonious or likely if a particular branch length were 0? Here a statistical technique called bootstrapping can help (Felsenstein 1985). In bootstrapping, a computer creates a new dataset from the existing one by repeated sampling. If there are 300 base pairs of sequence in a study, for example, the computer begins the bootstrapping process by randomly selecting one of the sites and using it as the first entry in a new dataset. Then it randomly selects another site, which becomes the second data point in the new dataset. (There is a 1/300 chance that this second point will be the same site as the first.) The computer keeps resampling like this until the new dataset has 300 base pairs of data, representing a random selection of the original data. This new dataset is then used to estimate the phylogeny. By repeating this process many times, the investigator can say that particular branches occur in 50%, 80%, or 100% of the trees estimated from the resampled datasets. The more times a branch occurs in the bootstrapped estimates, the more confidence we have that the branch actually exists. If bootstrap support for a particular branch is low, say under 50%, authors frequently conclude that the branch length is probably zero and collapse the branch into a polytomy in the published tree.

Scaling Branch Lengths for Time

A carefully reconstructed phylogeny describes the order of branching events during the history of a clade. But when did these events occur? If we can place branching events in time, we can ask an entirely new suite of questions about the rate and pattern of evolution. Were speciation events clustered in time? Are they recent or ancient? Do they correlate with plate movements, mass extinction events, climate changes, or other processes that can trigger diversification?

There are two ways to estimate the timing of branching events: through dating fossils, and by correlating genetic distances with time. We examine each in turn.

Aligning Nodes with Fossil Dates

When extant species also occur in the fossil record, or when the origin and duration of a lineage is well-documented in the fossil record, branching points can be directly aligned with an absolute time scale. Luis Chiappe (1995) was able to use dates from the fossil record to scale his cladogram of Mesozoic era bird groups for time (Figure 10.8). Adding information on time enriches the interpretation of this phylogeny. It is clear that several of the major bird lineages disappeared 65 million years ago, during the mass extinction event at the Cretaceous-Tertiary boundary. Even more interesting is that the branching patterns and the chronology do not align closely in every case. Look again at the tree, and note that the early dates documented for *Archaeopteryx* and the group called Enantiornithes dictate that branches 1–4 have to be placed deep in time. But fossil material that old has not been found from the derived groups called Dromaeosauridae, Alvarezsauridae, and the genus *Iberomesornis*. The placement of branches 5–7 in the early Cretaceous is startling, given that fossil material that old has not yet been found. The point is that many of the branches on this tree have to be much longer than the current fossil record indicates (that is, they have to go further back in time). The cladogram thus makes a strong prediction: Older representatives of these fossil taxa will eventually be found. If extensive fossil collecting fails to turn up evidence for these taxa in the early Cretaceous,

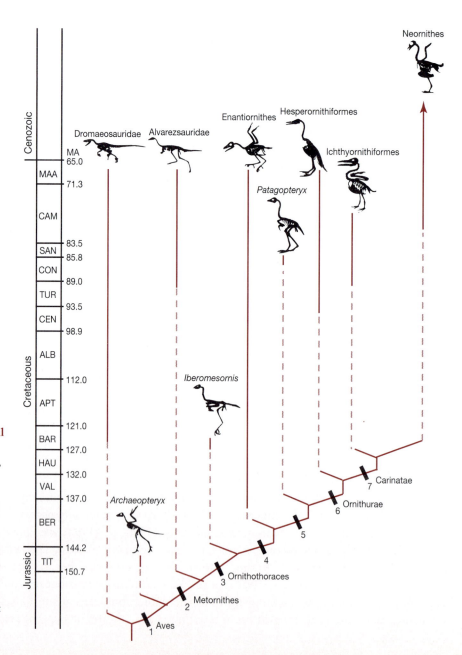

Figure 10.8 **A phylogeny of fossil birds, scaled for time** This is the same cladogram given in Figure 10.4, modified to reflect additional information on times of divergence and the duration of branches. The scale at the left of this diagram gives the era or period, stage, and absolute age of the events depicted in this phylogeny. The solid vertical branches leading to each taxon represent intervals where fossil representatives of a lineage have been found; the dashed bars represent intervals where fossil forms are yet to be discovered. From Chiappe (1995).

we will have grounds for rejecting at least some of the relationships that Chiappe has hypothesized.

The Molecular Clock

A remarkable claim emerged from early studies of molecular evolution: Sequences diverge with clock-like regularity. This hypothesis, called the molecular clock, originated with E. Zuckerkandl and Linus Pauling (1962). Zuckerkandl and Pauling compared the amino acid sequences of hemoglobin and cytochrome *c* in several different orders of mammals and found similar amounts of divergence in each species pair. Because the mammalian orders are thought to have split at about the same time, the amino acid substitutions observed must have occurred at roughly the same rate in the different lineages. The molecular clock hypothesis was promoted and elaborated throughout the 1970s and 80s in a series of papers from Allan Wilson's laboratory (see Wilson et al. 1977; Wilson et al. 1985; Wilson 1985). If molecular clocks exist and can be calibrated, they would provide a powerful tool for dating branching events that are poorly chronicled in the fossil record.

Vince Sarich and Allan Wilson (1967) used this approach to investigate the timing of the split between hominids and the great apes. They first measured the degree of divergence among several ape and monkey species in a blood protein called albumin. Then they scaled the amount of molecular differentiation for time by dividing by the well-dated divergence between Old and New World monkeys. Combining this calibration with their measurements of divergence in the albumin of humans and chimps produced an estimate for the date of the hominid-ape split at 5 million years ago (Ma). This was a dramatic departure from the estimate of 15 Ma provided by the fossil material available at the time. As we will see in Chapter 14, more recent fossil finds have largely corroborated the estimate made from molecular data.

Molecular clocks started out as an empirical phenomenon—a pattern that was observed in some datasets. But the neutral theory of molecular evolution also provides a theoretical basis for their existence. In Chapter 7 we reviewed the following expectation derived by Motoo Kimura: Base substitutions that are neutral with respect to natural selection become fixed at a rate equal to the mutation rate. If the mutation rate does not change over time, the neutral theory predicts that substitutions will occur in a clock-like fashion.

But in Chapter 7 we also reviewed several studies showing that certain gene sequences from humans and fruit flies—specifically, replacement substitutions in the antigen recognition site of the major histocompatibility complex (MHC) locus and the alcohol dehydrogenase gene—have *not* evolved in a neutral fashion. These sequences have undoubtedly undergone strong natural selection. In selecting genes to serve as molecular clocks, then, we want to focus on loci where strong positive selection is not operating. Using this criterion, three types of sequences become candidates for clock-like behavior:

- Sequences where most of the observed substitutions are silent changes in the third position of codons. If codon bias is low, silent site substitutions are expected to be largely neutral and thus clock-like.
- Untranslated or untranscribed sequences, like flanking regions and pseudogenes. Substitutions in these regions are not exposed to selection and should, like silent changes, occur in a clock-like fashion.

- Highly conserved genes where purifying selection is dominant. Recall that purifying selection refers to selection against replacement substitutions that worsen a protein's function. When purifying selection is very strong, only neutral or nearly neutral substitutions are likely to become fixed. The hypothesis being invoked here is that highly conserved genes have been under intense selection for so long that their function is optimized. As a consequence, advantageous mutations are rare to nonexistent.

Although molecular clocks have some empirical and theoretical support, it is crucial to recall (from Chapter 7) that substitution rates vary widely among sites, genes, and species. Rate variation will even be characteristic of completely neutral change. This is because neutral evolution depends on the mutation rate, which varies among genes and species (as we established in Chapter 4). There is also growing evidence that substitution rates vary among species as a function of generation time (see Martin and Palumbi 1993; Hillis et al. 1996). In sum, there is no such thing as a universal molecular clock. The most optimistic expectation is that a wide variety of "local" molecular clocks might exist, each specific to a particular lineage and sequence, and each ticking at a different rate.

How fast does a particular molecular clock run? There is only one way to scale a base substitution rate to time: by matching an observed genetic divergence with an absolute time of divergence documented in the fossil or geological record.

To illustrate how calibration is done in practice, we consider the mtDNA clock in animals. Wes Brown and colleagues (1979) estimated the amount of sequence divergence in the mtDNA of a variety of mammals, and plotted the data against estimates for time of divergence from the fossil record. The resulting graph, shown in Figure 10.9, makes two important points: Change in mtDNA is roughly linear, or clock-like, for about 10 million years. After that the rate of change falls off quickly. After about 40 million years, no additional genetic divergence occurs. The explanation for this pattern is that the vast majority of substitutions in mtDNA are silent substitutions in the third positions of codons. These occur in a clock-like manner until most or all available silent changes have taken place. At this point the molecule is "saturated." Additional substitutions represent multiple hits that do not change the overall amount of divergence. Figure 10.9 suggests that the mtDNA in mammals offers a reliable molecular clock for about 10 million years. During this interval, divergence averages about 2% per million years.

The mtDNA clock seems to move at about the same rate in at least some other animal groups as well. Bermingham and Lessios (1993), for example, estimated mtDNA sequence divergence in sister species of sea urchins found on either side of the Isthmus of Panama. Because the land bridge closed and separated these populations about three million years ago, the authors were able to calibrate the amount of sequence divergence for absolute time. The urchins provide an estimate of 1.8% to 2.2% sequence divergence per million years—remarkably close to that of mammals. Similar calibrations have been made in butterflies (Brower 1994) and geese (Shields and Wilson 1987).

Important exceptions to the 2% per million years calibration have cropped up, however. Andrew Martin and co-workers calibrated the mtDNA clock in sharks, which have an exceptionally complete and well-dated fossil record, and found that rates of change are seven to eight times slower than in mammals (Martin et

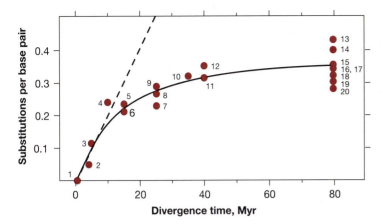

Figure 10.9 Sequence divergence in mammalian mtDNA as a function of time In this graph the average number of substitutions found in mtDNA is plotted on the ordinate, and the best estimate for time of divergence on the abscissa. Each data point represents a species pair as follows: 1 = mean difference among human populations, 2 = goat and sheep, 3 = human and chimpanzee, 4 = baboon and rhesus monkeys, 5 = guenon and baboon, 6 = guenon and rhesus, 7 = human and guenon, 8 = human and rhesus, 9 = human and baboon, 10 = rat and mouse, 11 = hamster and mouse, 12 = hamster and rat, 13−20 = various rodent-primate species pairs. From Brown et al. (1979).

al. 1992; Martin and Palumbi 1993). Their hypothesis is that rates of change in ectothermic vertebrates are markedly slower than in endotherms. Their results clearly invalidate the hypothesis of a universal mtDNA clock in animals. The general message is that clocks need to be calibrated for each sequence and in each taxonomic group where they are used (for clock calibrations in other genes, see DeSalle et al. 1987; Moran et al. 1993; Cano and Borucki 1995).

David Hillis and colleagues (1996) recently made the following recommendations regarding the use of molecular clocks:

Molecular clocks require careful calibration, preferably with multiple dates from the fossil record or from geological events.

1. Caution should be exercised in making references to dates in the fossil record. Paleontologists recognize that the quantity and precision of time estimates varies widely among taxa, based on the number of fossils available and their relationship to datable strata.

2. Because calibrations are made by comparing genetic divergence to times of divergence with a statistical technique called regression analysis, it should be possible to calculate a confidence interval on the estimated rate. A confidence interval describes the range of values we would expect to see through purely random variation about the true average value of the clock.

3. Because rates of molecular evolution vary among lineages, researchers should be extremely cautious about taking a calibration made for a particular taxonomic group and extrapolating it to other species.

Relative Rate Tests

As the preceding discussion shows, it can be difficult to calibrate a molecular clock and evaluate its validity. Further, we often need to know how rapidly sequence change has occurred in lineages that lack a fossil record entirely. Parsimony models and some of the algorithms for clustering distance data assume that rates of change have been constant or nearly constant across all of the branches in a phylogeny. How can we test this assumption of clock-like change?

An effective approach for assessing rate constancy is called the relative rate test (see Sarich and Wilson 1967; Li and Graur 1991). Here is how it works: Given the phylogeny pictured in Figure 10.10, we can measure the amount of genetic divergence that has taken place between species D_1 and O and D_2 and O. To check the assumption of rate constancy as D_1 and D_2 diverged, what we

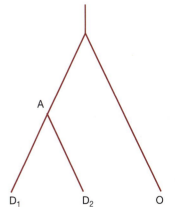

Figure 10.10 Relative rate tests In this hypothetical phylogeny, D stands for descendant species, A for ancestor, and O for outgroup.

want to know is this: Are the amounts of change that took place between A and D_1 and A and D_2 equal? In other words, is $A:D_1 - A:D_2$ zero? By inspection, it is clear that

$$\mathbf{D_1:O} = A:O + A:D_1, \text{ and}$$

$$\mathbf{D_2:O} = A:O + A:D_2.$$

We can measure the quantities in bold type. We want to know the difference between the italicized quantities. Subtracting, we get

$$D_1:O - D_2:O = A:O + A:D_1 - (A:O + A:D_2), \text{ or}$$

$$D_1:O - D_2:O = A:D_1 - A:D_2.$$

Relative rate tests can test the assumption of rate constancy without data from the fossil record.

Relative rate tests are usually reported in tables, with species pairs like D_1 and D_2 listed next to the difference in genetic distances $D_1:O - D_2:O$. When rate constancy occurs, these values should cluster near 0.

10.6 Using Phylogenies to Answer Questions

Research programs based on estimating and interpreting evolutionary trees are growing rapidly (Hillis 1997). In Chapter 1 we introduced the use of phylogenetic thinking by using a parsimony analysis to determine whether HIV jumped from humans to monkeys or from monkeys to humans. In Chapter 3 we showed how phylogenies can be used to control for shared history when making comparative tests. We did this by reviewing a study of seed dispersal systems in flowering plants. In the remainder of this chapter, our goal is to sample the diversity of applications for phylogenetic thinking. This discussion will set the stage for reviewing the history of life in Chapters 11 through 13.

Rates of Change: Radiation of Hawaiian Fruit Flies

Some of our most basic questions about the history of life concern when major events occurred, and how rapidly. In Chapter 12 we introduce events that have been dated from the fossil record. Here, we look at how a molecular clock was used to time a major episode in evolution.

The 500-plus species of fruit flies endemic to the Hawaiian archipelago rank as one of the most spectacular of all adaptive radiations. The oldest island, Kauai, is 5–6 million years old, implying that these speciation events have been extraordinarily rapid. But molecular clocks suggest a very different view.

To determine when flies first colonized the Hawaiian archipelago, Stephen Beverly and Allan Wilson (1985) estimated genetic divergences in a series of Hawaiian flies and outgroup species. Specifically, they studied the larval hemolymph protein (LHP) using a technique called microcomplement fixation. Microcomplement fixation measures how tightly antibodies from one species bind to the target protein in another species. Because the binding capacity of the antibody correlates closely with the identity of amino acid sequences between species, the technique offers a way of estimating the divergence of proteins that is cheaper and faster than amino acid sequencing.

Using phenetic methods with their genetic distance data, Beverly and Wilson were able to estimate the phylogeny of the Hawaiian flies and outgroup species (Figure 10.11). Employing a calibration for the LHP clock determined from

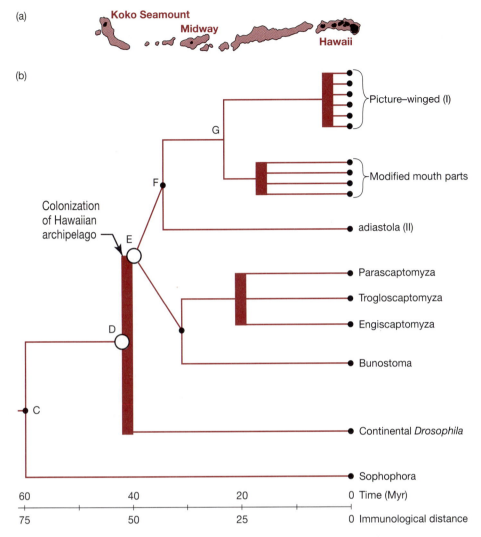

(a) Koko Seamount
Midway
Hawaii

(b)

Picture–winged (I)

G

Modified mouth parts

F

Colonization
of Hawaiian
archipelago

E

adiastola (II)

Parascaptomyza

Trogloscaptomyza

Engiscaptomyza

D

Bunostoma

C

Continental *Drosophila*

Sophophora

| 60 | | 40 | | 20 | | 0 | Time (Myr) |
| 75 | | 50 | | 25 | | 0 | Immunological distance |

Figure 10.11 Young islands, old flies Part (a) in this diagram shows the Hawaiian archipelago, scaled so that the age of the landforms corresponds to the absolute time scale at the bottom of the figure. Islands are shown in black and the extent of undersea structures is outlined.

The phylogeny of fruit flies, part (b), is based on clustering of immunological distance data from microcomplement fixation of LHP. The branch lengths are scaled so that a molecular clock calibration for divergence in LHP corresponds to the absolute time scale at the bottom of the figure. The "Picture-winged" through "Bunostoma" designations on the right refer to subgroups of Hawaiian fruit flies. The outgroup called Sophophora is represented by the common household fruit fly, *Drosophila melanogaster.* Node D, then, represents the ancestral population that first colonized the Hawaiian islands. From Beverly and Wilson (1985).

Note that several nodes are reported as polytomies (among species within the picture-winged group, for example). These branch points cannot be distinguished on the basis of immunological distance. Also note that the picture-winged group has radiated within the last 5 million years, solely within the current group of Hawaiian islands.

other fly groups (Beverly and Wilson 1984), they were able to date each of the major nodes on the tree. The analysis pointed to a startling conclusion: Hawaiian flies first colonized the archipelago some 42 Ma. This was long before the existing islands were formed.

How can this be? Recall, from our discussion of fly speciation in Chapter 9, that the Hawaiian islands originated from a hotspot under the Pacific plate. These volcanic islands drift to the north and west after they are formed and gradually erode, eventually becoming underwater mountains called seamounts. Many of the landforms in the chain have been dated using K-Ar techniques, confirming the old-to-new sequence. In Figure 10.11 this picture of island formation is matched to the fly phylogeny, the LHP divergence data, and the dates implied by the molecular clock.

Taken together, the geographical, phylogenetic, and clock data offer a logical resolution to the paradox of "young islands, old flies": *Drosophila* first colonized the Hawaiian chain when islands now in the Koko seamount were over the hotspot. As these older islands eroded and new islands were formed, the flies hopped from the old landforms to new. This "ancient origin" hypothesis for Hawaiian flies has now been confirmed by investigators using different gene sequences and independent clock calibrations (see DeSalle and Hunt 1987). Hawaiian crickets, in contrast, have radiated solely on the current group of islands (Shaw 1996).

There are several other examples in which molecular clocks have challenged traditional ideas about the timing of major evolutionary events. We already mentioned the dating of the hominid–great-ape split at 5–6 Ma, which was later confirmed by new fossil finds. In Chapter 12 we review fossil evidence that the radiation of metazoans, or multicellular animals, occurred about 560 Ma; for a molecular clock analysis that claims metazoans actually originated 1,200 Ma, see Wray et al. (1996) and Vermeij (1996).

Establishing the Sequence of Character Change: Swim Bladders in Fish

A basic way to use a phylogeny is to map a character of interest onto the tips and nodes and analyze the sequence and polarity (or direction) of change. As an example, we look at the evolution of the swim bladder in fish. Like the seed dispersal example in Chapter 3, this is an exercise in mapping traits onto a phylogeny to study the pattern of change. The organizing principle here is this: Once we have estimated a phylogeny based on characters selected as reliable indicators of history, we can use the tree to analyze change in other, independent traits. In this way, the phylogenetic approach gives us a rigorous way to study the evolution of adaptation.

Swim bladders, for example, are organs that help maintain neutral buoyancy in fish. Muscle and bone are heavier than water, so to avoid sinking constantly many fish maintain a gas-filled chamber called a swim bladder. These organs are homologous with the lungs found in fish and tetrapods. During development, both originate as outpocketings of the gut. Darwin recognized this homology and inferred that because fish evolved earlier than tetrapods, and because swim bladders are much more common in fish than are lungs, it follows that lungs must have been derived from swim bladders (see Gould 1993). But is this true?

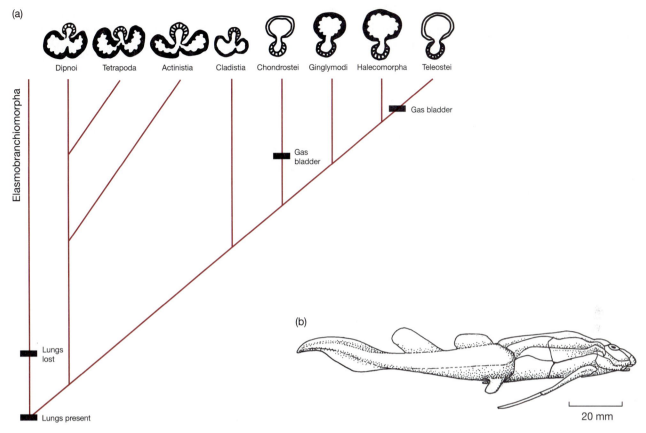

Figure 10.12 Evolutionary history of the lung and swim bladder   The cladogram in part (a) shows the evolutionary relationships of the major groups of extant vertebrates (Liem 1988). The Elasmobranchiomorpha are the sharks and rays; Dipnoi are lungfish; Tetrapoda are tetrapods (amphibians, reptiles, birds, and mammals); Actinistia include the "living fossil" fish called coelocanths; Cladistia are the bichirs (a family of armored fish from Africa); Chondrostei are a group of primitive ray-finned fish; Ginglymodi are the gars; Teleostei are the modern ray-finned fish. In numbers of species, teleosts are now far and away the dominant group of fish.

The outgroup in this phylogeny is not shown, but consists of a group of fossil forms called the placoderms. These were some of the first fish with jaws. They were armored, and radiated about 400 million years ago (Ma). A placoderm called *Bothriolepis* is illustrated in part (b).

The diagrams above each group in part (a)—the phylogeny—are cross-sections of the lungs or swim bladder, with the dorsal side on top and ventral position below. The sacs in heavy black outline are lungs, and the sacs with open outlines are swim bladders. The smaller sacs with hatching represent the gastrointestinal tract.

A cladogram of fish groups and tetrapods is shown in Figure 10.12a. Lungs are indicated as present at the base of the tree because they occurred in one of the earliest groups of jawed fish, the placoderms. Specimens of a placoderm common in rocks on Quebec's Gaspé Peninsula, called *Bothriolepis* (Figure 10.12b), show impressions that clearly indicate a well-developed pair of lungs opening from the pharynx (Liem 1988; Colbert and Morales 1991). Lungs or swim bladders also occur in every descendant group on the phylogeny save for the sharks and rays (elasmobranchs). Swim bladders, though, are found only in two descendant groups: the superorders Chondrostei (a group of primitive ray-finned fish) and Teleostei (the group of ray-finned fish that includes most modern forms).

The phylogenetic analysis, then, shows that lungs were present at the basal node, that lungs were lost completely in the lineage leading to elasmobranchs, and that lungs were transformed into swim bladders in two different lineages of fish. This last conclusion is bolstered by the observation that the swim bladders of chondrosteans and teleosts develop differently: In the former they originate as outpocketings from the stomach, and in the latter as outpocketings of the esophagus (Liem 1988).

Note that we have used two different principles to reach these conclusions: Outgroup analysis allowed us to map the presence of lungs onto the basal node, and a parsimony criterion made it logical to infer one complete loss of lungs and two transformations of lungs to swim bladders. Under parsimony we reject the alternative hypothesis that swim bladders evolved early and then reverted to lungs four times (in the lineages leading to the Dipnoi et al. clade, Cladistia, Ginglymodi, and Halecomorpha).

What misled Darwin and many other workers was the assumption that ancestral forms must have had swim bladders, because most extant and many fossil fish do. It is not necessarily true that widespread traits are ancestral. Interpreting the evolution of the swim bladders in a phylogenetic context made the actual direction of change clear.

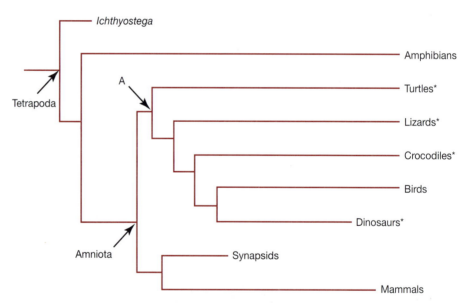

Figure 10.13 **Reptiles are a paraphyletic group** This cladogram shows the relationships among some major groups of terrestrial vertebrates (for information on the synapomorphies used to define this branching pattern, see Benton 1990a and Figures 10−2 and 10−18 in Pough et al. 1996). Tetrapoda is the name for the group of terrestrial, four-limbed vertebrates that evolved from fish; Amniota include all vertebrates with an amniotic, or membrane-bound, egg. *Ichthyostega* is one of the first tetrapods found in the fossil record, about 375 Ma. The branches shown here are scaled for time in an informal way; groups with living representatives are shown at the far right, and extinct groups are shown terminating in the correct sequence.

The groups with an asterisk are those traditionally considered as reptiles. The traditional vertebrate class called "Reptilia" is paraphyletic because it includes some, but not all, of the descendants of the ancestor at node A (see Figure 10.14).

Classifying the Diversity of Life

In Chapter 9 we reviewed the phylogenetic species concept, which considers every population that occupies a tip on an evolutionary tree as a different species. What does phylogenetic thinking have to say about how we should organize higher-order taxa like genera, families, orders, and classes?

Traditional classifications are based on a phenetic approach, in which organisms are grouped according to similar features. Reptiles, for example, are a phenetic group defined by shared primitive characteristics including internal fertilization, scaly skin, a four-chambered heart, and the membrane-bound, or amniotic, egg. Reptiles include the turtles, dinosaurs, crocodiles, snakes, and lizards. Birds share many of the same features, but have feathers as a unique and distinguishing characteristic. Because classification schemes have been phenetic, and because reptiles and birds are so distinct, reptiles and birds were traditionally considered as separate classes in the kingdom Animalia (Figure 10.13).

Phylogenetic schemes for classifying organisms, in contrast, are based on evolutionary relationships. Phylogenetic systematics argues that classification systems should be tree-based, with names and categories that reflect the actual sequence of branching events. Only monophyletic groups, which include all descendants of a common ancestor, are named (Figure 10.14).

Although phenetic and phylogenetic naming schemes frequently lead to the same conclusions, they can produce conflicts (see Schwenk 1994). The relationship between reptiles and birds that we just reviewed is a prominent example. The phylogeny of the dinosaurs, shown in Figure 10.15, provides additional detail. This tree clearly indicates that birds are a clade within the dinosaur radiation. Instead of considering reptiles and birds as separate classes, a cladistic classification would nest birds as a subgroup within the dinosaurs, which in turn are a subgroup of the Reptilia.

Phylogenetic systematics represents an important philosophical break from classical approaches to taxonomy (de Queiroz 1992; de Queiroz and Gauthier 1990, 1992; Simpson and Cracraft 1995). An increasing number of taxonomists and systematists are calling for a complete overhaul of the traditional phenetic scheme, with the goal of creating a phylogenetic classification (see Pennisi 1996). A phylogenetic naming scheme is already being implemented in prominent forums like GenBank, the online database of DNA sequences.

Before moving on, we should insert a semantic note. Early in the chapter we introduced the term *cladistics* in the context of a particular methodology for reconstructing phylogenies (based on identifying monophyletic groups via synapomorphies). The term also refers to the philosophy of classification we just introduced. In current usage, the terms *cladistic* and *phylogenetic* classification are synonymous. The term *phenetics* also has two meanings, parallel to the term *cladistics*. Phenetics refers both to the philosophy of classification we just referred to, based on naming groups according to their similarities, and to a family of phylogeny inference methods, based on statistical criteria for grouping distance data. For background on phenetic classification, see Sneath and Sokal (1973).

Coevolution

Coevolution is an umbrella term for interactions among species that result in reciprocal adaptation. We introduced this field of research in Chapter 8 when

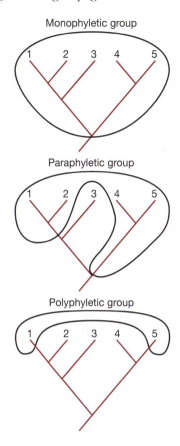

Figure 10.14 **Monophyletic, paraphyletic, and polyphyletic groups** Taxonomic groups can be monophyletic, paraphyletic, or polyphyletic. Monophyletic groups, or clades, contain all the descendants of a common ancestor. Paraphyletic groups contain some, but not all, the descendants of a common ancestor. A polyphyletic grouping does not contain the most recent common ancestor of all the taxa.

In a cladistic classification, only monophyletic groups are named. In contrast, traditional taxonomic schemes occasionally assign names to polyphyletic and paraphyletic groups.

In the 1960s and 1970s, advocates of phenetic and cladistic approaches engaged in a controversy over the best system for classifying the diversity of life.

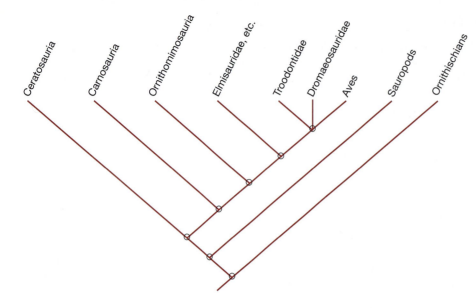

Figure 10.15 Phylogeny of some dinosaur groups This cladogram was reconstructed from a large catalog of skeletal characters (see Benton 1990a and citations therein for the complete list of synapomorphies identified at each circled node). The names across the top include both genera and families; "Aves" refers to the birds.

we discussed predation, mutualism, and parasitism. Here we introduce how phylogenetic thinking is used in coevolutionary studies. Our focal system is ants that farm fungi.

The 200 species of ants in the tribe Attini are the dominant herbivores in the New World tropics. The group includes the leaf-cutting ants, which dissect

Figure 10.16 Comparing phylogenies of fungi and their ant symbionts I   The left side of this figure shows the phylogeny of five species of fungi cultivated by ants, along with several free-living fungi. The free-living forms are shown in light type and the symbiotic forms in bold type. Other free-living forms included in the study are ancestral to the species shown here. This fact confirms that symbiosis is derived. The numbers indicate the percentage of bootstrapped trees that include given branches; branches that appeared in less than 50% of the bootstrapped replicates were collapsed into polytomies.

The right side of the figure shows the phylogeny of some fungal-farming ants. The topology is based on synapomorphies identified in the morphology of prepupal worker larva. The names in bold type are the names of ants that farm fungi; *Blepharidatta* serves as an outgroup.

The thick lines highlight the congruence between ants and the fungi they farm. From Hinkle et al. (1994).

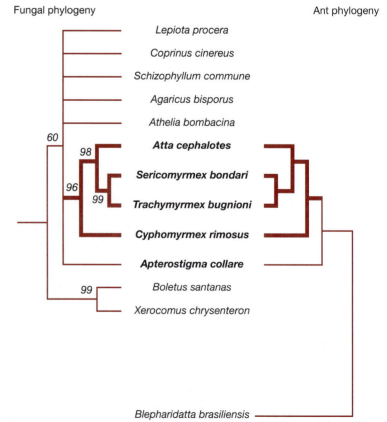

pieces from leaves and carry them to their nests. There they use the leaf material as a substrate for growing fungi in underground gardens. The fungi are harvested and serve as the primary food source for the ant colony. This symbiosis, or mutually beneficial relationship, is thought to be 50 million years old. This time estimate is based on the first appearance of attine ants in the fossil record. The relationship is also obligate: To the best of our knowledge, none of the symbiotic ant or fungal species can live without the other.

We would like to answer the following question about this association: If the symbiosis originated just once, did the ants and fungi subsequently evolve in

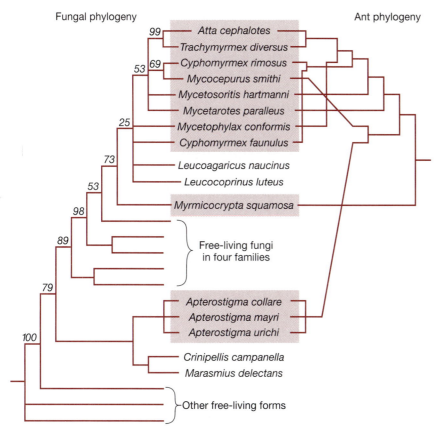

Figure 10.17 Comparing phylogenies of fungi and their ant symbionts II
The left side of this figure (Chapela et al. 1994) shows the phylogeny of fungi cultivated by ants, along with free-living fungi. The free-living forms are shown on a light background and the symbiotic forms in a shaded background. This is a strict consensus tree of the 22 equally most-parsimonious trees (meaning that only branches found in all 22 best trees are shown). The numbers indicate the percentage of bootstrapped phylogenies that include given branches.

The right side of the figure shows the phylogeny of fungal-farming ants, based on the same cladistic data used in Figure 10.16.

To interpret this (very complicated) diagram, begin by tracing the ant tree up from the base. Note that the first lineage to branch off leads to *Myrmicocrypta squamosa*. If cospeciation were occurring, the next branch off should have led to *Leucoprinus* or *Leucoagaricus*. Instead it leads to *Apterostigma*. This means that ants "domesticated" free-living fungal species more than once. Now read from the fungal side of the paired phylogenies. Note that *Leucoprinus* and *Leucoagaricus* are part of an ant-associated clade. This fact suggests that these fungal species escaped domestication and again became free-living.

tandem? That is, did the two groups "cospeciate"? If so, we would expect the phylogenies of the two groups to be congruent (match up branch for branch).

To answer this question, Gregory Hinkle and colleagues (1994) collected fungi from the nests of five different species of attine ants and sequenced over 1,800 base pairs from the coding region for the small subunit of their ribosomal RNA. They used these data to estimate the phylogeny of the attine fungi and several outgroups, using a parsimony criterion. When they matched the fungal phylogeny to a cladogram for the attine ants, they found a close correspondence: Branching patterns were identical in four of the five species compared (Figure 10.16, page 390). This is strong support for the idea that ants and fungi have cospeciated. The fifth fungus-farming ant species, however, poses a dilemma. The sequence data were unable to resolve the relationship of the fungus used by the ant *Apterostigma collare* relative to free-living and symbiotic forms. If this species branched earlier than the free-living forms in the polytomy shown in Figure 10.16, it breaks the pattern of cospeciation. If so, it means that ants acquired fungal symbionts more than once, or that some fungi have escaped from the farms to resume a free-living life-style.

A study performed by Ignacio Chapela et al. (1994) has helped clarify the history of mutualism. These researchers sequenced the 28S rRNA gene in 37 species of ant-associated and free-living fungi and matched the branching pattern inferred from these data to the same ant phylogeny shown in Figure 10.16. Their analysis clearly shows that the *Apterostigma* fungus split earlier than free-living forms in the polytomy we just examined (Figure 10.17, page 391). This confirms that strict cospeciation has not been universal in the coevolution of ants and fungi. Early in the evolution of ant-fungi symbiosis, some ants that farmed fungi switched species, picking up new, free-living fungi to "domesticate." Cospeciation has occurred only in the most recently evolved forms.

It is noteworthy that in at least some of these highly derived ant genera, like *Trachymyrmex* and *Atta,* queens that leave established nests to found new colonies carry a small ball of fungus with them, and use it to start the new garden. This behavior, which should lead to strict cospeciation, has not been found in the older lineages like *Apterostigma* and *Myrmicocrypta*.

Summary

Recent conceptual and technological advances have revolutionized our ability to estimate phylogenies accurately. Research in systematics is advancing rapidly, and phylogenetic thinking is beginning to pervade evolutionary biology.

The cladistic method for reconstructing evolutionary relationships is based on the identification of shared, derived traits, or synapomorphies. The first step in estimating a phylogeny is to select and measure characters that can be phylogenetically informative. The molecular or morphological characters employed in phylogeny inference have to be independent, homologous, variable among the taxa in the study, and resistant to homoplasy. The second step in phylogeny inference is to determine whether a parsimony, maximum likelihood, or minimum-distance ("phenetic") criterion is appropriate for selecting the best tree implied by a particular set of data. With each of these approaches, different models of character change can be utilized. Several different computer algorithms can be employed to search among the very large number of trees that are possible and evaluate them according to a parsimony, likelihood, or minimum-distance criterion.

Once an optimal tree has been found, nodes can sometimes be assigned times through fossil dating or the calibration of molecular clocks. Molecular clocks can exist because some types of gene sequences evolve in a neutral fashion. A variety of taxon and gene-specific clocks have now been calibrated using dates from the geological or fossil record.

Phylogenetic thinking has been applied to a wide variety of problems in evolutionary biology, from the transmission of HIV to systems for classifying the diversity of life. Informative uses of phylogenies include dating events that are poorly documented in the fossil record, studying the rate and pattern of evolution in characters other than those used in constructing the phylogeny, and studying coevolution.

Questions

1. Paddle-like limbs or flippers are found in numerous mammals, including manatees (Order Sirenia), seals (Order Carnivora), whales and dolphins (Order Cetacea), and duck-billed platypus (Order Monotremata). Examine Figure 10.18, which shows the relationships of these and other mammalian orders. Are all mammalian flippers and paddles a synapomorphy (derived from the same flipper-bearing common ancestor), or have they evolved independently via convergent evolution?

2. In what sense do the HIV-positive local individuals included in Figure 10.3 serve as a control?

3. Suppose you are studying four species whose true evolutionary relationships (unknown to you) are shown in Figure 10.19a. After studying many characters of these four species, you produce the unrooted tree shown in Figure 10.19b. Your next goal is to determine the *order* of evolutionary branching. That is, you want to root the tree. You choose an outgroup species that you think probably has ancestral states of the characters you have studied. However, due to convergent evolution, your outgroup actually shares many derived character states with species C. What will your rooted tree look like? How will it differ from the true tree?

4. Examine the three primate phylogenies shown in Figure 10.20. Figure 10.20a shows a detailed phylogeny for most Old World primates, Figure 10.20b emphasizes relationships of great apes and humans, and Figure 10.20c emphasizes relationships of Old World monkeys. Do the three phylogenies agree with each other? (That is, do they show the same relationships and the same order of branching?) Do they give different impressions of whether there was a "goal" of primate evolution, or of what the "highest" primate is? If so, what aspects of the figures give the impression of a goal, and is this truly reflected in the complete dataset? (Source: data for great apes from various sources listed in Chapter 14; data for macaques from Hayasaka, K., K. Fujii, and S. Horai. 1996. Molecular phylogeny of macaques: implications of nucleotide sequences from an 896-base pair region of mitochondrial DNA. *Molecular Biology and Evolution* 13:1044−1053.)

5. Suppose a cladistics expert is giving a seminar at the biology department at your school. While chatting with her after the seminar, you happen to mention that you

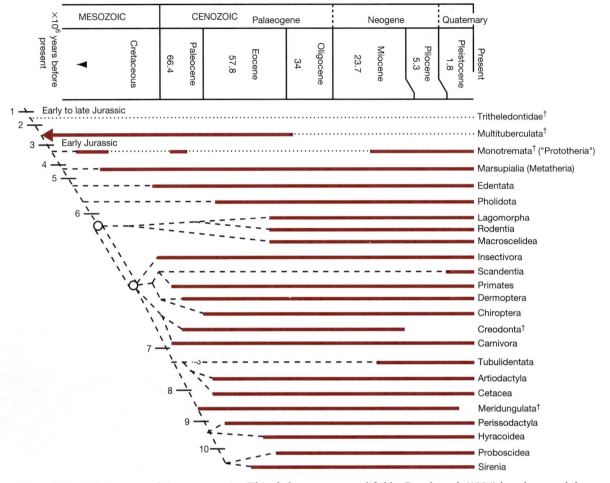

	MESOZOIC	CENOZOIC	Palaeogene			Neogene		Quaternary

Figure 10.18 **Phylogeny of the mammals**   This cladogram was modifed by Pough et al. (1996) based on work by Michael Novacek (1992, 1993). The names given along the right of the tree refer to mammalian orders. The numbers across branches refer to synapomorphies listed in Pough et al. The dark branches indicate the presence of fossil forms. The dashed lines indicate when the presence of an order has to be inferred.

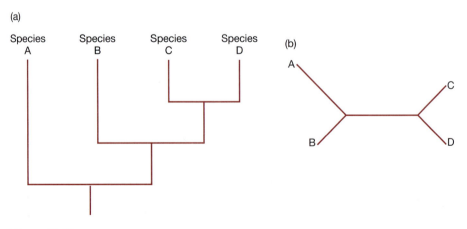

Figure 10.19

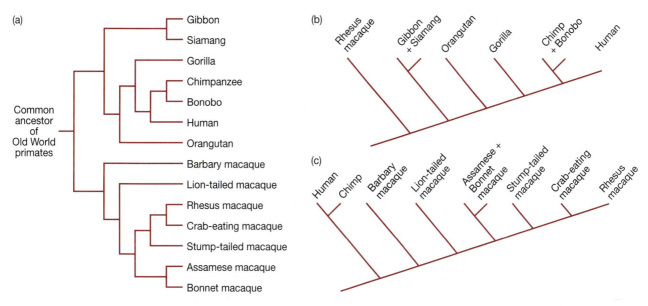

Figure 10.20   Phylogenies showing relationships of several Old World primates   (Branch lengths are not scaled.)

have a pet fish. She replies, "There's no such taxonomic group as "fish." "Fish" is a paraphyletic group. I don't know if you're talking about a pet shark, a pet guppy, or a pet cat." You remember (too late) that much of her seminar was about the importance of using taxonomic terms that refer only to monophyletic groups. Explain each of her statements. (Hint: refer to the relationships of "fishes" and tetrapods shown in Figure 10.12.)

6. For several centuries, biologists have used the Linnean system of classification, in which all species are assigned a spot in a hierarchy with exactly seven tiers (kingdom, phylum or division, class, order, family, genus, species). As explained in the text, some biologists have called for the abolition of the Linnean system and the institution of new names, such as "Tetrapoda" and "Amniota," for each major evolutionary branching event. In this new system, what would determine the number of tiers required to classify a species? Would different species have the same number of tiers? The Linnean system also gives extra weight to those evolutionary branching events that produced noticeable morphological innovations, such as the evolution of birds (see text). Do you think it is worthwhile to distinguish between those branching events that produced noticeable morphological innovation and those that did not?

7. The vagaries of fossilization and fossil discovery mean that there may easily have been many other bird species that lived at the same time as *Archeopteryx,* but that have not (yet) been discovered in the fossil record. In fact, it is now considered likely that *Archeopteryx* was not the direct ancestor of today's birds, but instead belonged to another branch of early birds, while the true ancestor of modern birds remains undiscovered (Figure 10.4). Similarly, other "transitional" fossils might not be the actual ancestors of modern species. In your opinion, does this mean that these "transitional fossils" do not offer useful information on the origin of today's species? In other words, is it necessary to find an actual direct ancestor of today's species in order to study the origin of modern traits, or is a fossil "great-uncle" good enough?

Exploring the Literature

8. Leading phylogenetic systematists are now maintaining World Wide Web sites containing finished trees for all types of life forms and/or copies of the raw datasets used to estimate phylogenies. For background on this effort and the addresses of some key sites, see Morell (1996) and the Web site associated with this text.

9. Phylogenetic studies of coevolution are a booming research area. For entrees to the literature on relationships between flowering plants and herbivorous insects, fungal-algal relationships in lichen, and the evolution of warning coloration in butterflies, see:

Brower, A.V.Z. 1996. Parallel race formation and the evolution of mimicry in *Heliconius* butterflies: A phylogenetic hypothesis from mitochondrial DNA sequences. *Evolution* 50:195–221.

Funk, D.J., D.J. Futuyma, G. Ortí, and A. Meyer. 1995. A history of host associations and evolutionary diversification for *Ophraella* (Coleoptera: Chrysomelidae): New evidence from mitochondrial DNA. *Evolution* 49:1008–1017.

Gargas, A., P.T. DePriest, M. Grube, and A. Tehler. 1995. Multiple origins of lichen symbioses in fungi suggested by SSU rDNA phylogeny. *Science* 268:1492–1495.

Miller, J.S. 1991. Host-plant associations among prominent moths. *BioScience* 42:50–57.

Mitter, C., B.D. Farrell, and D.J. Futuyma. 1991. Phylogenetic studies of insect/plant interactions: Insights into the genesis of diversity. *Trends in Ecology and Evolution* 6:290–293.

10. Mapping traits onto phylogenies has confirmed or challenged several traditional views of how traits evolved. The following papers offer a test of how social behavior evolved in wasps, insight into the evolution of endothermy in fish, an analysis of the evolution of self-fertilization in a family of flowering plants, and general reviews of the field.

Block, B.A., J.R. Finnerty, A.F.R. Stewart, and J. Kidd. 1993. Evolution of endothermy in fish: Mapping physiological traits on a molecular phylogeny. *Science* 260:210–214.

Brooks, D.R., D.A. McLennan, J.M. Carpenter, S.G. Weller, and J.A. Coddington. 1995. Systematics, ecology, and behavior. *BioScience* 45:687–695.

Carpenter, J.M. 1989. Testing scenarios: Wasp social behavior. *Cladistics* 5:131–144.

Kohn, J.R., S.W. Graham, B. Morton, J.J. Doyle, and S.C.H. Barrett. 1996. Reconstruction of the evolution of reproductive characters in Pontederiaceae using phylogenetic evidence from chloroplast DNA restriction-site variation. *Evolution* 50:1454–1469.

Maddison, D.R. 1994. Phylogenetic methods for inferring the evolutionary history and processes of change in discretely valued characters. *Annual Review of Entomology* 39:267–292.

Schultz, T.R., R.B. Cocroft, and G.A. Churchill. 1996. The reconstruction of ancestral character states. *Evolution* 50:504–511.

Citations

Benton, M.J. 1990a. Origin and interrelationships of dinosaurs. In D.B. Weishampel, P. Dodson, and H. Osmolka, eds. *The Dinosauria* (Berkeley: University California Press), 11–30.

Benton, M.J. 1990b. Reptiles. In K.J. McNamara, ed. *Evolutionary Trends* (Tucson: University of Arizona Press), 279–300.

Benton, M.J. 1990c. Phylogeny of the major tetrapod groups: Morphological data and divergence dates. *Journal of Molecular Evolution* 30:409–424.

Bermingham, E., and H.A. Lessios. 1993. Rate variation of protein and mitochondrial DNA evolution as revealed by sea urchins separated by the Isthmus of Panama. *Proceedings of the National Academy of Sciences, USA* 90:2734–2738.

Beverly, S.M., and A.C. Wilson. 1984. Molecular evolution in *Drosophila* and the higher Diptera II. A time scale for fly evolution. *Journal of Molecular Evolution* 21:1–13.

Beverly, S.M., and A.C. Wilson. 1985. Ancient origin for Hawaiian Drosophilinae inferred from protein comparisons. *Proceedings of the National Academy of Sciences, USA* 82:4753–4757.

Brower, A.V.Z. 1994. Rapid morphological radiation and convergence among races of the butterfly *Heliconius erato,* inferred from patterns of mitochondrial DNA evolution. *Proceedings of the National Academy of Sciences, USA* 91:6491–6495.

Brown, W.M., M. George, Jr., and A.C. Wilson. 1979. Rapid evolution of animal mitochondrial DNA. *Proceedings of the National Academy of Sciences, USA* 76:1967–1971.

Cann, R.L., M. Stoneking, and A.C. Wilson. 1987. Mitochondrial DNA and human evolution. *Nature* 325:31–36.

Cano, R.J., and M.K. Borucki. 1995. Revival and identification of bacterial spores in 25- to 40-million-year-old Dominican amber. *Science* 268:1060–1064.

Chapela, I.H., S.A. Rehner, T.R. Schultz, and U.G. Mueller. 1994. Evolutionary history of the symbiosis between fungus-growing ants and their fungi. *Science* 266:1691–1694.

Chiappe, L.M. 1995. The first 85 million years of avian evolution. *Nature* 378:349–355.

Colbert, E.H., and M. Morales. 1991. *Evolution of the Vertebrates* (New York: Wiley-Liss).

DeBry, R.W., L.G. Abele, S.H. Weiss, M.D. Hill, M. Bouzas, E. Lorenzo, F. Graebnitz, and L. Resnick. 1993. Dental HIV transmission? *Nature* 361:691.

DeSalle, R., T. Freedman, E.M. Prager, and A.C. Wilson. 1987. Tempo and mode of sequence evolution in mitochondrial DNA of Hawaiian *Drosophila. Journal of Molecular Evolution* 26:157–164.

DeSalle, R., and J.A. Hunt. 1987. Molecular evolution in Hawaiian drosphilids. *Trends in Ecology and Evolution* 2:212–215.

Dixon, M.T., and D.M. Hillis. 1993. Ribosomal RNA structure: Compensatory mutations and implications for phylogenetic analysis. *Molecular Biology and Evolution* 10:256–267.

Doyle, J.J., J.L. Doyle, and A.H.D. Brown. 1990. A chloroplast-DNA phylogeny of the wild perennial relatives of soybean (*Glycine* subgenus *Glycine*): congruence with morphological and crossing groups. *Evolution* 44:371–389.

Eldredge, N., and J. Cracraft. 1980. *Phylogenetic Patterns and the Evolutionary Process* (New York: Columbia University Press).

Felsenstein, J. 1978. Cases in which parsimony or compatibility methods will be positively misleading. *Systematic Zoology* 27:401–410.

Felsenstein, J. 1983. Parsimony in systematics: Biological and statistical issues. *Annual Review of Ecology and Systematics* 14:313–333.

Felsenstein, J. 1985. Confidence limits on phylogenies: An approach using the bootstrap. *Evolution* 39:783–791.

Felsenstein, J. 1988. Phylogenies and quantitative characters. *Annual Review of Ecology and Systematics* 19:445–471.

Felsenstein, J. 1993. PHYLIP (Phylogeny Inference Package) version 3.5c. Distributed by the author (Seattle: Department of Genetics, University of Washington).

Gould, S.J. 1993. *Eight Little Piggies* (New York: W.W. Norton).

Hennig, W. 1979. *Phylogenetic Systematics* (Urbana: University of Illinois Press).

Hillis, D.M. 1997. Biology recapitulates phylogeny. *Science* 276:218–219.

Hillis, D.M., J.P. Huelsenbeck, and C.W. Cunningham. 1994. Application and accuracy of molecular phylogenies. *Science* 264:671–677.

Hillis, D.M., B.K. Mable, and C. Moritz. 1996. Applications of molecular systematics: The state of the field and a look to the future. In D.M. Hillis, C. Moritz, and B.K. Mable, eds. *Molecular Systematics* (Sunderland, MA: Sinauer).

Hinkle, G., J.K. Wetterer, T.R. Schultz, and M.L. Sogin. 1994. Phylogeny of the attine ant fungi based on analysis of small subunit ribosomal RNA gene sequences. *Science* 266:1695–1697.

Huelsenbeck, J.P., and D.M. Hillis. 1993. Success of phylogenetic methods in the four-taxon case. *Systematic Biology* 42:247–264.

Huelsenbeck, J.P., and B. Rannala. 1997. Phylogenetic methods come of age: Testing hypotheses in an evolutionary context. *Science* 276:227–232.

Kimura, M. 1980. A simple method for estimating evolutionary rates of base substitution through comparative studies of nucleotide sequences. *Journal of Molecular Evolution* 16:111–120.

Kreitman, M. 1983. Nucleotide polymorphism at the alcohol dehydrogenase locus of *Drosophila melanogaster. Nature* 304:412–417.

Lauder, G.V., and K.F. Liem. 1983. The evolution and interrelationships of actinopterygian fishes. *Bulletin of the Museum of Comparative Zoology* 150:95–197.

Li, W.-H., and D. Graur. 1991. *Molecular Evolution* (Sunderland, MA: Sinauer).

Liem, K.F. 1988. Form and function of lungs: The evolution of air breathing mechanisms. *American Zoologist* 28:739–759.

Maddison, D.R. 1991. African origin of human mitochondrial DNA reexamined. *Systematic Zoology* 40:355–377.

Maddison, W.P., M.J. Donoghue, and D.R. Maddison. 1984. Outgroup comparison and parsimony. *Systematic Zoology* 33:83–103.

Martin, A.P., G.J.P. Naylor, and S.R. Palumbi. 1992. Rates of mitochondrial DNA evolution in sharks are slow compared with mammals. *Nature* 357:153–155.

Martin, A.P., and S.R. Palumbi. 1993. Body size, metabolic rate, generation time, and the molecular clock. *Proceedings of the National Academy of Sciences, USA* 90:4087–4091.

Mayden, R.L., and E.O. Wiley. 1992. The fundamentals of phylogenetic systematics. In R.L. Mayden, ed. *Systematics, Historical Ecology, and North American Freshwater Fishes* (Stanford, CA: Stanford University Press), 114–185.

Moran, N.A., M.A. Munson, P. Baumann, and H. Ishikawa. 1993. A molecular clock in endosymbiotic bacteria is calibrated using the insect hosts. *Proceedings of the Royal Society of London,* Series B 253:167–171.

Morell, V. 1996. Web-crawling up the tree of life. *Science* 273:568–570.

Nelson, G., and N. Platnick. 1981. *Systematics and Biogeography: Cladistics and Vicariance* (New York: Columbia University Press).

Novacek, M.J. 1992. Mammalian phylogeny: Shaking the tree. *Nature* 356:121–125.

Novacek, M.J. 1993. Reflections on higher mammalian phylogenetics. *Journal of Mammalian Evolution* 1:3–30.

Ou, C.-Y., Carol A. Ciesielski, G. Myers, C.I. Bandea, C.-C. Luo, B.T.M. Korber, J.I. Mullins, G. Schochetman, R.L. Berkelman, A.N. Economou, J.J. Witte, L.J. Furman, G.A. Satten, K.A. MacInnes, J.W. Curran, H.W. Jaffe, Laboratory Investigation Group, and Epidemiologic Investigation Group. 1992. Molecular epidemiology of HIV transmission in a dental practice. *Science* 256:1165–1171.

Pennisi, E. 1996. Evolutionary and systematic biologists converge. *Science* 273:181–182.

Pough, F.H., J.B Heiser, and W.N. McFarland. 1996. *Vertebrate Life* (Upper Saddle River, NJ: Prentice Hall).

de Queiroz, K. 1992. Phylogenetic definitions and taxonomic philosophy. *Biology and Philosophy* 7:295–313.

de Queiroz, K., and J. Gauthier. 1990. Phylogeny as a central principle in taxonomy: Phylogenetic definitions of taxon names. *Systematic Zoology* 39:307–322.

de Queiroz, K., and J. Gauthier. 1992. Phylogenetic taxonomy. *Annual Review of Ecology and Systematics* 23:449–480.

Sarich, V., and A.C. Wilson. 1967. Immunological time scale for hominid evolution. *Science* 158:1200–1203.

Schwenk, K. 1994. Comparative biology and the importance of cladistic classification: A case study from the sensory biology of squamate reptiles. *Biological Journal of the Linnean Society* 52:69–82.

Shaw, K.L. 1996. Sequential radiations and patterns of speciation in the Hawaiian cricket genus *Laupala* inferred from DNA sequences. *Evolution* 50:237–255.

Shields, G.F., and A.C. Wilson. 1987. Calibration of mitochondrial DNA evolution in geese. *Journal of Molecular Evolution* 24:212–217.

Simpson, B.B., and J. Cracraft. 1995. Systematics: The science of biodiversity. *BioScience* 45:670–672.

Sneath, P.H.A., and R.R. Sokal. 1973. *Numerical Taxonomy: The Principles and Practice of Numerical Classification* (San Francisco: Freeman).

Swofford, D.L., and W.P. Maddison. 1992. Parsimony, character state reconstructions, and evolutionary inferences. In R.L. Mayden, ed. *Systematics, Historical Ecology, and North American Freshwater Fishes* (Stanford CA: Stanford University Press), 186–223.

Swofford, D.L., and G.J. Olsen. 1990. Phylogeny reconstruction. In D.M. Hills and C. Moritz, eds. *Molecular Systematics* (Sunderland, MA: Sinauer), 411–501.

Swofford, D.L., G.J. Olsen, P.J. Waddell, and D.M. Hillis. 1996. Phylogenetic Inference. In D.M. Hillis, C. Moritz, and B.K. Mable, eds. *Molecular Systematics* (Sunderland, MA: Sinauer), 407–514.

Templeton, A.R., S.B. Hedges, S. Kumar, K. Tamura, and M. Stoneking. 1991. Human origins and analysis of mitochondrial DNA sequences. *Science* 255:737–739.

Vermeij, G.J. 1996. Animal origins. *Science* 274:525–526.

Wakely, J. 1996. The excess of transitions among nucleotide substitutions: New methods of estimating transition bias underscore its significance. *Trends in Ecology and Evolution* 11:158–163.

Wiley, E.O. 1981. *Phylogenetics: The Theory and Practice of Phylogenetic Systematics* (New York: Wiley).

Wilson, A.C. 1985. The molecular basis of evolution. *Scientific American* 253:164–173.

Wilson, A.C., S.S. Carlson, and T.J. White. 1977. Biochemical evolution. *Annual Review of Biochemistry* 46:573–639.

Wilson, A.C., R.L. Cann, S.M. Carr, M. George, U.B. Gyllensten, K.M. Helm-Bychowski, R.G. Higuchi, S.R. Palumbi, E.M. Prager, R.D. Sage, and M. Stoneking. 1985. Mitochondrial DNA and two perspectives in evolutionary genetics. *Biological Journal of the Linnean Society* 26:375–400.

Wray, G.A., J.S. Levington, and L.H. Shapiro. 1996. Molecular evidence for deep Precambrian divergences among metazoan phyla. *Science* 274:568–573.

Zuckerkandl, E., and L. Pauling. 1962. Molecular disease, evolution and genic heterogeneity. In M. Kash and B. Pullman, eds. *Horizons in Biochemistry* (New York: Academic Press).

The Origin of Life, and Precambrian Evolution

In the final chapter of *On the Origin of Species,* Charles Darwin wrote:

> [A]ll living things have much in common, in their chemical composition, their germinal vesicles, their cellular structure, and their laws of growth and reproduction. . . . Therefore I should infer . . . that probably all the organic beings which have ever lived on this earth have descended from some one primordial form, into which life was first breathed. (1859, page 484)

Nearly 14 decades have passed since Darwin's tentative conclusion. In that time geneticists have discovered the rules of heredity, identified the genetic material, and learned to read and decipher the coded messages in the genes. Biochemists have mapped the steps of cellular metabolism from photosynthesis to the Krebs cycle, and identified the reactants, products, cofactors, and enzymes that run the engines of life. Cell biologists have peered into the tiniest internal compartments and examined the machinery of growth, reproduction, and cell-cell communication. Everything biologists have learned supports Darwin's inference of the unity of life. There is but one evolutionary tree.

This chapter is about the "one primordial form" and one of its descendants, the most recent common ancestor of all living things (Figure 11.1). We focus our attention on the most recent common ancestor, sometimes referred to as the cenancestor (Fitch and Upper 1987), but our inquiry takes us back in life's history to a time before the primordial form, and forward to the present. About the cenancestor, we ask four questions:

What was it?

When did it live?

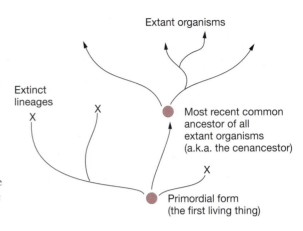

Figure 11.1   **Cartoon of the tree of life**   The first living organism presumably had several descendant lineages, all but one of which eventually died out. The most recent common ancestor of all living things is the organism whose immediate descendants diverged to found the lineages that ultimately became all extant organisms. The whole-life phylogeny drawn here does not include the viruses, whose position on the tree is unclear.

Where did it come from?

By what route did its descendants evolve into today's orchids, ants, mushrooms, amoebae, and bacteria?

The questions are straightforward, but they are among the most difficult to answer in all biology. They concern events that happened in deep time, and most of the direct evidence we might ideally rely on is lost. In recent years researchers have made remarkable progress, however, and some of the answers are beginning to emerge.

11.1   What Was the Most Recent Common Ancestor of All Living Things?

In principle, the fossil record could allow us to identify the cenancestor. In practice, the fossil record is too incomplete.

The first place we might look in trying to identify the common ancestor is the fossil record. In principle, a complete fossil record would allow us to trace lines of descent from present-day organisms all the way back to the cenancestor. However, it does not appear that the fossil record so far assembled can take us that deep into the past. The oldest fossils definitively established as those of living organisms are 3.465 billion years old (Schopf 1993; see also Schopf and Packer 1987; Schopf 1992). These fossils, which come from a rock formation called the Apex chert in Western Australia, show simple cells growing in short filaments (Figure 11.2). Their discoverer, J. William Schopf (1993), believes these organisms belonged to the cyanobacteria, a group of photosynthesizing bacteria formerly known as the blue-green algae. This identification is tentative. It is based on the morphology of the fossil cells and on analysis of the isotopic composition of the organic carbon in the rocks, neither of which is taxonomically definitive. If Schopf's identification of the Apex chert fossils is correct, then the fossils probably represent organisms that are already fairly high up the evolutionary tree, well above the most recent common ancestor (see Figure 11.5). Furthermore, the fossil record for times earlier than 2.5 billion years ago is too spotty to allow paleontologists to trace lines of evolutionary descent from present-day organisms back to the Apex chert fossils (Altermann and Schopf 1995). As a result, we have no direct way of knowing whether the organisms recorded in the Apex chert represent extinct or living branches of the tree of life. If we

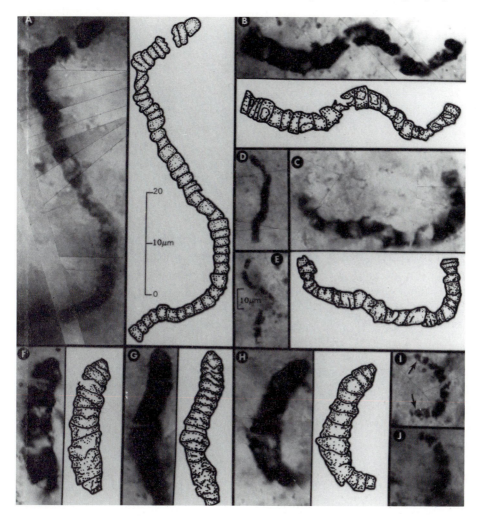

Figure 11.2 **The oldest known fossils of living organisms** Most photographs of fossils are accompanied by an interpretive drawing showing the fossil's morphology. The fossils show filaments composed of individual cells lined up like beads on a string. Specimens A, B, C, D, and E belong to the species *Primaevifilum amoenum;* F, G, H, I, and J belong to *P. conicoterminatum*. Reproduced by permission from J. William Schopf, Microfossils of the early Archean Apex chert: New evidence of the antiquity of life, *Science* 260: 640–646, 1993. © 1993 by the American Association for the Advancement of Science.

want to discover the characteristics of the most recent common ancestor, we must use methods other than examination of the fossil record.

Finding the Common Ancestor Without Fossils

If we knew the phylogeny, or evolutionary tree, of all living things, we could use it to infer the characteristics of the cenancestor. The general logic for making such inferences is illustrated in Figure 11.3, which uses the tetrapod vertebrates as an example. The figure shows an estimate of the phylogeny of all living tetrapods, with an arrow indicating their most recent common ancestor. What was this ancestor like? We can infer from this phylogeny that the common ancestor of the tetrapods had legs, even though some of its descendants (the caecilians, the snakes, and some of the lizards) subsequently lost them. An alternative scenario is that the last common ancestor of the tetrapods was a legless animal (a fish perhaps), and that legs evolved independently in the lineages leading to frogs and salamanders, turtles, at least two groups of lizards, the crocodiles and birds, and the mammals. This alternative is less plausible than a common ancestor with legs, for at least four reasons: (1) the alternative requires more evolutionary transi-

A phylogeny of all extant organisms would allow us to infer the characteristics of the cenancestor.

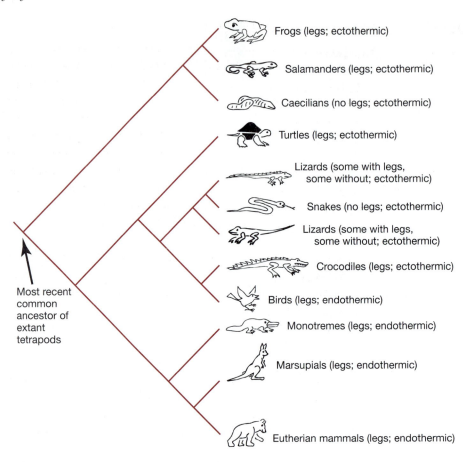

Figure 11.3 **An estimate of the phylogeny of the tetrapod vertebrates** Redrawn from Pough et al. (1996).

tions between the legless and legged states; (2) independent losses of a complex structure are more plausible than independent originations; (3) the anatomical similarities of the legs of the tetrapods that have them strongly suggests a common evolutionary origin; and (4) many of the legless tetrapods retain vestiges of legs. We leave it as an exercise for the reader to infer whether the most recent common ancestor of the tetrapods was an endotherm or an ectotherm.

What we need, then, in our search for the most recent common ancestor of all extant organisms, is an estimate of the phylogeny of all living things. The first attempts to construct such a phylogeny were based on the morphologies of organisms (see reviews in Woese 1991; Doolittle and Brown 1994). The morphological approach was productive for biologists interested in the branches of the tree of life that contain eukaryotes. Morphology was, indeed, the basis of the phylogeny of the tetrapods in Figure 11.3. The fact that the monotremes, marsupials, and eutherians have hair, for example, indicates that these mammal groups are all more closely related to each other than any of them is to any other tetrapod. The morphological approach led only to frustration, however, for biologists interested in the branches of the universal phylogeny containing prokaryotes. Prokaryotes lack sufficient structural diversity to allow the reconstruction of morphology-based evolutionary trees.

When biologists developed methods for reading the sequences of amino acids in proteins, and the sequences of nucleotides in DNA and RNA, a new technique for estimating phylogenies quickly became established (Zuckerkandl and Pauling 1965). Some of the details of this technique are devilish (see Chapter 10), but the basic idea is straightforward. Imagine that we have a group of species, all carrying in their genomes a particular gene. We can read the sequence of nucleotides in this gene in each of the species, then compare the sequences. If species are closely related, their sequences ought to be fairly similar. If species occupy distant branches on the evolutionary tree, then their sequences ought to be less similar. As a result, we can use the relative similarity of the sequences of species to infer their evolutionary relationships. We place species with more similar sequences on neighboring branches of the evolutionary tree, and species with less similar sequences on more distant branches.

Comparison of molecular sequences provides a method for estimating the phylogeny of all extant organisms.

Such a tree is shown in Figure 11.4; it is an estimate of the tetrapod phylogeny based on amino acid sequences of an eye lens protein called αA-crystallin. Although it includes fewer taxa, this tree agrees with the morphology-based tree in Figure 11.3. The first branch point in the tree separates the amphibians from the rest of the tetrapods. The next group to split off is the mammals; next are turtles; then lizards. Finally, the crocodiles and birds are each other's closest living relatives.

We should emphasize that no method of estimating phylogenies is perfect. Sequence-based trees using different genes often disagree among themselves. For example, when S. Blair Hedges (1994) used nucleotide sequences for three genes on the mitochondrial chromosome to estimate the phylogeny of the tetrapods, the result was a tree that placed the lizards closer to the turtles than to the crocodiles and birds (see also Hedges et al. 1990; Hedges and Maxson 1991, 1992; Marshall 1992; Eernisse and Kluge 1993; Huelsenbeck and Bull 1996). Morphology-based trees often conflict as well. The most accurate estimates of any phylogeny are those that combine data from many different sources. When Eernisse and Kluge (1993) used such an approach for the tetrapods, they got a tree that agrees with the morphology-based tree in Figure 11.3.

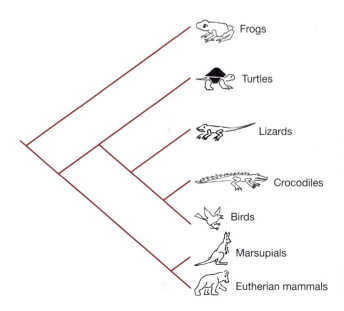

Frogs

Turtles

Lizards

Crocodiles

Birds

Marsupials

Eutherian mammals

Figure 11.4 An estimate of the phylogeny of the tetrapod vertebrates based on the protein αA-crystallin Gert-Jan Caspers and colleagues (1996) isolated mRNA from turtles, reverse-transcribed it to cDNA, and then sequenced the gene for the eye lens protein αA-crystallin. From the nucleotide sequence in the cDNA, the researchers inferred the amino acid sequence in the protein. Using their own sequence for turtles, along with sequences for other species obtained from databases, the researchers determined that the best estimate of the tetrapod phylogeny is the one shown here. Redrawn from Caspers et al. (1996).

The Phylogeny of All Living Things

The challenge in using sequence data to estimate the evolutionary tree for all living things is to find a gene that shows recognizable sequence similarities even between species that are as distantly related as *Escherichia coli* and *Homo sapiens* (Woese 1991). We need a gene that is present in all organisms, and that encodes a product whose function is essential and thus subject to strong stabilizing selection. Without strong stabilizing selection, billions of years of genetic drift will have obliterated any recognizable similarities in the sequences of distantly related organisms. Additionally, the function of the gene must have remained the same in all organisms. This is because when a gene product's function shifts in some species but not in others, then selection on the new function can cause a rapid divergence in nucleotide sequence that makes species look more distantly related than they actually are.

The first reliable estimates of the universal phylogeny were based on sequences of small subunit ribosomal RNAs.

One gene that meets all the criteria for use in reconstructing the universal phylogeny is the gene that codes for the small-subunit ribosomal RNA (Woese and Fox 1977; Woese 1991). All organisms have ribosomes, and in all organisms the ribosomes have a similar composition, including both ribosomal RNA (rRNA) and protein. All ribosomes have a similar tertiary structure, including small and large subunits. In all organisms the function of the ribosomes is the same: They are the machines responsible for translation. Translation is so vital, and organisms are under such strong natural selection to maintain it, that the ribosomal RNAs of humans and their intestinal bacteria show recognizable similarities in nucleotide sequence, even though humans and bacteria last shared a common ancestor billions of years ago. The small subunit rRNA was the molecule chosen by Carl R. Woese, the chief pioneer of the use of molecular sequences in estimating the universal phylogeny (Fox et al. 1977; Woese and Fox 1977; see also Doolittle and Brown 1994). The small subunit rRNA remains the single most informative resource for whole-life phylogenies.

An estimate of the universal phylogeny, based on sequences of the small subunit rRNA, appears in Figure 11.5. Before using this tree to infer the characteristics of the most recent common ancestor, we need to discuss the phylogeny's general shape. The whole-life rRNA phylogeny has prompted a dramatic revision of our traditional view of the organization of life, because it reveals that the five-kingdom system of classification (Whittaker 1969) bears only a limited resemblance to actual evolutionary relationships (Woese et al. 1990; for a contrary view, see Margulis 1996).

The rRNA universal phylogeny contradicts the traditional view of the tree of life.

The prokaryotes, for example, which are all grouped in the kingdom Monera in the traditional classification, occupy two of the three main branches of the rRNA phylogeny. One of these two branches, the Bacteria, includes virtually all of the well-known prokaryotes. The gram positive bacteria, for instance, include *Mycobacterium tuberculosis,* the pathogen that causes tuberculosis. The purple bacteria include *E. coli.* (The purple bacteria are so named because some of them are purple and photosynthetic, although *E. coli* is neither.) The cyanobacteria, all of which are photosynthetic, include *Nostoc,* an organism often seen in introductory biology labs.

The other prokaryote branch, the Archaea, is not as well known. Many of the Archaea live in physiologically harsh environments, are difficult to grow in cul-

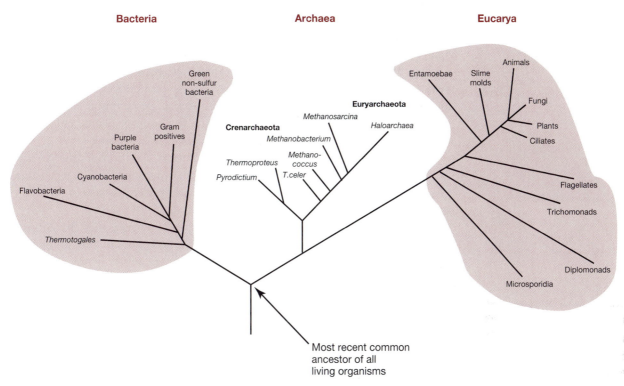

***Figure 11.5* An estimate of the phylogeny of all living organisms**   This tree is based on the analysis of nucleotide sequences of small subunit rRNAs. From Woese (1996).

ture, and were discovered only recently (see Madigan and Marrs 1997). Most of the Crenarchaeota, for example, are hyperthermophiles, living in hot springs at temperatures as high as 110°C. Many of the Euryarchaeota are anaerobic methane producers. Another group in the Euryarchaeota, the Haloarchaea, are highly salt dependent, and thus referred to as extreme halophiles.

Because of their prokaryotic cell structure, the Archaea were originally considered bacteria. When Woese and colleagues discovered that these organisms were only distantly related to the rest of the bacteria, they renamed them the archaebacteria (Fox et al. 1977; Woese and Fox 1977). Eventually biologists realized that, as the phylogeny in Figure 11.5 shows, the archaebacteria are in fact more closely related to the eukaryotes than they are to the true bacteria (see Bult et al. 1996; Olsen and Woese 1996). In recognition of this, Woese and colleagues (1990) proposed the new classification used in Figure 11.5. Woese and colleagues dropped the *bacteria* from "archaebacteria," renaming this group the Archaea. The most inclusive taxonomic units in the new classification are three domains corresponding to the three main branches on the tree of life: the Bacteria, the Archaea, and the Eucarya. Woese and colleagues proposed that the two fundamental branches of the Archaea, the Crenarchaeota and the Euryarchaeota, be designated as kingdoms.

Woese et al. (1990) declined to offer a detailed proposal on how to divide the Eucarya into kingdoms. The Protista, a single kingdom in the traditional classification, are scattered across several fundamental limbs on the eukaryotic branch of the tree of life. The diplomonads, for example, which include the intestinal parasite *Giardia lamblia,* represent one of the deepest branches of the Eucarya. They are well separated from such other protists as the flagellates, which include *Euglena,* and the ciliates, which include *Paramecium.* If we want our kingdoms to be natural evolutionary groups, they should be monophyletic. That is, each kingdom should include all the descendants of a single common ancestor. Unless we want the kingdom Protista to include the animals, plants, and fungi, it will have to be disbanded and replaced by several new kingdoms.

The remaining three kingdoms in the traditional classification, the Animals, Plants, and Fungi, require only minor revision. To make the Fungi a natural group, for example, the cellular slime molds (such as *Dictyostelium,* a favorite of developmental biologists) will have to be removed.

The universal rRNA phylogeny demonstrates, however, that the Animals, Plants, and Fungi, the kingdoms that have absorbed most of the attention of evolutionary biologists (and represent most of the examples in this book), are mere twigs on the tip of one branch of the tree of life. The multicellular, macroscopic organisms in these three kingdoms are newcomers on the evolutionary scene, with a relatively recent common ancestor. For genes shared among all organisms, like the gene for the small subunit rRNA, Animals, Plants, and Fungi appear to contribute less than 10% of the genetic diversity on Earth (Olsen and Woese 1996).

Inferring the Characteristics of the Cenancestor

Now that we have a universal phylogeny, what does it tell us about the last common ancestor of all extant organisms? As indicated by the arrow in Figure 11.5, the universal ancestor was the species whose descendants diverged to become the bacteria on one side, and the Archaea/Eucarya on the other (see Box 11.1). Before discussing particular traits, we note that when we look at the distribution of complex traits on the universal phylogeny and try to infer whether the cenancestor had them, we may encounter five different patterns (Figure 11.6). If a trait is universal, then the simplest evolutionary scenario is that the trait existed in the cenancestor (that is, the trait arose only once, in an ancestor of the cenancestor; Figure 11.6a). If a trait occurs in the Bacteria and the Archaea, but not in the Eucarya, then the simplest scenario is that the trait existed in the cenancestor and was lost on the line that led to the Eucarya (Figure 11.6b). Likewise, if a trait occurs in the Bacteria and the Eucarya, but not in the Archaea, then the simplest scenario is that the trait existed in the cenancestor and was lost on the line that led to the Archaea (Figure 11.6c). If a trait occurs in the Archaea and the Eucarya, but not in the Bacteria, then the simplest scenario is that the trait did not exist in the cenancestor, and that the trait arose only once, in the line that led to the Archaea and the Eucarya (Figure 11.6d). Finally, if a trait is unique to a particular domain, then the simplest scenario is that the trait did not exist in the cenancestor, and that the trait arose only once, in the line that led to the domain that has it (Figure 11.6e). The key to discovering the nature of the cenancestor thus lies in knowing how the basic traits of organisms are distributed across the three domains of life.

We provide a detailed analysis of trait distributions in the following paragraphs, but the essential message is this: The most recent common ancestor was

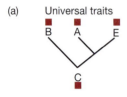

(a)   Universal traits

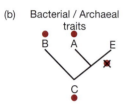

(b)   Bacterial / Archaeal traits

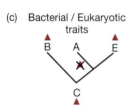

(c)   Bacterial / Eukaryotic traits

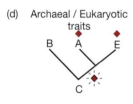

(d)   Archaeal / Eukaryotic traits

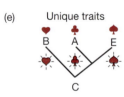

(e)   Unique traits

Figure 11.6 **Five possible distributions of traits among the three domains of life**   The first appearance of a symbol on a tree represents the origin of the trait represented by the symbol. A crossed-out symbol represents the loss of a trait.

rather like a modern bacterium. It used DNA as its genetic material, and had many elements of the modern machinery for replication, transcription, and translation. It also had a complex metabolism, with many elements of modern biochemical pathways, and may even have been an autotroph.

The cenancestor used DNA. All extant organisms store genetic information in DNA, which suggests that the common ancestor did the same (Figure 11.6a). An alternative possibility is that the cenancestor stored its genetic information in some other molecule, such as RNA, but that storage in DNA was favored so strongly by natural selection that a conversion from RNA storage to DNA storage occurred independently in more than one domain. Use of DNA by the cenancestor appears more likely than this scenario of convergent evolution. One clue is that the DNA-dependent RNA polymerases used in transcription show strong similarities across all three domains. This suggests that the cenancestor had a DNA-dependent RNA polymerase. The existence of a DNA-dependent RNA polymerase implies DNA (Benner et al. 1989). Likewise, DNA polymerases found in all three domains show enough similarities to suggest that the cenancestor had a DNA polymerase too (see the next paragraph). Where there was a DNA polymerase, there was probably DNA.

Although the cenancestor probably stored genetic information in DNA, its handling of DNA replication was apparently somewhat rudimentary. The archaean microbe *Methanococcus jannaschii* has only one DNA polymerase gene (Bult et al. 1996). This gene has a nucleotide sequence recognizably similar to the genes for three eukaryotic DNA polymerases, for bacterial DNA polymerase II, and for the DNA polymerases of several other archaeans. In other words, homologs of the *M. jannaschii* polymerase gene occur in all three domains (Figure 11.6a). This suggests that the cenancestor had a version of the polymerase gene, and thus that the cenancestor used an enzymatic protein for DNA replication.

Based on the universal phylogeny, we can infer that the cenancestor was a remarkably modern organism.

However, no central components of the DNA replication machinery have retained their precise functions across all three domains (Bult et al. 1996; Olsen and Woese 1996). For example, bacteria replicate their genomes with DNA polymerase III, whereas the archaean *M. jannaschii* lacks this enzyme altogether and presumably replicates its genome with its homolog of DNA polymerase II. The bacteria that have a version of DNA polymerase II use it for functions other than genome replication, including replication of episomes (plasmids that can insert themselves into the bacterial chromosome), and DNA repair (Foster et al. 1995; Tomer et al. 1996). Some bacteria, including two whose genomes have been completely sequenced, do not have DNA polymerase II at all (Fleischmann et al. 1995; Fraser et al. 1995; Bult et al. 1996). Thus each domain's DNA replication machinery has many unique features (Figure 11.6e). Overall, the archaean machinery resembles the eukaryotic machinery more than it does the bacterial machinery (Figure 11.6d; Olsen and Woese 1996). We can conclude that the cenancestor had a DNA polymerase enzyme, but beyond that we can infer little about how the cenancestor replicated and repaired its DNA.

We can infer somewhat more about the last common ancestor's machinery for transcription (Benner et al. 1989; Bult et al. 1996; Olsen and Woese 1996). All organisms perform transcription with a DNA-dependent RNA polymerase. The bacterial RNA polymerase has three types of subunit, and these bacterial subunits have recognizable homologs in all organisms (Figure 11.6a). Thus the

BOX 11.1 Rooting the tree of life

Readers familiar with the estimation of phylogenies from molecular sequence data (see Chapter 10) may have noticed that when we said the tree in Figure 11.5 identifies the most recent common ancestor of all extant organisms, we ignored an important issue: How was the universal phylogeny rooted? Comparison of the sequences of a gene (like the small subunit rRNA) that is found in the Bacteria, Archaea, and Eucarya can tell us that the archaeal and eukaryotic versions of the gene are more similar to each other than either is to the bacterial version (Figure 11.7a). But this analysis does not tell us the order of the branchings during evolutionary history. The sequence differences diagrammed in Figure 11.7a are equally compatible with three different evolutionary trees (Figure 11.7b, c, and d). How do we decide which is most likely to be correct?

If we were dealing with some subset of the extant organisms, we could solve the tree rooting problem by including an outgroup. If we were trying to root a tree that included the amphibians, the birds, and the mammals, for example, we could do so by including the fish as the outgroup (Figure 11.8). The point at which the outgroup connects to the tree becomes the root for the clade that includes the amphibians, birds, and mammals. The root defines the branching order as the tetrapod vertebrates diverged from their last common ancestor and from each other. This simple outgroup method does not work, however, for rooting the universal phylogeny. By definition, the evolutionary tree of all extant organisms has no outgroup.

In 1989 two teams, one led by Johann Gogarten and the other by Naoyuki Iwabe, overcame this problem (Gogarten et al. 1989; Iwabe et al. 1989; for a review of the Iwabe team's work, see Doolittle and Brown 1994). The trick, suggested by Robert Schwartz and Margaret Dayhoff (1978), was to use a pair of genes that exists in all organisms and is derived from a gene duplication. For any such gene pair, the duplication event must have happened in the most recent common ancestor of all living things, or one of its ancestors. One of the genes derived from the duplication can serve as the outgroup for the tree constructed from the sequence similarities of the other gene (Figure 11.9). Both teams concluded that the rooting shown in Figure 11.7b (and Figure 11.5) is the correct one.

Two other teams have used the same technique to root the universal phylogeny, using larger datasets and addressing criticisms of the Gogarten and Iwabe papers. James R. Brown and W. Ford Doolittle (1995) used genes for aminoacyl-tRNA synthetase enzymes. Aminoacyl-tRNA synthetases are the enzymes that charge each tRNA molecule with the appropriate amino acid before translation. The aminoacyl-tRNA synthetase genes for valine, leucine, and isoleucine (among others) share sequence similarities within and among organisms that indicate that the enzyme genes originated as pre-cenancestor duplications of a single ancestral gene (Nagel and Doolittle 1991). Brown and Doolittle (1995) used gene sequences for the isoleucine aminoacyl-tRNA synthetase to estimate the universal phylogeny, and rooted it by including as outgroups the leucine and valine aminoacyl-tRNA synthetase genes from a variety of bacteria and eukaryotes. Brown and Doolittle's tree shows the same root as the Gogarten and Iwabe trees: The first branch

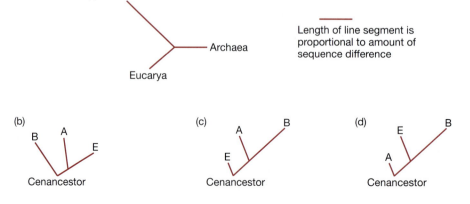

Figure 11.7 Rooting the tree of life (a) The sequences of the Archaea and the Eucarya are more similar to each other than either is to the Bacteria. (b, c, and d) The three trees compatible with the sequence similarities (B = Bacteria; A = Archaea; E = Eucarya).

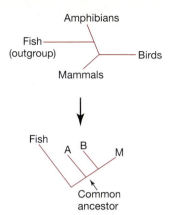

Figure 11.8 Rooting the tree of tetrapods A phylogeny for the tetrapod vertebrates can be rooted by adding fish as an outgroup.

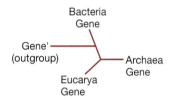

Figure 11.9 A gene duplication present in all organisms can root the universal phylogeny Gene′ arose as a duplicate of Gene; Gene′ serves as the outgroup for the phylogeny of Gene.

point on the tree of life separates the Bacteria on one limb from the Archaea/Eucarya on the other (Figure 11.10). Sandra Baldauf, Jeffrey Palmer, and W. Ford Doolittle (1996) used genes for translation elongation factors. Again, they found that the universal root was between the Bacteria and the Archaea/Eucarya.

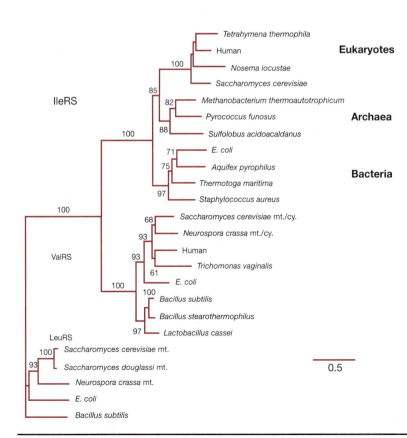

Figure 11.10 A rooted estimate of the universal phylogeny The top portion of this phylogeny is based on sequence similarities among the isoleucine aminoacyl-tRNA synthetase (IleRS) genes of organisms representing all three domains (the Bacteria, the Archaea, and the Eucarya). Sequences of the valine (ValRS) and leucine (LeuRS) aminoacyl-tRNA synthetase genes from a variety of bacteria and eukaryotes are the out-groups that root the tree. From Brown and Doolittle (1995).

cenancestor probably had these bacterial subunits too. The archaean *M. jannaschii* has genes for six additional subunit types, all with sequences recognizably similar to eukaryotic RNA polymerase subunits but not to bacterial subunits (Figure 11.6d). Thus the cenancestor probably did not have these six genes. The mechanism for transcription initiation is not universal; the archaean and eukaryotic systems are similar (Figure 11.6d), although the archaean system is much simpler. Thus we can infer that the cenancestor had an RNA polymerase like those of modern bacteria, but we cannot infer how the cenancestor initiated transcription.

The cenancestor had a highly elaborated machinery for translation (Nagel and Doolittle 1991; Doolittle and Brown 1994; Brown and Doolittle 1995; Olsen and Woese 1996; Bult et al. 1996). Translation is universally performed by two-subunit ribosomes composed of RNA and protein. The ribosomal RNAs used in all three domains have recognizable sequence similarities, as do the genes for many of the ribosomal proteins. Transfer RNAs are homologous across all three domains, and the genetic code is so similar in all organisms that it is nearly universal. Most of the enzymes responsible for charging the tRNAs with their amino acids (the aminoacyl–tRNA synthetases) are recognizably related across all three domains. In other words, nearly the entire apparatus for translation is universal (Figure 11.6a) and must have been present in the cenancestor. Among the few exceptions to this pattern are the translation initiation factors, which are similar in the Archaea and the Eucarya but distinctive in the Bacteria (Figure 11.6d), making it impossible to infer the mechanism of translation initiation used by the common ancestor.

The cenancestor must have possessed a complex and sophisticated metabolism. Many metabolic enzymes show strong similarities between the Bacteria and the Archaea (Figure 11.6b), including enzymes involved in amino acid, nucleotide, and coenzyme synthesis (Bult et al. 1996; Olsen and Woese 1996). Other enzymes with either universal or bacterial/archaeal distributions (Figures 11.6a and b) are involved in glycolysis, gluconeogenesis, and the citric acid cycle (Zillig 1991). Olsen and Woese (1996) note that some of the key enzymes involved in carbon fixation are similar in Bacteria and Archaea (Figure 11.6b), suggesting that the most recent common ancestor may have been an autotroph.

The organization of the genome in the cenancestor is important later in the chapter, when we consider the evolution of the eukaryotes. The Bacteria and Archaea have circular chromosomes (Figure 11.6b), whereas the Eucarya have linear chromosomes (Figure 11.6e). These patterns suggest that the cenancestor had a circular chromosome. Compared to other ways in which a small genome might be organized, however, a circular chromosome is not a particularly complex trait. Circular chromosomes may have evolved independently in the Bacteria and Archaea, leaving open the possibility that the cenancestor's genome had some other organization (Doolittle and Brown 1994; Olsen and Woese 1996).

Bacterial and archaeal genomes have many genes arranged in operons—clustered groups of genes with related functions that are regulated and transcribed as a unit. Eukaryotic genomes lack operons (Figure 11.6b; Zillig 1991; Doolittle and Brown 1994). The manifest efficiency of a genome organized in operons suggests that operons may be strongly favored by natural selection, and could have evolved independently in the Bacteria and the Archaea. However, at least some operons appear to date back to the common ancestor. (Doolittle and Brown 1994). For example, there are four different operons for ribosomal proteins, containing 4, 11, 8, and 3 genes, respectively, that occur in both the bacterium *E. coli* and in several archaeans. In every instance the genes in these operons are arranged in the same order on the chromosome. Because there is no known functional reason why the ribosomal protein genes need to be in any particular order, it seems reasonable to infer that these operons are derived from the cenancestor, rather than that they are the products of convergent evolution. Since operons were present in the cenancestor, we are challenged to explain their loss in the Eucarya. Olsen and Woese (1996) note that the most extensively studied archaean to date, *M. jannaschii,* has relatively few of its genes in operons. The extent to which the cenancestor had its genes organized in operons should become clearer with further study.

Another puzzling question about the organization of the cenancestor's genome concerns introns, intervening sequences of noncoding DNA inserted into the coding regions of genes. Introns, and the machinery for post-transcriptional processing needed to remove introns from mRNAs, are virtually unique to the Eucarya. This suggests that introns were absent in the cenancestor (Figure 11.6e). If this is true, then we must explain how and why introns were acquired by the Eucarya. We will return to this issue later.

Finally, note that many features of cell structure are unique to the Eucarya (Figure 11.6e). These include a nucleus, organelles, and a cytoskeleton. These complex features were almost certainly absent from the common ancestor. We will also return to the acquisition of organelles later in this chapter.

In summary, the most recent common ancestor of all extant organisms was a highly evolved, biologically sophisticated creature. It had a DNA genome, and was capable of complete information processing, including replication, transcription, and translation. Its metabolic capabilities were extensive; it may even have been an autotroph. Its genome was at least to some extent efficiently organized. In all, the cenancestor appears to have been rather like a modern bacterium.

Our Picture of the Cenancestor Will Continue to Improve

Our knowledge of the cenancestor depends crucially on genetic sequence data. Such data allow us to estimate the universal phylogeny. Furthermore, sequence data provide much of the information about the traits of organisms that, when placed on the universal phylogeny, allow us to make inferences about the common ancestor. As of this writing, the amount of sequence data we have is limited, but growing rapidly. Two trends in particular promise to yield many new insights.

First, our knowledge of the Archaea is increasing dramatically. As we mentioned earlier, many archaeans live in harsh and unusual environments. *M. jannaschii,* for example, lives anaerobically in deep sea hydrothermal vents at temperatures near 85°C and depths of at least 2,600 m (Jones et al. 1983). Most archaeans are, not surprisingly, difficult or impossible to grow in culture, and thus hard to study. In 1984, a team of biologists working in the laboratory of Norman Pace pioneered a new approach to studying the environmental distribution of the Archaea. The researchers extracted DNA directly from mud and

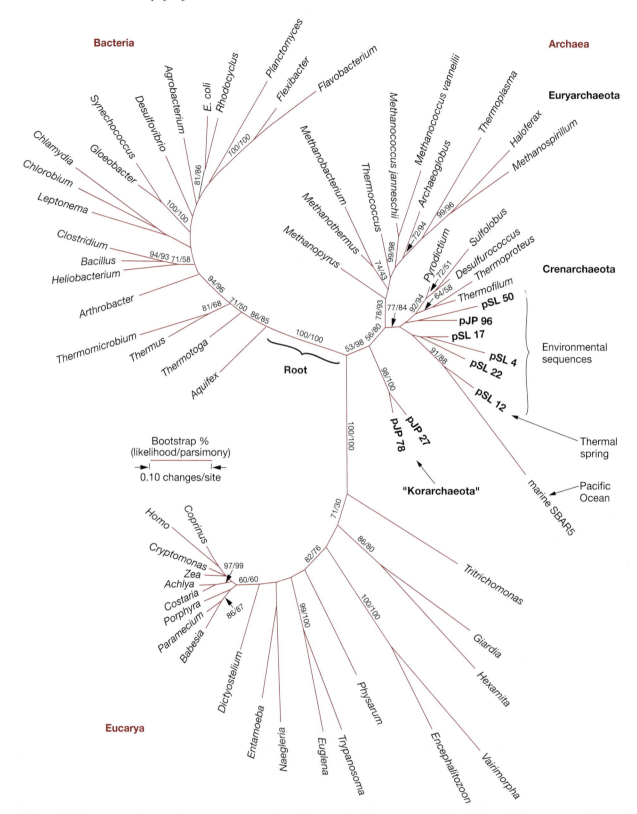

Figure 11.11 **An estimate of the phylogeny of all extant organisms based on rRNA sequences** Codes such as "pSL 22" designate organisms known only from their ribosomal RNA genes. From Barns et al. (1996).

water samples collected in nature, then amplified and sequenced it in the lab (Stahl et al. 1984). Following this approach, Edward DeLong and colleagues examined ribosomal RNA genes extracted from seawater collected in the Antarctic and off the coast of North America. DeLong and colleagues found many genes that were recognizable, based on their sequences, as belonging to previously unknown archaeans (DeLong 1992; DeLong et al. 1994). Susan Barns and colleagues (1994) likewise looked at rRNA genes extracted directly from mud in a hot spring in Yellowstone National Park. They also detected several rRNAs from previously unknown archaeans. Researchers in several laboratories are now pursuing similar studies (Service 1997).

The new Archaean sequences, plus additional new gene sequences from organisms in the other two domains, will improve our understanding of the universal phylogeny. Barns and colleagues (1996) used the new archaean rRNA sequences in the estimate of the whole-life phylogeny shown in Figure 11.11. This tree suggests the existence of a previously unknown kingdom of archaeans, the Korarchaeota. Also note that some of the deep branches in the Eucarya occur in a different order than in the tree in Figure 11.5.

Two of the deepest branch points on the tree of life appear to be well established. These are the fundamental split between the Bacteria versus the Archaea/Eucarya (see Box 11.1) and the deep split within the Archaea between the Euryarchaeota versus the Crenarchaeota (Figure 11.5 and Figure 11.11). Another fundamental branch point, the origin of the Eucarya, remains to be definitively established. The rRNA phylogenies in Figure 11.5 and Figure 11.11 show the split between the Archaea and the Eucarya as the second branch point on the universal phylogeny. In other words, these trees indicate that both the Archaea and the Eucarya are monophyletic. Other recent phylogeny estimates, such as one by Sandra Baldauf and colleagues (1996), show the Eucarya arising within the Archaea, as the sister group of the Crenarchaeota (Figure 11.12; see also Rivera and Lake 1992; Lake et al. 1996). As confidence in estimates of the universal phylogeny increases, confidence in inferences about the cenancestor will increase with it.

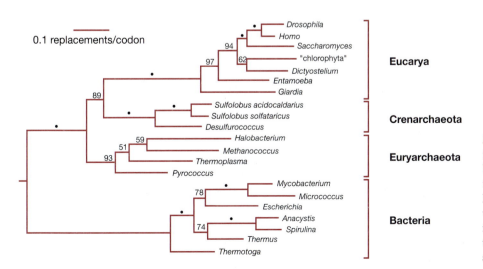

Figure 11.12 **An estimate of the universal phylogeny based on translation elongation factor gene sequences** Note that the Eucarya arise from within the Archaea. This configuration is sometimes referred to as the Eocyte tree. From Baldauf et al. (1996).

The second trend that will improve our understanding of both the universal phylogeny and the biology of the most recent common ancestor is the advent of whole-genome sequencing. As of this writing, the complete genomes of at least five organisms have been sequenced: three bacteria, *Haemophilus influenzae* (Fleischmann et al. 1995), *Mycoplasma genitalium* (Fraser et al. 1995), and *Escherichia coli* (see Wade 1997); one eukaryote, the yeast *Saccharomyces cerevisiae* (Bassett et al. 1996; Williams 1996); and one archaean, *Methanococcus jannaschii* (Bult et al. 1996). Efforts are underway to sequence dozens more (Wade 1997). The knowledge of trait distributions already gained from analyzing whole genomes has been invaluable in inferring the biology of the cenancestor. As additional complete genomes are sequenced, our picture of the cenancestor will continue to improve dramatically (Koonin and Mushegian 1996; Leipe 1996).

11.2 When Did the Cenancestor Live?

We do not have a fossil of the cenancestor that can be dated. We can, however, use a variety of geological and fossil evidence, together with the universal phylogeny shown in Figures 11.5 and 11.11, to put upper and lower limits on when the cenancestor could have lived.

The Earliest Possible Date for the Cenancestor

If we assume that the cenancestor lived on Earth, then the cenancestor could not have lived before the Earth existed and was inhabitable. The Earth is descended, along with the sun and other planets, from an interstellar dust cloud (Hughes 1989; Wetherill 1990). The cloud contained hydrogen and helium left over from the big bang, plus heavier elements ejected in the explosions of earlier generations of stars. These heavy elements were vital: They became major constituents of the Earth.

As the spinning interstellar dust cloud condensed under the force of gravity, it formed countless small planetesimals in orbit around a central mass. The central mass became the sun; its thermonuclear furnace was ignited by heat generated during its gravitational collapse. The orbiting planetesimals collided with each other, forming the planets by gradual accretion. The process of planetary accretion has slowed dramatically, but it is not yet finished. Planetesimals—meteors—continue to collide with the Earth and other planets. Rocks dating from the time of Earth's formation do not exist on the planet's surface, but radioisotopic dating of meteorites yields an estimated age for the solar system, and hence the Earth, of 4.5 to 4.6 billion years (see Badash 1989).

The newborn Earth remained uninhabitable for at least a few hundred million years. At first, it was simply much too hot. The collisions of the planetesimals that formed the Earth released enough heat to melt the entire planet (Wetherill 1990). Eventually the Earth's outer surface cooled and solidified to form the crust, and water vapor released from the planet's interior cooled and condensed to form the oceans. We know that at least some solid crust existed by about 4 billion years ago, because there are are 3.96-billion-year-old rocks in Arctic Canada (Bowring et al. 1989), and 4.2 to 4.3 billion-year-old zircon crystals embedded in younger sandstones in Australia (Myers and Williams 1985; Compston and Pidgeon 1986). We know that surface water existed by 3.85 billion years ago, because there are sedimentary rocks of that age at Akilia

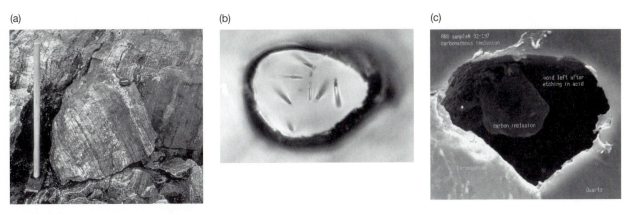

Figure 11.13 **The oldest known sedimentary rocks also contain the oldest known evidence suggesting life**
(a) A boulder from the banded iron formation on Akilia Island, Greenland. This rock is at least 3.85 billion years old. The dark bands are magnetite (an iron-containing mineral); the light bands are silicates (silicon-containing minerals). The rock hammer at the left edge of the photograph is about 40 cm tall (Allen P. Nutman, Australian National University/Scripps Institution of Oceanography).   (b) An apatite crystal from an Akilia Island sedimentary rock. The dark border around the crystal is made of carbonaceous material. The dark streaks within the crystal are tracks left by fission products released during the decay of radioactive elements imbedded in the crystal. The crystal is a little over 20 micrometers long (Scripps Institution of Oceanography).   (c) A photograph, taken with a scanning electron microscope, of the void left after an apatite crystal has been etched away with nitric acid. The void contains a speck of carbonaceous material (an inclusion) that was embedded in the apatite. The ratio of C^{12} to C^{13} in this carbonaceous speck suggests that the speck was produced by a living organism. (Stephen J. Mojzsis, Scripps Institution of Oceanography, University of California at San Diego).

Island, Greenland, and sedimentary rocks that are 3.8 billion years old at nearby Isua, Greenland (see Mojzsis et al. 1996).

The oldest sedimentary rocks contain the oldest evidence suggesting the presence of life. The rocks are of a type known as banded iron formations (Figure 11.13a). Geological processes have exposed the rocks to high temperatures and pressures, which have compacted the rocks and crystallized many of the minerals they contain. This transformation would have destroyed any microfossils the rocks might have originally harbored. The rocks do contain apatite crystals, however (Figure 11.13b). Apatites are calcium phosphate minerals. Apatites are produced by many microorganisms, and are also the main component of bones and teeth in vertebrates. Apatites can also be produced by nonbiological processes.

Following up on earlier work by Manfred Schidlowski (1988), Stephen Mojzsis, Gustaf Arrhenius, and their colleagues (1996) hypothesized that the apatite in the Akilia and Isua rocks was produced by organisms. They tested this hypothesis by examining specks of carbonaceous material embedded in the apatite crystals (Figure 11.13c). Carbon has two stable isotopes, C^{12} and C^{13}. When organisms capture and fix environmental carbon, during photosynthesis for example, they harvest C^{12} at a slightly higher rate than C^{13}. As a result, carbonaceous material produced by biological processes has a slightly higher ratio of C^{12} to C^{13} than does carbonaceous material produced by nonbiological processes. Mojzsis, Arrhenius, and colleagues used an ion microprobe to measure the ratio of C^{12} to C^{13} in the carbonaceous specks imbedded in Akilia and Isua apatite crystals. The ion microprobe knocks individual carbon atoms out of a sample; a mass spectrometer then weighs the atoms. The scientists also examined

The oldest evidence of life is in rocks that are 3.85 billion years old.

carbonaceous specks in apatite crystals from 3.25-billion-year-old sedimentary rocks from Pilbara, Australia. In all three cases, the scientists found carbon isotope ratios characteristic of life. Other workers in the field have greeted these results with cautious enthusiasm. John M. Hayes (1996), for example, calls for further evaluation of the suitability of the ion microprobe for measuring carbon isotope ratios, and notes that some as-yet-undiscovered prebiotic chemical reaction could explain the data. Hayes agrees with Mojzsis, Arrhenius, et al., however, that their evidence suggests that life was already established on Earth by 3.85 billion years ago.

Solid crust and oceans probably existed earlier than the 3.96 and 3.85 billion years ago demonstrated by the Canadian and Akilia Island rocks, but erosion, plate tectonics, and volcanic eruptions have obliterated all direct evidence. Even if crust and oceans did exist earlier, however, continued bombardment of the planet by large meteors probably would have prevented life from being established much earlier than 3.85 billion years ago. Large meteor impacts generate heat, create sun-blocking dust, and produce a blanket of debris. All of these effects, if intense enough, can kill organisms. Indeed, the mass extinction at the end of the Cretaceous that killed off the non-avian dinosaurs may have been caused by a large meteor impact (see Chapter 13). Figure 11.14 shows the history of very large impacts on the Earth and Moon. As time passed, and the largest planetesimals got swept up by the Earth and other planets, the sizes of the largest impacts decreased. Norman Sleep and colleagues (1989) estimated that the last impact with sufficient energy to vaporize the entire global ocean, and thereby kill any and all organisms then alive, probably happened between 4.44 and 3.8 billion years ago.

The first life on Earth thus appeared only a short time after the Earth became inhabitable.

Based on the Akilia sedimentary rocks, then, and on the history of large impacts, we can estimate that the earliest the cenancestor could possibly have lived was somewhere between 3.85 and about 4.4 billion years ago. We now consider the other side of the question: What is the most recent possible time at which the cenancestor could have lived?

The Latest Possible Date for the Cenancestor

The cenancestor could not have lived any more recently than the occurrence of any of the branch points above it on the universal phylogeny. Attempts have been made to calibrate the timing of the branch points using sequence data and molecular clocks (see, for example, Doolittle et al. 1996, but also Hasegawa et al. 1996). However, the most definitive information about branching times comes from fossils. The fossils useful in this regard are those that can be confidently identified as belonging to a particular group of organisms. If we can place a fossil on the whole-life phylogeny, then we know that the cenancestor is older than the fossil.

Fossils of single-celled organisms are identifiable as eukaryotes if they show sufficient structural complexity. The size of fossils also provides evidence of their identity: Any fossil larger than 10 μm in diameter is potentially a eukaryote, and any fossil larger than 60 μm in diameter is probably a eukaryote (Schopf 1992). The 590-million-year-old fossil in Figure 11.15a, for example, is clearly eukaryotic: It is about 250 μm in diameter and, unlike any known Archaean or Bac-

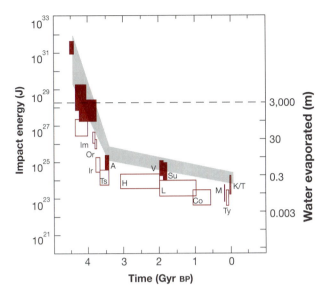

Figure 11.14   The history of large impacts on the Earth and Moon   The horizontal axis represents time (billions of years before present). The left vertical axis represents the energy of impact (joules). The right vertical axis shows the depth (m) to which the oceans would be vaporized by an impact with a given energy. The dashed line represents vaporization of the entire global ocean. Each box encloses the range of times during which a particular impact is estimated to have occurred, and the range of energies the impact is estimated to have fallen within. The open boxes are for the Moon; the filled boxes are for the Earth. Boxes with labels represent impacts documented by craters or other geological evidence. The unlabeled box at the upper left represents a large impact thought to be responsible for the formation of the Moon. The other unlabeled boxes are for hypothetical impacts. The gray band shows the largest impacts likely to have hit the Earth at any given time. Lunar craters: Im = Imbrium; Or = Orientale; Ir = Iridum; Ts = Tsiolkovski; H = Hausen; L = Langrenus; Co = Copernicus; Ty = Tycho. Terrestrial craters: A = Archaean spherule beds; V = Vredevort; Su = Sudbury; M = Manicougan; K/T = Cretaceous/Tertiary impact (crater located off the Yucatan peninsula). From Sleep et al. (1989).

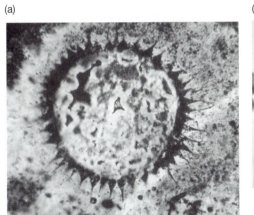

(a)

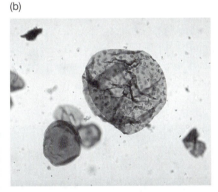

(b)

Figure 11.15   Fossils of single-celled Eucarya   (a) A spiny fossil from the Doushantuo Formation, China. This 590-million-year-old fossil represents either the preserved cell wall of a single-celled eukaryote, the reproductive cyst of a multicellular alga, or the egg case of an early animal. The fossil has a diameter of about 250 μm. From Knoll (1994); see also Knoll (1992).   (b) A structurally complex fossil from the Miroyedicha Formation, Siberia. This 850-to-950-million-year-old fossil is clearly eukaryotic, but like that shown in (a), its exact identification is unclear. It probably represents a single-celled organism. It has a diameter of about 40 μm. From Knoll (1994); see also Knoll (1992).   (c) Fossil cell from the Roper Group, Australia. This cell is 1.4 to 1.5 billion years old; it probably represents a eukaryote, but it lacks sufficient complexity for a definitive identification. The fossil is about 60 μm in diameter (Andrew H. Knoll, Harvard University).

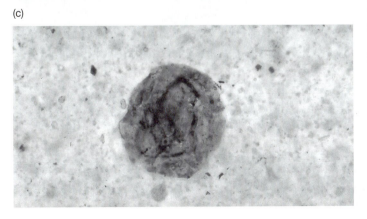

(c)

terium, it is covered with spikes. The 850-to-950-million-year-old fossil in Figure 11.15b is also eukaryotic: It is about 40 μm in diameter and covered with knobs. The 1.4-to-1.5-billion-year-old fossil in Figure 11.15c is about 60 μm in diameter, but simple in structure. Based on its size, it is probably, but not certainly, eukaryotic.

The oldest known fossils that are probably those of eukaryotes are 2.1 billion years old (Figure 11.16). Found by Tsu-Ming Han and Bruce Runnegar (1992) at the Empire Mine in Michigan, these fossils show a spiral-shaped organism similar to a more recent fossil named *Grypania spiralis*. *G. spiralis* is known from fossils in Montana, China, and India that range in age from 1.1 to 1.4 billion years old (see Han and Runnegar 1992). Because of its large size and structural complexity, paleontologists believe that *Grypania* was a eukaryote, probably an alga.

The first eukaryotes lived at least 2.1 billion years ago. The cenancestor must have lived earlier.

The extant algae occupy limbs of the universal phylogeny that are well above the common ancestor (Figure 11.17). The fossil algae from Michigan thus indicate that the cenancestor could not have lived more recently than 2.1 billion years ago, and probably lived substantially earlier than that.

Fossil cyanobacteria support a similar conclusion (Schopf 1994a). In a testament to the length of time that successful organisms can remain at least superficially unchanged, many fossil cyanobacteria are identifiable based on their structural similarity to extant forms. Each row of Figure 11.18 shows an extant species of cyanobacteria on the left, and a similar fossil form on the right. The fossils range in age from 850 million to 2 billion years old. The extant cyanobacteria occupy a limb of the universal phylogeny that is, like those occupied by the extant algae, several branch points above the last common ancestor (Figure 11.5). Again we can conclude that the cenancestor lived at least 2 billion years ago, and probably much earlier. If Schopf (1993) is correct in identifying

The first cyanobacteria lived at least 2 billion years ago, and possibly 3.5 billion years ago. The cenancestor must have lived earlier.

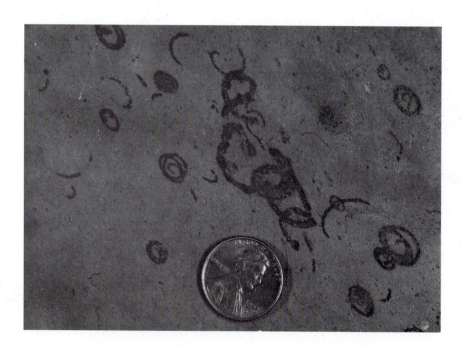

***Figure 11.16* 2.1-billion-year-old fossils from Michigan** Paleontologists believe these fossils represent eukaryotic algae. The penny is 18.5 mm in diameter (Tsu-Ming Han).

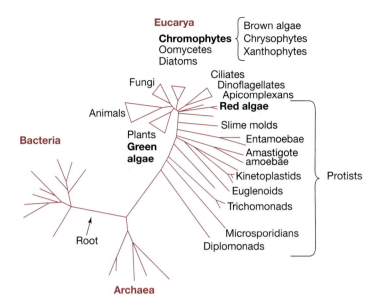

Figure 11.17   An estimate of the universal phylogeny, emphasizing the Eucarya   This tree is based on small-subunit rRNA sequences. Note the locations of the extant algae. From Sogin (1991).

the 3.465 billion-year-old Apex Chert fossils (Figure 11.2) as belonging to a diversity of cyanobacteria, then the cenancestor must have lived at least 3.5 billion years ago.

In summary, we can use the estimated universal phylogeny, along with geological and paleontological data, to bracket the time when the cenancestor could have lived. The earliest possible date is between 4.4 and 3.85 billion years ago; the most recent possible date is between 3.5 and 2 billion years ago.

11.3   Where Did the Cenancestor Come From?

The most recent common ancestor of all extant organisms was a descendant of the "one primordial form," the first self-replicating entity that could evolve by natural selection (Figure 11.1). But what was the primordial form, and how was it created by nonbiological processes? This is perhaps the most compelling puzzle we consider in this chapter. It is also, unfortunately, the puzzle for which we are farthest from a satisfactory solution.

We will probably never have direct fossil evidence of the nature of the first living thing (Schopf 1994b). The oldest known rocks from the inhabitable Earth contain evidence suggesting that life was already established by the time the rocks were formed (Figure 11.13). Even if we find rocks that date from the time of the primordial form, it is unlikely that they will contain any useful clues to the origin of life. The first self-replicating entity was probably chemical, not cellular, meaning that it had no cell walls to fossilize. Furthermore, the organic molecules that likely made up the primordial form are far too unstable to have remained preserved in rock for 4 billion years.

When we encountered a similar lack of direct evidence in our search for the most recent common ancestor, we turned to estimates of the universal phylogeny and the inferences we could draw from them. The cenancestor is essentially the end of the line for such inferences, however. Some biologists have

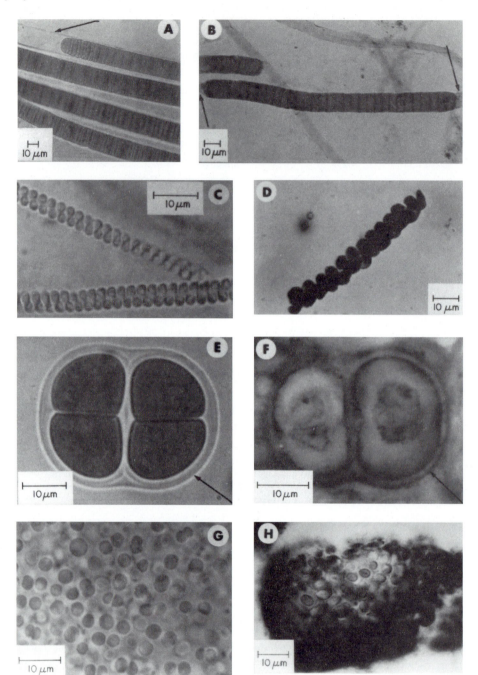

Figure 11.18 **Fossil cyanobacteria and their extant relatives** (a) *Lyngbya* (extant); (b) *Paleolyngbya* fossil from the 950-million-year-old Lakhanda Formation, Siberia. (c) *Spirulina* (extant); (d) *Heliconema* fossil from the 850-million-year-old Miroedikha Formation, Siberia. (e) *Gloeocapsa* (extant); (f) *Gloeodiniopsis* fossil from the 1.55-billion-year-old Satka Formation, Bashkiria. (g) *Entophysalis* (extant); (h) *Eoentophysalis* fossil from the 2 billion-year-old Belcher Group, Canada (J. William Schopf, University of California at Los Angeles) See Schopf (1994a).

attempted to use the features of existing organisms to extract information about ancestors of the cenancestor (Benner et al. 1989; Maizels and Weiner 1994). Nancy Maizels and Alan Weiner (1994), for example, argue that at least some viruses (which are not included in any phylogeny we have shown) are descended from predecessors of the last common ancestor, and thus provide clues about pre-cenancestor life (see also Keese and Gibbs 1993). But such efforts take us back to only a short time before the cenancestor; they do not approach the primordial form. We are left to make educated guesses, and to use experiments to sort among them for the most plausible scenarios.

We will assume that the primordial form originated on Earth (but see Box 11.2). We know that evolution before the cenancestor had to progress from a collection of nonliving chemicals to the primordial form to the cenancestor itself, and we know that the cenancestor was essentially a modern organism. Following the most widely accepted scenario, sometimes referred to as the Oparin-Haldane theory (see Lazcano and Miller 1996), we can break the spontaneous generation of the primordial form into a series of steps that occurred sequentially in the waters or moist clay of the young Earth:

1. Nonbiological processes synthesized organic molecules, such as amino acids and nucleotides, that would later serve as the building blocks of life. The accumulation of these organic molecules in solution created what is known as the pre-biotic soup.

2. The organic building blocks in the pre-biotic soup were assembled into biological polymers, such as proteins and nucleic acids.

3. Some combination of biological polymers was assembled into a self-replicating organism that fed off the organic molecules in the pre-biotic soup.

Once the first self-replicating organism appeared, its descendants acquired cellular form and the metabolic ability to synthesize their own organic building blocks. The actual route of evolution may, of course, have been less direct than this outline indicates, but with the outline at least we have a starting point. Now we can ask what evidence exists that the individual steps in the outline are feasible.

Nonbiological Synthesis of the Building Blocks of Life

The first step, the nonbiological synthesis of the building blocks of life, is fairly easy, at least under the right conditions. In 1953 Stanley L. Miller, then a graduate student, reported a simple and elegant experiment. He built an apparatus that boiled water and circulated the hot vapor through an atmosphere of methane (CH_4), ammonia (NH_3), and hydrogen (H_2), past an electric spark, and finally through a cooling jacket that condensed the vapor and directed it back into the boiling flask (Figure 11.20a, page 424). Miller let the apparatus run for a week; the water inside turned deep red and cloudy. Using paper chromatography, Miller identified the cause of the red color as a mixture of organic molecules, most notably the amino acids glycine, α-alanine, and β-alanine (Figure 11.20b). Since 1953, chemists working on the pre-biotic synthesis of organic molecules have, in similar experiments, documented the formation of a tremendous diversity of organic molecules, including amino acids, nucleotides, and sugars (see Fox and Dose 1972; Miller 1992).

Miller used methane, ammonia, and hydrogen as his atmosphere; at the time, this mixture was thought to model the atmosphere of the young Earth. The implication of Miller's result was that if lightning or UV radiation could have played the role that the spark did in his experiment, then the young Earth's oceans would have quickly become a rich pre-biotic soup.

The fundamental building blocks of life may have formed on the young Earth . . .

Box 11.2   The panspermia hypothesis

Although most contemporary specialists assume that life originated on Earth, there is an alternative: Life could have originated elsewhere and traveled to Earth. This suggestion, the panspermia hypothesis, begs the complaint that it merely shifts the problem of life's origin to some remote location where it is even harder to study. Francis Crick and Leslie Orgel (1973) point out, however, that such criticism is not only unfair, it could even prevent us from discovering the truth:

> For all we know there may be other types of planet on which the origin of life . . . is greatly more probable than on our own. For example, such a planet may possess a mineral, or compound, of crucial catalytic importance, which is rare on Earth.

The notion that Earth's primordial form evolved elsewhere and arrived here complete has become at least somewhat more attractive as evidence has mounted that life was established on Earth virtually as soon as it could have been, and that the cenancestor was biologically sophisticated and may have lived more than 3.5 billion years ago (see main text). On the other hand, Lazcano and Miller (1994) argue that life could have arisen, and evolved from primordial form to cenancestor, in as little as 10 million years. In any case, given that no one has definitively demonstrated the mechanism by which life originated on Earth, panspermia remains a possibility.

One version of the panspermia hypothesis is that life originated on another planet within our own solar system. Microbes could then have been dislodged from their home world by a meteor impact, carried through space on a chunk of debris, and dropped to Earth in a second meteor impact. Two crucial questions are (1) Do (or did) microbes exist on other planets in our solar system? and (2) Could they survive such a trip?

In an effort, among other things, to answer the first question, the United States landed two Viking spacecraft on Mars in 1976. The Viking landers carried three experiments designed to determine whether microbes were living in the Martian soil. All three involved attempts to detect gases released as by-products of metabolism. None yielded positive results. The results did not, however, rule out the possibility that life existed on Mars at other places or times.

David McKay and colleagues (1996) report evidence suggesting that life did indeed exist on Mars some 4 billion years ago. McKay's team studied a rock from Mars that had fallen as a meteorite onto Antarctica. On freshly chipped pieces of the rock, the team found globules of carbonate in close association with magnetite, iron sulfide, and organic molecules called polycyclic aromatic hydrocarbons. All of these chemicals can be produced by either biological or nonbiological processes. In addition, McKay and colleagues found objects on the Martian rock that resemble tiny bacteria (Figure 11.19). McKay et al. conclude that the most plausible explanation for their findings is that the objects are fossils of microbes that produced the minerals and chemicals. Other scientists are not convinced (see Anders et al. 1996; Kerr 1997a). A new series of expeditions to Mars has begun, and promises eventually to yield more decisive information about past or present life on the red planet.

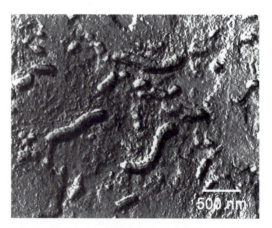

Figure 11.19   Fossils of Martian bacteria?
This scanning electron microscope image of a Martian meteorite shows objects that resemble tiny bacteria (NASA/Johnson Space Center).

There are other candidate locations for extraterrestrial life in the solar system. One is Jupiter's moon, Europa. Recent photographs taken by the spacecraft Galileo suggest that Europa has abundant liquid water and active volcanoes. Together, these offer the possibility that life could have evolved around hydrothermal vents in a liquid ocean hidden under Europa's icy surface (Belton et al. 1996; Kerr 1996, 1997b).

What about a microbe's chances of surviving a trip to Earth? For microbes on meteorites knocked loose from Mars, the trip to Earth would typically take several million years, but a lucky few would reach Earth in less than a year (Gladman et al. 1996; Gladman and Burns 1996). Spacefaring microbes would be exposed to cold, vacuum, and both UV and ionizing radiation. Two teams, Peter Weber and J. Mayo Greenberg (1985) and Klaus Dose and Anke Klein (1996), have measured the survival rates of spores of the bacterium *Bacillus subtilis* exposed to various combinations of these conditions; Jeff Secker and colleagues (1994) have made theoretical calculations. The consensus is that bare spores could not survive an interplanetary trip. With some sort of shield against radiation, however, spores would have some chance of making it. A shield could be provided by ice, rock, or carbon.

To our knowledge, only once has the long-term survival of microbes in space been directly tested. In November of 1969, astronauts from Apollo 12 recovered a camera from the unmanned lunar lander Surveyor 3, which had touched down on the Moon two and a half years earlier. Back on Earth, NASA scientists found that a piece of foam insulation inside the camera contained viable bacteria *(Streptococcus mitis)*. The microbes had apparently stowed away on Surveyor before it left Earth, and, shielded from radiation by the camera, survived their stay in the vacuum and cold of space (Mitchell and Ellis 1972).

A second possibility under panspermia is that life originated in another solar system and traveled to Earth through interstellar space. Spores embarking on such a voyage would need a force to accelerate them to sufficient velocity to escape the gravitational field of their home star. This force can be supplied by radiation pressure (Arrhenius 1908; Secker et al. 1994). Sailing on radiation pressure severely limits the mass of the shielding a spore can carry. Secker and colleagues suggest that spores encased in a film of carbonaceous material could both achieve escape velocity and survive the radiation they would encounter on a trip between stars.

Finally, Crick and Orgel (1973) suggest a third possibility, which they call directed panspermia: That Earth's founding microbes were sent here intentionally, aboard a spacecraft, by intelligent extraterrestrials bent on seeding the galaxy with life. Crick and Orgel argue that within the foreseeable future, it will probably be possible for us to launch such a mission. It is therefore at least conceivable that some other civilization actually did so 4 billion years ago.

Many atmospheric chemists now believe that the early atmosphere was dominated by carbon dioxide (CO_2) rather than methane, and molecular nitrogen (N_2) rather than ammonia (Kasting 1993). This conclusion is based on the mixture of gases released by contemporary volcanoes, and on improved knowledge of the chemical reactions that occur in the upper atmosphere. Which recipe for the early atmosphere is correct is important, because an atmosphere dominated by carbon dioxide and molecular nitrogen appears to be much less conducive to the formation of organic molecules. The composition of the early atmosphere remains a subject of debate (Lazcano and Miller 1996), and atmospheric chemists are looking for mechanisms by which organic molecules could have been synthesized even in relatively impermissive mixtures of gases (Kasting 1993).

Even if the early atmosphere did not allow the synthesis of organic molecules on Earth, there may have been another mechanism for the accumulation of a pre-biotic soup. We noted earlier that the young Earth experienced heavy

. . . or they may have been delivered to the Earth by meteors and comets.

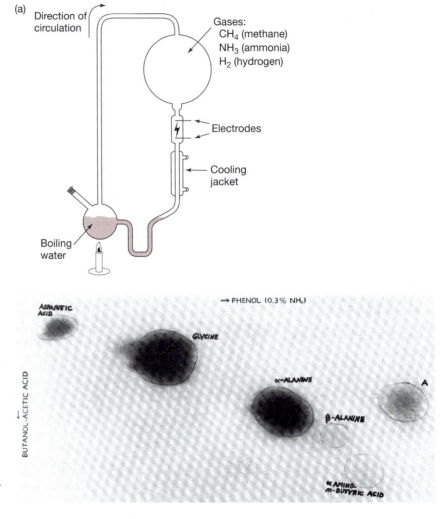

Figure 11.20 **Miller's 1953 experiment**   (a) The apparatus.   (b) The paper chromatograph identifying the molecules formed during the experiment (Stanley Miller, University of California at San Diego).

bombardment by meteors and comets (Figure 11.14). Carbonaceous chondrite meteorites, believed to be fragments of asteroids, contain an abundance of organic molecules (see Chyba et al. 1990; Lazcano and Miller 1996). The Murchison meteorite, for example, which fell in Australia in 1969, contained 18 amino acids, including six that occur in organisms (see Orgel 1973; Miller 1992). Many comets also contain a variety of organic molecules (Chyba et al. 1990; Cruikshank 1997). Thus the very same bombardment of the young Earth that made early conditions uncomfortable for life may, ironically, have supplied the ingredients for the construction of the primordial form.

There is at least one crucial difficulty with this hypothesis: When meteors and comets crash to Earth, friction with the atmosphere and collision with the ground generate tremendous heat (Anders 1989). This heat may destroy most or all of the organic molecules the meteors and comets carry. Edward Anders (1989) notes that very small incoming particles are slowed gently enough by the atmosphere to avoid incinerating all of their organics; he suggests that cometary dust may have been the primary source of the young Earth's organic molecules. Christopher Chyba (1990) and colleagues look instead to the possibility that the early atmosphere was dense with carbon dioxide. A dense CO_2 atmosphere may have provided a soft enough landing, even for large meteors and comets, for some of their organics to survive.

Luann Becker and colleagues (1996) provide direct evidence that at least some complex organic molecules survived the impact of the large meteor that created the Sudbury crater 1.85 billion years ago. The Sudbury crater, located in Canada, is the one marked "Su" in Figure 11.14. The meteor that produced it was larger than the one that may have caused the end-Cretaceous extinction of the non-avian dinosaurs. In rocks formed during the Sudbury impact, Becker et al. found fullerenes (C_{60} and C_{70}). Fullerenes, named after R. Buckminster Fuller, are large spherical organic molecules made of carbon atoms arranged in a pattern similar to that in a geodesic dome. The Sudbury meteor fell onto rocks that contain little carbon of their own. This indicates that the fullerenes could have gotten into the rocks formed during the impact in only one of two ways: (1) the fullerenes could have been created during the impact itself from carbon supplied by the meteor; (2) the fullerenes could have been in the meteor already and survived the impact.

Becker and colleagues knew that when fullerenes form, they can trap inside themselves atoms of certain gases, including helium. Any sample of helium gas contains a mixture of two isotopes, He^3 and He^4. The scientists reasoned that if the fullerenes had been formed during the impact, they would contain helium showing a mixture of isotopes characteristic of Earth's atmosphere, whereas if the fullerenes rode in on the meteor and survived the impact, they would contain helium with a mixture of isotopes characteristic of outer space. Becker et al. heated the fullerenes to break them open, then used a mass spectrometer to determine the masses of the helium atoms released. The released helium showed a ratio of isotopes different than that in the Earth's atmosphere, but similar to the ratios found in meteorites and extraterrestrial dust particles. In other words, the helium in the fullerenes was from outer space. This result demonstrates that the fullerenes were also from outer space.

Becker and colleagues note that the survival of fullerenes riding in the Sudbury meteor suggests that earlier estimates of the rates at which bombardment could have delivered intact organic molecules to the young Earth may have been overly pessimistic. Whether the mechanism that formed the pre-biotic soup was atmospheric chemistry, delivery by meteors and comets, or something else, it seems reasonable to conclude that the first step of the Oparin-Haldane theory is a plausible one.

The Assembly of Biological Polymers

The second step in the Oparin-Haldane theory, the formation of biological polymers from the building blocks in the pre-biotic soup, has presented a theoretical and practical challenge. The pre-biotic soup is an aqueous solution of organic building blocks, and although biological polymers can be readily synthesized in aqueous solution, they also break down by hydrolysis even as they are being built. This problem raises doubts that polymers sufficiently long to serve as the basis of a self-replicating primordial organism would ever have formed in a simple organic soup (Ferris et al. 1996).

James Ferris and colleagues (1996), extending a tradition that dates from the

The first organic polymers may have formed on the surface of clay crystals.

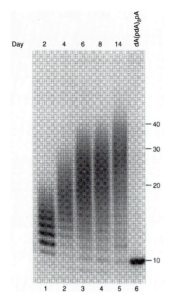

Figure 11.21 Synthesis of long nucleotide chains on clay   This electrophoresis gel has separated mixtures of polyadenylates by size. The right lane contains a single band, corresponding to nucleotide chains 10 bases long; this was the starting point for Ferris et al.'s (1996) experiment. The left lane contains the mixture of polynucleotides produced when 10-nucleotide polyadenylates were allowed to bind to montmorillonite, then given two successive baths with activated adenosine nucleotides. Each successive band represents a one-nucleotide difference in length. The leftmost lane thus contains polyadenylates ranging from 11 to 20 nucleotides in length. The second lane from the left shows the results of four successive baths with activated nucleotides, and so on. Reactions run without montmorillonite failed to produce elongated nucleotide chains. (James P. Ferris, Chemisty Dept., Rensselaer Polytechnic Institute, Troy, NY 12180).

1940s and 1950s (see Cairns-Smith et al. 1992), demonstrated a plausible mechanism to overcome the hydrolysis problem. Ferris et al. formulated a simple pre-biotic soup in the lab and added the common clay mineral montmorillonite. Montmorillonite is a naturally occurring aluminum-silicate clay to which organic molecules readily adhere. When polynucleotides stick to montmorillonite, the polynucleotides are much less vulnerable to hydrolysis. At the same time, the polynucleotides remain able to grow in length by addition of new bases to the end of the chain.

Ferris and colleagues started with polyadenylate nucleotide chains 10 nucleotides long, and let them bind to the montmorillonite. The scientists then added to the polyadenylate/clay solution a bath of activated adenosine nucleotides. The activated nucleotides reacted with the polyadenylates, adding themselves to the nucleotide chains. Ferris and colleagues then used a centrifuge to spin down the clay (and its attached nucleotide chains), poured off the spent solution, and added a fresh bath of activated nucleotides. By repeating this process, adding a fresh bath of activated nucleotides once each day, Ferris et al. synthesized polyadenylates over 40 nucleotides long (Figure 11.21).

Ferris et al. (1996) used an analogous procedure to grow polypeptides up to 55 amino acids long on the minerals illite and hydroxylapatite. The team asserts that their method models a mechanism by which biological polymers could have grown on the early Earth. Minerals in sediments that were repeatedly splashed with the pre-biotic soup, or continuously bathed by it, could have nursed the formation of polymers that were long enough to form a self-replicating primordial form. Thus the second step of the Oparin-Haldane theory appears to be plausible.

Self-Replication

The third step in the Oparin-Haldane theory still presents difficulties. The acquisition by a collection of organic molecules of the ability to self-replicate is arguably the point at which nonliving matter came to life. No one knows how a biological polymer, or a mixture of polymers, could have acquired the ability to self-replicate. Graham Cairns-Smith and colleagues (1992) suggest that the first self-replicating proto-organisms were actually the clay minerals themselves, rather than the polymers that grew on them. Imperfections in the structure of complex clay crystals could have served as a primitive sort of genetic information, with the imperfections repeated in layer after layer as the clay crystals grew. If an interaction between the clay crystals and the biological polymers that grew on them resulted in greater stability, more faithful replication, or more rapid growth, then the clay-polymer hybrid would have had a selective advantage. Evolution by natural selection would have begun. Eventually all the genetic and replication functions could have been taken over by the polymers themselves. Other scientists are skeptical of this hypothesis of genetic takeover. The skeptics appear content to leave to chance and time the formation of the first biological polymer capable of replicating itself.

Many biologists believe that an important early self-replicator was made largely or entirely of RNA. This idea is called the RNA-world hypothesis. Early versions of the RNA-world hypothesis envisioned a self-replicating RNA as the primordial form itself. This view has largely been abandoned; among other concerns, ribose and other sugars are too unstable, and there is no plausible nonbio-logical mechanism for the synthesis of sufficient quantities of ribonucleotides (Joyce et al. 1987; Larralde et al. 1995; Lazcano and Miller 1996). Nonetheless, many biologists think that self-replicating RNAs were in the direct line of de-scent from primordial form to cenancestor. This conclusion is based on two lines of reasoning (see James and Ellington 1995). First, when using the universal phy-logeny in making inferences back to the ancestors of the cenancestor, the clues that we can find seem to point to RNA. The most highly conserved and univer-sal component of the information processing machinery in cells, for example, is the apparatus for translation. This apparatus, while it incorporates proteins, is built on a frame made of RNA. Second, RNA molecules can both store genetic in-formation and, like proteins, act as catalysts. It is thus conceivable that an RNA molecule could, functioning as a ribozyme, catalyze its own replication.

The hypothesis that an RNA molecule could replicate itself, serving as a simple proto-organism, is testable: If the hypothesis is correct, then we should be able make a self-replicating RNA molecule in the lab. As of this writing, biochemists are racing to do so. David Bartel, Jack Szostak, and Eric Ekland, for example, are using test-tube evolution-by-natural-selection to create ribozymes capable of synthesizing RNA (Bartel and Szostak 1993; Ekland et al. 1995).

Figure 11.22 shows the selection scheme that Bartel and Szostak (1993) used to make ribozymes that can catalyze the formation of a phosphodiester bond to link a pair of adjacent RNA nucleotides. Bartel and Szostak started with a large pool of RNA polynucleotides. Every polynucleotide had the same sequence on its 5′ and 3′ ends (represented by lines), plus a unique 220-nucleotide stretch of random sequence in the middle (represented by the box labeled "Random 220"). Figure 11.22 follows two molecules that are members of this RNA pool: Random 220 A (left column) and Random 220 B (right column).

Bartel and Szostak bound the pool RNAs to agarose beads by means of a base-pairing interaction on their 3′ ends. The scientists then bathed the pool RNAs in a solution containing many copies of a specific substrate polynu-cleotide (row 1). This short RNA molecule had on its 5′ end a sequence of nu-cleotides forming a tag, whose function will soon become clear. On its 3′ end, the substrate RNA had a sequence of nucleotides complimentary to the free end of the pool RNA molecules. The substrate molecules quickly became bound, by base-pairing hydrogen bonds, to the pool RNAs (row 2).

This annealing brought into adjacent position the triphosphate group (PPP) on the 5′ end of the pool RNA and the hydroxyl group (OH) on the 3′ end of the substrate RNA. If, by chance, the 220-base stretch of random sequence in a pool RNA molecule had some ability to catalyze the formation of phosphodi-ester bonds, then it catalyzed the formation of such a bond between the sub-strate and pool RNA molecules. In row 3, Random 220 A has catalyzed such a reaction, liberating a diphosphate molecule, whereas Random 220 B has not.

Bartel and Szostak then rinsed the pool RNAs under conditions that washed away any substrate RNAs not covalently bound (by phosphodiester bonds) to pool RNAs, and liberated the pool RNAs from their agarose beads (row 4). Random 220 A still has its substrate (with tag); Random 220 B does not.

Finally, the scientists ran the pool RNAs through an affinity column (row 5). The affinity column caught hold of the tag sequence on the substrate RNA by base-pairing. The column thus captured any pool RNA whose 220-base stretch

The identity of the first self-repli-cating molecules is unknown, but biochemists appear to be close to demonstrating that an early self-replicator could have been made of RNA.

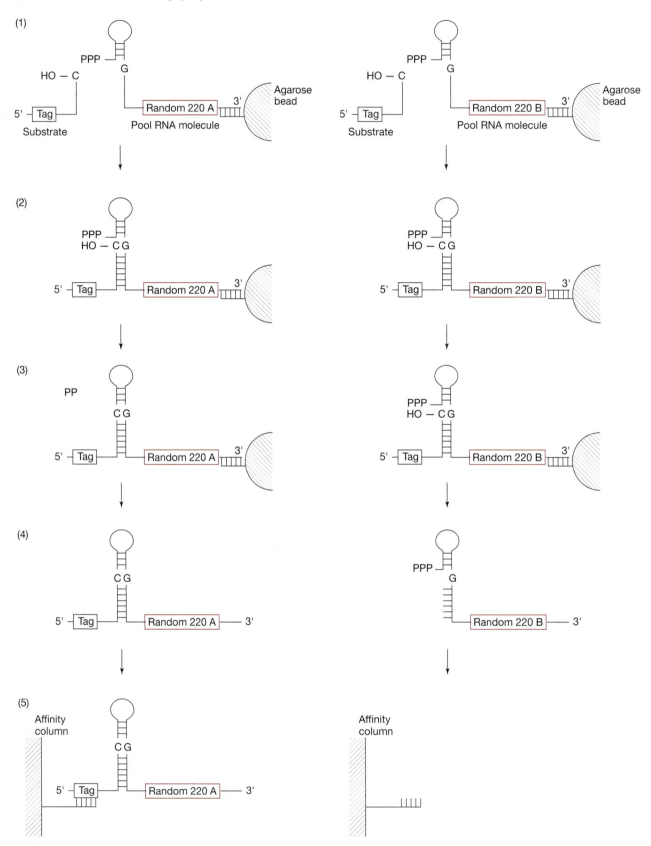

***Figure 11.22*   In vitro selection scheme for identifying ribosomes that can synthesize RNA**   See text for explanation. After Bartel and Szostak (1993).

of random sequence had catalytic activity (like Random 220 A), and let pass any pool RNA whose 220-base sequence did not.

Now Bartel and Szostak released the captured pool RNAs from the affinity column, made many copies of each by a process that allowed some mutations, and repeated the whole process again. Notice that Bartel and Szostak's protocol gives to the pool RNAs all the properties necessary and sufficient for evolution by natural selection. The RNAs have reproduction with heritability (via the copying process), variation (due to mutation), and differential survival (in the affinity column). The RNAs most likely to survive from one generation to the next are the ones that are most efficient at catalyzing phosphodiester bonds. After ten rounds of selection, the RNA pool had evolved ribozymes that could catalyze the formation of phosphodiester bonds at a rate several orders of magnitude faster than such bonds form without a catalyst (Figure 11.23).

More recently, Ekland and Bartel (1996) reported that they had used a similar scheme to evolve a catalytic RNA that can add up to six nucleotides to a growing RNA chain. This ribozyme uses template RNA and nucleoside triphosphates, and catalyzes RNA polymerization by the same chemical reaction promoted by the protein-based RNA polymerase enzymes used by living organisms. Laboratory-evolved ribozymes are not yet capable of self-replication, but it appears that biochemists may one day create ribozymes that are.

As Gerald Joyce (1996) put it, "Once an RNA enzyme with RNA replicase activity is in hand, the dreaming stops and the fun begins." If given an organic soup to live in, a population of self-replicating RNAs should be able to evolve on its own by mutation and natural selection. The population would not require the generation-by-generation management practiced by Bartel and Szostak. Would a species of self-replicating RNA evolve a DNA genome with DNA replication and transcription? Would it invent proteins and translation? Would its machinery be anything like the machinery in naturally evolved organisms? Would one of its descendants resemble our reconstruction of the cenancestor? Perhaps one day we will know the answers.

It is a long way from a self-replicating RNA molecule to the cenancestor, and many questions remain. How, for example, did the earliest organisms acquire cellular form? One potential answer has come from the work of Sidney Fox and colleagues, who found that mixtures of polyamino acids in water or salt solution spontaneously organize themselves into microspheres with properties reminiscent of living cells (see Fox and Dose 1972; Fox 1988; Fox 1991). How did metabolism arise? One possible solution comes from Claudia Huber and Günter Wächtershäuser (1997; see also Crabtree 1997), who suggest that metabolism may have originated in chemical reactions catalyzed by iron nickel sulfides at deep-sea hydrothermal vents.

The research programs spawned by the Oparin-Haldane theory of the origin of life have been successful. The continuous chain of descent from nonliving matter to primordial form to cenancestor has yet to be persuasively reassembled, but many of the individual links have been rendered plausible.

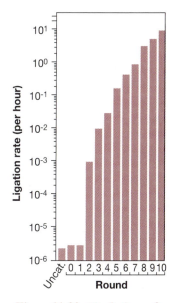

Figure 11.23 **Evolution of catalytic ability in a laboratory population of ribozymes** The graph shows the average rate at which the members of Bartel and Szostak's (1993) RNA pool catalyzed the formation of phosphodiester bonds (ligation rate) as a function of round of selection. Note the logarithmic scale on the vertical axis. Over the course of the experiment, the catalytic activity of the molecules in the RNA pool increased by several orders of magnitude. From Bartel and Szostak (1993).

11.4 How Did the Cenancestor's Descendants Evolve into Contemporary Organisms?

If we accept the reconstruction of the most recent common ancestor outlined in the first section of this chapter, then modern Bacteria and Archaea appear to resemble the cenancestor in many respects. Evolution from the cenancestor to the extant Bacteria and Archaea appears to have been a process of gradual refinement. The Eucarya, on the other hand, are dramatically different from the cenancestor. Among other things, eukaryotes have complex cells containing a variety of organelles, and dramatically more complicated genomes, in which the coding regions of genes are frequently interrupted by intervening sequences of noncoding DNA (introns). In addition, many eukaryotes are multicellular, with differentiated cells and tissues. We consider the evolutionary origin of organelles and introns here; we will discuss the evolution of multicellularity in Chapter 12.

The Origin of Organelles

The organelles whose evolution is best understood are mitochondria and chloroplasts. Most Eucarya have mitochondria and many also have chloroplasts, but some early-branching eukaryotes, such as *Giardia lamblia* (Figure 11.11), have neither (Sogin et al. 1989). Mitochondria and chloroplasts thus do not appear to be defining characteristics of the Eucarya; instead, they arose within the eukaryotic branch of the tree of life (but see Palmer 1997).

Superficially, mitochondria and chloroplasts resemble simple bacteria. With the discovery that mitochondria and chloroplasts have their own chromosomes, and that these chromosomes are small circular DNA molecules similar to those of bacteria, biologists began to take seriously an old idea that had been dismissed by all but a persistent few. This idea is that mitochondria and chloroplasts originated as bacteria that lived as internal symbionts of early eukaryotic cells (Margulis 1970; 1993). The definitive test of this hypothesis was to sequence genes from mitochondria and chloroplasts, such as the genes for the small subunit ribosomal RNA, then determine their position on the universal phylogeny. If the organelles arose via symbiosis, then their rRNA genes should branch from within the Bacteria; if, instead, the organelles arose independently within the Eucarya, then their rRNA genes should branch from within the Eucarya. The results are in, and they prove that Lynn Margulis was correct in championing the endosymbiosis theory (Figure 11.24).

The acquisition of organelles via symbiosis was a key early event in evolution of the eukaryotic lineage. Mitochondria and chloroplasts are the descendants of free-living bacteria.

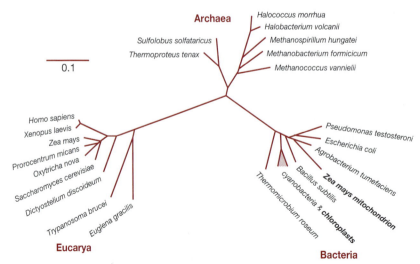

Figure 11.24 An estimate of the universal phylogeny, showing the locations of the mitochondria and chloroplasts   The tree is based on sequences of small-subunit rRNA genes. The mitochondria are represented by that of *Zea mays* (corn). The chloroplasts are represented by a variety of species, all of which branch within the shaded region. The scale bar represents the number of point mutations fixed per nucleotide. From Giovannoni et al. (1988).

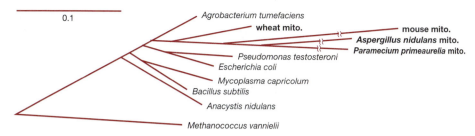

Figure 11.25 Phylogeny of the purple bacteria and their relatives, with four mitochondria This tree is based on sequences of small-subunit rRNA genes. The out-group, *Methanococcus vannielii,* is an Archaean. The scale bar represents estimated number of mutations per nucleotide. The branches leading to the mitochondria of the mouse, *Aspergillus,* and *Paramecium* are shortened by 0.4, 0.2, and 0.2 units. From Yang et al. (1985).

Mitochondria are closely related to the purple bacteria (Yang et al. 1985). Figure 11.25 shows an estimated phylogeny of three purple bacteria and some of their relatives. The purple bacteria are at the top: *Agrobacterium tumefaciens; Pseudomonas testosteroni;* and *E. coli.* Also included are the mitochondria of four eukaryotes, each representing a different traditional kingdom: wheat; mouse; *Aspergillus nidulans* (a fungus); and *Paramecium primeaurelia* (a protist). The mitochondria are all more closely related to each other than any is to any other organism, and the mitochondrial branch is within the purple bacteria. In an evolutionary sense, the mitochondria *are* purple bacteria. Yang et al. (1985) note that the branch of the purple bacteria that holds the mitochondria also includes a variety of other species, including the rhizobacteria, the agrobacteria, and the rickettsias, that typically form close associations with eukaryotic cells. Mitochondria are the descendants of such an association, one that became intimate to the point of mutual dependence.

Chloroplasts and other photosynthetic organelles are cyanobacteria (Giovannoni et al. 1988). Figure 11.26 shows an estimate of the phylogeny of the cyanobacteria. A representative chloroplast (liverwort) and another type of

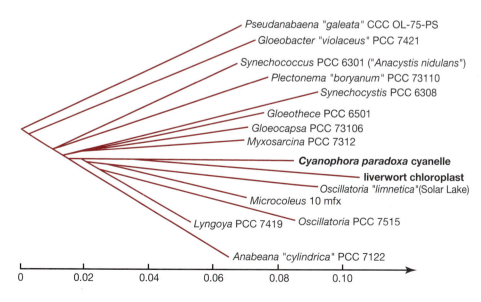

Figure 11.26 Phylogeny of the cyanobacteria, with a representative chloroplast and a cyanelle This tree is based on sequences of small-subunit rRNA genes. The chloroplasts are represented by liverwort; the cyanelle is a photosynthetic organelle from the flagellate *Cyanophora paradoxa.* The scale at the bottom shows the number of point mutations fixed per nucleotide. From Giovannoni et al. (1988).

photosynthetic organelle (a cyanelle) from a flagellate are each others' closest relatives. Their branch is located solidly within the cyanobacteria.

· Some eukaryotes that acquired photosynthetic organelles were later acquired themselves, *as* photosynthetic organelles. *Cryptomonas* Φ is an alga whose chloroplasts have two pairs of envelope membranes instead of the more typical single pair. Inside the inner membrane pair the *Cryptomonas* chloroplast has a typical circular chloroplast chromosome; between the inner and outer membrane pairs it also has a small nucleus-like organelle, called the nucleomorph. The nucleomorph makes a functional ribosome, which remains between the membrane pairs. Susan Douglas and colleagues (1991) sequenced the small–subunit ribosomal RNA from both the nucleomorph and the nucleus, then placed these sequences on a phylogeny of the eukaryotes (Figure 11.27). Both rRNAs are clearly of eukaryotic origin, but they are not closely related to each other. The implication is that the outer chloroplast membrane pair and the nucleomorph are vestiges of a eukaryotic ancestor. This ancestor, itself host to a chloroplast, became the endosymbiont of another eukaryotic host.

The acquisition of symbionts has clearly been an important factor in eukaryotic evolution.

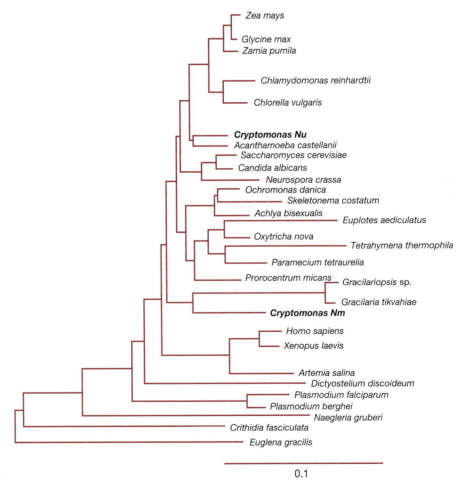

Figure 11.27 **Phylogeny of the eukaryotes, with *Cryptomonas*** The *Cryptomonas* rRNAs are shown in boldface: Nu = nuclear rRNA; Nm = nucleomorph rRNA. The scale bar represents the number of mutations per nucleotide. From Douglas et al. (1991).

The Origin of Introns

Comparing the genome organization of the Bacteria, Archaea, and Eucarya, it is tempting to conclude that the cenancestor's genome was more like that of the Eucarya (Doolittle and Brown 1994). Eukaryotes have large unwieldy genomes. Functionally related genes are often scattered about apparently at random. Genes are riddled with apparently nonfunctional introns. Bacteria and Archaea, in contrast, have small, streamlined genomes. Functionally related genes are often arranged in jointly transcribed operons behind a single promoter. Genes have virtually no introns. The bacterial/archaeal genome organization looks like it could well be an evolutionary refinement of the eukaryotic plan.

Introns are a prominent feature of eukaryotic genomes, but are virtually absent in bacterial and archaeal genomes.

When we plotted genome organization onto the universal phylogeny (section 11.1), however, we concluded that the simplest explanation was that the traits unique to the Eucarya, such as introns, arose within the Eucarya. The alternative is that introns were present in the cenancestor, and were subsequently and independently lost in the Bacteria and Archaea. Given the manifest advantage of an efficiently organized genome, this latter scenario appears plausible too.

Proponents of the introns-early theory, which is also known as the exon theory of genes, trace introns back to a time before the cenancestor (see Mattick 1994; Long et al. 1995a; 1995b). They envision a population of early organisms that had modular genomes. The modules, or exons, were relatively short, each encoding a single functional unit for a protein, rather than an entire protein. A functional unit might be a membrane-spanning domain, or a catalytic domain. The modules were connected to each other by linker regions that recombined easily with one another. This arrangement allowed the rapid creation of new genes via the shuffling of modules, a process analogous to the rapid creation of different movies via alternative splicings of a set of semi-independent scenes.

Some evidence suggests that introns derive from the earliest life forms.

Proponents of the introns-early theory explain the phylogenetic distribution of introns as follows. Although a modular genome is advantageous, in that it allows rapid evolution of new protein functions, it comes at the cost of inefficiency. All those linker regions take time and resources to replicate. In organisms that are selected primarily for efficiency of reproduction, like the Bacteria and the Archaea, the linkers have virtually all been lost. In organisms that are selected for innovation and complexity, like the multicellular eukaryotes, the linkers have been retained; they survive as today's introns.

Consistent with the introns-early theory, molecular biologists have documented examples of exon shuffling via recombination within introns. For instance, Manyuan Long et al. (1996) showed that such an event has occurred between two plant genes, one for a protein called cytosolic glyceraldehyde-3-phosphate dehydrogenase, the other for a protein called cytochrome c1 precursor. Three terminal exons from the former gene have been moved to the latter, giving the latter gene a new function.

Manyuan Long et al. (1995a; 1995b) offer additional evidence in support of the introns-early theory. They analyzed a large dataset of intron-bearing gene sequences, looking at intron phase and exon symmetry. An intron has a phase of 0 if it is inserted between codons and a phase of 1 or 2 if it is inserted within a codon (Figure 11.28a). An exon is symmetrical if the introns on either side are in the same phase (Figure 11.28b). Long et al. found higher proportions of introns

(a) Intron phases:

Phase 0: AAA –**INTRON**– GCG CAG TGA GGT

Phase 1: AAA G–**INTRON**–CG CAG TGA GGT

Phase 2: AAA GC–**INTRON**–G CAG TGA GGT

(b) Exon symmetry:

Figure 11.28 **Intron phase and exon symmetry**
Groups of three nucleotides represent codons. See text
for additional explanation.

Symmetrical exon: AAA –INTRON– **GCG CAG TGA** –INTRON–GGT

Asymmetrical exon: AAA –INTRON– **GCG CAG TGA G**–INTRON–GT

in phase 0, and higher proportions of symmetrical exons, than would be expected if introns had been inserted into the genome at random. They interpret this result as consistent with the introns-early theory, because exon shuffling works best if introns are in phase with each other and exons are symmetrical.

Other evidence suggests that introns are derived from selfish genes that recently invaded eukaryotic genomes.

Proponents of the introns-late, or insertional theory of introns, see the origin of introns in selfish transposable genes that inserted themselves into host chromosomes (see Palmer and Logsdon, 1991; Mattick 1994). They explain the phylogenetic distribution of introns as follows. In Bacteria and Archaeans, transcription and translation are linked: The ribosome attaches to the mRNA while the mRNA is still being transcribed. This linkage means that introns would disrupt protein synthesis, and would be strongly selected against. In eukaryotes, on the other hand, transcription and translation are separated. This provides the opportunity for post-transcriptional processing of mRNAs, making introns less disruptive and less of a selective disadvantage. Eukaryotes are thus inherently more tolerant of introns than are Bacteria and Archaeans.

Consistent with the introns-late theory, molecular biologists have strong circumstantial evidence that at least some introns can insert themselves into new locations. Copertino and Hallick (1993), for example, have found several twintrons in *Euglena*. Twintrons are introns embedded within introns. When a twintron is spliced out of a messenger RNA, it happens in two steps: first the internal intron gets spliced out; then the external intron. Twintrons are most easily explained by the insertion of the internal intron into the external intron.

A detailed look at the phylogenetic distribution of introns also appears to favor the introns-late theory (Figure 11.29). The fact that introns occur only in the most recently derived lineages of eukaryotes argues for their recent acquisition. The alternative is that thousands of introns were retained in all the common ancestors up to the most recent one shared by *Entamoeba* and *Dictyostelium,* then lost separately and independently in the Archaeans and six early-branching lineages of eukaryotes. Furthermore, introns are present in those mitochondrial and chloroplast genes that have moved to the nucleus, but absent in the genes that have remained in the organelle. This pattern is easily explained under the introns-late theory, but not readily reconciled with the introns-early theory.

The origin of introns remains the subject of contentious debate. Some early proponents of the introns-early theory, such as W. Ford Doolittle, have recently concluded that the theory is untenable (Doolittle and Brown 1994; Stoltzfus et al. 1994). On the other hand, a caveat concerning Palmer and Logsdon's evidence for the introns-late theory is that the lower branches of eukaryote

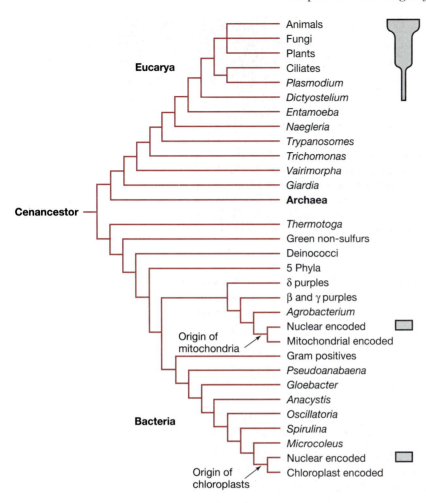

Figure 11.29 **The phylogenetic distribution of introns** The upper branch holds both the Archaea and the Eucarya; the lower branch holds the Bacteria. Nuclear-encoded genes of mitochondria and chloroplasts are genes that originated in the organelle genome but moved to the nucleus after the formation of the endosymbiotic relationship. The shaded bars represent the presence of spliceosome-dependent introns (the spliceosome is a nuclear organelle that functions in post-transcriptional processing of mRNA). The width of the shaded bars represents the relative density of introns in the genome. From Palmer and Logsdon (1991).

phylogeny shown in Figure 11.29 may require revision (see Palmer 1997). If any of the intron-free taxa represented on the lower branches in the figure is actually located among the intron-bearing branches, then interpretation of the distribution of introns will be more complicated. The current explosion in the availability of sequence data should eventually produce a decisive conclusion.

Summary

The tree of life has three main branches: the Bacteria, the Archaea, and the Eucarya. The first branch point divides the Bacteria from the other two groups. Whether the Eucarya branch off from the Archaea at the base of the Archaea or within one of the archaean sub-branches has not been definitively established.

All extant organisms share fundamental features of replication, transcription, and translation. This indicates that the most recent common ancestor of all extant organisms (the cenancestor) had a more or less complete apparatus for storing and managing genetic information. Other trait distributions across the universal phylogeny suggest that the cenancestor also had a fairly complete metabolism. In all, the cenancestor appears to have been similar to a modern bacterium.

At the earliest, the cenancestor could have appeared on Earth shortly after the planet became inhabitable, some 4.4 to 3.85 billion years ago. Based on the dates of fossil eukaryotes and cyanobacteria, both of which are descendants of the cenancestor, the cenancestor could not have lived more recently than 2 billion years ago. Possibly, it could not have lived more recently than 3.5 billion years ago.

The origin of the cenancestor's ultimate ancestor, the primordial life form, remains a mystery. However, a number of plausible steps that may have led up to the primordial form have been worked out. Organic building blocks could have been synthesized nonbiologically in the atmosphere, or delivered to Earth by comets and meteors. These building blocks could have been assembled into long biological polymers on the surface of mineral clays. One of these biological polymers may have become the primordial form by achieving the ability to replicate itself, and thereby the ability to evolve by natural selection.

The process of evolution from cenancestor to modern Bacteria and Archaea was largely one of refinement. The evolution of the Eucarya, in contrast, involved major innovations. Eucarya obtained mitochondria and chloroplasts by acquiring symbiotic bacteria. Whether introns represent a new acquisition or a hold-over from life before the cenancestor remains unresolved.

Questions

1. The genesis of life is sometimes said to have required four things: energy, concentration, protection, and catalysis (for example, Cowen 1995). Explain why each of these four things was necessary for the generation of the primordial form.

2. It has often been thought that life could develop on Earth only because Earth is just the right distance from the sun. Any closer, and the Earth would have been too hot (like Mercury or Venus); any farther, and there would not have been sufficient solar energy for the evolution of living things. Recently, communities of organisms have been found in deep-sea vents on Earth. These communities seem to get all of their energy from the vents rather than the sun. That is, the vent communities derive energy from the inner heat of the Earth (which is provided ultimately by radioactivity). How does this discovery inform consideration of whether life might exist on other planets or moons that are not at "the right distance" from the sun?

3. Suppose you are trekking through remote Greenland on a day off from your summer job at a scientific camp, and you find an unusual layer of sedimentary rocks that is not mapped on your geological charts. You suspect these rocks might be even older than the 3.85 billion-year-old rocks from Akilia Island. What would you do to determine whether these rocks have any evidence of ancient life? What results would show that life was indeed present before 3.85 billion years ago?

4. Imagine an extremely primitive organism that has very primitive ribosomes with no proteins. Would it be possible to place this organism on the phylogenies in Figure 11.5 or Figure 11.11? Why or why not? Is it conceivable that there are some as-yet-undiscovered primitive organisms that cannot be placed on these phylogenies? How would the discovery of such an organism affect our reconstruction of the cenancestor?

5. When biologists worked out details of DNA replication in bacteria and eukaryotes, many researchers were surprised to discover that there are several different DNA polymerases, each with a different role. The machinery for replication seemed enor-

mously complex, and every piece seemed essential if the whole system was to function at all. Many people found it hard to imagine how such a complex system of interdependent parts could have evolved by natural selection. Does the discovery of organisms with only one DNA polymerase (such as *Methanococcus jannaschii*) offer new insight into the evolution of replication? Why or why not?

6. Giovannoni et al.'s (1988) study on chloroplasts showed that plant chloroplast DNA is very closely related to cyanelle DNA from a flagellate (Figure 11.26). Look back at Figure 11.11 and find the flagellates (such as *Euglena*) and the plants (such as *Zea*, the corn plant). Do you think it is plausible that flagellates and plants might have acquired their related endosymbionts independently? If chloroplast/cyanelle acquisition took place only once, where on the eukaryote tree (that is, at which node) did that event occur? What other questions does your inference raise? How could you answer them?

7. Suppose Yang et al.'s (1985) study on mitochondria (Figure 11.25) had shown that mouse mitochondria, *Aspergillus* mitochondria, and *Paramecium* mitochondria were nested within the purple bacteria but were not each other's closest relatives. For example, suppose mouse mtDNA and *E. coli* DNA were more similar to each other than either was to any other mtDNA, while *Aspergillus* mtDNA and *Bacillus* DNA were more similar to each other than either was to any other mtDNA. What would these (hypothetical) results imply about the evolutionary history of mitochondria? How about if the study had shown that mouse mtDNA was very similar to the mouse nuclear DNA?

8. In the experiment diagramed in Figure 11.22, why was it important for the researchers to include a tag on the $5'$ end of the small substrate RNAs?

9. Many people have attempted to calculate the chance that "advanced" extraterrestrial life exists on other planets. By "advanced," they typically mean highly intelligent lifeforms that have a technological civilization with radio communication, such as has existed on Earth in this century. One of the fundamental uncertainties in these calculations is the probability that any life at all will evolve on a planet. On Earth, how soon after the Earth became habitable did life appear? How long did it take until eukaryotes appeared? How long until intelligent life appeared? In your opinion, do the answers indicate that the evolution of life (of any kind) on other earthlike planets is probable or improbable? How about the evolution of intelligent life? How about advanced civilization?

10. Leslie Orgel admitted to John Horgan (1991) that he and Crick intended their directed panspermia hypothesis as "sort of a joke." In their 1973 paper, however, Crick and Orgel treat the idea seriously enough to consider biological patterns that might serve as evidence. They point out, for example, that:

> It is a little surprising that organisms with somewhat different [genetic] codes do not exist. The universality of the code follows naturally from an "infective" theory of the origins of life. Life on Earth would represent a clone derived from a single extraterrestrial organism.

Since 1973, biologists have discovered that the genetic code is not universal, and that organisms with "somewhat different codes" do in fact exist. Our mitochondria, for example, use a code slightly different from that used by our nuclei (see Ayala and Kiger 1984). Many ciliates also have deviant codes (see Cohen and Adoutte 1995). How strongly does the discovery that the genetic code is not universal refute the directed panspermia hypothesis? How strongly does it refute other versions of panspermia? Explain your reasoning. Can you think of other kinds of evidence that could (or do) either support or refute some version of panspermia?

11. Are the intron-early and the introns-late theories mutually exclusive? Or is it possible to believe both that early organisms had modular genomes and that present-day introns represent recently inserted selfish DNA?

Exploring the Literature

12. Many of the fields discussed in this chapter are progressing rapidly. Search Current Contents, the Science Citation Index, or another online database for the most recent papers by the researchers whose work we have cited. What has changed since this writing?

13. During their first 2 billion years on earth, organisms wrought dramatic changes in the chemical composition of the atmosphere and oceans. For an introduction, see:

    Schopf, J.W. 1992. The oldest fossils and what they mean. In J.W. Schopf, ed. *Major events in the history of life* (Boston: Jones and Bartlett Publishers), 29–63.

14. Why are there not more very deep branches on the universal phylogeny? For a hypothesis, see:

    Gogarten-Boekels, M., E. Hilario, and J.P. Gogarten. 1995. The effects of heavy meteorite bombardment on the early evolution: The emergence of the three domains of life. *Origins of Life and Evolution of the Biosphere* 25:251–264.

15. Zhu Shixing and Chen Huineng recently reported 1.7 billion-year-old fossils that they interpreted as representing multicellular algae with differentiated tissues:

    Zhu Shixing and Chen Huineng. 1995. Megascopic multicellular organisms from the 1700-million-year-old Tuanshanzi Formation in the Jixian area, North China. *Science* 270:620–622.

    Examine Zhu and Chen's photographs. How convincing is their evidence?

16. The organisms on the deepest eukaryotic branches in Figure 11.5 were long thought to lack mitochondria. For a review of recent evidence suggesting that this belief is mistaken, see:

    Palmer, J.D. 1997. Organelle genomes: Going, going, gone! *Science* 275:790–791.

17. For another example of a eukaryote acquiring a symbiont, then being acquired as a symbiont, see:

    Köhler, S., C.F. Delwiche, P.W. Denny, L.G. Tilney, P. Webster, R.J.M. Wilson, J.D. Palmer, and D.S. Roos. 1997. A plastid of probable green algal origin in apicomplexan parasites. *Science* 275:1485–1489.

Citations

Altermann, W., and J.W. Schopf. 1995. Microfossils from the Neoarchean Campbell Group, Griqualand West Sequence of the Transvaal Supergroup, and their paleoenvironmental and evolutionary implications. *Precambrian Research* 75:65–90.

Anders, E. 1989. Pre-biotic organic matter from comets and asteroids. *Nature* 342:255–257.

Anders, E., C.K. Shearer, J.J. Papike, J.F. Bell, S.J. Clemett, R.N. Zare, D.S. McKay, K.L. Thomas-Keprta, C.S. Romanek, E.K. Gibson, Jr., and H. Vali. 1996. Evaluating the evidence for past life on Mars. *Science* 274:2119–2125.

Arrhenius, S. 1908. *Worlds in the Making* (New York: Harper and Row).

Ayala, F.J., and J.A. Kiger, Jr. 1984. *Modern Genetics* (Menlo Park, CA: Benjamin/Cummings).

Badash, L. 1989. The Age-of-the-Earth Debate. *Scientific American* 261:90–96.

Baldauf, S.L., J.D. Palmer, and W.F. Doolittle. 1996. The root of the universal tree and the origin of eukaryotes based on elongation factor phylogeny. *Proceedings of the National Academy of Sciences, USA* 93:7749–7754.

Barns, S.M., R.E. Fundyga, M.W. Jeffries, and N.R. Pace.

1994. Remarkable archaeal diversity detected in a Yellowstone National Park hot spring environment. *Proceedings of the National Academy of Sciences, USA* 91:1609–1613.

Barns, S.M., C.F. Delwiche, J.D. Palmer, and N.R. Pace. 1996. Perspectives on archaeal diversity, thermophily and monophyly from environmental rRNA sequences. *Proceedings of the National Academy of Sciences, USA* 93:9188–9193.

Bartel, D.P., and J.W. Szostak. 1993. Isolation of new ribozymes from a large pool of random sequences. *Science* 261:1411–1418.

Bassett, D.E., Jr., M.A. Basrai, C. Connelly, K.M. Hyland, K. Kitagawa, et al. 1996. Exploiting the complete yeast genome sequence. *Current Opinion in Genetics and Development* 6:763–766.

Becker, L., R.J. Poreda, and J. L. Bada. 1996. Extraterrestrial helium trapped in fullerenes in the Sudbury impact structure. *Science* 272:249–252.

Belton, M.J.S., et al. 1996. Galileo's first images of Jupiter and the Galilean satellites. *Science* 274:377–385.

Benner, S.A., A.D. Ellington, and A. Tauer. 1989. Modern metabolism as a palimpsest of the RNA world. *Proceedings of the National Academy of Sciences, USA* 86:7054–7058.

Bowring, S.A., I.S. Williams, and W. Compston. 1989. 3.96 Ga gneisses from the Slave province, Northwest Territories, Canada. *Geology* 17:971–975.

Brown, J.R., and W.F. Doolittle. 1995. Root of the universal tree of life based on ancient aminoacyl-tRNA synthetase gene duplications. *Proceedings of the National Academy of Sciences, USA* 92:2441–2445.

Bult, C.J., et al. 1996. Complete genome sequence of the methanogenic archaeon, *Methanococcus jannaschii*. *Science* 273:1058–1073.

Cairns-Smith, A.G., A.J. Hall, and M.J. Russell. 1992. Mineral theories of the origin of life and an iron sulfide example. *Origins of Life and Evolution of the Biosphere* 22:161–180.

Caspers, G.-J., G.-J. Reinders, J.A.M. Leunissen, J. Wattel, and W.W. de Jong. 1996. Protein sequences indicate that turtles branched off from the amniote tree after mammals. *Journal of Molecular Evolution* 42:580–586.

Chyba, C.F., P.J. Thomas, L. Brookshaw, and C. Sagan. 1990. Cometary delivery of organic molecules to the early Earth. *Science* 249:366–373.

Cohen, J., and A. Adoutte. 1995. Why does the genetic code deviate so easily in ciliates? *Biology of the Cell* 85:105–108.

Compston, W., and R.T. Pidgeon. 1986. Jack Hills, evidence of more very old detrital zircons in Western Australia. *Nature* 321:766–769.

Copertino, D.W., and R.B. Hallick. 1993. Group II and group III introns of twintrons: potential relationships with nuclear pre-mRNA introns. *Trends in Biochemical Science* 18:467–471.

Cowen, R. 1995. *History of Life*, 2nd ed. (Cambridge: Blackwell Scientific Publications).

Crabtree, R.H. 1997. Where smokers rule. *Science* 276:222.

Crick, F.H.C., and L.E. Orgel. 1973. Directed panspermia. *Icarus* 19:341–346.

Cruikshank, D.P. 1997. Stardust memories. *Science* 275:1895–1896.

Darwin, C. 1859. *On the Origin of Species by Means of Natural Selection, Or the Preservation of the Favoured Races in the Struggle for Life* (London: John Murray).

DeLong, E.F. 1992. Archaea in coastal marine environments. *Proceedings of the National Academy of Sciences, USA* 89:5685–5689.

DeLong, E.F., K.Y. Wu, B.B. Prézelin, and R.V.M. Jovine. 1994. High abundance of Archaea in Antarctic marine picoplankton. *Nature* 371:695–697.

Doolittle, W.F., and J.R. Brown. 1994. Tempo, mode, the progenote, and the universal root. *Proceedings of the National Academy of Sciences, USA* 91:6721–6728.

Doolittle, R.F., D.-F. Feng, S. Tsang, G. Cho, and E. Little. 1996. Determining divergence times of the major kingdoms of living organisms with a protein clock. *Science* 271:470–477.

Douglas, S.E., C.A. Murphy, D.F. Spencer, and M.W. Gray. 1991. Cryptomonad algae are evolutionary chimaeras of two phylogenetically distinct unicellular eukaryotes. *Nature* 350:148–151.

Dose, K., and A. Klein. 1996. Response of *Bacillus subtilis* spores to dehydration and UV irradiation at extremely low temperatures. *Origins of Life and Evolution of the Biosphere* 26:47–59.

Eernisse, D.J., and A.G. Kluge. 1993. Taxonomic congruence versus total evidence, and amniote phylogeny inferred from fossils, molecules, and morphology. *Molecular Biology and Evolution* 10:1170–1195.

Ekland, E.H., J.W. Szostak, and D.P. Bartel. 1995. Structurally complex and highly active RNA ligases derived from random RNA sequences. *Science* 269:364–370.

Ekland, E.H., and D.P. Bartel. 1996. RNA-catalysed RNA polymerization using nucleoside triphosphates. *Nature* 382:373–376.

Ferris, J.R., A.R. Hill, Jr., R. Liu, and L.E. Orgel. 1996. Synthesis of long prebiotic oligomers on mineral surfaces. *Nature* 381:59–61.

Fitch, W.M., and K. Upper. 1987. The phylogeny of tRNA sequences provides evidence for ambiguity reduction in the origin of the genetic code. *Cold Spring Harbor Symposia on Quantitative Biology* 52:759–767.

Fleischmann, R.D., et al. 1995. Whole-genome random sequencing and assembly of *Haemophilus influenzae* Rd. *Science* 269:496–512.

Foster, P.L., G. Gudmundsson, J.M. Trimarchi, H. Cai,

and M.F. Goodman. 1995. Proofreading-defective DNA polymerase II increases adaptive mutation in *Escherichia coli*. *Proceedings of the National Academy of Sciences, USA* 92:7951–7955.

Fox, G.E., L.J. Magrum, W.E. Balch, R.S. Wolfe, and C.R. Woese. 1977. Classification of methanogenic bacteria by 16S ribosomal RNA characterization. *Proceedings of the National Academy of Sciences, USA* 74:4537–4541.

Fox, S.W. 1988. *The Emergence of Life: Darwinian Evolution from the Inside* (New York: Basic Books).

Fox, S.W. 1991. Synthesis of life in the lab? Defining a protoliving system. *Quarterly Review of Biology* 66:181–185.

Fox, S.W., and K. Dose. 1972. *Molecular Evolution and the Origin of Life* (San Francisco: W.H. Freeman).

Fraser, C.M., et al. 1995. The minimal gene complement of *Mycoplasma genitalium*. *Science* 270:397–403.

Giovannoni, S.J., S. Turner, G.J. Olsen, S. Barns, D.J. Lane, and N.R. Pace. 1988. Evolutionary relationships among cyanobacteria and green chloroplasts. *Journal of Bacteriology* 170:3584–3592.

Gladman, B.J., and J.A. Burns. 1996. Mars meteorite transfer: Simulation. *Science* 274: Letters.

Gladman, B.J., J.A. Burns, M. Duncan, P. Lee, and H.F. Levison. 1996. The exchange of impact ejecta between terrestrial planets. *Science* 271:1387–1392.

Gogarten, J.P., et al. 1989. Evolution of the vacuolar H+-ATPase: Implications for the origin of eukaryotes. *Proceedings of the National Academy of Sciences, USA* 86:6661–6665.

Han, T.-M., and B. Runnegar. 1992. Megascopic eukaryotic algae from the 2.1-billion-year-old Negaunee Iron-Formation, Michigan. *Science* 257:232–235.

Hasegawa, M., W.M. Fitch, J.P. Gogarten, L. Olendzenski, E. Hilario, C. Simon, K.E. Holsinger, R.F. Doolittle, D.-F. Feng, S. Tsang, G. Cho, and E. Little. 1996. Dating the cenancester of organisms. *Science* 274:1750–1753.

Hayes, J.M. 1996. The earliset memories of life on Earth. *Nature* 384:21–22.

Hedges, S.B. 1994. Molecular evidence for the origin of birds. *Proceedings of the National Academy of Sciences, USA* 91:2621–2624.

Hedges, S.B., and L.R. Maxson. 1991. Pancreatic polypeptide and the sister group of birds. *Molecular Biology and Evolution* 8:888–891.

Hedges, S.B., and L.R. Maxson. 1992. 18S rRNA sequences and amniote phylogeny: Reply to Marshall. *Molecular Biology and Evolution* 9:374–377.

Hedges, S.B., K.D. Moberg, and L.R. Maxson. 1990. Tetrapod phylogeny inferred from 18S and 28S ribosomal RNA sequences and a review of the evidence for amniote relationships. *Molecular Biology and Evolution* 7:607–633.

Horgan, J. 1991. In the beginning . . . *Scientific American* 264(2):116–125.

Huber, C., and G.Wächtershäuser. 1997. Activated acetic acid by carbon fixation on (Fe,Ni)S under primordial conditions. *Science* 276:245–247.

Huelsenbeck, J.P., and J.J. Bull. 1996. A likelihood ratio test to detect conflicting phylogenetic signal. *Systematic Biology* 45:92–98.

Hughes, D.W. 1989. Evolution of the universe, stars and planets. In K. Allen and D. Briggs, eds. *Evolution and the Fossil Record* (Washington, DC: Smithsonian Institution Press). 1–25.

Iwabe, N., K.-i. Kuma, M. Hasegawa, S. Osawa, and T. Miyata. 1989. Evolutionary relationship of archaebacteria, eubacteria, and eukaryotes inferred from phylogenetic trees of duplicated genes. *Proceedings of the National Academy of Sciences, USA* 86:9355–9359.

James, K.D., and A.D. Ellington. 1995. The search for missing links between self-replicating nucleic acids and the RNA world. *Origins of Life and Evolution of the Biosphere* 25:515–530.

Jones, W.J., J.A. Leigh, F. Mayer, C.R. Woese, and R.S. Wolfe. 1983. *Methanococcus jannaschii* sp. nov., an extremely thermophilic methanogen from a submarine hydrothermal vent. *Archives of Microbiology* 136:254–261.

Joyce, G.F. 1996. Ribozymes: Building the RNA world. *Current Biology* 6:965–967.

Joyce, G.F., A. Schwartz, S.L. Miller, and L. Orgel. 1987. The case for an ancestral genetic system involving simple analogues of the nucleotides. *Proceedings of the National Academy of Sciences, USA* 84:4398–4402.

Kasting, J.F. 1993. Earth's early atmosphere. *Science* 259:920–926.

Keese, P., and A. Gibbs. 1993. Plant viruses: Master explorers of evolutionary space. *Current Opinion in Genetics and Development* 3:873–877.

Kerr, R.A. 1996. Galileo turns geology upside down on Jupiter's icy moons. *Science* 274:341.

Kerr, R.A. 1997a. Martian "microbes" cover their tracks. *Science* 276:30–31.

Kerr, R.A. 1997b. An ocean emerges on Europa. *Science* 276:355.

Knoll, A.H. 1992. The early evolution of eukaryotes: A geological perspective. *Science* 256:622–627.

Knoll, A.H. 1994. Proterozoic and Early Cambrian protists: Evidence for accelerating evolutionary tempo. *Proceedings of the National Academy of Sciences, USA* 91:6743–6750.

Koonin, E.V., and A.R. Mushegian. 1996. Complete genome sequences of cellular life forms: Glimpses of theoretical evolutionary genomics. *Current Opinion in Genetics and Development* 6:757–762.

Lake, J.A., M. Rivera, R. Haselkorn, W.J. Buikema, S.W. Golomb, G.J. Olsen, C.R. Woese, O. White, and J.C. Venter. 1996. *Methanococcus* genome. *Science* 274:901–903.

Larralde, R., M.P. Robertson, and S.L. Miller. 1995. Rates of decomposition of ribose and other sugars: Implications for chemical evolution. *Proceedings of the National Academy of Sciences, USA* 92:8158−8160.

Lazcano, A., and S.L. Miller. 1994. How long did it take for life to begin and evolve to cyanobacteria? *Journal of Molecular Evolution* 39:546−554.

Lazcano, A., and S.L. Miller. 1996. The origin and early evolution of life: Prebiotic chemistry, the pre-RNA world, and time. *Cell* 85:793−798.

Leipe, D.D. 1996. Biodiversity, genomes, and DNA sequence databases. *Current Opinion in Genetics and Development* 6:686−691.

Long, M., S.J. De Souza, and W. Gilbert. 1995a. Evolution of the intron-exon structure of eukaryotic genes. *Current Opinion in Genetics and Development* 5:774−778.

Long, M., C. Rosenberg, and W. Gilbert. 1995b. Intron phase correlations and the evolution of the intron/exon structure of genes. *Proceedings of the National Academy of Sciences, USA* 92:12495−12499.

Long, M., S.J. De Souza, C. Rosenberg, and W. Gilbert. 1996. Exon shuffling and the origin of the mitochondrial targeting function in plant cytochrome c1 precursor. *Proceedings of the National Academy of Sciences, USA* 93:7727−7731.

Madigan, M.T., and B.L. Marrs. 1997. Extremophiles. *Scientific American* 276(4):82−87.

Maizels, N., and A.M. Weiner. 1994. Phylogeny from function: Evidence from the molecular fossil record that tRNA originated in replication, not translation. *Proceedings of the National Academy of Sciences, USA* 91:6729−6734.

Margulis, L. 1970. *Origin of Eukaryotic Cells* (New Haven, CT: Yale University Press).

Margulis, L. 1993. *Symbiosis in Cell Evolution* (New York: Freeman).

Margulis, L. 1996. Archaeal-eubacterial mergers in the origin of Eukarya: Phylogenetic classification of life. *Proceedings of the National Academy of Sciences, USA* 93:1071−1076.

Marshall, C.R. 1992. Substitution bias, weighted parsimony, and amniote phylogeny as inferred from 18S rRNA sequences. *Molecular Biology and Evolution* 9:370−373.

Mattick, J.S. 1994. Introns: Evolution and function. *Current Opinion in Genetics and Development* 4:823−831.

McKay, D.S., et al. 1996. Search for past life on Mars: Possible relic biogenic activity in martian meteorite ALH84001. *Science* 273:924−930.

Miller, S.L. 1953. A production of amino acids under possible primitive Earth conditions. *Science* 117:528−529.

Miller, S.L. 1992. The prebiotic synthesis of organic compounds as a step toward the origin of life. In J.W. Schopf, ed. *Major Events in the History of Life* (Boston: Jones and Bartlett), 1−28.

Mitchell, F.J., and W.L. Ellis. 1972. Surveyor 3: Bacterium isolated from lunar-retrieved television camera. In *Analysis of Surveyor 3 Material and Photographs Returned by Apollo 12* (Washington, DC: National Aeronautics and Space Administration), 239−248.

Mojzsis, S.J., G. Arrhenius, K.D. McKeegan, T.M. Harrison, A.P. Nutman, and C.R.L. Friend. 1996. Evidence for life on Earth before 3,800 million years ago. *Nature* 384:55−59.

Myers, J.S. and I.R. Williams. 1985. Early Precambrian crustal evolution at Mount Narryer, Western Australia. *Precambrian Research* 27:153−163.

Nagel, G.M., and R.F. Doolittle. 1991. Evolution and relatedness in two aminoacyl-tRNA synthetase families. *Proceedings of the National Academy of Sciences, USA* 88:8121−8125.

Olsen, G.J., and C.R. Woese. 1996. Lessons from an Archaeal genome: What are we learning from *Methanococcus jannaschii*? *Trends in Genetics* 12:377−379.

Orgel, L.E. 1973. *The Origins of Life: Molecules and Natural Selection* (New York: John Wiley & Sons).

Palmer, J.D. 1997. Organelle genomes: Going, going, gone! *Science* 275:790−791.

Palmer, J.D., and J.M. Logsdon, Jr. 1991. The recent origins of introns. *Current Opinion in Genetics and Development* 1:470−477.

Pough, F.H., J.B. Heiser, and W.N. McFarland. 1996. *Vertebrate Life* (Upper Saddle River, NJ: Prentice Hall).

Rivera, M. C., and J.A. Lake. 1992. Evidence that eukaryotes and eocyte prokaryotes are immediate relatives. *Science* 257:74−76.

Schidlowski, M. 1988. A 3,800-million-year isotopic record of life from carbon in sedimentary rocks. *Nature* 333:313−318.

Schopf, J.W. 1992. The oldest fossils and what they mean. In J.W. Schopf, ed. *Major Events in the History of Life* (Boston: Jones and Bartlett), 29−63.

Schopf, J.W. 1993. Microfossils of the early Archean apex chert: New evidence of the antiquity of life. *Science* 260:640−646.

Schopf, J.W. 1994a. Disparate rates, differing fates: Tempo and mode of evolution changed from the precambrian to the phanerozoic. *Proceedings of the National Academy of Sciences, USA* 91:6735−6742.

Schopf, J.W. 1994b. The early evolution of life: Solution to Darwin's dilemma. *Trends in Ecology and Evolution* 9:375−378.

Schopf, J.W., and B.M. Packer. 1987. Early Archean (3.3-billion to 3.5-billion-year-old) microfossils from Warrawoona Group, Australia. *Science* 237:70−73.

Schwartz, R.M., and M.O. Dayhoff. 1978. Origins of prokaryotes, eukaryotes, mitochondria, and chloroplasts. *Science* 199:395−403.

Secker, J., J. Lepock, and P. Wesson. 1994. Damage due to

ultraviolet and ionizing radiation during the ejection of shielded micro-organisms from the vicinity of 1 M main sequence and red giant stars. *Astrophysics and Space Science* 219:1–28.

Service, R.F. 1997. Microbiologists explore life's rich, hidden kingdoms. *Science* 275:1740–1742.

Sleep, N.H., K.J. Zahnle, J.F. Kasting, and H.J. Morowitz. 1989. Annihilation of ecosystems by large asteroid impacts on the early Earth. *Nature* 342:139–142.

Sogin, M.L. 1991. Early evolution and the origin of eukaryotes. *Current Opinion in Genetics and Development* 1:457–463.

Sogin, M.L., J.H. Gunderson, H.J. Elwood, R.A. Alonso, and D.A. Peattie. 1989. Phylogenetic meaning of the kingdom concept: An unusual ribosomal RNA from *Giardia lamblia*. *Science* 243:75–77.

Stahl, D.A., D.J. Lane, G.J. Olsen, and N.R. Pace. 1984. Analysis of hydrothermal vent-associated symbionts by ribosomal RNA sequences. *Science* 224:409–411.

Stoltzfus, A., D.F. Spencer, M. Zucker, J.M. Logsdon, and W.F. Doolittle. 1994. Testing the exon theory of genes: The evidence from protein structure. *Science* 265:202–207.

Tomer, G., O. Cohen-Fix, M. O'Donnel, M. Goodman, and Z. Livneh. 1996. Reconstitution of repair-gap UV mutagenesis with purified proteins from *Escherichia coli:* A role for DNA polymerases III and II. *Proceedings of the National Academy of Sciences, USA* 93:1376–1380.

Wade, N. 1997. Thinking small paying off big in gene quest. *The New York Times* February 3: A1; A14.

Weber, P., and J.M. Greenberg. 1985. Can spores survive in interstellar space? *Nature* 316:403–407.

Wetherill, G.W. 1990. Formation of the Earth. *Annual Review of Earth and Planetary Science* 18:205–256.

Whittaker, R.H. 1969. New concepts of kingdoms of organisms. *Science* 163:150–160.

Williams, N. 1996. Yeast genome sequence ferments new research. *Science* 272:481.

Woese, C.R. 1991. The use of ribosomal RNA in reconstructing evolutionary relationships among bacteria. In R.K. Selander, A.G. Clark, and T.S. Whittam, eds. *Evolution at the Molecular Level* (Sunderland, MA: Sinauer), 1–24.

Woese, C.R. 1996. Phylogenetic trees: Wither microbiology? *Current Biology* 6:1060–1063.

Woese, C.R., and G.E. Fox. 1977. Phylogenetic structure of the prokaryotic domain: The primary kingdoms. *Proceedings of the National Academy of Sciences, USA* 74:5088–5090.

Woese, C.R., O. Kandler, and M.L. Wheelis. 1990. Towards a natural system of organisms: Proposal for the domains Archaea, Bacteria, and Eucarya. *Proceedings of the National Academy of Sciences, USA* 87:4576–4579.

Yang, D., Y. Oyaizu, H. Oyaizu, G.J. Olsen, and C.R. Woese. 1985. Mitochondrial origins. *Proceedings of the National Academy of Sciences, USA* 82:4443–4447.

Zillig, W. 1991. Comparative biochemistry of Archaea and Bacteria. *Current Opinion in Genetics and Development* 1:544–551.

Zuckerkandl, E., and L. Pauling. 1965. Molecules as documents of evolutionary history. *Journal of Theoretical Biology* 8:357–366.

The Cambrian Explosion and Beyond

Once the fundamental life processes of DNA replication, protein synthesis, respiration, and cell division had evolved, a spectacular diversification of life ensued. Breakthroughs like the evolution of photosynthesis and the nuclear membrane occurred. These events, along with others we reviewed in Chapter 11, spanned some 3.2 billion years and created the deep branches on the tree of life. During this interval all organisms, with the exception of some red, brown, and green algae, were unicellular.

The first multicellular animals do not appear in the fossil record until about 565 million years ago (Ma). But then, over a span of just 40 million years, virtually every major phylum of animals evolved. This period is called the Cambrian explosion. Although it represents a mere 0.8% of the Earth's history, it ranks as one of the great events in the history of life. Our goal in this chapter is to introduce this and other turning points that have taken place over the past 543 million years. This interval in earth history is called the Phanerozoic ("visible life") eon.

Our focus is not as much on what happened during the elaboration of multicellular life as it is on how and why. Which genes were involved in the diversification and elaboration of metazoan body plans? Do we know anything about the loci involved in creating innovations like the jointed limbs of arthropods and the flower? Can we reconstruct intermediate stages in adaptive breakthroughs like the evolution of flight? What patterns can we detect in the history of adaptation and diversification?

These are the types of questions driving current research. Some are starting to be answered through a remarkable fusion of paleontology, molecular genetics, and developmental biology. But we start with the basics: a look at how paleontologists read the fossil record and document the history of life.

12.1   The Nature of the Fossil Record

In Chapter 2 we introduced the geological time scale, established by paleontologists in the early 19th century. We also looked at how 20th-century geochronologists are using radioactive isotopes to estimate the absolute age of each eon, era, period, and epoch. We now elaborate on how life has changed through time by reviewing the process of fossilization, examining the strengths and weaknesses of the fossil record, and presenting a timeline of major events in evolution.

How Organic Remains Fossilize

A fossil is any trace left by an organism that lived in the past. Fossils are enormously diverse, but four general categories can be defined by method of formation (see following list). There are two important issues to focus on here: Which part of the organism is preserved and available for study? What kinds of habitats produce fossils?

- **Compression and impression fossils** (Figure 12.1) can result when organic material is buried in water- or wind-borne sediment before it decomposes. Under the weight of sand, mud, ash, or other particles deposited above, a structure can leave an impression in the material below. The resulting fossil is analogous to the record left by footprints in mud or leaves in wet concrete. In anoxic settings, the deposits can compress the cells into a black, carbonaceous deposit on the substrate. If the sediment is extremely fine-grained, compressions often record details of external structure as fine as the stomata and guard cells on plant leaves.

- **Permineralized fossils** (Figure 12.2) can form when structures are buried in sediments and dissolved minerals precipitate in the cells. This process, which is analogous to the way a microscopist embeds tissues with resin before sectioning, can preserve details of internal structure.

- **Casts and molds** (Figure 12.3) originate when remains decay or when shells are dissolved after being buried in sediment. Molds consist of unfilled spaces, while casts form when new material infiltrates the space, fills it, and hardens into rock. This process is analogous to the lost-wax casting technique used by sculptors. Molds and casts can preserve information about external and internal surfaces.

- **Unaltered remains** (Figure 12.4) can be preserved in environments that discourage loss from weathering, consumption by scavenging animals, and decomposition by bacteria and fungi. How long can organic material remain unaltered? Human cadavers from the Iron Age, buried in the highly acidic environment of peat bogs, have been recovered with some flesh still intact. Woolly mammoths dug out of permafrost can be so well preserved that paleontologists have eaten them. Dried but otherwise unaltered dung from giant

(a)

(b)

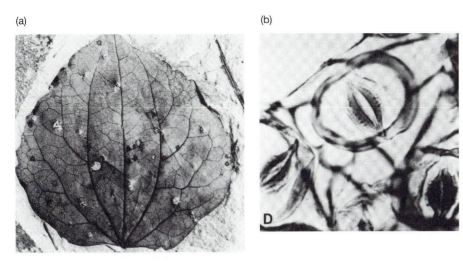

Figure 12.1 **Compression and impression fossils**   These are two-dimensional fossils, usually found by splitting sedimentary rocks along the bedding plane.   (a) A compression fossil of a Paleocene leaf found near Alberta, Canada.   (b) A close-up from the leaf pictured in (a), showing stomata and guard cells. Reproduced by permission from W. N. Stewart, *Paleobotany and the Evolution of Plants,* page 12, Fig. 2.5A and 2.5D. New York: Cambridge University Press, 1983. © 1983 Cambridge University Press.

(a)

(b)

Figure 12.2 **Permineralized fossils**   Over time, buried organic material can be replaced with various water-soluble minerals, including silicates and calcium carbonates. Permineralized fossils are usually found in rock outcrops after they have been partially exposed by natural weathering.   (a) The skull of *Tyrannosaurus Rex,* a predatory dinosaur (Francis Gohier/Photo Researchers, Inc.).   (b) Petrified wood (John D. Cunningham/Visuals Unlimited).

Figure 12.3 **Casts and molds**   These three-dimensional fossils are usually found in rock outcrops or quarries. This is a cast of a horsetail stem from the Carboniferous. Reproduced by permission from W. N. Stewart, *Paleobotany and the Evolution of Plants,* page 10, Fig. 2.4C. New York: Cambridge University Press, 1983. © 1983 Cambridge University Press.

Figure 12.4 **Unaltered remains**   A variety of unusual environmental conditions can preserve organic material for long periods. This is a winged male termite, preserved in amber, from the Upper Cretaceous of Canada (John Koivula/Science Source/Photo Researchers, Inc.).

ground sloths, which lived in the Pleistocene, can be found in protected, desiccating environments like desert caves. Viscous plant resins can harden into amber, preserving the insects trapped inside so well that wing veins are visible. Paleobotanists have driven nails into 100-million-year-old logs recovered from oil-saturated tar sands.

Though spectacular, intact remains are so rare that they represent only a small fraction of the total fossil record. Compression, impression, casting, molding, and permineralization are far more common. All of these processes depend on two key factors: resistance to decay, burial (usually in a water-saturated sediment), and lack of oxygen. As a result, the fossil record consists primarily of hard structures left in depositional environments like deltas, beaches, floodplains, marshes, lakeshores, and sea floors. The database is dominated by shells from marine invertebrates, shark and mammal teeth, leaves and other plant organs associated with coal deposits, and microfossils like pollen and foraminiferans.

It is not hard to understand why these structures predominate. Marine bivalves that burrow are automatically buried in saturated sediments after death. Because enamel is one of the highest-density substances known in nature, teeth decay slowly and have more time to fossilize. Trees that grow in floodplain forests drop their leaves and seeds onto substrates that are frequently washed over with sediment, and the saturated soils of swamps are routinely anaerobic and permit only slow decomposition. The coats of pollen and spores are made of sporopollenin, a highly resistant molecule. These structures rain into lakes and marshes, settle to the bottom, and are buried in sediment. Shelly organisms in the marine plankton drift to the ocean floor after death and are likewise buried.

The upshot is that fossilization preserves a small but well-defined sample of the biosphere. Soft tissues and organisms from upland habitats are rarely preserved; pollen, spores, and hard tissues from aquatic organisms are more common.

Strengths and Weaknesses of the Fossil Record

There are three main types of bias in the fossil record: geographic, taxonomic, and temporal. We have already mentioned the propensity for fossils to come from lowland and marine habitats. To appreciate the taxonomic bias more fully, consider this: Marine organisms dominate the fossil record but make up only 10% of extant species. A full two-thirds of animal phyla living today lack any sort of mineralized hard parts, like bone or shell, that are amenable to fossilization. Critical parts of plants, including reproductive structures such as flowers, are seldom preserved. The temporal bias results because the Earth's crust is constantly being recycled. When tectonic plates subduct or mountains erode, their fossils go with them. As a result, old rocks are rarer than new rocks, and our ability to sample ancient life forms adequately decreases.

It is important to realize, however, that these types of sampling issues are by no means unique to paleontology. Advances in developmental genetics depend on the generality of a few model systems like *Drosophila melanogaster,* the roundworm *Caenorhabditis elegans,* the annual flower *Arabidopsis thaliana,* corn *(Zea mays),* and zebrafish. Most work in molecular genetics is done on a few phages, the filamentous fungus *Neurospora crassa,* and *E. coli.* The vast majority of research in behavioral ecology is done on birds and mammals; community ecology has historically focused on upland habitats in North America and Europe.

The important point is to recognize that like any source of data, the fossil record has characteristics that limit the types of information that can be retrieved and how broadly the data can be interpreted. The goal is to recognize the constraints and work creatively within them.

With these caveats in place, we begin our intensive use of the fossil record with a broad look at the sequence of events during the Phanerozoic.

Life Through Time: An Overview

The geologic time scale is a hierarchy divided into eons, eras, periods, epochs, and stages. Each named interval is defined by a suite of diagnostic fossils. When the scale was first being formulated, in the early 1800s, the intervals were arranged by relative age only. It was only much later, after the discovery of radioisotopes and the development of accurate dating techniques, that absolute times were assigned to each interval. Consequently, levels in the hierarchy are not equivalent in terms of time. The Paleozoic era lasted 292 million years and the Mesozoic era 186 million, for example. Also, the geologic time scale is a work in progress. Estimates for the absolute ages in the scale are constantly improving as dating techniques become more sophisticated and more rocks are sampled.

Figure 12.5 presents time lines for the three eras that make up the Phanerozoic. The eon begins with the Cambrian explosion and ends in the present. Its three component eras are the Paleozoic (ancient life), Mesozoic (middle life), and Cenozoic (recent life). In addition to offering a compact overview of the history of multicellular life, the diagrams in Figure 12.5 should inspire questions. For example, each of the fossil "firsts" mapped along the top of the time lines represents a suite of new traits. What are these? Why did selection favor them? What was the ancestral state, and what intermediate forms occurred during the course of their evolution? What genetic changes were responsible for these turning points in evolution? Note, too, that many of these events led to the creation of new orders, classes, and phyla.

Given enough time and space, we could explore these questions for any of the events diagrammed on the time lines. Because our goal is to introduce how research in contemporary paleontology is done and to illustrate the most important concepts in the field, we have selected a few to review in detail.

(a)

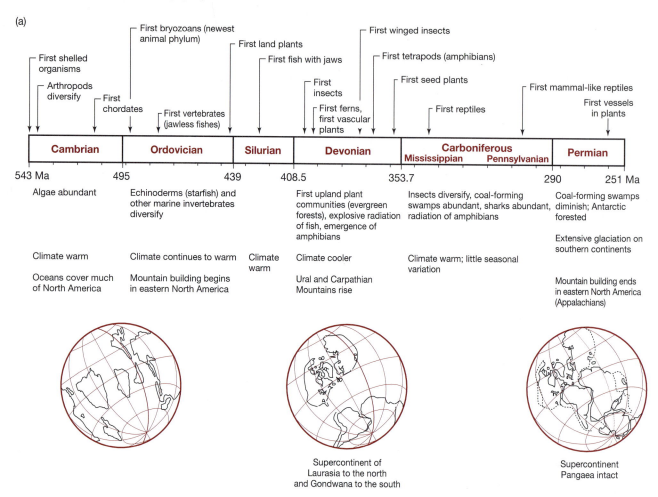

Figure 12.5 **The Phanerozoic eon**   The diagrams show a selection of events from the three eras that make up the Phanero-zoic. We have included the first appearances for life forms discussed in this and other chapters, the names of periods or epochs within the era, absolute ages assigned by radioactive dating, notes on prominent plant communities, the climate and important geological events, and the estimated positions of major land masses. References used in compiling this figure include Niklas et al. 1983; Taylor and Hickey 1990; Niklas 1994; Li et al. 1996; Taylor and Taylor 1993; Irving 1977; Harland et al. 1990; Erwin, D. 1996, pers. comm.   (a) The Paleozoic, or "ancient life" era. The Paleozoic begins with the radiation of metazoans and ends with a mass extinction event at the end of the Permian. The time bar is labeled with the names of periods and subperiods within the Paleozoic, but omits the names of epochs and stages.   (b) The Mesozoic, or "middle life" era. The Mesozoic, sometimes nick-named the age of reptiles, begins after the end–Permian extinction event and ends with the extinction of the dinosaurs and other groups at the Cretaceous-Tertiary boundary. The time bar is labeled with the names of periods within the Mesozoic, but omits the names of epochs and stages.   (c) The Cenozoic, or "recent life" era. The Cenozoic is divided into the Tertiary and Quater-nary periods. The Tertiary includes the Paleocene, Eocene, Oligocene, Miocene, and Pliocene epochs. The Quaternary includes the Pleistocene and Holocene epochs. The Cenozoic is sometimes nicknamed the age of mammals.

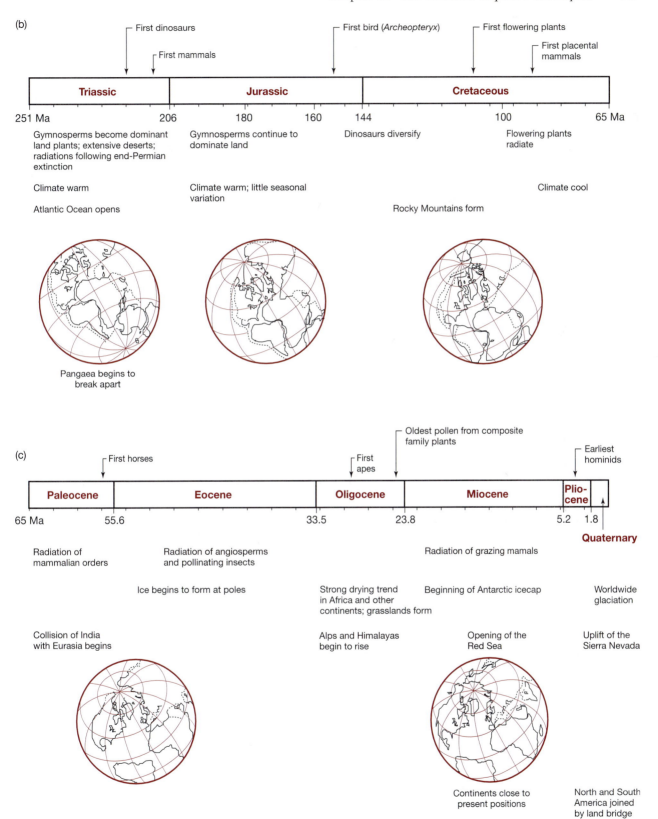

(b)

First dinosaurs

First mammals

First bird (*Archeopteryx*)

First flowering plants

First placental mammals

Triassic	Jurassic	Cretaceous

251 Ma   206   180   160   144   100   65 Ma

Gymnosperms become dominant land plants; extensive deserts; radiations following end-Permian extinction

Gymnosperms continue to dominate land

Dinosaurs diversify

Flowering plants radiate

Climate warm

Climate warm; little seasonal variation

Climate cool

Atlantic Ocean opens

Rocky Mountains form

Pangaea begins to break apart

(c)

First horses

First apes

Oldest pollen from composite family plants

Earliest hominids

Paleocene	Eocene	Oligocene	Miocene	Plio-cene	

65 Ma   55.6   33.5   23.8   5.2   1.8

Quaternary

Radiation of mammalian orders

Radiation of angiosperms and pollinating insects

Radiation of grazing mamals

Ice begins to form at poles

Strong drying trend in Africa and other continents; grasslands form

Beginning of Antarctic icecap

Worldwide glaciation

Collision of India with Eurasia begins

Alps and Himalayas begin to rise

Opening of the Red Sea

Uplift of the Sierra Nevada

Continents close to present positions

North and South America joined by land bridge

12.2   The Cambrian Explosion

All but one of the 35 metazoan phyla currently recognized by taxonomists appear in the fossil record over a span of just 40 million years. This is the blink of an eye, geologically speaking. The origin and rapid diversification of these multicellular animals ranks as one of the great events in the history of life. To appreciate the full drama of this episode, we need to review the changes that occurred. This means delving into a little classical embryology.

Diversification of Animal Body Plans

Metazoan phyla are distinguished by how they develop as embryos and how their adult body plan is organized. There are three key divisions:

- **Radiata and Bilateria** (Figure 12.6).   Radiata are animals with two embryonic tissue types and radial symmetry, while Bilateria have three embryonic tissues and bilateral symmetry. The tissue types present in both groups are the ectoderm (outer skin) and endoderm (inner skin). Ectodermal cells produce the adult skin and nervous system, and endodermal cells produce the gut and associated organs. The third embryonic tissue is the mesoderm (middle skin). In bilaterians these cells develop into the gonads, heart, connective tissues, and blood.

- **Coelomates, pseudocoelomates, and acoelomates** (Figure 12.7). Coelomates have a true coelom. This is a fluid-filled cavity in the body derived from mesoderm and lined with mesodermal cells called the peritoneum. Coeloms create a "tube within a tube" body plan that provides space for the arrangement of internal organs. Because fluid-filled cavities are required for hydrostatic skeletons to operate, coeloms also triggered improvements in swimming and crawling ability. In contrast, acoelomates have no body cavity. Their mesoderm forms a solid mass between the body wall and the gut. Pseudocoelomates have cavities in their bodies that are not derived from mesoderm and are not lined completely with mesoderm.

- **Protostomes and deuterostomes** (Figure 12.8).   Although both of these groups have three embryonic tissue types, true coeloms, and bilaterally symmetric body plans, they differ in how their embryos undergo a key developmental event called cleavage. Cleavage refers to the initial cell divisions that take place after the egg is fertilized. In protostomes cleavage is spiral, meaning that the cells divide at an acute angle to the axis of the embryo (Figure 12.8a). In deuterostomes cleavage is radial, meaning that the cells divide perpendicular or parallel to one another (Figure 12.8b).

Protostomes and deuterostomes also differ in the way they undergo gastrulation. This is the mass movement of cells that rearranges cells after cleavage and defines the ectoderm, endoderm, and mesoderm. In protostomes gastrulation forms the mouth region first; in deuterostomes gastrulation forms the mouth region after the anal region forms. These differences inspired the use of the Greek roots *proto* (first), *deutero* (second), and *stoma* (mouth).

The Cambrian explosion produced other major morphological innovations, including the first segmented body plans, shells, external skeletons, appendages,

(a)

(b)

Figure 12.6 **Radial and bilateral symmetry** (a) In radially symmetric animals, like this sea star, body parts are arranged around a central axis (Andrew J. Martinez/Photo Researchers, Inc.). (b) In bilaterally symmetric animals, like this crab, body parts are arranged on either side of a central axis (Tom McHugh/Photo Researchers, Inc.).

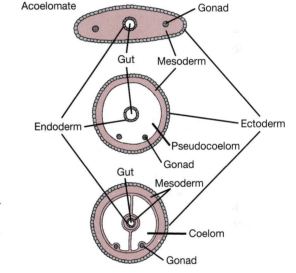

Figure 12.7 **Coelomate, acoelomate, and pseudocoelomate body plans** Coeloms can be formed in one of two ways. In schyzo-coelous development, found in protostomes, the body cavity forms when a hollow develops in a solid block of mesodermal cells. In enterocoelous development, found in deuterostomes, the body cavity forms from pouches extending out from the gut.

(a)

(b)

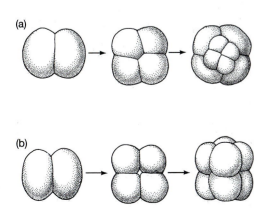

Figure 12.8 **Cleavage patterns in metazoans** (a) Spiral cleavage. (b) Radial cleavage.

(a)   (b)   (c)

Figure 12.9 **Ediacaran faunas** (a) A radially symmetric organism, a jellyfish, called *Brachina delicata.* The scale bar represents 2 cm. (b) A bilaterally symmetric sea pen, called *Charniodiscus arboreus.* The scale bar represents 5 cm. [Photos (a) and (b) from Simon Conway Morris, University of Cambridge] (c) A possible segmented worm, *Dickinsonia costata.* The scale bar represents 2 cm. Reproduced by permission from Fedonkin, Sun, and Bengtson, *Early Life on Earth.* New York: Columbia University Press.

and notochords. An enormous amount of evolutionary change occurred. How do we know all this? Snapshots offered by two beautifully preserved faunas, dated about 40 million years apart, tell the story.

The Ediacaran and Burgess Shale Faunas

The first unequivocal evidence for multicellular animals in the fossil record comes from the Ediacaran faunas (Conway Morris 1989; Weiguo 1994). These specimens were first found in the 1940s in the Ediacara Hills of south Australia; similar fossils have now been found at some 20 sites around the world. The earliest members of this fauna are dated at or around 565 Ma, placing them at the start of the Vendian period in the Proterozoic (early life) era, and the youngest at 544 Ma. These latest Precambrian faunas were entirely soft-bodied and are preserved primarily as compression and impression fossils (Figure 12.9). The partial or full-body specimens that are present are often difficult to identify, but most experts now agree that sponges, jellyfish, and comb jellies are represented (Conway Morris 1989; Weiguo 1994). Many of the specimens are traces, meaning that they are the remnants of burrows, fecal pellets, or tracks. Trace fossils can be difficult to interpret, but at least some must have been made by bilaterally symmetric organisms.

The second, and slightly younger, snapshot is provided by the Burgess Shale fauna (Figure 12.10), discovered early in this century near the town of Field, British Columbia. The Burgess Shale, dated to 520–515 Ma, and the Chengjiang biota from Yunnan Province China (525–520 Ma) are arguably the most spectacular fossil deposits ever found (Gould 1989; Conway Morris 1989, 1993; Briggs 1991; Weiguo 1994; Erwin et al. 1997). They are primarily impression and compression fossils, but are extraordinary for the detail they preserve and the story they tell. The specimens include a wide diversity of unusual arthropods, including trilobites, but there are also segmented worms, worm-like priapulids, a possible chordate named *Pikaia,* and molluscs. There is little if any overlap between species found in the Ediacaran and Burgess Shale deposits (Conway Morris 1989).

To appreciate just how much change took place in the interval between the two faunas, examine the metazoan phylogeny diagrammed in Figure 12.11. These relationships were estimated using data from 18s rRNA sequences. The

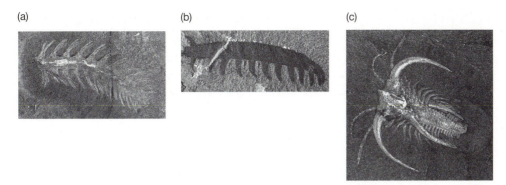

(a)   (b)   (c)

***Figure 12.10* Burgess shale faunas**   The Burgess Shale and Chengjiang faunas are extraordinary not only for the diversity of arthropods and other hard-shelled animals, but for the number of soft-bodied organisms that are preserved. In other localities from the same time period, only the shelly organisms fossilized. But shelled organisms represent a mere 20% of the diversity recorded in the Burgess Shale and Chengjiang deposits. Note that animals with well-developed segmentation, heads, and limbs are present (Chip Clark/National Museum of Natural History/Smithsonian Institution).

branching events that split the Radiata and Bilateria, protostomes and deuterostomes, and coelomates and acoelomates are mapped on the tree, as are the origins of multicellularity, tissues, segmented body plans, exoskeletons, and the notochord. These events, which define the universe of basic body plans and early developmental sequences found in animals, are recorded in these two early fossil assemblages. Just 40 million years of evolution created a suite of body plans that has persisted for well over 500 million years.

In addition, a significant portion of the Burgess Shale and Chengjiang faunas is made up of organisms whose body plans are so unusual that they cannot be placed in extant phyla. Taxonomists group these animals in a miscellaneous bin called the problematica (Figure 12.12). If we imposed the criteria we use today to classify these organisms, at least some might be assigned to new phyla represented by single species (Conway Morris 1989; Briggs et al. 1992). The message of the problematica is that there are many more ways to put an animal together than are exhibited by extant phyla. The fact that none of these clades survived more than a few million years suggests that natural selection rapidly sorted out a restricted set of developmental sequences and body plans that worked well.

A wide variety of animal body plans evolved during the Cambrian explosion. A subset of these plans defines the phyla living today.

What Caused the Cambrian Explosion?

Although we have emphasized the variety of body plans, cell types, and developmental patterns that evolved, the Cambrian explosion was also ecological in nature. Appreciating this point is a key to understanding why the event occurred. Most of the Ediacaran metazoans are either sessile filter feeders or predators that floated high in the water column and fed on planktonic organisms. But the Burgess Shale fauna introduces a huge variety of benthic and pelagic predators, filter feeders, grazers, scavengers, and detritivores who burrowed, walked, floated, clung, and swam. The Cambrian explosion filled many of the niches present in shallow marine habitats. Also, the explosion was clearly powered, in part, by entirely new modes of locomotion. What events triggered this morphological and ecological radiation?

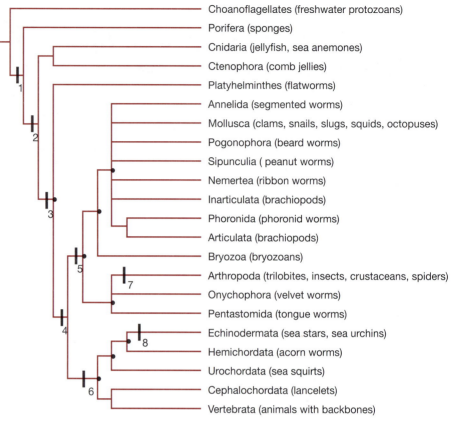

Figure 12.11   A phylogeny of metazoans   This tree, estimated from small-subunit (18s) rRNA sequences, is adapted from Halanych et al. (1995), Valentine et al. (1996), and Raff (1996). The branch lengths are not drawn to scale. All of the names refer to a phylum except for the following: Articulata and Inarticulata are classes in the phylum Brachiopoda, and Urochordata, Cephalochordata, and Vertebrata are subphyla in the phylum Chordata. The filled circles indicate branching events that occurred during the Cambrian explosion. The dark bars indicate important changes in body plans: (1) Multicellularity, radial symmetry; (2) tissues, differentiation of endoderm and ectoderm; (3) differentiation of endoderm, ectoderm, and mesoderm, development of anterior-posterior axis, and bilateral symmetry; (4) coelom; (5) protostome pattern of development; (6) deuterostome pattern of development; (7) exoskeleton; (8) radial, pentameral (five-formed) symmetry.

Figure 12.12   **Problematica**
Some 60% of the Burgess Shale fauna represents *incertae sedis,* or problematica—organisms of unknown affinity (Conway Morris 1989). The body plans of these organisms are so unusual that they cannot be placed in extant phyla (Chip Clark/National Museum of Natural History/Smithsonian Institution).

(a)

(b)

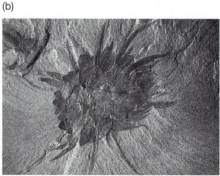

Environmental Factors

Rising oxygen concentrations in seawater, due to the increase of photosynthetic algae in the Proterozoic, was clearly a key to the origin of multicellularity (Valentine 1994). Increased availability of oxygen makes larger size and higher metabolic rates possible. Large size is a prerequisite for the evolution of tissues and higher metabolic rates are required for powered movement. Both of these traits are recorded in the Ediacaran faunas.

The origin of hard parts, another hallmark of the radiation, is documented in other late Precambrian fossils. Why did shells evolve? Stefan Bengtson and Yue Zhao (1992) addressed this question by analyzing the first mineralized exoskeletons in the fossil record. These are found in a widespread, tube-forming, 1- to 2-cm-long species called *Cloudina hartmannae*. Bengtson and Zhao examined 524 *Cloudina* tubes excavated in Shaanxi Province, China, and found conspicuous rounded holes, 40 to 400 μm in diameter, in 17 of them (Figure 12.13). Some of the holes penetrate only the outer layers of the tubes, but most go all the way through. Bengtson and Zhao suggest that the holes were drilled by predators. There is, for example, a positive correlation between the size of the hole and the width of the tube at the hole. This result implies that predators were selecting tubes by size, with larger predators attacking larger prey. This work supports an important hypothesis: that predation set up selection gradients favoring mineralized shells.

Genetic Factors

Predation pressure may have favored increased body size, new modes of locomotion, and mineralized skeletons, but a response to selection only occurred because genetic variability existed. The massive amount of morphological evolution we have just documented requires a significant increase in the amount of information contained in the genome. What genetic changes allowed the Cambrian explosion to occur?

For information content and processing, perhaps the most basic challenge posed by multicellularity is creating a system for arranging cells in three-dimensional space. Cells must be identified by their location—in relation to other cells and in relation to time—for cleavage and gastrulation to occur; for symmetry, segmentation, and body cavities to be organized properly; and for the correct differentiation of muscle, gut, and other tissues.

The genes that carry this information are called homeotic loci. The most important homeotic genes in animals are called the *HOM* loci in invertebrates and the *Hox* loci in vertebrates. Determining how these gene complexes originated and changed through time is currently one of the most active research topics in developmental and evolutionary biology.

HOM/Hox genes have been found in all major animal phyla except for sponges, and share three key traits (Carroll 1995):

- They are organized in gene complexes. This means that related genes are found in close proximity on the chromosome, and suggests that the original genes were elaborated by duplication events. Every phylum or class that has been surveyed shows a unique pattern of gene duplication or loss compared to other groups (see Kenyon and Wang 1991; Krumlauf 1994; McGinnis and Kuziora 1994; Balavoine and Telford 1995; Valentine et al. 1996).

Figure 12.13 Evidence for predatorial borings in the first exoskeletons This is a fossilized skeleton of *Cloudina,* showing a hole made by predators. The white scale bars represent 100 μm (Stefan Bengtson. Reproduced by permission from *Science.* © American Association for the Advancement of Science).

- There is a perfect correlation between the 5′–3′ order of genes along the chromosome and the anterior-posterior location of gene products in the embryo (Lewis 1978). Genes located at the 5′ end of the complex are expressed in the head region of the embryo, while genes located at the 3′ end are expressed toward the posterior of the embryo. Genes in the 5′ end of *Hox* complexes are also expressed earlier in development than genes located downstream. This phenomenon is called temporal and spatial colinearity (Krumlauf 1994), and is unique to *HOM* and *Hox* genes. Why it occurs is a complete mystery.

- Each locus within the complex contains a highly conserved 180-bp sequence called the homeobox. These bases code for a DNA binding motif. The discovery of the homeobox (McGinnis et al. 1984a, b) confirmed that *HOM* and *Hox* gene products are regulatory proteins that bind to DNA and control the transcription of other genes.

What do *HOM/Hox* loci regulate? The answer to this question came through experiments that produced fruit flies lacking specific *HOM* genes (Lewis 1978; McGinnis and Kuziora 1994). In *Drosophila,* homeotic genes are found in two main clusters called the *bithorax* and *Antennopedia* complexes. Mutations in *bithorax* complex genes tend to cause defects in the posterior half of embryos, while mutations in *Antennopedia* genes affect the anterior. The key observation is that flies missing one or more of these gene products can produce segment-specific appendages in the wrong place. By mixing and matching mutations in the two complexes, researchers have been able to produce adult flies with legs growing from the head region, or with four pairs of wings instead of two (Figure 12.14). The cells that produced the misplaced antennae or wings acted as if they misunderstood their location. This means that rather than speci-

(a)

(b)

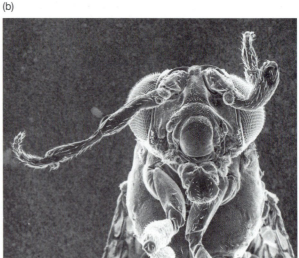

Figure 12.14 **Homeotic mutants in *Drosophila*** (a) Mutations in the *Hom* complex loci *bithorax (bx), posterior bithorax (pbx),* and *anterobithorax (abx)* produce four-winged flies. The phenotype occurs because appendages that should appear on thoracic segment 2 also appear on thoracic segment 3. In effect, the mutations have changed the identity of segment 3 to equal that of segment 2 (E.B. Lewis, California Institute of Technology).   (b) Mutations in the *Antennopedia (antp)* locus can produce adult flies with legs growing from the head. For limb development, the identity of the head segment has been changed to that of a thoracic segment (Dr. F.R. Turner, Biology Department, Indiana University).

fying a particular structure, gene products from *HOM/Hox* loci demarcate relative positions in the embryo. That is, instead of signaling "make wing," a *HOM/Hox* protein indicates "this is thoracic segment 2." Other genes, presumably those regulated by the protein products of *HOM/Hox* loci, make the structure specified for each location. In sum, *HOM/Hox* genes regulate the fate of cells by specifying where they are in time and space.

When did these loci originate? An early hypothesis was that they coincided with the evolution of segmentation during the Cambrian explosion. This was inspired by the observation that homoetic genes tend to be expressed in a segment-specific manner in flies. But homologs to the *HOM/Hox* genes have now been found in green plants, fungi, the roundworm *Caenorhabditis elegans,* and other nonsegmented animals, including jellyfish (see Kenyon and Wang 1991). These results push the origin of the *HOM/Hox* complex back in time, perhaps to the origin of multicellularity in the Precambrian.

James Valentine and co-workers (1996; see also Erwin et al. 1997) have synthesized data on the origin and elaboration of homeotic genes by mapping the presence and absence of *HOM/Hox* loci onto the metazoan phylogeny. Using a parsimony criterion, they inferred which genes in the *HOM/Hox* complex have been gained or lost at key branching points near the base of the tree. Their analysis (Figure 12.15) confirms that the loci called *labial (lab), proboscipedea (pb), Deformed (Dfd), Antennapedia (Antp),* and *Abdominal-B (Abd-B)* are ancestral to many other genes in the complex; that the entire *Hox* complex was duplicated several times in the lineage leading to vertebrates; and that the number of loci within the complex correlates roughly with the overall complexity of metazoan body plans. These observations support the hypothesis that the duplication and

Loci in homeotic gene complexes code for regulatory proteins that control the transcription of other genes. The effect of homeotic genes is to demarcate positions within embryos.

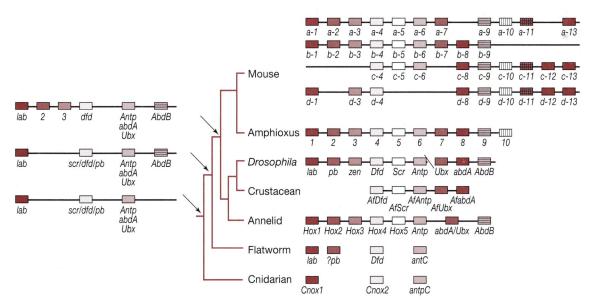

Figure 12.15 HOM/Hox genes from various metazoan phyla   The boxes on the right indicate which homeotic loci have been found in each of the taxa on the tree. Homologous loci are lined up vertically. Using these presence/absence data, Valentine et al. (1996) used parsimony to infer which loci were present at branch points early in the metazoan radiation. These loci are indicated at the left, with arrows pointing to the appropriate node. Note that Amphioxus, a lancelet in the subphylum Cephalochordata, is the most "primitive" living chordate.

elaboration of genes in the *HOM/Hox* complex was a genetic innovation that helped make the Cambrian explosion possible.

Current Research on the Cambrian Explosion

Research on the radiation is now focusing on six questions:

- When did the *HOM/Hox* genes first appear, and from what loci are they derived? Additional surveys of these loci, in sponges and other groups near the base of the radiation, are underway.

- Does the presence or absence of certain loci in the *HOM/Hox* complex correlate with specific changes in body plans, like the origin of segmentation? To answer this question, we need to know much more about which genes are found in which taxa. Even if we could establish that certain correlations exist, establishing a causative role for the homeotic loci will depend on continued advances in the study of their function in *Drosophila, Caenorhabditis,* and other model organisms.

- Do different alleles at homeotic loci produce variation in morphology within species? This variation could be acted upon by natural selection and drift. Research on intra- and interspecific polymorphism in homeotic genes is only beginning. Recent results show that measurable variation is present, however, and raises the possibility of measuring selection gradients on homeotic loci (Gibson and Hogness 1996; see also Tautz 1996).

- Can we improve our understanding of the metazoan phylogeny? If we can obtain sequence data from phyla that have not been studied yet, and include information from genes other than 18s rRNA, the resulting tree would be more complete and better resolved. This would allow us to refine and extend the analysis of *HOM/Hox* evolution illustrated in Figure 12.15.

- Can we get a better picture of what key groups like annelids, molluscs, and arthropods looked like, close to the time they originated? Interpreting phylogenies based on molecular data is difficult unless we have well-dated, well-preserved specimens to map onto the key branches and nodes. Collection, preparation, and interpretation of fossils from the Vendian to Middle Cambrian is ongoing.

- Just how "explosive" was this early radiation of metazoans? Based on a molecular clock analysis of seven genes, Gregory Wray and colleagues (1996) estimated the divergence between vertebrates and invertebrates at about a billion years ago—almost twice as long as the time of divergence documented in the fossil record. Although their calibration is controversial, many biologists concede that at least some of the major metazoan phyla may have originated long before they were documented in Ediacaran or Burgess Shale faunas.

To summarize, synthesizing phylogenetic, fossil, and molecular genetic data is proving to be a powerful new approach to understanding the great events in evolution. Having introduced this style of evolutionary analysis to explore the origin of multicellularity and the diversification of complex body plans, we can refine our focus and explore the developmental and genetic mechanisms responsible for a specific new trait, using the tetrapod limb and the flower as case studies.

12.3 Innovations: Developmental Genetics and Macroevolutionary Change

Explaining the origin of adaptations is a classical problem in paleontology. In Chapter 2 we discussed the Giant Panda's thumb and noted that it evolved through elaboration of a wrist bone, presumably in response to natural selection for increased efficiency in stripping the leaves off bamboo. In Chapter 3 we investigated the middle ear in mammals, which was probably modified in response to natural selection for the ability to hear high-pitched sounds. With the ear example, we introduced the logical sequence that evolutionary biologists use to study morphological innovations: establishing homologies with ancestral characters and documenting change in those characters through time. In this section we introduce a third step: identifying the genetic changes that underlie morphological novelties.

The Tetrapod Limb

The terrestrial vertebrates include the amphibians, reptiles, birds, and mammals. The signature adaptation of this lineage, called the Tetrapoda, is the limb. This structure distinguished the early tetrapods from fish and allowed them to crawl about on land. Later this innovation was modified into an enormous variety of shapes and sizes, with functions ranging from digging to flying. But as we noted in Chapter 2, tetrapod limbs are variations on the same theme (look back at Figure 2.1b). From frogs to foxes, the number and arrangement of limb bones is similar (Figure 12.16).

What Structure Gave Rise to the Limb?

Establishing homologies between the tetrapod limb and ancestral forms is the key to answering this question. Cladistic analyses show that the sister group of the tetrapods is a lineage of lobe-finned fishes from the late Devonian (Ahlberg and Milner 1994; Ahlberg 1995). In particular, a group of fish called the Panderichthyidae share several synapomorphies with early tetrapods like *Acanthostega*—the Devonian tetrapod we introduced in Chapter 3. The panderichthyids were large predators in shallow, fresh-water habitats, and undoubtedly used their limbs to walk or pull themselves along much as modern lungfish do. As Figure 12.17 illustrates, there are numerous structural homologies between the fins of lobe-finned fish and the limbs of tetrapods. The phylogenetic and morphological analyses support the hypothesis that the tetrapod limb is derived from the fins of lobe-finned fish.

What Developmental and Genetic Changes Were Responsible for Its Origin and Subsequent Modification?

To answer this question we need to understand some developmental genetics. We start by investigating the mechanisms that create the common ground plan of the tetrapod limb, then consider how these mechanisms first appeared at the base of the tetrapod radiation.

Evidence from developmental biology backs up our claim that tetrapod limbs share a common ground plan: Every tetrapod investigated to date shares the

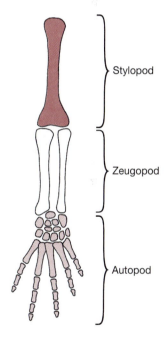

Figure 12.16 **The tetrapod limb** This diagram shows the basic elements of the tetrapod limb. There is a single proximal element, or stylopod (called the humerus in the forelimb and femur in the hindlimb) followed by an element with two bones (radius and ulna or tibia and fibula), a complex of small elements (carpus or wrist, tarsus or ankle), and the digits (fingers or toes).

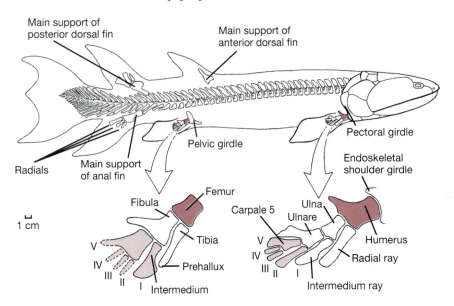

Figure 12.17   Fin bones in lobe-finned fish from the Devonian
Eusthanopteron, the genus pictured here, had fins containing bones in the typical tetrapod limb arrangement. In the detailed drawing of the fins, dashed lines indicate elements that have not yet been found in fossilized specimens. These are inferred to have existed from the position and structure of adjoining elements. Modified from Carroll (1988) and Jarvik (1980).

Tetrapod limbs have a common ground plan. This results from a shared developmental program.

same basic features of limb development (Hinchcliffe 1989; Gilbert 1991). Tetrapod limbs originate with a bud formed from mesodermal cells (Figure 12.18a). At the tip of this bud is a structure called the apical ectodermal ridge, or AER (Figure 12.18b, c). Cells in the AER secrete a molecule that keeps the underlying mesodermal cells in a growing and undifferentiated state. This population of mesodermal cells is called the progress zone. The progress zone grows outward and defines the principal axis of the developing limb (Figure 12.18c).

At the base of the bud in Figure 12.18c is a group of cells called the zone of polarizing activity (ZPA). A molecule secreted from the ZPA diffuses into the surrounding tissue and establishes a gradient that supplies positional information to cells in the structure. This is critical, because the limb develops in a four-dimensional system consisting of time and the three spatial coordinates. The spatial dimensions are defined in relation to the main body orientation: anterior to

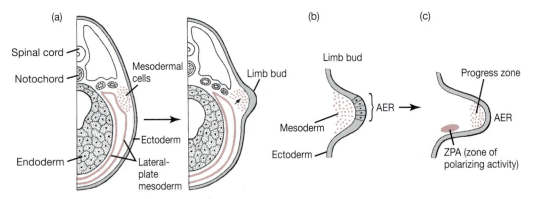

Figure 12.18   The developing limb bud   (a) In tetrapods, the limb bud originates from cells that migrate from the lateral-plate mesoderm early in development.   (b) These cells induce the formation of a structure called the apical ectodermal ridge (AER).   (c) If the AER is removed experimentally, cells in the progress zone stop their rapid division and differentiate, so only part of the limb is formed.

posterior, dorsal to ventral, and proximal to distal (Figure 12.19). The concentrations of molecules diffusing from the AER and ZPA help set up a coordinate system and tell each cell where it is in the four-dimensional space.

These molecules are, in fact, now well characterized and take our recognition of homology among limbs to the gene level. The products of genes called fibroblast growth factors (specifically, proteins called FGF-2 and -4) are found in the AER. Application of FGF-2, even when the AER is removed, leads to normal limb development in chicks (Fallon et al. 1995; Cohn et al 1995). This is strong evidence that FGF-2 is the molecule we referred to before as the AER product maintaining the progress zone. Similarly, the product of a gene called sonic hedgehog *(shh)* localizes to the ZPA, and grafting *shh*-expressing cells onto limb buds produces a ZPA (Riddle et al. 1993; Tabin 1995). In addition, expression of a recently discovered gene called *Wnt7a* causes cells in developing limbs to form dorsal structures in mice (Yang and Niswander 1995; Parr and McMahon 1995). In sum, though we do not yet have a candidate clock molecule, we are zeroing in on the genes responsible for specifying the other three coordinates in the limb system. And they are old genes: Homologs of the *Wnt* and hedgehog gene families are expressed in the legs of developing insects (see Martin 1995).

Equally important is the recent discovery that loci in the *Hox* family are critical to limb development. Earlier we noted how loci in the *Hox* complex are expressed along the embryo's anterior to posterior axis in the same order that they are found along the chromosome. This linear sequence of *Hox* genes is also expressed in the developing tetrapod limb—specifically in the progress zone (Duboule 1994). As we will detail later, *Hox* gene expression in the limb is regulated by FGF and *Shh*. *Hox* loci are the position-determining genes in the limb, just as they are along the main body axis (Cohn et al. 1995; Tabin 1995).

This brief introduction to the developmental genetics of the vertebrate limb carries two messages. First, shared developmental and genetic pathways underlie the structural homology of amphibian, reptile, bird, and mammal limbs. Second, changes in the timing or level of expression in the pattern-forming genes we have reviewed—fibroblast growth factor loci, *shh, Wnt,* or the *Hox* complex—could be responsible for adaptive changes in the limb. The evolution of longer limbs, for example, might be a function of variation in FGF gene expression and how long the progress zone is maintained.

Just this type of genetic connection has been made concerning the origin of a prominent part of the tetrapod limb: the hand and foot. The lobe-finned ancestors of the tetrapods, like extant fish, do not have feet or hands (see Figure 12.17); tetrapods do. A recent study by Paolo Sordino and co-workers (1995; see also Nelson and Tabin 1995; Sordino and Duboule 1996) suggests a genetic mechanism for the change. Sordino et al. sequenced *shh* and members of the *Hox* gene family in zebrafish *(Danio rerio),* labeled copies of the genes with radioactive atoms, then hybridized the DNA with various stages of developing fins. The pattern of expression of *shh* and the various *Hox* loci, when contrasted with results for the same genes in a tetrapod (the mouse), is striking. Early in development, gene expression in tetrapods and fish is similar. *Hox* transcripts, for example, localize to the posterior margin of both the elongating fin and limb buds (Figure 12.20a, b). But in tetrapods, a switch then occurs. Later in development, *Hox* gene transcripts are found in the anterior and distal parts of the

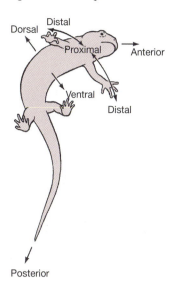

Figure 12.19 **Developmental and morphological axes** Dorsal = toward the back, ventral = toward the belly, anterior = toward the front, posterior = toward the rear, proximal = toward the main body axis, distal = away from the main body axis.

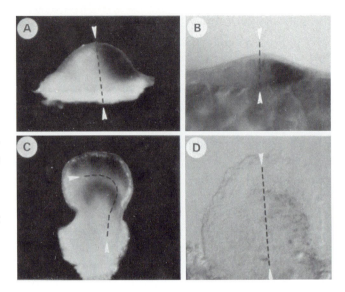

***Figure 12.20* Expression of pattern-forming genes in fish and tetrapod limbs**   (a) In early limb buds in mice, *Hoxd-11* expression (indicated by the dark splotch) is limited to the posterior half.   (b) Early limb buds in zebrafish show the same pattern of *Hoxd-11* expression.   (c) In later mouse embryos, a second wave of *Hoxd-11* expression is oriented anterior to posterior, in the most distal portion of the developing limb.   (d) In contrast, in late fish embryos, *Hoxd-11* expression has decreased and is still restricted to the posterior margin. In each of the photographs, the dotted lines indicate Sordino et al.'s (1995) proposal for the axis of development (Prof. Denis Duboule, Université de Genève, Geneva, Switzerland).

mouse limb bud, and there is a late expression of *shh* (Figure 12.20c). Why is this interesting? Because it does not happen in fish. There is no expression of *Hox* or *shh* in the late limb bud of zebrafish (Figure 12.20d); instead, the ectoderm folds and development stops after formation of the tibia and fibula (a structure called the zeugopod in fish). The late expression of *shh* and *Hox* is an evolutionary novelty and may have produced tissues that became the first hands and feet in the history of life. Hands and feet are *not* homologous to any part of a lobefin's fin. They are evolutionary add-ons.

This pattern of expression of *Hox* genes and *shh*—specifically the change from the early restriction on the posterior margin to the late anterior and distal expression—also supports a recent hypothesis about the axis of development in the tetrapod limb. For years morphologists had assumed that the axis of development, as specified by the progress zone and the sequence of bones formed, ran straight through the hand and foot; there were long debates about exactly which elements this straight axis ran through. (This is not trivial, because identifying the axis is important to hypotheses about patterns of digit reduction among tetrapods.) But in 1986, Neil Shubin and Pere Alberch proposed that the principal axis of limb development in tetrapods is not straight, as it obviously is in *Eusthenopteron* and extant fish, but bent (Figure 12.21). In a spectacular verification of the Shubin–Alberch hypothesis, Sordino et al.'s work suggests a mechanism for the bending: the switch in the pattern of expression of *shh* and *Hox*.

What is the regulatory change that produced the developmental switch, the change in the direction of the progress zone, and the resulting foot/hand structure? We do not yet know. Research is continuing.

Our exploration of the genetic changes responsible for macroevolution continues with another type of limb—one that evolved independently from the limbs of tetrapods.

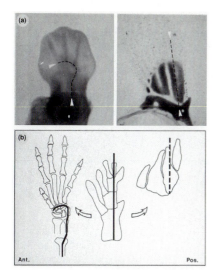

Figure 12.21 **The axis of development in the vertebrate limb** (a) These photographs show mouse (left) and fish (right) limbs at the stage of cartilage condensation. The proposed axis of development is indicated by dashed lines. (b) Adult limbs in mouse (left), *Eusthenopteron* (middle), and zebrafish (right), with the proposed axis of development indicated (Prof. Denis Duboule, Université de Genève, Geneva, Switzerland).

The Arthropod Limb

Arthropods are far and away the most diverse animal phylum (Brusca and Brusca 1990). Over a million species have been described, and perhaps 50 times that many are still to be named. Members of the group are distinguished by having an exoskeleton (which necessitates growth by molting), a body organized into cephalic (head), thoracic, and abdominal regions or tagma, pairs of appendages on each body segment, and an open circulatory system. Although estimating the phylogeny of arthropods is notoriously difficult and controversial (see Budd 1996), one recent classification identifies four subphyla:

- Chelicerates include scorpions, spiders, mites, horseshoe crabs, and a variety of extinct groups (Figure 12.22a). Prominent among the extinct forms are the eurypterids, or sea spiders, which occasionally reached three meters in length.
- Uniramians (Figure 12.22b) encompass the centipedes and millipedes (or myriapods) and insects.
- Trilobites (Figure 12.22c) were abundant in marine environments throughout the Paleozoic but went extinct at the end of the Permian.
- Crustaceans (Figure 12.22d) are the shrimp, copepods, barnacles, crabs, lobsters, crayfish, and pill bugs.

The initial diversification of arthropods occurred during the Cambrian explosion, but many of the classes, orders, and families originated and diversified later. Insects do not show up in the fossil record until early in the Devonian, for example. This is long after the phylum's initial radiation. As we saw in Chapter 3, insects then underwent a dramatic adaptive radiation in the Triassic and Jurassic that was correlated with the evolution of seed plants. In sum, the diversification of arthropods is marked by a series of radiations, beginning in the Precambrian and continuing to the present.

Like tetrapods, the hallmark adaptation of arthropods, and the feature most responsible for the diversification of the group, is the limb. Arthropod appendages

(a) (b)

(c) (d)

Figure 12.22 **The arthropod radiation**   Note the diversity of form and function in the limbs of these representative taxa.   (a) Chelicerates (Bill Johnson/Visuals Unlimited).   (b) Uniramians (Tom McHugh/Photo Researchers, Inc.). (c) Trilobites (A. Graffham/Visuals Unlimited).   (d) Crustaceans (Scott Johnson/Animals Animals/Earth Scenes).

come in two basic forms: uniramous (one branch) and biramous (two branch). Chelicerates and Uniramians have unbranched appendages while trilobite and crustacean limbs have two elements (Figure 12.23). In addition, several crustacean groups have multiply branched or phyllopodous (leafy feet) appendages that are used for swimming. The position, form, and function of biramous, uniramous, and phyllopodous limbs can all vary enormously. Arthropod limbs can be located on the cephalic, thoracic, and/or abdominal segments, and can be used for swimming, walking, breathing, fighting, or foraging. The ecological diversification of arthropods goes hand in hand with the elaboration of this morphological innovation.

What do we know about the genes involved in the evolution of arthropod appendages? Research on *Drosophila melanogaster* has established three major types of genetic control over limb formation in arthropods:

- The decision whether to make a limb depends on a gene called *wingless (wg)*. Mutant flies that lack *wg* fail to make any limb primordia at all. *Wg* is expressed in the anterior section of limb primordia; the protein products of a locus called *engrailed (en)* show up in the posterior. These observations suggest

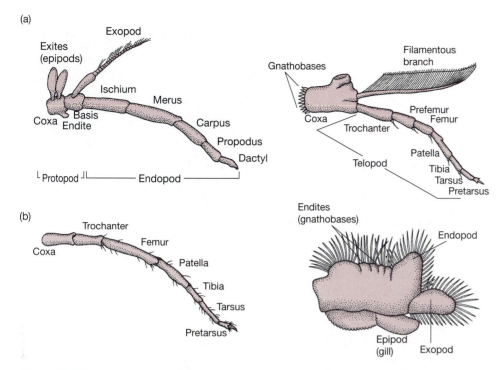

Figure 12.23 **Uniramous and biramous limbs of arthropods** This figure shows just a few of the many variations on the jointed limbs of arthropods (along with some of the terminology required to categorize the variation). A major distinction among arthropod limbs is whether they are biramous (a) or uniramous (b). Classes within the phylum tend to have one type of limb or the other (except that that some crustaceans have secondarily derived uniramous limbs). Modified from Brusca and Brusca (1990).

that the two loci may be responsible for defining the anterior-posterior axis of the early limb.

- The decision to extend the limb primordium distally seems to hinge on the expression of a gene called *Distal-less (Dll)*. This is the first gene activated specifically in limb primordia.

- The decision on which type of limb will develop is controlled by homeotic genes. This is not surprising, because the fate of a limb primordium should depend on its position in the embryo, and homeotics specify location.

What do these observations have to do with change in limb morphology through time, and the radiation of arthropods? There are hints in the *Drosophila* literature that changes in these genes correlate with significant evolutionary events. Specifically, abnormal expression of *Dll* can lead to branched limbs in flies. This result suggests that changes in the regulation of the locus could have played a role in the evolution of uniramous, biramous, and phyllopodous appendages. Variation in when or where *Dll* is turned on or off might lead to new or modified limb outgrowths.

Recent work by Grace Panganiban and colleagues (1995; see also Roush 1995) has extended our understanding of *Dll*'s role in the evolution of the arthropod limb. The researchers succeeded in making an antibody to the *Dll*

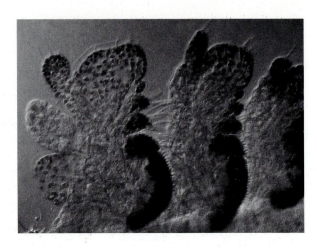

Figure 12.24   *Distal-less* is expressed in each branch of crustacean phyllopodous limbs   The dark staining grains indicate the protein product of *Dll*. In both the first and second thoracic segments of the common brine shrimp *(Artemia fransci- cana)* pictured here, marked T1 and T2, all of the limb branches (indicated by the letters and arrowheads) contain *Dll* product. The scale bar represents 0.1 mm. Reproduced by permission from G. Panganiban et al., The development of crustacean limbs . . . , *Science.* © American Association for the Advancement of Science.

protein product, and used it to stain early embryos of the common brine shrimp *Artemia franscicana* and the opossum shrimp *Mysidopsis bahia. Artemia* has phyllopodous appendages on its thorax that are used for swimming. *Mysidopsis* has biramous limbs: antennae on its head, walking legs on its thorax, and pleopods used for swimming on its abdomen. In *Artemia, Dll* is expressed in each of the outgrowths present in phyllopods (Figure 12.24). In *Mysidopsis,* the anti-*Dll* antibody localizes to each branch of the biramous appendages (Figure 12.25).

Dll seems to be regulated in different ways in different body regions of *Mysidopsis,* however. In the thorax, the two limb branches grow out from two independent clusters of cells, each of which expresses *Dll*. That is, the expression of the gene varies spatially. In the abdomen, the branches of pleopods form sequentially. The gene is turned on at two different times in the same location. This means that *Dll* is regulated temporally. In both cases there is a close correlation between when and where *Dll* is turned on or off and where limbs are produced. Just as it does in flies, *Dll* seems to signal "grow limb here" wherever it is expressed. Changes in when and where the gene is expressed appear to affect the branching of limbs.

Mutations that change the timing or location of distal-less *expression could change the morphology of crustacean limbs.*

We are still a long way, however, from understanding the genetics of macroevolutionary change in arthropod limbs. Two of the more urgent tasks now are to find out if *Distal-less* functions in similar ways in close relatives of the Arthropoda like the Onychophora, Tardigrada, and Pentastomida (Figure 12.11), and to expand the fossil record of these groups in the Vendian and Cambrian. The phylum Onychophora, or velvet worms, are burrowers restricted to the humid tropics. They have small, unbranched, saclike limbs called lobopods. The Tardigrada, or water bears, are tiny animals that live in moss or soil and lumber about on unbranched, clawed legs. The Pentastomida, or tongue worms, are all parasitic and have tiny lobe-like legs with claws that allow them to cling to respiratory passages in their vertebrate hosts. Based on the work done to date, we can predict that *Distal-less* will be found in all three of these "worms with legs," but not in true worms like annelids. We can also predict that the earliest fossils in these forms, and in the ancestors to the arthropods, will have the same sort of unbranched, saclike legs.

The Flower

We close our look at "molecular paleontology" by examining the origins of to-day's dominant plants. The radiation of terrestrial life began in the Silurian period, when algae first made the transition to land. There have been four major radiations of terrestrial plants since then: early groups called the Prilophyta and Rhyniophyta, ferns, early seed plants without flowers, and angiosperms. Each of these radiations was associated with one or more evolutionary innovations:

• Cuticle, a waxy covering over leaves, cut down on water loss through transpiration and allowed early land plants to tolerate drying. The evolution of stomata and guard cells provided breaks in the cuticle and a mechanism for acquiring carbon dioxide from the atmosphere instead of from water.

• Vascular tissues permitted land plants to grow beyond small size, and set the stage for a radiation of ferns and other large Devonian plants (Niklas 1994). These cells provide a mechanism for conducting water along the potential gradient created by uptake at the roots and loss at the leaves. This process is much more efficient than osmosis.

• Pollen and seeds are multicellular structures with a protective coat, and resist drying much better than sperm and spores. Pollen can be transported by wind while sperm are restricted to swimming; seeds can remain viable for years while spores often dry out and die within weeks. Gymnosperms (pines, junipers, and allies) were among the first plants with pollen and seeds, and radiated throughout the Triassic and Jurassic.

All of the features we have just reviewed have a common theme: They allowed plants to be less dependent on wet environments for growth and reproduction. The latest radiation, though, and the one we will examine in detail, is associated not with adaptation to drying but with a new reproductive structure.

The flowering plants, or Anthophyta, are a phylum of the plant kingdom. With an estimated 200,000 to 300,000 living species, the angiosperms are far and away the most species-rich plant group in the history of life. The innovation that occurred at the base of this radiation is a reproductive structure consisting of four organs: the calyx, corolla, stamens, and pistil (Figure 12.26). The calyx is a whorl of leaf-like protective structures, called sepals, that encloses the developing flower bud. The corolla includes all of the petals. Stamens are the male organ, and the pistil is the female organ that includes the ovules. All of these organs are thought to have evolved from leaf-like branching structures; possible steps in the evolution of the pistil and stamen, from leaves that bore female and male sporangia, are diagrammed in Figure 12.27.

Understanding the genetic changes responsible for the original modification of the flower is an active area of research. Not surprisingly, it turns out that flower morphology depends on the expression of homeotic genes. These loci are like the homeotic genes of animals, in two ways: They specify which organs appear in different locations, and they include a DNA-binding region called the MADS domain analogous to the homeodomain of *HOM/Hox* loci. A key experimental finding is that when certain combinations of loss-of-function mutations occur in these genes, floral organs are replaced by leaf-like structures

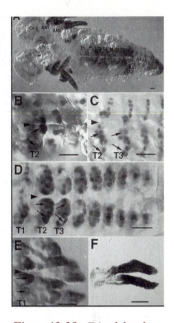

Figure 12.25 *Distal-less* is expressed in both branches of biramous limbs in crustaceans The dark staining grains in these photos indicate the protein product of *Dll*. The photos show how biramous limbs in the first and second thoracic segments (T1 and T2) of the opossum shrimp *Mysidopsis bahia* develop through time. The arrows highlight the way that *Dll* is localized in each branch of the biramous limb. The scale bars represent 0.1 mm. Reproduced by permission from G. Panganiban et al., *Science.* © American Association for the Advancement of Science.

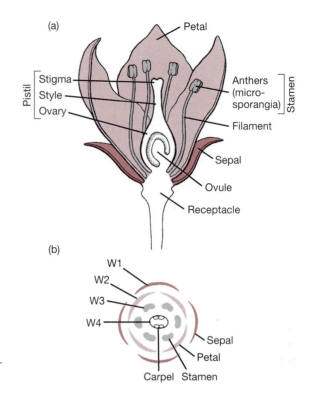

Figure 12.26   Parts of a flower   (a) A long section through an idealized flower;   (b) A top-down view showing how the four major organs are arranged in whorls (W1 through W4).

If homeotic or other types of mutants resemble ancestral forms, they demonstrate a plausible genetic mechanism for macroevolutionary change.

(Figure 12.28). The mutant appears to phenocopy the ancestral state of the flower (see Figure 12.29). This result is reminiscent of the mutations leading to the four-winged fruit fly. Furthermore, other combinations of mutations in homeotic loci lead to flowers lacking one or more organs: sepals, petals, stamens, or carpels (reviewed in Weigel and Meyerowitz 1994). Are these loci general position-determining genes, similar to homeotics in animals, or are they flower-specific genes that originated at the same time angiosperms did? We simply do not know. If homologous loci are found in gymnosperms or other ancestral groups, it would mean that they predate the evolution of flowers. A result like this would argue that they are not "the flower genes," but instead are active in establishing the general pattern of reproductive structures.

Definitive flower-specific genes have, however, recently been cloned from *Arabidopsis thaliana* (an annual in the mustard family) and the snapdragon *Antirrhinum majus*. These loci affect key aspects of floral morphology, such as the

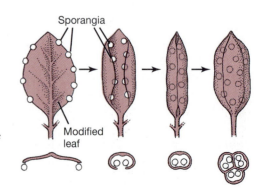

Figure 12.27   Steps in the evolution of the carpel and stamen
This a hypothesis for the steps leading to a flowering plant structure called the carpel. The carpel encloses the ovules; a pistil can contain one or more carpels. The rightmost arrow indicates that three carpels have fused to form the typical three-chambered ovary.

Figure 12.28 **Null mutants in floral homeotic genes**   (a) A wild-type flower of *Arabidopsis thaliana* (Elliot M. Meyerowitz, California Institute of Technology). (b) In a triple mutant lacking the protein product of the *AP2, PI,* and *AG* loci, sepals, petals, stamens, and ovules are replaced by leaf-like structures (M.P. Running/Elliot M. Meyerowitz, California Institute of Technology).

symmetry of the flower and the arrangement of flowers on the stalk (the inflorescence; see Coen and Nugent 1994). For example, mutations in the *cycloidea (cyc)* locus of *Antirrhinum* turn irregular flowers, with one plane of symmetry, into regular flowers, with more than one plane of symmetry (Figure 12.30). We have good fossil and phylogenetic evidence that regular flowers evolved first, and that irregular shapes are derived. This observation suggests that *cyc* originated during the angiosperm radiation, and may have contributed to the evolution of an entirely new floral morphology.

12.4 Transitional Forms

To a large degree, the focus in this chapter has been how innovations like complex body plans, jointed limbs, and the flower characterize the diversification of multicellular life. One of the profound ironies of evolutionary biology, however, is that these great events occur so quickly that they seldom leave hard (fossil) evidence of how and why they happened. What transitional steps occurred, and what was the selective advantage of each stage? The fossil record contains few smoking guns. The most notable gaps in the record involve intermediate forms.

Fortunately, a few of these gaps are closing. Here we review two research programs that are gradually uncovering the steps in the evolution of complex traits.

Figure 12.29 **The earliest flower** (Leo J. Hickey, Yale University).

Figure 12.30 **The *cyc* locus in snapdragons affects floral symmetry**   The flower at left is the normal phenotype of *Antirrhinum.* This is called an irregular flower, because it has a single plane of symmetry. The *Antirrhinum* flower (right) is the product of a mutation in *cyc.* It is regular, meaning that it has many planes of symmetry. Reproduced by permission from *Nature* 353: 31–37 (1991), Fig. 7, and from E.S. Coen and J.M. Nugent, Evolution of flowers and inflorescences (1994).

The Origin of Insect Flight

The flying insects, a subclass of the class Insecta, make up an estimated two-thirds of all multicellular species living today (Hammond 1992). The first insects appear in the fossil record early in the Devonian; and the first winged forms, or ptery-gotes, at 330−400 Ma. But there are no transitional or intermediate stages. Wings appear fully formed, similar in structure to contemporary insect wings used in flight. How could such a complex structure, requiring coordinated development and use of muscle, nerve, and circulatory tissues, evolve out of nowhere?

The short answer, of course, is that it did not evolve out of nowhere. Early insects had two candidate structures that selection could have elaborated into wings: paired, probably immovable "winglets" on thoracic segments, and mobile plates that covered the gills of aquatic forms (Kukalova-Peck 1978). Neither of these structures was used for flight when it first appeared. What function did the structures have, and how would it have preadapted insects for powered flight?

Although "few events in the Animal Kingdom have been the subject of as many hypotheses as the origin of insect wings" (Kukalova-Peck 1978, p. 57), attention is currently focused on two: the use of proto-wings in thermoregulation and in skimming across water surfaces. Both hypotheses are attractive because they suggest a selective gradient for increasingly large, moveable wings long before they had anything to do with flight.

Joel Kingsolver and M.A.R. Koehl (1985) are primarily responsible for elaborating and testing the thermoregulation theory. The idea here is that proto-wings functioned as solar panels, helping ectothermic insects collect heat and become active earlier in the day than nonwinged insects. To test the plausibility of the hypothesis, Kingsolver and Koehl constructed epoxy models of flightless Paleozoic insects, inserted thermocouples capable of recording temperature in the "body," and attached artificial wings of various sizes. When they put the models into a wind tunnel with strain gauges capable of measuring lift and drag, they found that short wings had almost no aerodynamic effect. Wings did not start to generate significant amounts of lift until they were at least a centimeter long. Short wings were, however, highly effective at raising the temperature of the model's body when a lamp was placed in the tunnel. Based on these results, Kingsolver and Koehn proposed that selection for heat gain could have acted on the thoracic lobes found in fossil forms, favoring elongation and possibly even the evolution of an articulation and movement. This latter change is conceivable because many contemporary insects actively position their wings to increase heat gain.

Originally, the selective advantage of the insect wing may have been to promote heat gain or to aid in skimming over water surfaces.

James Marden and Melissa Kramer (1994, 1995; see also Thomas and Norberg 1996) originated and developed an alternative called the skimming theory for the origin of insect wings. This hypothesis was inspired by watching contemporary stoneflies sail across water. Stoneflies are flightless insects that resemble fossil forms found in the Carboniferous. In late winter or early spring in northern North America, stonefly adults emerge from their aquatic nymph stage, raise their wings in response to wind, and sail across the water surface to the riverbank, where they climb up and spend the rest of their adult life on land. Marden and Kramer propose that individuals in a semi-aquatic ancestor of flying insects raised their gill plates to facilitate sailing. These individuals subsequently underwent strong directional selection for larger gill plates, and eventually their gill plates were transformed into lightweight wings. The skimming theory is more plausible than the thermoregulatory hypothesis in one important respect: The starting point for evolution is a structure already supplied with musculature, an articulation, and nerve enervation. These are all components required for flapping flight.

The two hypotheses are not mutually exclusive, and Marden and Kramer (1995) point out that flight may have originated several times in insects. Perhaps selection on thermoregulation was responsible for the initial elaboration of thoracic and abdominal lobes, with skimming later selecting for elaboration of selected thoracic lobes or gill plates.

Work on the topic continues. The skimming hypothesis, for example, makes an interesting prediction that can be tested by reexamining existing fossils or scouring late Devonian or Carboniferous outcrops for additional specimens. Stoneflies have hairs on their wings and legs. The hairs probably function to decrease wetting and increase buoyancy on the water surface. If selection for sailing was important early in insect evolution, we should be able to find these hairs on the earliest wingless forms.

Feathers and Flight

Although we have been stressing that morphological change is often extremely rapid in geological time, the fossil record does have several spectacular examples of intermediate forms. The stem-tetrapods *Acanthostega* and *Ichthyostega*, which we introduced earlier, were semiterrestrial organisms that are clearly transitional between fish and amphibians. When we examine human evolution in Chapter 14, we will introduce *Australopithecus afarensis*, a species with a combination of ancestral characteristics present in great apes and derived traits retained or elaborated in the hominids. Perhaps the most famous of all intermediate forms, though, is the first bird ever found in the fossil record: *Archaeopteryx lithographica* (Figure 12.31).

A total of seven specimens of *Archaeopteryx* have now been found and dated to the late Jurassic, about 150 Ma. Although the presence of feathers clearly identifies them as birds, the skeleton is so similar to the theropod dinosaur *Compsognathus* that one of the *Archaeopteryx* fossils was misidentified as *Compsognathus* when it was first found (Briggs 1991). Only when it was reexamined, and the presence of faint feather traces confirmed, was the specimen identified correctly.

Archaeopteryx has a large series of traits found in its dinosaur ancestors or bird descendants, but not both. Some of the more prominent "primitive" traits are its teeth, extensive tail vertebrae, claws, and lack of a calcified sternum, or breastbone. Its signature derived trait is feathers. The placement, number, and morphology of *Archaeopteryx* feathers are indistinguishable from those of modern birds. But where did feathers come from? And what about endothermy, the other outstanding derived character of birds? Could *Archaeopteryx* regulate its body temperature by expending metabolic energy? Finally, how did flight evolve? We will consider these questions in turn.

Several lines of evidence suggest that feathers may have evolved from scales. They are made of a similar protein, called keratin, and develop from similar embryonic structures. Extant birds actually have scales on their toes that form in the same way that scales on contemporary reptiles do. Given that scales probably

(a)

(b)

(c)

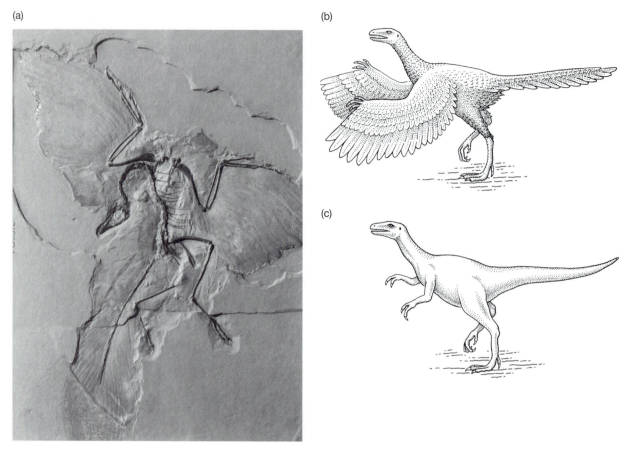

Figure 12.31 Archaeopteryx, **the first bird**   (a) A fossil from the limestone quarry in Eichstätt, Germany. Notice the breast area. Contemporary birds have a huge breastbone with a keel that provides surface area for the attachment of flight muscles (Neg. No. 325097, shot from Sanford Bird Hall, Courtesy Dept. Library Services, American Museum of Natural History). (b) An artist's reconstruction of how *Archaeopteryx* may have looked in life. Note the prominent claws on the forelimbs. In extant birds, only juveniles of a South American species called the Hoatzin have claws on their wings (they use them to climb trees).   (c) An artist's reconstruction of how the small dinosaur *Compsognathus* may have looked in life.

represent the ancestral state, what sort of selection pressure led to their transformation into feathers? For years the leading hypothesis has been that the earliest feathers evolved to provide insulation. This hypothesis got a potential boost recently, with reports from China of a meter-long bipedal dinosaur fossil, similar to *Compsognathus,* with impressions along its head, spine, and upper back that resemble feathers (Browne 1996; Monastersky 1996). The fossil has been tentatively dated to 120–140 Ma, however. This is long after *Archaeopteryx* lived. If the presence of feathers and the dating are confirmed, the find would confirm that feathers were not used for flight in all of the early bird-like lineages. Instead, they could have provided insulation or been used in display.

The new fossil also brings added urgency to a hot topic in dinosaur biology: Did selection favor insulation because the animals were endothermic? Although some indirect data support the idea that dinosaurs maintained high metabolic levels, the most direct studies we have suggest that many were ectotherms. This conclusion comes from research into a structure called the respiratory turbinate. This is a bony or cartilaginous tube in the front part of the nose, present in over 99% of living birds and mammals. Endotherms have respiratory turbinates for a simple reason: They breathe so quickly, because of their high metabolic rates, that they are susceptible to massive water loss. The tube increases the surface area of the nasal passage and leads to more efficient retention of heat and water from exhaled air.

John Ruben and colleagues (1996) measured the size of nasal passages in a wide variety of living mammals, birds, and reptiles, and found that they defined two statistically different distributions: The nasal passages of endotherms were significantly bigger than those of ectotherms (Figure 12.32). Ruben's group then did computed axial topography (CAT) scans on several extremely well preserved dinosaur skulls from the late Cretaceous, and found that their nasal passages were relatively small (Figure 12.32). This suggests that they lacked respiratory turbinates and implies that they were ectothermic. These dinosaurs lived 80 Ma after *Archaeopteryx,* however, so the result does not preclude the idea that at least some earlier dinosaur lineages were endothermic. Also, the sample sizes in the study are small, so the result will need to be confirmed by follow-up work.

We simply do not know about the metabolic status of *Archaeopteryx* itself, but the best evidence we have argues that this bird was not capable of sustained flapping flight. There was no place on the skeleton to attach a large flight muscle (called the supracoracoideus), meaning that *Archaeopteryx* probably could not generate a powered downstroke (Rayner 1989). Its feathers, however, may have made it an efficient glider. But did it glide from the ground up or from the trees down?

Support for the trees-down theory comes from a variety of sources. Gliding is an efficient form of locomotion, and theoretical studies have shown that gliders can start to produce lift with even slight flapping movements (Norberg 1985). Gliding from trees has evolved in a number of vertebrate lineages, and often involves specialized structures. Examples include rodents like flying squirrels, which parachute with the aid of flaps of skin between their fore and hindlimbs, and lizards in the genus *Draco,* which have extendible membranes supported by several pairs of ribs. These examples suggest that it is not especially difficult for selection to favor increasingly specialized structures, like feathers, to aid in gliding. A problem for the trees-down theory, though, is that the lagoon-like environment where the *Archaeopteryx* specimens were fossilized did not contain trees. The closest things to tree-like forms were 3-meter-tall stem conifers, which had a growth habit something like that of modern cacti (Norberg 1990). Proponents of the theory counter that these and other tree-like

It is still not clear whether dinosaurs were endothermic. If they were not, it is doubtful that feathers evolved because of their insulating value.

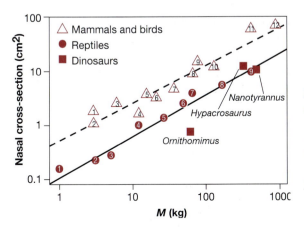

Figure 12.32 **Dinosaurs have nasal passages typical of ectotherms** Each triangle and circle on the plots represents the average value for a species. *M* on the abscissa stands for average body mass of the species. The lines through the data points are the best-fit lines produced by regression analyses of the mammal and reptile data. The three data points for dinosaurs represent single specimens; their body masses were estimated from their lengths. These points were not included in the regression line calculated for reptiles. From Ruben et al. (1996).

plants were abundant, and that the birds may also have been capable of limited flapping flight.

A recent paper by Alan Feduccia (1993) offers more direct evidence in favor of the trees-down proposal. Feduccia measured the geometry of claws from a variety of extant birds, and compared the curvature of claws from birds that spend most of their time perching in trees, climbing on tree trunks, or running on the ground. Perching birds have claws intermediate between ground-dwellers, which have relatively flat claws, and climbers, which have strongly curved claws. The curvature of the foot claws on *Archaeopteryx* places them firmly among the perching birds in Feduccia's data set, while the curvature of the claws on their hand are more similar to the climbers. Feduccia concludes that *Archaeopteryx* was definitely arboreal.

Current work on the evolution of avian flight is focused on understanding the function and distribution of early feathers, and on exploring the distribution of endothermy in early birds and their dinosaur ancestors. But already, *Archaeopteryx* is one of the best studied of all major transitional forms in the history of life.

12.5 Macroevolutionary Patterns

Understanding the great innovations in evolution is fundamental. That is why we spent most of Chapter 11 and much of this one on subjects like the origin of DNA synthesis, endosymbiosis, multicellularity, and breakthrough adaptations like the arthropod limb and the flower. But documenting evolution's "greatest hits" is only part of the historical biology portfolio. Searching for broad patterns in the fossil record is an equally important research program. The microevolutionary processes of mutation, natural selection, drift, and gene flow have produced an almost overwhelming diversity of life forms. The composition of this biota has changed radically over vast reaches of time. What patterns can we discern to make sense of all this variation?

The literature on macroevolutionary patterns is enormous, and we can only touch on a few of the major results. We start by summarizing a pattern that we have already explored in some detail: adaptive radiation in response to a morphological innovation or ecological opportunity.

Adaptive Radiations

Adaptive radiations are characterized by the rapid diversification of a clade into a wide variety of ecological niches (Figure 12.33). We have introduced several examples in earlier chapters: the Galápagos finches, the insect families, Hawaiian *Drosophila,* and Rift Lake cichlids.

What factors trigger adaptive radiations? Why do only certain lineages diversify broadly and rapidly? The answers vary from time to time and clade to clade.

Ecological opportunity was a key factor in several of the examples we have reviewed. The ancestors of the Rift Lake cichlids, Hawaiian *Drosophila,* and Galápagos finches colonized a lake or island habitat that had few competitors and a wide variety of resources to exploit. These circumstances created conditions favorable to rapid diversification and speciation. But why these particular populations colonized the lake or island was largely a matter of blind luck. Sim-

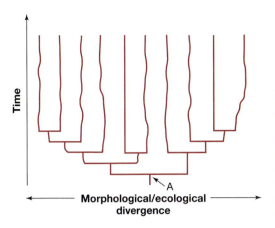

Time

← Morphological/ecological →
divergence

Figure 12.33 **Adaptive radiation** This diagram shows the branching pattern produced by a hypothetical adaptive radiation. Time is plotted on the vertical axis, and morphological or ecological differentiation on the horizontal axis. Note that the initial radiation is rapid, producing lineages with widely divergent forms. When phylogenetic trees are estimated for lineages that have undergone adaptive radiation, the early branches on the tree are often extremely short and later branches much longer. The pattern is "bushy" or "stemmy." The event that triggered the radiation occured at node A.

ilarly, in Chapter 13 we will examine the aftermath of a mass extinction at the end of the Cretaceous period. Mammals diversified rapidly after the dinosaurs went extinct. The leading hypothesis for why they did so was because they lacked competitors, not because of superior adaptations.

Finally, many adaptive radiations are correlated with morphological innovations. The diversification of arthropods is a prime example. The variety of ecological niches occupied by members of the phylum is remarkable, and closely associated with modifications and elaborations of their jointed limbs.

Throughout the chapter we have emphasized that many of the great events in evolutionary history have, like adaptive radiations, been rapid. This would seem to conflict with the classical view of Darwinism presented in Chapter 2: that gradual change occurs in populations through time, based on slight variations favored by natural selection. Is there a conflict here? If so, how do we resolve it? This is the issue we take up next.

Punctuated Equilibrium

Perhaps the most prominent pattern in the history of life is that new morphospecies appear in the fossil record suddenly. Even major morphological innovations seem to arise instantaneously in geological time. Transitional forms do occur, as we have seen, but they are relatively rare.

Darwin (1859) was well aware of this fact and considered it a problem for his theory. Because his ideas were presented in opposition to the Theory of Special Creation, which predicts the instantaneous creation of new forms, Darwin repeatedly emphasized the gradual nature of evolution by natural selection. He attributed the sudden appearance of new taxa to the incompleteness of the fossil record, and predicted that as specimen collections grew, the sudden jumps that were observed would be filled in by forms showing gradual transitions between morphospecies. For a century thereafter, most paleontologists followed his lead.

In 1972, however, Niles Eldredge and Stephen Jay Gould published an influential paper that accepted the sudden appearance of new forms as fact, and proposed an explanation. They called their proposal the theory of punctuated equilibrium. Like most scientific theories, punctuated equilibrium consists of two components: a pattern, and a process that explains the pattern. As originally formulated, the theory made two major claims:

- The dominant pattern in the history of life is that rapid morphological change occurs during the formation of new species. Rapid change at speciation is then followed by long periods of no change, or stasis, in that lineage (Figure 12.34). As a result, the vast majority of evolutionary change occurs during speciation events.

- This pattern results from the process of peripatric speciation.

At the time, peripatric speciation was the leading hypothesis for how speciation occurs. According to a model formulated by Ernst Mayr (1963), the vast majority of speciation events begin when a small population becomes isolated on the periphery of a species' range. Mayr proposed that genetic drift would lead to radical changes in allele frequencies in the small population, and lead to what he termed a genetic revolution. As we reviewed in Chapter 9, the hypothesis was that a drift-induced reorganization of gene frequencies would produce rapid genetic change, and lead to reproductive isolation if the ancestral and derived populations later came into contact.

If a small population with a limited range changes rapidly, it is unlikely that any of the transitional forms will fossilize. Only later, if the new species expands in numbers and endures through time, is it probable that fossil traces will be left. Because of this, the Mayr model seemed a perfect fit with the observation that new morphospecies appear suddenly in the fossil record. The process component of punctuated equilibrium has been controversial, however, because population genetic models have shown that genetic revolutions due to drift are extremely unlikely (Lande 1980; Barton and Charlesworth 1984).

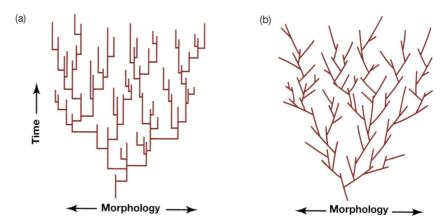

Figure 12.34 Patterns of morphological change under punctuated equilibrium and phyletic gradualism   When time is plotted against morphology, two extreme patterns are possible along with many intermediate or mixed patterns. Diagram (a) shows the extreme called punctuated equilibrium: All morphological variation occurs at the speciation (branching) event, with no change occurring for the rest of each species' life span. Diagram (b) shows the other extreme, called phyletic gradualism: All morphological change occurs steadily and gradually and is completely unrelated to speciation events. Eldredge and Gould (1972) said the punctuated pattern dominates the history of life; Darwin (1859; see his Figure 1) said that the gradual pattern does.

Partly because of this critique, Gould and Eldredge (1993) reformulated the theory of punctuated equilibrium. They noted that Mayr's hypothesis is closely tied to Sewall Wright's (1932; 1988) concept of the adaptive landscape. We introduced adaptive landscapes in Chapter 5 by examining the frequencies of hemoglobin alleles in humans, in environments where malaria is present. In Figure 5.33, we plotted the average fitness of a population as a function of the allele frequencies present, and showed that several adaptive peaks existed. Different combinations of hemoglobin alleles put the population on different adaptive peaks. These peaks are separated by troughs of lower average fitness. We also showed how drift or gene flow could allow a population to cross a trough and shift from one adaptive peak to another.

Wright's metaphor can be extended to consider the entire genome instead of just a single locus. In this version, each genotype represents a point on the fitness landscape. Because populations contain distinct combinations of genotypes, they occupy dissimilar regions of the fitness landscape. Populations can, in fact, occupy discrete fitness peaks. One population might occupy a fitness peak defined by a warmer, dryer habitat; another might occupy a peak defined by a cooler, moister habitat. Over time, gene flow occurs between these populations, creating intermediate genotypes. This keeps the populations from differentiating.

Douglas Futuyma (1989), however, pointed out that these populations on different fitness peaks become isolated when speciation occurs. This keeps the populations from fusing over time and creates an opportunity for rapid divergence. In Futuyma's view, a speciation event "captures" variation that exists among populations within species. This is variation that otherwise would have been homogenized over time by gene flow, and lost. Gould and Eldredge (1993) have proposed that Futuyma's mechanism may be responsible for the pattern component of punctuated equilibrium.

This brings us back to the question that motivated Eldredge and Gould's original paper: Do the data support the claim that punctuation is the predominant pattern of change through time?

Before looking at some tests of the theory, we need to clarify several points. The first concerns time and scales of observation. Biologists and paleontologists often talk past one another when they talk about time. The fossil record rarely allows researchers to resolve events that occurred on time scales much less than a million years in length. When a researcher is studying the history of a lineage over 10 or 20 millions years, as paleontologists routinely do, a couple hundred thousand years is trivial. Consequently, paleontologists say that speciation occurs "instantaneously" in geological time (see Gould 1982). But to a biologist measuring selection on an extant population, a long-term study lasts a decade or less. From this perspective, a hundred thousand years is not instantaneous—it is overwhelming.

Furthermore, different patterns of change can result from measuring evolution on different time scales. As an example, think back to the long-term study Richard Lenski and co-workers did with *E. coli,* introduced in Chapter 5. In that experiment, cell size and relative fitness both increased rapidly as favorable mutations occurred and swept through the populations. Figure 12.35 shows two graphs that Lenski's group generated from exactly the same set of data. In one, the data are plotted every 100 generations, and the pattern of change looks

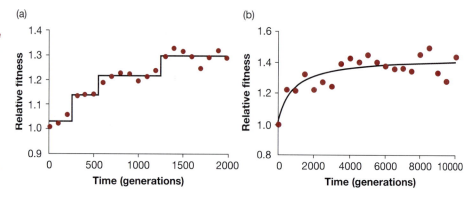

Figure 12.35 **Gradual and punctuated patterns from the same dataset** The data points show the fitness of descendant versus ancestral populations of *E. coli* during evolution in glucose-limited medium (Lenski and Travisano 1994). The solid lines describe the function that best fits the data when points are plotted every 100 generations (a) or 500 generations (b).

stepped. In the other, the data are plotted at intervals of 500 generations, and the pattern looks smooth. The message here is that we need to be cautious about interpreting patterns sensitive to this scaling effect. In the case of distinguishing punctuated (stepped) from gradualist (smooth) patterns in the fossil record, the finer the time scale, the better.

The second point of clarification concerns rules for testing the theory. Remember the goal: We want to follow changes in morphology in speciating clades through time, and determine (1) whether change occurs in conjunction with speciation events or independently, and (2) whether rapid change is followed by stasis or continued change. As critics of the theory have emphasized, testing for punctuated vs. gradualist patterns rigorously is exceptionally difficult. This is because the theory of punctuated equilibrium can become tautological if we are not careful. We define fossil species on the basis of morphology, so it might be trivial to observe a strong correlation between speciation and morphological change. To avoid circularity, an acceptable test requires that

- The phylogeny of the clade is known, so we can identify which species are ancestral and which descendant.
- Ancestral species survive long enough to co-occur with the new species in the fossil record.

The second criterion is critical. If it is not fulfilled, it is impossible to know whether the new morphospecies is indeed a product of a splitting event, or whether it is the result of rapid evolution in the ancestral form without speciation taking place. This second possibility is called phyletic transformation, or anagenesis.

These are demanding criteria, especially when compounded with other difficult practical issues: the problem of misidentifying cryptic species in the fossil record, the need for analyzing change at the level of species, the requirement of sampling at frequent time intervals, and the necessity of sampling multiple localities to distinguish normal, within-species geographic variation from authentically different morphospecies.

There are relatively few fossil series available, anywhere in the world, that meet these stringent requirements. Several do, however. We next look in detail at two studies that tested the theory of punctuated equilibrium rigorously. After reviewing these, we will close our look at the theory by trying to reach a tentative conclusion about which of the two patterns predominates.

Gradual Phyletic Change in Marine Plankton

Fossils from microscopic marine organisms called foraminifera offer some of the best datasets available for testing the theory of punctuated equilibrium. Foraminiferans are planktonic (surface-dwelling) or benthic (bottom-dwelling) protozoans, many of which have hard, calcareous shells that fossilize readily. Because they settle to the ocean floor after death, they leave a more or less continuous record that can either be recovered from sediment cores or sampled in marine strata that have been uplifted and exposed.

Richard Hunter and colleagues (1988) studied eight million years of morphologic change and speciation in a series of foram species extracted from cores drilled in the Gulf of Mexico. Starting in the Paleocene, about 66 Ma, the cores provide an almost continuous record of evolution in this lineage. Four well-defined morphospecies occur sequentially, with ancestral forms co-occurring with descendants in every case (Figure 12.36a).

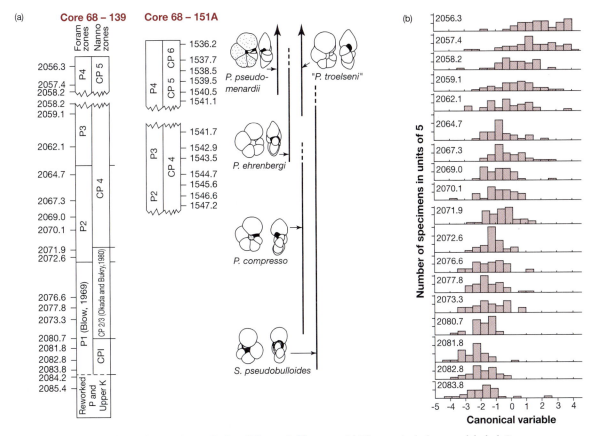

Figure 12.36 Gradual change in a clade of foraminiferans (a) The vertical elements labeled Core represent the sediment cores analyzed by Hunter et al. (1988). Strata at the bottom are the oldest. The diagram at the right shows the four species that were analyzed in the study, aligned with the strata where they were found. Note that *S. pseudobulloides* coexists with all the of the descendant species.   (b) These histograms show the distributions of canonical scores. These data summarize the amount of morphological change in forams sampled at various strata in the sediment cores.

Hunter's group measured 11 characters on each fossil shell, then used a statistical technique called discriminant function analysis to measure the overall morphological difference between the species that started and ended the interval they sampled (the species called *Subbotina pseudobulloides* and *Planorotalites pseudomenardii* in Figure 12.36a). Then for each shell sampled and measured in the strata between these two species, the researchers computed how different it was, in discriminant function, from the values for the ancestral and descendent species. This distance is called a canonical score. When canonical scores are plotted through time (as you move up the core) a strong pattern of continuous, gradual change results (Figure 12.36b). There are no sharp breaks at species boundaries, nor are there periods of stasis. This means that this lineage of foraminiferans evolved through phyletic gradualism and not punctuated equilibrium.

Studies that have looked at change through time in forams from other localities or periods have generally found the same pattern, whether the investigators looked at single traits at a time or a large suite of characteristics, as in the Hunter et al. study. Gradualism seems to dominate in these taxa.

Punctuated Change in Bryozoans

In Chapter 9 we introduced the phylum of marine invertebrates called the Bryozoa. Our focus then was on experiments performed by Jeremy Jackson and Alan Cheetham (1990, 1994a) on a subgroup called the cheilostome bryozoans. Their research showed that the criteria used to differentiate fossil morphospecies also work for genetically differentiated species alive today. This phylum is a good group for testing the theory of punctuated equilibrium, for two reasons: We are confident that species designations can be made rigorously, and they are abundant in the fossil record of the past 100 million years.

Cheetham (1986) and Jackson and Cheetham (1994b) performed a high-resolution analysis of speciation and morphologic change in cheilostomes from the Caribbean, starting in the Miocene and ending with living taxa. They began the study by defining 19 morphospecies in the genus *Stylopoma*, based on a discriminant function analysis of 15 skeletal characters. Seven of these species still have living representatives, so they could check the validity of the morphospecies designations. Recall that Jackson and Cheetham did this by surveying allozymes in the living species and confirming that almost all morphospecies pairs were distinguished either by unique alleles or by significant differences in the frequencies of shared alleles. Then they estimated the phylogeny of the 19 morphospecies from differences in skeletal characters, and scaled the tree so that the branch points and branch tips lined up with the dates of first and last appearance for fossil forms. They did a similar analysis for 19 living and extinct morphospecies in the genus *Metrarabdotos*.

The trees generated by the study are pictured in Figure 12.37. In both genera, species clearly appear in a punctuated mode. The phylogenies show an unequivocal pattern of rapid morphological change, followed by stasis. The fact that ancestral and descendant species co-occur defends the idea that morphologic change was strongly associated with speciation events. This is an almost flawless example of punctuated equilibrium.

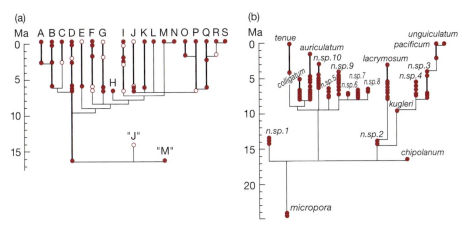

Figure 12.37 **Punctuated change in cheilostome Bryozoa** These phylogenies, for 19 living and fossil morphospecies in genus *Stylopoma* (a) and 19 living and fossil morphospecies in the genus *Metrarabdotos* (b), were estimated from differences in skeletal characters (Jackson and Cheetham 1994). Each dot indicates a population that was sampled. None of the populations in the study showed skeletal traits that were intermediate between species, and the characteristics of species were stable through time. As a result, the pattern of change is strongly punctuated.

What Is the Relative Frequency of Gradualist and Punctuated Patterns?

As many participants in the punctuated equilibrium debate have recognized, the argument is really about relative frequency. No one denies that both patterns have occurred in the history of life. The question is, which one predominates?

Doug Erwin and Robert Anstey (1995a, b) recently reviewed a total of 58 studies conducted to test the theory of punctuated equilibrium. The studies represent a wide variety of taxa and time periods. Although they varied in their ability to meet the strict criteria we listed, their sheer number may compensate somewhat. Erwin and Anstey's conclusion? "Paleontological evidence overwhelmingly supports a view that speciation is sometimes gradual and sometimes punctuated, and that no one mode characterizes this very complicated process in the history of life." Further, Erwin and Anstey noted that a quarter of the studies reported a third pattern: gradualism and stasis.

Although partisans in the debate might be disappointed in this conclusion, it does suggest new avenues for research. For example, Is it possible that different types of organisms exhibit different patterns of change? Hunter et al. (1988) suggested that a "tentative consensus" is emerging among researchers who have worked on the problem: that gradualist patterns tend to predominate in forams, radiolarians, and other microscopic marine forms, while punctuated patterns occur more often in macroscopic fossils such as marine arthropods, bivalves, corals, and bryozoans. If so, why? Research continues.

Why Does Stasis Occur?

One of Eldredge and Gould's prominent claims about the fossil record is that "stasis is data." That is, lack of change is a pattern that needs to be explained. We have run into several examples of stasis in the course of examining other issues in the text. Jackson and Cheetham's study of bryozoans, which showed that virtually no change occurred in these sessile invertebrates over millions of years, comes to mind. The long-term laboratory experiment showing dramatic change in the cell size of *E. coli* over time also demonstrates stasis (Lenski and Travisano 1994; Elena et al. 1996). In these experimental populations, both cell size and relative fitness stopped changing after about 1,500 generations (Figure 12.37). The question is, why?

To organize our thinking about this issue, we return to the metaphor of the adaptive landscape. In this context there are two general explanations for stasis: Either the peak that a population occupies does not change over time, or the population simply cannot get off an existing peak.

Populations can remain on the same peak through time as a result of stabilizing selection. Genotypes that are off the peak experience lower relative fitness and tend to be eliminated by selection. A famous example is human birthweight. Very small and very large babies survive significantly less well than do average-sized babies (Figure 12.38). As a result, we expect little or no change to occur in the average birthweight of humans over time. Populations are sitting on a fitness peak, and stasis results.

But suppose selection suddenly favored larger or smaller babies, meaning that a new fitness peak has emerged nearby. Would human populations move to it? The answer is not necessarily. Evolution does not always occur, even when it "should." We have already discussed several examples of this phenomenon. In Chapter 2 we saw that the width of finch beaks on the Galápagos did not change over time, even though narrow beaks worked better at cracking hard seeds. This lack of change was due to positive genetic correlations with a related character, beak depth, being selected in the opposite direction. In Chapter 5 we saw that the recessive allele for hemoglobin, called C, has not increased in human populations, even though it should be favored in malarial environments. Finally, in some instances there may simply not be enough genetic variation in a population to generate genotypes that cross the fitness trough and end up on the adjacent, higher peak.

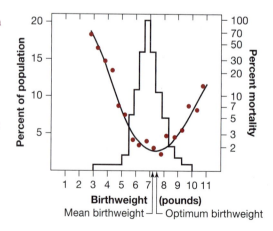

Figure 12.38 **Stabilizing selection on human birthweight** The histogram shows the distribution of birthweights in humans, graphed as the percentage of the population of newborns. The curve that is superimposed shows the percent mortality as a function of birthweight. Notice that very light and very heavy newborns suffer high mortality. Also note that the mean birthweight in the histogram is almost exactly the same as the optimum calculated from the mortality data.

Figure 12.39 **"Living fossils"** Here are just a few of numerous examples of contemporary and fossil species that are extremely similar. (a) Contemporary stromatolite forming bacteria from Australia (top: Francois Gohier/Photo Researchers, Inc.) and 1.8 Ba fossil forms (bottom: Biological Photo Service) from the Great Slave Lake area of Canada. (b) Leaves from a living gingko tree (top: Hugh Spencer/Photo Researchers, Inc.) and 40 Ma impression fossils made by gingko leaves (bottom: Sinclair Stammers/Science Photo Library/Photo Researchers, Inc.). We put the term "living fossils" in quotes because it is an oxymoron.

Listing possible explanations like this is an important first step in any scientific analysis, but what we really want to do is perform some tests and discover which possibilities actually occur in nature. To date, studies of stasis have concentrated on the so-called living fossils. These are species or clades showing little or no measurable morphological change over extended periods (Figure 12.39). Horseshoe crabs are a spectacular example. The extant species, in the genus *Limulus,* are virtually identical to fossil species in a different family that existed 150 Ma. While these horseshoe crab lineages stayed virtually unchanged, the entire radiation of birds, mammals, and flowering plants took place.

Have these species failed to change simply because they lack genetic variation? John Avise and colleagues (1994) answered this question by sequencing several genes in the mitochondrial DNA of horseshoe crabs, and comparing the amount of genetic divergence they found to a previously published study of genetic distances in another arthropod clade: the king and hermit crabs

(a)

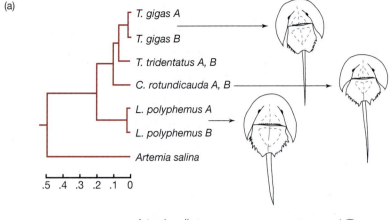

(b)

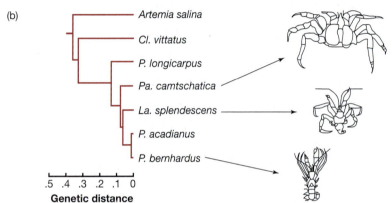

Genetic distance

Figure 12.40 **Genetic and morphological change in two arthropod clades** The lengths of each branch on these phylogenies represents a genetic distance, measured as the percent difference in 16S rRNA sequences in mtDNA. The scale is the same for both trees. The fairy shrimp *Artemia salina* was used as the outgroup to root each of these trees. Slightly more genetic diverence in 16S rRNA sequences has occured in the horseshoe crab clade (a), even though much more morphological divergence has occured in the clade that includes hermit crabs and allies (b). From Avise et al. (1994).

(Cunningham et al. 1992). The result is striking: The horseshoe crabs show just as much genetic divergence as the king-hermit crab clade, even though far less morphological change has occurred (Figure 12.40). This is strong evidence that stasis is not from a lack of genetic variability.

What about some of the other possibilities? Steve Stanley and Xiangning Yang (1987) looked at bivalve species that have shown remarkably little change over the past 15 million years, and discovered an interesting pattern. When they mapped change in 24 different shell characters over this interval, they found that even though most showed little net change within species, many had undergone large fluctuations, or what Stanley and Yang called "zigzag evolution" (Figure 12.41). These clam populations probably tracked changes in the fitness surface over time. But because the fitness peak tended to fluctuate about a mean value, we perceive stasis as a result. This phenomenon has also been called habitat tracking, or dynamic stasis.

As a consequence of studies like these, the current view is that there is no single and general explanation for the low rates of morphological change that occur in particular lineages. Stasis is best tested and explained case by case (see Eldredge 1984; Stanley 1984; Thomson 1986).

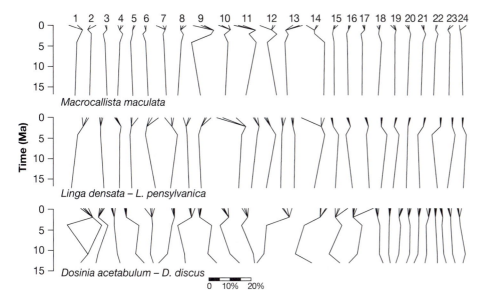

Figure 12.41 **Zigzag evolution in Pliocene bivalves results in stasis** This diagram shows how 24 different morphological characters change through time in three lineages. The horizontal axis plots percent change in shell morphology between each time interval sampled in the study. Several populations are plotted for the Recent. From Stanley and Yang (1987).

Summary

The most efficient fossilization processes are compression, impression, casting, and permineralization. Because these events depend on the rapid burial of organic remains in water-saturated sediments, the fossil record is dominated by organisms with hard parts that lived in lowland or shallow-water marine environments. Thanks to new fossil finds and increasingly high-resolution dating techniques, the geological record of life on earth is steadily improving.

The Cambrian explosion was characterized by the rapid evolution of diverse metazoan body plans and the filling of ecological niches in shallow-water marine communities. The event is associated with the elaboration of homeotic gene complexes. Changes in the timing and spatial expression of loci in the *HOM/Hox* complexes and other genes may have been important in the elaboration of the tetrapod and arthropod limbs. Preliminary evidence suggests that the diversification of angiosperms may also be associated with the origin of homeotic loci. Several genes involved in the elaboration of flower symmetry and the morphology of inflorescences have been identified.

Even though new morphological traits often evolve too rapidly to leave a detailed fossil record, researchers have been able to reconstruct intermediate stages in the evolution of insect and bird flight. Insect wings probably evolved in response to selection for efficient sailing across water surfaces. Bird feathers most likely evolved for their insulation value in dinosaurs, while flapping flight in birds probably started as an elaboration of gliding from trees.

The search for macroevolutionary patterns is a prominent theme in paleobiology. One classical pattern is called adaptive radiation. This is the rapid diversification of a clade into a wide variety of niches. Adaptive radiations can be triggered by ecological opportunities, key adaptations, or chance events like colonizing a new habitat or surviving a mass extinction. Punctuated equilibrium is another prominent pattern in evolutionary history. It is characterized by rapid morphological change at speciation, followed by extended periods of stasis. Stasis is widespread in the fossil record and can result from stabilizing selection, habitat tracking (zig-zag evolution), or lack of genetic variation. Punctuated, gradualist, and intermediate patterns have all been documented in the fossil record.

Questions

1. Terrestrial fossils from a particular time (say, 230 million years ago) are patchily distributed. Instead of being evenly distributed over the continents in a continuous thin layer, they often occur in narrow strips or pockets a few miles wide. Why is this?

2. Most fossils of Mesozoic birds are from marine diving birds. Relatively few terrestrial species are known. Does this mean that most Mesozoic birds were, in fact, marine diving birds? Explain your reasoning.

3. Recall that Wray et al. (1996) concluded from molecular data that the Metazoan phyla diverged hundreds of millions of years before the Cambrian explosion. If this analysis is correct, does it tell us anything about when homeotic genes first arose, or when the Metazoan phyla acquired their classic body plans?

4. Listed below are some hypothetical fossils. For each one, describe whether discovery of such a fossil would strengthen or weaken the popular hypothesis that birds are derived from a theropod dinosaur, and why.

   a bird-like theropod that pre-dates *Archaeopteryx* by 10 million years
   a true bird that pre-dates all dinosaurs by 50 million years
   an early reptile that is not a dinosaur and that has feathers

5. One way to examine the abilities of extinct animals is to develop models of them and then test the abilities of the models. We have seen an example of this in studies of insect flight. Researchers have also used computer simulations, scale models of extinct animals, and even remote-control robots. With these ideas in mind, outline some experiments to test the flying abilities of an extinct pterosaur (flying reptile) or the running abilities of a large sauropod dinosaur.

6. A common phenomenon in the fossil record is the appearance of a new group which gradually becomes more common, until it has taken over the dominant role in its ecosystem from a previous group. Some classic examples are the brachiopods with hinged shells "taking over" a dominant role in the oceans from brachiopods with unhinged shells; angiosperm plants "taking over" the dominant land-plant role from gymnosperms; and diapsid reptiles (a large group including dinosaurs, lizards, and birds) "taking over" the dominant large-vertebrate role from the synapsids (the lineage that, much later, produced the mammals). In these cases, it is tempting to conclude that the new lineage was competitively superior to the old lineage. Why might this assumption be wrong? That is, what are some other explanations that might account for the decrease of an old group and the near-simultaneous increase of a new group?

7. Now let's look at one of these "takeovers" in a little more detail. The rise of angiosperms over gymnosperms is widely thought to be due to the competitive superiority of angiosperms. However, just what gave the angiosperms the edge has been debated. Some of the hypotheses are: 1) angiosperms have a better method of pollination; 2) angiosperms have a better method of water transport; and 3) angiosperms have more efficient leaves and can grow faster. Are any of these theories testable? Remember that both angiosperms and gymnosperms are still alive today, and thus are available for experiments.

8. Figure 12.35 illustrates changes in the pattern of evolution that are solely a product

of scale of measurement. The text says that testing the theory of punctuated equilibrium, the finer the time scale the better. Suppose a study plotted morphological change in a lineage on two scales: every million years and every 10,000 years. Which time scale would be more likely to show punctuated change, and which would be more likely to show gradual change? Explain.

Exploring the Literature

9. Devonian tetrapods had different numbers of digits on their fore- and hindlimbs. Most later tetrapods have five digits on both limbs. To investigate the evolution of the pentadactyl limb, see:

Coates, M.I., and J.A. Clack. 1990. Polydactyly in the earliest known tetrapod limbs. *Nature* 347:66–69.

Gould, S.J. 1990. *Eight Little Piggies.* (New York: W.W. Norton), Essay 4.

10. The theory of punctuated equilibrium has been intensely controversial since it was proposed over 25 years ago. To begin a study of how scientific debates are conducted and how theories evolve over time, see:

Eldredge, N., S.J. Gould, J.A. Coyne, and B. Charlesworth. 1997. On punctuated equilibria. *Science* 276:338–341.

Citations

Ahlberg, P.E. 1995. *Elginerpeton panchei* and the earliest tetrapod clade. *Nature* 373:420–425.

Ahlberg, P.E., and A.R. Milner. 1994. The origin and early diversification of tetrapods. *Nature* 368:507–518.

Avise, J.C., W.S. Nelson, and H. Sugita. 1994. A speciational history of "living fossils": Molecular evolutionary patterns in horseshoe crabs. *Evolution* 48:1986–2001.

Balavoine, G., and M. J. Telford. 1995. Identification of planarian homeobox sequences indicates the antiquity of most *Hox*/homeotic gene subclasses. *Proceedings of the National Academy of Sciences, USA* 92:7227–7231.

Barton, N.H., and B. Charlesworth. 1984. Genetic revolutions, founder effects, and speciation. *Annual Review of Ecology and Systematics* 15:133–164.

Bengtson, S., and Y. Zhao. 1992. Predatorial borings in Late Precambrian mineralized exoskeletons. *Science* 257:367–369.

Briggs, D.E.G. 1991. Extraordinary fossils. *American Scientist* 79:130–141.

Briggs, D.E.G., R.A. Fortey, and M.A. Wills. 1992. Morphological disparity in the Cambrian. *Science* 256:1670–1673.

Browne, M. 1996. Feathery fossil hints dinosaur-bird link. *New York Times,* October 19, page 1.

Brusca, R.C., and G.J. Brusca. 1990. *Invertebrates* (Sunderland, MA: Sinauer).

Budd, G.E. 1996. Progress and problems in arthropod phylogeny. *Trends in Ecology and Evolution* 11:356–358.

Carroll, R.L. 1988. *Vertebrate Paleontology and Evolution* (New York: W.H. Freeman).

Carroll, S.B. 1995. Homeotic genes and the evolution of the chordates. *Nature* 376:479–485.

Cheetham, A.H. 1986. Tempo of evolution in a Neogene bryozoan: rates of morphologic change within and across species boundaries. *Paleobiology* 12:190–202.

Coen, E.S., and J.M. Nugent. 1994. Evolution of flowers and inflorescences. *Development Supplement:*107–116.

Cohn, M.J., J.C. Izpisúa-Belmonte, H. Abud, J.K. Heath, and C. Tickle. 1995. Fibroblast growth factors induce additional limb development from the flank of chick embryos. *Cell* 80:739–746.

Conway Morris, S. 1989. Burgess shale faunas and the Cambrian explosion. *Science* 246:339–346.

Conway Morris, S. 1993. The fossil record and the early evolution of the Metazoa. *Nature* 361:219–225.

Cunningham, C.W., N.W. Blackstone, and L.W. Buss. 1992. Evolution of king crabs from hermit crab ancestors. *Nature* 355:539–542.

Darwin, C. 1859. *The Origin of Species* (London: John Murray).

Duboule, D. 1994. How to make a limb. *Science* 266:575–576.

Edwards, J.L. 1989. Two perspectives on the evolution of the tetrapod limb. *American Zoologist* 29:235–254.

Eldredge, N. 1984. Simpson's inverse: Bradytely and the phenomenon of living fossils. In N. Eldredge and S.M. Stanley, eds. *Living Fossils* (Berlin: Springer-Verlag), 272–277.

Eldredge, N., and S.J. Gould. 1972. Punctuated equilibria: An alternative to phyletic gradualism. In T.J.M. Schopf, ed. *Models in Paleobiology* (San Francisco: Freeman, Cooper & Company), 82–115.

Elena, S.F., V.S. Cooper, and R.E. Lenski. 1996. Punctuated evolution caused by selection of rare beneficial mutations. *Science* 272:1802–1804.

Erwin, D.H. 1996. The geologic history of diversity. In R.C.

Szaro and D.W. Johnston, eds. *Biodiversity in managed landscapes* (New York: Oxford University Press), 3–16.

Erwin, D.H., and R.L. Anstey. 1995a. Introduction. In D.H. Erwin and R.L. Anstey, eds. *New Approaches to Speciation in the Fossil Record* (New York: Columbia University Press), 1–8.

Erwin, D.H., and R.L. Anstey. 1995b. Speciation in the fossil record. In D.H. Erwin and R.L. Anstey, eds. *New Approaches to Speciation in the Fossil Record* (New York: Columbia University Press), 11–38.

Erwin, D., J.Valentine, and D. Jablonski. 1997. The origin of animal body plans. *American Scientist* 85:126–137.

Fallon, J.F., A. Lopez, M.A. Ros, M.P. Savage, B.B. Olwin, and B.K. Simandi. 1995. FGF-2: Apical ectodermal ridge growth signal for chick limb development. *Science* 264:104–106.

Feduccia, A. 1993. Evidence from claw geometry indicating arboreal habits of *Archaeopteryx*. *Science* 259:790–793.

Futuyma, D.J. 1989. Macroevolutionary consequences of speciation: Inferences from phytophagous insects. In D. Otte and J.A. Endler. *Speciation and its Consequences* (Sunderland, MA: Sinauer), 557–578.

Gibson, G., and D.S. Hogness. 1996. Effect of polymorphism in the *Drosophila* regulatory gene *Ultrabithorax* on homeotic stability. *Science* 271:200–203.

Gilbert, S.F. 1991. *Developmental Biology* (Sunderland, MA: Sinauer).

Gould, S.J. 1982. The meaning of punctuated equilibrium and its role in validating a hierarchical approach to macroevolution. In R. Milkman, ed. *Perspectives on Evolution* (Sunderland, MA: Sinauer), 83–104.

Gould, S.J. 1989. *Wonderful Life* (New York: W.W. Norton).

Gould, S.J., and N. Eldredge. 1993. Punctuated equilibrium comes of age. *Nature* 366:223–227.

Halanych, K.M., J.D. Bacheller, A.M.A. Aguinaldo, S.M. Liva, D.M. Hillis, and J.A. Lake. 1995. Evidence from 18S ribosomal DNA that the lophophorates are protostome animals. *Science* 267:1641–1643.

Hammond, P.M. 1992. Species inventory. In B. Groombridge, ed. *Global Biodiversity* (London: Chapman and Hall), 17–39.

Harland, W.B., R.L. Armstrong, A.V. Cox, L.E. Craig, A.G. Smith, and D.G. Smith. *A Geologic Time Scale 1989.* (Cambridge: Cambridge University Press).

Hinchcliffe, J.R. 1989. Reconstructing the archetype: Innovation and conservatism in the evolution and development of the pentadactyl limb. In D.B. Wake and G. Roth, eds., *Complex Organismal Functions: Integration and Evolution in Vertebrates* (Chichester, England: Wiley & Sons), 171–189.

Hinchcliffe, J.R., and D.R. Johnson. 1980. *The Development of the Vertebrate Limb* (Oxford: Clarendon Press).

Hunter, R.S.T., A.J. Arnold, and W.C. Parker. 1988. Evolu-

tion and homeomorphy in the development of the Paleocene *Planorotalites pseudomenardii* and the Miocene *Globorotalia (Glororotalia) maragritae* lineages. *Micropaleontology* 34:181–192.

Irving, E. 1977. Drift of the major continental blocks since the Devonian. *Nature* 270:304–309.

Jackson, J.B.C., and A.H. Cheetham. 1990. Evolutionary significance of morphospecies: A test with cheilostome bryozoa. *Science* 248:579–583.

Jackson, J.B.C., and A.H. Cheetham. 1994a. Phylogeny reconstruction and the tempo of speciation in the cheilostome Bryozoa. *Paleobiology* 20:407–423.

Jackson, J.B.C., and A.H. Cheetham. 1994b. Phylogeny reconstruction and the tempo of speciation in the cheilostome Bryozoa. *Paleobiology* 20:407–423.

Jarvik, E. 1980. *Basic Structure and Evolution of Vertebrates,* vol. 2 (London: Academic Press).

Kenyon, C., and B. Wang. 1991. A cluster of Antennapedia-class homeobox genes in a nonsegmented animal. *Science* 253:516–517.

Kingsolver, J.G., and M.A.R. Koehl. 1985. Aerodynamics, thermoregulation, and the evolution of insect wings: Differential scaling and evolutionary change. *Evolution* 39:488–504.

Krumlauf, R. 1994. *Hox* genes in vertebrate development. *Cell* 78:191–201.

Kukalova-Peck, J. 1978. Origin and evolution of insect wings and their relation to metamorphosis, as documented by the fossil record. *Journal of Morphology* 156:53–126.

Lande, R. 1980. Genetic variation and phenotypic evolution during allopatric speciation. *American Naturalist* 116:463–479.

Lenski, R.E., and M. Travisano. 1994. Dynamics of adaptation and diversification: A 10,000-generation experiment with bacterial populations. *Proceedings of the National Academy of Sciences, USA* 91:6808–6814.

Lewis, E.B. 1978. A gene complex controlling segmentation in *Drosophila*. *Nature* 276:565–570.

Li, H., E.L. Taylor, and T.N. Taylor. 1996. Permian vessel elements. *Science* 271:188–189.

Marden, J.H., and M.G. Kramer. 1994. Surface-skimming stoneflies: A possible intermediate stage in insect flight evolution. *Science* 266:427–430.

Marden, J.H., and M.G. Kramer. 1995. Locomotor performance of insects with rudimentary wings. *Nature* 377:332–334.

Martin, G.R. 1995. Why thumbs are up. *Nature* 374:410–411.

Mayr, E. 1963. *Animal Species and Evolution* (Cambridge, MA: Harvard University Press).

McGinnis, W., M.S. Levine, E. Hafen, A. Kuroiwa, and W.J. Gehring. 1984a. A conserved DNA sequence in homoeotic genes of the *Drosophila* Antennapedia and bithorax complexes. *Nature* 308:428–433.

McGinnis, W., R.L. Garber, J. Wirz, A. Kuroiwa, and W.J. Gehring. 1984b. A homologous protein-coding sequence in *Drosophila* homeotic genes and its conservation in other metazoans. *Cell* 37:403–408.

McGinnis, W., and M. Kuziora. 1994. The molecular architects of body design. *Scientific American* 270:58–66.

Monastersky, R. 1996. Hints of a downy dinosaur in China. *Science News* 150:260.

Nelson, C.E., and C. Tabin. 1995. Footnote on limb evolution. *Nature* 375:630–631.

Niklas, K.J. 1994. One giant step for life. *Natural History* 6:22–25.

Niklas, K.J., B.H. Tiffney, and A.H. Knoll. 1983. Patterns in vascular land plant diversification. *Nature* 303:614–616.

Norberg, U.M. 1985. Evolution of vertebrate flight: An aerodynamic model for the transition from gliding to active flight. *American Naturalist* 126:303–327.

Norberg, U.M. 1990. *Vertebrate Flight* (Berlin: Springer-Verlag).

Panganiban, G., A. Sebring, L. Nagy, and S. Carroll. 1995. The development of crustacean limbs and the evolution of arthropods. *Science* 270:1363–1366.

Parr, B.A., and A.P. McMahon. 1995. Dorsalizing signal *Wnt-7a* required for normal polarity of D-V and A-P axes of mouse limb. *Nature* 374:350–353.

Raff, R.A. 1996. *The Shape of Life* (Chicago: University of Chicago Press).

Rayner, J.M.V. 1989. Vertebrate flight and the origin of flying vertebrates. In K.C. Allen and D.E.G. Briggs, eds., *Evolution and the Fossil Record* (London: Bellhaven Press), 188–217.

Riddle, R.D., R.L. Johnson, E. Laufer, and C. Tabin. 1993. *Sonic hedgehog* mediates the polarizing activity of the ZPA. *Cell* 75:1401–1416.

Roush, W. 1995. Gene ties arthropods together. *Science* 270:1297.

Ruben, J.A., W.J. Hillenius, N.R. Geist, A. Leitch, T.D. Jones, P.J. Currie, J.R. Horner, and G. Espe III. 1996. The metabolic status of some late Cretaceous dinosaurs. *Science* 273:1204–1207.

Shubin, N.H., and P. Alberch. 1986. A morphogenetic approach to the origin and basic organization of the tetrapod limb. In M.K. Hecht, B. Wallace, and G.T. Prance eds., *Evolutionary Biology,* vol. 20. (New York: Plenum Press), 319–387.

Sordino, P., F. van der Hoeven, and D. Duboule. 1995. *Hox* gene expression in teleost fins and the origin of vertebrate digits. *Nature* 375:678–681.

Sordino, P., and D. Duboule. 1996. A molecular approach to the evolution of vertebrate paired appendages. *Trends in Ecology and Evolution* 11:114–119.

Stanley, S.M. 1984. Does bradytely exist? In N. Eldredge and S.M. Stanley, eds. *Living Fossils* (Berlin: Springer-Verlag), 278–280.

Stanley, S.M., and X. Yang. 1987. Approximate evolutionary stasis for bivalve morphology over millions of years: A multivariate, multilineage study. *Paleobiology* 13:113–139.

Stewart, W.N., and G.W. Rothwell. 1993. *Paleobotany and the Evolution of Plants* (Cambridge: Cambridge University Press).

Tabin, C. 1995. The initiation of the limb bud: growth factors, *Hox* genes, and retinoids. *Cell* 80:671–674.

Tautz, D. 1996. Selector genes, polymorphisms, and evolution. *Science* 271:160–161.

Taylor, D.W., and L.J. Hickey. 1990. An Aptian plant with attached leaves and flowers: Implications for angiosperm origin. *Science* 247:702–704.

Taylor, T.N., and E.L. Taylor. 1993. *The Biology and Evolution of Fossil Plants.* (Englewood Cliffs, NJ: Prentice Hall).

Taylor, E.L., T.N. Taylor, N.R. Cúneo. 1992. The present is not the key to the past: A polar forest from the Permian of Antarctica. *Science* 257:1675–1677.

Thomas, A.L.R., and R.A. Norberg. 1996. Skimming the surface—the origin of flight in insects? *Trends in Ecology and Evolution* 11:187–188.

Thomson, K.S. 1986. Living fossils. *Paleobiology* 12:495–498.

Valentine, J.W. 1994. The Cambrian explosion. In S. Bengtson, ed. *Early Life on Earth,* Nobel Symposium No. 84 (New York: Columbia University Press), 401–411.

Valentine, J.W., D.H. Erwin, and D. Jablonski. 1996. Developmental evolution of metazoan bodyplans: The fossil evidence. *Developmental Biology* 173:373–381.

Weigel, D., and E.M. Meyerowitz. 1994. The ABC's of floral homeotic genes. *Cell* 78:203–209.

Weiguo, S. 1994. Early multicellular fossils. In S. Bengtson, ed. *Early Life on Earth,* Nobel Symposium No. 84 (New York: Columbia University Press), 358–369.

Wray, G.A., J.S. Levinton, and L.H. Shapiro. 1996. Molecular evidence for deep Precambrian divergences among metazoan phyla. *Science* 274:568–573.

Wright, S. 1932. The roles of mutation, inbreeding, crossbreeding, and selection in evolution. *Proceedings of the 6th International Congress of Genetics* 1:356–366.

Wright, S. 1988. Surfaces of selective value revisited. *American Naturalist* 131:115–123.

Yang, Y., and L. Niswander. 1995. Interaction between the signaling molecules *WNT7a* and *SHH* during vertebrate limb development: dorsal signals regulate anteroposterior patterning. *Cell* 80:939–947.

Mass Extinctions and Their Consequences

In earlier chapters we reviewed the events and processes responsible for the rapid diversification of lineages. The African cichlids, Hawaiian flies, and Galápagos finches are all examples of a small population colonizing a new habitat and initiating a dramatic radiation. In the Cambrian explosion and the arthropod, angiosperm, insect, and bird radiations, extensive diversification was associated with one or more morphological innovations. Now we examine a new, equally important force for explaining the diversification of life through time: a phenomenon known as mass extinctions.

We start with an overview of background and mass extinction events during the Phanerozoic and proceed with three case studies, each illustrating a different mechanism responsible for a biological cataclysm. We close the chapter by asking whether a mass extinction event is now underway, and examining the role of chance in the evolution and history of life.

13.1 Background Versus Mass Extinction

Although research on extinction "is still at a reconnaissance level" (Raup 1994: 6758), several patterns are clear. Perhaps the most striking is that the global extinction rate has been anything but constant through time. Instead, there have been periods of particularly intense extinction. Consider a plot David Raup (1991, 1994) constructed by calculating, for each one-million-year interval over the last 543 million years, the percent of taxa that went extinct in that interval (Figure 13.1). The histogram has a pronounced right skew, created by a few particularly large events. The most extreme of these intense periods, called the Big Five, are commonly referred to as mass extinctions and are currently the focus of intense research. We are emphasizing them in this chapter because their consequences appear to be qualitatively different from the effects of "background

Figure 13.1 Distribution of extinction intensities during the Phanerozoic David Raup (1994) broke the 543-million-year fossil record for the Phanerozoic into one-million-year intervals, and calculated the percentage of existing species that went extinct during each of these intervals. The long right-hand tail of the distribution is created by intervals in which over 60% of the species alive went extinct in the span of a million years.

Many of the larger events plotted in this figure were recognized early in the 19th century and were used to define boundaries for the eras, periods, and epochs that make up the geologic time scale.

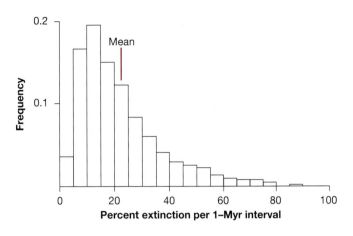

During background times, the probability that a taxon will go extinct at any time is constant . . .

extinctions." But as Raup notes, the combined species kill of the Big Five amounts to perhaps 4% of all extinctions recorded in the history of life.

What about the other extinction events in Raup's histogram? Background extinctions form the other 96% of events we would like to understand.

Several interesting patterns have been resolved from data on background extinctions. First, within any radiation, the likelihood of subclades going extinct is constant and independent of how long the taxa have been in existence. Leigh Van Valen (1973) discovered this when he plotted simple survivorship curves for a wide variety of fossil groups. Survivorship curves show the proportion of an original sample that survives for a particular amount of time. For fossil taxa, Van Valen plotted the number of species, genera, or families from an order or phylum of fossil animals that survived for different intervals. He put the number surviving on a logarithmic scale, so the slope of the curve at any point equaled the probability of going extinct at that time. Virtually every plot he constructed, from many different fossil groups and eras, produced a straight line. This means that the probability of subgroups going extinct was constant over the life span of the larger clade. The data in Figure 13.2 are typical. Note that the slopes of the lines vary from taxon to taxon, meaning that rates of extinction vary dramatically between lineages. In sum, during background times extinction rates are constant within clades but highly variable across clades.

Second, in marine organisms, extinction rates vary with how far larvae disperse after eggs are fertilized and begin development. David Jablonski (1986a) came to this conclusion by studying extinction patterns in bivalve (clams and mussels) and gastropod (slugs and snails) species from the Gulf of Mexico and

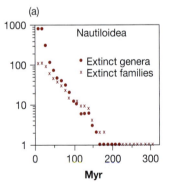

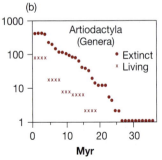

Figure 13.2 Survivorship curves The first step in constructing these curves is to select a random sample of taxa from the fossil record of a particular clade—say a family or class. The taxa included in the sample can come from any time period. The logarithm of the number of genera or families in the clade that survive for different intervals is then plotted. The curves reproduced here are typical (Van Valen 1973): (a) is for genera and families of fossil marine invertebrates called nautiloids, and (b) is for genera in the deer family.

Figure 13.3 The duration of marine bivalve species depends on larval life-styles
Planktrotrophs are species with larvae that spend at least some time floating in the plankton. Nonplanktotrophs are species with larvae that develop directly from the egg. The number of species plotted in each histogram is given by n, and the mean duration by M; Myr stands for million years. These data are from Jablonski (1986a) .

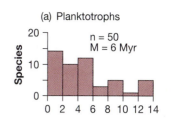

(a) Planktotrophs

n = 50
M = 6 Myr

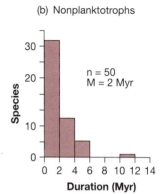

(b) Nonplanktotrophs

n = 50
M = 2 Myr

Atlantic coastal plain region, over the last 16 million years of the Cretaceous period. He found that marine invertebrate species with a planktonic larval stage survived longer, on average, than species whose young develop directly from the egg (Figure 13.3). In living species, planktonic larvae are carried on currents and often disperse long distances. This gives them greater colonizing ability, which might reduce the frequency of extinction. Populations with this life history also tend to have larger ranges. Indeed, Jablonski confirmed that geographic range also influences extinction rates: species with large ranges survived longer than those with more limited ranges (Figure 13.4). Taxa found in small areas are less likely to survive sea-level changes, new predation or disease regimes, and other stresses that can lead to extinction.

13.2 The Big Five: Extinction Events in the Phanerozoic

In contrast to background extinctions, mass extinctions are global in extent, involve a broad range of organisms, and are rapid relative to the average duration of taxa that are wiped out (Jablonski 1995). It is difficult to define mass extinction more precisely than this, however. As Raup's analysis makes clear, mass extinctions simply represent the tail of a continuous distribution of extinction events over time (Figure 13.1).

The five events commonly recognized and studied under the rubric of mass extinctions (Figure 13.5) are the terminal-Ordovician (ca. 440 Ma), late-Devonian (ca. 365 Ma), end-Permian (250 Ma), end-Triassic (ca. 215 Ma), and Cretaceous-Tertiary, or K–T (65 Ma). (Note that the Cretaceous is routinely symbolized with a K to distinguish it from other eras and periods that start with C.)

. . . and species with large geographic ranges have lower extinction rates.

The five largest extinctions in the Phanerozoic are known as the Big Five.

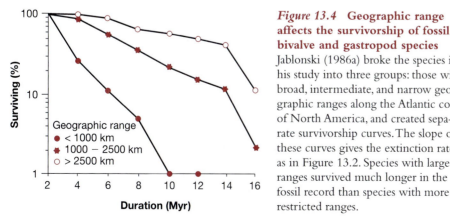

Figure 13.4 Geographic range affects the survivorship of fossil bivalve and gastropod species
Jablonski (1986a) broke the species in his study into three groups: those with broad, intermediate, and narrow geographic ranges along the Atlantic coast of North America, and created separate survivorship curves. The slope of these curves gives the extinction rate, as in Figure 13.2. Species with large ranges survived much longer in the fossil record than species with more restricted ranges.

Figure 13.5 **Patterns of extinctions of families through time**   M.J. Benton (1995) plotted the percent of families that died out during each stratigraphic stage in the fossil record of the past 510 million years. Because stages have different lengths, the percentages are not directly comparable in terms of extinctions per million years. Still, the diagram shows the distribution of mass extinction events throughout the Phanerozoic. The Big Five extinction events are circled. The Big Five were first defined by Raup and Sepkoski (1982).

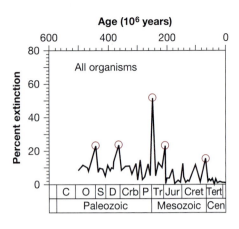

We will review two important issues in research on mass extinctions before exploring the greatest extinction event of all time, the end–Permian, in detail.

Dating Extinctions: The Signor–Lipps Effect

In 1982 P.W. Signor and J.H. Lipps pointed out that the fossil record tends to "smear" the dates of extinction events, making mass extinctions appear much more gradual than they may actually have been. This is because paleontologists have to estimate the duration of an extinction event by compiling the last records of taxa that were alive at the time. Simply due to the vagaries of sampling and variability in the probability of fossilization, these last records will be staggered. This is true even if the species being studied actually went extinct at exactly the same time. Rare or localized species, for example, tend to have last records much earlier than the actual date of extinction. Abundant and widespread organisms with hard shells are more likely to have last records closer to the actual event.

Another sampling effect influences our understanding of how rapidly mass extinction events have occurred. When paleontologists are fortunate enough to find more or less continuous strata that straddle the extinction event, the traditional approach has been to study the interval by collecting fossils every meter or so. It was not until the mid- to late-1980s that researchers began more intensive sampling, at intervals of 1 to 10 centimeters. Though it involves a tremendous increase in effort, the finer resolution of these datasets has vastly improved our estimates of how long mass extinction events take. An outstanding example is Charles Marshall and Peter Ward's (1996) study of mollusc fossils across the K-T boundary. Using refined sampling intervals and statistical testing of the Signor-Lipps effect, Martin and Ward show that the K-T extinction of inoceramid bivalves and ammonites was much more sudden than previously reported.

Estimating Species Losses Via Rarefaction

Although paleontologists recognize that species are the fundamental evolutionary unit, they are rarely able to use them as the level of analysis in research. Gaps in the fossil record usually create an overwhelming amount of sampling bias at the species level. These gaps occur because fossilization does not always record a species at a particular time and place, even though the organisms were present. Datasets based on species tend to be so noisy that finding patterns and drawing conclusions is problematic.

This noise is dramatically reduced, however, at higher taxonomic levels. Because genera or familes are less likely to drop out of the fossil record due to sampling problems, most paleontologists consider them a more reliable level of analysis. The downside of this approach is that genera, families, and orders are named on the basis of subjective judgments by one or more experts on the group. As a result, a genus of gymnosperms is not comparable, in any real biological sense, to a genus of clams. Fortunately this fact does not obscure the questions we are trying to answer in many cases, and genera and families can be appropriate proxies for species. In extinction studies, for example, counting the genera or families that have been wiped out should let us estimate the relative severity of different events.

But to illustrate the biological impact of an extinction event, we often want to report how many species were eliminated. If we cannot count them directly, how can we estimate this quantity from data on genera and families?

To do this, David Raup borrowed a technique from ecology called rarefaction. Raup (1979) used data from extant echinoderms (sea stars, sea urchins, sand dollars, and sea cucumbers) to answer the following question: If a random 10%, or 20%, or 30% of the species alive today were to go extinct, how many of the living genera, families, and orders would be wiped out? When Raup did these calculations, he came up with the curves reproduced in Figure 13.6. Given two important assumptions—that mass extinctions remove species randomly, and that species of fossil taxa are assigned to genera, families, and orders in the same way as the taxa used in the rarefaction analysis (Jablonski 1995)— we can use these curves to estimate the number of species that went extinct. Because these assumptions will not strictly be met, rarefaction estimates have to be interpreted cautiously.

Given this introduction to the distribution of extinction events and some issues in extinction research, we are ready to examine the scope and cause of a mass extinction event in detail.

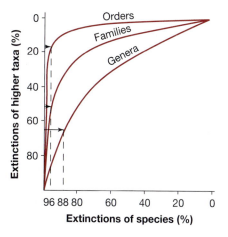

Figure 13.6 **Rarefaction curves** Assuming that extinction occurs randomly, rarefaction curves plot the percentage of higher-level taxa that are extinguished as more species are eliminated. David Raup (1979) generated these three curves using the number of species present in genera, families, and orders of living echinoderms in a database of 894 species. Note that almost all the species in a phylum have to go extinct for orders to begin dropping out. A more recent analysis, which combined data from 12 higher taxa of marine animals, yields a similar curve for the relationship between genus and species extinctions (Stanley and Yang 1994). This increases our confidence that it is legitimate to apply this curve to nonechinoderms.

13.3   End-Permian: The Mother of Mass Extinctions

"The extinctions that brought the Paleozoic era to a close . . . constituted the most severe biotic crisis in the history of animal life" (Stanley and Yang 1994: 1340). Some observers consider the end-Permian extinction one of the four most significant events in the history of life—on a par with the origin of life, the evolution of the eukaryotic cell, and the Cambrian explosion (Erwin 1993). It is the closest multicellular life has ever come to total extermination.

The mass extinction at the end of the Permian came close to exterminating multicellular life.

How bad was it? The best data we have indicate that over 50% of all families went extinct during the event and, based on rarefaction, perhaps as many as 90% to 96% of all species. The end-Permian includes the only mass extinction event recorded in the history of the insects (with 8 of 27 orders wiped out), the extermination of 21 of 27 reptile families, the elimination of 6 of 9 amphibian families, and the loss of over 70% of genera of marine invertebrates, including all reef-building corals. It is the only time in history that an order of Foraminifera has gone extinct (Erwin 1993). In one major fossil plant locality from Australia, up to 97% of the species were eliminated (Retallack 1995). In Europe, conifers were wiped out at the family level (Visscher et al. 1996). The end-Permian was a biological disaster of almost unimaginable proportions.

Selectivity

Although remarkably widespread, the end-Permian extinction was uneven in intensity. Among marine invertebrates, for example, groups with narrow geographic distributions were hit much harder than taxa with extensive ranges. Only 35% of genera that were broadly distributed went extinct, versus 93% of genera that were restricted to the margins of particular ocean basins (Figure 13.7). This is the same pattern that occurs during background times.

There were also taxonomic biases among marine invertebrates. Marine arthropods suffered few losses, though the last of the trilobites were wiped out. Inarticulate brachiopods, which were not common, were virtually unaffected; the articulate brachiopods, a dominant form living on the substrate of Permian oceans, were devastated (Figure 13.8). The articulate brachiopods, in fact, never again approached their former importance in distribution, abundance, or diversity. Instead, clams and snails took over that niche in shallow-water marine environments.

What Caused the End-Permian Extinctions?

We turn now to an unresolved issue that is currently the focus of intense research: What caused the greatest extinction event of all time? This is a challenging question. Any mechanism that is proposed has to explain the size, timing, duration, and specificity of the kill.

The Geological Setting: Pangea

The Permian was the only time during the Phanerozoic that the continents were assembled into a single land mass. This supercontinent, called Pangea (Figure 13.9), had formed by the early Permian and did not begin to break up until the mid-Triassic, when the Atlantic Ocean started to form.

(a)

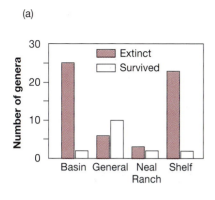

(b)

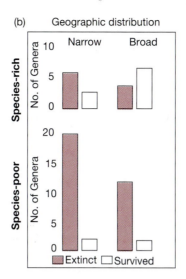

Figure 13.7 **Broadly distributed genera of marine invertebrates survived the end–Permian extinction best** Doug Erwin (1989) generated these bar charts based on an intensive study of marine genera found in Permian-Triassic deposits of the American Southwest. (a) Surviving genera outnumbered extinct genera only among habitat generalists. *Basin* refers to genera found only in ocean basin deposits, *General* to genera found in a variety of habitat types, *Neal Ranch* to genera found only in the shallow-water habitats of the Neal Ranch formation, and *Shelf* to genera found only in deposits typical of ocean-shelf environments. (b) Surviving genera outnumbered extinct genera only if they were species-rich and broadly distributed. Species-rich genera are defined as those with four or more species. Erwin categorized genera as narrowly distributed if they were found only in the American Southwest, and broadly distributed if they were also found at end-Permian localities in Malaysia that have been intensively sampled.

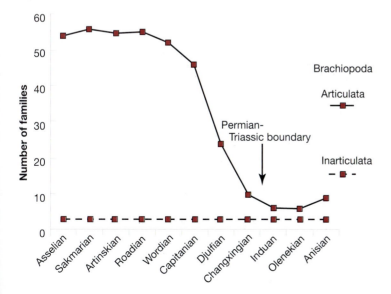

Figure 13.8 **Articulate, but not inarticulate, Brachiopods die off** The articulate brachiopods, a class in the phylum Brachiopoda, are morphologically similar to clams but depend on a different filter-feeding mechanism (called the lophophore). During the end-Permian extinction this group was almost extinguished. This graph (Erwin 1993) plots the number of families of articulate and inarticulate brachiopods through the course of the extinction event.

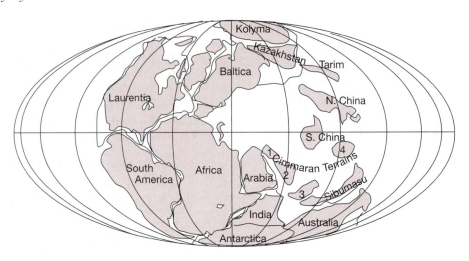

Figure 13.9 **Pangea** This is how the continents were positioned at the end of the Permian, about 251 Ma. Note that north and south China were island continents in the only ocean, called the Panthalassic. The remaining continental plates were joined into a single land mass.

(a)

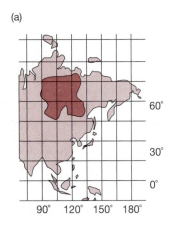

(b)

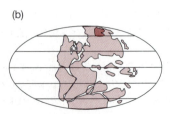

Figure 13.10 **The Siberian flood basalts** (a) The extent of Siberian flood basalts in present-day Asia. (b) Their position at time of formation.

The formation of Pangea did not, however, cause the extinctions directly. The supercontinent existed for tens of millions of years on either side of the cataclysm. But the formation of Pangea did create the setting for some of the most extraordinary geological events in history.

One of the most striking of these was the Siberian flood basalts, also called the Siberian traps. Flood basalts are outpourings of magma that flow across the surface instead of forming volcanoes. The Siberian flood basalts are 400m to 3,000m thick and cover 1.5 million km^2 of northeast Asia. The resulting formation represents as much as 3 million km^3 of rock (Figure 13.10). This was the largest outpouring of magma ever to occur on a continent during the Phanerozoic; it occurred in a series of events over 600,000 years, starting 250 Ma. Recent argon-argon dating work has shown that the flows spanned the end-Permian extinctions (Renne et al. 1995).

These basalt flows added tremendous quantities of heat, carbon dioxide, sulfur dioxide, and rock to the crust and atmosphere. But the connection between the flows and the extinctions, especially in the oceans, is anything but direct. What we do know is that the end-Permian was a time of massive change in several key aspects of the Earth's life support system: sea-levels, ocean chemistry, climate, and atmospheric chemistry. We now examine each of these changes in turn.

Sea-Level Changes

Several extinction events in earth history have been contemporaneous with a major regression in the extent of the oceans. It is difficult to point to sea-level change as a causative agent, however, because many such changes have occurred without accompanying extinctions (Erwin 1993). During the most recent glacial age, for example, the vast quantities of water locked up in continental ice sheets caused sea levels to drop. These regressions were 60 to 150 meters in extent. This is enough to reduce the amount of total ocean-shelf habitat from 50% to as little as 10% of current levels. But there were no mass extinctions of marine invertebrates during the Pleistocene. Instead, the organisms responded by adjusting their ranges, which tend to be broad in the along-shore dimension and narrow in the onshore-offshore direction. As sea level retreated, organism distributions simply tracked the onshore-offshore shifts in appropriate habitat (Valentine and Jablonski 1991).

Several studies have confirmed that global sea level regressed gradually throughout the mid-Permian, with a dramatic, rapid drop at the end of the period. This sharp decrease was followed by an equally rapid and dramatic transgression, or sea-level rise, at the start of the Triassic. Some of the best evidence is presented by W.T. Holser and M. Magaritz (1987), who reviewed data from 68 sedimentary basins scattered over the continents. At these localities the strata date to the interval spanning the extinction event. Holser and Magaritz looked at the rocks that made up each formation and, for each geological stage over the interval, determined the areal extents of strata that formed in marine versus nonmarine environments, based on the fossils present. Figure 13.11, which summarizes their work, shows how much of the continents were flooded with seawater from the beginning of the Permian to the mid-Triassic. The coverage of the continents dropped from an estimated 40% in the early Permian to about 10% at the time of extinction. This confirms a major regression event, and a massive loss of shallow-water habitat (see also Schopf 1979).

A second possible source of evidence for a large drop-and-rise event comes from the study of carbon isotopes. Carbon exists in three forms: ^{14}C, ^{13}C, and ^{12}C. ^{14}C is too unstable to be used for studies in deep time, but changes in the ratio of ^{13}C to ^{12}C found in old rocks can be informative. This is because ^{12}C is preferentially taken up during photosynthesis and incorporated into organic matter.

Why might changes in $^{13}C/^{12}C$ ratios indicate a major sea-level regression? In the ocean, carbon is deposited in one of two reservoirs: an organic reservoir made up of oil deposits and gas hydrates (this is methane locked in ice; huge deposits of gas hydrates are found in offshore sediments along the outer continental margins and in permafrost), and an inorganic reservoir consisting of carbonate rocks. Thus, a high $^{13}C/^{12}C$ ratio in carbonate rocks of marine origin can indicate a period of high productivity, when large quantities of ^{12}C were being deposited in the organic reservoir. A low ratio can indicate a period in which

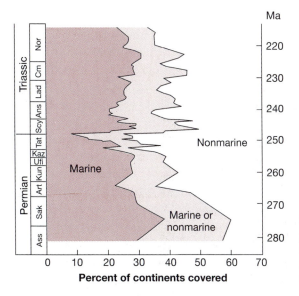

Figure 13.11 **Sea-level changes at the end-Permian** This diagram shows how much of the continents were flooded with seawater from the beginning of the Permian to the mid-Triassic (Holser and Magaritz 1987). The "marine or nonmarine" category represents rocks whose origin is not clear.

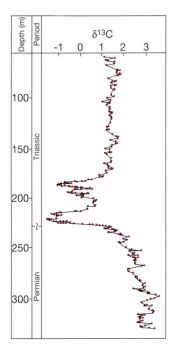

Figure 13.12 **The carbon spike** Detailed carbon-isotope analyses for the interval spanning the end-Permian have now been done at about eight localities around the globe. This pattern, which is fairly typical, is from the Carnic Alps in Austria. By convention, $^{13}C/^{12}C$ ratios are reported as a standardized quantity called $\delta^{13}C$. This is calculated as the percentage difference between $^{13}C/^{12}C$ of the sample and the ratio observed in a certain shell from the Pee Dee formation of South Carolina. Because they are standardized in this way, $\delta^{13}C$ values are comparable from period to period. For a more detailed explanation of carbon ratios and citations illustrating how this work is done, see Erwin (1993).

For perhaps as long as 20 million years, the water at the bottom of the Panthalassic Ocean contained little or no oxygen.

one of two things was happening: Little organic carbon was being deposited, or buried organic carbon was being oxidized and released to the environment as CO_2. Studies of carbon isotope ratios from various localities around the world indicate that there was dramatic swing from high to low ratios at or near the Permo-Triassic boundary (Figure 13.12).

What caused the carbon spike? CO_2 released from the Siberian traps probably contributed to the swing, but this cannot be the whole story. Vast quantities of light carbon are required to produce a drop of this magnitude. So much, in fact, that a major sea-level regression might be involved (Erwin 1993, 1995). The logic here is that much of the Earth's ^{12}C is buried in the form of oil, coal, and gas hydrates. If part of this buried ^{12}C were liberated from ocean sediments, it could contribute to the low $^{13}C/^{12}C$ ratios recorded in the marine rocks sampled in Figure 13.12. One mechanism for release is the oxidation of offshore gas hydrates. These hydrocarbon deposits are exposed during major sea-level regressions and released by weathering. But a dramatic drop in the primary productivity of the oceans would also increase the fraction of ^{12}C deposited in the inorganic reservoir. The carbon spike is compatible with a major sea-level regression event, but other explanations are possible.

Based on the arguments we made earlier, though, it is unlikely that the swing in sea level was a major killing agent in itself. The regression was important because it may have played a role in changing the chemistry of water in shallow marine habitats, which in turn could have led to mass extinctions.

Changes in Ocean Chemistry

The plates that make up the sea floor are constantly being subducted below the continental plates. As a result, few examples of pelagic rocks from deep time exist. Fortunately we have two good samples of rocks from deep-water environments at the Permo-Triassic boundary. These are from Japan and British Columbia, Canada, and are remarkable for what they imply about ocean chemistry in the end-Permian (Isozaki 1994, 1997; see also Knoll et al. 1996). The sediments are chert: a hard, compact rock that Native Americans favored for making projectile points. Chert is made up of silica (SiO_2). Some cherts develop in deposits of siliceous skeletons from sponges and other marine planktonic organisms. The end-Permian cherts are extraordinary because they contain abundant pyrite (FeS_2). Pyrite, in turn, forms when bacteria reduce sulfate to sulfide in iron-containing mud. Why is all this interesting? Because pyrite is a product of sulfur-reducing bacteria, it forms only in the absence of oxygen. The pyritic cherts strongly suggest that the bottom of the Panthalassic was anoxic across the Permian-Triassic boundary.

Today the oceans are well oxygenated in shallow-water, continental shelf habitats as well as deep-water environments underlain by ocean crust. Why would the deep-water environments have turned anoxic during the Permian? The leading hypothesis is that the formation of Pangea disrupted the circulation of seawater, which led to widespread stagnation. Then the aerobic decay of organic debris, in the form of dead plankton raining down from the ocean surface, would eventually use up the available oxygen and lead to anaerobic conditions. A.H. Knoll and colleagues (1996) have suggested that anoxic conditions would also lead to extraordinarily high concentrations of dissolved CO_2 in the deep oceans, because sulfate-reducing bacteria oxidize carbon to CO_2.

In sum, the data indicate that the ocean bottom took on some characteristics of bog water.

For aquatic organisms, contact with anoxic water and high CO_2 concentrations can lead to suffocation or hypercapnia (poisoning due to increased carbon dioxide in the blood). Both are deadly. But how could anoxic, carbon-dioxide-rich water in the deep ocean rise to the surface and spread onto the continental shelves, where the bulk of macroscopic marine organisms live?

One of the more remarkable properties of water is that it reaches its highest density at 4°C. If a global cooling event was extensive enough to cool parts of the ocean surface to 4°C, or to create continental glaciers that contributed quantities of meltwater to the surface, then the 4° water would sink and force water from the bottom of the ocean to the surface. This phenomenon, known as an overturn, occurs every spring and fall in high-latitude lakes.

A.H. Knoll and colleagues (1996) propose that an overturn occurred in the Panthalassic Ocean late in the Permian, flooding what remained of the continental shelves with quantities of anoxic, carbon-dioxide-enriched water. Alternatively, as the volume of anoxic water at the bottom of the Panthallasic gradually increased, it could simply have accreted onto the shelves (Isozaki 1997).

Wignall and Twitchett (1996) recently found evidence supporting widespread anoxia in the form of sedimentary rocks enriched with uranous fluoride, a compound that is deposited in reducing environments. In the strata they studied, the elevated uranium signal showed up in shallow-water marine sediments contemporaneous with the extinction event. Normal amounts of uranium deposition, indicating oxygenated waters, returned by the early Triassic.

Although the issue is still hotly debated, an anoxia/hypercapnia kill in shallow marine habitats remains an interesting possibility. We turn now to a different issue: What about the terrestrial extinctions?

Changes in Atmospheric Chemistry and Climate

The formation of Pangea made highly seasonal continental climates an almost global phenomenon. There is also abundant evidence that strong warming and drying trends occurred throughout the Late Permian: Large coal deposits formed at high latitudes, huge salt-rich evaporite basins were created where inland seas dried up, tropical reefs were extensive, and the distribution of plant and animal fossils indicative of hot, arid climates increased.

There is some evidence for widespread continental glaciation at the end of the Permian, however. Glacial boulders have been found in Australian and northeast Asian sediments dated to the Late Permian (Stanley 1988a, 1988b). If these continental glaciers are confirmed, they may furnish a source of cold water for the ocean overturn event hypothesized by Knoll and co-workers.

What could have caused this cooling after the long period of warming? Here the Siberian flood basalts may come into play. In many volcanic events, sulfur-containing aerosols like SO_2 are emitted in huge quantities with the magma. These molecules are remarkable for their ability to backscatter and adsorb incoming solar radiation. Even historically recent, relatively small-scale eruptions have added enough sulfur dioxide to the atmosphere to produce a short-term "volcanic winter." The Siberian flood basalts were huge and long-lasting by comparison, with potentially profound effects on climate. The sheer mass of the flood basalts could also have been responsible for cooling, simply by raising the

A dramatic swing from warmer to cooler climates may have occurred at the end-Permian.

altitude of the northern continent. Even a small increment in elevation may have been enough to trigger glacier formation (Erwin 1993).

Counteracting these rapid cooling processes, however, is a longer-term factor: The massive amounts of carbon dioxide released with the magmas or from an ocean overturn would have created a significant greenhouse effect, and contributed to a long warming trend. If the flood basalts were associated closely in time with the release of gas hydrates during the end-Permian regression, the additional carbon release from this enormous reservoir—and its subsequent oxidation—would have added even more carbon dioxide to the atmosphere, resulting in still more extensive warming.

The Multiple-Causation Hypothesis

If this discussion has confirmed our earlier claim, that the end-Permian was a time of unprecedented change in physical systems, it has been a success. Researchers in the field tend to search for a single, underlying causative mechanism for the end-Permian cataclysm—one capable of uniting the geological, atmospheric, and oceanographic changes with the biological responses on both land and sea. But to date we have not been able to tease apart all of the interconnecting physical events and link their occurrence in time and space with the timing, cause, and extent of the extinctions. As a result, the best point of view on causation may well be a conservative one, embodied by what has been called the multiple-causation or "world-went-to-hell" hypothesis (Erwin 1993, 1994, 1995). This idea holds that the near-simultaneous occurrence of extreme climate change, a major sea-level regression, and ocean anoxia combined to create the kill (Figure 13.13). These events, and the extinctions that resulted, occurred in less than a million years.

Further complicating the situation, there is increasing evidence that—in some regions at least—the end-Permian extinction represents not one but two events, both short in duration and separated by about five million years (Stanley and Yang 1994). If the spacing between episodes is confirmed by further re-

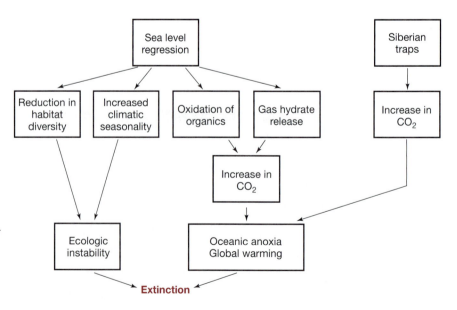

Figure 13.13 **The multiple-causation hypothesis for the end-Permian extinction**　This diagram indicates that a large-scale drop in sea level, combined with huge volumes of flood basalts from the Siberian traps, could set off a chain reaction that ended with profound ocean anoxia and general ecological instability.

search on the timing of the extinction, it will place new demands on all of the causal mechanisms proposed. We will need to figure out who went extinct, when and why, twice.

One thing *is* clear about the end-Permian extinction: It was not caused by an extraterrestrial object that slammed into the Earth. None of the evidence we review in section 13.4 for the K–T extinction has been confirmed at the Permo-Triassic boundary.

13.4 Cretaceous–Tertiary: High-Impact Extinction

The impact hypothesis for the extinction at the K–T boundary (Alvarez et al. 1980) fired the imaginations of scientists and the lay public. The idea that a huge object had hit the Earth and caused widespread extinctions, including the demise of the dinosaurs, provoked intense debate and research. In just ten years after its publication, the hypothesis had inspired over 2,000 scientific papers and innumerable accounts in the popular press (Glen 1990).

We organize our review of the K–T extinction by looking first at the evidence for the impact event, considering the killing mechanisms that could have resulted, and investigating who died. In section 13.5 we examine evidence that impact events, and mass extinctions, may have occurred at regular intervals during earth history.

Evidence for the Impact Event

The discovery of anomalous concentrations of iridium in sediments that were laid down at the K–T boundary (Figure 13.14) was the first clue that an asteroid, comet, or other body (the general term is bolide) hit the Earth 65 Ma. The

Figure 13.14 **Clay lenses at the K-T boundary** This photograph shows a dark band of clay sandwiched between limestones laid down over the Cretaceous-Tertiary boundary. These rocks, which were formed in a deep-sea environment, are found today in Gubbio, Italy. The clay layer was laid down at the K-T boundary. Because limestone is made up of calcareous shells of marine invertebrates, the interruption of limestone formation and deposition of clay represents the period immediately after the extinction event. Few marine invertebrates were present in the ocean at this time (Alessandro Montanari/Geological Observatory of Coldigioco).

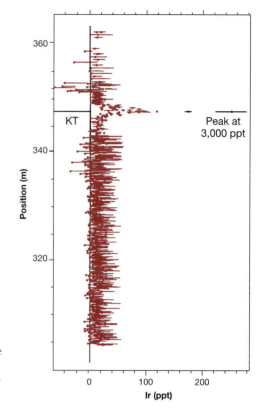

Figure 13.15   **The iridium anomaly at the K-T boundary**   This graph shows the concentration of the element iridium, in parts per trillion (ppt), found in the strata pictured in Figure 13.14. The large spike is in the clay layer. Note that the peak concentration is off the scale, far to the right. From Alvarez et al. (1990).

element iridium is rare in the Earth's crust, but abundant in meteorites and other extra-terrestrial objects. Figure 13.15 shows a typical iridium spike found in strata that were laid down over the Cretaceous-Tertiary boundary. At last count, 95 different sampling localities from all over the globe had been found with iridium anomalies dated to the K-T boundary (Glen 1990). Based on estimates for the amount of iridium needed to produce the anomalies and the density of iridium in typical meteorites, Alvarez et al. (1980) suggested that the bolide was on the order of 10 km wide. It was, quite literally, the size of a mountain. It would also have been intensely hot, from friction with the atmosphere.

The discovery of two unusual minerals in K-T boundary layers provide additional support for the hypothesis. Shocked quartz particles (Figure 13.16)

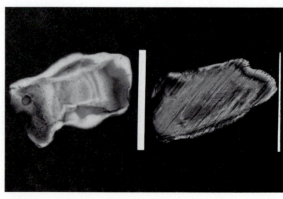

Figure 13.16   **Shocked quartz**   Small quartz grains (1–2mm across) with parallel planes called lamellae are routinely found near meteorite strikes. The deformation is thought to be caused by the shock of impact. A shocked quartz grain is shown on the right, and a normal grain on the left (Glen A. Izett/U.S. Geological Survey, U.S. Department of the Interior).

Figure 13.17 **Microtektites** Microtektites are spherical or teardrop-shaped particles of glass associated with impact sites. They are thought to originate as particles of rock melted by the heat of impact, and either cooled in place or splashed into the atmosphere. The tektites pictured here have been sectioned to show the interior (Glen A. Izett/U.S. Geological Survey, U.S. Department of the Interior).

had been found only on the margins of well-documented meteorite impact craters until they were discovered at K-T boundary sites (Glen 1990). The other unusual structures are tiny glass particles called microtektites (Figure 13.17). Microtektites can have a variety of mineral compositions depending on the source rock, but all originate as grains melted by the heat of bolide impact. If the melted particles are ejected from the crash site instead of being cooled in place, they are often teardrop- or dumbbell-shaped. This is a result of solidifying in flight.

The discovery of abundant shocked quartz and microtektites in K-T boundary layers from Haiti and other localities in the Caribbean helped investigators narrow the search for the crater (Florentin and Sen 1990; see also McKinnon 1992; Alvarez et al. 1995). Then, in the early 1990s, a series of papers on magnetic and gravitational anomalies confirmed the existence of a crater 180 kilometers in diameter, centered near a town called Chicxulub (cheek-soo-LOOB) in the northwest part of Mexico's Yucatán peninsula (Figure 13.18). Subsequent dating work confirmed that microtektites from the wall of the crater, recovered from cores drilled in the ocean floor, were 65.06 ± 0.18 million years old (Swisher et al. 1992). This is an almost exact match to dates for glasses ejected from the site and recovered in the Haitian K-T boundary layer.

The discovery of the crater was the long-sought smoking gun. It solidified a consensus among paleontologists, physicists, geologists, and astronomers that a large bolide struck the Earth 65 Ma. The existence of the impact is no longer controversial; the consequences of the impact are.

Evidence for a bolide impact at the end-Cretaceous is now overwhelming.

Killing Mechanisms

The 10-km bolide that struck the ocean would have produced a series of events capable of affecting climate and atmospheric and oceanic chemistry all over the globe. The ocean floor near the impact site at Chicxulub consisted of carbonates, including large beds of anhydrite ($CaSO_4$), over a granitic basement layer. The distribution of shocked quartz and microtektites, far to the north and west

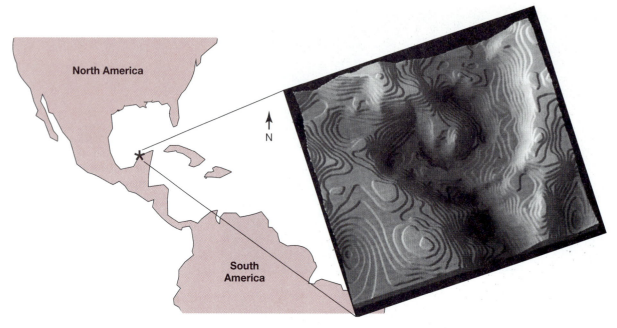

***Figure 13.18* Location and shape of the Chicxulub crater** Gravitational and magnetic anomalies outline the rim and walls of a 180-km diameter crater, buried beneath sediments near Chicxulub on the Yucatán peninsula. The inset is a relief map of the impact site, showing gravity field density (Schultz and D'Hondt 1996)(Peter H. Schultz, Brown University).

of Chicxulub, confirms that a large quantity of this material was ejected from the site, and that significant amounts were melted or vaporized by the heat generated at impact.

What consequences did the ejected material have? Vaporization of anhydrite and seawater would have contributed an enormous influx of SO_2 and water vapor to the atmosphere. Although the source of the sulfur is different than at the end–Permian, the results would have been the same. The molecules would react to form H_2SO_4, with acid rain resulting. Sulfur dioxide is also a strong scatterer of solar radiation in the visible spectrum, which would lead to global cooling (McKinnon 1992). The cooling effect would have been enhanced by dust-sized carbonate, granitic, and other particles. These were ejected in quantities large enough to block incoming solar radiation. Further, soot and fullerene deposits at numerous K–T boundary localities (for example, Wolbach et al. 1988; Heymann et al. 1994) suggest that widespread wildfires occurred, perhaps triggered by the fireball of hot gas and particulates expelled from the impact site. The ash and soot produced by the fires would have added to a widespread smog layer and the accentuation of global cooling. All these data indicate that the Earth became cold and dark soon after the impact.

A variety of models, reviewed by Glen (1990), suggest that the force of the impact was also sufficient to trigger massive earthquakes, perhaps as large as magnitude 13 on the Richter scale, and to set off volcanoes. The second largest flood basalts of the Phanerozoic, the Deccan Traps of India, were contemporaneous with the K–T extinctions, though no causal connection has yet been established to directly tie this event to the impact. Widespread volcanism would have added sulfur dioxide, carbon dioxide, and ash to the atmosphere. The ash and sulfur dioxide would accentuate global cooling, while carbon dioxide would contribute to a longer-term greenhouse effect and warming.

Finally, the impact would have created an enormous tidal wave, or tsunami, in the Atlantic. If the bolide was indeed 10 km wide, models suggest that the wave produced by the strike would have been as large as 4 km high. The mountain of rock made a mountain-size splash. What evidence do we have that a tsunami of this size actually occurred? Joanne Bourgeois and colleagues (1988) discovered a huge sandstone deposit along the Brazos River in Texas, which has been mapped throughout northeast Mexico. It is 300 km long and several meters thick, and is now interpreted by most geologists as a product of the rapid and massive deposition typical of tsunamis (see Kerr 1994). In the Haitian boundary layer there is a thick jumble of coarse- and fine-grained particles sandwiched between the iridium-enriched clay layer above and extensive tektite deposits below. Florentin and Sen (1990) interpret this middle stratum as the product of tsunami-induced mixing and deposition. This occurred after the initial splash of microtektites but before the fallout of iridium-enriched particulates from the atmosphere.

The upshot is that the bolide strike would have affected the world's oceans in two ways. Globally, the primary productivity of phytoplankton would have been dramatically reduced by the cooling and darkening of the atmosphere. Locally, temperature regimes and chemical gradients in the Atlantic would have been disrupted by the largest tsunami ever recorded.

The impact had profound effects on both marine and terrestrial ecosystems.

In sum, between acid rain, widespread wildfires, intense cooling, extensive darkness, and an enormous tsunami, there is no shortage of plausible killing mechanisms for both terrestrial and marine environments. The question now becomes: Are there any patterns indicating which groups died out? If so, do these patterns tell us anything about which of the possible kill mechanisms were most important?

Extent and Selectivity of Extinctions

To date, our best estimate is that 60% to 80% of all species went extinct at the end-Cretaceous (Jablonski 1991). As in the end-Permian extinction, however, the losses were not distributed evenly across taxa. Among vertebrates, for example, amphibians, crocodilians, mammals, and turtles were little affected. Prominent terrestrial groups like the dinosaurs and pterosaurs (flying reptiles) were wiped out, and large-bodied marine reptiles like the ichthyosaurs, plesiosaurs, and mosasaurs disappeared. Only one order of birds, a shorebird-like group, survived (Feduccia 1995; but see Cooper and Penney 1997), while insects escaped virtually unscathed. Among marine invertebrates the ammonites and the rudists (a group of clams) were obliterated. Marine plankton became so scarce that micropaleontologists describe a "plankton line" at the K-T boundary. At some localities in North America, more than 35% of land plant species went extinct (Schultz and D'Hondt 1996). Pollen and spore deposits dated to just after the boundary show a prominent "fern spike" at these localities, implying that forest communities were dramatically reduced and replaced by widespread stands of ferns (Nichols et al. 1992; Sweet and Braman 1992).

What does this imply about the kill mechanism? It has traditionally been argued that the K-T extinction was size selective, with large-bodied organisms suffering most. The logic was that an extended period of cold and dark after the impact would have affected large animals and plants the most, because of their higher nutritional requirements. But the analyses done to date have shown no difference in extinction rates between small- and large-bodied forms of marine bivalves and gastropods (Figure 13.19). And even though a definitive study has not been done for either marine or terrestrial vertebrates, Jablonski (1996) and others have pointed out that large crocodilians survived while small-bodied and juvenile dinosaurs did not.

Further, William Clemens and L. Gayle Nelms (1993) have found fossils from a variety of dinosaur species on the north slope of Alaska. This locality lay well within the Arctic Circle at the time of the extinction. The collections made by Clemens and Nelms include juveniles of some species, suggesting that they were hatched in the north. If so, it is unlikely that these individuals grew fast enough to migrate overland to warmer climates in the fall. It is much more sensible to argue that they were year-round residents of cold, dark climates. If some dinosaurs could tolerate cold, it suggests that an impact-induced period of cold and dark is not sufficient to explain the selectivity of the extinction. Could the taxonomic selectivity of the vertebrate losses be related to hibernation or other adaptations that protect individuals from long food shortages, rather than to body size? We simply do not know. The selectivity of the vertebrate extinctions is unresolved, and still the focus of vigorous debate and research.

Although research on plant extinctions is still gathering momentum, an interesting pattern may be starting to emerge: Losses seem to be much more severe in North America. The reason may be that species there occupied the splash zone. Peter Shultz and Steven D'Hondt (1996) have proposed that the impact occurred at an oblique angle, with the bolide approaching North America from the southeast and ejecting material and heat predominantly to the north and west.

We have made more progress in understanding the selectivity of the marine invertebrate extinctions, largely because the fossil record is so much more exten-

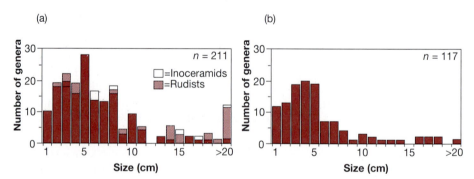

Figure 13.19 **Bivalve extinctions at K–T were not size dependent**   The body size of bivalve genera is plotted on the x-axis of these histograms. Body size was estimated as the average of shell lengths for species in each genus; *n* is the total number of genera. Groups that went extinct at the K–T boundary are shown in (a) and survivors in (b). When the group of unusually large bivalves called rudists are excluded from the analysis, there is no significant difference between victims and survivors (Jablonski and Raup 1995; see also Jablonski 1986a; 1996).

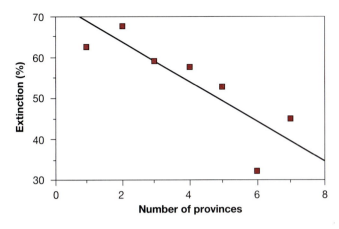

Figure 13.20 **Marine bivalve genera with wide geographic ranges were less likely to go extinct at the K-T boundary** On this graph, the percent of bivalve genera that went extinct at the K-T boundary is plotted as a function of the number of biogeographic provinces occupied by those genera. (Biogeographic provinces are regions of the world that share similar floras and faunas.) There is a statistically significant trend for more localized genera to be wiped out during the mass extinction. For additional data on this pattern, see Jablonski 1986a.

sive (Jablonski 1986a, 1991). For example, the probability that a bivalve genus survived the K-T boundary had nothing to do with whether it lived by burrowing or in exposed positions, or whether it lived close to shore or far offshore, or whether it occupied tropical or polar latitudes (Raup and Jablonski 1993; Jablonski and Raup 1995). The outstanding pattern that *has* emerged in studies of bivalves and gastropods is that genera with wide geographic ranges were less susceptible to being eliminated than genera with narrower ranges (Figure 13.20). This is exactly the same result observed for the end-Permian extinction (Figure 13.7), and for bivalves and gastropod species undergoing background extinction in the interval just before the bolide impact at the end-Cretaceous (Figure 13.4). Wide geographic ranges clearly buffer marine invertebrate clades against extinction. To date, this pattern represents the most robust result to emerge from the study of both background and mass extinctions.

Extinctions of marine invertebrates at K-T were selective: Genera with broad geographic ranges survived better.

13.5 Are Mass Extinctions Periodic?

To complete our look at impact-induced mass extinction, turn back to Figure 13.5 and note that the spikes indicating high-intensity extinction episodes seem to occur at regular intervals through the Phanerozoic. In 1982 David Raup and Jack Sepkoski saw this pattern and, inspired by the impact theory for the K-T extinction, suggested: (1) that a major extinction event has occurred every 26 million years on average (Figure 13.21), and (2) that the most likely cause of periodicity was bolide impacts.

To their credit, Raup and Sepkoski have been careful to emphasize the tentative nature of the hypothesis, given the uncertainties in identifying and dating major extinction events. Still, the idea was immediately and intensely controversial, and remains so today (to get started with this literature, see Raup and Sepkoski 1984, 1986; Sepkoski 1989; Rampino et al. 1995). Is the pattern real? That is, is there a statistically significant regularity to mass extinctions? If so, it would be the most astonishing pattern ever recognized in the history of life. Immediately we would want to find out what astronomical or other process could be responsible for clock-like catastrophes.

There are a series of important questions about both the proposed pattern and the process, however:

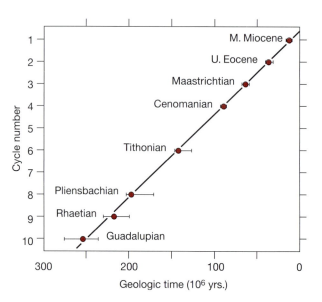

***Figure 13.21* Evidence for periodicity of mass extinctions?** Extinction events from the end-Permian to the present are plotted against time in this graph. The horizontal lines through each data point indicate the earliest and latest dates proposed for each extinction event. In this figure, extinction events are identified by geological stage or epoch instead of period (so, for example, the K-T event is denoted Maastrichtian). Note that the extinction events that should be numbered 5 and 7 are missing. The straight line describes a periodicity of 26 million years. From Raup (1987).

Are We Counting the Number of Mass Extinction Events Correctly?

While most research on mass extinctions has focused on the Big Five, the periodic pattern depends on a series of smaller events. Are these lower peaks real, or could they simply represent normal variation in extinction intensity? In other words, how many peaks qualify as mass extinction events statistically? Raup and Sepkoski (1986) determined that eight peaks in the fossil record for marine families were significantly larger than background rates, when the background rate was defined as the extinction intensity during the geological stage just before and after the peak. But using an expanded and updated database that included nonmarine forms, Benton (1995; see also Rampino et al. 1995) was able to identify only six of the same peaks in family-level extinctions. The two doubtful peaks are the most recent: those hypothesized to have occurred in the upper-Eocene and mid-Miocene. Until larger databases are available and resolved at the level of genera, which should allow us to confirm or reject the two most recent events, the conservative position is to acknowledge that six to eight statistically significant mass extinctions have occurred over the past 250 million years.

Is the Dating of Mass Extinction Events Precise and Robust Enough to Make the Pattern Believable?

Dating mass extinctions precisely is difficult for two reasons: the Signor-Lipps effect, and because only a small fraction of fossil localities can be dated using radioisotopes. As a result, the date that a species, genus, or family disappeared from the fossil record is usually recorded at the level of a geological stage. Geological stages, in turn, can span from two to over 10 million years. Furthermore, the absolute ages assigned to geological stages are constantly being revised and improved. All this uncertainty makes it difficult to fit a periodic waveform to the record of mass extinctions statistically (Raup and Sepkoski 1986) and test whether the periodic pattern is real or illusory.

Another dating issue has emerged recently: Several catastrophes recorded in Figure 13.5, including the end-Permian, end-Devonian, and end-Triassic, may actually represent two independent periods of intensified extinction spaced closely in time (Benton 1991; Wang 1992; Weems 1992; Stanley and Yang 1994). That is, as fossil collections grow and dating becomes more precise and reliable, investigators are finding that these events may have occurred in two different geological stages, not one. If these consecutive-stage events are confirmed, it means that additional peaks have occurred. The claim of 26-million-year intervals would be seriously weakened.

Have Other Impact Events Been Confirmed That Correlate with Mass Extinctions?

The answer to this question is a qualified yes. The end-Devonian mass extinction, the second-largest marine extinction, has been tied to impact events. Evidence in the form of microtektites, shocked quartz, and appropriately dated craters in Sweden, south China, and possibly Quebec and Chad, argues that several bolides hit Gondwanaland approximately 367 Ma (Claeys et al. 1992; Wang 1992; Kerr 1996). But this is the only other mass extinction event besides the K-T that is associated with bolide strikes; the other Big Five extinctions are associated with climate change, dramatic sea-level regressions, or unknown causes (Erwin 1996).

Ironically, the end-Devonian event has not been included in the periodicity analyses, even though it is most closely tied to an impact event. As a result, there is little evidence that the data points in Figure 13.21 correspond to collisions with extraterrestrial objects, as hypothesized by Raup and Sepkoski.

Have Any Astronomical Events Been Identified That Could Produce Regular Showers of Comets or Asteroids Capable of Striking the Earth?

Periodic or cyclical phenomena imply a common, underlying cause. Raup and Sepkoski's original idea was that if some sort of orbiting body or collection of objects passes close to the earth every 25 to 30 million years, it would make collisions much more likely than at other times. Although astronomers and astrophysicists have been keenly interested in the hypothesis and have evaluated a variety of asteroid belts and other structures for appropriate orbits, no convincing candidates have yet been found. To date, the answer to the question is no.

In sum, the evidence for the periodicity hypothesis is not particularly strong. Even if further work confirms the pattern, by finding statistically significant and regularly spaced peaks in the extinction of genera, we are still left without a plausible process. For most scientists, patterns in nature are not especially convincing or interesting until a causative agent is confirmed.

13.6 Recent: The Human Meteor

Concern about widespread extinction is on the minds of people from all walks of life, from grade school children to heads of state. But despite celebrity examples like the dodo, passenger pigeon, and Carolina parakeet, is anything of special evolutionary significance going on now? That is, are we currently experiencing or contemplating an event anything like the Big Five in scale and speed?

To answer this question, we examine well-documented extinction events that have occurred during the current geological period: the 1.8 million years that have passed since the Tertiary ended and the Quaternary began.

Pleistocene Megafauna

During the Pleistocene epoch a series of glacial periods, averaging about 100,000 years, alternated with interglacials lasting 10,000 to 20,000 years (see Porter 1989). Large areas of North America and northern Europe were covered with ice at the maximum extent of the glaciations. Figure 13.22 shows the peak of the continental glaciers during the most recent cycle. These ice sheets had retreated to the Arctic by about 10,500 years before present (ybp), when the present epoch, the Holocene, began.

The most recent of these glacial periods saw the extinction of numerous large, terrestrial mammals in the Americas, Eurasia, and Australia (Figure 13.23). The extinctions are concentrated in the last glacial interval, from 100,000 to 10,000 ybp. The losses are particularly striking for species averaging an adult weight of 44 kg (100 lb) or more.

For example, in North America 38 of 54, or 70%, of the large-bodied species disappeared (Stuart 1991). Some of these, like the lion and the horse, survived elsewhere in the world. But others were endemic: an extinct camel and musk ox, the mastodon and woolly mammoth, the saber-toothed cat, a beaver weighing 150 kg, the American cheetah, two species of tapir, and the giant ground sloth. An extinction of similar scale occurred in South America, where 46 of 58 (80%) of the large-bodied genera were wiped out (Martin 1984). Although a relatively small number of appropriately dated digs are available from Australia, the data that do exist suggest that 19 of 22 genera (86%) of giant marsupials were lost (Martin 1984).

In northern Eurasia, however, the impact on large mammals was substantially less severe. Stuart (1991), for example, estimates that just seven of 24 genera

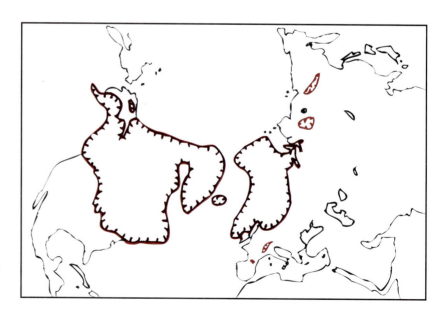

Figure 13.22 **Distribution of ice sheets 18,000 ybp** This figure shows a glacial maximum in the late Pleistocene. From Stuart (1991).

Figure 13.23 **Extinct Pleistocene megafauna** The animals illustrated here are the cow-sized *Nesodon* of South America (top), giant kangaroo of Australia (middle left), woolly mammoth of Eurasia and North America (middle right) and giant ground sloth of North America (bottom).

(29%) went extinct. Species that were lost completely include the straight-tusked elephant, woolly mammoth, woolly rhinoceros, cave bear, giant deer (or "Irish elk"), and species of bison and rhinoceros. The hippopotamus, spotted hyena, lion, and musk ox were extinguished in the region but survived elsewhere. The impact in Africa was even less pronounced, and probably indistinguishable from background rates. Counting large-bodied forms distributed south of the Sahara, only two of 44 genera (4.5%) were lost at the Pleistocene-Holocene boundary (Stuart 1991, based on data in Klein 1984). Over the entire 90,000-year interval, seven of 49 (14%) of the large-bodied genera in Africa went extinct (Martin 1984). The contemporary African fauna remains essentially unchanged from forms present in the late Pleistocene.

Because it was not global in extent, the event does not qualify as a mass extinction. It was also much too restricted taxonomically. We noted earlier that the lack of extinctions among marine invertebrates during the Pleistocene was remarkable, given that glaciers tied so much water up in ice. Sea-level regressions of over 100m occurred repeatedly as glaciation waxed and waned (Valentine and Jablonski 1991). But insects, plants, and birds were also virtually untouched.

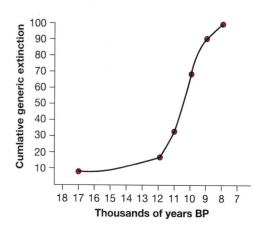

Figure 13.24 **Extinctions of North American mammal genera were clustered in time**   The ordinate shows the percentage of 23 North American mammal genera that went extinct during the last 100,000 years, accumulated over time. The abscissa records absolute time, estimated from radiocarbon dates. BP stands for before present. There was a sharp increase in extinctions between 11,000 and 9,500 years before present (Mead and Meltzer 1984).

The striking thing about the end-Pleistocene extinction, then, is not its size but its selectivity. Among mammals, large forms were lost at much greater rates than small species (Martin 1984). We have already reviewed the geographic selectivity: the Americas and Australia were hit much harder than northern Eurasia, and Africa and south Asia were affected little if at all. Variability also occurred in timing. Losses in the Americas were clustered tightly in time (Figure 13.24), while extinctions were spread over much longer intervals in Europe and Australia.

What caused the extinction of Pleistocene megafauna? Several hypotheses have been advanced since the event was first recognized in the late 19th century, but two dominate the debate today (Grayson 1984). The climate-change hypothesis contends that extinctions were caused by loss of open, "mammoth-steppe" habitats. This was due to the expansion of forests at the start of the Holocene. In contrast, the overkill hypothesis states that predation by humans was the causative agent.

Climate change, overhunting by humans, or both, could have been responsible for the extinction of Pleistocene megafauna.

There are a number of problems with both hypotheses (Stuart 1991). The leading objection to the climate-shift hypothesis is that changes in temperature and habitat types equivalent to those at the end-Pleistocene had occurred repeatedly in the preceding 900,000 years, without being accompanied by waves of extinctions. Questions about the overkill hypothesis are equally serious. The losses in Australia and northern Eurasia, for example, postdate the arrival of humans by tens of thousands of years. Several extinctions were of species not likely to have been prey items for human hunters, and it is extremely rare to find extinct forms in archaeological sites. There is also a plausibility issue: Could hunters using paleolithic technologies extinguish so many species?

Distinguishing between the two hypotheses is probably impossible. In the Americas, for example, the extinctions, the arrival of humans, and climate change were all contemporaneous. This knowledge makes it impossible to tease apart the effects of the two processes. The current consensus in the field, then, is that both climate change and predation by humans were important agents in the extinction of Pleistocene megafauna. Both stresses were rapid, and both acted selectively on large species.

Polynesian Avifauna

David Steadman (1995) has amassed compelling evidence that an important extinction event has just occurred among birds. Steadman estimates that 2,000 avian species have been extinguished over the past two millennia in the Pacific region alone, as a result of human colonization of islands. Because slightly over 9,000 species of birds exist today, Steadman's work means that the clade called Aves has recently lost almost 20% of its species.

The evidence for this claim comes from archaeological digs throughout the Pacific islands, conducted over the past 20 years by Steadman, Storrs Olsen, Helen James, and colleagues. For example:

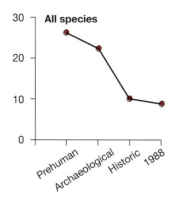

Figure 13.25 **Extinction of forest birds on the island of 'Eua** This graph plots the number of species present on an island in the south Pacific at four different time intervals. *Archaeological* indicates data from cultures present on the island before contact with Europeans *(Historic)*. From Steadman (1995).

- Sixty bird species endemic to the Hawaiian archipelago went extinct after the arrival of settlers there about 1,500 ybp. These species begin dropping out of the fossil record soon after the first fire pits, middens (trash piles), and tools appear in the digs.
- In New Zealand, 44 bird species went extinct after human colonization but before historical times. The losses included eight species of gigantic flightless birds called moas.
- On the island of 'Eua in the Kingdom of Tonga, only six of the 27 land birds represented in the prehuman fossil record are still extant (Figure 13.25).
- On each of the seven best-studied islands in central Polynesia, where research has been intense enough to recover and identify at least 300 bones and thus provide a reasonable sample of the fossil avifauna, at least 20 endemic species or populations were wiped out after human settlement.

The total losses are staggering. Extrapolating based on data from the best-studied sites, Steadman suggests that a minimum of 10 species or populations have been lost on each of Oceania's 800 major islands. This estimate is probably conservative (see Balmford 1996). Consider, for example, the fossil record for the small, forest-dwelling, flightless birds called rails. One to four endemic species have been recovered as fossils from each of the 19 islands where enough research has been done to uncover at least 50 land-bird bones. Extrapolating from this well-studied subset implies that there may have been 2,000 species of rail in the Pacific alone. But just four species are left in Oceania. As Steadman (1995: 1127) has written: "only the bones remain as evidence of one of the most spectacular examples of avian speciation."

Several possible agents of human-caused extinction have been documented in the digs (Olson and James 1982; James et al. 1987; Steadman 1995). In Hawaii, for example, bird bones have been found that are split or charred by fire. These observations suggest direct predation by humans. The presence of pig, dog, and rat remains in the human-associated deposits indicates that these animals were imported by the colonists. Because most islands in the world lacked mammalian predators before the arrival of people, the introduction of rats and dogs was potentially devastating for ground-nesting birds and for the many species, like rails, that had evolved flightlessness. Habitat destruction was also a factor. Records from early European visitors to Hawaii indicate that slash-and-burn agriculture, or permanent irrigated fields

Archaeological evidence suggests that humans were responsible for the extinction of at least 2,000 bird species on islands in the Pacific.

for the cultivation of taro, had virtually wiped out the lowland forests of Hawaii before the 1700s.

As a control for the hypothesis that humans were the causative agent, Steadman offers the Galápagos archipelago. The Galápagos lacked any permanent human settlement until 1535 and had a tiny population until 1800. Archaeological and paleontological work has been extensive on the five largest islands in the group: Over 500,000 vertebrate bones have been excavated. The extensive fossil record documents the loss of only three populations in the 4,000 to 8,000 years preceding the arrival of humans. Over 20 taxa, however, have been eliminated since humans arrived.

Is a Mass Extinction Event Underway?

As impressive (and discouraging) as the data are for Pleistocene megafauna and recent Polynesian birds, they do not approach the intensity or geographic scope of the Big Five, when between 50% and 90% of species were lost. What about more recent rates? The data in Table 13.1 indicate only modest numbers and rates of extinctions documented since 1600. Is concern about an impending mass extinction overblown?

To answer this question, it is important to note that the majority of the losses represented in Table 13.1 occurred on islands. These island extinctions, in turn, usually resulted from the introduction of non-native predators or competitors. This introduction process has probably peaked in intensity, and should be less important in the future.

TABLE 13.1 Extinction events in the past 400 years

Smith et al. (1993), who compiled these data, point out that they should be interpreted with caution. The numbers reported here are most accurate for well-studied groups like birds and mammals, but are almost certainly gross underestimates for poorly studied groups like insects and plants. Note that the category called "threatened" lumps together species considered by the International Union for the Conservation of Nature (IUCN) as probably extinct, endangered (likely to go extinct if present trends continue), or vulnerable (likely to become endangered if present trends continue).

		No. of species certified extinct since 1600	No. of species listed as threatened	Approximate total of recorded, extant species	Percentage extinct	Percentage threatened
Animals	Molluscs	191	354	10^5	0.2	0.4
	Crustaceans	4	126	4×10^3	0.01	3
	Insects	61	873	10^6	0.005	0.07
	Vertebrates	229	2,212	4.7×10^4	0.5	5
	Fishes	29	452	2.4×10^4	0.1	2
	Amphibians	2	59	3×10^3	0.07	2
	Reptiles	23	167	6×10^3	0.4	3
	Birds	116	1,029	9.5×10^3	1	11
	Mammals	59	505	4.5×10^3	1	11
	Total	485	3,565	1.4×10^6	0.04	0.3
Plants	Gymnosperms	2	242	758	0.3	32
	Dicotyledons	120	17,474	1.9×10^5	0.2	9
	Monocotyledons	462	4,421	5.2×10^4	0.2	9
	Palms	4	925	2820	0.1	33
	Total	584	22,137	2.4×10^5	0.3	9

Instead, current concern is focused on a different agent of extinction: habitat loss due to expanding human populations. The current human population is 5.7 billion and is growing at the rate of 1.6% per year (Figure 13.26). If this rate continues, world population will double in 43 years (Cohen 1995). The point is that a new extinction agent has emerged, and will grow in intensity unless human population growth declines rapidly.

Do we have any data to use in projecting the impact of human-induced habitat destruction as an agent of mass extinction? One approach (see Smith et al. 1993) is to look at extinctions that have occurred in the very recent past. These events are much more likely to be due to habitat destruction than the extinctions recorded in Table 13.1. For example, from 1986 to 1990, 15 vertebrates were added to the list of species formally considered extinct. If this rate were sustained into the future, it would take just 7,000 years to eliminate half of the 45,000 vertebrate species that have been named. If similar losses occurred among other taxonomic groups, this would clearly qualify as a mass extinction event comparable to the Big Five. The estimate is crude, however. Habitat destruction should eliminate particularly rare or localized species first. This fact suggests that the rate of 15 in four years is too high to be sustained, unless habitat destruction continues to accelerate.

Fortunately, two additional approaches can be employed to provide independent estimates of the extinction threat. The first method simply assumes that all species currently listed as threatened will go extinct in the next 100 years. A continued rate of loss over longer time intervals can then be made, based on this snapshot. The second method is based on estimates for the number of species occupying habitats of specified size (these are called species-area relationships). The rate of habitat loss, documented in satellite photographs taken over time, is then used to estimate the number of species likely to go extinct. Both types of analysis suggest that extinctions will occur at rates 100 to 1,000 times background levels (Pimm et al. 1995). These results are consistent with the estimate made by extrapolating from extremely recent extinction rates.

Although a paucity of data on the existing number of species and rates of recent loss makes these analyses very imprecise, their overall message is clear: A mass extinction event is probably inevitable if present trends in human population growth and habitat destruction continue.

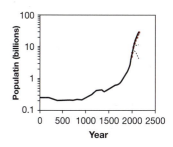

Figure 13.26 **Human population growth** The solid line shows the world human population from A.D. 1 to 1990. The three lines graphed from 1990 to 2150 represent projections made by the United Nations based on high, medium, and low growth parameters. From Cohen (1995).

13.7 Consequences of Mass Extinctions

Earlier in the chapter, we stated that background and mass extinctions are considered as separate processes because their consequences are qualitatively different. We now need to defend that statement.

The basic dichotomy can be stated as follows: Background extinctions result from natural selection, while mass extinctions result from chance (Raup 1986; Jablonski 1986b). Stated another way, background extinctions are based on changes in the fitnesses of individuals within populations; mass extinctions eliminate individuals irrespective of their fitness. This is because mass extinctions result from short-term exposure to unique and extreme environmental conditions. These events represent rapid and massive changes in the fitness land-scape, triggered by changes in physical conditions.

As a result, mass extinction is a drift-like evolutionary process. Survivors are not better adapted than the victims. They are just luckier, or more widely distributed.

Mass extinctions prune the tree of life in a largely random fashion.

Recovery and Replacement

What did the biosphere look like after each of the Big Five? After the end-Permian extinction, stromatolites become extraordinarily common in the marine fossil record (Erwin 1993). Stromatolites are sedimentary structures formed by bacteria and cyanobacteria. Today they are found only in extreme, biologically depauperate environments like high-salinity lagoons. Because of this, paleontologists consider them a "disaster form." On land, a strong fungal spike is observed in the post-extinction fossil record of the end-Permian, in numerous and widely spaced localities (Visscher et al. 1996). The fungal spike suggests that an enormous amount of rotting vegetation was available. The plant fossil record indicates that almost four million years passed before coniferous forests began to be reestablished.

Indications of low-diversity ecosystems also occur in the fossil record of the early Tertiary. We have already referred to the fern spike in the North American fossil record. This pronounced increase in fungal, fern, and lycopod (club moss) spores also occurs just across the K-T boundary of Australia (Retallack 1995). The ferns that dominate are from groups that are considered ecological opportunists. In contemporary environments, these types of species colonize blast zones created by volcanoes and landscapes devastated by extraordinarily hot forest fires.

Abundant stromatolites in shallow marine habitats and increases in ferns and fungi in terrestrial environments point to the same conclusion: After a mass extinction, ecosystems are radically simplified. The diversity and abundance of organisms are low.

Detailed records from specific taxonomic groups, traced across mass extinction boundaries, provide additional details about the recovery process. Figure 13.27, for example, illustrates the record of brachiopod species across the K-T boundary at a locality in Denmark. After the extinction event, brachiopods completely disappeared from the fossil record for over three million years. Some

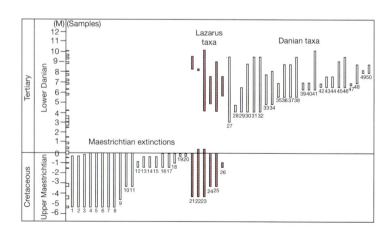

Figure 13.27   Brachiopods across the K-T boundary   Each vertical bar in this diagram represents a species. The Upper Maestrichtian (also spelled Maastrichtian) is the last portion of the last stage in the Cretaceous period; the Lower Danian is the first portion of the first stage in the Tertiary period. The time scale along the vertical axis is in millions of years; the horizontal line at time 0, labeled "Maestrichtian extinctions," indicates the K-T boundary. From Raup (1986).

of the species that reappear are new; others had existed before the catastrophe. The latter are called Lazarus taxa, and are found in a wide variety of taxonomic groups. Why do Lazarus taxa occur? The consensus interpretation is that they represent species whose populations were devastated, but not eliminated, by the extinction event. These reduced populations are thought to have existed for long periods of time in small geographic areas, called refugia, before they finally spread and recolonized parts of their former range. This recovery process is exceedingly slow on an ecological time scale. Raup (1994) maintains that most ecosystems take 5 to 10 million years to regain their former levels of diversity and abundance after a mass extinction event.

Without a doubt, the most important fact about mass extinctions is that the communities produced by recovery are often radically different than those that dominated before. Here are a few examples that drive this point home:

- After the end–Permian mass extinction, shallow-water marine communities came to be dominated by bivalve and gastropod molluscs instead of by brachiopods and bryozoans. Molluscs continue to dominate these environments today. On land, pteridophytes and primitive gymnosperms were replaced by advanced gymnosperms. Early insect groups, called the Paleoptera (old wings), had a simple hinge attaching the wings to the body so that the wings could not be folded over the back, as in modern dragonflies and mayflies. These taxa were almost completely replaced by neopterous (new winged) forms, which can fold their wings over their backs. The Neoptera now make up 98% of the modern insect fauna.

- A large suite of early tetrapod groups, including the mammal-like reptiles, were wiped out at the end–Triassic and replaced by turtles, crocodilians, dinosaurs, pterosaurs (flying reptiles), and mammals (Benton 1991).

- Mammals and birds radiated rapidly after the demise of the dinosaurs at the K–T boundary. Most of the current mammalian and bird orders had become established just 5 to 10 million years after the extinction event. Mammals and birds are still the most abundant, diverse, and widely distributed terrestrial vertebrates today.

A major consequence of mass extinctions, then, is that rapid radiations follow, and that these radiations lead to a sometimes dramatic reordering in the distribution and abundance of species (Jablonski 1989). In a metaphorical sense, the extinction event left vacant adaptive peaks, or empty ecological space. Frequently, the populations that climb these peaks do not come from the lineages that occupied the peak before.

A key point here is that all of the radiations we have just mentioned occurred in an ecological vacuum. Clams and snails, for example, did not replace brachiopods and bryozoans in the Triassic because they were competitively superior. Rather, brachiopods and bryozoans were wiped out, and *then* clams and snails radiated.

Another major consequence of mass extinctions is the loss of evolutionary potential. When mass extinctions eliminate certain adaptations, they also eliminate the possibility of further elaboration of those traits as well as the possibility that new innovations might be created from those starting points (Jablonski

Recoveries from mass extinctions are slow, and often result in radically different plant and animal communities.

1996). In a fundamentally important sense, mass extinctions change the direction of history.

The Role of Chance in Evolution

Because they are responsible for such thorough turnover in dominant life forms, mass extinctions are an important force in explaining the diversification of life through time. It is doubtful whether many of the major changes observed over the past 250 million years would have taken place had mass extinctions not occurred. The Big Five were defining moments in the history of life.

It is important to recognize, though, that these defining moments were largely random events. Just as mutation and drift introduce a strong random component into the process of adaptation, mass extinctions introduce chance into the process of diversification. This is because mass extinctions are a sampling process analogous to genetic drift. Instead of sampling allele frequencies, mass extinctions sample species and lineages. The only pattern we have observed in the sampling process is that clades with large geographic ranges are less likely to be wiped out.

The punchline? Chance plays a large role in the processes responsible for adaptation and diversity.

Catastrophes in Earth History

Research on mass extinctions has helped alter the perspective of evolutionary biologists in subtle but important ways. The classical Darwinian world view, formalized in the Modern Synthesis, was that evolution proceeded in slow, almost imperceptible steps. Research on adaptive radiations and the debate over the theory of punctuated equilibrium has produced a broader view: Some of the most important evolutionary events occur instantaneously in geological time (too quickly, in fact, to be documented in the fossil record). Research on mass extinctions now suggests that turning points in the history of life can also occur rapidly in ecological time. That is, truly catastrophic rates of change can and do occur.

Combined with an increased emphasis on the role of chance in directing change, recognition of extremely rapid evolution has helped make Darwinism a more pluralistic world view than originally conceived, with a broader and more dynamic agenda for research. This research program is increasingly interdisciplinary, phylogenetically based, and oriented toward solving real-world problems. Our agenda in Part IV is to sample just a few of the many topics inspiring contemporary research.

Summary

When rates of extinction per million years are plotted through the Phanerozoic, a continuous distribution of extinction intensities results. The five most intense extinction events are designated as mass extinctions and are commonly distinguished from background extinctions. During background times species go extinct due to natural selection, while mass extinctions eliminate species based on chance exposure to extreme environmental conditions. During both background and mass extinctions, widespread species are less likely to go extinct.

Levels of species diversity are slow to recover after a mass extinction, and formerly widespread and abundant species are often replaced by species in distantly related lineages.

The mass extinction at the end-Permian was the greatest crisis in the history of multicellular life. A major sea-level regression, ocean anoxia, the eruption of the Siberian traps, and rapid climate change all played a role in the event. The K-T extinction, in contrast, was caused by a bolide that slammed into the Earth near Mexico's Yucatán peninsula. The impact produced a variety of killing mechanisms including acid rain, widespread wildfires, intense cooling, extensive darkness, and an enormous tsunami in the Atlantic Ocean.

During the Pleistocene and recent times, two prominent extinction events have been documented: loss of large mammals in North America and Europe and loss of bird species on Polynesian islands. Although dramatic, these events were too local to qualify as mass extinctions. But current projections for species loss due to rapid habitat destruction indicate that a mass extinction event may now be underway.

Research on mass extinctions has altered our perspective on the history of life. Chance extinctions have played a large role in the diversification of life through time, and rates of speciation and extinction can be extremely rapid.

Questions

1. One of the (many) mysteries of the K-T extinction is the different fate of ammonites and nautiloids. These were mollusks with buoyant, chambered shells that lived in open-water habitats. Ammonites went extinct during the K-T extinction, but some nautiloids survived. The two groups had different reproductive strategies. Ammonites are thought to have produced many free-swimming young each year which fed near the ocean surface and grew rapidly. In contrast, a female nautilus produces just a few large eggs each year, each of which rests quietly in the depths for up to a year before hatching into a small, slow-growing nautilus. Based on these different reproductive strategies, suggest a possible hypothesis for why the nautiloids, but not the ammonites, might have been able to survive an asteroid impact. See the following reference for further discussion of the biology and history of ammonites and nautiloids:

   Ward, P. 1992. *On Methuselah's Trail: Living Fossils and the Great Extinctions.* (New York: W.H. Freeman & Co.).

2. The Biscay coast of France and Spain has especially well-preserved K-T strata. Yet until several years ago, no good ammonite specimens had yet been found near the K-T boundary, despite many years of study by two different teams. The paleontologists on these teams were well aware that ammonite fossils might exist, but might not have been found simply due to limited searching time and limited personnel. Would this limited sampling make the extinction of ammonites appear to be earlier or later than it actually was? If ammonites had been more common, how would that affect the chance of discovering ammonite fossils, and the resulting apparent time of extinction? How about if ammonites had been less common?

   As David Raup (1991) describes, in the late 1980s there happened to be a paleontology meeting near the Biscay Coast, and many of the paleontologists made a field trip to see these K-T boundary strata. The leader of one of the Biscay research teams took advantage of this opportunity by offering a prize for a good ammonite near the K-T boundary. The paleontologists fanned out and began combing the rocks, and one person discovered a lovely ammonite (and claimed the prize) after just a few hours of searching. What does this story illustrate about the need for fine-scale sampling? For more insight into the importance of high-intensity sampling see Marshall and Ward (1996).

3. In 1996, Gregory Retallack announced that he had found shocked quartz crystals dating from the same time as the end-Permian extinction. What is the implication of this finding? Other geologists point out no one has found any evidence of elevated iridium from these strata (despite much searching). What is the significance of the lack of iridium?

4. $^{13}C/^{12}C$ ratios can change due to several quite different causes. Review the studies of the end-Permian extinction and explain how each of the following different events can cause changes in $^{13}C/^{12}C$ ratios. What other evidence can clarify which event(s) actually occurred to cause the observed changes in $^{13}C/^{12}C$ ratios?

   Sea level regression
   A drop in primary productivity
   Volcanic activity

5. This chapter discussed many possible causes of the end-Permian and the K-T mass extinctions. As a review, examine the following list of possible causes of mass extinction. For the end-Permian extinction, what evidence is there for or against the occurrence of each event on the list? What is the proposed chain of causation—that is, which events are suspected to have caused some of the other events?

   Asteroid impact
   Drop in sea level (regression)
   Release of CO_2 stores from deep ocean waters
   Reduced O_2 levels in shallow oceans
   Continental glaciations
   Extremely high volcanic activity
   Acid rain
   Extensive wildfires

6. Because the data presented in Figure 13.16 are simple counts of genera, they do not control for the effects of phylogeny—an issue we warned about in Chapters 3 and 9. Jablonski and Raup tried to mitigate this problem by excluding rudists from their analysis. The rudistid clams have unusually large body size and happened to go extinct at or near the boundary. Note that it changed their result—with rudists included, there is a statistically significant trend for large-bodied taxa to be extinguished at the K-T boundary. Was removing the rudists justified? If you think that removing rudists was justified, was this removal sufficient to control for phylogenetic effects? Do you still believe the claim that body size did not affect the probability that bivalve genera would go extinct at the K-T boundary?

7. Norman Kitz (1996) has proposed the rather outlandish idea that mass extinctions occur whenever our solar system crosses the plane of our galaxy, exposing life on earth to unusually high levels of cosmic radiation. Kitz suggests that the high cosmic radiation caused high mutation rates, killing off many species but also causing some lucky species to develop evolutionary innovations. Is this hypothesis testable? (Two ways of approaching this question are to ask "Is each step in the proposed chain of events plausible?" and then to ask "Is there any evidence that the events actually happened at the same time as mass extinctions of the past?") In your opinion, are the end-Permian and K-T extinctions likely to have been caused by this occurrence? Why or why not? The original paper is:

   Kitz, N. 1996. Radiation, mutation, and extinction. *American Journal of Science* 296:954–965.

8. The popular press frequently portrays evolution on earth as leading inexorably to the evolution of intelligent life, namely humans. Yet it seems clear from the fossil record that the K–T mass extinction was necessary to open the door for subsequent mammalian radiations. Speculate on what the dominant terrestrial vertebrates might look like today if the asteroid had not struck the Earth at the end of the Cretaceous. Would humans have evolved? Might intelligent life have evolved?

9. Suppose you are talking to a friend about extinctions, and you mention that humans are known to have caused thousands of extinctions in the last few millenia. Your friend responds "So? Extinction is natural. Species have always gone extinct. So it's really not something we need to worry about." Is your friend correct that extinction is "natural"? If so, is the current rate of extinction typical? Is your friend correct that if extinctions are natural, then they are never a problem for the dominant lifeforms on Earth?

Exploring the Literature

10. A controversy erupted over the claim that the end-Permian marine extinctions were caused by an ocean turnover event that flushed water rich in carbon dioxide onto the continental shelves. To examine an interesting exchange of letters that promote competing explanations for the mother of all mass extinctions, see:

Martin, R.E., G.J. Vermeij, D. Dorritie, K. Caldeira, M.R. Rampino, A.H. Knoll, R.K Bambach, D. Canfield, J.P Grotzinger, P.B. Wignall, and R.J. Twitchett. 1996. Late Permian extinctions. *Science* 274:1549–1552.

Citations

Alvarez, L.W., W. Alvarez, F. Asaro, and H.V. Michel. 1980. Extraterrestrial cause for the Cretaceous-Tertiary extinction. *Science* 208:1095–1108.

Alvarez, W., F. Asaro, and A. Montanari. 1990. Iridium profile for 10 million years across the Cretaceous-Tertiary boundary at Gubbio (Italy). *Science* 250:1700–1702.

Alvarez, W., P. Claeys, and S.W. Kieffer. 1995. Emplacement of Cretaceous-Tertiary boundary shocked quartz from Chicxulub crater. *Science* 269:930–935.

Balmford, A. 1996. Extinction filters and current resilience: The significance of past selection pressures for conservation biology. *Trends in Ecology and Evolution* 11:193–196.

Benton, M.J. 1991. What really happened in the Late Triassic? *Historical Biology* 5:263–278.

Benton, M.J. 1995. Diversification and extinction in the history of life. *Science* 268:52–58.

Bourgeois, J., T.A. Hansen, P.L. Wiberg, and E.G. Kauffman. 1988. A tsunami deposit at the Cretaceous-Tertiary boundary in Texas. *Science* 241:567–570.

Claeys, P., J.-G. Casier, and S.V. Margolis. 1992. Microtektites and mass extinctions: Evidence for a late Devonian asteroid impact. *Science* 257:1102–1104.

Clemens, W.A., and L.G. Nelms. 1993. Paleoecological implications of Alaskan terrestrial vertebrate fauna in latest Cretaceous time at high paleolatitudes. *Geology* 21:503–506.

Cohen, J.E. 1995. Population growth and earth's human carrying capacity. *Science* 269:341–346.

Cooper, A., and D. Penny. 1997. Mass survival of birds across the Cretaceous-Tertiary boundary: Molecular evidence. *Science* 275:1109–1113.

Erwin, D.H. 1989. Regional paleoecology of Permian gastropod genera, southwestern United States and the end-Permian mass extinction. *Palaios* 4:424–438.

Erwin, D.H. 1993. *The Great Paleozoic Crisis: Life and Death in the Permian* (New York: Columbia University Press).

Erwin, D.H. 1994. The Permo-Triassic extinction. *Nature* 367:231–236.

Erwin, D.H. 1995. Permian global bio-events. In O.H. Walliser, ed. *Global Events and Event Stratigraphy* (Berlin: Springer), 251–264.

Erwin, D.H. 1996. The geologic history of diversity. In R.C. Szaro and D.W. Johnston, eds. *Biodiversity in Managed Landscapes* (New York: Oxford University Press), 3–16.

Feduccia, A. 1995. Explosive evolution in tertiary birds and mammals. *Science* 267:637–638.

Florentin, J-M.R.M., and G. Sen. 1990. Impacts, tsunamis, and the Haitian Cretaceous-Tertiary boundary layer. *Science* 252:1690–1693.

Glen, W. 1990. What killed the dinosaurs? *American Scientist* 78:354–370.

Grayson, D.K. 1984. Nineteenth-century explanations of Pleistocene extinctions: A review and analysis. In P.S. Martin and R.G. Klein, eds. *Quarternary Extinctions* (Tucson: University of Arizona Press), 5–39.

Heymann, D., L.P. F. Chibante, R.R. Brooks, W.S. Wolbach, and R.E. Smalley. 1994. Fullerenes in the Cretaceous-Tertiary boundary layer. *Science* 265:645–647.

Holser, W.T., and M. Magaritz. 1987. Events near the Permian-Triassic boundary. *Modern Geology* 11:155–180.

Isozaki, Y. 1994. Superanoxia across the Permo-Triassic boundary: Record in accreted deep-sea pelagic chert in Japan. In A.F. Embry, B. Beauchamp, and D.L. Glass, eds. *Pangea: Global Environments and Resources,* Memoir 17 (Canadian Society of Petroleum Geologists), 805–812.

Isozaki, Y. 1997. Permo-Triassic boundary superanoxia and stratified superocean: Records from lost deep sea. *Science* 276:235–238.

Jablonski, D. 1986a. Background and mass extinctions: The alternation of evolutionary regimes. *Science* 231:129–329.

Jablonski, D. 1986b. Evolutionary consequences of mass extinction. In D.M. Raup and D. Jablonski, eds. *Patterns and Processes in the History of Life* (Berlin: Springer-Verlag), 313–329.

Jablonski, D. 1989. The biology of mass extinction: A palaeontological view. *Philosophical Transactions of the Royal Society of London,* Series B 325:357–368.

Jablonski, D. 1991. Extinctions: A paleontological perspective. *Science* 253:754–757.

Jablonski, D. 1995. Extinctions in the fossil record. In J.H. Lawton and R.M. May, eds., *Extinction Rates* (Oxford: Oxford University Press), 25–44.

Jablonski, D. 1996. Body size and macroevolution. In D. Jablonski, D.H. Erwin, and J.H. Lipps, eds. *Evolutionary Paleobiology* (Chicago: University of Chicago Press), 256–289.

Jablonski, D., and D.M. Raup. 1995. Selectivity of end-Cretaceous marine bivalve extinctions. *Science* 268:389–391.

James, H.F., T.W. Stafford, Jr., D.W. Steadman, S.L. Olson, P.S. Martin, A.J.T. Jull, and P.C. McCoy. 1987. Radiocarbon dates on bones of extinct birds from Hawaii. *Proceedings of the National Academy of Sciences, USA* 84:2350–2354.

Kerr, R.A. 1994. Testing an ancient impact's punch. *Science* 263:1371–1372.

Kerr, R.A. 1996. Impact craters all in a row? *Science* 272:33.

Klein, R.G. 1984. Mammalian extinctions and stone age people in Africa. In P.S. Martin and R.G. Klein, eds. *Quaternary Extinctions* (Tucson: University of Arizona Press), 553–573.

Knoll, A.H., R.K. Bambach, D.E. Canfield, and J.P. Grotzinger. 1996. Comparative earth history and late Permian mass extinction. *Science* 273:452–457.

Marshall, C.R., and P.D. Ward. 1996. Sudden and gradual extinctions in the latest Cretaceous of western European Tethys. *Science* 274:1360–1363.

Martin, P.S. 1984. Prehistoric overkill: The global model. In P.S. Martin and R.G. Klein, eds. *Quarternary Extinctions* (Tucson: University of Arizona Press), 354–403.

McKinnon, W.B. 1992. Killer acid at the K/T boundary. *Nature* 357:15–16.

Mead, J.I., and D.J. Meltzer. 1984. North American Late Quaternary extinctions and the radiocarbon record. In P.S. Martin and R.G. Klein, eds. *Quarternary Extinctions* (Tucson: University of Arizona Press), 440–450.

Nichols, D.J., J.L. Brown, M.A. Attrep, Jr., and C.J. Orth. 1992. A new Cretaceous-Tertiary boundary locality in the western Powder River basin, Wyoming: Biological and geological implications. *Cretaceous Research* 13:3–30.

Olson, S.L., and H.F. James. 1982. Prodromus of the fossil avifauna of the Hawaiian islands. *Smithsonian Contributions in Zoology* 365:1–59.

Pimm, S.L., G.J. Russell, J.L. Gittleman, and T.M. Brooks. 1995. The future of biodiversity. *Science* 269:347–350.

Porter, S.C. 1989. Some geological implications of average Quaternary glacial conditions. *Quaternary Research* 32:245–261.

Rampino, M.R., B.M. Haggerty, and M.J. Benton. 1995. Mass extinctions and periodicity. *Science* 269:617–619.

Raup, D.M. 1979. Size of the Permo-Triassic bottleneck and its evolutionary consequences. *Science* 206:217–218.

Raup, D.M. 1986. Biological extinction in Earth history. *Science* 231:1528–1533.

Raup, D.M. 1991. A kill curve for Phanerozoic marine species. *Paleobiology* 17:37–48.

Raup, D.M. 1994. The role of extinction in evolution. *Proceedings of the National Academy of Sciences, USA* 91:6758–6763.

Raup, D.M., and D. Jablonski. 1993. Geography of end-Cretaceous marine bivalve extinctions. *Science* 260:971–973.

Raup, D.M., and J.J. Sepkoski, Jr. 1982. Mass extinction in the marine fossil record. *Science* 215:1501–1503.

Raup, D.M., and J.J. Sepkoski, Jr. 1984. Periodicity of extinctions in the geologic past. *Proceedings of the National Academy of Sciences, USA* 81:801–805.

Raup, D.M., and J.J. Sepkoski, Jr. 1986. Periodic extinction of families and genera. *Science* 231:833–836.

Renne, P.R., Z. Zichao, M.A. Richards, M.T. Black, and A.R. Basu. 1995. Synchrony and causal relations between Permian-Triassic boundary crises and Siberian flood volcanism. *Science* 269:1413–1416.

Retallack, G.J. 1995. Permian-Triassic life crisis on land. *Science* 267:77–80.

Schopf, T.J.M. 1979. The role of biogeographic provinces in regulating marine faunal diversity through geologic time. In J. Gray and A.J. Boucot, eds. *Historical Biogeography, Plate Tectonics, and the Changing Environment* (Corvallis, OR: Oregon State University Press), 449–457.

Schultz, P.H., and S. D'Hondt. 1996. Cretaceous-Tertiary (Chicxulub) impact angle and its consequences. *Geology* 24:963–967.

Sepkoski, J.J., Jr. 1989. Periodicity in extinction and the problem of catastrophism in the history of life. *Journal of the Geological Society of London* 146:7–19.

Signor, P.W. III, and J.H. Lipps. 1982. Sampling bias, gradual extinction patterns, and catastrophes in the fossil record. In L.T. Silver and P.H. Schultz, eds. *Geological Implications of Impacts of Large Asteroids and Comets on the Earth* (Geological Society Special Paper 190), 291–296.

Smith, F.D.M., R.M. May, R. Pellew, T.H. Johnson, and K.S. Walter. 1993. Estimating extinction rates. *Nature* 364:494–496.

Stanley, S.M. 1988a. Climatic cooling and mass extinction of Paleozoic reef communities. *Palaios* 3:228–232.

Stanley, S.M. 1988b. Paleozoic mass extinctions: Shared patterns suggest global cooling as a common cause. *American Journal of Science* 288:334–352.

Stanley, S.M., and X. Yang. 1994. A double mass extinction at the end of the Paleozoic era. *Science* 266:1340–1344.

Steadman, D.W. 1995. Prehistoric extinctions of Pacific island birds: Biodiversity meets zooarchaeology. *Science* 267:1123–1131.

Stuart, A.J. 1991. Mammalian extinctions in the late Pleistocene of northern Eurasia and North America. *Biological Review* 66:453–562.

Sweet, A.R., and D.R. Braman. 1992. The K/T boundary and contiguous strata in western Canada: Interactions between paleoenvironments and palynological assemblages. *Cretaceous Research* 13:31–79.

Swisher, C.C. III, J.M. Grajales-Nishimura, A. Montanari, S.V. Margolis, P. Claeys, W. Alvarez, P. Renne, E. Cedillo-Pardo, F. J-M. R. Maurrasse, G.H. Curtis, J. Smit, and M.O. McWilliams. 1992. Coeval ^{40}Ar/^{39}Ar ages of 65.0 million years ago from Chicxulub crater melt rock and Cretaceous-Tertiary boundary tektites. *Science* 257:954–958.

Valentine, J.W., and D. Jablonski. 1991. Biologic effects of sea level change: The Pleistocene test. *Journal of Geophysical Research* 96:6873–6878.

Van Valen, L. 1973. A new evolutionary law. *Evolutionary Theory* 1:1–30.

Visscher, H., H. Brinkhuis, D.L. Dilcher, W.C. Elsik, Y. Eshet, C.V. Looy, M.R. Rampino, and A. Traverse. 1996. The terminal Paleozoic fungal event: Evidence of terrestrial ecosystem destabilization and collapse. *Proceedings of the National Academy of Sciences, USA* 93:2155–2158.

Wang, K. 1992. Glassy microspherules (microtektites) from an Upper Devonian limestone. *Science* 256:1547–1550.

Weems, R.E. 1992. The terminal Triassic catastrophic extinction event in perspective—a review of Carboniferous through early Jurassic terrestrial vertebrate extinction patterns. *Palaeogeography, Palaeoclimatology, and Palaeoecology* 94:1–29.

Wignall, P.B. and R.J. Twitchett. 1996. Oceanic anoxia and the end Permian mass extinction. *Science* 272:1155–1158.

Wolbach, W.S., I. Gilmour, E. Anders, C.J. Orth, and R.R. Brooks. 1988. Global fire at the Cretaceous-Tertiary boundary. *Nature* 334:665–669.

PART IV

CURRENT RESEARCH— A SAMPLER

In Parts I, II, and III we explored the methods of evolutionary analysis, the mechanisms of evolution, and the history of life on Earth. In doing so we developed a toolkit of knowledge, concepts, and techniques for asking and answering questions about organismal diversity and adaptation. In Part IV we use this toolkit to investigate issues that drive current research in evolutionary biology. The topics we cover represent but a small sample of the questions under active investigation. We have chosen them in the hopes of highlighting research areas most relevant to readers' personal and professional lives. Although our emphasis is on current studies, many of the chapters also include reviews of classical research. This is because many of the questions we consider have held the attention of evolutionary biologists for decades.

Chapter 14 surveys recent advances in answering a question that has been a fundamental motivator for evolutionary biologists since the field's beginnings: Where do humans come from? Chapter 15 addresses classical and recent work on a question Darwin posed: Why are the females and males within a species often strikingly different? Chapter 16 considers social behavior, focusing in particular on an evolutionary paradox: How could natural selection favor the apparently self-sacrificing behavior exhibited by individuals in a variety of species? Among the questions posed in Chapter 17 is: Why do organisms age and die? Chapter 18 explores applications of evolutionary thinking to human behavior, human health, and our species' evolutionary future. Finally, Chapter 19 considers the diversity of life on Earth, the extent to which that diversity is under threat from human activities, and the ways in which an evolutionary analysis can inform efforts to mitigate the biodiversity crisis.

Human Evolution

The first printing of *On the Origin of Species* sold out on 24 November 1859, the very day it was published. Among the profound implications that made the book such a sensation was what it told its readers about themselves. Although Darwin saw this as clearly as anyone, his only explicit treatment of human evolution was a single paragraph in the last chapter, in which his strongest claim was that "Light will be thrown on the origin of man and his history" (Darwin 1859, page 488).

Not until 12 years later did Darwin reveal the depth and breadth of his thinking about humans. In 1871 he published a two-volume work, *The Descent of Man, and Selection in Relation to Sex*. In the introduction Darwin explained his initial reticence on the subject of human evolution: "During many years I collected notes on the origin or descent of man, without any intention of publishing on the subject, but rather with the determination not to publish, as I thought that I should thus only add to the prejudices against my views" (Darwin 1871, page 1).

Darwin's apprehensions were well-founded: The human implications of evolutionary biology have been and remain the focus of heated controversy. In 1925 Tennessee schoolteacher John T. Scopes was convicted of violating a new state law prohibiting the teaching of evolution (see Chapter 2). The Scopes case was popularly known as the Monkey Trial, indicating that for many observers the central issue at stake was the origin of the human species. In 1995 the Alabama state board of education ruled that all textbooks discussing evolution must carry a disclaimer that admonishes readers to consider evolution as theory, not fact. National Public Radio reporter Debbie Elliot interviewed members of the school board about their decision (National Public Radio 1995). Again, the origin of our species was a key issue. Boardmember Stephanie Bell, for example, told Elliot, "Most people do not believe that we evolved from apes. . . . "

In this chapter we explore research on the evolutionary history of our species. We start by reviewing attempts to determine the evolutionary relationships among humans and the extant apes. Then we consider the fossil evidence bearing on the course of human evolution following the split between our lineage and the lineage of our closest living relatives. Next we look at fossil and molecular evidence on the emergence of *Homo sapiens.* Finally, we consider the evolutionary origins of some of our species' defining characteristics, including tool use and language. Our exploration illustrates that the subject of human evolution generates controversies within the scientific community that, while different in focus, are as heated as those it generates among the lay public.

14.1 Relationships Among Humans and the Extant Apes

Humans *(Homo sapiens)* belong to the primate taxon Catarrhini, which includes the Old World monkeys, such as the baboons and macaques, and the Hominoidea (Figure 14.1). The Hominoidea, also refered to as the hominoids, include the gibbons *(Hylobates)* of southeast Asia, and the Hominidae. The Hominidae (Wilson and Reeder 1993), or hominids, include the orangutan *(Pongo pygmaeus),* also of southeast Asia, and three African species: the gorilla *(Gorilla gorilla),* the common chimpanzee *(Pan troglodytes),* and the bonobo, or pygmy chimpanzee *(Pan paniscus).* The gibbons, orangutan, gorilla, and chimpanzees are

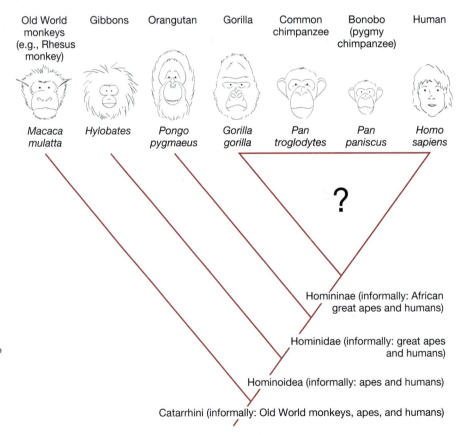

Figure 14.1 **Phylogeny of the Catarrhini** This evolutionary tree shows the relationships among the Old World monkeys, represented by a macaque, and the apes and humans. Among the apes, the gibbons branch off first, followed by the orangutan. The evolutionary relationships among the gorilla, the two chimpanzees, and humans (triangle with question mark) have been the subject of considerable dispute. The classification used here follows that in Pough et al. (1996).

refered to informally as the apes. The orangutan, gorilla, and chimpanzees are refered to as the great apes.

Humans Belong to the Same Clade as the Apes

Scientists universally agree that humans belong with the apes in the Hominoidea. Humans share with the apes numerous derived characteristics (synapomorphies). These evolutionary innovations distinguish the hominoids from the rest of the Catarrhini and indicate that the hominoids are descended from a common ancestor (see Chapter 10). The shared derived traits of the hominoids include taillessness, a more erect posture, greater flexibility of the hips and ankles, increased flexibility of the wrist and thumb, and changes in the structure and use of the arm and shoulder (Andrews 1992; see also Groves 1986; Andrews and Martin 1987). In addition to this morphological evidence, the molecular analyses described later in this chapter unequivocally demonstrate that humans are hominoids.

Humans Belong to the Same Clade as the African Great Apes

Figure 14.1 includes a reconstruction of the phylogenetic relationships among the hominoids. This reconstruction places humans with the great apes in the Hominidae. Furthermore, it places humans with the African great apes in the Homininae. This reconstruction has raised dispute. Jeffrey Schwartz (1984), for example, suggested on morphological grounds that humans and orangutans might be closest relatives. Kluge's (1983) analysis indicated that humans might not be hominids at all, but rather the sister group of a clade containing orangutans, gorillas, and the two chimpanzees.

But in recent years, as more data have been collected and analyzed, scientists in all fields have accepted the tree in Figure 14.1. Cladistic analyses of morphology support the tree. Humans and the African great apes share a number of derived traits that distinguish them from the rest of the apes. These synapomorphies include enlarged brow ridges, shortened but stout canine teeth, changes in the front of the upper jaw (premaxilla), fusion of certain bones in the wrist, enlarged ovaries and mammary glands, changes in muscular anatomy, and reduced hairiness (Ward and Kimbel 1983; Groves 1986; Andrews and Martin 1987; Andrews 1992; Begun et al. 1997).

Molecular analyses concur; they have indicated a close relationship between humans and the African great apes since the beginnings of modern molecular systematics in the 1960s. In 1967, for example, Vincent Sarich and Allan Wilson published the results of an immunological study of the evolutionary relationships among the apes. Working in the era before DNA sequencing, Sarich and Wilson used an indirect method to determine the relative similarities of proteins from different species. Sarich and Wilson took purified serum albumin, a blood protein, from humans, and injected it into rabbits. After giving the rabbits time to make antibodies against the human albumin protein, Sarich and Wilson took blood serum from the rabbits (which contained rabbit antihuman antibodies). The researchers mixed the rabbit serum with purified serum albumin from a variety of apes and Old World monkeys. The scientists used the strength of the

Morphological and molecular analyses demonstrate that humans are closely related to gorillas and chimpanzees.

TABLE 14.1 Similarities among the serum albumins of various apes
The numbers reflect the relative strength of the immune reaction between serum albumin from the animals listed in the first column and rabbit antibodies raised against serum albumin from the animals listed at the top of the remaining three columns. The numbers indicate similarity between the serum albumin proteins of the animals in question: 0 means no similarity; 1 means identical. The second column, for example, indicates that gorilla albumin (0.92) is more similar to human albumin than is gibbon albumin (0.78). The numbers in the table are not all entirely consistent with each other. Calculated from data in Sarich and Wilson (1967).

Serum albumin from:	Relative reaction with rabbit . . .		
	. . . Antihuman antibodies	. . . Antichimp	. . . Antigibbon
Human	1.00	0.92	0.78
Chimpanzee	0.88	1.00	0.71
Bonobo	0.88	1.00	0.71
Gorilla	0.92	0.85	0.76
Orangutan	0.82	0.81	0.78
Siamang	0.77	0.80	0.93
Gibbon	0.78	0.80	1.00
Mean of 6 Old World monkeys	0.41	0.45	0.44

immune reaction between, for example, rabbit antihuman antibodies and chimpanzee serum albumin, as a measure of the similarity between chimpanzee and human serum albumin (Table 14.1).

Sarich and Wilson assumed that the similarity of two species' serum albumin proteins reflects the species' evolutionary kinship. This assumption allowed Sarich and Wilson to use their data to estimate the phylogeny of the hominoids (Figure 14.2). The phylogeny shows that humans are closely related to gorillas

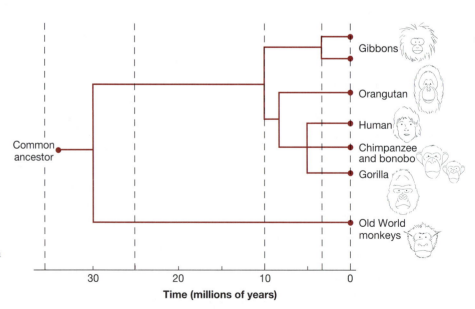

Figure 14.2 **Sarich and Wilson's phylogeny of the Hominoidea**   The time line on the bottom is in millions of years before present. From Sarich and Wilson (1967).

and the two chimpanzees. Sarich and Wilson put a time line on their phylogeny by assuming that (1) serum albumin evolves at a constant rate; and (2) the split between the apes and the Old World monkeys occurred 30 million years ago (this latter assumption is based on the fossil record). The time line suggests that humans and the African great apes shared a common ancestor about 5 million years ago, which is much more recently than scientists had previously suspected (see Lowenstein and Zihlman 1988). We will review other, more recent molecular phylogenies shortly; all are consistent with Sarich and Wilson's conclusions in showing a close kinship between humans and the African great apes.

The phylogenies in Figure 14.1 and Figure 14.2 show that humans, gorillas, and the two chimpanzees are closest relatives, but they do not resolve the evolutionary relationships among these four species. The true phylogeny for humans, gorillas, and the two chimpanzees could be any one of the four trees shown in Figure 14.3. It is probably safe to say that more scientists have invested more effort in attempting to determine which of these trees is correct than has been invested in any other species-level problem in the history of systematics.

Humans, Gorillas, Chimpanzees, and Bonobos: Morphological Evidence

Paleontologists have attempted to solve the human/African ape puzzle by performing cladistic analyses of morphology. These researchers point out several features shared by gorillas and the two chimpanzees, but lacking in humans. These traits include characteristics of tooth enamel (Martin 1985; but see Beynon et al. 1991), and especially skeletal traits associated with knucklewalking (Andrews and Martin 1987). Knucklewalking is a derived trait in the African great apes. Gorillas and both species of chimpanzee are knucklewalkers, whereas humans are not. The simplest explanation for the distribution of knucklewalking is that the human lineage diverged first from the lineage that would later produce the gorilla and the two chimpanzees (Figure 14.3c). This scenario requires only one appearance of knucklewalking, in a common ancestor of the gorilla and chimpanzees,

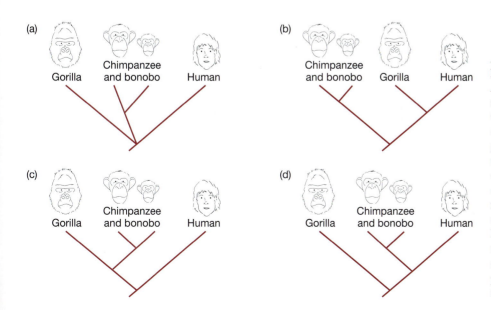

Figure 14.3 **Possible phylogenies of humans and the African great apes** The figure shows four possible resolutions of the evolutionary relationships among humans and the African great apes. All assume that the two species of chimpanzee are closest relatives. The true tree could have: (a) A genuine simultaneous three-way split (trichotomy); (b) Gorillas and humans as closest relatives; (c) Gorillas and chimpanzees as closest relatives; or (d) Chimpanzees and humans as closest relatives.

and no losses. However, the scenario does require that several traits shared only by humans and the two chimpanzees be interpreted as either ancestral traits that were lost in gorillas, or as convergent traits that evolved independently in humans and the chimpanzees. These traits include features of the teeth, skull, and limbs, delayed sexual maturity, and prominent labia minora in females and a pendulous scrotum in males (Groves 1986; Begun 1992).

Cladistic reconstructions of evolutionary trees depend on accurate identification of which traits are ancestral and which are derived (see Chapter 10). David R. Begun classified characteristics of the skulls of the great apes by including in his analysis an extinct European ape, called *Dryopithecus*. *Dryopithecus* is known only from fossils about 10 million years old (Begun 1992; see also Begun 1995). *Dryopithecus* shares several cranial traits with gorillas that are absent in the two chimpanzees and humans. These traits might previously have been classified as uniquely derived in gorillas, but given their presense in *Dryopithecus* the traits now appear to be ancestral. This, in turn, means that some of the traits thought to be ancestral or convergent in humans and chimpanzees now appear to be derived. When Begun reconstructed the ape evolutionary tree with the new classification of traits, he concluded that humans and chimpanzees are closest relatives (Figure 14.3d). This implies either that (1) the most recent common ancestor of humans, gorillas, and the chimpanzees was a knucklewalker, and that knucklewalking was subsequently lost in the human lineage; or (2) that knucklewalking evolved independently in gorillas and the chimpanzees. It also implies that the delayed sexual maturity and shared genital anatomy of the humans and chimpanzees need have evolved only once.

David Dean and Eric Delson (1992) are not convinced by Begun's reasoning. They argue that some of the skull features that Begun believes are shared derived traits in humans and chimpanzees may be ancestral or convergent, and that knucklewalking may not be so readily evolved or lost as Begun's phylogeny requires. Dean and Delson consider the phylogeny of humans, gorillas, and chimpanzees to be an as-yet-unresolved trichotomy (Figure 14.3a; see also Andrews 1992).

A recent morphological analysis suggests that humans and chimpanzees are closest relatives, but this conclusion is controversial.

Humans, Gorillas, Chimpanzees, and Bonobos: Molecular Evidence

Molecular biologists have attempted to resolve the human/African great ape phylogeny by analyzing DNA sequences. These efforts are producing a consensus that humans and the chimpanzees are closest relatives (Figure 14.3d). The evidence is still controversial, however. Figure 14.4 shows three estimates of the phylogeny based on different parts of the genome. The tree in Figure 14.4a is based on the mitochondrial genome, which is inherited only along the maternal line. The tree in Figure 14.4b is based on a gene on the Y chromosome, which is inherited only along the paternal line. The tree in Figure 14.4c is based on autosomal nuclear genes, which are inherited both maternally and paternally. All three phylogenies strongly support the version of the human/African great ape phylogeny in which humans and the chimpanzees are closest relatives (Figure 14.3d).

Recent molecular analyses indicate that humans and chimpanzees are closest relatives . . .

Jonathan Marks (1994) objects to this interpretation of the molecular data. Marks emphasizes that phylogenies such as those in Figure 14.4 are actually gene

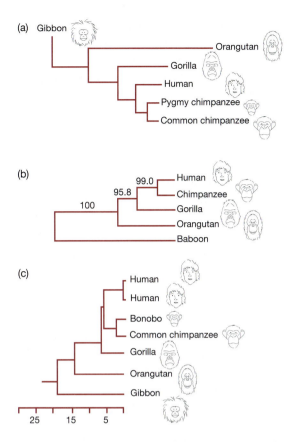

Figure 14.4   Three molecular phylogenies of humans and the African great apes
(a) A phylogeny based on mitochondrial DNA. Satoshi Horai and colleagues (1992) analyzed variation in the sequence of a 4,700-base-pair-long stretch of the mitochondrial
genome, including genes for 11 transfer RNAs and 6 proteins. They estimated the phylogeny
using the maximum likelihood method (see Chapter 10). Statistically, this tree, in which humans and chimpanzees are closest relatives, is about 10^{23} times as likely as the alternative tree
in which chimpanzees and gorillas are closest relatives. Horai and colleagues estimate that
the gorillas branched off the tree 7.7 $\pm$ 0.7 million years ago, and that humans and chimpanzees diverged 4.7 $\pm$ 0.5 million years ago. From Horai et al. (1992).   (b) A phylogeny
based on a Y-linked gene. Heui-Soo Kim and Osamu Takenaka (1996) analyzed variation in
the sequence of the gene for testis-specific protein Y, a protein expressed only in the testes.
Kim and Takenaka estimated the phylogeny by the maximum parsimony method (see Chapter 10). The tree has three nodes; the numbers on the branches indicate the percentage of
bootstrap replicates in which the most parsimonious tree contained each node. These results
provide strong statistical support for the hypothesis that humans and chimpanzees are closest
relatives. From Kim and Takenaka (1996).   (c) A phylogeny based on autosomal genes.
Morris Goodman and colleagues (1994) analyzed sequence variation in noncoding regions
of the β-globin gene cluster. They estimated the phylogeny by the maximum parsimony
method (see Chapter 10). They also performed a maximum likelihood analysis, and found
strong statistical support for the hypothesis that humans and chimpanzees are closest relatives. The scale bar below the phylogeny shows the estimated times for the branch points, in
millions of years ago. From Goodman et al. (1994).

. . . *but the phylogenies of genes and the phylogenies of species are not necessarily the same.*

trees, not species trees. Marks points out that if an ancestral species was genetically variable for the gene under study, then the gene tree estimated from sequence data may not be the same as the true species tree. Figure 14.5 illustrates the reasoning. If different descendent species lose different ancestral alleles, and if we sequence only one allele from each descendant species, then we can end up reconstructing only a portion of the original gene tree. This portion may imply a different branching pattern than the true species tree. Marks notes that his own study of chromosome morphology (1993), and Djian and Green's study of the involucrin gene (1989), examined variation within as well as between species. Both studies suggest that gorillas and the chimpanzees are closest relatives (Figure 14.3c). In light of the difference in conclusions between these studies and studies such as those shown in Figure 14.4, Marks asserts that the human/African great ape phylogeny is still an unresolved trichotomy (Figure 14.3a).

Maryellen Ruvolo and colleagues (1994) address Marks' objection. Ruvolo and colleagues reconstruct a mitochondrial gene tree using several samples each

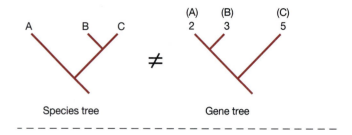

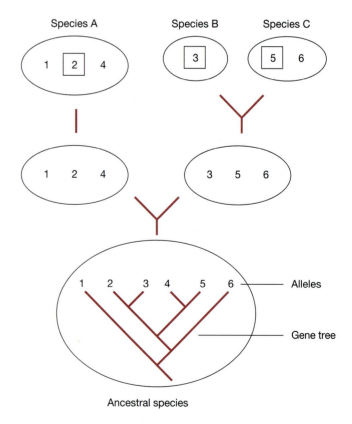

Figure 14.5 **Gene trees versus species trees**   Read the figure from the bottom to the top. The oval at the bottom represents an ancestral species containing six alleles for a gene. The alleles are related to each other as shown by the gene phylogeny inside the oval: Each allele was derived from its ancestors via a series of mutations. Moving up the figure, a speciation event produces two sister species from the ancestral species. By selection or drift, one species loses alleles 3, 5, and 6, while the other species loses alleles 1, 2, and 4. Moving up the figure, another speciation event occurs, followed by the loss of alleles in its descendant species. We now have: species A, containing alleles 1, 2, and 4; species B, containing allele 3; and species C, containing alleles 5 and 6. Now imagine that we sequence a single allele from each species (boxes) and reconstruct a phylogeny. In the true species tree, species B and C are closest relatives, but in the gene tree alleles 2 and 3 are closest relatives. The gene tree and the species tree show different branching patterns. Note that if we had sampled more extensively from each species and found all six alleles, we would have realized that the gene tree was a misleading guide for estimating the species phylogeny, because the alleles from species A, for example, would not cluster together. After Ruvolo (1994).

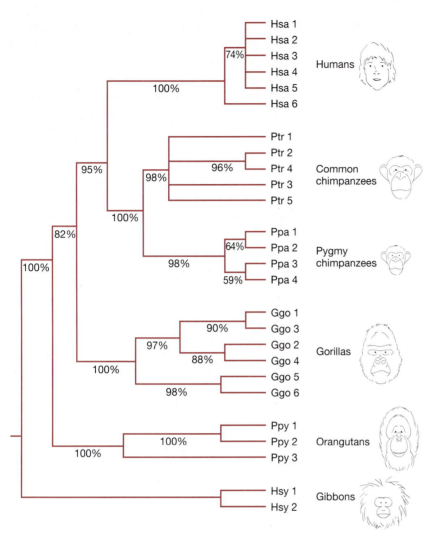

Figure 14.6 **Phylogeny of mitochondrial cytochrome oxidase II alleles in humans and the African great apes** Ruvolo and colleagues estimated this tree using the maximum parsimony method; the numbers on the branches indicate the percentage of times each branch was present in the most parsimonious trees from 1,000 bootstrap replicates (see Chapter 10). Note that the branch leading to the split between humans and chimpanzees has a bootstrap value of 95%. This represents strong support for the hypothesis that humans and chimpanzees are closest relatives. From Ruvolo et al. (1994).

from all of the ape species (Figure 14.6). If the gene tree were different from the species tree, then different haplotypes from a single species might fail to cluster together. For example, the human haplotype Hsa 1 might be more closely related to one of the chimpanzee haplotypes than to the other human haplotypes. In fact, in Ruvolo et al.'s tree all the haplotypes cluster together by species. This suggests that the gene tree is a reliable estimate of the species tree, and that humans and the chimpanzees are indeed closest relatives.

However, Ruvolo et al.'s test is not definitive. Had the mitochondrial sequences failed to cluster by species, it would have proved that the gene tree and the species tree are different. But failure to prove that the gene and species trees are different does not prove that they are the same. Look back at Figure 14.5 and consider the following scenario. Imagine that after the second speciation event, species A, B, and C all go through bottlenecks in which the number of individuals in each species is small. Now imagine that genetic drift eliminates alleles 1 and 4 in species A, and allele 6 in species C. Genetic diversity has been reduced to a single allele in each species. Now imagine that all three species

A more thorough molecular analysis supports the hypothesis that humans and chimpanzees are closest relatives . . .

recover from their bottlenecks, and accumulate new genetic diversity. In species A this diversity accumulates as a result of mutations in allele 2; in species B this diversity accumulates as a result of mutations in allele 3; in species C this diversity accumulates as a result of mutations in allele 5. Now we reconstruct the phylogeny. No matter how many alleles we include from each species, the alleles will cluster by species, and the gene tree and species tree will be different.

A way around this problem is to reconstruct the phylogeny using several independent genes, and see if the independent phylogenies agree. Ruvolo (1995) reviews and tallies the independent datasets of DNA sequences that are informative about the human/African great ape phylogeny. Ruvolo points out that if Marks is correct, and there is a real three-way split of the branches leading to humans, gorillas, and the two chimpanzees, then there should be equal numbers of independent datasets supporting a human-chimpanzee pairing, a human-gorilla pairing, and a chimpanzee-gorilla pairing. Ruvolo counts all mitochondrial DNA studies as a single dataset, because all mitochondrial genes are linked and are thus not independent of each other. Likewise, any groups of nuclear genes that are near each other on the same chromosome count as a single dataset, because the genes are linked. Ruvolo counts 10 independent datasets: 8 show humans and the chimpanzees as closest relatives; 2 show gorillas and the chimpanzees as closest relatives; none show humans and gorillas as closest relatives. Ruvolo calculates that under Marks' hypothesis of a true trichotomy, this distribution of results has a probability of only 0.0034. Ruvolo concludes that the molecular phylogeny data reject the trichotomous tree, and favor the tree in which humans and the chimpanzees are closest relatives (Figure 14.3d). For more of the controversy concerning the molecular evidence, see the exchange between Ruvolo (1994; 1995) and Green and Djian (1995), Marks (1994; 1995), and Rogers and Comuzzie (1995).

. . . as does a combined analysis of several molecular datasets.

14.2 The Recent Ancestry of Humans

According to the evidence presented in Figure 14.4a and c, humans and chimpanzees last shared an ancestor about 5 million years ago. What has been the pattern of evolution leading from that common ancestor to ourselves? Has the history of our lineage been one of steady transformation of a single lineage, finally culminating in *Homo sapiens,* or have there been repeated splits and extinctions in our recent evolutionary tree?

The Fossil Evidence

The fossil record includes a diversity of hominids that lived after the human lineage and the chimpanzee lineage separated.

Fossils provide the only data available for answering these questions. The fossil database for hominids is frustratingly sparse, but steadily improving (see Tattersall 1995; Johanson et al. 1996; Tattersall 1997). Illustrations of some of the key specimens appear in Figures 14.7 through 14.10.

Paleoanthropologists disagree about the most appropriate names for many of these specimens. We use the names used by Johanson et al. (1996) in the belief that they will be the names most familiar to readers. In several cases we note alternative names. Likewise, paleoanthropologists disagree about the number of species represented by the specimens in the figures (see Tattersall 1986; 1992). For example, the specimens of *Homo habilis* and *Homo rudolfensis* in Figure 14.9b and c are both from Koobi Fora, Kenya, and are both about 1.9 million years old. Some researchers consider them to be variants of the same species *(Homo habilis)*, whereas others consider them different species. As with the names, we have followed the classification used by Johanson et al. (1996). The time ranges noted in the figures are also those given in Johanson et al. (1996); they differ somewhat from the estimates of other researchers, including the estimates given by Strait et al. (1997) and used in Figure 14.11.

Figure 14.7 shows examples of the gracile australopithecines and *Ardipithecus.* The species depicted in Figure 14.7a and b, *Australopithecus africanus* and *Australopithecus afarensis,* had skulls with small braincases (400 to just over 500 cm³) and relatively large, projecting faces (Johanson et al. 1996). The females grew to heights of about 1.1 meter (3′7″), whereas the males were some 1.4 to 1.5 meters tall (4′7″ to 4′11″). Both species walked on two legs. Evidence for their erect posture includes the position of the foramen magnum, the hole in the base of the skull through which the spinal cord connects to the brain. In *A. africanus* and *A. afarensis* the foramen magnum faces downward as it does in our own skulls, instead of backward as it does in quadrupeds. Other evidence for bipedal

(a) Name: ***Australopithecus africanus***
Specimen: Sts 5
Age: 2.5 million years
Found by: Robert Broom and John T. Robinson
Location: Sterkfontein, South Africa
Color photo: Johanson et al.
   (1996) pages 3; 135

Species Time Range: ~2.4–2.8 Ma

(b) Name: ***Australopithecus afarensis***
Also known as: *Praeanthropus africanus*
Specimen: Reconstruction from fragments
Color photo of same species:
   Johanson et al. (1996) page 129

Species Time Range: ~3.0–3.9 Ma

(c) Name: ***Australopithecus anamensis***
Specimen: KNM-KP 29281
Age: 4.1 million years
Found by: Peter Nzube
Location: Kanapoi, Kenya
Color photo:
   Johanson et al. (1996) page 123

Species Time Range: ~3.9–4.2 Ma

(d) Name: ***Ardipithecus ramidus***
Originally named as: *Australopithecus ramidus*
Specimen: ARA-VP-1/128
Age: 4.4 million years
Found by: T. Assebework
Location: Aramis, Ethiopia
Color photo of same species:
   Johanson et al. (1996) page 116

Species Time Range: ~4.4 Ma

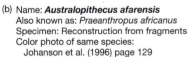

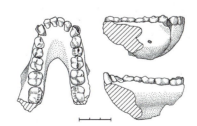

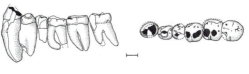

Figure 14.7 **Gracile australopithecines and *Ardipithecus***
(a) By Don McGranaghan, page 70 in Tattersall (1995). Scale bar = 1 cm.   (b) By Don McGranaghan, page 146 in Tattersall (1995). Scale bar = 1 cm. (c) After Figure 1a in Leakey et al. (1995). Scale unit: 1 cm. (d) After Figure 3b in White et al. (1994); see also White et al. (1995). Scale bar = 1 cm.

locomotion in these species comes from the structure of other bones, and from the fossilized footprints at Laetoli, Tanzania of a pair of *A. afarensis* that walked side-by-side through fresh ash from the Sadiman volcano 3.6 million years ago (Stern and Susman 1983; White and Suwa 1987).

The two species depicted in Figure 14.7c and d, *Australopithecus anamensis* and *Ardipithecus ramidus,* are less well known. The structure and size of a tibia from *A. anamensis* indicates that its owner was a biped somewhat larger than *A. afarensis* (Leakey et al. 1995). *Ardipithecus ramidus* has skeletal features that suggest it was a biped, but its discoverers are withholding final judgement until they can complete a more thorough analysis (White et al. 1994; Johanson et al. 1996).

Figure 14.8 shows examples of the robust australopithecines. Like the gracile australopithecines, these species had relatively small braincases and large faces. Unlike the gracile australopithecines, they had enormous cheek teeth, robust jaws, and massive jaw muscles, sometimes anchored to a bony crest running along the centerline on the top of the skull (Johanson et al. 1996). These adaptations for powerful chewing have given one of the species, *A. boisei,* the nickname "nutcracker man." The robust australopithecines were about the same size as the gracile forms, and all were bipeds.

Figure 14.9 shows examples of early members of the genus *Homo,* all from Africa. Compared to the australopithecines, *Homo* species have larger braincases and relatively smaller faces (Johanson et al. 1996). Braincase volume in the genus *Homo* ranges from around 600 cm^3 in *H. habilis* to some 2,000 cm^3 in ex-

(a) Name: **Australopithecus robustus**
Also known as: *Paranthropus robustus*
Specimen: SK 48
Age: 1.5–2.0 million years
Found by: Fourie
Location: Swartkrans, South Africa
Color photo: Johanson et al.
  (1996) pages 108; 150

Species Time Range: ~1.0–2.0 Ma

(b) Name: **Australopithecus boisei**
Also known as: *Paranthropus boisei*
Specimen: KNM-ER 406
Age: 1.7 million years
Found by: Richard Leakey and H. Mutua
Location: Koobi Fora, Kenya
Color photo: Johanson et al.
  (1996) pages 54; 159; 160

Species Time Range: ~1.4–2.3 Ma

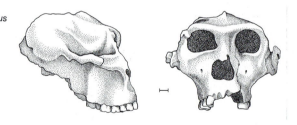

Figure 14.8 Robust australopithecines (a) After pages 108 and 150 in Johanson et al. (1996). Scale unit = 1 cm.
(b) By Don McGranaghan, page 131 in Tattersall (1995). Scale bar = 1 cm. (c) By Don McGranaghan, page 195 in Tattersall (1995). Scale bar = 1 cm.

(c) Name: **Australopithecus aethiopicus**
Also known as: *Paranthropus aethiopicus*
Specimen: KNM-WT 17000 (Black Skull)
Age: 2.5 million years
Found by: Alan C. Walker
Location: Lake Turkana, Kenya
Color photo: Johanson et al.
  (1996) pages 153; 154

Species Time Range: ~1.9–2.7 Ma

(a) Name: **Homo ergaster**
    Also known as: (African) *Homo erectus*
    Specimen: KNM-ER 3733
    Age: 1.75 million years
    Found by: Bernard Ngeneo
    Location: Koobi Fora, Kenya
    Color photo: Johanson et al.
      (1996) pages 180; 181

    Species Time Range: ~1.5–1.8 Ma

(b) Name: **Homo habilis**
    Specimen: KNM-ER 1813
    Age: 1.9 million years
    Found by: Kamoya Kimeu
    Location: Koobi Fora, Kenya
    Color photo: Johanson et al.
      (1996) pages 6; 175

    Species Time Range: ~1.6–1.9 Ma

(c) Name: **Homo rudolfensis**
    Also known as: *Homo habilis*
    Specimen: KNM-ER 1470
    Age: 1.8-1.9 million years
    Found by: Bernard Ngeneo
    Location: Koobi Fora, Kenya
    Color photo: Johanson et al.
      (1996) pages 178; 179

    Species Time Range: ~1.8–2.4 Ma

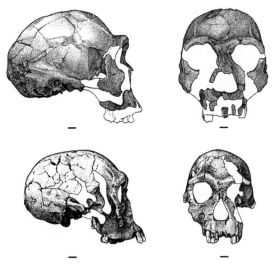
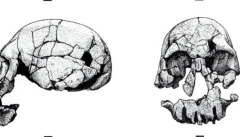

Figure 14.9 **Early humans**
(a) By Don McGranaghan, page 138 in Tattersall (1995). Scale bar = 1 cm.   (b) By Don Mc-Granaghan, page 134 in Tattersall (1995). Scale bar = 1 cm.
(c) By Don McGranaghan, page 133 in Tattersall (1995). Scale bar = 1 cm.

amples of *H. sapiens.* In present-day humans, braincase volume averages about 1,200 cm^3. The teeth and jaws of *Homo* species are smaller than those of the australopithecines. Members of the genus *Homo* tend to be taller than the australopithecines, with longer legs. The difference in size between females and males within species tends to be less pronounced in *Homo* than in *Australopithecus.*

Figure 14.10 shows examples of more recent members of the genus *Homo* from Africa, Europe, and Java. Modern *H. sapiens,* like Cro-Magnon I, whose skull appears in Figure 14.10a, differ from earlier forms in a variety of traits (Johanson et al. 1996). Modern humans have very large braincases (Cro-Magnon I's is over 1,600 cm^3, substantially larger than the present-day average). Associated with their large braincases, modern humans have high, steep foreheads. They also have relatively short, flat, vertical faces, and prominent noses. Cro-Magnon I was a man who died in middle age about 30,000 years ago. His skeleton was found in a single, prepared grave along with those of two other adult men, an adult woman, and an infant. The group had been buried with an assortment of animal bones, jewelry, and stone tools.

Interpreting the Fossil Evidence

Paleoanthropologists using fossils to reconstruct evolutionary history employ a two-step process (Strait et al. 1997). First, the researchers use a cladistic analysis to estimate the evolutionary relationships among the various fossil species. Then they make educated guesses about which fossil species represent ancestors that

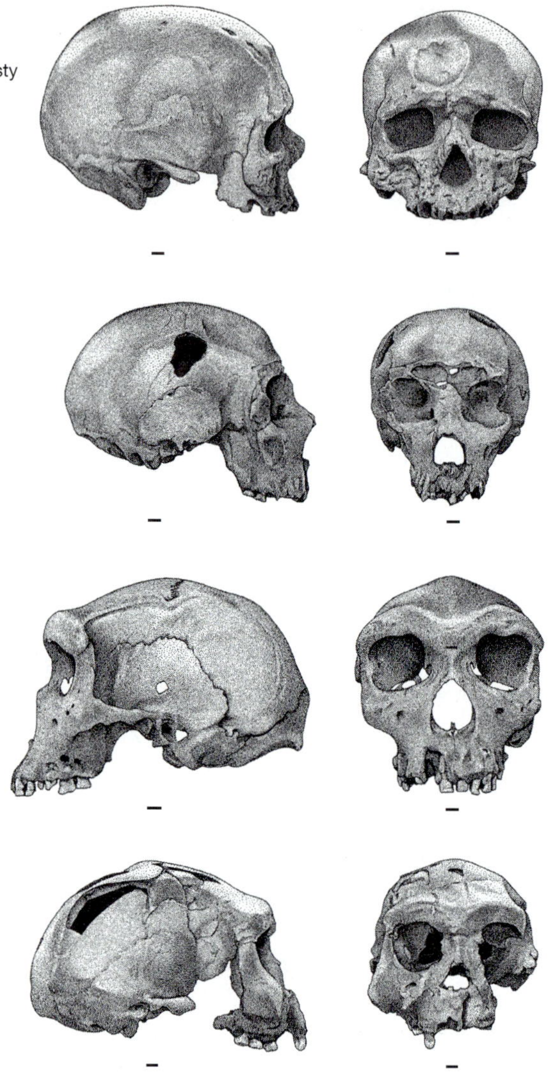

(a) Name: ***Homo sapiens***
Specimen: Cro-Magnon I
Age: 30,000 to 32,000 years
Found by: Louis Lartet and Henry Christy
Location: Abri Cro-Magnon,
      Les Eyzies, France
Color photo: Johanson et al.
      (1996) pages 245; 246

Species Time Range: ~0.1 Ma–Present

(b) Name: ***Homo neanderthalensis***
Specimen: Saccopastore 1
Age: ~120,000 years
Found by: Mario Grazioli
Location: Saccopastore quarry,
      Rome, Italy
Color photo: Johanson et al.
      (1996) pages 213; 214

Species Time Range: ~0.03–0.3 Ma

(c) Name: ***Homo heidelbergensis***
Specimen: Broken Hill 1
Age: ~300,000 years
Found by: Tom Zwigelaar
Location: Kabwe, Zambia
Color photo: Johanson et al.
      (1996) pages 209; 210

Species Time Range: ~0.2–0.6 Ma

(d) Name: ***Homo erectus***
Specimen: Sangiran 17
Age: ~800,000 years
Found by: Mr. Towikromo
Location: Sangiran, Java, Indonesia
Color photo: Johanson et al.
      (1996) pages 192; 193

Species Time Range: ~0.4–1.2 Ma

Figure 14.10   **Recent humans**
(a) By Don McGranaghan, page 25 in Tattersall (1995). Scale bar = 1 cm.   (b) By Don Mc-Granaghan, page 83 in Tattersall (1995). Scale bar = 1 cm. (c) By Don McGranaghan, page 54 in Tattersall (1995). Scale bar = 1 cm.   (d) By Don Mc-Granaghan, page 172 in Tattersall (1995). Scale bar = 1 cm.

lived at the branch points of the cladogram, and which fossil species represent extinct side branches.

The results of one such study, by David S. Strait and colleagues (1997), appear in Figure 14.11. All of the species included, with the exception of *H. sapiens,* occur exclusively, or almost exclusively, in Africa. The first of our recent relatives to leave Africa were members of the genus *Homo.* They were probably descended from *H. ergaster.* We consider *H. ergaster* and its relationship to *H. sapiens* in the next section.

Figure 14.11 shows Strait et al.'s cladogram (a), and two hypotheses about what the cladogram tells us concerning the phylogenetic relationships among the various species (b and c). The cladogram is based on a variety of skull and tooth characters. Note that in the cladogram, the lengths of the branches are meaningless; the only information encoded in the cladogram is in the order of branching. The hypothesized phylogenies differ in the actual lengths ascribed to the branches in the cladogram. For example, in Figure 14.11b, the branch lead-

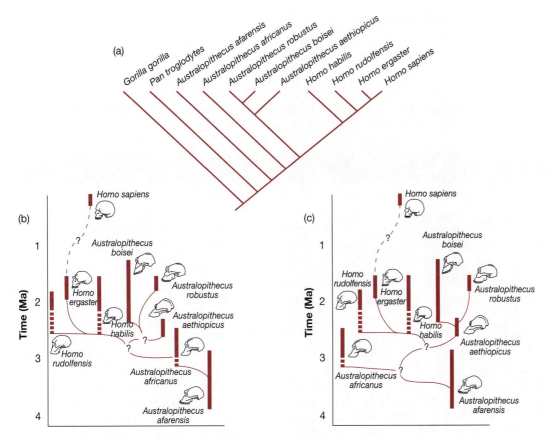

Figure 14.11 Phylogenies of *Homo sapiens* and its recent ancestors and extinct relatives
In both b and c, the heavy vertical bars indicate the known range of times over which each species existed, whereas the heavy dashes represent the suspected range of times over which each species existed. The transition represented by the dashed line and question mark is the subject of debate. We discuss the evidence on this transition in section 14.3.
(a) A cladogram of three extant hominids (the gorilla, the common chimpanzee, and humans), and several extinct hominids known only from fossils.
(b) A hypothesis about the ancestor/descendant relationships implied by the cladogram in (a).

• *A. afarensis* is the ancestor of *A. africanus,* which in turn is the ancestor of an as yet undiscovered species.

• This undiscovered species was the ancestor of sister clades, one containing the robust australopithecines and the other containing the genus *Homo.*

• The three species of robust australopithecines (*A. aethiopicus, A. robustus,* and *A. boisei*) are sister taxa whose most recent common ancestor is yet to be discovered.

• The three early species of *Homo* are essentially sister taxa, although early *H. habilis* may be the ancestor of the other two *Homo* lineages.

(c) A different hypothesis about the ancestor/descendant relationships implied by the cladogram in (a).

• *A. afarensis* is the ancestor of an as yet undiscovered species. This undiscovered species was the ancestor of *A. africanus,* which is an extinct lineage.

• The undiscovered species was also a common ancestor of the sister clades containing the robust australopithecines and *Homo,* but not their most recent common ancestor; their most recent common ancestor also remains to be discovered.

• *A. aethiopicus* is the common ancestor of the other two species of robust australopithecine.

• The three early species of *Homo* are essentially sister taxa, although early *H. habilis* may be the ancestor of the other two *Homo* lineages.

From Strait et al. (1997), with modifications to the scientific names.

ing to *Australopithecus aethiopicus* is relatively long, so that *A. aethiopicus* is a sister species to *A. boisei* and *A. robustus.* In contrast, in Figure 14.11c, the length of the branch leading to *A. aethiopicus* is zero, so that *A. aethiopicus* is the common ancestor of *A. boisei* and *A. robustus.*

The hypothesized phylogenies in Figure 14.11b and c are testable, because they predict the existence of specific fossils that have not yet been discovered. For example, the phylogeny in Figure 14.11b predicts that paleontologists will one day find a new species of *Australopithecus,* between 2.6 and 3 million years old, that is the common ancestor of the three currently known robust australopithecines (*A. aethiopicus, A. robustus,* and *A. boisei*). This ancestral robust australopithecine should have all of the shared derived traits that define the robust group, but few if any of the unique traits that distinguish the three known species. The phylogeny in Figure 14.11c, in contrast, predicts that no such generalized robust australopithecine will be found, because the phylogeny asserts that the immediate ancestor of the group is already known.

Both of the hypothesized phylogenies in Figure 14.11b and c suggest that early *H. habilis* was the ancestor of the genus *Homo,* and thus one of our own direct ancestors. Another possibility is that there existed an additional, and as yet undiscovered, species of *Homo* that is the common ancestor to *H. rudolfensis, H. ergaster,* and *H. habilis.* In November 1994, a team of paleontologists lead by William H. Kimbel, Robert C. Walter, and Donald C. Johanson found a 2.3-million-year-old fossil jaw, together with a collection of simple stone tools, at Hadar, Ethiopia. The jaw may prove to be that of the heretofore undiscovered common ancestor of all members of the genus *Homo* (Gibbons 1996; Kimbel et al. 1996).

As the two different evolutionary trees offered by Strait and colleagues suggest, paleontologists have not reached consensus on the correct phylogeny for *Homo, Australopithecus,* and their kin. Bernard Wood (1992) and Henry M. McHenry (1994), for example, propose phylogenies that are somewhat different than those suggested by Strait and colleagues (1997). Researchers disagree about which characters represent shared derived traits, and which represent ancestral features. Two recently discovered fossil species may prove helpful in determining trait polarities. *Australopithecus anamensis* (Figure 14.7c), discovered in Kenya by Meave G. Leakey and colleagues (1995), lived 3.9 to 4.2 million years ago. *Ardipithecus ramidus* (Figure 14.7d), discovered in Ethiopia by Tim D. White and colleagues (1994; 1995), lived 4.4 million years ago. So far these two species are incompletely known; it may not be possible to confidently infer their precise relationships to the species included in Figure 14.11 until paleontologists find more complete skulls (Strait et al. 1997).

The phylogenetic relationships among the species of Australopithecus *and* Homo *have not been definitively established.*

Some Answers

Regardless of differences in the details, all recent estimates of the phylogeny of *Homo, Australopithecus,* and their relatives give the same general answers to the questions we posed at the beginning of this section. The pattern of evolution leading from our common ancestor with the chimpanzees to ourselves has been anything but simple. Frequent speciation produced a diversity of species. Throughout most of the last four million years, multiple species, as many as five at a time, have coexisted in Africa. For example, specimen KNM-ER 406 (Figure 14.8b) and specimen KNM–ER 3733 (Figure 14.9a) clearly represent different species. Both were found at Koobi Fora, Kenya, in sediments of nearly the same age. We *Homo sapiens* are the lone survivors of an otherwise extinct radiation of bipedal African hominids.

14.3   The Origin of the Species *Homo sapiens*

Figure 14.11 represents the origin of our own species with only a dashed line and a question mark. The figure does not include three of the four recent *Homo* species shown in Figure 14.10. The question mark and the omissions reflect the considerable uncertainty over the origin of *Homo sapiens.*

Controversies Over the Origin of Modern Humans

Paleoanthropologists are split on the taxonomic status of *H. ergaster* (Figure 14.9a) and *H. erectus* (Figure 14.10d). Some researchers consider these two forms to be regional variants of a single species *(H. erectus),* whereas others consider *H. erectus* to be a distinct Asian species descended from the African species *H. ergaster.* Likewise, some researchers consider *H. neanderthalensis* (Figure 14.10b) and *H. heidelbergensis* (Figure 14.10c) to be regional variants of transitional forms between *H. erectus* and modern *H. sapiens.* Others consider them to be distinct species, with *H. heidelbergensis* descended from *H. ergaster,* and *H. neanderthalensis* descended from *H. heidelbergensis* (see Tattersall 1997). Paleoanthropologists generally agree that modern humans are the descendants of some or all of the populations in the *H. ergaster/erectus* group. However, how and where the transition from *H. ergaster/erectus* to *H. sapiens* took place is a matter of debate.

The oldest examples of *H. ergaster/erectus* appear in the fossil record nearly simultaneously at Koobi Fora in Africa, at Dmanisi in Caucasus (Eastern Europe), at Longgupo Cave in China, and at Sangiran and Mojokerto in Java—all 1.6 to 1.9 million years ago (Gibbons 1994; Swisher et al. 1994; Gabunia and Vekua 1995; Huang Wanpo et al. 1995; Wood and Turner 1995). Because its immediate ancestors and closest relatives appeared to be confined to Africa, most paleontologists had assumed that *H. erectus* evolved in Africa, then moved to Asia. The fossils at Longgupo Cave, China, however, are similar enough to African *H. habilis* and *H. ergaster* to suggest that *H. erectus* may have evolved in Asia from earlier migrants (Huang Wanpo et al. 1995). Either way, prior to 2 million years ago the ancestors of our species within the genus *Homo* almost certainly lived in Africa.

Anatomically modern *H. sapiens* first appear in the fossil record about 100,000 years ago in Africa and Israel, and somewhat later throughout Europe and Asia (Stringer 1988; Valladas et al. 1988; Aiello 1993). The range of hypotheses concerning the evolutionary transition from *H. ergaster/erectus* to *H. sapiens* is illustrated in Figure 14.12. At one extreme, the African replacement (or out of Africa) model (Figure 14.12a) posits that *H. sapiens* evolved in Africa, then migrated to Europe and Asia, replacing *H. erectus* and *H. neanderthalensis* without interbreeding. At the other extreme, the candelabra model (Figure 14.12d) holds that *H. sapiens* evolved independently in Europe, Africa, and Asia, without gene flow between regions. Between the extremes are hypotheses that postulate different combinations of migration, gene flow, and local evolutionary transition

The origin of modern Homo sapiens *is controversial.*

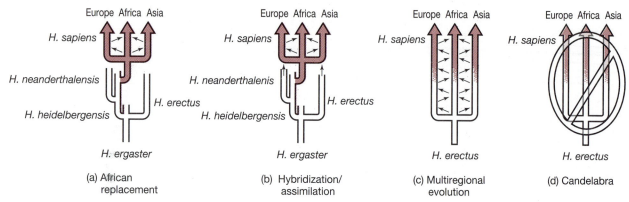

Figure 14.12 **Hypotheses concerning the transition from *Homo ergaster/erectus* to *Homo sapiens***   The open portions of the phylogenies represent various archaic forms of *Homo,* including *H. ergaster, H. erectus, H. heidelbergensis,* and *H. neanderthalensis.* The filled portions represent modern *H. sapiens.* The small arrows represent gene flow. Note that specimens identified as *H. heidelbergensis* have been found in Europe and Africa, and specimens identified as *H. neanderthalensis* have been found in Europe and the Middle East.   (a) The African replacement model. According to this model, modern *H. sapiens* evolved in Africa, then migrated to Europe and Asia. *H. sapiens* replaced the local forms without hybridization. No genes from these earlier forms persist in modern human populations.   (b) The hybridization and assimilation model. According to this model, modern *H. sapiens* evolved in Africa, then migrated to Europe and Asia. *H. sapiens* largely replaced the local populations, but there was hybridization between the newcomers and the established residents. As a result, some genes from the archaic local populations were assimilated and persist in modern human populations.   (c) The multiregional evolution model. According to this model, *H. sapiens* evolved concurrently in Europe, Africa, and Asia, with sufficient gene flow among populations to maintain their continuity as a single species. Gene pools of all present-day human populations are derived from a mixture of distant and local archaic populations.   (d) The candelabra model: *H. sapiens* evolved independently in Europe, Africa, and Asia, without gene flow among populations. All genes in present-day European and Asian populations are derived from local archaic populations. The model names and characterizations we use are based on those used by Aiello (1993), Ayala et al. (1994), and Tattersall (1997). Not all authors would agree with our characterizations and names. Frayer et al. (1993), for example, apparently consider both models (b) and (c) to be variations of the multiregional evolution model they favor.

from *H. ergaster/erectus* to *H. sapiens.* These intermediates are the hybridization and assimilation model (Figure 14.12b), and the multiregional origin model (Figure 14.12c).

At stake in the debate over these models is the nature and antiquity of the present-day geographic races of humans. If the African replacement model is correct, then present-day racial variation is the result of recent geographic differentiation that occurred within the last 100,000 to 200,000 years, after anatomically modern *H. sapiens* emerged from Africa. If one of the intermediate models is correct, then present-day racial variation represents some mixture of recent and ancient geographic differentiation. If the candelabra model is correct, then present-day racial variation derives from geographic differentiation among *H. ergaster/erectus* populations, and may be as much as 1.5 to 2 million years old.

The candelabra model has been widely and thoroughly rejected by scientists in all fields (see Frayer et al. 1993; Ayala et al. 1994). It is flatly implausible that the same single descendant species, *H. sapiens,* could emerge in parallel in three different regions with no gene flow to maintain its continuity (see Chapters 5 and 9). However, this rejection of the candelabra model is the beginning and end of consensus in the field. Arguments over the remaining three models are based on archaeological and paleontological evidence, and on genetic analyses. In much of the following discussion, we focus on distinguishing between the remaining extremes (the African replacement model of Figure 14.12a versus the multiregional evolution model of Figure 14.12c), but it is useful to keep in mind that these two models fall at the ends of a continuum of possibilities.

African Replacement Versus Multiregional Evolution: Archaeological and Paleontological Evidence

David Frayer and colleagues (1993) use archaeological and paleontological data to argue against the African replacement model (Figure 14.12a) and in favor of some alternative (Figure 14.12b or c). The researchers note that the African replacement model holds that long-established populations of one or more tool-using, hunter-gatherer species (*H. erectus* and other archaic forms of *Homo*) were supplanted wholesale throughout Europe and Asia by populations of another tool-using, hunter-gatherer species (modern *H. sapiens* emerging from Africa). It is hard to imagine how this could have happened except by direct competition between the invaders and the established residents. It is implausible that modern *H. sapiens* could have been such a relentlessly superior competitor without a substantial technological advantage in the form of better tools or weapons. Thus, Frayer and colleagues conclude, the African replacement model predicts that the archaeological record will show evidence of abrupt changes in the level of technology in Europe and Asia as modern *H. sapiens* replaced archaic *Homo*. In fact, the researchers say, there is no evidence of any such abrupt technological changes.

Frayer and colleagues (1993) also argue that the African replacement model predicts that fossils of *Homo* populations in any given non-African region should show distinct changes in morphology when modern *H. sapiens* migrating from Africa replaced local archaic *Homo*. To refute this prediction, Frayer and colleagues point to distinctive traits of regional populations that have persisted from the distant past to the present. One-million-year-old fossils of *H. erectus* from Java, for example, have a straighter, more prominent browridge than their contemporaries elsewhere in the world. This strong browridge remains a distinctive feature of present-day Australian aborigines, whose ancestors arrived by boat from Java some 60,000 years ago. Likewise, many present-day Asians have shovel-shaped upper front teeth, a trait that characterizes virtually all fossil specimens of Asian *H. erectus* and *H. sapiens*. (For other examples of the continuity of distinctive regional traits, see Thorne and Wolpoff 1981; Li Tianyuan and Etler 1992; Frayer et al. 1993). On these and other grounds, Frayer and colleagues reject the African replacement model.

Daniel Lieberman (1995) also takes a paleontological approach, but uses a cladistic analysis of fossil and modern skulls in an attempt to distinguish between the African replacement model (Figure 14.12a) and the multiregional evolution model (Figure 14.12c). The two models predict strikingly different cladograms. Under the African replacement model, modern humans from all regions and archaic forms from Africa should be closest relatives (Figure 14.13a). In contrast, under the multiregional evolution model, the modern humans in each region should be closest relatives of the local archaic forms (Figure 14.13b). To determine which model is closer to the truth, we need an accurate cladogram that includes modern and archaic humans from all regions.

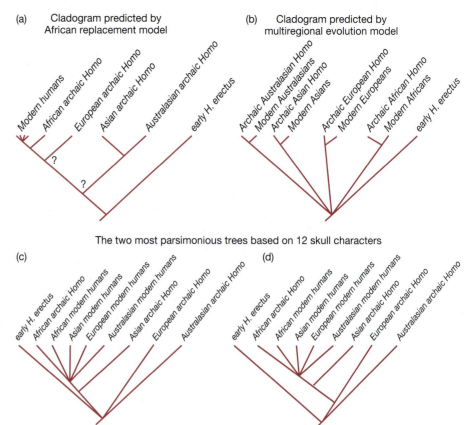

Figure 14.13 A cladistic test of the African replacement model versus the multiregional evolution model
(a) The African replacement model predicts that all modern humans will be more closely related to each other than any is to any archaic species, and that among the archaic species, those from Africa will be the most closely related to modern humans   (b) In contrast, the multiregional evolution model predicts that the archaic and modern humans in each region will be each other's closest relatives. (c and d) In both trees from an empirical test, modern humans from all regions and archaic humans from Africa are closest relatives. From Lieberman (1995).

Recent morphological analyses, while not definitive, suggest that modern humans evolved in Africa, then replaced archaic humans elsewhere.

Lieberman examines 33 skull characters. Some of these are characters that other researchers have used to link all modern humans to archaic Africans; others are characters that other researchers have used to link modern humans in a particular region to the local archaic species. Of the 33 characters, only 12 prove appropriate for inclusion in a cladistic analysis. Lieberman stresses that a dataset of only 12 characters is too small to produce a reliable conclusion. However, to the extent that the 12 characters are informative, they tend to support the African replacement model rather than the multiregional evolution model. The characters yield two most-parsimonious cladograms. In both of these trees the archaic Africans and all modern humans are closest relatives (Figure 14.13c and d).

Diane Waddle (1994) uses an alternative statistical approach to evaluate predictions similar to those examined by Lieberman. Waddle, too, concludes that the available data support the African replacement model. If Lieberman and Waddle are correct, then the examples described by Frayer and colleagues of apparent long-term continuity of regionally distinctive traits must be the result of convergent evolution in *H. erectus* and *H. sapiens.*

African Replacement Versus Multiregional Evolution: Molecular Evidence

In principle, we could take Lieberman's approach with DNA sequence data as well. If we had sequences of genes from both modern and archaic humans in all regions, we could estimate their phylogeny and see whether it most closely matches the tree predicted by the African replacement model or the tree predicted by the multiregional evolution model. Recently, a team led by Svante Pääbo recovered a sequence of mitochondrial DNA from the skeleton of a *Homo neanderthalensis* that lived in Germany some 30,000 to 100,000 years ago (Krings et al. 1997). Placing this sequence on a phylogeny, the researchers found that modern humans from Europe, Africa, Asia, America, Australia, and Oceania are all more closely related to each other than any is to the archaic European. This result is consistent with the African replacement model (Figure 14.12a). However, the result is not definitive. It is based on a single archaic individual from only one geographic region, and it leaves open the possibility that modern Europeans inherited nuclear genes, if not mitochondrial genes, from archaic Europeans. Unfortunately, retrieving DNA from fossils is difficult, and may prove impossible for bones older than 100,000 years.

Working with DNA sequences of present-day humans only, it is more difficult to design tests that distinguish the African replacement model from the multiregional evolution model (see, for example, the exchange between Sarah Tishkoff and colleagues [1996a] and Milford Wolpoff [1996]). The trouble is that from a genetic perspective, the two models are identical in most respects. Both describe a species originating in Africa and spreading throughout Europe and Asia, then differentiating into regionally distinct populations that nonetheless remain connected by gene flow (Figure 14.12a and c). The only difference is that under the African origin model this process began 200,000 years ago or less, whereas under the multiregional evolution model it began some 1.8 million years ago. This means that any genetic patterns that might allow us to distinguish between the two models will involve quantitative differences rather than qualitative differences. Table 14.2 lists four criteria that molecular geneticists have used in efforts to distinguish between the African replacement versus multiregional evolution models. We will refer to the table in the paragraphs that follow.

Linda Vigilant and colleagues (1991) attempt to distinguish between the African replacement model and the multiregional evolution model by analyzing sequence data for mitochondrial DNA. The researchers sequenced a piece of the noncoding mitochondrial DNA from 189 people from a diversity of geographic regions, then used a parsimony technique to estimate the evolutionary tree linking the 189 mitochondrial sequences (see also earlier work from the same lab in Cann et al. 1987). Recall that mitochondrial DNA carries a record of direct maternal ancestry. By direct maternal ancestry we mean a person's mother, mother's mother, and so on. (A person's mother's father is part of the person's indirect maternal ancestry). Vigilant et al.'s mitochondrial evolutionary tree traces the direct maternal ancestries of the 189 people back to a point at which they all converge to a single woman. Because the deepest branches in the mitochondrial phylogeny involve splits within African lineages, Vigilant and colleagues conclude that this woman lived in Africa (Figure 14.14). (Note that this woman, often called by the misleading name Mitochondrial Eve, is not, in her generation, the sole female ancestor of the 189 present-day people. The 189 present-day people undoubtedly have a large number of indirect female ancestors who were contemporaries of their common direct maternal ancestor, and from whom they inherited many of their nuclear genes. See Ayala et al. 1994; Ayala 1995.)

Vigilant et al.'s conclusion that the common ancestor of all mitochondrial DNAs lived in Africa proves to be poorly grounded statistically (see Chapter 10;

TABLE 14.2 Genetic predictions distinguishing the African replacement versus multiregional evolution models

Each of the criteria in the first column is a category of data we might use to distinguish between the African replacement model and the multiregional evolution model. The next two columns predict the patterns in each type of data under each model. The last column gives reasons why the distinctions implied by the predictions are not definitive. See text for more details.

| | *Predictions* | | |
Criteria	African replacement	Multiregional evolution	Caveat
1. Location of ancestor of neutral alleles	Mostly Africa	Random	African origin of *H. ergaster/erectus* may bias the location of alleles toward Africa even under multiregional evolution.
2. Divergence time of African vs. non-African populations	200,000 years or less	1 million years or more	Gene flow among regional populations can reduce the apparent age of population divergence under multiregional evolution.
3. Genetic diversity	Genetic diversity is greater in Africa.	Genetic diversity is roughly equal in all regions.	African origin of *H. ergaster/erectus* and gene flow or selection may lead to greater diversity in Africa, even under multiregional evolution.
4. Sets of neutral alleles	Alleles present in Europe and Asia are subsets of those in Africa	Each region has some unique alleles; no region's alleles are a subset of another region's.	African origin of *H. ergaster/erectus* may mean that alleles present in Europe and Asia are subsets of those in Africa, even under multiregional evolution.

Hedges et al. 1992; Templeton 1992; 1993). Nonetheless, the conclusion may be correct (see Zischler et al. 1995). If we assume for the sake of argument that the common ancestor of all present-day human mitochondrial DNAs did, in fact, live in Africa, does that help us decide, by criterion 1 of Table 14.2, between the African replacement model and the multiregional evolution model? No, because both models include a common ancestry for all present-day humans that traces back to Africa. The key question is, when did the common ancestor of all human mitochondrial DNAs live?

Vigilant and colleagues estimate that the common ancestor of all present-day mitochondrial DNAs lived between 166,000 and 249,000 years ago. The researchers arrive at this figure by (1) assuming that mutations accumulate in mi-

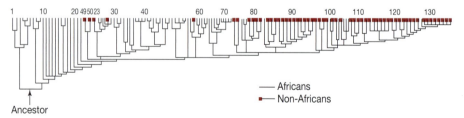

Figure 14.14 **Vigilant et al.'s (1991) evolutionary tree of human mitochondrial DNA** Vigilant et al. rooted the tree by using a chimpanzee sequence as the out-group. Note that the deepest branches in the tree are all among Africans. From Templeton (1993).

tochondrial DNAs at a constant rate; (2) comparing human sequences to chimpanzee sequences; and (3) using accepted estimates for the divergence time between humans and chimpanzees to callibrate the molecular clock (see Chapter 10). Ruvolo et al. (1993) use similar data and reasoning to estimate that the common ancestor of all human mitochondrial DNAs lived 129,000 to 536,000 years ago. The most accurate estimate, because it is based on sequences of entire mitochondrial genomes, is probably that of Horai et al. (1995): 125,000 to 161,000 years ago.

At first glance these dates, which are consistent with the African replacement hypothesis, appear to refute the multiregional origin hypothesis by criterion 2 of Table 14.2. The dates suggest that non-African populations of humans diverged from African populations no more than a few hundred thousand years ago, and certainly nothing like a million years ago. Examination of Figure 14.15, however, shows that this is not necessarily the case. It is true that species cannot diverge any earlier than the divergence of any of their alleles, but popu-

Recent molecular analyses, while not definitive, suggest that modern humans evolved in Africa, then replaced archaic humans elsewhere.

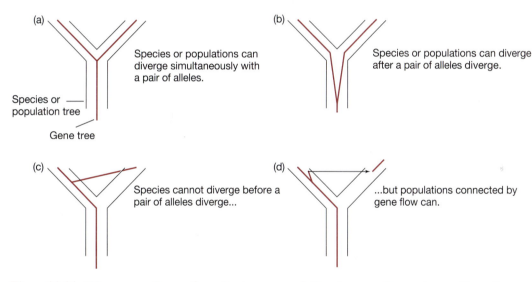

Figure 14.15 Divergence times of species trees, population trees, and gene trees   These figures illustrate a hypothetical gene phylogeny within a species or population phylogeny.   (a) In this scenario, a mutation creates a new allele to produce a split in the gene tree at the same time the species or population splits into two. One allele is then lost by drift or selection in each descendant species or population. The result is a gene tree that is exactly congruent with the species or population tree. If we use a molecular clock to estimate the divergence time for the species or populations, we will get the right answer.   (b) Here, a mutation creates a new allele, producing a split in the gene tree. Some time later, the species or population splits into two. One allele is then lost by drift or selection in each descendant species or population. If we use a molecular clock for the gene tree to estimate the divergence time for the species or populations, the species or populations will appear to have diverged earlier than they actually did.   (c) First, a species splits into two. Sometime later, a mutation creates a new allele, producing a split in the gene tree. Finally one of the alleles moves from one species to the other. This last step is impossible; thus a molecular clock will not make a split between species appear more recent than it was.   (d) A scenario like that in (c) is however, possible for populations. First, a population splits into two. Some time later, a mutation creates a new allele, producing a split in the gene tree. Then a migrant carries the new allele to the other population. Finally, the new allele is lost in one population, and the ancestral allele is lost in the other population. If we use a molecular clock for the gene tree to estimate the divergence time for the population tree, the populations will appear to have diverged more recently than they actually did.

lations connected by gene flow can. It is possible that the mitochondrial clock, which is effectively based on a single gene, makes the split between African and non-African populations look more recent than it actually was.

What we need to do is look at many loci at once, and see whether, taken together, they tell the same story of a recent divergence between African and non-African populations. A. M. Bowcock and colleagues (1994) look at 30 nuclear microsatellite loci from people in each of 14 populations. Microsatellite loci are places in the genome where a short string of nucleotides, usually 2–5 bases long, is tandemly repeated. The number of repeats at any given locus is highly variable among individuals, meaning that each microsatellite locus has many alleles. Bowcock and colleagues calculate multilocus genetic distances among the 14 populations based on the allele frequencies at each of the 30 microsatellite loci. The researchers then use the genetic distances among populations to estimate the population phylogenetic tree (Figure 14.16).

On Bowcock et al.'s phylogenetic tree, geographically neighboring populations cluster together. Furthermore, the deepest branch point separates African from non-African populations. Analyzing the same data, D. B. Goldstein and colleagues (1995) estimate that the split between African and non-African populations occurred 75,000 to 287,000 years ago. This date, which is consistent with the African replacement model, makes a more persuasive case than the mitochondrial clock date that the multiregional evolution model can be rejected by criterion 2 of Table 14.2. It is still possible under the multiregional evolution model to argue that there was enough gene flow to make the population split look more shallow than it was. But if there was that much gene flow, then it becomes hard to explain how any regional differentiation of characters could be maintained for a million years or more (Nei 1995).

Finally, we consider a study by Sarah Tishkoff and colleagues (1996b). These researchers examined allelic variation at a locus on chromosome 12 that is the site of a short-tandem-repeat polymorphism. This is a region of noncoding

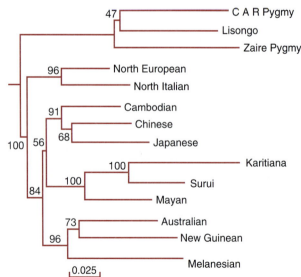

Figure 14.16   Phylogenetic tree for 14 human populations based on allele frequencies at 30 microsatellite loci   The number at a node indicates the percentage of times that the node was present in 100 bootstrap replicates (see Chapter 10). The deepest split in the tree was present in 100% of the bootstrap replicates, indicating strong statistical support for the conclusion that African (C A R Pygmy, Lisongo, Zaire Pygmy) versus non-African is the most fundamental division in the population phylogeny. From Bowcock et al. (1994).

DNA in which the sequence TTTTC is repeated between 4 and 15 times, producing a total of 12 alleles. Tishkoff and colleagues determined the genotypes of more than 1,600 people from seven different geographic regions. Figure 14.17 shows plots of the allele frequencies in each of the seven regions.

The African populations show much greater allelic diversity than non-African populations. This pattern is consistent with the African replacement model. If non-African populations were founded by small bands of people migrating out of Africa, then non-African populations should have reduced genetic diversity because of the founder effect (see Chapter 5).

Notice also that the graphs in Figure 14.17 are arranged by travel distance from sub-Saharan Africa, with the closest regions near the bottom and the most distant regions near the top. Moving up the figure from sub-Saharan Africa to northeast Africa, then to the Middle East, to Europe, and to Asia and beyond, each region shows a set of alleles that is a subset of those present in the region below. Again this is consistent with the African replacement model. It is what we would expect if each more-distant region were founded by a small band of people picking up from where their ancestors had settled and moving on. Not only is the pattern of allelic diversity consistent with the African replacement model, but it also tends to refute the multiregional evolution model by criteria 3 and 4 of Table 14.2. This refutation is not definitive, however, because the multiregional evolution model postulates the same pattern of migration and settlement, just earlier in time.

Tishkoff and colleagues can estimate when the founders of the non-African populations left Africa. They use a method based on linkage disequilibrium between the short-tandem-repeat locus and a second locus nearby (see Box 14.1). Tishkoff and colleagues estimate that the founders of the non-African populations left Africa not more than 102,000 to 450,000 years ago. These dates are consistent with the African replacement model, and tend to refute the multiregional evolution model by criterion 2 of Table 14.2.

Present-Day Geographic Variation Among Human Populations Appears to Be of Recent Origin

The balance of evidence we have reviewed appears to favor the African replacement model for the origin of *H. sapiens*. None of the tests are definitive, so some form of intermediate model (Figure 14.12b) cannot be ruled out. But taken together, the genetic data and at least some of the morphological data suggest that (1) all present-day people are descended from African ancestors, and (2) all present-day non-African people are descended from *H. sapiens* ancestors who left Africa within the last few hundred thousand years. Present-day differences among races must have arisen since then.

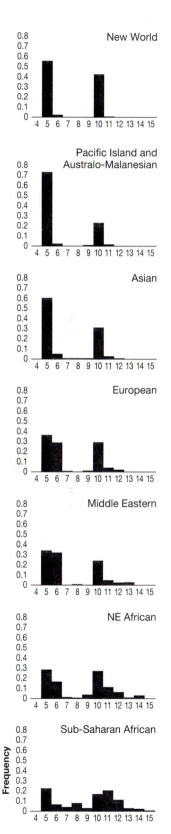

Figure 14.17 **Genetic diversity at a single locus among the people of seven geographic regions** Each plot shows, for the people of a particular geographic region, the frequencies of the various alleles (numbered 4–15) at a short-tandem-repeat locus on chromosome 12. Plotted from tables in Tishkoff et al. (1996b).

Box 14.1 Using linkage disequilibrium to date the divergence between African and non–African populations

Near the short-tandem-repeat locus on chromosome 12 is another locus, also noncoding, with a nucleotide sequence known as an Alu element. This Alu locus has two alleles: the ancestral, or Alu(+) allele; and a derived, Alu(−), allele with a 256–base-pair deletion. Gorillas and chimpanzees lack the Alu(−) allele, so it probably arose in the human lineage after the split between the human lineage and the chimpanzee lineage.

The deletion mutation that created the Alu(−) allele probably occurred only once, in Africa, most likely in a chromosome that carried the six-repeat allele at the short-tandem-repeat locus (Tishkoff et al. 1996b). Upon its appearance in the population, the Alu(−) allele was in linkage disequilibrium with the short-tandem-repeat allele (see Chapter 6). The only kind of Alu(−) gamete in the population had the genotype short-tandem-repeat-6-Alu(−).

Recall from Chapter 6 that sexual reproduction in a population reduces linkage disequilibrium. After the Alu(-) allele appeared in Africa, mutations at the short-tandem-repeat locus and genetic recombination between it and the Alu locus created a great variety of other genotypic combinations on chromosome 12 (Figure 14.18, African populations). The sub-Saharan African population is near linkage equilibrium at the two loci.

When the first migrants left Africa, it appears that they carried with them at appreciable frequency only three 2-locus genotypes: 5-Alu(+); 10-Alu(+); and 6-Alu(−). In other words, genetic drift in the form of the founder effect put the migrant populations in linkage disequilibrium. The time since the migrants' departure from Africa has not been sufficient for mutation and recombination to create new two-locus genotypes and replenish gametic diversity to the levels seen in Africa. In other words, the non-African populations are still in linkage disequilibrium, a population-genetic legacy from their ancient ancestors who left Africa. Using estimates of the rates of mutation and recombination, both of which affect the rate at which populations approach linkage equilibrium (see Chapter 6), Tishkoff and colleagues estimate that the founders of the non-African populations left Africa not more than 102,000 to 450,000 years ago (see also Pritchard and Feldman 1996; Risch et al. 1996).

14.4   The Evolution of Uniquely Human Traits

Humans have a number of traits unique among extant primates: We walk bipedally, we have very large brains, we manufacture and use complex tools, and we use language. We discussed bipedal locomotion and brain size briefly in section 14.2. Here we consider evidence on the origin of tools and language.

Which of Our Ancestors Made and Used Stone Tools?

Chimpanzees make and use simple tools. They strip stems and twigs of leaves and use the resulting tools to fish termites out of termite mounds; they use rocks and sticks to hammer open nuts; they use leaves as umbrellas. Other animals use tools as well. One species of Darwin's finch, the woodpecker finch *(Camarhynchus pallidus),* uses cactus spines to extract insects from bark. So making and using tools is not, in itself, unique to humans. What is unique to humans is making and using complex tools.

The earliest uniquely complex tools that appear in the archaeological record are sharp-edged stone flakes and hand-held chopping tools (Figure 14.19). A

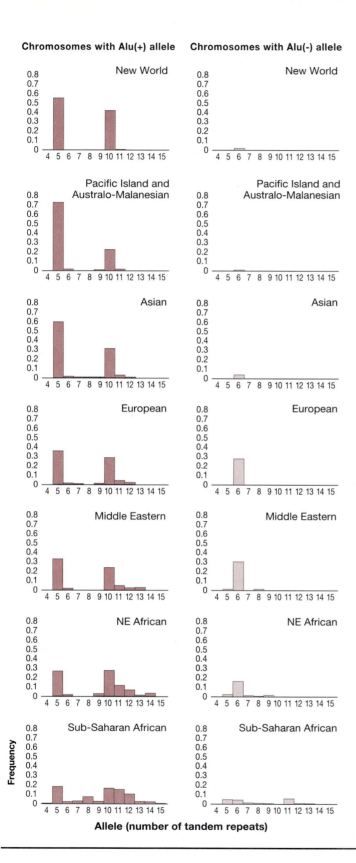

Figure 14.18 **Genetic diversity at two linked loci in the people of seven geographic regions** These graphs represent the same populations and the same alleles represented in Figure 14.17. Near the short-tandem-repeat locus is an Alu element polymorphic for a deletion. There are two alleles at this locus: Alu(+) and Alu(−). The left column shows allele frequencies at the short-tandem-repeat locus among chromosomes that carry the Alu(+) allele. The right column shows allele frequencies at the short-tandem-repeat locus among chromosomes that carry the Alu(−) allele. If a population is at linkage equilibrium for the short-tandem-repeat locus and the Alu locus, then the shape of the distributions of alleles in the two columns will be the same (but not necessarily the height of the distributions). In sub-Saharan Africa, the distribution of alleles is roughly the same for Alu(+) and Alu(−) chromosomes. This pattern indicates that the short-tandem-repeat-locus and the Alu locus are near linkage equilibrium in this population. In the people of other regions, the distribution of alleles is dramatically different in Alu(+) versus Alu(−) chromosomes, indicating that the short-tandem-repeat locus and the Alu locus are in linkage disequilibrium. Plotted from tables in Tishkoff et al. (1996b).

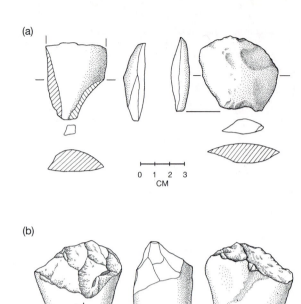

(a)

0 1 2 3
CM

(b)

0 1 2 3
CM

Figure 14.19 **Oldowan stone tools from Hadar, Ethiopia**
These 2.3-million-year-old stone tools are among the oldest known.
(a) Two sharp-edged flakes. Each is shown from various sides.   (b) A
hand-held chopper, shown from three sides. After Kimbel et al. 1996.

stone knapper making such tools began by selecting an appropriate cobble from
a river bed, preferably one of fine-grained volcanic rock (Johanson et al. 1996).
He or she then struck the cobble with a second rock to chip off flakes. The
flakes themselves were usable as cutting tools. Chipping numerous flakes off a
cobble in an appropriate pattern produced a chopper. Tools of this style are said
to belong to the Oldowan industrial complex, because they were first discovered at Olduvai Gorge, Tanzania. Archaeologists have learned firsthand that
making Oldowan-style stone tools requires skill and experience.

The oldest known Oldowan tools are from Gona, Ethiopia. Based on the
ages of the strata just above and below the sediment layer that contains the
tools, Sileshi Semaw and colleagues (1997) established that the tools are 2.5 to
2.6 million years old. Who were the stone knappers who made them?

An obvious answer is, some early member of the genus *Homo*. The trouble
with this answer is that we have no definitive evidence that any species of *Homo*
had appeared by 2.5 million years ago. The oldest reliably dated *Homo* fossil is a
2.3-million-year-old upper jaw (maxilla) from Hadar, Ethiopia (section 14.2;
Gibbons 1996; Kimbel et al. 1996). Which species this fossil represents is unclear; it could be *H. habilis* (Figure 14.9b), *H. rudolfensis* (Figure 14.9c), or some
heretofore unknown species.

Circumstantial evidence certainly suggests that the Hadar fossil may represent
the same species that made the 2.5-million-year-old Gona tools. Hadar is geographically close to Gona, 2.3 million years ago is geologically close to 2.5 million years ago, and the Hadar fossil was found near 34 Oldowan tools. It is entirely possible that 2.5-million-year-old *Homo* fossils will eventually be found at
Gona, and that these early *Homo* were the Gona stone knappers. On the other
hand, as Bernard Wood (1997) points out, circumstantial evidence is not proof.

*The earliest known stone tools pre-
date the earliest known* Homo
specimens.

If other hominids were present at the same time and place, then they are suspects too. This is true even at Hadar, where the 2.3-million-year-old *Homo* jaw was found near Oldowan tools.

Wood argues that we should take into account an observation made by Semaw at al.: that the Oldowan industrial complex persisted unchanged for about a million years, from 2.5 to 1.5 million years ago. This was a period of substantial evolutionary change in the genus *Homo*. Wood finds it puzzling that *Homo* should have experienced little cultural evolution in its toolmaking over a time span during which it experienced two or more speciation events and dramatic morphological evolution. Wood suggests instead that we consider another group of suspects as the makers of the Gona tools: the robust australopithecines (Figure 14.8). Members of this group existed in the same part of Africa over approximately the same time span as the Oldowan industrial complex.

What other evidence is there concerning the toolmaking abilities of early *Homo* and the robust australopithecines? Randall L. Susman (1994) makes an argument based on the anatomy of opposable thumbs. He starts by comparing the bones and muscles of the thumb in humans versus chimpanzees (Figure 14.20). Humans have three muscles that chimpanzees lack. Associated with these extra muscles, humans have thicker metacarpals with broader heads (Figure 14.21a). These differences in thumb anatomy make the human hand more adept at precision grasping than the chimpanzee hand. Susman argues that the modified anatomy of the human thumb evolved in response to selection pressures associated with the manufacture and use of complex tools.

Susman then compares the relative thickness of the thumb metacarpals in humans and chimpanzees to that in a variety of hominid fossils (Figure 14.21b). *H. neanderthalensis, H. erectus,* and *A. robustus* resemble *H. sapiens* in having thumb metacarpals with broad heads for their length. *A. afarensis,* a gracile australopithecine that disappears from the fossil record before Oldowan tools appear, is like the chimpanzees in having thumb metacarpals with narrow heads

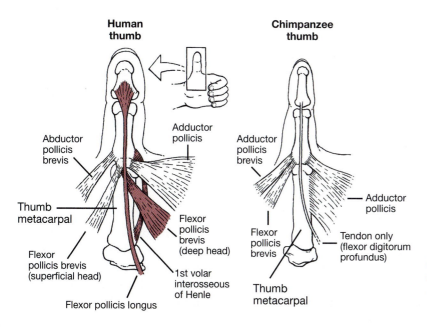

Figure 14.20 **Human versus chimpanzee thumbs** The three muscles in a human thumb that chimpanzees lack are indicated by shading. After Susman (1994).

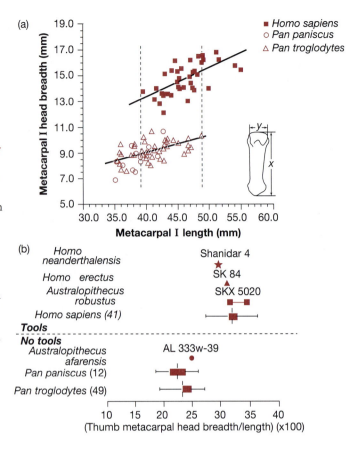

Figure 14.21   Thumb metacarpal bones in a variety of hominids (a) This graph plots the breadth of the thumb metacarpal head (the end of the bone pointing out along the thumb) against the length of the thumb metacarpal for a sample of humans, bonobos, and common chimpanzees. Longer metacarpals have broader heads, but human metacarpals have broader heads for their size than chimpanzee metacarpals. In other words, the ratio of metacarpal head breadth to length is greater in humans. (b) This plot shows the range of ratios (metacarpal head breadth to length) for several hominid fossils (the labels on the data points are the specimen numbers) and samples of present-day humans, bonobos, and common chimpanzees. For *Australopithecus robustus,* the ratio is reported as an estimated range because the bone in question is not quite complete. Species above the dashed line are associated (at least temporally) with manufactured stone tools in the archaeological/fossil record and/or the present. Species below the dashed line are not. From Susman (1994) with some changes in the species names.

Morphological analyses suggest that both early Homo *species and the robust australopithecines could have been tool makers.*

for their length. Susman asserts that we can use the thumb metacarpals to diagnose whether extinct hominids were makers and users of stone tools. Susman concludes than both *H. erectus* and *A. robustus* were toolmakers. Susman's argument has been controversial (McGrew et al. 1995).

None of the evidence we have discussed establishes for certain whether the Oldowan tools at Gona were made by a species of *Homo* or a robust australopithecine. Instead, the evidence argues that they could have been made by either or both. If we accept Susman's conclusion that the robust australopithecines were toolmakers but the gracile australopithecines were not, and if we accept one of the phylogenies of Strait et al. (Figure 14.11), then we must make one of two inferences: (1) The manufacture and use of complex stone tools originated in an undiscovered common ancestor of *Homo* and the robust australopithecines, or (2) the manufacture and use of complex stone tools originated independently in at least two hominid lineages.

Which of Our Ancestors Had Language?

If the history of hominid tool use is murky, the history of hominid language is even murkier. Like tool use, language is a behavior. Because behaviors do not fossilize, we have no direct evidence of their history. We are left to examine circumstantial evidence in the archaeological and fossil record. Before the invention of writing, language left even less circumstantial evidence than tool use.

Language is a complex adaptation located in the neural circuitry of the brain. The vocabulary and particular grammatical rules of any given language are transmitted culturally, but the capacity for language and a fundamental grammar are, in present-day humans, both innate and universal (see Pinker 1994). Among the evidence for this assertion is the observation that communities of deaf children, if isolated from native signers, invent their own signed languages from scratch. By the end of two generations of transmission to young children within the new deaf culture, these new sign languages develop all the hallmarks of genuine language. They have standardized vocabulary and grammar, and their fluent users can efficiently communicate the full range of human ideas and emotions. Each of these new sign languages is unique, but all reflect the same universal grammar that linguists have identified in spoken languages.

Many of the brain's language circuits are concentrated in an area called the perisylvian cortex, usually in the brain's left hemisphere (see Pinker 1994). These language circuits include Broca's area and Wernicke's area. Homologous structures exist in the brains of monkeys (Galaburda and Pandya 1982). The monkey homologues of Wernicke's area function in the recognition of sounds, including monkey calls. The monkey homologues of Broca's area function in controlling the muscles of the face, tongue, mouth, and larynx. However, neither of these structures plays a role in the production of monkey vocal calls. Instead, vocal calls are generated by circuits in the brain stem and limbic system. These same structures control nonlinguistic vocalizations in humans, such as laughing, sobbing, and shouting in pain. Thus the human language organ appears to be a derived modification of neural circuits common to all primates. But among extant species the nature of this modification, its specialization for linguistic communication, appears to be unique to humans. The implication is that the language organ, as such, evolved after our lineage split from the lineage of chimpanzees and bonobos.

Spoken language also relies on derived modifications of the larynx that are unique to humans. In modern human newborns, and in other mammals, the larynx is high enough in the throat that it can rise to form a seal with the back opening of the nasal cavity (see Pinker 1994). This allows air to bypass the mouth and throat on its way from the nose to the lungs, and prevents the infant from choking on accidentally inhaled food or water. When human babies are about three months old, the larynx descends to a lower position in the throat. This clears more space in which the tongue can move, changes the shape of a pair of resonating chambers, and makes it possible for humans to articulate a much greater diversity of vowel sounds than, for example, chimpanzees.

How far back in our evolutionary lineage can we trace the existence of language, and what evidence can we use? William Noble and Iain Davidson (1991) assert that the only reliable source of evidence is the archaeological record. In their view, the hallmark of language is the use of arbitrary symbols, standardized within a culture, to represent objects and ideas. To find language, then, we should look for such symbols in the archaeological record. The first unequivocally arbitrary symbols occur in cave paintings in Germany and France that are about 32,000 years old. Even Noble and Davidson cannot quite accept that language is as recent an innovation as that. They note that *Homo sapiens* had colonized Australia by 40,000 years ago, and confess that they cannot imagine how a group of people could build boats and cross the open ocean without language to help them conceive and coordinate the expedition. However, Noble and Davidson hold the line at about 40,000 years. This would imply that *H. sapiens* is the only species ever to use language.

Other support for the view that language is a recent innovation of *H. sapiens* came from anatomical studies of the skulls of *H. neanderthalensis*. These studies were attempts to reconstruct the neanderthal larynx. Analyses of the shape of the base of the skull, and the position of its muscle attachment sites, were used to argue that neanderthals had an undescended larynx that would have limited their ability to articulate vowels. This, it was argued, would have prevented them from developing language.

Early studies suggested that neanderthals could not talk . . .

B. Arensburg and colleagues (1989; 1990) refuted these arguments with a rare paleontological find. A 60,000-year-old neanderthal skeleton from Israel included an intact hyoid bone (Figure 14.22). The hyoid bone is the only free-floating bone in the body. It is located in the larynx and serves as the anchor for throat muscles that, in humans, are important in speaking. Arensburg et al. made a detailed analysis of the hyoid bone, and compared it to the hyoids of chimpanzees and present-day humans. They found the neanderthal hyoid to be dramatically different from that of chimpanzees, and virtually identical to that of present-day humans. Based on this hyoid bone, Arensburg and colleagues suggest that neanderthals had a descended larynx. Given that a descended larynx entails an increased risk of choking, it is hard to imagine why it would evolve unless it carried a substantial benefit. An obvious candidate for such a benefit is an improved ability to speak.

. . . but this conclusion was contradicted by the discovery of a key neanderthal bone.

If we accept the proposition that neanderthals could talk, can we trace language back any further in our lineage? Given that the language organ is in the brain, what can we glean from the fossil record about the brains of our ancestors? Following David Pilbeam and Stephen Jay Gould (1974), Figure 14.23 plots brain size as a function of body size in extant great apes, australopithecines, and three species of *Homo*. Inspection of the figure indicates that there has been something dramatically different about our brains since the first emergence of our genus. Not only do *Homo* species have larger brains for their size,

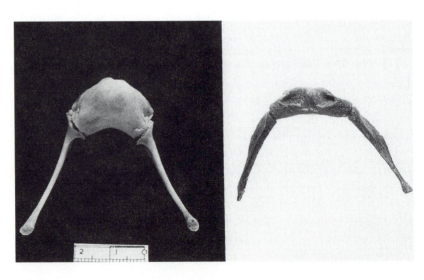

Figure 14.22 **Hyoid bones from *Homo neanderthalensis* and a common chimpanzee** The neanderthal hyoid is on the right. The scale bar, which applies to both hyoids, is in cm. Reproduced by permission from Arensburg, Schepartz, Tillier, Vandermeersch, and Rak, 1990. A reappraisal of the anatomical basis for speech in Middle Paleolithic hominids. *American Journal of Physical Anthropology* 83: 137–146. © American Association of Physical Anthropologists.

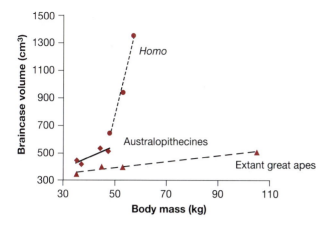

Figure 14.23 **Brain size versus body size in a variety of hominoids** Data points represent species averages, with best-fit lines. In all three groups, species with larger brains have larger bodies. Australopithecines have larger brains for their size than extant great apes. *Homo* species have larger brains for their size, as well as a dramatically different relationship between brain size and body size. The extant great apes are the bonobo, common chimpanzee, orangutan, and gorilla; the australo-pithecines are *A. afarensis, A. africanus, A. boisei,* and *A. robustus;* the *Homo* species are *H. habilis, H. ergaster/erectus,* and *H. sapiens.* The data are from Tobias (1987) and Pilbeam and Gould (1974). See also McHenry (1994). After Pilbeam and Gould (1974).

but brain size increases much more steeply with body size in *Homo* than in its hominid relatives. Like a descended larynx, large brains come at a cost. They require a great deal of energy to maintain, and they generate considerable heat. What benefits could have compensated for these costs? Two possibilities are an increased capacity to make and use tools, and an increased capacity for language.

Phillip Tobias (1987) examined castes of the insides of the braincases of specimens of *H. habilis.* In addition to their sheer size, these endocasts revealed the existence of derived structural traits unique to our genus. Among them are clear enlargments of Broca's and Wernicke's areas. By his own account, this dis-covery converted Tobias from a skeptic to an advocate of the hypothesis that language first emerged, at least in rudimentary form, in *H. habilis.* We do not know for certain, and perhaps never will, but language may be as much as 2 million years old.

An analysis of the skulls of Homo habilis *suggest that they may have had at least rudimentary language.*

Summary

As Darwin predicted, an evolutionary perspective throws light on the origin and nature of humans. Humans are relatives of the great apes. Although the evi-dence is conflicting, morphological and molecular studies suggest that our clos-est living relatives are the chimpanzees. Our most recent common ancestor with the chimpanzees lived about 5 million years ago.

Following its split from the chimpanzee lineage, our own lineage gave rise to several species of bipedal African hominids. Fossils provide strong evidence for the coexistence of at least two of these species, and perhaps as many as five. We are the sole survivors of this evolutionary radiation.

The first members of genus *Homo* left Africa nearly 2 million years ago. Whether these populations ultimately contributed genes to present-day popula-tions of humans is the subject of debate. No definitive tests have been per-formed, but the balance of the evidence suggests that all present-day non-African populations are descended from a more recent wave of emigrants that left Africa within the last 200,000 years. This implies that present-day geo-graphic variation among populations of humans is of relatively recent origin.

Among the derived traits unique to our species are the manufacture and use of complex tools and the capacity for language. Because behavior does not fossilize, researchers have to rely on circumstantial evidence to reconstruct the history of these traits. Tool use appeared at least 2.5 million years ago. It may even have arisen independently in an early species of *Homo* and a robust australopithecine. The evidence on language is more tenuous, but it suggests that language may have emerged nearly as early as tool use.

Questions

1. Explain why Ruvolo et al. (1995) thought it was important to look at several nuclear genes, and not just the mtDNA genes, to study the relationships of humans and the great apes.

2. Suppose humans and chimpanzees are indeed each other's closest relatives. Is it then accurate to say that humans evolved from chimpanzees? How about that chimpanzees evolved from humans?

3. In a recent study of the phylogeny of Old World monkeys (Hayasaka et al. 1996), the three individual Rhesus macaques that were studied did not form a monophyletic group. Instead, the mtDNA of one of the Rhesus macaques was more similar to the mtDNA of Japanese and Taiwanese macaques (which are different species) than it was to the other Rhesus macaques. How might this have happened? (There are at least two possibilities.) How does this relate to Jonathan Marks' objection to the conclusions many researchers draw from the human-ape molecular phylogenies (Figures 14.4 and 14.6)?

4. What's in a name? Jared Diamond (1992), who believes that humans and chimpanzees are closest relatives (Figure 14.3d), suggests that if we follow the naming traditions of cladistic taxonomy, then humans, chimpanzees, and bonobos should all be considered members of a single genus. Diamond proposes calling these species, respectively, *Homo sapiens, Homo troglodytes,* and *Homo paniscus.* Jonathan Marks objects to Diamond's taxonomic reasoning. Marks advocates a human/great ape phylogeny in which either gorillas and chimpanzees are closest relatives (Figure 14.3c; Marks 1993), or there is an unresolved trichotomy (Figure 14.4a; Marks 1994; 1995). Concerning the nature of humans and apes, Marks (1994) asserts:

   > [P]opular works tell us that we are not merely *genetically* apes, but that we are *literally* apes (e.g., Diamond 1992). Sometimes there is profundity in absurdity, but I don't think this is one of those times. It merely reflects the paraphyletic nature of the category "apes"—humans are apes, but only in the same sense that pigeons are reptiles and horses are fish. . . . Focusing on the genetic relations obscures biologically significant patterns of phenotypic divergence.

   Do you think humans, chimpanzees, and bonobos should all be classified as members of the same genus? Is there more at stake in the disagreement between Diamond and Marks than just latin names? If so, what?

5. Jared Diamond finds ethical dilemmas in the close kinship between humans and chimpanzees:

   > [I]t's considered acceptable to exhibit caged apes in zoos, but it's not acceptable to do the same with humans. I wonder how the public will feel when the identifying label on the chimp cage in the zoo reads *"Homo troglodytes."* (Diamond 1992, page 29)

Diamond finds the use of chimpanzees in medical research even more problematic. The scientific justification for the use of chimpanzees is that chimpanzee physiology is extremely similar to human physiology, so chimpanzees are the best substitute for human subjects. Diamond notes that jails are a very rough analog to zoos, in the sense that they represent conditions under which we do consider it acceptable to keep people in cages without their consent (if not to display them). But there is no human analog to research on chimpanzees: There are no conditions under which we consider it acceptable to do medical experiments on humans without their consent. Is it ethically justified to keep animals in zoos? To use animals in medical research? Does the phylogenetic relationship between ourselves and the animals in question matter? If so, how and why?

6. Look at the two different phylogenies illustrated in Figure 14.11b and c. Suppose you are a hominid paleontologist who wants to figure out which of these trees (if either) is correct. What strata would you choose to search for hominid fossils? If you are lucky enough to find a hominid skull, what features of the skull could tell you which tree is correct? If you find no fossils at all, can you conclude anything about the two trees?

7. One of the most heated aspects of human racial politics is the contention that human races are genetically distinct. How do the African replacement versus multiregional evolution models address this issue? That is, which model predicts that human races are more genetically different from each other?

8. Several genetic studies show that African versus non-African is the most fundamental division in the phylogeny of present-day human populations. Some people might conclude from this data that modern African people are in some sense "primitive." What is the logical flaw in this thinking?

9. Jared Diamond and others have pointed out that different ethnic groups within Africa are more diverse than are the ethnic groups on all other continents taken together. What does this imply about the common U.S. practice of categorizing people into "African," "Caucasian," "Asian," and "Native American"?

10. Recent work by C. Swisher and colleagues (1996) indicates that *Homo erectus* may have persisted in Java until 53,000 years ago at Sambungmachan and until 27,000 years ago at Ngandong. If correct, these dates imply that *H. erectus* and *H. sapiens* coexisted in Java. Will this finding help settle the debate over the out of Africa model versus the multiregional origin model? Why or why not?

11. For the sake of argument, adopt the proposition implied by Wood (1997) that early species of *Homo* did not participate in the production of Oldowan stone tools. What other puzzles must we now solve? Note that after the invention of Oldowan tools, the next advance in toolmaking is marked in the archaeological record by the appearance of Acheulean tools. Acheulean tools are substantially more sophisticated than Oldowan tools (Johanson et al. 1996). Acheulean tools appear about 1.4 million years ago, and persist until less than 200,000 years ago.

12. Derek Bickerton (1995) and Charles Catania (1995) object to the suggestion that *Homo habilis* had language. Bickerton writes, "If *H. habilis* already had all the necessary ingredients for language, what happened during the next 2 1/2 million years?" And Catania writes, "I . . . [deduce] that our hominid ancestors should have taken over the world 950,000 years ago if a million years ago they were really like us in their language competence. But they did not, so they were not; if they had been, those years would have been historic instead of prehistoric." How do you think Phillip Tobias would respond to Bickerton and Catania? Who do you think is right?

Exploring the Literature

13. Read Susman's (1994) paper on opposable thumbs and tool use. What weaknesses can you identify in Susman's argument? What additional data would you like to see? If you were Susman, how would you respond to these critiques? Read McGrew et al. (1995) to see if the critiques and responses therein are similar to your own.

    McGrew, W.C., M.W. Hamrick, S.E. Inouye, J.C. Ohman, M. Slanina, G. Baker, R.P. Mensforth, and R.L. Susman. 1995. Thumbs, tools, and early humans. *Science* 268: 586–589.

    Susman, R.L. 1994. Fossil evidence for early hominid tool use. *Science* 265: 1570–1573.

14. There were a couple of hints in this chapter that present-day humans have smaller bodies and brains than some of our recent ancestors. For documentation that this is indeed the case, see:

    Gibbons, A. 1997. Bone sizes trace the decline of man (and woman). *Science* 276:896–897.

    Ruff, C.B., E. Trinkaus, and T.W. Holliday. 1997. Body mass and encephalization in Pleistocene *Homo. Nature* 387:173–176.

15. We noted that bipedal locomotion appears early in the *Australopithecus/Homo* lineage, but we did not discuss the adaptive significance of this trait. Researchers have offered a variety of hypotheses, some more intuitively plausible than others. For a start on this literature, see:

    Chaplin, G., N.G. Jablonski, and N.T. Cable. 1994. Physiology, thermoregulation, and bipedalism. *Journal of Human Evolution* 27:497–510.

    Hunt, E.D. 1994. The evolution of human bipedality: ecology and functional morphology. *Journal of Human Evolution* 26:183–199.

    Jablonski, N.G., and G. Chaplin. 1993. Origin of habitual terrestrial bipedalism in the ancestor of the Hominidae. *Journal of Human Evolution* 24:259–280.

    Wheeler, P.E. 1994. The foraging times of bipedal and quadrupedal hominids in open equatorial environments (a reply to Chaplin, Jablonski & Cable, 1994). *Journal of Human Evolution* 27:511–517.

    Wheeler, P.E. 1994. The thermoregulatory advantages of heat storage and shade-seeking behavior to hominids foraging in equatorial savannah environments. *Journal of Human Evolution* 26:339–350.

16. The recent phylogeny of *Homo* may continue to get more complex. Recently E. Carbonell and colleagues (1995) reported 780,000-year-old fossils from Spain that do not readily fit into any previously defined species:

    Carbonell, E., J.M. Bermúdez de Castro, J.L. Arsuaga, J.C. Díez, A. Rosas, G. Cuenca-Bescós, R. Sala, M. Mosquera, and X.P. Rodríguez. 1995. Lower pleistocene hominids and artifacts from Atapuerca-TD6 (Spain). *Science* 269:826–830.

Citations

Aiello, L.C. 1993. The fossil evidence for modern human origins in Africa: A revised review. *American Anthropologist* 95:73–96.

Andrews, P. 1992. Evolution and environment in the Hominoidea. *Nature* 360:641–646.

Andrews, P., and L. Martin. 1987. Cladistic relationships of extant and fossil hominoids. *Journal of Human Evolution* 16:101–118.

Arensburg, B., L.A. Schepartz, A.M. Tillier, B. Vandermeersch, and Y. Rak. 1990. A reappraisal of the anatomical basis for speech in Middle Paleolithic hominids. *American Journal of Physical Anthropology* 83:137–146.

Arensburg, B., A.M. Tillier, B. Vandermeersch, H. Duday, L.A. Schepartz, and Y. Rak. 1989. A middle paleolithic human hyoid bone. *Nature* 338:758–760.

Ayala, F.J. 1995. The myth of Eve: Molecular biology and human origins. *Science* 270:1930–1936.

Ayala, F.J., A. Escalante, C. O'hUigin, and J. Klein. 1994. Molecular genetics of speciation and human origins. *Proceedings of the National Academy of Sciences, USA* 91:6787–6794.

Begun, D.R. 1992. Miocene fossil hominids and the chimp-human clade. *Science* 257:1929–1933.

Begun, D.R. 1995. Late Miocene European orang-utans, gorillas, humans, or none of the above? *Journal of Human Evolution* 29:169–180.

Begun, D.R., C.V. Ward, and M.D. Rose, eds. 1997. *Function, Phylogeny, and Fossils: Miocene Hominid Evolution and Adaptations* (New York: Plenum).

Beynon, A.D., M.C. Dean, and D.J. Reid. 1991. On thick and thin enamel in hominoids. *American Journal of Physical Anthropology* 86:295–309.

Bickerton, Derek. 1995. Finding the true place of *Homo habilis* in language evolution. Open peer commentary in Wilkins and Wakefield (1995).

Bowcock, A.M., A. Ruiz-Linares, J. Tomfohrde, E. Minch, J.R. Kidd, and L.L. Cavalli-Sforza. 1994. High resolution of human evolutionary trees with polymorphic microsatellites. *Nature* 368:455–457.

Cann, R.L., M. Stoneking, and A.C. Wilson. 1987. Mitochondrial DNA and human evolution. *Nature* 325:31–36.

Catania, A.C. 1995. Single words, multiple words, and the functions of language. Open peer commentary in Wilkins and Wakefield (1995).

Darwin, C. 1859. *On the Origin of Species by Means of Natural Selection, Or the Preservation of the Favoured Races in the Struggle for Life* (London: John Murray).

Darwin, C. 1871. *The Descent of Man, and Selection in Relation to Sex* (London: John Murray).

Dean, D., and E. Delson. 1992. Second gorilla or third chimp? *Nature* 359:676–677.

Diamond, J. 1992. *The Third Chimpanzee* (New York: HarperCollins).

Djian, P., and H. Green. 1989. Vectorial expansion of the involucrin gene and the relatedness of the hominoids. *Proceedings of the National Academy of Sciences, USA* 86:8447–8451.

Frayer, D.W., M.H. Wolpoff, A.G. Thorne, F.H. Smith, and G.G. Pope. 1993. Theories of modern human origins: The paleontological test. *American Anthropologist* 95:14–50.

Gabunia, L., and A. Vekua. 1995. A plio-pleistocene hominid from Dmanisi, East Georgia, Caucasus. *Nature* 373:509–512.

Galaburda, A.M., and D.N. Pandya. 1982. Role of architectonics and connections in the study of primate brain evolution. In E. Armstrong and D. Falk, eds. *Primate Brain Evolution* (New York: Plenum).

Gibbons, A. 1994. Rewriting—and redating—prehistory. *Science* 263:1087–1088.

Gibbons, A. 1996. A rare glimpse of an early human face. *Science* 274:1298.

Goldstein, D.B., A. Ruiz Linares, L.L. Cavalli-Sforza, and M.W. Feldman. 1995. Genetic absolute dating based on microsatellites and the origin of modern humans. *Proceedings of the National Academy of Sciences, USA* 92:6723–6727.

Goodman, M., W.J. Bailey, K. Hayasaka, M.J. Stanhope, J. Slightom, and J. Czelusniak. 1994. Molecular evidence on primate phylogeny from DNA sequences. *American Journal of Physical Anthropology* 94:3–24.

Green, H., and P. Djian. 1995. The involucrin gene and Hominoid relationships. *American Journal of Physical Anthropology* 98:213–216.

Groves, C.P. 1986. Systematics of the great apes. In D.R. Swindler and J. Erwin, eds. *Comparative Primate Biology, Volume 1: Systematics, Evolution, and Anatomy* (New York: Alan R. Liss, Inc.), 187–217.

Hayasaka, K., K. Fujii, and S. Horai. 1996. Molecular phylogeny of macaques: Implications of nucleotide sequences from an 896-base pair region of mitochondrial DNA. *Molecular Biology and Evolution* 13:1044–1053.

Hedges, S.B., S. Kumar, K. Tamura, and M. Stoneking. 1992. Human origins and analysis of mitochondrial DNA sequences. *Science* 255:737–739.

Horai, S., Y. Satta, K. Hayasaka, R. Kondo, T. Inoue, T. Ishida, S. Hayashi, and N. Takahata. 1992. Man's place in the Hominoidea revealed by mitochondrial DNA genealogy. *Journal of Molecular Evolution* 35:32–43.

Horai, S., K. Hayasaka, R. Kondo, K. Tsugane, and N. Takahata. 1995. Recent African origin of modern humans revealed by complete sequences of hominoid mitochondrial DNAs. *Proceedings of the National Academy of Sciences, USA* 92:532–536.

Huang Wanpo, R. Ciochon, G. Yúmin, R. Larick, F. Qiren, H. Schwarcz, C. Yonge, J. de Vos, and W. Rink. 1995. Early *Homo* and associated artefacts from Asia. *Nature* 378:275–278.

Johanson, D.C., B. Edgar, and D. Brill. 1996. *From Lucy to Language* (New York: Simon & Schuster Editions).

Kim, H.-S., and O. Takenaka. 1996. A comparison of TSPY genes from Y-chromosomal DNA of the great apes and humans: Sequence, evolution, and phylogeny. *American Journal of Physical Anthropology* 100:301–309.

Kimbel, W.H., R.C. Walter, D.C. Johanson, K.E. Reed, J.L. Aronson, Z. Assefa, C.W. Marean, G.G. Eck, R. Robe, E. Hovers, Y. Rak, C. Vondra, T. Yemane, D. York, Y. Chen, N.M. Evensen, and P.E. Smith. 1996. Late Pliocene

Homo and Oldowan tools from the Hadar Formation (Kada Hadar member), Ethiopia. *Journal of Human Evolution* 31:549–561.

Kluge, A.G. 1983. Cladistics and the classification of the great apes. In R.L. Ciochon and R.S. Corruccini, eds. *New Interpretations of Ape and Human Ancestry* (New York: Plenum Press), 151–177.

Krings, M., A. Stone, R.W. Schmitz, H. Krainitzki, M. Stoneking, and S. Pääbo. 1997. Neandertal DNA sequences and the origin of modern humans. *Cell* 90: 19–30.

Leakey, M.G., C.S. Feibel, I. McDougall, and A. Walker. 1995. New four-million-year-old hominid species from Kanapoi and Allia Bay, Kenya. *Nature* 376:565–571.

Lieberman, D.E. 1995. Testing hypotheses about recent human evolution from skulls: Integrating morphology, function, development, and phylogeny. *Current Anthropology* 36:159–197.

Li Tianyuan and D.A. Etler. 1992. New middle pleistocene hominid crania from Yunxian in China. *Nature* 357:404–407.

Lowenstein, J., and A. Zihlman. 1988. The invisible ape. *New Scientist* 3 (December):56–59.

Marks, J. 1993. Hominoid heterochromatin: Terminal C-bands as a complex genetic trait linking chimpanzee and gorilla. *American Journal of Physical Anthropology* 90:237–246.

Marks, J. 1994. Blood will tell (won't it?): A century of molecular discourse in anthropological systematics. *American Journal of Physical Anthropology* 94:59–79.

Marks, J. 1995. Learning to live with a trichotomy. *American Journal of Physical Anthropology* 98:211–213.

Martin, L. 1985. Significance of enamel thickness in hominoid evolution. *Nature* 314:260–263.

McGrew, W.C., M.W. Hamrick, S.E. Inouye, J.C. Ohman, M. Slanina, G. Baker, R.P. Mensforth, and R.L. Susman. 1995. Thumbs, tools, and early humans. *Science* 268:586–589.

McHenry, H. 1994. Tempo and mode in human evolution. *Proceedings of the National Academy of Sciences, USA* 91:6780–6784.

National Public Radio. 1995. Evolution disclaimer to be placed in Alabama textbooks. Transcript 1747, segment 13.

Nei, M. 1995. Genetic support for the out-of-Africa theory of human evolution. *Proceedings of the National Academy of Sciences, USA* 92:6720–6722.

Noble, W., and I. Davidson. 1991. The evolutionary emergence of modern human behavior: Language and its archaeology. *Man* 26:223–253.

Pilbeam, D., and S.J. Gould. 1974. Size and scaling in human evolution. *Science* 186:892–901.

Pinker, S. 1994. *The Language Instinct* (New York: Harper-Collins).

Pough, F.H., J.B. Heiser, and W.N. McFarland. 1996. *Vertebrate Life* (Upper Saddle River, NJ: Prentice Hall).

Pritchard, J.K., and M.W. Feldman. 1996. Genetic data and the African origin of humans. *Science* 274:1548.

Rak, Y. 1983. *The Australopithecine Face* (New York: Academic Press).

Risch, N., K.K. Kidd, and S.A. Tishkoff. 1996. Genetic data and the African origin of humans—reply. *Science* 274:1548–1549.

Rogers, J., and A.G. Comuzzie. 1995. When is ancient polymorphism a potential problem for molecular phylogenetics? *American Journal of Physical Anthropology* 98:216–218.

Ruvolo, M. 1994. Molecular evolutionary processes and conflicting gene trees: The Hominoid case. *American Journal of Physical Anthropology* 94:89–113.

Ruvolo, M. 1995. Seeing the forest and the trees: Replies to Marks; Rogers and Commuzzie; Green and Djian. *American Journal of Physical Anthropology* 98:218–232.

Ruvolo, M., S. Zehr, M. von Dornum, D. Pan, B. Chang, and J. Lin. 1993. Mitochondrial COII sequences and modern human origins. *Molecular Biology and Evolution* 10:1115–1135.

Ruvolo, M., D. Pan, S. Zehr, T. Goldberg, T.R. Disotell, and M. von Dornum. 1994. Gene trees and hominoid phylogeny. *Proceedings of the National Academy of Sciences, USA* 91:8900–8904.

Sarich, V.M., and A.C. Wilson. 1967. Immunological time scale for Hominid evolution. *Science* 158:1200–1203.

Schwartz, J.H. 1984. The evolutionary relationships of man and orang-utans. *Nature* 308:501–505.

Semaw, S., P. Renne, J.W.K. Harris, C.S. Feibel, R L. Bernor, N. Fesseha, and K. Mowbray. 1997. 2.5-million-year-old stone tools from Gona, Ethiopia. *Nature* 385:333–336.

Strait, D.S., F.E. Grine, and M.A. Moniz. 1997. A reappraisal of early hominid phylogeny. *Journal of Human Evolution* 32:17–82.

Stern, J.T., and R.L. Susman. 1983. The locomotor anatomy of *Australopithecus afarensis*. *American Journal of Physical Anthropology* 60:279–317.

Stringer, C. 1988. The dates of Eden. *Nature* 331:565–566.

Susman, R.L. 1994. Fossil evidence for early hominid tool use. *Science* 265:1570–1573.

Swisher, C.C., III, G.H. Curtis, T. Jacob, A.G. Getty, A. Suprijo, and Widiasmoro. 1994. Age of the earliest known hominids in Java, Indonesia. *Science* 263:1118–1121.

Swisher, C.C., III, W.J. Rink, S.C. Antón, H.P. Schwarcz, G.H. Curtis, A. Suprijo, and Widiasmoro. 1996. Latest *Homo erectus* of Java: Potential contemporaneity with *Homo sapiens* in Southeast Asia. *Science* 274:1870–1874.

Tattersall, I. 1986. Species recognition in human paleontology. *Journal of Human Evolution* 15:165–175.

Tattersall, I. 1992. Species concepts and species identification in human evolution. *Journal of Human Evolution* 22:341–349.

Tattersall, I. 1995. *The Fossil Trail* (Oxford: Oxford University Press).

Tattersall, I. 1997. Out of Africa again . . . and again? *Scientific American* 276 (April):60–67.

Templeton, A.R. 1992. Human origins and analysis of mitochondrial DNA sequences. *Science* 255:737.

Templeton, A.R. 1993. The "Eve" hypothesis: A genetic critique and reanalysis. *American Anthropologist* 95:51–72.

Thorne, A.G., and M.H. Wolpoff. 1981. Regional continuity in Australasian pleistocene hominid evolution. *American Journal of Physical Anthropology* 55:337–349.

Tishkoff, S.A., E. Dietzsch, W. Speed, A.J. Pakstis, J.R. Kidd, K. Cheung, B. Bonné-Tamir, A.S. Santachiara-Benerecetti, P. Moral, M. Krings, S. Pääbo, E. Watson, N. Risch, T. Jenkins, and K.K. Kidd. 1996b. Global patterns of linkage disequilibrium at the CD4 locus and modern human origins. *Science* 271:1380–1387.

Tishkoff, S.A., K.K. Kidd, and N. Risch. 1996a. Interpretations of multiregional evolution—reply. *Science* 274:706–707.

Tobias, P.V. 1987. The brain of *Homo habilis:* A new level of organization in cerebral evolution. *Journal of Human Evolution* 16:741–761.

Valladas, H., J.L. Reyss, J.L. Joron, G. Valladas, O. Bar-Yosef, and B. Vandermeersch. 1988. Thermoluminescence dating of Mousterian "Proto-Cro-Magnon" remains from Israel and the origin of modern man. *Nature* 331:614–616.

Vigilant, L., M. Stoneking, H. Harpending, K. Hawkes, and A.C. Wilson. 1991. African populations and the evolution of human mitochondrial DNA. *Science* 253:1503–1507.

Waddle, D.M. 1994. Matrix correlation tests support a single origin for modern humans. *Nature* 368:452–454.

Ward, S.C., and W.H. Kimbel. 1983. Subnasal alveolar morphology and the systematic position of *Sivapithecus. American Journal of Physical Anthropology* 61:157–171.

White, T.D., and G. Suwa. 1987. Hominid footprints at Laetoli: Facts and interpretations. *American Journal of Physical Anthropology* 72:485–514.

White, T.D., G. Suwa, and B. Asfaw. 1994. *Australopithecus ramidus,* a new species of early hominid from Aramis, Ethiopia. *Nature* 371:306–312.

White, T.D., G. Suwa, and B. Asfaw. 1995. Corrigendum: *Australopithecus ramidus,* a new species of early hominid from Aramis, Ethiopia. *Nature* 375:88.

Wilkins, W.K., and J. Wakefield. 1995. Brain evolution and neurolinguistic preconditions. *Behavioral and Brain Sciences* 18:161–226.

Wilson, D.E., and D.M. Reeder, eds. 1993. *Mammal species of the world,* 2nd edition (Washington: Smithsonian Institution Press).

Wolpoff, M.H. 1996. Interpretations of multiregional evolution. *Science* 274:704–706.

Wood, B. 1992. Origin and evolution of the genus *Homo. Nature* 355:783–790.

Wood, B. 1997. The oldest whodunnit in the world. *Nature* 385:292–293.

Wood, B., and A. Turner. 1995. Out of Africa and into Asia. *Nature* 378:239–240.

Zischler, H., H. Geisert, A. von Haeseler, and S. Pääbo. 1995. A nuclear "fossil" of the mitochondrial D-loop and the origin of modern humans. *Nature* 378:489.

Sexual Selection

The females and males in many species are strikingly different from each other. In marine iguanas, for example, the largest females are about half the size of the largest males. In túngara frogs the females are bigger than the males, by about 10%. In long-tailed widow birds the impressive difference is not in size, but feathers: The males are jet-black, carry tailfeathers several times the length of their own bodies, and have yellow and red shoulder patches; the females are colored a cryptic brown, have short tail feathers, and have no shoulder patches. In some species of pipefish the females have blue stripes and skin folds on their bellies, which the males lack. In stalk-eyed flies, both sexes wear their eyes on the ends of long thin stalks, but males have longer eyestalks than females. The differences between the sexes also often include behavior. In marine iguanas the males are territorial during the breeding season, while the females remain gregarious; in túngara frogs only the males make the mating call that gives the species its name.

In humans, too, females and males are conspicuously different. The differences include not just the obvious and essential ones like our genitalia and reproductive organs, but also facial structure, vocal tone, distribution of body fat and body hair, and body size. The difference in body size between women and men is documented in Figure 15.1.

A difference between the males and females of a species is called a sexual dimorphism. Sexual dimorphism occurs in a great variety of organisms (Figure 15.2). In this chapter we ask why. In previous chapters, we have explained many features of organisms with the theory of evolution by natural selection. The goal has been to discover whether the features in question are adaptive, and if so then how they improve the survival or fecundity of the individuals that possess them. But differences between the sexes of a species are often hard to explain with the theory of evolution by natural selection.

A difference between the sexes is called a sexual dimorphism.

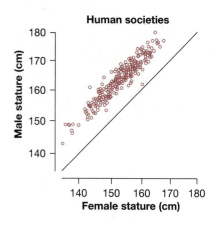

Human societies

Figure 15.1 **Women and men differ in height**   For each of more than 200 human societies, the average height of the men is plotted against the average height of the women. The diagonal line shows where the points would fall if men and women were of equal height. People vary widely in height from society to society: In the shortest society, the average man is about 143 cm tall (about 4 feet 8 inches) and the average woman about 135 cm ($\sim 4'5''$); in the tallest society, the average man is about 180 cm tall ($\sim 5'11''$) and the average woman about 165 cm ($\sim 5'5''$). But in every society the average man is taller than the average woman, usually by about 10%. From Rogers and Mukherjee 1992.

For example, compare the well-documented example of the evolution of beak size in Darwin's finches (Chapter 2) to a hypothetical story we might tell about the evolution of long tail feathers in male long-tailed widow birds (Figure 15.3). Two problems arise. First, when the finches get hit by a drought and small, soft seeds become rare, big beaks are as useful to the females as to the males. But if long tail feathers can improve the survival or fecundity of a widow bird, then why do only the males have them? Second, how could those enor-

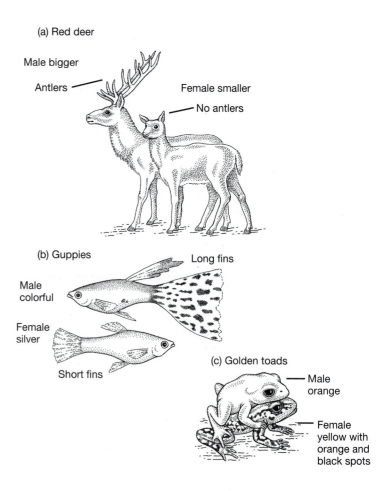

Figure 15.2 **The differences between males and females (the sexual dimorphism) in red deer *(Cervus elaphus)*, guppies *(Poecilia reticulata)*, and golden toads *(Bufo periglenes)***

Figure 15.3 **The sexual dimorphism in long-tailed widow birds** *(Euplectes progne)* The male is black with long tail feathers and red and yellow shoulder patches; the female is cryptic brown.

mously long tail feathers improve survival or fecundity? The long tail feathers probably make the male widow birds easier for predators to find and catch. Furthermore, growing long tail feathers requires a considerable investment of energy. Any energy spent on feathers is energy that cannot be spent on making offspring. It appears that the theory of evolution by natural selection can explain neither why male and female widow birds are different, nor why the unusual trait, long tail feathers, exists at all.

Sex provides a solution to the puzzle of sexual dimorphism. First consider life without sex. For organisms that reproduce asexually, getting genes into the next generation is fairly straightforward. The two main challenges are surviving long enough to reproduce, and then reproducing. Sex complicates life by adding a third major challenge: finding a member of the opposite sex and persuading him or her to cooperate.

Charles Darwin recognized that individuals vary not only in their success at surviving and reproducing, but also in their success at persuading members of the opposite sex to mate. Considering birds, for example, Darwin wrote, "inasmuch as the act of courtship appears to be with many birds a prolonged and tedious affair, so it occasionally happens that certain males and females do not succeed during the proper season, in exciting each other's love, and consequently do not pair" (1871, page 107). Darwin applied the label natural selection to differences among individuals in survival and reproduction. Differences among individuals in success at getting mates he called sexual selection.

Our goal in this chapter is to explore how the theory of evolution by sexual selection does what the theory of evolution by natural selection cannot: Explain the frequent existence of conspicuous differences between females and males, particularly when those differences involve traits that seem likely to impair survival. We first review classical work that elucidated the precise mechanism through which sexual reproduction creates different selection pressures for females versus males. Then we consider recent research on the evolutionary consequences of these differing selection pressures in different species.

Sexual dimorphism is an evolutionary puzzle, because natural selection cannot explain it.

15.1 The Asymmetry of Sex

In this section we argue that sexual reproduction creates different selection pressures for females versus males. The logic we develop to support this conclusion was clearly articulated by A. J. Bateman (1948) and refined by Robert Trivers (1972). It hinges on a crucial fact: Eggs (or pregnancies) are more expensive than ejaculates. In more general terms, females typically make a larger parental investment in each offspring than males. This investment can take the form of energy, time, or both.

The parental investments of female and male orangutans are typical for mammals. Adult orangutans of opposite sex favor each others' company only for the purpose of mating (Nowak 1991). After a brief tryst, including a copulation that lasts about 15 minutes, the male and female go their separate ways. If pregnancy results, then the mother, who weighs about 40 kilograms, will carry the fetus for 8 months, give birth to a $1\frac{3}{4}$-kilogram baby, nurse it for about $3\frac{1}{2}$ years, and continue to protect it until it reaches the age of 7 or 8. For the father, who weighs about 70 kilograms, the beginning and end of parental investment is a few grams of semen, which he can replace in a matter of hours or days. Females that provide substantial parental care and males that provide none whatsoever are the rule in more than 90% of mammal species (Woodroffe and Vincent 1994). Because female mammals provide such intensive parental care, mammals are a somewhat extreme example of disparity in parental investment.

In most animal species, neither parent cares for the young: They just lay the eggs and leave them. But in these species, too, the females usually make a larger investment in each offspring than the males. Eggs are typically large and yolky, with a big supply of stored energy and nutrients. Think of a sea turtle's eggs, some of which are as large as hen's eggs. Most sperm, on the other hand, are little more than DNA with a propeller. Even when a single ejaculate delivers hundreds of millions of sperm, the ejaculate seldom represents more than a fraction of the investment contained in a clutch of eggs.

The key to explaining sexual dimorphism is in recognizing that sexual reproduction imposes different selection pressures on females versus males.

The fact that eggs are more expensive than ejaculates yields the prediction of a profound difference in what limits the lifetime reproductive success of females versus males. A female's maximum potential reproductive success is relatively small, and her realized reproductive success is likely to be limited more by the number of eggs she can make (or pregnancies she can carry) than by the number of males she can convince to mate with her. In contrast, a male's maximum potential reproductive success is relatively large, and his realized reproductive success is likely to be limited more by the number of females he can convince to mate with him than by the number of ejaculates he can make. In other words, we predict that access to females will be a limiting resource for males, but access to males will not be a limiting resource for females.

The Asymmetry of Sex Demonstrated in Fruit Flies

A.J. Bateman (1948) tested this prediction in laboratory populations of the fruitfly, *Drosophila melanogaster*. Bateman set up small populations of flies in bottles. Each population consisted of three virgin males and three virgin females. Each fly was heterozygous for a unique dominant genetic mutation: one that gave the fly curly wings, for example, or hairless patches on parts of its body. Bateman let the flies mate with each other. He then raised their offspring to adulthood. He was able to identify the parents of each offspring by looking at the mutations the offspring inherited. In this way, Bateman could figure out which females had mated with which males, and vice versa. He could also count the number of offspring each parent produced. Note that the males and females living together in a bottle had the same number of potential mates. If the factors that limit reproductive success are different for the two sexes, then the data for the two sexes should show different relationships between reproductive success and number of actual mates. Figure 15.4 gives details of the experiment for one of Bateman's bottle populations.

Bateman conducted the experiment shown in Figure 15.4 a dozen times. He also conducted a dozen replicates of the reciprocal experiment, swapping the genetic markers carried by the males and females. The combined results appear in Figure 15.5. They confirm the prediction that the two sexes differ in the factors limiting reproductive success.

For males, reproductive success increased in direct proportion with number of mates (Figure 15.5a). Males also showed considerable variation in number of mates, with many males in the extreme groups with zero mates and three mates (Figure 15.5b). These two patterns combined to produce considerable variation among males in reproductive success: Males with three mates were big winners; males with no mates were big losers (Figure 15.5c).

For females, on the other hand, reproductive success did not substantially increase with more than one mate (Figure 15.5a). Further, females showed less

Because females typically invest more in each offspring than males, a female's reproductive success is limited by the number of eggs she can make. In contrast, male's reproductive success is limited by the number of females he can mate with.

(a) Six flies placed in a bottle and allowed to mate among themselves:

	Sex	Genetic Marker	Abbreviation	Description
1	Female	Curly-Lobe	CyL	Curly wings and small eyes
2	Female	Curly	Cy	Curly wings
3	Female	Microcephalus	Mc	Eyes tiny or absent
4	Male	Stubble	Sb	Short bristles
5	Male	Plum	Pm	Brown eyes
6	Male	Hairless	H	Parts of body lack hairs

(b) Their offspring, scored by the markers they carry:

		Markers that identify the mother:				
			CyL	Cy	Mc	None	Total
Markers that idenfity the father:	Sb		13	0	0	16	29
	Pm		10	12	15	68	105
	H		7	29	0	41	77
	None		60	38	40	110	248
	Total		90	79	55	235	459

(c) Mating and reproductive success of parents inferred from offspring:

	Sex	Abbreviation	No. of Mates	Identity of Mates	No. of Identifiable Offspring
1	Female	CyL	3	Sb; Pm; H	90
2	Female	Cy	2	Pm; H	79
3	Female	Mc	1	Pm	55
4	Male	Sb	1	CyL	29
5	Male	Pm	3	CyL; Cy; Mc	105
6	Male	H	2	CyL; Cy	77

(d) Reproductive success versus number of mates:

(e) Variance in reproductive success:

Males: 985   Females: 214

Figure 15.4   One of Bateman's laboratory populations of *Drosophila melanogaster*   (a) Each of the six parents is heterozygous for a dominant genetic marker.   (b) Approximately half of each parent's offspring are identifiable because they inherited the parent's marker.   (c) The results from (b) allowed Bateman to determine the number of mates and relative number of offspring for each parent.   (d) The results for this population are consistent with the prediction that the reproductive success of males depends more strongly on number of mates than does the reproductive success of females.   (e) Furthermore, variation in reproductive success was much higher for males than for females. Here, variation is characterized by the variance. To calculate the variance of a list of numbers, first calculate the mean. Then take the difference between each number and the mean. These differences are called the deviations from the mean. Now take each deviation and square it. The mean of the squared deviations is the variance. The larger the variation among the numbers in a list, the larger the variance. After Bateman (1948).

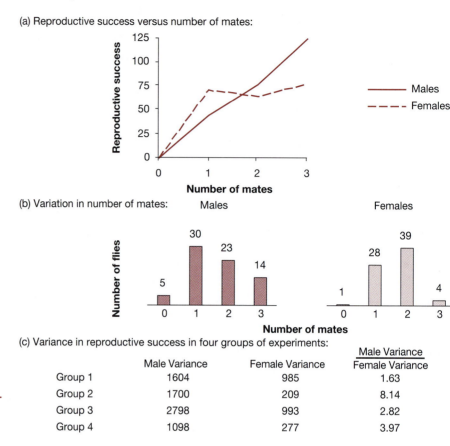

(a) Reproductive success versus number of mates:

(b) Variation in number of mates: Males   Females

(c) Variance in reproductive success in four groups of experiments:

	Male Variance	Female Variance	Male Variance / Female Variance
Group 1	1604	985	1.63
Group 2	1700	209	8.14
Group 3	2798	993	2.82
Group 4	1098	277	3.97

***Figure 15.5* The combined results from Bateman's experiments**   After Bateman (1948).

variation in number of mates, with most females having one or two (Figure 15.5b). These two patterns combined to produce relatively little variation among females in reproductive success: There were few big winners and few big losers (Figure 15.5c).

Access to mates indeed proved to be a limiting resource for males, but not for females. Sexual selection—variation in fitness due to variation in success at getting mates—was much stronger in males than in females. Evolutionary biologists believe that this pattern is quite general (although not universal).

Behavioral Consequences of the Asymmetry of Sex

The asymmetry between males and females in what limits reproductive success allows us to predict differences in the mating behavior of the two sexes:

Sexual selection theory predicts that males will compete with each other over access to mates, and that females will be choosy.

- Males should be competitive. If the fitness of males is limited by access to females, then we predict that males will compete among themselves over opportunities to mate.
- Females should be choosy. If the fitness of females is not limited by opportunities to mate, but any given mating may involve the commitment by the female of a large investment in offspring, then we predict that females will be choosy about who they mate with.

Male-male competition for mates and female choosiness can play out in two ways. First, in species in which males can directly monopolize access to females, males typically fight with each other over such monopolies. The females mate with the winners. This form of sexual selection is called intrasexual selection, because the key event that determines reproductive success involves interactions among the members of a single sex (the males). Second, in species in which males cannot directly control access to females, the males advertise for mates. The females then choose among them. This form of sexual selection is called intersexual selection, because the key event that determines reproductive success involves an interaction between members of the two sexes (females choose males).

Many readers will have noticed that our treatment of the asymmetry of sex has been full of crass generalizations. We want to emphasize that expectations of male-male competition and female choosiness are not based on anything inherent to maleness or femaleness per se, but on the observation that females commonly invest more per offspring than males. This investment pattern is broken in a great variety of species. For species in which males care for their offspring, males may invest as much or more in each offspring as females. Species with male parental care include humans, many fish, about 5% of the frogs, and over 90% of the birds. For animals in which males actually do invest more per offspring than females, and are thus a limiting resource for the females, we predict that females will compete with each other over access to males, and that males will be choosy. We will return to such animals toward the end of this chapter, to see if they are exceptions that can prove the rules of sexual selection. For now, however, we will focus on sexual selection by male-male competition and female choice.

15.2 Male-Male Competition: Intrasexual Selection

Sexual selection by male-male competition occurs when individual males can monopolize access to females. This monopolization of females can happen through direct control of the females themselves, or through control of some resource important to females. In this section, we consider examples of recent research into three forms of male-male competition: outright combat, sperm competition, and infanticide.

Combat

Outright combat is the most obvious form of male-male competition for mates. Intrasexual selection involving male-male combat over access to mates can favor morphological traits including large body size, weaponry, and armor. Male-male combat also selects for effective tactics.

Our example comes from the marine iguanas *(Amblyrhynchus cristatus)* of the Galápagos Islands (Figure 15.6). Marine iguanas have a life-style unique among the lizards. They make their living by basking on rocks near the ocean to warm themselves, grazing on algae in the intertidal zone, then basking again to aid digestion. As we mentioned earlier, the males get much larger than the females (Figure 15.7a).

Male-male competition can take the form of combat over access to females.

The marine iguana sexual size dimorphism is an excellent example for the study of sexual selection, because we know a great deal about how marine iguana size is affected by natural selection (Wikelski et al. 1997; Wikelski and Trillmich 1997). Martin Wikelski and Fritz Trillmich documented natural selection on iguana body size by monitoring the survival of marked individuals on two islands over one to two years (Figure 15.7b). Natural selection was much harsher on Genovesa than on Santa Fé, but it was stabilizing on both islands, with medium-sized iguanas surviving at higher rates than either small or large iguanas.

Potential agents of this natural selection on body size are few. Marine iguanas do not compete with other species for food and have virtually no predators. Other than reproduction, about all the iguanas have to contend with is competing among themselves for food.

Wikelski and colleagues showed that food is, in fact, a limiting resource for marine iguanas (Wikelski et al. 1997). The researchers measured the relationship between body size and the amount of energy, in the form of algae, that an individual iguana can harvest each day. The larger an iguana gets, the more energy it can harvest. The researchers also calculated, based on figures in the literature, how much energy an iguana uses each day on metabolism. The larger an iguana gets, the more energy it uses. Wikelski and colleagues discovered that the amount of energy an iguana uses each day increases faster with body size than does the amount of energy it can harvest. This means that below a threshold size, an iguana can harvest more energy than it needs for metabolism; but above the threshold, it cannot (Figure 15.7c).

The maximum sustainable body size varies from island to island and year to year. Marine iguanas graze at low tide, and during each low tide they graze the algae right down to the rocks. The amount of energy an individual of a given size can harvest therefore depends on how fast the algae have grown since the last low tide. The growth rate of algae varies from island to island and from year to year. However, the maximum sustainable iguana body size is consistently larger on Santa Fé than on Genovesa.

Now compare Figure 15.7a with Figure 15.7b. The largest females in each population are near the optimal body size for survival, but the largest males are

Figure 15.6 **A male Galápagos marine iguana** (Carlo Dani and Ingrid Jeske/Animals Animals/Earth Scenes)

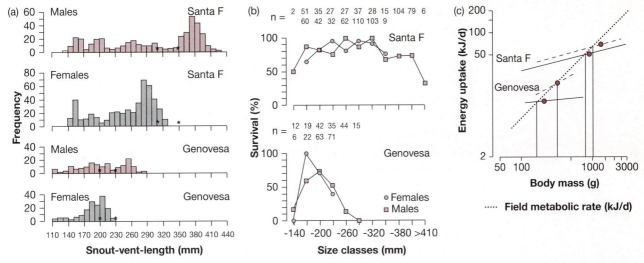

Figure 15.7 Natural selection on body size in marine iguanas (a) Histograms showing the size distributions of male and female marine iguanas on two different Galápagos Islands, Genovesa and Santa Fé. Iguanas on Santa Fé get much larger than iguanas on Genovesa; but in both populations, males get larger than females. See (c) for explanation of asterisks. From Wikelski et al. (1997).   (b) Survival rates of marked individuals of different sizes (snout-vent length, mm) from March 1991 to March 1992 on Genovesa, and from February 1990 to February 1992 on Santa Fé. In both sexes on both islands, medium-sized individuals survived at higher rates than small or large individuals. The sample sizes, or number of individuals in each group, are given by n. From Wikelski and Trillmich (1997).   (c) Energy balance for marine iguanas. The dotted line shows the amount of energy in kilojoules (kJ) iguanas use each day on metabolism as a function of body mass. The dashed and solid lines show the amount of energy iguanas can harvest each day as a function of body mass; the dashed lines are from a good year (1992–1993), and the solid lines are from a bad year (the El Niño of 1991–1992). The intersections of dashed and solid lines with the dotted line, marked by vertical lines, are predicted thresholds for positive versus negative energy balance. For example, the top dashed line represents the amount of energy an iguana can harvest each day during a good year on Santa Fé. The point at which this dashed line intersects the dotted line indicates the body mass (about 1,000g) at which an iguana can harvest exactly the amount of energy it burns each day on metabolism. Iguanas weighing 1,000g will neither gain nor lose weight. Iguanas weighing less than 1,000g can harvest more energy than they use; they will grow. Iguanas weighing more than 1,000g use more energy than they can harvest; they will lose weight. The large dots mark the body masses above which marked iguanas monitored by Wikelski and colleagues were actually observed to be losing weight. These sizes are also marked by asterisks in (a). From Wikelski et al. (1997).

much larger than the optimal body size. The large body size of male marine iguanas is thus an evolutionary puzzle: We cannot explain it by natural selection, because Wikelski and Trillmich have shown that natural selection acts against it. It is exactly the kind of puzzle for which Darwin invoked sexual selection.

As we discussed earlier, a crucial issue in sexual selection is the relative investment per offspring made by females versus males. In marine iguanas the parental investment by females is much larger. Each female digs a nest on a beach away from the basking and feeding areas, buries her eggs, guards the nest for a few days, and then abandons it (Rauch 1988). Males provide no parental care at all. So parental investment by females consists mostly of producing eggs, and parental investment by males consists entirely of producing ejaculates. Females lay a single clutch of one to six eggs each year, into which they put about 20% of their body mass (Rauch 1985; Rauch 1988; Wikelski and Trillmich 1997). Compared to this, the cost of the single ejaculate needed to fertilize all the eggs in a clutch is paltry. This difference in investment suggests that the maximum potential reproductive success of males is much higher than that of females.

Number of mates will be a limiting resource for the lifetime reproductive success of males but not females.

The iguanas' mating behavior is consistent with these inferences. Females copulate only once each reproductive season. Martin Wikelski, Silke Bäurle, and their field assistants followed several dozen marked females on Genovesa. The researchers watched the females from dawn to dusk every day during the entire month-long mating season in 1992–1993 and 1993–1994 (Wikelski and Bäurle 1996). They also watched the marked females from dawn to dusk every day during the subsequent nesting seasons. Every marked female that dug a nest and laid eggs had been seen copulating, but no marked female had been seen copulating more than once. Male iguanas, in contrast, attempt to copulate many times with many different females. But the opportunity to copulate with females is a privilege a male iguana has to fight for (Figure 15.8).

Prior to the mating season each year, male iguanas stake out territories on the rocks where females bask between feeding bouts. In small, densely packed territories (Figure 15.9a), males attempt to claim and hold ground by ousting male interlopers. Confrontations begin with head-bobbing threats, and escalate to chases and head pushing. If neither male backs down, fights can end with bites leaving serious injuries on the head, neck, flanks, and legs (Trillmich 1983). While a male holds a territory, he has a more or less exclusive right to mate with any receptive females that happen to be there, typically females that use the territory as their basking site (Rauch 1985). Because only some males manage to claim territories, and some males manage to maintain their claims for longer than others, there is extreme variation among males in the number of females they copulate with (Figure 15.9b).

Because claiming and holding a territory involves combat with other males, bigger males tend to win. In the iguana colony that Krisztina Trillmich (1983) studied on Camaaño Islet, the male that got 45 copulations (Figure 15.9b), far more than any other male, was iguana 59 (his territory is shown in Figure 15.9a). His neighbor, iguana 65, was the second most successful with 10 copulations. Both of their territories were females' favorite early-morning and late-afternoon basking places. Trillmich reported that iguana 59 was the largest male

Male marine iguanas fight over territories where females congregate. Large iguanas win more fights, and thus get to copulate with more females. This pattern of sexual selection has led to the evolution of large body size in males.

Figure 15.8 Male marine iguanas in combat (Martin Wikelski, University of Washington)

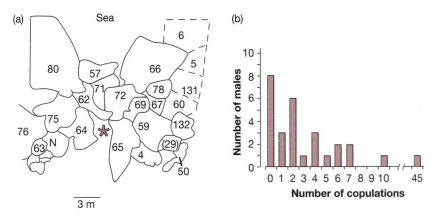

Figure 15.9 Mating success in male marine iguanas   (a) A cluster of iguana mating territories on Camaaño Islet, Galápagos. Lines show boundaries of mating territories on 16 January 1978; numbers identify territory owners. As the scale bar for this map shows, mating territories are only a few square meters in size. The asterisk indicates where Krisztina Trillmich sat to watch the iguanas. (Camaaño Islet has only 880m of shoreline and supports a population of nearly 2,000 iguanas.) From Trillmich (1983).   (b) Histogram showing variation in number of copulations obtained by male iguanas on mating territories shown in (a). Note the break in the horizontal scale; the most successful male, iguana 59, got more than four times as many copulations as any of his rivals. The histogram includes only males that claimed a territory for at least a short time during the mating season. From Trillmich (1983).

in the colony; that to claim his territory he had to eject four other males who tried to take it; and that during his tenure he lost parts of it to four neighboring males who were pushing their territories in from the sides. Wikelski and co-workers studied iguana colonies on Genovesa and Santa Fé (Wikelski et al. 1996; Wikelski and Trillmich 1997). Consistent with Krisztina Trillmich's observations, these researchers found that the mean size of males that got to copulate was significantly larger than the mean size of all males that tried to copulate (Table 15.1).

If we assume that body size is heritable in marine iguanas, then we have variation, heritability, and differential mating success. These are the elements of evolution by sexual selection. We thus have an explanation for why male marine iguanas get so much bigger than the optimal size for survival. Male iguanas get big because bigger males get more mates and pass on more of their big-male genes.

Male-male combat, analogous to that in marine iguanas, happens in a great variety of species (Figure 15.10). When mating opportunities are a limiting resource for males, and when males can monopolize either the females themselves, or some resource that is vital to the females and thus sure to attract them, males fight among themselves for access to the females or the resource. In addition to large body size, this kind of sexual selection leads to the evolution of other traits that are assets in combat, such as weaponry and armor. Male-male combat can also lead to the evolution of alternative male mating strategies (see Box 15.1).

TABLE 15.1 Sexual selection differentials for male body size in marine iguana colonies on Santa Fé and Genovesa

Body size is given as snout-vent length (SVL). The standardized selection differential (see Chapter 6) is the difference between the average body size of all males that copulated at least once and the average body size of all males that tried to copulate, expressed in standard deviations of the distribution of body sizes of all males that tried to copulate. (The standard deviation is the square root of the variance.) Both standardized selection differentials are positive (P < 0.05), indicating that males that got to copulate were larger on average than males that tried to copulate. From Wikelski and Trillmich (1997).

	N	Average size (SVL)	Standard deviation	Standardized selection differential
Santa Fé				
Males that copulated	253	401	13	0.42
All males that tried to	343	390	26	
Genovesa				
Males that copulated	25	243	26	0.77
All males that tried to	147	227	21	

Sperm Competition

Male-male competition can take the form of sperm competition.

Male-male competition does not necessarily stop when copulation is over. The real determinant of a male's mating success is not whether he copulates, but whether his sperm actually fertilize eggs. If an animal has internal fertilization, and if a female mates with two or more different males within a short period of time, then the sperm from the males will be in a race to the eggs. Indeed, females may produce litters or clutches in which different offspring are fathered

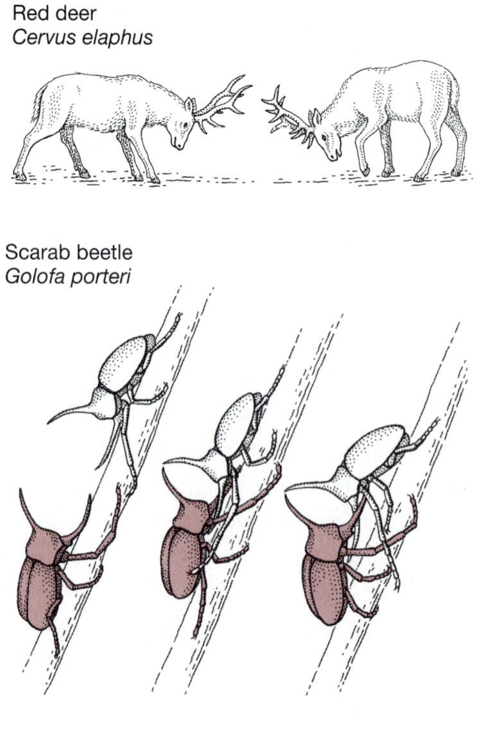

Red deer
Cervus elaphus

Scarab beetle
Golofa porteri

Figure 15.10 **Sexual selection in red deer and scarab beetles** In red deer, *Cervus elaphus,* males fight over control of good grazing areas, and over the groups of females that come to feed there. Males able to control access to bigger groups of females for longer periods of time father more offspring. As a result of selection on fighting ability, males have evolved large body size (twice that of the females) and antlers. See Clutton-Brock (1985). Similar factors are apparently responsible for the long horns and front legs of the males in the scarab beetle, *Golofa porteri.* Male beetles use their horns and legs to pry other males from the plants on which the beetles feed, presumably to monopolize access to females that are feeding there too. After Anderson (1994).

Box 15.1 Alternative male mating strategies

Victory in male–male combat typically goes to the large, strong, and well armed. But what about the smaller males? Is their only chance at fitness to survive until they are large enough to win fights? Often small males attempt to mate by employing alternative strategies.

In marine iguanas, small adult males are ousted from the mating territories on the basking grounds. But many do not give up; they continue trying to get females to copulate with them. The small males are not terribly successful, but they do get about 5% of the matings in the colony (Wikelski et al. 1996). Needless to say, small males attempting to mate with females are often harassed by other males. This happens to large territorial males too, but it happens more often to small males. Furthermore, copulations by small males are more likely to be disrupted before the male has time to ejaculate (Figure 15.11).

The small males solve the problem of disrupted copulations by ejaculating ahead of time (Wikelski and Bäurle 1996). They use the stimulation of an attempted copulation, or even of seeing a female pass by, to induce ejaculation. The males then store the ejaculate in their cloacal pouches. If he gets a chance to mate, a small male transfers his stored ejaculate to the female at the beginning of copulation. Wikelski and Bäurle examined the cloacae of a dozen females caught immediately after copulations that had lasted less than three minutes. None of these females had copulated earlier that mating season, but 10 of the 12 females had old ejaculates in their cloacae that must have been transferred during the short copulation.

The sperm in these old ejaculates were viable. From dawn to dusk every day for about a month, Wikelski and Bäurle watched five of the females until they laid their eggs. None of the five copulated again, but all laid fertilized eggs.

Prior ejaculation appears to be a strategy practiced more often by small nonterritorial males than by large territorial males. Wikelski and Bäurle caught 13 nonterritorial and 13 territorial males at random; 85% of the nonterritorial males had stored ejaculates in their cloacal pouches, versus only 38% of the territorial males ($P < 0.05$). This difference is unlikely to result from more frequent copulation by territorial males, because even territorial males copulate only about once every six days (Wikelski et al. 1996).

Alternative, or sneaky male, mating strategies have also evolved in a variety of other species. In coho salmon, *Oncorhynchus kisutch,* for example, males return from the sea to spawn at two different ages (Gross 1984; Gross 1985; Gross 1991). One group, called hooknoses, returns at 18 months. They

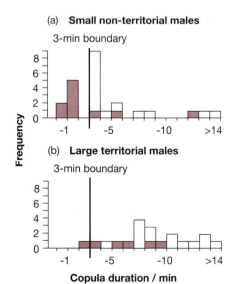

Figure 15.11 Duration of copulations by male marine iguanas Histograms showing the distribution of copulation durations for   (a) 24 small nonterritorial males and   (b) 20 large territorial males. Filled areas indicate copulations that were disrupted by other males. The heavy vertical line at 3 minutes marks the approximate amount of time a male must copulate before he can ejaculate. Large territorial males had copulations that were significantly longer ($P < 0.001$) and less likely to be disrupted before the 3-minute boundary ($P < 0.05$). From Wikelski and Bäurle (1996).

Box 15.1 *(continued)*

are large, armed with enlarged hooked jaws, and ar-mored with cartilaginous deposits along their backs. The other group, called jacks, returns at 6 months. They are small, poorly armed, and poorly armored.

When a female coho is ready to mate, she digs a nest and then lays her eggs in it. As she prepares the nest, males congregate. The males use one of two strategies in trying to fertilize the female's eggs (Figure 15.12). Some males fight for a position close to the female. These fighters quickly sort themselves out by size. When the female lays her eggs, the males spawn over them in order. The first male to spawn fertilizes the most eggs. Other males do not fight for position, but instead look for a hiding place near the female. When the female lays her eggs, these sneak-ers attempt to dart out and spawn over the eggs.

Among hooknoses, those that adopt the fighting strategy are more successful. Among jacks, those that adopt the sneaky strategy are more successful. The relative fitness of hooknoses versus jacks depends, in part, on the frequency of each type of male in the breeding population.

There is an important distinction between the iguana example and the coho example. In marine iguanas, the small nonterritorial males appear to be making the best of a bad situation while they grow to a large enough size to successfully fight for a ter-ritory. In coho, a male irreversibly becomes either a hooknose or a jack. Which strategy a male coho pursues depends on a mixture of environmental and genetic factors.

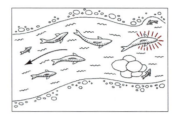

Figure 15.12 **Alternative mating strategies in coho salmon**   This figure shows a coho mating group. The large fish at the right (upstream) is a female that has built a nest and is ready to lay her eggs. Down-stream from the female are five males that have opted for the fighting strategy. These males, four hooknoses and a jack, have sorted themselves out by size. Two other jacks have opted for the sneaky strategy. They have found hiding places near the fe-male, one behind a rock, the other in a shallow. After Gross (1991).

by different males. Batches of offspring with multiple fathers have been docu-mented in a great variety of animals, including squirrels (Boellstorff et al. 1994), bears (Schenk and Kovacs 1995), red-winged blackbirds (Gibbs et al. 1990), many other birds, lizards (Olsson et al. 1994), and spiders (Watson 1991). It hap-pens in humans too; Smith (1984) reviews reports of over a dozen sets of twins with clearly different fathers.

Given sperm competition, what traits contribute to victory? One useful trait might simply be the production of large ejaculates containing many sperm. If sperm competition is something of a lottery, then the more tickets a male buys, the better his chances of winning. This hypothesis has been tested by Matthew Gage (1991) with the Mediterranean fruit fly, *Ceratitis capitata*. Gage's experi-ment was based on the observation that, although ejaculates are cheaper than eggs, they are not free (see, for example, Nakatsuru and Kramer 1982). Gage rea-soned that if male Mediterranean fruit flies are subject to any constraints on sperm production, they might benefit from conserving their sperm, using during each copulation only the minimum number necessary to ensure complete fertilization of the female's eggs. But if larger ejaculates contribute to victory in sperm competition, males whose sperm are at risk of competition should release more sperm during copulation than males whose sperm are not at risk. If the number of sperm released is unimportant to the outcome of competition, then males should release the same number of sperm regardless of the risk of competition.

Gage raised and mated male medflies under two sets of conditions. One group of 20 he raised by themselves and allowed to mate in private; the other group of 20 he raised in the company of another male and allowed to mate in the presence of that second male. Immediately after each mating, Gage dissected the females and counted the number of sperm the males had released. Males raised and mated in the presence of a potential rival ejaculated during copulation more than $2\frac{1}{2}$ times as many sperm (average $\pm$ standard error = 3520 $\pm$ 417) as males raised and mated in isolation (1379 $\pm$ 241), a highly significant difference (P < 0.0001). Gage's interpretation was that large ejaculates do contribute to victory in sperm competition, and that male medflies dispense their sperm to balance the twin priorities of ensuring successful fertilization and conserving sperm.

When male medflies are at risk of sperm competition, they produce larger ejaculates.

In addition to large ejaculates, sperm competition has apparently led to various other adaptations such as males prolonging copulation, depositing a copulatory plug, applying pheromones that reduce the female's attractiveness, and using special structures on the copulatory organ to scoop out sperm left by rivals (Gilbert 1976; Sillén-Tullberg 1981; Thornhill and Alcock 1983).

Infanticide

In some species of mammals, competition between males continues even beyond conception. One example, discovered by B.C.R. Bertram (1975) and also studied by Craig Packer and Anne Pusey (reviewed in Packer et al. 1988), happens in lions. The basic social unit of lions is the pride. The core of a pride is a group of closely related females—mothers, daughters, sisters, nieces, aunts, and so on—and their cubs. Also in the pride is a small group of adult males; two or three is a typical number. The males are usually related to each other, but not to the adult females. This system is maintained because females reaching sexual maturity stay in the pride they were born to, whereas newly mature males move to another pride.

The move for young adult males from one pride to another is no stroll in the park. The adult males already resident in the new pride resist the invaders. That is why males stay with their other male kin: Each group, the residents and the newcomers, is a coalition that together fights the other group, sometimes violently, over the right to live in the pride. If the residents win, they stay in the pride and the newcomers search for a different pride to take over. If the residents lose they are evicted, and the newcomers have exclusive access to the pride's females. Exclusive, that is, until another coalition of younger, stronger, or more numerous males comes along and kicks *them* out. Pusey and Packer found that the average time a coalition of males holds a pride is a little over two years. Because residence in a pride is the key to reproductive success in lions, males in a victorious coalition quickly begin trying to father cubs. One impediment to quick fatherhood, however, is the presence of still-nursing cubs fathered

Figure 15.13 **Lion infanticide**   In this photo by George B. Schaller, a male lion has just killed another male's cub, which it now carries in its mouth.

Male-male competition can take the form of infanticide. By killing other males' cubs, male lions gain more opportunities to mate.

by males of the previous coalition. That is because females do not return to breeding condition until after their cubs are weaned.

How can the males overcome this problem? They frequently employ the obvious, if grisly, solution: They kill any cubs in the pride that are not weaned (Figure 15.13). Packer and Pusey have shown that this strategy causes the cubs' mothers to return to breeding condition an average of 8 months earlier than they otherwise would. Infanticide by males is the cause of about 25% of all cub deaths in the first year of life, and over 10% of all lion mortality.

Infanticide improves the males' reproductive prospects, but is obviously detrimental to the reproductive success of the females. The females have two options for making the best of their own interests in this bad situation (Packer and Pusey 1983). One is to defend their cubs from infanticidal males, which females often do, occasionally at the cost of their own lives. Nonetheless, Packer and Pusey report that young cubs rarely survive more than two months in the presence of a new coalition of males. The females' other tactic is to spontaneously abort any pregnancies in progress when a new coalition gains residence in the pride. This cuts the females' losses: They do not waste energy and time on cubs that would be killed anyway shortly after birth. With this shift in focus to female reproductive strategy, we leave the subject of male-male conflict and move to the other side of sexual selection: female choice.

15.3   Female Choice

In a great variety of species access to females is limiting to males, but the males are unable to monopolize either the females themselves or any resource vital to the females. In many such species, the males advertise for mates and the females choose among them. This form of sexual selection leads to the evolution of elaborate courtship displays by males.

When males cannot monopolize access to females, they often advertise for mates. Although biologists were long skeptical that females discriminate among the advertising males, female choice is now well established.

The existence of active discrimination among males by females is widely accepted today. However, female choice remained controversial for many decades after Darwin asserted its importance in 1871 in *The Descent of Man, and Selection in Relation to Sex*. Evolutionary biologists thought that female discrimination was limited to choosing a male of the right species (see Trivers 1985). Beyond species identification, courtship displays functioned primarily in overcoming a general female reluctance to mate. Once ready to mate, a female would accept any male of her own species. We begin this section by describing two recent experiments that demonstrate conclusively that females are in fact selective, actively choosing particular males from among the many available.

We then consider the functions of female choice. Why should females prefer males with some traits and not others? One theory suggests that in some cases the traits preferred by females are arbitrary, evolving by a mechanism called runaway selection. Various other theories offer reasons why females choose males with particular kinds of traits. Potential benefits to the choosy female include directly acquiring resources offered by the male, avoiding hybridization (the traditional theory), and acquiring good genes for her offspring. Alternatively, females may prefer particular traits because of preexisting sensory biases.

Are Female Barn Swallows Choosy?

Our first example of an experiment demonstrating active female choice comes from the work of Anders Møller on barn swallows. Barn swallows, *Hirundo rustica,* are small insect-eating birds that breed in colonies of up to about 80 individuals. The swallows Møller (1988) studied breed in Denmark during spring and summer, and spend the European winter in Africa. Upon arriving in a Danish breeding colony, each male swallow sets up a territory a few square meters in size. He then tries to attract a mate by displaying his tail while perching and flying. Each female visits several males, then chooses one to pair with. Upon pairing, the male and female together build a mud nest in the male's territory and raise one or, if they have time before summer's end, two clutches of young. The female incubates the eggs by herself, but both male and female feed the chicks.

At first glance, barn swallows may not seem promising subjects for a study of sexual selection. The fact that the males help care for the young should tend to equalize parental investment by the two sexes, and the fact that the swallows appear to mate monogamously means that neither sex should be in short supply for the other. Indeed, there is relatively little sexual dimorphism; females and males are similar in size and appearance. The biggest difference is that the outermost tail feathers, which are elongated in both sexes, are about 15% longer in males than in females (Figure 15.14).

Three factors suggest, however, that sexual selection may be at work in barn swallows. First, even in a monogamous species in which both sexes care for the young, males and females may vary in quality, both as parents and as donors of genes to offspring. The members of both sexes should thus benefit, in higher reproductive success, by trying to identify and attract the best mate possible. Second, although barn swallows appear to be monogamous, many in fact are not. Males sometimes solicit copulations with females other than their pair-mates. When these offers of extra-pair copulation are accepted by the females, as occasionally they are, the successful male may benefit by fathering offspring that some other male will help to raise. Third, the one trait that does differ between the sexes—the length of the long outer tail feathers—is precisely the trait that males show off when advertising for mates. Møller hypothesized that sexual selection does indeed occur in barn swallows, through the mechanism of female choice, and that females prefer to mate with males displaying longer tail feathers.

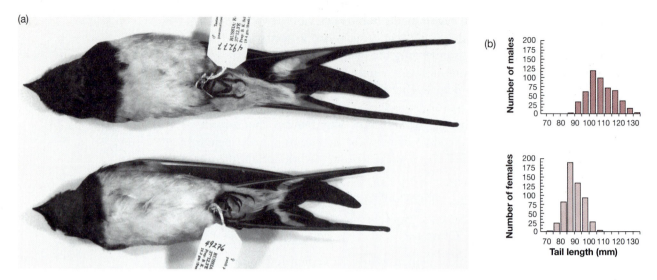

Figure 15.14 Sexual dimorphism in barn swallows (a) In male (top) and female (bottom) barn swallows, the outermost tail feathers are elongated. (Museum specimens courtesy of University of Washington Burke Museum; Jon Herron)   (b) The distribution of tail streamer length in male and female barn swallows. The distributions overlap, but the average male has longer tail feathers than the average female ($P < 0.0001$). From Møller (1991).

Møller captured and color-banded 44 males that had established territories but not yet attracted mates. He divided them at random into four groups of 11, and altered the tail feathers of each group as follows:

- **Shortened tail feathers.** Møller clipped about 2 cm out of the middle of each outer tail feather, then reattached the feather tips to the bases with superglue.
- **Mock-altered (control I).** Møller clipped the tail feathers, then glued them back together. This did not change the length of the feathers, but it otherwise subjected the birds to the same handling, clipping, and gluing as the shortened and lengthened groups.
- **Unaltered (control II).** Møller captured and banded these birds, but did nothing to their tail feathers.
- **Elongated tail feathers.** Møller added, by clipping and gluing, the 2 cm of feather removed from the shortened group into the middles of the tail feathers of these birds.

Møller then released the birds back into their colony. He predicted that if females prefer males with longer tail feathers, then the group of males with elongated tails would attract mates sooner, and fledge more young, than either control I males or control II males. The control males should, in turn, be more successful than the shortened males. If, on the other hand, the females have no preferences based on tail feathers, then there should be no differences in success among the groups.

The results appear in Figure 15.15. The elongated males, on average, attracted mates more quickly than the control males, and control males attracted mates more quickly than shortened males (Figure 15.15a). Among the advantages of attracting a mate quickly is that the male and his partner can get an earlier start on rearing a clutch of chicks. An earlier start means that the parents are more likely to have time to raise a second clutch before the summer ends (Figure 15.15b). Finally, the advantage of raising two clutches is that the parents fledge more chicks in one summer (Figure 15.15c).

Just as Møller predicted, females prefer pair-mates with longer tail feathers. For males, being one of the more desirable mates results in higher reproductive success. We mentioned earlier that barn swallows sometimes engage in extra-pair copulations; elongated males were winners there too (see Box 15.2).

Female barn swallows prefer to mate with the males whose tail feathers are longest.

Are Female Stalk-Eyed Flies Choosy?

The stalk-eyed flies of Southeast Asia carry their eyes on the ends of long thin appendages. In both sexes bigger flies have longer eyestalks, but males have longer eyestalks for their size than females (Figure 15.17, page 590). By day the flies are solitary, and forage for rotting plants. In the evening, however, the flies congregate beneath overhanging stream banks, where they cling in small groups to exposed root hairs and spend the night (Figure 15.18). At dawn and dusk, the flies roosting together on a root hair often mate with each other. Neither sex cares for the young, so a female's investment in each offspring is larger than a male's. Not surprisingly, males attempt to evict each other from the root-hair roosts in order to be the only male in the group at mating time. In male-male confrontations, the male with longer eyestalks typically wins, so male-male competition may partially explain the evolution of eyestalks (Burkhardt and de la Motte 1983; Burkhardt and de la Motte 1987). Gerald Wilkinson and Paul Reillo (1994) suspected, however, that female choice has also played a role.

Wilkinson and Reillo's test of their hypothesis is based on an idea first elaborated by Ronald Fisher in 1915, and perhaps even traceable to a remark made by T.H. Morgan in 1903 (reviewed in Anderson 1994). The idea is worth explaining in some detail, because it illustrates some useful concepts in evolutionary

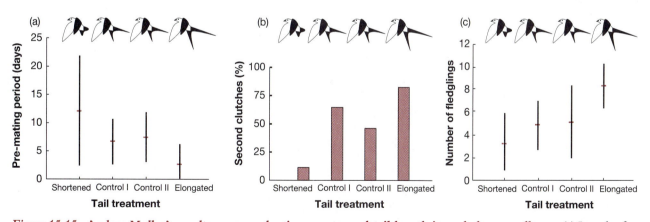

Figure 15.15 **Anders Møller's results on reproductive success and tail length in male barn swallows** (a) Length of time required by males to attract a mate. Horizontal ticks represent the average for each group; the vertical lines represent the standard deviation (a statistical measure of the amount of variation within each group). There were significant differences among groups (P < 0.01). (b) Differences in time required to attract a mate carried over into which males had time to raise a second clutch of chicks. Bar height indicates the percentage of males in each group who, with their mates, raised two clutches. Again there was significant variation among groups (P < 0.02). (c) Differences in second-clutch success carried over into reproductive success—the number of chicks each male had fledged by summer's end. The meanings of ticks and lines are the same as in (a). There was significant variation among groups (P < 0.001). Elongated males fledged more chicks than control males, who in turn fledged more chicks than shortened males. From Møller (1988).

genetics. Using stalk-eyed flies as an example, imagine that both males and females are variable, males in the lengths of their eyestalks and females in their mating preferences for stalk length. These two patterns of variation should combine to produce assortative mating, in which the females that prefer the longest stalks mate with the longest-stalked males, and the females that prefer the shortest stalks mate with the shortest-stalked males (Figure 15.19a, page 591). Assume, furthermore, that in both sexes the variation is heritable; that is, that at least part of the variation in stalk length and preference is due to variation in genes (see Chapter 6). If so, then offspring that receive from their fathers genes for long eyestalks tend to also receive from their mothers genes for a preference for long-stalked males. In other words, if the assortative mating persists for some generations, then it will establish a genetic correlation (linkage disequilibrium)

Box 15.2 Extra-pair copulations

Barn swallows are socially monogamous. Females and males form pairs, build nests, and rear young together. Social monogamy does not necessarily entail sexual monogamy, however. While conducting his experiment on female choice, Anders Møller (1988) saw both male and female barn swallows copulate with individuals other than their social mates. It was the males who solicited extra-pair copulations; the females sometimes accepted these advances and sometimes rejected them.

Female barn swallows had opportunities to be choosy whenever they were solicited by males for extra-pair copulations. Møller was able to watch the birds closely enough to estimate the rates at which the male swallows attempted to copulate with females other than their pair-mates, the rates at which the females they solicited accepted these copulations, and the rates at which the study males' pair-mates copulated with other males. The males in the four study groups showed no differences in the rate at which they attempted to gain extra-pair copulations. Nor did their pair-mates differ in the rates at

which they were solicited. The males in the groups did differ, however, in the rates at which the extra-pair females they propositioned actually accepted them for copulation (Table 15.2). Furthermore, the males' pair-mates differed in the rates at which they accepted extra-pair copulations with other males (Table 15.2) Apparently females who had to settle for less desirable short-tailed males attempted to compensate by copulating out-of-pair with more desirable long-tailed males. Thanks to these females, the long-tailed males won again, at the expense of the short-tails.

Barn swallows are somewhat unusual in that a careful observer can see enough copulations to estimate the rate of extra-pair copulations directly. Over the last several years, however, biologists have developed methods of genetic analysis that enable them to indirectly estimate the rate of extra-pair copulations (Figure 15.16). Through the use of such techniques, biologists have discovered that many socially monogamous birds engage in frequent extra-pair copulations.

TABLE 15.2 Extra-pair copulations in Møller's barn swallow experiment

The numbers reported are rates, measured as extra-pair copulations per hour. P values give statistical significance of variation among groups. From Møller (1988).

	Male Tail Treatment				
Extra-pair copulations	Shortened tails	Control I	Control II	Lengthened tails	P
By males	0	0	0	0.040	<0.001
By their social pair-mates	0.036	0.014	0.017	0	<0.01

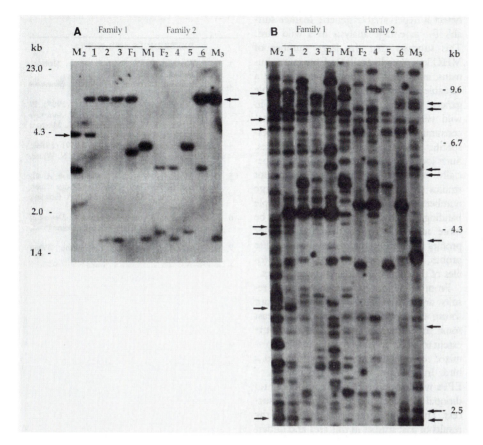

Figure 15.16   **Genetic analyses demonstrating extra-pair copulations in red-winged blackbirds**   (a) A paternity analysis using a traditional restriction-fragment length polymorphism. Each lane in this electrophoresis gel contains DNA that has been extracted from an individual bird, cut with a restriction enzyme, and labeled with a probe that recognizes a sequence of DNA that occurs at a single locus. Bands on the gel are inherited as simple Mendelian alleles. Individual M_1 (center lane) is an adult male red-wing who had two mates on his territory, F_1 and F_2. M_2 and M_3 are adult male neighbors of M_1, F_1, and F_2. The numbers 1, 2, and 3 represent chicks in the nest shared by M_1 and F_1. Chick 1 has a band in its lane (arrow) that is present in neither its mother (F_1) nor its social father (M_1). This band is present, however, in M_2. We can infer that F_1 had an extra-pair copulation with M_2. Chick 6 has also a band in its lane (arrow) that is present in neither its mother (F_2) nor its social father (M_1). This band is present, however, in M_3. We can infer that F_2 had an extra-pair copulation with M_3.   (b) A paternity analysis of the same families using DNA fingerprints. Each lane contains DNA that has been extracted from an individual bird, cut with a restriction enzyme, and labeled with a probe that recognizes a sequence of DNA that occurs at many loci. Bands on the gel are inherited as simple Mendelian alleles, although we do not know which band corresponds to an allele at which locus. The DNA fingerprints confirm the same cases of extra-pair copulation inferred in (a). Reproduced by permission from H.L. Gibbs, 1990. Realized reproductive success of polygynous red-winged blackbirds revealed by DNA markers. *Science* 250:1394−96, Dec. 7, 1990, p. 1395, Fig. 1. © American Association for the Advancement of Science.

between the stalk-length genes and the preference genes (see Chapter 6). If we were to take a group of males, mate each with a number of randomly chosen females, and then examine their sons and daughters, we would find that sons with long eyestalks tend to have sisters with a preference for long-stalked-males (Figure 15.19b). This correlation means that if we conduct artificial selection on

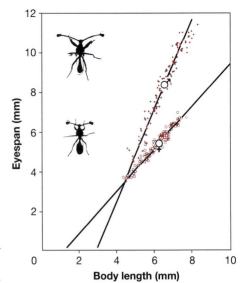

Figure 15.17   The relationship between eyespan and body length in males (crosses) and females (circles) of the Malaysian stalk-eyed fly, ***Cyrtodiopsis whitei***   The male and female symbols show the position of the average member of each sex. The illustrations show a large male (top) and a large female, drawn $1\frac{1}{2}$ times their actual size. From Burkhardt and de la Motte (1988).

Figure 15.18   A group of Malaysian stalk-eyed flies ***(Cyrtodiopsis whitei)*** **gathered on a root hair to spend the night**   The largest fly is a male; the others are females. (Gerald Wilkinson, University of Maryland)

Female stalk-eyed flies prefer to mate the males whose eyestalks are longest — except in populations selected for short eyestalks.

the stalk lengths of the males (and only on the males) in a population, we should get a correlated evolutionary response in the preferences of the females (Figure 15.19c). It was this final proposition that Wilkinson and Reillo tested.

Wilkinson and Reillo collected stalk-eyed flies *(Cyrtodiopsis dalmanni)* in Malaysia and used them to establish three laboratory populations. In each population the researchers separated the males and females immediately after the adults flies emerged from their pupae, and kept them apart for two to three months. Wilkinson and Reillo then chose breeders for the next generation of each population as follows. For the control line, they used 10 males and 25 females picked at random. For the long-selected line, they used the 10 males with longest eyestalks from a pool of 50 males picked at random, and 25 females picked at random. For the short-selected line, they used the 10 males with shortest eyestalks from a pool of 50 males picked at random, and 25 females picked at random. After 13 generations the populations had diverged substantially in eyestalk length. Wilkinson and Reillo then performed paired-choice tests to assay female preferences in each population.

In each test, Wilkinson and Reillo placed two males in a cage with five females. The males had the same body size, but one was from the long-selected line, and thus had eyestalks that were long, and the other was from the short-selected line, and thus had eyestalks that were short (but still longer than the eyestalks of females from any line). The two males were separated by a clear plastic barrier, and each had his own artificial root hair on which to roost. In the center of the plastic barrier was a hole, just large enough to allow the females to pass back and forth, but too small for the males to fit through with their longer eyestalks. Wilkinson and Reillo then watched to see which male attracted more females. The scientists performed 15 to 25 tests for each of the three lines. In both the control and the long-selected lines, more females chose to roost for the night with the long-stalked male. In the short-selected line, however, more females chose to roost for the night with the short-stalked male (Figure 15.20). Artificial selection for short eyestalks in males had changed the mating preferences of females.

(a) Variation in eyestalks and preferences should lead to assortative mating:

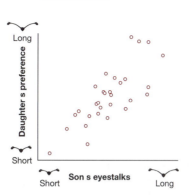

(b) Assortative mating should produce genetic correlations between sons and daughters within families:

(c) Selection on male eyestalks should produce a response in female preference:

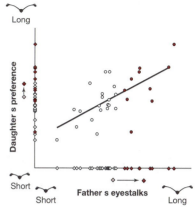

Figure 15.19 **The reasoning behind Wilkinson and Reillo's (1994) test of female choice in stalk-eyed flies** (a) Assortative mating. Females with different tastes choose among males with different eyestalk lengths. If both traits are heritable, then offspring receiving genes for long eyestalks also tend to receive genes for "long" tastes.   (b) A genetic correlation between male stalk length and female preference. Each point represents the average value of the offspring of a male mated with each of a number of randomly chosen females. Males whose sons have long eyestalks tend to also have daughters that seek long eyestalks in their mates. After figure 4.6 in Arnold (1983).   (c) Female preference evolves as a correlated response to selection on male stalk length. Each circle represents the stalklength and preference, respectively, of a father-daughter pair. The fathers are also represented by diamonds on the horizontal axis, and the daughters by diamonds on the vertical axis. If we select the longest-stalked males as breeders (filled diamonds on the horizontal axis, and filled circles), we should see a response in the daughters. The arrows indicate the selection differential and predicted response (see Chapter 6). The light-filled diamond below the horizontal axis marks the average of all fathers in the population, and the dark-filled diamond marks the average of selected fathers. The light-filled diamond to the left of the vertical axis marks the average of all daughters, and the dark-filled diamond marks the average of daughters of selected males. After Falconer (1989).

This result neatly accomplishes several things at once:

- It demonstrates that female stalk-eyed flies are choosy.
- It demonstrates that both male eyestalk length and female preference are heritable.
- It is consistent with Fisher's 80-year-old prediction that sexual selection by female choice produces genetic correlations between male traits and female preferences.
- It clearly illustrates that selection on one trait can produce an evolutionary response in another trait (see Chapter 6).
- It is consistent with Wilkinson and Reillo's hypothesis that sexual selection in the form of female choice is at least partly responsible for the evolution and maintenance of long eyestalk length in males.

The scenario for the evolution of long eyestalks is that in the past the majority of the females preferred long eyestalks, so the males with long eyestalks left more offspring. As Fisher first noted, this can create a positive feedback loop, because as the experiment showed, selection on males for longer eyestalks produces a correlated response in female preferences. Each generation's males have longer eyestalks than their fathers had, but each generation's females prefer longer stalks than their mothers did. Under the right circumstances, this positive-feedback loop can result in the runaway evolution of ever-longer eyestalks

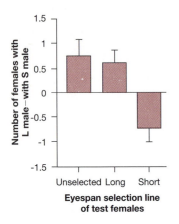

Figure 15.20 The results of Wilkinson and Reillo's paired choice tests for female preference The height of each bar represents the average value (± standard error), for a number of trials, of the difference between the number of females that preferred the long-stalked male and the number that preferred the short-stalked male. Positive values indicate that more females preferred long-stalked males. Unselected females preferred long-stalked males (P = 0.0033), as did females from the long-selected line (P = 0.005). Females from the short-selected line preferred short-stalked males (P = 0.023). From Wilkinson and Reillo (1994).

Theory suggests that female preferences can be completely arbitrary. Alternatively, females may benefit from their choosiness.

Choosy female hanging flies get food from their mates.

(see Fisher 1958; Lande 1981; Arnold 1983). In other words, it is at least theoretically possible that females prefer long eyestalks not because this preference carries any intrinsic fitness advantage, but simply because a small arbitrary preference, once established, led to runaway selection for ever more extreme preferences and ever longer eyestalks.

The possibility of runaway selection raises a new question: Why are the males' eyestalks not even longer than they are now—twice the length of the flies' bodies, or three times? One hypothesis is that if the eyestalks were any longer they would be a serious impediment to survival. As far as we know, this hypothesis has not been tested in stalk-eyed flies.

Female choice, as illustrated by barn swallows and stalk-eyed flies, is thought to be the selective force responsible for the evolution of a great variety of male advertisement displays—from the wing-beat love song of the fruitfly, to the peeping and croaking of frogs, to the gaudy tail feathers of the peacock. Some male displays are loud and clear; others, like those of barn swallows, are more subtle. It is curious that an extra two centimeters added to two tail feathers should make male barn swallows so much more attractive to females as to dramatically affect the males' reproductive success. Why should the females care about such a small difference? And for that matter, why should females care about any of the advertisements, even the loud ones, that males use to attract mates? We have just noted the theoretical possibility of runaway selection, in which there is, in fact, no good reason at all. We now consider several other explanations for female choice.

Choosy Females May Benefit Directly Through the Acquisition of Resources

In many species the males provide food, parental care, or some other resource that is directly beneficial to the female, and females prefer males offering bigger gifts. Such is the case in the hangingfly *(Bittacus apicalis),* studied by Randy Thornhill (1976). Hangingflies live in the woods of eastern North America, where they hunt for other insects. After a male catches an insect, he hangs from a twig and releases a pheromone to attract females. When a female approaches, the male presents his prey. If she accepts it the pair copulates while she eats (Figure 15.21a). The larger the prey, the longer it takes her to eat it, and the longer the pair copulates (Figure 15.21b). The longer the pair copulates, the more sperm the female accepts from the male (Figure 15.21c). If she finishes her meal in less than 20 minutes, the female breaks off the copulation and flies away looking for another male and another meal. The female's preference for males bearing large gifts benefits her in two ways: (1) it provides her with more nutrients, allowing her to lay more eggs; and (2) it saves her from the need to hunt for herself. Hunting is dangerous, and males die in spider webs at more than twice the rate of females. The males behave in accord with the same kind of economic analysis: If the female is still eating after accepting all the sperm she can, the male grabs his gift back and flies off to look for a second female to share it with.

Choosy Females May Get Better Genes for Their Offspring

Another potential benefit for choosy females is that males with the best advertisement displays may be genetically better in general. One popular version of this hypothesis, proposed by William Hamilton and Marlene Zuk (1982), sug-

(a)

(b)

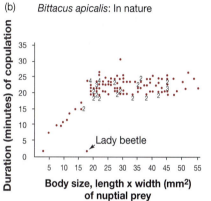

Bittacus apicalis: In nature

(c)

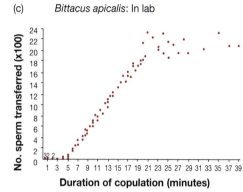

Bittacus apicalis: In lab

Figure 15.21 Courtship and mating in hangingflies
(a) A female (right) copulates with a male while eating a blowfly he has captured and presented to her. (Randy Thornhill, University of New Mexico). (b) The larger the gift the male presents to the female, the longer the pair copulates. Copulation ends after about 20 minutes, even if the female is still eating. A lady beetle presented by one male is an exception to the general pattern. Even though the beetle was fairly large, the female hangingfly rejected it and broke off copulation almost immediately. From Thornhill (1976). (c) The longer a pair copulates, the more sperm the female allows the male to transfer. The male must present a gift that takes at least 5 minutes to eat, or the female accepts no sperm. From Thornhill (1976).

gests that display structures such as long feathers and bright colors are expensive to make, and that only the healthiest males, those most resistant to parasites and diseases, will be able to develop them fully.

Anders Møller (1990) tested this hypothesis in barn swallows. Barn swallows are hosts to parasitic mites, and Møller has shown that heavy infestations with mites are detrimental to the birds. Heavily infested adults are less likely to be able to produce a second clutch during a breeding season, for example, and heavily infested chicks grow more slowly. Møller measured the tail feathers of a number of males, and counted the parasitic mites on their chicks. Some chicks had hatched and lived in their biological father's nests, and some chicks Møller had moved to other nests to be raised by foster parents. Males with longer tails had biological offspring infested with fewer mites, whether the offspring grew up in the males' own nests or in foster nests (compare Figure 15.22a and b), but there was no relationship between the tail lengths of males and the mite loads of their foster offspring (Figure 15.22c). A female barn swallow can thus reap at least two benefits by choosing as her mate a male with the longest possible tail. First, the male himself will be infested with fewer parasites, and is therefore less likely to infect the female and her offspring. Second, the female's offspring will inherit their father's resistance genes, and consequently be healthier and more likely to survive.

Choosy female swallows get mite-resistant genes for their offspring.

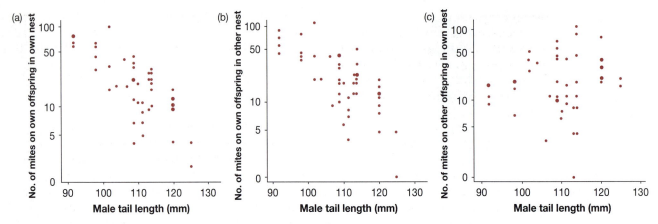

Figure 15.22   The relationship between the length of a male barn swallow's tail feathers and the mite loads carried by his offspring   (a) When offspring grow up in their biological fathers' nests, fathers with longer tail feathers have offspring infested with fewer mites (P < 0.001).   (b) When offspring are fostered in other males' nests, biological fathers with longer tail feathers still have biological offspring infested with fewer mites (P < 0.001).   (c) When offspring are fostered in other males' nests, there is no relationship between the tail-feather length of foster fathers and the mite loads of their foster offspring. From Møller (1990).

Choosy Females May Avoid Hybridizing

Another function of female choosiness may be to ensure that the female mates with a male of the same species. This seems to be the case in fireflies. In these insects (which are actually beetles, not flies), the males advertise for mates at night by flying about flashing their lights on and off in a pattern unique to their species (see Gould 1982). When waiting females see the right code, they respond with a species-specific flash of their own, and the males approach to proceed with courtship. Readers who live in an area with fireflies may be able to attract males themselves by using a penlight to mimic female responses. Males of several different species may be flying on a given night in any given area. By responding only to males flashing the right species-specific code, female fireflies avoid hybridizing. If hybrids are less fit, then females that avoid hybridizing will have more fit offspring.

Choosy female fireflies get higher fitness by avoiding hybridizing.

In the firefly species *Photuris versicolor,* females use the code-flashing system for another purpose as well (Lloyd 1981). Like a biologist with a penlight, these females respond to males of other species, with the codes that females of the other species would give. When the males approach expecting to mate, the *P. versicolor* females eat them (Figure 15.23).

Choosy Females May Have Preexisting Sensory Biases

Females use their sensory organs and nervous systems for many other things than just discriminating among potential mates. It is possible that selection for such abilities as avoiding predators, finding food, and identifying members of the same species may result in sensory biases that make females particularly responsive to certain cues (see Enquist and Arak 1993). This may in turn select on males to display those cues, even if the cues would otherwise have no relation to mating or fitness. In other words, the preexisting bias hypothesis holds that female preferences evolve first and male traits follow. Alexandra Basolo tested

Figure 15.23 **A female firefly, *Photuris versicolor,* eats a male firefly of another species** The female (top) has lured the male by responding to his flashes in the same way a female of his own species would have. Reproduced by permission from J. E. Lloyd, Aggresive mimicry in Photuris fireflies: signal repertoires by femmes fatales. *Science* 187: 452–3, 7 February 1975. © 1975 by the American Association for the Advancement of Science.

this hypothesis in fish of the genus *Xiphophorus,* which includes the platyfish and swordtails, some of which are common aquarium fish (Basolo 1990a, 1990b; da Silva et al. 1991; Basolo 1995a).

Basolo's test involved mapping two traits onto an evolutionary tree to infer their points of origin: male swords and the female preference for swords. A reconstruction of the phylogeny of the swordtail-platyfish genus suggests that swords are a derived character (Figure 15.24). The outgroup is swordless, as are the two platyfish clades that branch most deeply (a and b in Figure 15.24). The simplest interpretation is that elongated, colored swords evolved on the branch immediately preceding the split between clades c and d, and that the common ancestor of clades b and c + d (arrow in Figure 15.24) was swordless (for more details, see da Silva et al. 1991).

Basolo used paired-choice tests to assess the preferences of females in one sworded species, *X. helleri* (clade d), and two swordless species, *X. maculatus* and *X. variatus* (both in clade b). These tests were straightforward for the sworded species, and showed that females prefer males with longer swords (Basolo 1990b). Testing the preferences of females in the two swordless species was trickier, precisely because the males in these species lack swords. Basolo's solution was to surgically attach plastic swords to the male's tails, using colored swords for the experimental males and transparent swords for the controls. The sword-added males continued to swim, feed, and court females. Females in both swordless species preferred the males with colored swords (Basolo 1990a; Basolo 1995a; see also Haines and Gould 1994). When added to the phylogeny in Figure 15.24, these results (marked with asterisks) suggest that the female preference for swords dates back at least to the branch preceding the split between clades b and c + d. In other words, the common ancestor of clades b and c + d (arrow) appears to have been a species in which the females preferred males with swords even though none of the males had swords. This conclusion is

Choosy female swordfish may simply be exercising a pre-existing sensory bias.

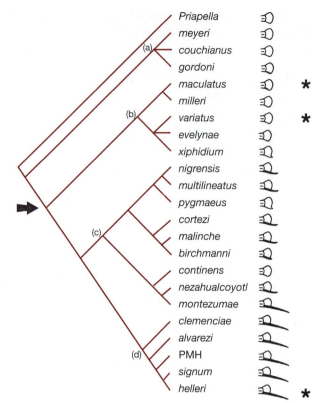

Figure 15.24 Estimated phylogeny of the swordtail-platy-fish genus *Xiphophorus*   The outgroup is the sister genus *Priapella* (top). The illustrations show the tail fins of typical males in each species. The asterisks mark the species in which Basolo (1990a, 1990b, 1995a) and Haines and Gould (1994) have tested female preferences; females in all three species prefer males with swords. This phylogeny suggests that the common ancestor of clades b and c + d (arrow) was a species in which males lacked swords but females nonetheless preferred swords. From Basolo (1995a).

consistent with the hypothesis that male swords evolved in response to a preexisting sensory bias in the females.

Basolo's analysis generated some controversy. The estimated phylogeny she used was based on a mixture of morphological and biochemical characters. Axel Meyer and colleagues (1994) used DNA sequence data to reconstruct an independent estimate of the *Xiphophorus* phylogeny. Their reconstruction of the phylogeny was substantially different from the one used by Basolo. Depending on the criteria used to decide whether each species' tail includes a structure worthy of the name *sword,* Meyer and colleagues' phylogeny can be interpreted as showing either that the female preference for swords evolved before male swords, or that the female preference for swords evolved after male swords (Wiens and Morris 1996).

Basolo's response was to assay female preferences in the genus *Priapella,* which served as the out-group in both the phylogeny Basolo used and the phylogeny reconstructed by Meyer and colleagues (Basolo 1995b, 1996). Female *Priapella* prefer males with swords. Regardless of which phylogeny is used, this result suggests that the female preference for swords evolved first, and that male swords evolved afterward in some, but not all *Xiphophorus.*

15.4 Diversity in Sex Roles

In all of the examples we have presented so far, the crucial fact explaining the roles taken by each sex is that access to mates limits the reproductive success of males more than it limits the reproductive success of females. This pattern is widespread, but it is by no means universal (Arnold and Duvall 1994). The pipefish species *Nerophis ophidion* and *Syngnathus typhle,* studied by Gunilla Rosenqvist, Anders Berglund, and their colleagues, provide a counterexample. These pipefish, which live in eelgrass beds, are relatives of the seahorses (Figure 15.25). As in the seahorses, males provide all the parental care. In *N. ophidion,* the male has a brood patch on his belly; in *S. typhle,* the male has a brood pouch. In both species the female lays her eggs directly onto or into the male's brood structure. The male supplies the eggs with both oxygen and nutrients until they hatch.

Although the extensive parental care provided by male pipefish requires energy, the pivotal currency for pipefish reproduction is not energy but time

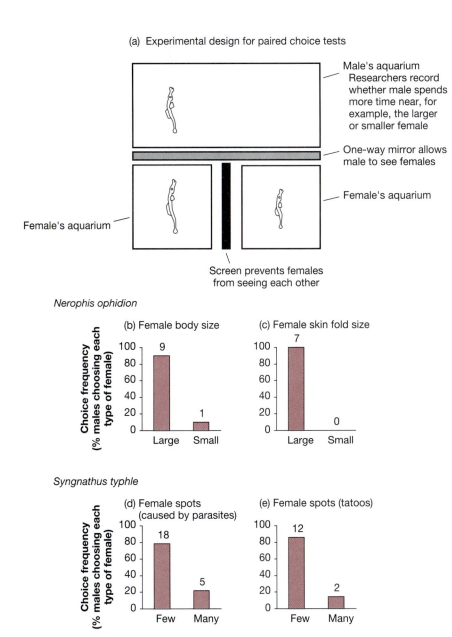

(a) Experimental design for paired choice tests

Male's aquarium
Researchers record whether male spends more time near, for example, the larger or smaller female

One-way mirror allows male to see females

Female's aquarium

Female's aquarium

Screen prevents females from seeing each other

Nerophis ophidion

(b) Female body size

Choice frequency (% males choosing each type of female)

9 — Large
1 — Small

(c) Female skin fold size

7 — Large
0 — Small

Syngnathus typhle

(d) Female spots (caused by parasites)

Choice frequency (% males choosing each type of female)

18 — Few
5 — Many

(e) Female spots (tatoos)

12 — Few
2 — Many

Figure 15.25 **Male choice in pipefish** (a) In paired-choice tests, researchers place a male pipefish in an aquarium from which he can see two females. The researchers infer which female the male would prefer as a mate from where he spends more of his time. In (b–d) the numbers above the bars indicate the number of males tested. After Rosenqvist and Johansson (1995). (b) Given a choice between large or small females, male pipefish prefer large females (P = 0.022). Replotted from Rosenqvist (1990). (c) Given a choice between females with large or small skin folds, male pipefish prefer large-folded females (P = 0.016). Replotted from Rosenqvist (1990). (d) Given a choice between females with many black spots (caused by a parasite) or females with few black spots, male pipefish prefer females with few spots (P < 0.05). Replotted from Rosenqvist and Johansson (1995). (e) Males continue to prefer females with few spots even when the spots are tatooed onto parasite-free females (P < 0.01). Replotted from Rosenqvist and Johansson (1995).

(Berglund et al. 1989). Females of both *N. ophidion* and *S. typhle* can make eggs faster than males can rear them to hatching. As a result, access to male brood space limits female reproductive success. If the theory of sexual selection we have developed is correct, then in these pipefish the females should compete with each other over access to males, and the males should be choosy.

In *N. ophidion* the females are larger than the males, and have dark blue stripes and skin folds on their bellies, both of which the males lack. These traits appear to function primarily as advertisements for attracting mates. For example, females develop skin folds during the breeding season and lose them after, and in captivity females develop skin folds only when males are present (Rosenqvist 1990). In paired-choice tests (Figure 15.25a), *N. ophidion* males are choosy, preferring larger females (Figure 15.25b) and females with larger skin folds (Figure 15.25c).

In *S. typhle* the males and females are similar to each other in size and appearance. Nonetheless, the females compete with each other over access to males (Berglund 1991), and the males are choosy (Rosenqvist and Johansson 1995). In paired-choice tests (Figure 15.25a), male *S. typhle* prefer females showing fewer of the black spots that indicate infection with a parasitic worm, whether the black spots were actually caused by parasites (Figure 15.25d) or were tattooed onto the females (Figure 15.25e). This choosiness benefits the males directly, because females with fewer parasites lay more eggs for the male to fertilize.

The mating behavior of pipefish males and females is thus consistent with the theory of sexual selection. Other examples of "role-reversed" species whose behavior appears to support the theory include moorhens (Petrie 1983) and some species of katydids (Gwynne 1981; Gwynne and Simmons 1990).

15.5 Sexual Selection in Plants

Many of the ideas we have developed about sexual selection in the context of animal mating can be applied as well to plants (Bateman 1948; Willson 1979). In plants, mating involves the movement of pollen from one individual to another. Reproductive investment per seed may be more expensive for the seed parent, which must produce a fruit, than for the pollen donor, which must only make pollen. When pollen is transported by animals, access to mates is a function of access to pollinators. Based on the principles of sexual selection in animals, we can hypothesize that access to pollinators limits the reproductive success of pollen donors to a greater extent than it limits the reproductive success of seed parents. Maureen Stanton and colleagues (1986) tested this hypothesis in wild radish *(Raphanus raphanistrum)*.

Wild radish is a self-incompatible annual herb that is pollinated by a variety of insects, including honey bees, bumble bees, and butterflies. Many natural populations of wild radish contain a mixture of white-flowered and yellow-flowered individuals. Flower color is determined by a single locus: white (W) is dominant to yellow (w). Stanton and colleagues set up a study population with eight homozygous white plants (WW) and eight yellow plants (ww). The scientists monitored the number of pollinator visits to plants of each color, then measured reproductive success through female and male function.

Measuring reproductive success through female function was easy: The researchers just counted the number of fruits produced by each plant of each color. Measuring reproductive success through male function was harder; in fact, it was not possible at the level of individual plants. Note, however, that a yellow seed parent (*ww*) will produce yellow offspring (*ww*) if it mated with a yellow pollen donor (*ww*), but white offspring (*Ww*) if it mated with a white pollen donor (*WW*). Thus by rearing the seeds produced by the yellow seed parents and noting the color of their flowers, Stanton and colleagues could compare the population-level reproductive success of white versus yellow pollen donors. The relative reproductive success of pollen donors through yellow seed parents should be a reasonable estimate of the pollen donors' relative reproductive success through seed parents of both colors. The scientists repeated their experiment three times.

As Stanton and colleagues expected from previous research, the yellow-flowered plants got about $\frac{3}{4}$ of the pollinator visits (Figure 15.26a). If reproductive success is limited by the number of pollinator visits, then the yellow–flowered plants should also have gotten about $\frac{3}{4}$ of the reproductive success. This was true for reproductive success through pollen donation (Figure 15.26c), but not for reproductive success through seed production (Figure 15.26b). Reproductive success through seed production was simply proportional to the number of plants of each type. These results are consistent with the typical pattern in animals: The reproductive success of males is more limited by access to mates than is the reproductive success of females. The results also suggest that the evolution of showy flowers that attract pollinators has been driven more by their effect on male reproductive success than on female reproductive success (Stanton et al. 1986).

If it is true in general that the number of pollinator visits is more important to male reproductive success than to female reproductive success, then in animal-pollinated plant species with separate male and female flowers, the male flowers should be more attractive. Lynda Delph (1996) and colleagues tested this hypothesis with a survey of animal- and wind-pollinated plants, including both dioecious species (separate male and female individuals) and monoecious species (separate male and female flowers on the same individual).

Delph and her co-authors first noted that the showiest parts of a flower, the petals and sepals that together form the perianth, serve not only to attract pollinators but also to protect the reproductive structures when the flower is developing in the bud. If protection were the only function of the perianth, then the sex that has the bigger reproductive parts should always have the bigger perianth. This was indeed the case in all 11 wind-pollinated species Delph and colleagues measured (Figure 15.27a, right). If, however, pollinator attraction is also important, and more important to males than to females, then there should be a substantial number of species in which the female flowers have bigger reproductive parts but the male flowers have bigger perianths. This was the case in 29% of the 42 animal-pollinated plants Delph and colleagues measured (Figure 15.27a, left). Furthermore, in species that are sexually dimorphic, male function tends to draw a greater investment in number of flowers per inflorescence and in strength of floral odor, although not in quantity of nectar (Figure 15.27b). These results are consistent with the hypothesis that sexual

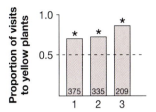

(a) Pollinator discrimination

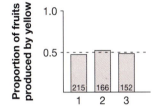

(b) Maternal function

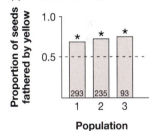

(c) Paternal function

Population

Figure 15.26 Reproductive success through pollen donation is more stongly affected by the number of pollinator visits than is reproductive success through the production of seeds The numbers inside each bar represent the number of pollinator visits (a), the number of fruits (b), and the number of seeds (c) examined. Bars marked with an asterisk have heights significantly different from 0.5 (P < 0.0001). (a) In populations with equal numbers of white and yellow flowers, the yellow-flowered plants got most of the pollinator visits. (b) In spite of the inequality in pollinator visits, white- and yellow-flowered plants produced equal numbers of fruits. (c) The majority of the seeds, however, were fathered by yellow plants. From Stanton et al. (1986).

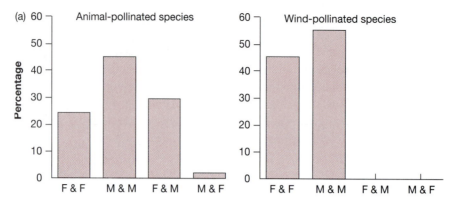

F&F: Female flower has larger reproductive parts and the larger perianth.

M&M: Male flower has larger reproductive parts and the larger perianth.

F&M: Female flower has larger reproductive parts, but male has the larger perianth.

M&F: Male flower has larger reproductive parts, but female has the larger perianth.

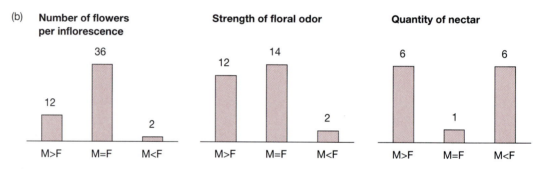

M>F: Male investment is greater than female investment.

M=F: Male investment and female investment are equal.

M<F: Male investment is less than female investment.

***Figure 15.27*   Patterns of sexual dimorphism in plants with separate male and female flowers**
(a) In all 11 wind-pollinated species measured (right), the sex with larger reproductive parts has the larger perianth. In 29% of the 42 animal-pollinated species measured (left), the female has larger reproductive parts but the male has the larger perianth—a pattern significantly different from that for wind-pollinated species (P < 0.001). From Delph et al. (1996).   (b) When animal-pollinated plants have flowers that are sexually dimorphic for investment in pollinator attraction, investment by males tends to be larger for two of the three traits studied (P < 0.01; P = 0.01; N.S.). Drawn from data in Delph et al. (1996).

selection via pollinator attraction is often stronger for male flowers than for female flowers.

15.6 Sexual Dimorphism in Body Size in Humans

One example of sexual dimorphism that we cited at the beginning of this chapter was for body size in humans (Figure 15.1). We now ask whether the sexual dimorphism in humans is the result of sexual selection. It is a difficult question to answer, because sexual selection concerns mating behavior. The evolutionary significance of human behavior is hard to study, for at least two reasons:

- Human behavior is driven by a complex combination of culture and biology. Studies based on the behavior of people in any one culture provide no means of disentangling these two influences. Cross-cultural studies can identify universal traits or broad patterns of behavior, either of which may warrant biological explanations. Cultural diversity is rapidly declining, however, and some biologists feel that it is no longer possible to do a genuine cross-cultural study.

- Ethical and practical considerations prohibit most of the kinds of experiments we might conduct on individuals of other species. This means that most studies of human behavior are observational. Observational studies can identify correlations between variables, but they offer little evidence of cause and effect.

Human behavior is inherently fascinating, however, and we therefore proceed cautiously in briefly considering the question of sexual selection and body size in humans.

Comparative studies of variation among primate species suggest that sexual selection may play a role in the evolution and maintenance of differences in body size between males and females. Across the 53 species of primates shown in Figure 15.28, polygynous species are more sexually dimorphic in body size than are monogamous species. A polygynous species is one in which at least some males mate with more than one female. Polygyny implies more intense competition among males for opportunities to mate, because (assuming a sex ratio of 1:1) for every male that gets more than one mate, there is another male that gets no mate. Note, however, that while this graph shows an association between mating system and the degree of sexual dimorphism and suggests a causal connection, it actually proves neither that the mating system is what drives the evolution of sexual dimorphism in primates in general, nor that sexual selection is responsible for the dimorphism in any particular species. What is needed for establishing cause and effect is research on forces of selection within individual species.

The most basic knowledge of human reproductive biology indicates that the opportunity for sexual selection is greater in men than in women, and data from a single culture will suffice for illustration. Research by Monique Borgerhoff Mulder (1988) on the Kipsigis people of southwestern Kenya revealed that the men with highest reproductive success had upwards of 25 children, while

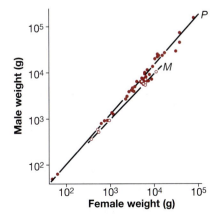

Figure 15.28 **The ratio of male weight to female weight is larger in polygynous than in monogamous primates** The plot shows average male body weight versus average female body weight for 53 species of primates. Filled dots and the best-fit line marked *P* represent 42 polygynous species; the slope of the best-fit line is significantly greater than 1 (P < 0.01). Open dots and the best-fit line marked *M* represent 11 monogamous species; the slope of the best-fit line is not significantly different from 1. From Leutenegger (1978).

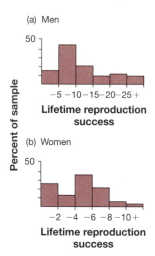

Figure 15.29   Variation in lifetime reproductive success among Kipsigis men and women   (a) For the men, the height of each bar represents the percentage of men who had 0 to 5 children, 6 to 10 children, and so on. (b) For the women, the height of each bar represents the percentage of women who had 0 to 2 children, 3 or 4 children, 5 or 6 children, and so on. Some of the men had more than 25 children; few of the women had more than 10. From Borgerhoff Mulder (1988).

the most prolific women rarely had more than 10 (Figure 15.29). In Kipsigis culture, it appears that the reproductive success of men was limited by mating opportunities to a greater extent than was the reproductive success of women. But is there any evidence that reproductive competition, either via male-male interactions or female choice, selects for larger body size in men?

Steven Gaulin and James Boster (1992) addressed this question with a comparative study of marriage systems across cultures. The marriage systems of human cultures range from polyandry (some women have more than one husband) to monogamy to polygyny (some men have more than one wife). Gaulin and Boster reasoned that if sexual selection on males is responsible for the sexual dimorphism in human body size, then polygynous cultures, whose marriage systems are presumed to entail a greater degree of male-male competition, should show more extreme sexual dimorphism than monogamous or polyandrous cultures. Gaulin and Boster also divided their sample of societies according to whether they were socially stratified or egalitarian, on the presumption that male-male competition is more intense in stratified societies than in egalitarian societies. The researchers drew on a widely used anthropological database called the Ethnographic Atlas (Murdock 1967), and found 118 societies whose marriage systems and degree of stratification were documented, and for which records existed on the heights of at least 20 men and 20 women. For each society they calculated the degree of sexual dimorphism as the average male height divided by the average female height; results appear in Table 15.3.

The data do not support the sexual selection hypothesis. There is no significant difference between monogamous and polyandrous societies versus polygynous societies (compare the column means). Neither is there a significant difference between stratified versus nonstratified societies (compare the row means). This result is perhaps not surprising; as Gaulin and Boster themselves realized, human marriage systems may be too transient to expect them to produce any appreciable evolutionary changes in the degree of sexual dimorphism (Rogers and Mukherjee 1992).

Gaulin and Boster would have ended their analysis with this conclusion except for an unexpected result: There is a statistically significant difference ($P < 0.014$) in the effect of stratification within marriage systems. Statisticians

TABLE 15.3 Marriage system, social stratification, and sexual size dimorphism in humans

The top number in each box is the average degree of sexual dimorphism (average male height/average femae height). In parentheses is the number of societies.

	Monogamous and polyandrous	Polygynous	Row means
Nonstratified	1.070 (13)	1.074 (59)	1.073 (72)
Stratified	1.079 (12)	1.068 (34)	1.071 (46)
Column Means	1.074 (25)	1.072 (93)	1.073 (118)

call this an interaction effect: Within monogamous and polyandrous marriage systems, stratified cultures show greater sexual dimorphism, while within polygynous marriage systems, nonstratified cultures show greater sexual dimorphism. Gaulin and Boster have no explanation for this pattern.

The evolutionary significance of sexual size dimorphism in humans is unresolved. What is really needed to evaluate the sexual selection hypothesis is data from a number of cultures on the relationship between body size, number of mates, survival, and reproductive success for both women and men. Preferably, the data would come from hunter-gatherers, whose members live the life-style ancestral for our species. The most technically challenging of these factors to measure is the reproductive success of men. Modern techniques for genetic analysis have made it feasible, in principle, to collect such data (Figure 15.16). The research remains to be done.

It is unclear whether sexual selection has played a role in maintaining the human sexual dimorphism in body size.

15.7 When Sexual Selection and Natural Selection Collide: Túngara Frogs

We noted earlier that Darwin was careful to distinguish between natural selection and sexual selection. Natural selection is variation among individuals in survival and fecundity; sexual selection is variation among individuals of the same sex in the number of mates obtained. Because natural selection and sexual selection both lead to variation in lifetime reproductive success, the ultimate cause of evolutionary change, it is sometimes hard to see why Darwin was at such pains to keep them separate. The reason is that natural selection and sexual selection are often at odds with each other, pushing in opposite directions on the evolution of a single trait. Darwin needed sexual selection to explain the persistence of traits that seem to be opposed by natural selection. We close our chapter on sexual selection with a final example illustrating this point, drawn

Sexual selection and natural selection often oppose each other.

Figure 15.30 A frog-eating bat catching a túngara frog (Merlin D. Tuttle/Bat Conservation International/Photo Researchers, Inc.)

from research in Panama on the calling behavior of male túngara frogs by Michael Ryan and his colleagues (reviewed in Ryan 1985).

Male túngaras congregate in pools to call for mates. When females are ready to lay eggs, they come to the pools, listen to various males, and then approach one to mate. The males' advertisement call has two parts, which Ryan calls the "whine" and the "chuck." The call always starts with a whine, to which the males add from 0 to 6 chucks. In choice tests, females approach calls that contain chucks in preference to calls that contain no chucks. One reason for this appears to be that the chucks give the female information about the size of the calling male: Larger males make chucks of lower pitch, and females prefer to mate with larger males. Larger males fertilize a higher proportion of the females' eggs during mating. This is probably for the simple reason that it is easier for larger males to simultaneously clasp the female (from the back just below the "armpits") and to reach around behind her to fertilize the eggs as she lays them. Male túngaras, incidentally, have unusually deep voices for their size.

Sexual selection, by female choice, favors advertisement calls that contain chucks. Why, then, do the males not always add chucks to their calls? Because somebody besides female túngaras may be listening too (Figure 15.30, page 603). In choice tests on frog-eating bats, Ryan, Merlin Tuttle, and Stanley Rand (1982) found that, just like female túngaras, bats are more attracted to calls with chucks. This is probably because chucks are lower in pitch and thus easier to localize. Natural selection by frog-eating bats favors advertisement calls without chucks.

So, when to whine and when to whine-chuck-chuck? Túngaras tend to call in choruses: All are quiet for a while, then one male calls, then lots of males call, then all are quiet again. Nobody likes to be bat bait alone. A chorus of calls can also be triggered by a biologist with a tape recorder. Rand and Ryan (1981) recorded the advertisements of males calling spontaneously, and those of males calling in response to a taped whine. Their results appear in Figure 15.31. Males calling by themselves are presumably noticeable enough that there is insufficient reason for them to indulge in chucks. Males competing for attention in the chorus, however, risk chucks to stand out. Besides, if a bat does attack, there is always the chance that it will eat the competition instead.

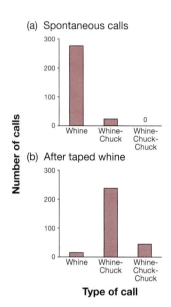

Figure 15.31 The kinds of calls given by male túngara frogs calling spontaneously versus calling in response to a taped whine When frogs call spontaneously (a), they almost always give a simple whine. When frogs call in response to a taped whine (b), they almost always give a whine-chuck. The difference in behavior is statistically significant (P < 0.005). Plotted from data in Table 1 of Rand and Ryan (1981).

Summary

Sexual dimorphism, a difference in form or behavior between females and males, is common. The difference often involves traits, like the enormous tail feathers of the peacock, that appear to be opposed by natural selection. To explain these puzzling traits, Darwin invoked sexual selection. Sexual selection is differential reproductive success resulting from variation in mating success.

Mating success is often a more important determinant of fitness for one sex than for the other. Usually, but by no means always, it is males whose reproductive success is limited by mating opportunities with females, and females whose reproductive success is limited by resources rather than matings. The members of the sex experiencing strong sexual selection typically compete among themselves over access to mates. This competition may involve direct combat, gamete competition, infanticide, or advertisement.

The members of the sex whose reproductive success is limited by resources rather than matings is typically choosy. This choosiness may provide the chooser with direct or indirect benefits, such as food or better genes for its offspring, or it may be the result of a preexisting sensory bias.

Sexual selection is distinct from natural selection in that the two forces often are at odds, as when a conspicuous mating display increases both the probability of finding a mate and the probability of being eaten by a predator.

Questions

1. The graphs in Figure 15.32 show the variation in lifetime reproductive success of male versus female elephant seals (Le Boeuf and Reiter 1988). Note that the scales on the horizontal axes are different. Why is the variation in reproductive success so much more extreme in males than females? Draw a graph showing your hypothesis for the relationship between number of mates and reproductive success for male and female elephant seals. Why are male elephant seals four times as large as female elephant seals? Why are the males not even bigger than that?

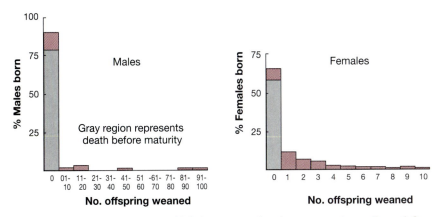

Figure 15.32 **Distributions of lifetime reproductive success in male and female elephant seals**

2. What sex is the sage grouse in Figure 15.33? What is it doing, and why? Do you think this individual provides parental care? Why or why not? What else would you like to know about this species?

3. We used long-tailed widow birds as an example of a dimorphic species at the beginning of this chapter (Figure 15.3). Suggest a hypothesis for why male widow birds have such long tails. Design an experiment to test your hypothesis. (Hint: Anders Møller's experiments on barn swallows were inspired by Malte Anderson's research on widow birds.) Look up Malte Anderson's paper on widow birds (Anderson 1982) and see if he designed the same experiment you did. Harder: The bar graphs in Anderson's Figure 1b give a somewhat biased impression of his (nonetheless valid) results. Looking at Anderson's Figure 1a, can you identify the bias and explain what caused it? To Anderson's great credit, he also included his raw data in the graphic. Use the raw data to draw an unbiased graphic representing Anderson's results.

4. Male butterflies and moths commonly drink from puddles, a behavior known as puddling. Scott Smedley and Thomas Eisner (1996) report a detailed physiological analysis of puddling in the moth *Gluphisia septentrionis*. A male *G. septentrionis* may

Figure 15.33 **Sage grouse**

puddle for hours at a time. He rapidly processes huge amounts of water, extracting the sodium and expelling the excess liquid in anal jets (see Smedley and Eisner's paper for a dramatic photo). The male moth will later give his harvest of sodium to a female during mating. The female will then put much of the sodium into her eggs. Speculate on the role this gift plays in the moth's mating ritual, and in the courtship roles taken by the male and the female. How would you test your ideas?

5. Males in many species often attempt to mate with strikingly inappropriate partners. Ryan (1985), for example, describes male túngara frogs clasping other males. Some orchids mimic female wasps and are pollinated by amorous male wasps—who have to be fooled twice for the strategy to work. Would a female túngara or a female wasp ever make the same mistake? Why or why not? More general answers—applicable to a wider range of species—are better.

6. Do you think there is any association in humans between infection with parasites and physical appearance? The scatterplot in Figure 15.34 shows the relationship between the importance of attractiveness in mate choice (as reported by subjects responding to a questionnaire) and the prevalence of six species of parasites in 29 cultures (Gangestad 1993; Gangestad and Buss 1993). (Note that statistical techniques have been used to remove the effects of latitude, geographic region, and mean income.) What is the pattern in the graph? Does this pattern make sense from an evolutionary perspective? What do women gain by choosing an attractive mate? What do men gain by choosing an attractive mate? Can you offer a cultural explanation that accounts for the pattern?

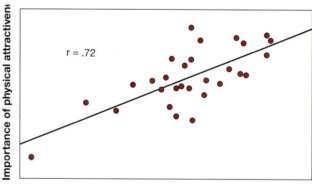

Figure 15.34 Importance of physical attractiveness in mate choice versus parasite prevalence in 29 human cultures

7. In many katydids, the male delivers his sperm to the female in a large spermatophore that contains nutrients the female eats (for a photo, see Gwynne 1981). The female uses these nutrients in the production of eggs. Darryl Gwynne and L.W. Simmons (1990) studied the behavior of caged populations of an Australian katydid under low food (control) and high food (extra) conditions. Some of their results are graphed in Figure 15.35 (The graph shown the results from four sets of replicate cages; calling males = average number of males calling at any given time; matings/female = average number of times each female mated; % reject by M = fraction of the time a female approached a male for mating and was rejected; % reject by F = fraction of the time a female approached a male for mating but then rejected him before copulating; % with F-F comp = fraction of matings in which one or more females were seen fighting over the male). Use sexual selection theory to explain these results. When were the females choosy and the males competitive? When were the males choosy and the females competitive? Why?

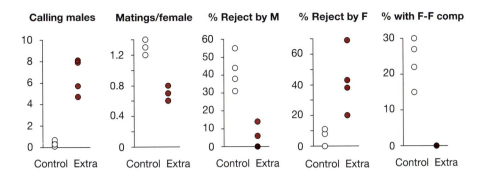

Figure 15.35 Behavior of male and female katydids under control versus extra-food conditions

8. In some species of deep sea anglerfish, the male lives as a symbiont permanently attached to the female (see Gould 1983, essay 1). The male is tiny compared to the female. Many of the male's organs, including the eyes, are reduced. Others, such as the jaws and teeth, are modified for attachment to the female. The circulatory systems of the two sexes are fused, and the male receives all of his nutrition from the female via the shared bloodstream. Often, two or more males are attached to a single female. What are the costs and benefits of the male's symbiotic habit for the male? For the female? What limits the lifetime reproductive success of each sex—the ability to gather resources, or the ability to find mates? Do you think that the male's symbiotic habit evolved as a result of sexual selection, or natural selection? (It may be helpful to break the male symbiotic syndrome into separate features, such as staying with a single female for life, physical attachment to the female, reduction in body size, and nutritional dependence on the female.)

Exploring the Literature

9. If a single insemination provides all the sperm necessary to fertilize an entire clutch of eggs, then what do females gain by engaging in extra-pair copulations? For hypotheses and tests, see:

Kempenaers, B., G.R. Verheyn, M. van den Broeck, T. Burke, C. van Broeckhoven, and A.A. Dhondt. 1992. Extra-pair paternity results from female preference for high-quality males in the blue tit. *Nature* 357:494–496.

Madsen, T., R. Shine, J. Loman, and T. Hakansson. 1992. Why do female adders copulate so frequently? *Nature* 355:440–441.

Gray, E.M. 1996. Female control of offspring paternity in a western population of red-winged blackbirds *(Agelaius phoeniceus). Behavioral Ecology and Sociobiology* 38:267–278.

Gray, E.M. 1997. Do female red-winged blackbirds benefit genetically from seeking extra-pair copulations? *Animal Behavior* 53:605–623.

Gray, E.M. 1997. Female red-winged blackbirds accrue material benefits from copulating with extra-pair males. *Animal Behavior* 53:625–629.

10. Why do females sometimes copulate more than once with the same male?

Petrie, M. 1992. Copulation frequency in birds: Why do females copulate more than once with the same male? *Animal Behavior* 44:790–792.

11. The testes in many mammals, including humans, are positioned in a scrotum outside the abdominal cavity. This dangerous arrangement has long defied adequate evolutionary explanation. One hypothesis is that the evolution of scrotal testes was driven by sperm competition. See:

Freeman, S. 1990. The evolution of the scrotum: A new hypothesis. *Journal of Theoretical Biology* 145:429–445.

12. For choosy females, symmetry is often an important criterion. Why?

Møller, A.P. 1992. Female swallow preference for symmetrical male sexual ornaments. *Nature* 357:238–240.

Thornhill, R. 1992. Fluctuating asymmetry and the mating system of the Japanese scorpionfly, *Panorpa japonica*. *Animal Behavior* 44:867–879.

13. Peacocks are among the most famous animals with an elaborate male mating display. For recent research on sexual selection in peacocks, see:

Manning, J.T., and M.A. Hartley. 1991. Symmetry and ornamentation are correlated in the peacock's train. *Animal Behavior* 42:1020–1021.

Petrie, M. 1992. Peacocks with low mating success are more likely to suffer predation. *Animal Behavior* 44:585–586.

Petrie, M., T. Halliday, and C. Sanders. 1991. Peahens prefer peacocks with elaborate trains. *Animal Behavior* 41:323–331.

Citations

Alcock, J. 1993. *Animal Behavior: An Evolutionary Approach* (Sunderland, MA: Sinauer).

Anderson, M. 1982. Female choice selects for extreme tail length in a widowbird. *Nature* 299: 818–820.

Anderson, M. 1994. *Sexual Selection* (Princeton, NJ: Princeton University Press).

Arnold, S.J. 1983. Sexual selection: The interface of theory and empiricism. In P. Bateson, ed. *Mate Choice* (Cambridge: Cambridge University Press), 67–107.

Arnold, S.J., and D. Duvall. 1994. Animal mating systems: A synthesis based on selection theory. *American Naturalist* 143:317–348.

Basolo, A.L. 1990a. Female preference predates the evolution of the sword in swordtail fish. *Science* 250:808–811.

Basolo, A.L. 1990b. Female preference for male sword length in the green swordtail, *Xiphophorus helleri* (Pisces: Poeciliidae). *Animal Behavior* 40:332–338.

Basolo, A.L. 1995a. A further examination of a pre-existing bias favoring a sword in the genus *Xiphophorus*. *Animal Behavior* 50:365–375.

Basolo, A.L. 1995b. Phylogenetic evidence for the role of a pre-existing bias in sexual selection. *Proceedings of the Royal Society of London,* Series B 259:307–311.

Basolo, A.L. 1996. The phylogenetic distribution of a female preference. *Systematic Biology* 45:290–307.

Bateman, A.J. 1948. Intra-sexual selection in *Drosophila*. *Heredity* 2:349–368.

Berglund, A. 1991. Egg competition in a sex-role reversed pipefish: Subdominant females trade reproduction for growth. *Evolution* 45:770–774.

Berglund, A., G. Rosenqvist, and I. Svensson. 1989. Reproductive success of females limited by males in two pipefish species. *American Naturalist* 133:506–516.

Bertram, C.R. 1975. Social factors influencing reproduction in wild lions. *Journal of Zoology* 177:463–482.

Boellstorff, D.E., D.H. Owings, M.C.T. Penedo, and M.J. Hersek. 1994. Reproductive behavior and multiple paternity of California ground squirrels. *Animal Behavior* 47:1057–1064.

Borgerhoff Mulder, M. 1988. Reproductive success in three Kipsigis cohorts. In T.H. Clutton-Brock, ed. *Reproductive Success* (Chicago: University of Chicago Press), 419–435.

Burkhardt, D., and I. de la Motte. 1983. How stalk-eyed flies eye stalk-eyed flies: Observations and measurements of the eyes of *Cyrtodiopsis whitei* (Dopsidae, Diptera). *Journal of Comparative Physiology* 151:407–421.

Burkhardt, D., and I. de la Motte. 1987. Physiological, behavioural, and morphometric data elucidate the evolutive significance of stalked eyes in Diopsidae (Diptera). *Entomologia Generalis* 12:221–233.

Burkhardt, D., and I. de la Motte. 1988. Big "antlers" are favored: Female choice in stalk-eyed flies (Diptera, Insecta), field collected harems and laboratory experiments. *Journal of Comparative Physiology A* 162:649–652.

Clutton-Brock, T.H. 1985. Reproductive Success in Red Deer. *Scientific American* 252(February): 86–92.

da Silva, J., S.T. Winquist, D.M. Weary, A.J. Inman, D.J. Mountjoy, E.A. Krebs, and A.L. Basolo. 1991. Male swords and female preferences. *Science* 253: 1426–1427.

Darwin, C. 1871. *The Descent of Man, and Selection in Relation to Sex* (London: John Murray).

Delph, L.F., L.F. Galloway, and M.L. Stanton. 1996. Sexual dimorphism in flower size. *American Naturalist* 148:299–320.

Enquist, M., and A. Arak. 1993. Selection of exaggerated male traits by female aesthetic senses. *Nature* 361:446–448.

Falconer, D.S. 1989. *Introduction to Quantitative Genetics.* (New York: John Wiley & Sons).

Fisher, R.A. 1958. *The Genetical Theory of Natural Selection,* 2nd ed. (New York: Dover).

Gage, M.J.G. 1991. Risk of sperm competition directly affects ejaculate size in the Mediterranean fruit fly. *Animal Behavior* 42:1036–1037.

Gangestad, S.W. 1993. Sexual selection and physical attractiveness: Implications for mating dynamics. *Human Nature* 4:205–235.

Gangestad, S.W., and D.M. Buss. 1993. Pathogen prevalence and human mate preferences. *Ethology and Sociobiology* 14:89–96.

Gaulin, S.J.C., and J.S. Boster. 1992. Human marriage systems and sexual dimorphism in stature. *American Journal of Physical Anthropology* 89:467–475.

Gibbs, H.L., P.J. Weatherhead, P.T. Boag, B.N. White, L.M. Tabak, and D.J. Hoysak. 1990. Realized reproductive success of polygynous red-winged blackbirds revealed by DNA markers. *Science* 250:1394–1397.

Gilbert, L.E. 1976. Postmating female odor in *Heliconius* butterflies: A male-contributed anti-aphrodesiac? *Science* 193:419–420.

Gould, J.L. 1982. *Ethology: The Mechanisms and Evolution of Behavior* (New York: W.W. Norton & Company).

Gould, S.J. 1983. *Hen's Teeth and Horse's Toes* (New York: W.W. Norton & Company).

Gross, M.R. 1984. Sunfish, salmon, and the evolution of alternative reproductive strategies and tactics in fishes. In G.W. Potts and R.J. Wootton, eds. *Fish Reproduction: Strategies and Tactics* (London: Academic Press), 55–75.

Gross, M.R. 1985. Disruptive selection for alternative live histories in salmon. *Nature* 313:47–48.

Gross, M.R. 1991. Salmon breeding behavior and life history evolution in changing environments. *Ecology* 72:1180–1186.

Gwynne, D.T. 1981. Sexual difference theory: Mormon crickets show role reversal in mate choice. *Science* 213:779.

Gwynne, D.T., and L.W. Simmons. 1990. Experimental reversal of courtship roles in an insect. *Nature* 346:172–174.

Haines, S.E., and J.L. Gould. 1994. Female platys prefer long tails. *Nature* 370:512.

Hamilton, W.D., and M. Zuk. 1982. Heritable true fitness and bright birds: A role for parasites? *Science* 218:384–387.

Lande, R. 1981. Models of speciation by sexual selection on polygenic traits. *Proceedings of the National Academy of Sciences, USA* 78:3721–3725.

Le Boeuf, B.J., and J. Reiter. 1988. Lifetime reproductive success in northern elephant seals. In T.H. Cutton-Brock, ed. *Reproductive Success* (Chicago: University of Chicago Press), 344–362.

Leutenegger, W. 1978. Scaling of sexual dimorphism in body size and breeding system in primates. *Nature* 272:610–611.

Lloyd, J.E. 1981. Mimicry in the sexual signals of fireflies. *Scientific American* 245(July):138–145.

Meyer, A., J.M. Morrissey, and M. Schartl. 1994. Recurrent origin of a sexually selected trait in *Xiphophorus* fishes inferred from a molecular phylogeny. *Nature* 368:539–542.

Møller, A.P. 1988. Female choice selects for male sexual tail ornaments in the monogamous swallow. *Nature* 332:640–642.

Møller, A.P. 1990. Effects of a haematophagous mite on the barn swallow *(Hirundo rustica):* A test of the Hamilton and Zuk hypothesis. *Evolution* 44:771–784.

Møller, A.P. 1991. Sexual selection in the monogamous barn swallow *(Hirundo rustica).* I. Determinants of tail ornament size. *Evolution* 45:1823–1836.

Murdock, G.P. 1967. *Ethnographic Atlas* (Pittsburgh: University of Pittsburgh Press).

Nakatsuru, K., and D.L. Kramer. 1982. Is sperm cheap? Limited male fertility and female choice in the lemon tetra (Pisces, Characidae). *Science* 216:753–755.

Nowak, R.M. 1991. *Walker's Mammals of the World* (Baltimore: Johns Hopkins University Press).

Olsson, M., A. Gullberg, and H. Tegelstrom. 1994. Sperm competition in the sand lizard, *Lacerta agilis. Animal Behavior* 48:193–200.

Packer, C., L. Herbst, A.E. Pusey, J.D. Bygott, J.P. Hanby, S.J. Cairns, and M. Borgerhoff Mulder. 1988. Reproductive success of lions. In T.H. Clutton-Brock, ed. *Reproductive Success: Studies of Individual Variation in Contrasting Breeding Systems* (Chicago: University of Chicago Press), 263–283.

Packer, C., and A.E. Pusey. 1983. Adaptations of female lions to infanticide by incoming males. *American Naturalist* 121:716–728.

Petrie, M. 1983. Female moorhens compete for small fat males. *Science* 220:413–415.

Rand, A.S., and M.J. Ryan. 1981. The adaptive significance of a complex vocal repertoire in a neotropical frog. *Zeitschrift für Tierpsychologie* 57:209–214.

Rauch, N. 1985. Female habitat choice as a determinant of the reproductive success of the territorial male marine iguana *(Amblyrhynchus cristatus). Behavioral Ecology and Sociobiology* 16:125–134.

Rauch, N. 1988. Competition of marine iguana females *Amblyrhynchus cristatus* for egg-laying sites. *Behavior* 107:91–106.

Rogers, A.R., and A. Mukherjee. 1992. Quantitative genetics of sexual dimorphism in human body size. *Evolution* 46:226–234.

Rosenqvist, G. 1990. Male mate choice and female-female competition for mates in the pipefish *Nerophis ophidion*. *Animal Behavior* 39:1110–1115.

Rosenqvist, G., and K. Johansson. 1995. Male avoidance of parasitized females explained by direct benefits in a pipefish. *Animal Behavior* 49:1039–1045.

Ryan, M.J. 1985. *The Túngara Frog: A Study in Sexual Selection and Communication* (Chicago: University of Chicago Press).

Ryan, M.J., M.D. Tuttle, and A.S. Rand. 1982. Bat predation and sexual advertisement in a neotropical frog. *American Naturalist* 119:136–139.

Schenk, A., and K.M. Kovacs. 1995. Multiple mating between black bears revealed by DNA fingerprinting. *Animal Behavior* 50:1483–1490.

Sillén-Tullberg, B. 1981. Prolonged copulation: A male "post-copulatory" strategy in a promiscuous species, *Lygaeus equestris* (Heteroptera: Lygaeidae). *Behavioral Ecology and Sociobiology* 9:283–289.

Smedley, S.R., and T. Eisner. 1996. Sodium: A male moth's gift to its offspring. *Proceedings of the National Academy of Sciences, USA* 93:809–813.

Smith, R.L. 1984. Human sperm competition. In R.L. Smith, ed. *Sperm Competition and the Evolution of Animal Mating Systems* (Orlando: Academic Press), 601–659.

Stanton, M.L., A.A. Snow, and S.N. Handel. 1986. Floral evolution: Attractiveness to pollinators increases male fitness. *Science* 232:1625–1627.

Thornhill, R. 1976. Sexual selection and nuptial feeding behavior in *Bittacus apicalis* (Insecta: Mecoptera). *American Naturalist* 110:529–548.

Thornhill, R., and J. Alcock. 1983. *The Evolution of Insect Mating Systems* (Cambridge, MA: Harvard University Press).

Trillmich, K.G.K. 1983. The mating system of the marine iguana *(Amblyrhynchus cristatus)*. *Zeitschrift für Tierpsychologie* 63:141–172.

Trivers, R.L. 1972. Parental investment and sexual selection. In B. Campbell, ed. *Sexual Selection and the Descent of Man 1871–1971* (Chicago: Aldine), 136–179.

Trivers, R. 1985. *Social Evolution* (Menlo Park, CA: Benjamin/Cummings).

Watson, P.J. 1991. Multiple paternity as genetic bet-hedging in female sierra dome spiders, *Linyphia litigiosa* (Linyphiidae). *Animal Behavior* 41:343–360.

Wiens, J.J., and M.R. Morris. 1996. Character definitions, sexual selection, and the evolution of swordtails. *American Naturalist* 147:866–869.

Wikelski, M., and S. Bäurle. 1996. Pre-copulatory ejaculation solves time constraints during copulations in marine iguanas. *Proceedings of the Royal Society of London, Series B* 263:439–444.

Wikelski, M., C. Carbone, and F. Trillmich. 1996. Lekking in marine iguanas: Female grouping and male reproductive strategies. *Animal Behavior* 52:581–596.

Wikelski, M., V. Carrillo, and F. Trillmich. 1997. Energetic limits to body size in a grazing reptile, the Galapagos marine iguana. *Ecology:* in press.

Wikelski, M., and F. Trillmich. 1997. Body size and sexual size dimorphism in marine iguanas fluctuate as result of opposing natural and sexual selection: An island comparison. *Evolution* 51:922–936.

Wilkinson, G.S., and P.R. Reillo. 1994. Female choice response to artificial selection on an exaggerated male trait in a stalk-eyed fly. *Proceedings of the Royal Society of London, Series B* 255:1–6.

Willson, M.F. 1979. Sexual selection in plants. *American Naturalist* 113:777–790.

Woodroffe, R., and A. Vincent. 1994. Mother's little helpers: Patterns of male care in mammals. *Trends in Ecology and Evolution* 9:294–297.

Social Behavior

Social interactions create the possibility for conflict or cooperation. Consider two American crows *(Corvus brachyrhynchos)* patrolling the edge of their adjacent nesting territories. If one moves across the established boundary its action may trigger aggressive calls, a flight chase, or even physical combat. But if a hawk flies by, the two antagonists will cooperate in chasing the predator away. Later in the day these same individuals may spend considerable time and effort feeding the young birds in their respective nests, even though the nestlings are not their own offspring.

When and why do these individuals cooperate with each other, and why do they help the parents of the nestlings? What conditions lead to conflicts with each other and with the parents on the territory, and how are these conflicts resolved? These are the types of questions we address in this chapter.

In fitness terms, an interaction between individuals has four possible outcomes (Table 16.1). Cooperation (or mutualism) is the term for actions that result in fitness gains for both participants. Altruism represents cases in which the individual instigating the action pays a fitness cost and the individual on the receiving end benefits. Selfishness is the opposite: The actor gains and the recipient loses. Spite is the term for behavior that results in fitness losses for both participants.

Understanding the evolution of these four interactions is made simpler because there are no clear-cut examples of spite in nature (Keller et al. 1994). It is obvious why spite has not evolved: A gene that results in fitness losses for both actor and recipient would quickly be eliminated by natural selection. But altruism would seem equally difficult to explain, and examples abound, from the crows who help at the nest to a human who dives into a river and saves a drowning child. Our primary focus in this chapter is on this problem: Why does altruism exist in nature?

16.1   Kin Selection and the Evolution of Altruism

Altruism is a central paradox of Darwinism. It would seem impossible for natural selection to favor a gene that results in behavior benefiting other individuals at the expense of the individual bearing the allele. For Darwin (1859: 257), the apparent existence of altruism presented a "special difficulty, which at first appeared to me insuperable, and actually fatal to my whole theory." Fortunately he was able to hint at a resolution to the paradox: Selection could favor traits that result in decreased personal fitness if they increase the survival and reproductive success of close relatives. Over a hundred years passed, however, before this result was formalized and widely applied.

Inclusive Fitness

In 1964, a young graduate student named William Hamilton developed a genetic model showing that a gene that favors altruistic behavior could spread under certain conditions. The key parameter in Hamilton's formulation is the coefficient of relationship, r. This is the probability that the homologous alleles in two individuals are identical by descent (Box 16.1). The parameter is closely related to F, the coefficient of inbreeding, which we introduced in Chapter 5. F is the probability that homologous alleles in the same individual are identical by descent.

Given r, the coefficient of relatedness between the actor and the recipient, Hamilton's rule states that a gene for altruistic behavior will spread if

$$Br - C > 0,$$

where B is the benefit to the recipient and C is the cost to the actor. Both B and C are measured in units of surviving offspring. This simple law means that altruism is more likely to spread when the benefits to the recipient are great, the cost to the actor is low, and the participants are closely related.

Inclusive fitness consists of direct fitness due to personal reproduction and indirect fitness due to additional reproduction by relatives. Behavior that results in indirect fitness gains is favored by kin selection.

To generalize this result, Hamilton created the concept of inclusive fitness. He pointed out that an individual's fitness can be partitioned into two components, which he called direct and indirect. Direct fitness is due to personal reproduction. Indirect fitness is due to additional reproduction by relatives that is made possible by the individual's actions (this means reproductive success above and beyond what those relatives would have achieved on their own). Natural selection that favors the spread of alleles increasing the indirect component of fitness is given a special name: kin selection. As we will see, most instances of altruism in nature are the result of kin selection.

Robert Trivers (1985:47) called Hamilton's rule and the concept of inclusive fitness "the most important advance in evolutionary theory since the work of

TABLE 16.1 Types of social interactions
The "actor" in any social interaction affects the recipient of the action as well as itself. The costs and benefits of interactions are measured in units of surviving offspring (fitness).

	Actor benefits	Actor is harmed
Recipient benefits	Cooperative	Altruistic
Recipient is harmed	Selfish	Spiteful

BOX 16.1 Calculating coefficients of relatedness

Calculating r, the coefficient of relatedness, requires a pedigree that includes the actor and the recipient. The researcher then performs a path analysis. Starting with the actor, all paths of descent are traced through the pedigree to the recipient. For example, half-siblings share one parent and have two genealogical connections, as indicated in Figure 16.1a. Parents contribute half their genes to each offspring, so the probability that genes are identical by descent (ibd) in each step in the path is $\frac{1}{2}$. Put another way, the probability that a particular allele was transmitted from parent to actor is $\frac{1}{2}$. The probability that the same allele was transmitted from parent to recipient is $\frac{1}{2}$. The probability that this same allele was transmitted to both the actor *and* the recipient (meaning that the alleles in actor and recipient are ibd) is the product of these two independent probabilities, or $\frac{1}{4}$.

Full siblings, on the other hand, share genes inherited from both parents. To calculate r when actor and recipient are full-sibs, we have to add the probabilities that genes are ibd through each path in the pedigree. In this case, we add the probability that genes are ibd through the mother to the probability that they are ibd through the father (Figure 16.1b). This is $\frac{1}{4} + \frac{1}{4} = \frac{1}{2}$.

Using this protocol results in the following coefficients:

- Parent to offspring, $\frac{1}{2}$
- Grandparent to grandchild, $\frac{1}{4}$
- Aunt or uncle to niece or nephew, $\frac{1}{4}$
- First-cousins, $\frac{1}{8}$

The analyses we have just performed assume that no inbreeding has occurred. If the population is inbred then coefficients will be higher. But when studying populations in the field, investigators usually have no data on inbreeding and have to assume that individuals are completely outbred. On this basis, coefficients of relationship that are reported in the literature should be considered minimal estimates. Another uncertainty in calculating r's comes in assigning paternity in pedigrees. As we indicated in Chapter 15, extra-pair copulations are common in many species. If paternity is assigned on the basis of male-female pairing relationships and extra-pair copulations go undetected, estimates of r may be inflated.

When constructing genealogies is impractical, coefficients of relatedness can be estimated directly from genetic data (Queller and Goodnight 1989). Microsatellite loci are proving to be extremely useful markers for calculating r in a wide variety of social insects (see Evans 1993; Hamaguchi et al. 1993; Queller et al. 1993).

(a) Half-siblings (b) Full-siblings

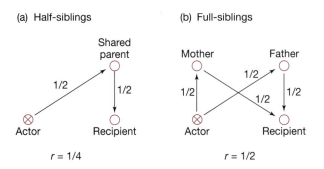

(c) Cousins

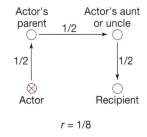

Figure 16.1 **Path analysis with pedigrees** The arrows describe paths by which genes can be identical by descent. The text explains how these paths are used to calculate r, the coefficient of relatedness.

Charles Darwin and Gregor Mendel." To see why, we will apply the theory by venturing to the Sierra Nevada of California and observing an intensely social mammal: Belding's ground squirrel *(Spermophilus beldingi)*.

Alarm Calling in Belding's Ground Squirrels

A classical application of inclusive fitness theory has been to explain alarm calling in birds and mammals. When flocks or herds are stalked by a predator, prey individuals that notice the intruder sometimes give loud, high-pitched calls.

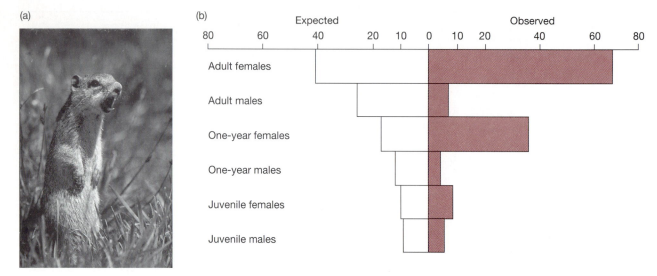

***Figure 16.2* In ground squirrels, most alarm calling is done by females**   (a) This female Belding's ground squirrel is giving an alarm call. (Richard R. Hansen/Photo Researchers, Inc.)   (b) This bar chart reports the observed and expected frequencies of alarm calling by different sex and age classes of Belding's ground squirrels, based on 102 encounters with predatory mammals. The expected values are calculated by assuming that individuals call randomly—that is, in proportion to the number of times they are present when the predator approaches. The observed and expected values are significantly different (P = 0.001). From Sherman (1977).

These warnings alert nearby individuals and allow them to flee or dive for cover. It also exposes the calling individual to danger. In Belding's ground squirrels, 13% of callers are stalked or chased by predators. Only 5% of non-calling individuals are (Sherman 1977).

Paul Sherman (1977, 1980) studied patterns in alarm calls given by Belding's ground squirrels to determine why such seemingly altruistic behavior evolved. Belding's ground squirrels are rodents that breed in colonies established in alpine meadows. Males disperse far from the natal burrow, while female offspring tend to remain and breed close by. As a result, females in proximity tend to be closely related. Because Sherman had individually marked many individuals over the course of the study, he was able to construct pedigrees and calculate coefficients of relationship for most members of the colony.

When Sherman compiled data on which squirrels called at the approach of a weasel, coyote, or badger, two outstanding patterns emerged: Females were much more likely to call than males (Figure 16.2), and females with close relatives within earshot were much more likely to call than females without close relatives nearby (Figure 16.3). These data strongly support the hypothesis that kin selection is responsible for the evolution of alarm calling. Sherman (1981) has also been able to show that mothers, daughters, and sisters are much more likely to cooperate when chasing trespassing squirrels off their territory than are more distant kin or non-kin (Figure 16.4).

The data show that altruistic behavior is not dispensed randomly. It is directed at close relatives and should result in indirect fitness gains.

Individuals are more likely to give alarm calls when close relatives are nearby.

White-Fronted Bee-Eaters

Another classical system for studying kin selection in vertebrates is helping behavior in birds (see Brown 1987; Stacey and Koenig 1990). In species from a

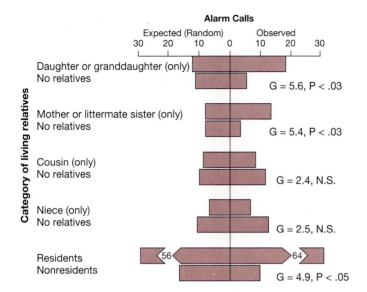

Figure 16.3 in right column:

Figure 16.3 **Female ground squirrels are more likely to give alarm calls when close kin are nearby** This bar chart summarizes data from 119 cases in which a mammalian predator approached a ground squirrel colony. Each paired comparison in the figure represents the occasions when at least one female in each category was present, but no others. The expected values are computed as in Figure 16.2. The "G" values reported here are a modified form of the χ^2 statistic we introduced in Chapter 5. "N.S." stands for not significant. From Sherman (1980).

wide variety of bird families, young that are old enough to breed on their own will instead remain and help their parents rear their brothers, sisters, or half-siblings. Helpers assist with nest building, nest defense, and/or food delivery to incubating parents and the fledglings.

Helping at the nest is usually found in species where breeding opportunities are extremely restricted, either because habitats are saturated with established breeding pairs or because suitable nest sites are difficult to obtain. In these cases, gaining direct fitness is almost impossible for young pairs. Gaining indirect fitness by helping becomes a best-of-a-bad-job strategy.

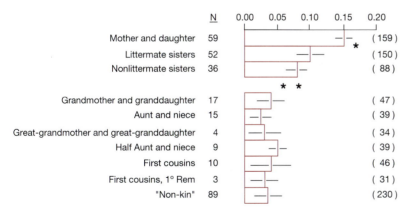

Figure 16.4 **Closely related female ground squirrels are more likely to cooperate than distant kin** This bar chart reports how frequently territory-owning females were joined by different categories of relatives and nonrelatives in chasing away trespassing ground squirrels. The number in parentheses indicates the number of chases that occurred when both types of individuals were present. "N" gives the number of different dyads of each kind that were observed. The horizontal bars give the standard deviations of the frequencies. There is a significant difference in the frequency of cooperation between the top three and the bottom seven categories of kinship (P < 0.01), but no significant difference among the seven kinship categories at the bottom (P > 0.09). From Sherman (1981).

Figure 16.5   **White-fronted bee-eaters**   The individual in the middle is performing a wing-waving display, and may be soliciting a feeding. (Gerard Lacz/Animals Animals/Earth Scenes)

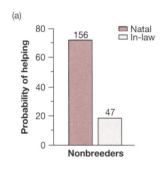

(a)

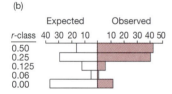

(b)

Figure 16.6   **In bee-eaters, helpers assist close relatives** (a) Bee-eater clans often contain nonbreeders that have paired with members of the clan. Their *r* with the offspring being raised that season is 0. This bar chart shows that they are much less likely to help than are clan members (P < 0.01).   (b) In this bar chart, the expected probability of helping is calculated by assuming that helpers assist clan members randomly, in proportion to the *r*'s of nestlings in clan nests. A G-test rejects the null hypothesis that helping is directed randomly with respect to kinship (P < 0.01). From Emlen and Wrege (1988).

Steve Emlen, Peter Wrege, and Natalie Demong have completed an intensive study of helping behavior in the white-fronted bee-eater *(Merops bullockoides)*. This colonial species, native to East and central Africa, breeds in nesting chambers excavated in sandy riverbanks (Figure 16.5). The 40-450 individuals in each colony are subdivided into groups of 3 to 17 birds, each of which defends a feeding territory up to seven kilometers away. These clans can include several sets of parents and offspring.

Many year-old bee-eaters stay to help at the nest during what would otherwise be their first breeding season. Clan members are related, so helpers usually have a choice of nestlings with different degrees of kinship as recipients of their helping behavior (Hegner et al. 1982; Emlen et al. 1995). This choice is a key point: Because kinship varies among the potential recipients of altruistic behavior, white-fronted bee-eaters are an excellent species for researchers to use in testing theories about kin selection.

After marking large numbers of individuals and working out genealogies over an eight-year study period, Emlen and Wrege (1988, 1991) found that bee-eaters obey Hamilton's rule. They determined, for example, that the coefficient of relatedness with recipients has a strong effect on whether a nonbreeding member of the clan helps (Figure 16.6a). Further, nonbreeders actively decide to help the most closely related individuals available (Figure 16.6b). That is, when young with different coefficients of relationship are being reared within their clan, helpers almost always chose to help those with the highest *r*. Their assistance is an enormous benefit to parents. More than half of bee-eater young die of starvation before leaving the nest. On average, the presence of each helper results in an additional 0.47 offspring being reared to fledging (Figure 16.7). For young birds, helping at the nest results in clear benefits for inclusive fitness.

Kin Recognition

The results we just reviewed suggest that individuals have accurate mechanisms for assessing their degree of kinship with conspecifics. This phenomenon, called kin recognition, has been divided into two broad categories: direct and indirect (Hepper 1991; Pfennig and Sherman 1995).

Indirect kin recognition is based on cues like the timing or location of interactions. Many species of adult birds rely on indirect kin recognition when their chicks are young, and will feed any young bird that appears in their nest.

Direct kin recognition, in contrast, is based on specific chemical, vocal, or other cues. Spadefoot toads provide a dramatic example (Pfennig et al. 1993). These amphibians, which live in small ponds in the deserts of the American Southwest, exist in two morphs: omnivorous and carnivorous. Whether a spadefoot tadpole becomes an omnivore or carnivore is environmentally determined. In the lab, brothers and sisters that are fed plant material or detritus become omnivores, while those fed fairy shrimp become carnivores. The carnivorous morph is highly cannibalistic. When carnivorous tadpoles are allowed to become hungry and then offered a choice between an unfamiliar omnivore sibling and an unfamiliar omnivore nonsibling, they strongly prefer to eat nonsiblings. The mechanism of choice is informative: They take siblings and nonsiblings into their mouths with equal probability, but tend to eat nonsiblings and spit out siblings. This suggests that spadefoot toads use chemical cues to recognize kin.

The spadefoot toad example leads us to an active area of research: mechanisms of kin recognition based on chemical cues and the senses of taste or smell. Specifically, there is increasing interest in the role played by genetic polymorphism in the major histocompatibility complex (MHC) in creating the so-called recognition loci. In Chapter 5 we introduced the MHC and its role in self/non-self recognition by cells in the immune system. Although loci in the MHC clearly evolved to function in disease prevention, polymorphism is so extensive that non-kin share very few alleles. As a result, these genes can serve as reliable markers of kinship (see Brown and Eklund 1994). To use the vocabulary we introduced in Chapter 2, the MHC may represent a preadaptation for recognizing kin.

In Chapters 5 and 14 we reviewed experiments suggesting that mice and humans can assess how similar conspecifics are in MHC genotype by smell, and then use this information as a criterion for choosing a mate. In those experiments mice and humans appeared to make mate choices based on degree of dissimilarity in MHC. But could similarity in MHC provide a reliable cue of kinship and offer a criterion for dispensing altruistic behavior?

Jo Manning and colleagues (1992) set out to answer this question using a semi-wild population of house mice *(Mus musculus domesticus)*. A kin recognition system requires three components: production of the signal, recognition of the signal by conspecifics, and action based on that recognition. Previous work had shown that glycoproteins coded for by MHC loci are released in the urine of mice, and that mice can distinguish these molecules by smell. Mice are, for example, able to distinguish full siblings from half-siblings on the basis of their MHC genotypes. But do mice dispense altruistic behavior accordingly?

House mice form communal nests and nurse each others' pups. Because individuals could take advantage of this cooperative system by contributing less than their fair share of milk, Manning et al. predicted that mothers would prefer to place their young in nests containing close relatives. The logic here is that close kin should be less likely to cheat on one another because of the cost to their indirect fitness. Through a program of controlled breeding in which wild-caught mice were crossed with laboratory strains, Manning et al. created a population

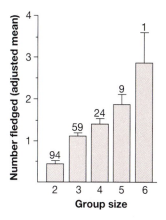

Figure 16.7 Fitness gains due to helping In bee-eaters the addition of 1–4 helpers dramatically increases fledging success (the number of young that leave the nest). Sample sizes for each group size are indicated. The vertical and horizontal lines above each bar represent the standard errors. From Emlen and Wrege (1991).

of semi-wild mice with known MHC genotypes. This population was allowed to establish itself in the semi-natural conditions of a large barn. The researchers then recorded where mothers in this population placed their newborn pups after birth. The null hypothesis was that mothers would choose to rear their offspring randomly with respect to the MHC genotypes present in the communal nests available at the time. Contrary to the null expectation, mothers showed a strong preference for rearing their young in nests containing offspring with similar MHC genotypes. This result confirms a role for MHC as a signal used in direct kin recognition, and shows that mice are capable of dispensing altruistic behavior on the basis of MHC genotypes.

Parent-Offspring Conflict

Conflicts arise when the fitness interests of close relatives do not coincide.

Although kin selection can clearly lead to close cooperation among individuals, it is not true that close relatives always cooperate. Even near kin can be involved in conflicts when the costs and benefits of altruism change or degrees of relatedness are not symmetrical. Robert Trivers (1974) was the first to point out that kin as close as parents and offspring are *expected* to disagree about each others' fitness interests. Conflicts over the amount of parental investment should be especially sharp in birds and mammals, simply because their parental care is so extensive.

Weaning Conflict

Weaning conflict is a well-documented example of parent-offspring strife. Aggressive and avoidance behaviors are common toward the end of nursing in a wide variety of mammals. Mothers will ignore or actively push young away when they attempt to nurse, and offspring will retaliate by screaming or by attacking their mothers (Figure 16.8).

(a)   (b)

Figure 16.8   Weaning conflict   (a) The infant langur monkey on the left has just attempted to nurse from its mother, at the right. The mother refused to nurse. In response, the infant is screaming at her. Reproduced by permission from Trivers, *Social Evolution,* page 147, Fig. 7-2a. Menlo Park, CA: Benjamin Cummings Publishing Co., 1985.   (b) The infant then dashes across the branch and slaps its mother. (Sarah Blaffer Hrdy/Anthro-Photo)

The key to explaining weaning conflict is to recognize that the fitness interests of parents and offspring are not symmetrical. Offspring are related to themselves with $r = 1.0$, but parents are related to their offspring with $r = 0.5$. Further, parents are equally related to all their offspring and are expected to equalize their investment in each. Siblings, in contrast, are related by 1.0 to themselves but 0.5 to each other. The theory of evolution by natural selection predicts that each offspring will demand an unequal amount of parental investment for themselves.

Figure 16.9 shows that when these asymmetries are applied to nursing, conflicts arise. At the start of nursing, the benefit to the offspring is high relative to the cost to the parent (Figure 16.9a). As nursing proceeds, however, this ratio declines. Young grow and demand more milk, which increases the cost of care. At the same time they are increasingly able to find their own food, which decreases the benefit. Natural selection should favor mothers who stop providing milk when the benefit-to-cost ratio reaches 1 (this is time p in Figure 16.9a). From the mother's perspective, this is when weaning should occur. Offspring, on the other hand, devalue their mother's cost of providing care. They do this because the "savings" that a parent achieves through weaning will be invested in brothers or sisters with $r = \frac{1}{2}$, instead of in themselves with $r = 1.0$. Natural selection should favor offspring who try to coerce continued parental investment until the benefit-cost ratio is $\frac{1}{2}$ (this is time o on Figure 16.9a). The period between time p and time o defines the interval of weaning conflict. Avoidance and aggressive behavior should be observed throughout this period. If mating systems are such that mothers routinely remate and produce half-siblings, the

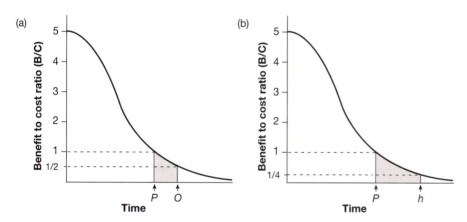

Figure 16.9 Changes in the costs and benefits of parental care and asymmetries in relationship produce parent-offspring conflict   (a) This graph illustrates why parent-offspring conflict occurs. The *y*-axis plots the benefit-to-cost ratio (B/C) for an act of parental care such as providing milk. Benefit is measured in terms of increased survival by the offspring receiving the care, while cost is measured as decreased production of additional offspring by the parent. Time is plotted along the *x*-axis. The curve drawn here is hypothetical; its shape will vary from species to species. See the text for an explanation of how this curve is interpreted.   (b) This is the same graph as in (a), modified to illustrate how the period of parent-offspring conflict is extended when parents produce half-siblings instead of full siblings. From Trivers (1985).

period of weaning conflict will extend to time *h*, when $B/C = r = \frac{1}{4}$ (Figure 16.9b). For field studies that confirm weaning conflict, see Trivers (1985, Chapter 7); for a theoretical treatment of other types of parent-offspring strife, see Godfrey (1995).

Harassment in White-fronted Bee-eaters

Another dramatic example of parent-offspring conflict occurs in the white-fronted bee-eaters introduced earlier. Steve Emlen and Peter Wrege (1992) have collected data suggesting that parents sometimes coerce sons into helping to raise their siblings. They do this by harassing sons who are trying to raise their own young.

A variety of harassment behaviors are observed at bee-eater colonies. Individuals chase resident birds off their territory, physically prevent the transfer of food during courtship feeding, or repeatedly visit nests that are not their own before egg laying or hatching. During the course of their study, Emlen and Wrege observed 47 cases of harassment. Over 90% of the instigators were male and over 70% were older than the targeted individual. In 58% of the episodes, the instigator and victim were close genetic kin. In fact, far from being targeted randomly among active pairs in the colony, statistical tests show that harassment behavior is *preferentially* directed at close kin ($P < 0.01$; χ^2 test).

Emlen and Wrege interpret this behavior by proposing that instigators are actively trying to break up the nesting attempts of close kin. Further, they suggest that instigators do this to recruit the targeted individuals as helpers at their own (the instigator's) nest.

What evidence do Emlen and Wrege present to support this hypothesis? In 16 of the 47 harassment episodes observed, the behavior actually resulted in recruitment: The harassed individuals abandoned their own nesting attempts and helped at the nest of the instigator. Of these successful events, 69% involved a parent and offspring and 62% involved a father and son. The risk of being recruited is clearly highest for younger males and for males with close genetic relatives breeding within their clan (Figure 16.10).

These data beg the question of why sons do not resist harassment more effectively. Emlen and Wrege suggest that harassment can be successful because sons are equally related to their own offspring and to their siblings. Parents, in contrast, are motivated to harass because they are more closely related to their own offspring ($r = \frac{1}{2}$) than they are to their grandchildren ($r = \frac{1}{4}$). We have already mentioned that on average, each helper is responsible for an additional 0.47 offspring being raised. In comparison, each parent at an unrecruited nest unaided by relatives is able to raise 0.51 offspring. This means that for a first-time breeder, the fitness payoff from breeding on its own is only slightly greater than the fitness payoff from helping. The payoffs are close enough to suggest that parents can change the bottom line of the fitness accounting. Perhaps harassing a son tips the balance by increasing his cost of rearing young. Then helping becomes a more favorable strategy for the son than raising his own young.

To summarize, Emlen and Wrege's data imply that bee-eater fathers recognize sons and preferentially coerce them into serving the father's reproductive interests. We can state the general message of studies reviewed in this section as

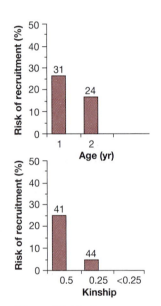

Figure 16.10 Bee-eaters recruit helpers who are younger and closely related Emlen and Wrege (1992) considered each paired male in the colony who had an equally aged or older male present in its clan, and calculated the percentage of these individuals who were recruited as helpers after experiencing harassment. This probability is plotted on the y-axis of these two bar charts. Different age and kinship categories of targeted individuals are plotted on the *x*-axis. Here kinship represents the *r*-value between the targeted individual and the offspring in the instigator's nest.

follows: Kin selection can result in either cooperative behavior or conflicts, depending on the costs and benefits involved and the degree of kinship between actor and recipient.

16.2   Evolution of Eusociality

Darwin (1859) recognized that social insects represent the epitome of altruism, and thus a special challenge to the theory of evolution by natural selection. Many worker ants and bees, for example, do not reproduce at all. They are helpers at the nests of their parents, for life. This is an extreme form of reproductive altruism.

Eusociality (true sociality) is used to describe social systems with three characteristics (Michener 1969; Wilson 1971; Alexander et al. 1991): (1) overlap in generations between parents and their offspring, (2) cooperative brood care, and (3) specialized castes of non-reproductive individuals. Eusocial species are found in snapping shrimp (Duffy 1996), a variety of insect orders (Table 16.2), and one family of rodents (the mammal family Bathyergidae, or mole-rats).

As an entree to the extensive literature on eusociality, in this section we consider how reproductive altruism evolved in two very different groups: the Hymenoptera (ants, bees, and wasps) and mole-rats.

Haplodiploidy and Eusocial Hymenoptera

The Hymenoptera represent the pinnacle of social evolution. A single ant colony may number millions of individuals, each appearing to function more like a cell in a superorganism than an individual pursuing its own reproductive interests. Worker, soldier, and reproductive castes, which seem analogous to tissues in a body, can be identified on the basis of their morphology and the tasks they perform. But unlike cells and tissues, individuals in the colony are not genetically identical. What factors laid the groundwork for such extensive altruism? Why is eusociality so widespread in Hymenoptera?

William Hamilton (1972) proposed that the unusual genetic system of ants, wasps, and bees predisposes them to eusociality. Hymenoptera have an unusual form of sex determination called a haplodiploid system: Males are haploid and females are diploid. Males develop from unfertilized eggs; females develop from fertilized eggs. As a result of this system female ants, bees, and wasps are more closely related to their sisters than they are to their own offspring. This follows because sisters share all of the genes they inherited from their father, which is half their genome, and half the genes they inherited from their mother (the colony's queen), which is the other half of their genome. The probability that homologous genes in hymenopteran sisters are identical by descent is thus $(1 \times \frac{1}{2}) + (\frac{1}{2} \times \frac{1}{2}) = \frac{3}{4}$. To their own offspring, however, females are related by the usual r of $\frac{1}{2}$.

In haplodiploid species, females are more closely related to their sisters than they are to their own offspring.

Before we discuss the implications of this fact, we should examine an important assumption in our calculation: that all female workers in the colony have the same father. This may not be true. Multiple mating is common in social Hymenoptera. Honeybee queens, for example, mate an average of 17.25 times before founding a colony (Page and Metcalf 1982). But because queens selectively store sperm in an organ called the spermatheca and control its release during

TABLE 16.2 Sociality in insects

(a) Insects exhibit degrees of sociality ranging from solitary to eusocial. This table summarizes the classes and key characteristics of social organization in insects.

	Continued care of young	Cooperative brood care	Reproductive division of labour	Colonies with at least two adult generations	Egg-layers morphologically differentiated
Solitary	−	−	−	No colonies	−
Subsocial	+	−	−	−	−
Parasocial					
Communal	+	−	−	−	−
Quasisocial	+	+	−	±	−
Semisocial	+	+	+	−	−
Eurosocial					
Primitive	+	+	+	+	−
Advanced	+	+	+	+	+

(b) This table summarizes the taxonomic distribution of eusociality in insects.

Order	Family	Subfamily	Eusocial species
Hymenoptera	Anthophoridae (carpenter bees)		In 7 genera
	Apidae	Apinae (honey-bees)	Six highly eusocial species
		Bombinae (bumble-bees)	300 primitively eusocial species
		Euglossinae (orchid bees)	None
		Meliponinae (stingless bees)	200 eusocial species
	Halictidae (sweat bees)		In 6 genera
	Sphecidae (sphecoid wasps)		In 1 genus
	Vespidae (paper wasps, yellow jackets)	Polistinae	Over 500 species, all eusocial
		Stenogastrinae	Some primitively eusocial species
		Vespinae	Ca. 80 species, all eusocial
	Formicidae (ants)	11 subfamilies	Over 8,800 described species, all are eusocial or descended from eusocial species
	Many other families		None
Isoptera (termites)	9 families		All species (over 2,288) are eusocial
Homoptera (plant bugs)	Pemphigidae		Sterile soldiers found in 6 genera
Coleoptera (beetles)	Curculionidae		*Austroplatypus incompertus*
Thysanoptera (thrips)	Phlaeothripidae		Subfertile soldiers are found in *Oncothrips*

Source: From Crozier and Pamilo (1996).

egg production, multiple paternity may be less widespread than multiple mating (Bourke and Franks 1995; Crozier and Pamilo 1996). Allozyme markers can be used to estimate the number of different fathers among colony members; studies in ants, wasps, and bees support the hypothesis that multiple paternity is rare (Boomsma and Ratnieks 1996).

Thus, the general expression for the coefficient of relationship between sisters in haplodiploid species is $\frac{1}{4} + (\frac{1}{2})g$, where g specifies the coefficient of relationship between their fathers. If their fathers are unrelated, then $g = 0$ and r between sisters is $\frac{1}{4}$. If sisters have the same father, then $g = 1$ and r is $\frac{3}{4}$.

When sisters have the same father, their high coefficient of relatedness suggests that the benefits of acting as a worker will often outweigh its costs. When this is true, females will maximize their inclusive fitness by acting as workers rather than as reproductives (Hamilton 1972). Specifically, their genes will increase in the population faster when they invest in the production of sisters rather than producing their own offspring. This is the haplodiploidy hypothesis for the evolution of eusociality in Hymenoptera.

The haplodiploidy hypothesis also predicts that workers will preferentially invest in sisters over brothers. This leads to an important prediction: Because their r with sisters is $\frac{3}{4}$ and only $\frac{1}{4}$ with brothers, workers should favor a $3:1$ female-biased sex ratio in the reproductive offspring produced by the colony (Trivers and Hare 1976). Queens, in contrast, are equally related to their sons and daughters and should favor a $1:1$ sex ratio in the reproductives produced (see Box 16.2). This leads us to one of the most hotly debated issues about eusociality in Hymenoptera: Who wins? Do queens or workers control the sex ratio? Answering this question gets to the heart of a broader issue: Do hymenopteran colonies primarily serve the reproductive interests of queens or of workers? The asymmetry in relationship produces a conflict of interests among members of the same colony.

Table 16.3 contains data on sex ratios, computed as the proportion of reproductive effort devoted to producing males versus reproductive females. These estimates were collected from a variety of ant species in which colonies have a single queen. In general, values are close to the predicted ratio of 0.25. This implies that workers are in control of colony reproduction and that the haplodiploidy hypothesis is correct. Data on levels of investment in males and females are notoriously difficult to collect, however, and the scatter in the data is considerable. Long-term studies have also documented year-to-year differences in sex allocation, suggesting that data from one-year studies may not be accurate. Finally, because the phylogeny of the ants is largely unknown, researchers have been unable to identify independent data points and perform rigorous statistical tests of between-species patterns. All of these issues complicate our ability to interpret the results, and advise caution. But given these qualifications, most researchers are now willing to concede that the studies done to date tend to support the reality of female-biased sex allocation. In the tug-of-war over the fitness interests, workers appear to have the upper hand over queens (for recent reviews, see Bourke and Franks 1995; Crozier and Pamilo 1996).

An even stronger consensus in the field, however, is that haplodiploidy supplies only a predisposition for eusociality. Many haplodiploid species are not eusocial, and many eusocial species are not haplodiploid. For example,

---

BOX 16.2   The evolution of the sex ratio

In many species the sex ratio at hatching, germination, or birth is 1:1. Ronald Fisher (1930) explained why this should be so. Fisher pointed out that if one sex is in short supply in a population, then a gene that leads to the production of the rarer sex will be favored. This is because individuals of the rarer sex will have more than one mate on average when they mature, simply because in a sexual species every individual has one mother and one father. Members of the rarer sex will experience increased reproductive success relative to individuals of the more common sex. Indeed, whenever the sex ratio varies from 1:1, selection favoring the rarer sex will exist until the ratio returns to unity. Fisher's explanation is a classic example of frequency-dependent selection—a concept we introduced in Chapter 5 with the example of left- and right-handed scale-eating fish.

Fisher's argument is based on an important assumption: that parents invest equally in each sex. When one sex is more costly than the other, parents should adjust the sex ratio to even out the investment in each. For this reason evolutionary biologists distinguish the numerical sex ratio from the investment sex ratio and speak, in general terms, about the issue of sex allocation (Charnov 1982).

Robert Trivers and Dan Willard (1973) came up with an important extension to Fisher's model. They suggested that when females who are in good physiological condition are better able to care for their young, and when differences in the condition of young are sustained into adulthood, then females in good condition should preferentially invest in male offspring. This is because differences in condition affect male reproductive success (RS) more than female RS (see Chapter 15). This prediction, called condition-dependent sex allocation, has been confirmed in a wide variety of mammals, including humans (for examples, see Clutton-Brock et al. 1984; Betzig and Turke 1986).

A third prominent result in sex-ratio theory is due to William Hamilton (1967). In insects that lay their eggs in fruit or other insects, the young often hatch, develop, and mate inside the host. Frequently, hosts are parasitized by a single female. Given this situation, Hamilton realized that selection should favor females that produce only enough males to ensure fertilization of their sisters, resulting in a sex ratio with a strong female bias. This phenomenon, known as local mate competition (LMC), has been observed in a variety of parasitic insects (for examples, see Hamilton 1967; Werren 1980, 1984).

---

nonreproductive casts are found in many termite species, even though they are diploid and have a normal chromosomal system of sex determination. To use formal terminology, haplodiploidy is neither a necessary nor sufficient condition for the evolution of eusociality. Further, if the identification of families and tribes in the taxonomy of Hymenoptera accurately reflects phylogeny, then eusociality has evolved at least 12 times independently in the order (Crozier and Pamilo 1996; also see Cameron 1993).

Haplodiploidy provides a predisposition to reproductive altruism, but ecological factors are also important.

Two ecological factors may add to the genetic predisposition caused by high coefficients of relatedness in Hymenoptera. These factors produce costs to independent breeding and benefits that favor reproductive altruism. The first is that safe, defensible nest sites near food supplies may be extremely rare (see Alexander et al. 1991). When this is true, the likelihood of an individual being able to breed on its own is low or nonexistent. Second, when predation rates are high but young are dependent on parental care for a long period, individuals who breed alone are unlikely to survive long enough to bring their young to independence (Queller 1989). To explain the evolution of eusociality, we clearly need to consider ecological factors that affect B and C as well as genetic factors that dictate r.

TABLE 16.3 Sex ratios observed in ant colonies

In this table, *C* represents the number of colonies studied and *m* the proportion of resources invested in males. This proportion is estimated as the dry weight of males divided by the dry weight of all colony individuals. Because males and females differ in size (males are smaller), calculating dry weights is a more informative index of sex allocation than simply counting the two sexes.

Species	C	m	Species	C	m
Acromyrmex octospinosus	10	0.25	*F. truncorum*	63	0.47
Aphaenogaster rudis	14	0.30	*Lasius alienus*	17	0.03
A. treatae	12	0.13	*L. flavus*	12	0.10
Apterostigma dentigerum	53	0.51	*L. niger*	9	0.61
Atta bisphaerica	5	0.28	*L. curvispinosus*	97	0.23
A. laevigata	6	0.26	*Myrmecina americana*	10	0.31
A. sexdens	7	0.39	*Myrmica schencki*	10	0.13
Camponotus ferrugineus	6	0.17	*Pogonomyrmex montanus*	35	0.40
C. herculeanus	1	0.32	*P. rugosus*	4	0.48
C. pennsylvanicus	12	0.16	*P. subnitidus*	7	0.40
Carebara vidus	7	0.08	*Prenolepis imparis*	12	0.24
Colobopsis nipponicus	20	0.25	*Pseudomyrmex belti*	1	0.40
Formica exsecta	15	0.73	*P. ferruginea*	1	0.21
F. fusca	29	0.59	*Rhytidoponera purpurea*	10	0.26
F. nitidiventris	19	0.14	*Stenamma brevicorne*	8	0.23
F. pratensis	35	0.71	*S. diecki*	10	0.27
F. rufa	32	0.46			

Source: From Crozier and Pamilo (1996).

To summarize, inclusive fitness theory inspired the haplodiploidy hypothesis for the evolution of eusociality in hymenopterans. This theory has been shown to provide an important predisposition to reproductive altruism. Research on the evolution of eusociality in Hymenoptera is now focused on a series of key issues:

- Why does the degree of sociality (Table 16.2) vary among some closely related species?
- Can phylogenetic analyses inform our understanding of how eusociality evolved?
- How do workers control the sex ratio? Liselotte Sundström and co-workers (1996) found that numerical sex ratios at egg laying were close to 1:1 in the ant colonies they studied, but become significantly female-biased by hatching. This implies that workers can determine the sex of eggs and selectively destroy them.
- How extensive is worker reproduction of male offspring? Sisters are related to their nephews with $r = \frac{3}{8}$ and should prefer to raise them to their brothers. How many of the males produced by colonies are the offspring of workers, and how many are the offspring of the queen?

Naked Mole-Rats

Naked mole-rats *(Heterocephalus glaber)* are one of the great oddities of the class Mammalia (Figure 16.11). They are neither moles nor rats, but are members of the family Bathyergidae, native to desert regions in the Horn of Africa. They eat tubers, live underground in colonies of 70–80 members, and construct tunnel systems up to 2 miles long by digging cooperatively in the fashion of a bucket brigade. Mole-rats are nearly hairless and ectothermic ("cold-blooded") and, like termites, can digest cellulose with the aid of specialized microorganisms in their intestines.

Naked mole-rats are also eusocial. All young are produced by a single queen and all fertilizations performed by a group of 2–3 reproductive males. As other members of the colony grow older and increase in size, their tasks change from tending young and working in the tunnels to specializing in colony defense. For unknown reasons, there is a slight male bias in the colony sex ratio, of 1.4:1. Naked mole-rats are diploid and have an XY chromosome system of sex determination.

The leading hypothesis to explain why naked mole-rats are eusocial centers on inbreeding. Analyses of microsatellite loci confirm that colonies are highly inbred (Reeve et al. 1990). Researchers studying colonies established in the lab have determined that approximately 85% of all matings are between parents and their offspring or between full siblings, and that the average coefficient of relationship among colony members is 0.81 (Sherman et al. 1992). These are among the highest coefficients ever recorded in animals.

Eusocial species that are not hap-lodiploid are often highly inbred.

Even extensive inbreeding does not mean that the reproductive interests of workers and reproductives are identical, however. Conflicts exist because workers are still more closely related to their own offspring than they are to their siblings and half-siblings. But queens are able to maintain control through physical dominance. If non-reproductives slow their pace of work, mole-rat queens push them. These head-to-head shoves are aggressive and can move a worker more than a meter backward through a tunnel. Shoves

***Figure 16.11* Naked mole-rats**   This photo shows a naked mole-rat queen lying on top of a pile of workers. For superb introductions to the biology of naked mole-rats, see Sherman et al. (1991), Honeycutt (1992), and Sherman et al. (1992). (Raymond A. Mendez/Animals Animals/Earth Scenes)

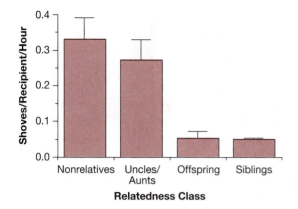

Figure 16.12 **Naked mole-rat queens preferentially shove non-relatives**   These data were collected from a captive colony of naked mole-rats. The bars indicate the average and standard errors of shoves given to different kin classes by a queen called three-bars. There is a statistically significant difference between shoving rates for nonrelatives and uncles/aunts versus the two closer kin classes. From Reeve and Sherman (1991).

are directed preferentially toward nonrelatives and toward relatives more distant than offspring and siblings of the queen (Figure 16.12). Workers respond to shoves by nearly doubling their work rate (Table 16.4). These data suggest that queens impose their reproductive interests on subordinates through intimidation.

Inbreeding has also been hypothesized as a key factor predisposing termites to eusociality (Bartz 1979; but see Pamilo 1984; Roisin 1994). But not all inbred species are eusocial. This means that inbreeding is just one of several factors that contributes to eusociality in naked mole-rats and termites.

Again, although we have stressed the unusual features of genetic systems present in eusocial species, ecological factors like extended parental care, group defense against predation, and severely constrained breeding opportunities are also important in explaining the evolution of reproductive altruism.

16.3   Reciprocal Altruism

Inclusive fitness theory has been remarkably successful in explaining a wide range of phenomena in social evolution. In many cases altruistic acts can be explained on the basis of Hamilton's rule, and conflicts can be understood by analyzing asymmetries in coefficients of relatedness and fitness payoffs. But the theory and data we have reviewed thus far concern only interactions among kin. What about the frequent occurrence of cooperation among unrelated individuals?

TABLE 16.4 Shoves from naked mole-rat queens motivate workers

Working in captive colonies, Hudson Reeve (1992) performed regular scans and recorded the activity of all individuals in the colony. The work level reported in this table represents the proportion of scans during which individuals were performing work, with standard errors given in parentheses. Shoving events by queens that took place in the nestbox or tunnels were also recorded. There is a statistically significant difference between work rates before and after shoves, for both types of shoves ($P < 0.01$).

Shove recipient's work level	Before shove	After shove
All shoves	0.14 (0.03)	0.25 (0.06)
Tunnel shoves only	0.34 (0.05)	0.58 (0.07)

Reciprocal altruism provides one theoretical framework for studying cooperation among non-kin. Robert Trivers (1971) created this concept when he proposed that individuals can be selected to dispense altruistic acts if equally valuable favors are later returned by the beneficiaries. Natural selection can favor altruistic behavior if the recipients reciprocate.

Two important conditions must be met for reciprocal altruism to evolve. First, selection can favor altruistic acts only if the cost to the actor is smaller than or equal to the benefit to the recipient. Alleles that lead to high-cost, low-benefit behavior cannot increase in the population even if this type of act is reciprocated. (As with kin selection, the costs and benefits of altruistic acts are measured by the numbers of surviving offspring.) Second, individuals that fail to reciprocate must be punished in some way. If they are not, then altruistic individuals will suffer fitness losses with no subsequent return. Alleles that lead to cheating behavior would increase in the population, and altruists would quickly be eliminated. As a result, the theory predicts that altruists will be selected to detect and punish cheaters by withholding any future benefits from them (Box 16.3).

Trivers pointed out that reciprocal altruism is most likely to evolve when

- Each individual repeatedly interacts with the same set of individuals (groups are stable).
- Many opportunities for altruism occur in an individual's lifetime.
- Potential altruists interact in symmetrical situations. This means that two interacting individuals are able to dispense roughly equivalent benefits at roughly equivalent costs.

Reciprocal altruism can evolve only under a restricted set of conditions.

Accordingly, we expect reciprocal altruism to be characteristic of long-lived social species with small group size, low rates of dispersal from the group, and a high degree of mutual dependence in group defense, foraging, or other activities. Reciprocal altruism should be less likely to evolve in species where strong dominance hierarchies are the rule. In these social systems, subordinate individuals are rarely able to provide benefits in return for altruistic acts dispensed by dominant individuals.

Based on these characteristics, Trivers (1971, 1985; see also Packer 1977) has suggested that reciprocal altruism is responsible for much of the cooperative behavior observed in primates like baboons, chimpanzees, and humans. Indeed, Trivers has proposed that human emotions like moralistic aggression, gratitude, guilt, and trust are adaptations that have evolved in response to selection for reciprocal altruism. He suggests that these emotions function as "score-keeping" mechanisms useful in moderating transactions among reciprocal altruists.

Studying reciprocal altruism in natural populations is difficult, however. It is likely that kin selection and reciprocal altruism interact and are mutually reinforcing in many social groups. This makes it difficult for researchers to disentangle the effect each type of selection has independently of the other. Also, the fitness effects of some altruistic actions can be difficult to quantify. When a young male baboon supports an older, unrelated male in his fight over access to a female, or when a young lioness participates in group defense of the territory, how do we quantify the fitness costs and benefits for each participant?

Box 16.3 Prisoner's dilemma: Analyzing cooperation and conflict using game theory

Robert Trivers (1971) recognized that a classical problem from the branch of mathematics called game theory closely simulates the problems faced by nonrelatives in making decisions about their interactions. The central idea in game theory is that the consequences of any move in a game are contingent: The result or payoff from an action depends on the move made by the opponent. When players in a game pursue contrasting strategies, game theory provides a way to quantify the outcomes and decide which strategy works best.

Game theory was invented in the 1940s to analyze contrasting strategies in games like poker and chess. Later the approach was applied by economists to a variety of problems in market economics and business competition. John Maynard Smith (1974, 1976, 1982) pioneered the use of game theory in evolutionary biology in analyses of animal contests. His work inspired a series of productive studies on the evolution of display behavior and combat (see Reichert 1978, 1979; Sigurjónsdóttir and Parker 1981; Hansen 1986; Hammerstein and Reichert 1988).

The game that Trivers employed to analyze cooperation is called Prisoner's Dilemma. Prisoner's Dilemma models the following situation: Two prisoners who have been charged as accomplices in a crime are locked in separate cells. The punishment they suffer depends on whether they cooperate with one another in maintaining their innocence or implicate the other in the crime. Each prisoner has to choose his strategy without knowing the other prisoner's choice. The payoffs to Player A in this game are as follows:

where $T > R > P > S$ and $R > (S + T)/2$. The highest payoff in the game comes when Players A and B both cooperate, but Player A does best when A defects and B cooperates. When Players A and B interact just once, the best strategy for each player is to defect.

What happens when the two players interact repeatedly? Robert Axelrod and William Hamilton (1981) performed a widely cited analysis of an iterated Prisoner's Dilemma. They invited game theorists from all over the world to submit strategies for players competing in a computerized simulation of the game. Each round in this tournament had the following payoffs: $R = 3$, $T = 5$, $S = 0$, and $P = 1$. Axelrod and Hamilton let each strategy play against all of the other strategies submitted, and computed the outcome of every one-on-one game over many interactions. Most of the theorists submitted complicated decision-making algorithms, but the winner was always the simplest strategy of all, called tit-for-tat. An individual playing tit-for-tat (TFT) starts by cooperating, then simply does whatever the opponent did in the previous round. This strategy has three prominent features: (1) it is never the first to defect; (2) it is provoked to immediate retaliation by defection; and (3) it is willing to cooperate again after just one act of retaliation for a defection.

In analyzing the outcome of games like this, researchers use the concept of an evolutionarily stable strategy (ESS). A strategy is an ESS if a population of individuals using it cannot be invaded by a rare mutant adopting a different strategy. Axelrod and Hamilton's tournament showed that TFT is an evo-

		Player B's Action	
		C **Cooperation**	**D** **Defection**
Player A's Action	**C** **Cooperation**	R (reward for cooperation)	S (sucker's payoff)
	D **Defection**	T (temptation to cheat and its payoff)	P (punishment for mutual defection)

Box 16.3 *(continued)*

---

lutionarily stable strategy with respect to other strategies employed in the tournament. Their result offers an explanation for the evolution of cooperative behavior in unrelated individuals, and laboratory experiments with guppies and sticklebacks suggest that animals may actually play TFT when they interact (see Milinski 1987; Dugatkin 1988; Milinski 1996).

Recently there has also been a great deal of interest in a Prisoner's Dilemma strategy called Pavlov, which is characterized as a "win-stay, lose-shift" algorithm. An individual playing Pavlov repeats its previous move if it was rewarded by R or T points, but switches if it received only P or S points. The strategy maintains an almost reflex-like response to the payoff, which inspired its name. A computerized tournament conducted by Martin Nowak and Karl Sigmund (1993) suggests that Pavlov can beat TFT and emerge as the ESS.

Carl Wedekind and Manfred Milinski (1996) followed up on this result by recruiting first-year biology students, who had no background in game theory, to participate in a Prisoner's Dilemma tournament. When players made their moves simultaneously, the winners were students who pursued a Pavlov-like strategy. When players made their moves alternately, the winning students pursued a strategy called generous tit-for-tat (GTFT). GTFT is identical to TFT, except that the player occasionally cooperates when the two interacting individuals are stuck in a round of mutual defections.

Game theoretical analyses of cooperation are now focused on the role of memory, when players make alternate moves and interact repeatedly. Another challenge will be gathering data on whether these strategies are actually pursued by individuals interacting in natural environments.

What are our chances of observing the return behavior, and quantifying its costs and benefits as well?

For these reasons it has been difficult for evolutionary biologists to document reciprocal altruism; difficult enough, in fact, to make many biologists suspect that the phenomenon is rare. Here we will examine one of the most robust studies done to date: food sharing in vampire bats. In this system the altruistic act is regurgitating a blood meal. The cost of altruism can be measured as an increase in the risk of starvation, and the benefit as a lowered risk of starvation.

Blood-Sharing in Vampire Bats

Gerald Wilkinson (1984) worked on a population of about 200 vampire bats *(Desmodus rotundus)* at a study site in Costa Rica (Figure 16.13). The basic social unit in this species consists of 8–12 adult females and their dependent offspring. Members of these groups frequently roost together in hollow trees during the day, though subgroups often move from tree to tree. (There were a total of 14 roosting trees at Wilkinson's study site.) As a result of this social structure, many individuals in the population associate with one another daily. But the degree of association between individuals varies widely. Wilkinson quantified the degree of association between each pair of bats in the population by counting the number of times they were seen together at a roost and dividing by the total number of times they were observed roosting.

Wilkinson was able to capture and individually mark almost all of the individuals at his study site over a period of four and a half years, and estimate coef-

Figure 16.13 **Vampire bats**   This photo shows a group of vampire bats roosting in a hollow tree. (Gerald Wilkinson, University of Maryland)

ficients of relatedness through path analysis of pedigrees. In the female group for which he had the most complete data, the average *r* between individuals was 0.11 (recall that cousins have an *r* of 0.125).

The combination of variability in association and relatedness raises interesting questions about the evolution of altruism. Vampire bats dispense altruistic behavior by regurgitating blood meals to one another. This food sharing is important because blood meals are difficult to obtain. The bats leave their roosts at night to search for large mammals—primarily horses and cattle—that can provide a meal. Prey are wary, however, and 33% of young bats and 7% of adults fail to feed on any given night. By studying weight loss in captive bats when food was withheld, Wilkinson was able to show that bats who go three consecutive nights without a meal are likely to starve to death.

Because the degree of relatedness and degree of association varied among individuals in the population, either kin selection or reciprocal altruism could operate in this system. Wilkinson was able to show that both occur. Over the course of the study, he witnessed 110 episodes of regurgitation. Seventy-seven of these were between mother and child and are simply examples of parental care. In 21 of the remaining 33 cases, Wilkinson knew both the *r* and degree of association between the actor and beneficiary and could examine the effect of both variables. Wilkinson discovered that both degree of relatedness and degree of association have a statistically significant effect in predicting the probability of regurgitation (Figure 16.14). Bats do not regurgitate blood meals to one another randomly. They are much more likely to regurgitate to relatives and to nonrelatives who are frequent roostmates.

To confirm that bats actually do reciprocate, Wilkinson held nine individuals in captivity, withheld food from a different individual each night for several weeks, and recorded who regurgitated to whom over the course of the experiment. Statistical tests rejected the null hypothesis that hungry individuals received blood randomly from cagemates. Instead, hungry individuals were much more likely to receive blood from an individual they had fed before. This confirms that vampire bats are reciprocal altruists.

Vampire bats reciprocate by sharing blood meals. They usually share with close relatives or nonrelatives who are nestmates and may later reciprocate.

Territory Defense in Lions

Lions (*Panthera leo;* Figure 16.15) are the only cat that lives in social groups. These groups, called prides, consist of 3–6 related females, their offspring, and a

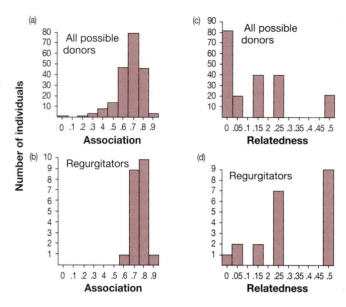

Figure 16.14   Association, relatedness, and altruism in vampire bats   These histograms plot the total number of bat-bat pairs at Gerald Wilkinson's study site versus their: (a) Degree of association, for all potential blood donors in the roost;   (b) Degree of association, for blood-sharing pairs who are not related as mother-offspring; (c) Coefficient of relatedness, for all potential blood donors in the roost;   (d) Coefficient of relatedness, for blood-sharing pairs who were not related as mother-offspring.

The visual impression in these figures is that regurgitators are more likely to be related *and* more likely to be roostmates than the general population. This is confirmed by a statistical procedure called a stepwise logistic regression. Both relatedness and association affect the probability of blood-sharing. From Wilkinson (1984).

coalition of males. Males are often related to one another but are unrelated to pride females (Packer and Pusey 1982; Packer et al. 1991). Males cooperate in defending themselves against other coalitions of males that attack and try to take over the pride. If invading males succeed, they quickly kill any young cubs. Infanticide in lions is adaptive because pride females respond to the loss of their young by coming back into estrus, copulating with the new males, and bearing their young. This increases the fitness of incoming males.

Females cooperate in defending their young against infanticidal attacks, in nursing young, in hunting prey that are difficult to capture (females do the vast majority of hunting in lions), and in defending the pride's territory against incursions by females in neighboring prides (Packer and Pusey 1983; Packer et al. 1990). Battles with intruding females are dangerous, especially if pride females do not cooperate in defense. Solitary lions are often killed in same-sex encounters.

To study cooperation during incursions by strange females who threaten the territory, Robert Heinsohn and Craig Packer (1995) placed speakers near the edge of pride territories and played tape recordings of roars given by unfamiliar

Figure 16.15   African lions   This female lion is moving across her territory toward the source of roars made by unfamiliar females. She is looking over her shoulder at a female member of her pride who is lagging behind. (Len Rue, Jr./Visuals Unlimited)

females. Pride females respond to these roars by approaching the speakers; they will even attack a stuffed female lion placed near the speaker. They also do not habituate to this stimulus—they continue responding when the experiment is repeated. As a result, Heinsohn and Packer were able to estimate the average response of females over multiple trials. They quantified these responses by measuring the time an individual took to reach the midpoint between the pride's original position and the speaker, calculating the difference between a focal individual's time-to-midpoint and the leader's time, recording the order within the group that each female reached the midpoint, and counting the number of glances a female made back at lagging pride members. They collected data on female responses in eight different prides.

Statistical analyses showed significant differences in the strength of responses among females within prides. Some females in each pride always led; others always lagged behind. Other females adopted conditional strategies: They would assist the lead female(s) more frequently when Heinsohn and Packer played tapes of more than one female roaring (that is, when the threat to the pride was greater). Still others would lag behind *more* when multiple roars were played. These differences were not correlated with the age or body size of females or with their coefficient of relationship with other pride members.

The dynamics of group defense by female lions cannot be explained by kin selection or reciprocal altruism.

These results are paradoxical in the context of both inclusive fitness and reciprocity theory. Why do the leaders tolerate the laggards? According to the theory we developed earlier, they should be punished. But leaders were never observed to threaten laggards or to withhold benefits. For example, they did not stop leading until laggards caught up. They glanced back, appearing to recognize that laggards were lagging, but continued approaching the speaker.

The simple answer is that we do not know why this combination of altruist and non-reciprocator strategies exists among lions. The leading hypothesis, which is still to be tested, is that laggard females make fitness benefits up to the leaders in other ways: through exceptional hunting prowess or milk production, for example. But clearly, social interactions among lions are substantially more complex than current theory would predict.

16.4 Behavioral Genetics

Throughout this chapter we have assumed that variation in behavior has a genetic basis. We have even used language such as "a gene that favors altruistic behavior." What evidence do we have for this claim? The patterns we have reviewed in this chapter imply that heritable variation exists in traits like the willingness to nurse, the inclination to help at the nest, and the ability to recognize kin on the basis of MHC genotype. But do we have any direct evidence for genetic variation in behavioral traits?

Answering this question is the focus of a discipline called behavioral genetics. The initial goal of this research program is to identify the loci involved in the expression of behavioral traits; the ultimate objective is to explain evolutionary changes in behavior in terms of base substitutions in these genes.

Before exploring this field further, however, we need to reemphasize points that we made in Chapters 6 and 8. First, the expression of behavioral traits, like the expression of morphological and physiological traits, is clearly the product of genetic *and* environmental factors. Second, variation in behavior among individuals in a population is frequently caused by condition-dependent decisions and not differences in genotype (this is an issue we will expand upon in Chapter 18). For example, as the bee-eater behavior summarized in this chapter shows, the same individual can act in radically different ways depending on its life situation. A white-fronted bee-eater may act in a selfish way when such tactics provide a fitness advantage, but change to altruistic or cooperative behavior when those tactics increase its ability to survive or reproduce. Behavioral biologists call these flexible, condition-dependent decisions "tactics" and distinguish them from behavioral "strategies" specified by different genotypes.

We reiterate these points because the fields of social biology and behavioral genetics have long been beset by a "nature-nurture" controversy (see Barlow and Silverberg 1980). Most evolutionary biologists now consider this debate to be unproductive. Behavioral flexibility is so pervasive, and evidence for joint environmental and genetic influence so well established, that it simply is not interesting to ask whether behavior is the product of genetics or the environment. A more productive approach has been to agree that organisms usually act in ways that optimize their fitness, ask whether variation in behavior is due to flexible tactics or alleles for different strategies, and investigate how genetic mechanisms are modified by environmental or cultural conditions.

The field of behavioral genetics has moved beyond the "nature-nurture" controversy.

Research is now proceeding rapidly within this conceptual framework. Because behavioral genetics is a relatively new field, our review focuses on the variety of approaches being employed to understand the genetic basis of social behavior.

Selection Studies

Artificial selection experiments are a classical approach to understanding the genetic basis of physiological and morphological traits. D.S. Falconer (1953, 1955), for example, used this technique to investigate variation in body size in mice. Falconer selected the smallest and largest individuals from a laboratory population, placed small individuals together and large individuals together, and allowed each subgroup to breed. He then selected the largest individuals from the larger population and the smallest individuals from the smaller population as breeders in the next generation. After repeating this procedure each generation for many generations, Falconer found that the small- and large-selected lines were significantly different from each other and from an unselected control population. Recall from Chapter 6 that $R = h^2S$ (the response to selection equals heritability times the selection differential). Because a measurable response (R) to selection (S) occurred, Falconer could conclude that heritable variation for the trait exists (that is, that $h^2 \neq 0$).

Ary Hoffmann (1988) has applied this experimental approach to the territorial behavior of *Drosophila melanogaster*. In fruit flies, mating and egg laying take place on food sources. Males defend these resources against other males when the food is in discrete patches. Hoffmann wanted to know if genetic variation exists in the ability of males to defend these territories.

To perform selection on territoriality, Hoffmann staged contests among male flies. He placed six males in cages that contained a small patch of food and observed the cages until one male had succeeded in defending the food source. Territorial behavior was confirmed when this male was seen forcing an intruder off the food by lunging or tussling. This male was then removed from the cage and allowed to mate with a female from the same selection line. Hoffmann replicated this procedure in 20 different cages every generation for over 20 generations. In control lines, males did not undergo selection on territoriality. Otherwise, control males and females were handled identically to individuals in the selection lines.

To test for a response to selection, Hoffmann staged contests between males from the selected and control lines. After 20 generations of selection, strong differences in territoriality had emerged: Males from selection lines won almost all of the staged contests (Table 16.5). Even though body size is the key variable in determining the outcome of contests in many animal species, Hoffmann was able to show that no changes in body size had occurred in the selected line when compared with controls. Rather, males from selection lines escalated territorial confrontations more often than control males, and were much more likely to win these escalated encounters. Based on these results, Hoffmann concluded that selection had elaborated a behavioral trait: the willingness to escalate and engage in physical combat. The experiment demonstrates that heritable variation exists in the territorial strategies pursued by fruit fly males.

It is important to recognize, however, that territorial behavior in fruit flies is a flexible tactic as well as a genetically based strategy. When cages contain evenly distributed food that cannot be defended, males are not territorial. Territorial behavior is induced by environments with a patchy distribution of resources.

Estimating the Heritability of Behavioral Traits

Estimating heritabilities directly is a second approach employed in behavioral genetics. This research program has been especially popular in studies of human behavior. For obvious reasons the cross-fostering and common garden

---

TABLE 16.5 Territorial behavior in fruit flies responds to selection
These data show the winners of contests between fruit fly males from two different selected (S1 and S2) and control (C1 and C2) lines, staged after 20 generations of selection. Ary Hoffmann placed two males from each line in a cage with a small source of food and recorded which male had established a territory on the food a day later. The Melbourne and Townsville lines were started from populations sampled at two different locations in Australia.

Lines compared	Melbourne lines		Townsville lines	
	Selection	Control	Selection	Control
S1 vs. C1	14	1	16	4
S1 vs. C2	17	2	18	1
S2 vs. C1	13	1	18	0
S2 vs. C2	16	1	16	3

Source: From Hoffmann (1988).

experiments we introduced earlier (Chapters 2 and 6), which are routinely used to estimate trait heritabilities in plants and nonhuman animals, are impossible to do with humans. Instead, researchers have relied on studies of twins to determine whether genetic variance for behavioral traits exists.

As we pointed out in Chapter 6, monozygotic (MZ) twins share their environment and all of their genes, while dizygotic (DZ) twins share their environment and half of their genes. If there is a difference in how similar MZ twins are compared to how similar DZ twins are, it should be primarily because MZ twins share twice as many of their genes. As a result, researchers can roughly estimate the heritability of a behavioral trait by

1. Plotting the values of that trait in MZ and in DZ twin pairs
2. Computing the correlation coefficients for each type of twin (this is a statistical measure of similarity within each class of twin)
3. Calculating the difference between the correlations computed for MZ and DZ twins
4. Doubling this difference

These estimates have to be interpreted cautiously, however, because MZ twins are likely to share more aspects of their environment than DZ twins (because their parents may dress them alike, for example). This similarity tends to inflate any observed differences between MZ and DZ twins and lead to an overestimate of heritabilities.

Heritabilities for several different human behavioral and cognitive traits have been estimated by this protocol (Figure 16.16). The values reported, which range from a low of 0.22 for memory to a high of 0.52 for general intelligence, are a close match to estimates made in studies of MZ twins that were separated soon after birth, placed in foster homes, and reared apart (Bouchard et al. 1990). These studies of MZ twins reared apart are the closest approximation we have to experiments done with other species, where related individuals are reared in randomized environments. The data from the two types of twin studies suggest that at least a small amount of genetic variation exists in complex personality and cognitive traits.

Several of the traits that have been investigated in the twin studies—particularly neuroticism, extraversion, verbal reasoning, and general intelligence—are involved in human social interactions. Earlier in the chapter, we mentioned the claim made by Robert Trivers that aspects of human emotions have been molded by natural selection through kin selection or reciprocal altruism. Although behavioral geneticists are still far from confirming or refuting this claim, the data we have reviewed here do support the hypothesis that aspects of human personality and cognitive skills can respond to natural selection.

As informative as they are, selection experiments and heritability studies are relatively imprecise. Their greatest value, in fact, may lie in identifying the suite of behavioral traits and syndromes that are the most heritable (Plomin et al. 1994). Traits with significant heritabilities then become attractive candidates for highly focused "gene hunts" conducted via QTL mapping, analyses of forward mutations, or sequencing studies. We turn now to these higher-resolution genetic techniques.

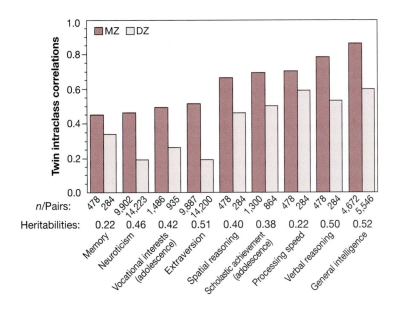

Figure 16.16 **Heritabilities in human personality and cognitive traits, estimated from twin studies** Robert Plomin and colleagues (1994) assembled this figure from data reported in the literature. The dark and light bars show the correlations among monozygotic and dizygotic twins, respectively, for each of the traits listed along the horizontal axis. Heritabilities are estimated as twice the difference in these correlations.

QTL Mapping

Quantitative trait loci (QTL) mapping, a technique we introduced in Chapter 9, is now being employed extensively in behavioral genetic studies. As an example of how the technique is applied, we review an effort to map loci that contribute to variation in the level of emotionality among mice.

In mice and rats, psychologists measure emotionality by observing an individual's activity when placed in a brightly lit, open environment. This is not a preferred habitat for rodents, presumably because individuals are more susceptible to predation in open areas, and individual mice and rats differ strongly in their response to it. Some individuals stay still and cower, while others move about and explore. Emotionality is defined as the correlation between an individual's willingness to move about in this open-field environment and its tendency to defecate. High emotionality is characterized by low movement and high defecation in the open-field habitat. Mouse and rat populations readily respond to selection for both high and low emotionality. Heritabilities are low to moderate: In mice the heritability of open-field activity has been estimated at 0.26 and defecation at 0.11. The two traits are highly correlated.

Understanding more about the genetics of emotionality is interesting because variation in the trait may have fitness consequences for rodents in the wild. Selection could favor different levels of emotionality among populations, depending on the nature of the vegetation in their habitat and predation pressure. Also, some psychologists hypothesize that emotionality in rodents may be related to human traits like susceptibility to anxiety and neuroticism.

Jonathon Flint and colleagues (1995) undertook a QTL mapping study of the trait. Their first step was to create inbred mouse strains from individuals selected for high and low emotionality scores. These populations were presumably homozygous for alleles contributing to high and low values. They crossed these two lines to create a population of F_1 heterozygotes, then backcrossed these F_1

hybrids to one of the two parental populations. They tested a total of 879 F_2 progeny in an open-field environment and scored their activity.

To search for associations between QTLs and level of activity in an open habitat, Flint et al. selected the 10% of F_2 individuals with the highest and lowest activity scores. They genotyped these mice by scoring the presence or absence of 84 microsatellite loci scattered throughout the mouse genome. The presence or absence of six microsatellite loci had a statistically significant association with open-field activity score. Three of these loci were also associated with level of defecation in the open-field environment and with the willingness of mice to explore two additional types of novel habitats (two different kinds of mazes).

Flint et al. maintain that their results show a relatively simple genetic basis for the complex trait called emotionality. Whether a mouse has or lacks these six QTLs accounts for a total of 31% of the phenotypic variation observed in the willingness of mice to explore an open-field environment. The implication is that selection or drift could act on these loci to change mouse behavior.

Though identifying QTLs allows us to estimate the number of loci involved in the expression of behavioral traits and to map their approximate location, it still leaves us fairly far from the actual genes themselves. To understand the mechanisms of gene action in behavior, we have to use techniques that target known loci.

Single-Gene Studies

Perhaps *the* classical approach to characterizing individual genes is studying the phenotypic effects of loss-of-function mutations. Forward mutations have traditionally been isolated by exposing a population to mutagens, screening F_1 offspring for mutant phenotypes, and then mapping the position of the locus responsible for each phenotype (this is done by crossing the mutants with individuals carrying known genetic markers). The general strategy is to learn about a gene's action by observing the phenotypes that result when the gene's product is missing.

A contemporary variant on this strategy has identified a locus in mice involved in parental care. Jennifer Brown and colleagues (1996) created a line of "knockout" mice missing the gene product from the *fosB* locus. In mice there are a total of four *fos* loci; all code for transcription factors that bind to DNA and influence the expression of other genes. Previous work has shown that members of the *fos* gene family are expressed in specific regions of the brain in response to environmental stimuli such as the presence of offspring. Beyond these observations, however, little was known about the function of these genes.

The knockout mutation created by Brown et al. was a gene created with recombinant techniques and inserted into embryos. By isolating proteins from transgenic animals, separating them on a gel, and probing them with antibodies to the *fosB* protein, Brown et al. were able to identify individuals that were homozygous for the mutant allele and confirm that these mice produced no *fosB* protein.

Mothers homozygous for the knockout alleles showed a striking change in their behavior: They did not care for their pups (Figure 16.17). As a result, the offspring of knockout mothers would die at age 1–2 days. In experiments

(a)

(b)

Figure 16.17 **Mouse mothers with *fosB* knockout genes do not care for their pups** Transgenic mice who do not produce the protein product of the *fosB* locus ignore their newly born pups. (a) Normal mothers make a nest, gather their pups into it, nurse them, and cover them to provide warmth. (b) Mothers who are homozygous for the knockout mutation do not show these behaviors. Reproduced by permission from J.R. Brown, H.Ye, R.T. Bronson, P. Dikkes, and M.E. Greenberg, 1996. A defect in nurturing in mice lacking the immediate early gene *fosB*. *Cell* 86(2): 297–309, July 26, 1996.

where pups were placed throughout the cage occupied by a mother, normal females quickly retrieved the young and gathered them into the nest. Knockout mothers approached and sniffed their pups but otherwise ignored them. They would not retrieve their young, did not build a nest, and did not nurse.

Brown et al. were able to show that the mutant females had normal mammary gland development and normal hormonal profiles. Rather than showing an anatomical or physiological defect, the phenotypic deficit was limited to their behavior. The research team concluded that normal *fosB* protein is a requirement for nurturing behavior.

Many questions remain about the mechanism of *fosB* gene action. Previous work indicates that the *fosB* gene may be transcribed in response to signals from olfactory receptors. If *fosB* is activated by the odor of young, what genes are then activated by the *fosB* protein? Do knockout mutants at these downstream loci produce the same phenotype as *fosB* knockouts? Further, do different alleles at the *fosB* locus affect the degree of care provided by females? If so, they would lead to genetic variation in the degree of parental care, and provide a target for the action of selection and drift.

The several genome sequencing projects now underway promise to supply a flood of data on the genes involved in sensory processing and behavioral responses. Knockout and other types of studies on gene function will continue to lend insight into the effects that specific loci have on social behavior. Combined with selection experiments and QTL analyses, these high-resolution techniques should dramatically increase our understanding of the genetic mechanisms behind the evolution of social behavior. But we are still far from being able to characterize the loci involved in the evolution of altruism, cooperation, and selfishness. Behavioral genetics has confirmed that at least a small amount of genetic variance exists in a wide variety of social traits. Beyond that, social behavior appears to be dominated by flexible decision-making and other gene-by-environment interactions.

Summary

When individuals interact, four outcomes are possible with respect to fitness: cooperation, altruism, selfishness, and spite. The evolution of altruism is one of the great paradoxes of evolutionary biology. This paradox was resolved by two important advances in evolutionary theory:

1. William Hamilton proved mathematically that a gene for altruism will spread when $Br - C > 0$, where B is the benefit to the recipient in units of surviving offspring, r is the coefficient of relatedness between actor and recipient, and C is the cost to the actor. When Hamilton's rule holds, kin selection results in altruistic behavior.
2. Robert Trivers developed the theory of reciprocal altruism. Altruism among unrelated individuals can evolve if the benefits of an altruistic act to the recipient are large and the cost to the actor is small, and the benefits are later returned to the actor by the recipient.

Kin selection and reciprocity theory have been successfully tested as explanations for alarm calling in ground squirrels, helping in birds, and food sharing in vampire bats. In conjunction with unusual genetic systems like haplodiploidy and extreme inbreeding, kin selection has also been important in explaining eusociality in Hymenoptera and naked mole-rats. Some cooperative behavior, such as group territorial defense in female lions, has been more difficult to explain in terms of kin selection or reciprocity.

Selfish behavior, like infanticide in lions and weaning conflict in mammals, occurs when the fitness interests of individuals conflict.

The success of kin selection theory confirms that social traits have responded to selection, and suggests that genetic variance exists for behavioral traits. The field of behavior genetics is devoted to exploring the extent and nature of this variation. Behavioral geneticists use selection and heritability studies to identify traits with significant genetic variance, then employ QTL mapping to assess the number and location of loci involved. Single-gene studies, often using knockout mutations, can uncover the specific function of loci influencing social behavior.

Questions

1. Suppose adult bee-eaters could raise only 0.3 more offspring with a helper than without a helper. Would you still expect male bee-eaters to "give in" to the harassment of their fathers, or would male bee-eaters tend to fight off their fathers? Explain your reasoning.
2. When a Thomson's gazelle detects a nearby stalking cheetah, the gazelle often begins bouncing up and down with a stiff-legged gait called "stotting." One hypothesis is that stotting, like the alarm calls of ground squirrels, has evolved because it may help alert the gazelle's kin to the presence of a predator. Caro (1986) reports that stotting does not seem to increase the gazelle's risk of being attacked. In fact, once a gazelle begins to stott, the cheetah often gives up the hunt. How is C (the cost of stotting) for a gazelle different from C (the cost of alarm calls) for a ground squirrel? Is it likely that stotting is an altruistic behavior? With this in mind, make a prediction about whether a gazelle will stott

when no other gazelles are around. Then look up Caro's paper to see if you are right.

> Caro, T.M. 1986. The function of stotting in Thomson's gazelles: Some tests of the hypotheses. *Animal Behavior* 34:663–684.

3. In some species it is common for young siblings to kill each other. The cubs of spotted hyenas begin fighting within moments of birth, and often one hyena cub dies. The mother hyena does not interfere. How could this behavior have evolved? For example, from the winning sibling's point of view, what must *B* (benefit of siblicide) be relative to *C* (cost of siblicide) to favor the evolution of siblicide? From the parent's point of view, what must *B* be, relative to *C*, for the parent to watch calmly rather than to interfere?

> See Frank (1997) for more information on the unusual social system of spotted hyenas, and Anderson and Ricklefs (1992) for some other examples of siblicide.

> Frank, Laurence G. 1997. Evolution of genital masculinization: Why do female hyenas have such a large "penis"? *Trends in Ecology and Evolution* 12:58–62.

> Anderson, D.J., and R.E. Ricklefs. 1992. Brood size and food provisioning in masked and blue-footed boobies. *Ecology* 73:1363–1374.

4. Which form of altruism—kin selection or reciprocal altruism—tends to favor the ability to recognize and remember different individuals?

5. The population geneticist J.B.S. Haldane was explaining kin selection to some friends in a pub (this incident actually occurred a decade or more before William Hamilton formalized the theory). As the story goes, he scribbled some calculations on an envelope and announced that he would be willing to die for two brothers or eight cousins. Explain his reasoning.

6. Suppose a honeybee father had some way of influencing the colony's behavior after his death. Would he "prefer" (evolutionarily) that more workers (daughters) than drones (sons) be produced, more drones than workers, or equal numbers of workers and drones? Why?

7. Look at Figure 16.9 on parent-offspring conflict. Explain, in general terms, why the behavior of females should evolve so that mothers start weaning when B/C falls below 1. (Hint: Consider the reproductive success of mothers who wean very early, and of mothers who wean very late.) If a mother could have only one litter of young in her lifetime, how would the period of weaning conflict change?

8. Do the knockout mice of Figure 16.17 demonstrate that the *fosB* gene is the only gene, or the most important gene, governing parental behavior in mice? Why or why not?

9. Suppose you showed Figure 16.16 to two friends. The first says "Aha, clear evidence that genes strongly affect human personality!" The second says "Aha, clear evidence that environment strongly affects human personality!" Which features of the graph have each of your friends observed? Are both statements correct?

Exploring the Literature

10. Throughout this chapter we concentrated on the fitness consequences of social interactions and paid little attention to the issue of why organisms live in groups in the first place. To learn how social living has been favored by factors such as a requirement for group defense against predators, benefits from group foraging, and a need for long-term care of dependent young, see:

Alexander, R.D. 1974. The evolution of social behavior. *Annual Review of Ecology and Systematics* 5:325–383.

Packer, C., D. Scheel, and A.E. Pusey. 1990. Why lions form groups: Food is not enough. *American Naturalist* 136:1–19.

11. Because this chapter emphasizes theories that explain why cooperative behavior can evolve when a fitness cost is involved, we spend little time on the evolution of mutualism. For an example of cooperative behavior that is *not* caused by kin selection or reciprocal altruism, see:

McDonald, D.B., and W.K. Potts. 1994. Cooperative display and relatedness among males in a lek-breeding bird. *Science* 266:1030–1032.

12. Evidence for associations between specific genes and behavior have been found in characteristics other than nurturing, including susceptibility to alcoholism and to aggression. To get started with the literature on these topics, see:

Cases, O., I. Seif, J. Grimsby, P. Gaspar, K. Chen, S. Pournin, U. Müller, M. Aguet, C. Babinet, J.C. Chih, and E. De Maeyer. 1995. Aggressive behavior and altered amounts of brain serotonin and norepinephrine in mice lack MAOA. *Science* 268:1763–1766.

Thomasson, H.R., and T-K. Li. 1993. How alcohol and aldehyde dehydrogenase genes modify alcohol drinking, alcohol flushing, and the risk for alcoholism. *Alcohol Health and Research World* 17:167–172.

Citations

Alexander, R.D., K.M. Noonan, and B.J. Crespi. 1991. The evolution of eusociality. In P.W. Sherman, J.U.M. Jarvis, and R.D. Alexander. *The Biology of the Naked Mole Rat* (Princeton, NJ: Princeton University Press), 3–44.

Axelrod, R., and W.D. Hamilton. 1981. The evolution of cooperation. *Science* 211:1390–1396.

Bartz, S.H. 1979. Evolution of eusociality in termites. *Proceedings of the National Academy of Sciences, USA* 76:5764–5768.

Barlow, G.W., and J. Silverberg, eds. 1980. *Sociobiology: Beyond Nature/Nurture?* (Washington, DC: AAAS; and Boulder, CO: Westview Press).

Betzig, L.L., and P.W. Turke. 1986. Parental investment by sex on Ifaluk. *Ethology and Sociobiology* 7:29–37.

Boomsma, J.J., and F.L.W. Ratnicks. 1996. Paternity in eusocial Hymenoptera. *Philosophical Transactions of the Royal Society of London* 351:947–975.

Bouchard, T.J. Jr., D.T. Lykken, M. McGue, N.L. Segal, and A. Tellegen. 1990. Sources of human psychological differences: The Minnesota study of twins reared apart. *Science* 250:223–228.

Bourke, A.F.G., and N.R. Franks. 1995. *Social Evolution in Ants* (Princeton, NJ: Princeton University Press).

Brown, J.L. 1987. *Helping and Communal Breeding in Birds* (Princeton, NJ: Princeton University Press).

Brown, J.L., and A. Eklund. 1994. Kin recognition and the major histocompatibility complex: An integrative review. *American Naturalist* 143:435–461.

Brown, J.R., H. Ye, R.T. Bronson, P. Dikkes, and M.E. Greenberg. 1996. A defect in nurturing in mice lacking the immediate early gene *fosB*. *Cell* 86:297–309.

Cameron, S.A. 1993. Multiple origins of advanced eusociality in bees inferred from mitochondrial DNA sequences. *Proceedings of the National Academy of Sciences, USA* 90:8687–8691.

Charnov, E.L. 1982. *The Theory of Sex Allocation* (Princeton: Princeton University Press).

Clutton-Brock, T.H., S.D. Albon, and F.E. Guinness. 1984. Maternal dominance, breeding success and birth sex ratios in red deer. *Nature* 308:358–360.

Crozier, R.H., and P. Pamilo. 1996. *Evolution of Social Insect Colonies* (Oxford University Press).

Darwin, C. 1859. *The Origin of Species* (London: John Murray).

Duffy, J.E. 1996. Eusociality in a coral-reef shrimp. *Nature* 381:512–514.

Dugatkin, L.A. 1988. Do guppies play tit for tat during predator inspection visits? *Behavioral Ecology and Sociobiology* 25:395–399.

Emlen, S.T., and P.H. Wrege. 1988. The role of kinship in helping decisions among white-fronted bee-eaters. *Behavioral Ecology and Sociobiology* 23:305–315.

Emlen, S.T., and P.H. Wrege. 1991. Breeding biology of white-fronted bee-eaters at Nakuru: The influence of helpers on breeder fitness. *Journal of Animal Ecology* 60:309–326.

Emlen, S.T., and P.H. Wrege. 1992. Parent-offspring conflict and the recruitment of helpers among bee-eaters. *Nature* 356:331–333.

Emlen, S.T., P.H. Wrege, and N.J. Demong. 1995. Making decisions in the family: An evolutionary perspective. *American Scientist* 83:148–157.

Evans, J. 1993. Parentage analysis in ant colonies using simple sequence repeat loci. *Molecular Ecology* 2:393–397.

Falconer, D.S. 1953. Selection for large and small size in mice. *Journal of Genetics* 51:470–501.

Falconer, D.S. 1955. Patterns of response in selection experiments in mice. *Cold Spring Harbor Symposium on Quantum Biology* 20:178–196.

Fisher, R.A. 1930. *The Genetical Theory of Natural Selection* (Oxford: Clarendon Press).

Flint, J., R. Corley, J.C. DeFries, D.W. Fulker, J.A. Gray, S. Miller, and A.C. Collins. 1995. A simple genetic basis for a complex psychological trait in laboratory mice. *Science* 269:1432–1435.

Godfrey, H.C.J. 1995. Evolutionary theory of parent-offspring conflict. *Nature* 376:133–138.

Hamaguchi, K., Y. Itô, and O. Takenaka. 1993. GT dinucleotide repeat polymorphisms in a polygynous ant, *Leptothorax spinosior* and their use for measurement of relatedness. *Naturwissenschaften* 80:179–181.

Hamilton, W.D. 1964. The genetical evolution of social behaviour. I. *Journal of Theoretical Biology* 7:1–16.

Hamilton, W.D. 1964. The genetical evolution of social behaviour. II. *Journal of Theoretical Biology* 7:17–52.

Hamilton, W.D. 1967. Extraordinary sex ratios. *Science* 156:477–488.

Hamilton, W.D. 1972. Altruism and related phenomena, mainly in the social insects. *Annual Review of Ecology and Systematics* 3:193–232.

Hammerstein, P., and S.E. Riechert. 1988. Payoffs and strategies in territorial contests: ESS analyses of two ecotypes of the spider *Agelenopsis aperta*. *Evolutionary Ecology* 2:115–138.

Hansen, A.J. 1986. Fighting behavior in bald eagles: A test of game theory. *Ecology* 67:787–797.

Hegner, R.E., S.T. Emlen, and N.J. Demong. 1982. Spatial organization of the white-fronted bee-eater. *Nature* 298:264–266.

Heinsohn, R., and C. Packer. 1995. Complex cooperative strategies in group-territorial African lions. *Science* 269:1260–1262.

Hepper, P.G., ed. 1991. *Kin Recognition* (Cambridge University Press).

Hoffmann, A.A. 1988. Heritable variation for territorial success in two *Drosophila melanogaster* populations. *Animal Behaviour* 36:1180–1189.

Honeycutt, R.L. 1992. Naked mole-rats. *American Scientist* 80:43–53.

Keller, L., M. Milinski, M. Frischknecht, N. Perrin, H. Richner, and F. Tripet. 1994. Spiteful animals still to be discovered. *Trends in Ecology and Evolutionary Biology* 9:103.

Manning, C.J., E.K. Wakeland, and W.K. Potts. 1992. Communal nesting patterns in mice implicate MHC genes in kin recognition. *Nature* 360:581–583.

Maynard Smith, J. 1974. The theory of games and the evolution of animal conflicts. *Journal of Theoretical Biology* 47:209–221.

Maynard Smith, J. 1976. Evolution and the theory of games. *American Scientist* 64:41–45.

Maynard Smith, J. 1982. *Evolution and the Theory of Games* (Cambridge University Press).

Michener, C.D. 1969. Comparative social behavior of bees. *Annual Review of Entomology* 14:299–342.

Milinski, M. 1987. Tit for tat and the evolution of cooperation in sticklebacks. *Nature* 325:433–435.

Milinski, M. 1996. By-product mutualism, tit-for-tat reciprocity and cooperative predator inspection: A reply to Connor. *Animal Behaviour* 51:458–461.

Nowak, M., and K. Sigmund. 1993. A strategy of win-stay, lose-shift that outperforms tit-for-tat in the Prisoner's Dilemma game. *Nature* 364:56–58.

Packer, C. 1977. Reciprocal altruism in *Papio anubis*. *Nature* 265:441–443.

Packer, C., and A.E. Pusey. 1982. Cooperation and competition within coalitions of male lions: Kin selection or game theory? *Nature* 296:740–742.

Packer, C., and A.E. Pusey. 1983. Adaptations of female lions to infanticide by incoming males. *American Naturalist* 121:716–728.

Packer, C., D. Scheel, and A.E. Pusey. 1990. Why lions form groups: Food is not enough. *American Naturalist* 136:1–19.

Packer, C., D.A. Gilbert, A.E. Pusey, and S.J. O'Brien. 1991. A molecular genetic analysis of kinship and cooperation in African lions. *Nature* 351:562–565.

Page, R.E. Jr., and R.A. Metcalf. 1982. Multiple mating, sperm utilization, and social evolution. *American Naturalist* 119:263–281.

Pamilo, P. 1984. Genetic relatedness and evolution of insect sociality. *Behavioral Ecology and Sociobiology* 15:241–248.

Pfennig, D.W., H.K. Reeve, and P.W. Sherman. 1993. Kin recognition and cannibalism in spadefoot toad tadpoles. *Animal Behaviour* 46:87–94.

Pfennig, D.W., and P.W. Sherman. 1995. Kin recognition. *Scientific American* 272:98–103.

Plomin, R., M.J. Owen, and P. McGuffin. 1994. The genetic basis of complex human behaviors. *Science* 264:1733–1739.

Queller, D.C. 1989. The evolution of eusociality: Reproductive head starts of workers. *Proceedings of the National Academy of Sciences, USA* 86:3224–3226.

Queller, D.C., and K.F. Goodnight. 1989. Estimating relatedness using genetic markers. *Evolution* 43:258–275.

Queller, D.C., J.E. Strassmann, and C.R. Hughes. 1993. Microsatellites and kinship. *Trends in Ecology and Evolution* 8:285–288.

Reeve, H.K. 1992. Queen activation of lazy workers in colonies of the eusocial naked mole-rat. *Nature* 358:147–149.

Reeve, H.K., D.F. Westneat, W.A. Noon, P.W. Sherman, and C.F. Aquadro. 1990. DNA "fingerprinting" reveals high levels of inbreeding in colonies of the eusocial naked mole-rat. *Proceedings of the National Academy of Sciences, USA* 87:2496–2500.

Richerson P.J., and R. Boyd. 1984. Natural selection and culture. *BioScience* 34:430–434.

Riechert, S.E. 1978. Games spiders play: Behavioral variability in territorial disputes. *Behavioral Ecology and Sociobiology* 3:135–162.

Riechert, S.E. 1979. Games spiders play II. Resource assessment strategies. *Behavioral Ecology and Sociobiology* 6:121–128.

Roisin, Y. 1994. Intragroup conflicts and the evolution of sterile castes in termites. *American Naturalist* 143:751–765.

Sherman, P.W. 1977. Nepotism and the evolution of alarm calls. *Science* 197:1246–1253.

Sherman, P.W. 1980. The limits of ground squirrel nepotism. In George W. Barlow and J. Silverberg, eds. *Sociobiology: Beyond Nature/Nurture?* (Washington, DC: AAAS; and Boulder, CO: Westview Press), 505–544.

Sherman, P.W. 1981. Kinship, demography, and Belding's ground squirrel nepotism. *Behavioral Ecology and Sociobiology* 8:251–259.

Sherman, P.W., J.U.M. Jarvis, and R.D. Alexander. 1991. *The Biology of the Naked Mole Rat* (Princeton, NJ: Princeton University Press).

Sherman, P.W., J.U.M. Jarvis, and S.H. Braude. 1992. Naked mole rats. *Scientific American* 267:72–78.

Sigurjónsdóttir, H., and G.A. Parker. 1981. Dung fly struggles: Evidence for assessment strategy. *Behavioral Ecology and Sociobiology* 8:219–230.

Stacey, P.B., and W.D Koenig, eds. 1990. *Cooperative Breeding in Birds: Long-term Studies of Ecology and Behaviour* (Cambridge University Press).

Sundström, L., M. Chapuisat, and L. Keller. 1996. Conditional manipulation of sex ratios by ant workers: A test of kin selection theory. *Science* 274:993–995.

Trivers, R.L. 1971. The evolution of reciprocal altruism. *Quarterly Review of Biology* 46:35–57.

Trivers, R.L. 1974. Parent-offspring conflict. *American Zoologist* 14:249–264.

Trivers, R.L. 1985. *Social Evolution* (Menlo Park, CA: Benjamin Cummings).

Trivers, R.L., and D.E. Willard. 1973. Natural selection of parental ability to vary the sex ratio of offspring. *Science* 179:90–92.

Trivers, R.L., and H. Hare. 1976. Haplodiploidy and the evolution of the social insects. *Science* 191:249–263.

Wedekind, C., and M. Milinski. 1996. Human cooperation in the simultaneous and the alternating Prisoner's Dilemma: Pavlov versus generous tit-for-tat. *Proceedings of the National Academy of Sciences, USA* 93:2686–2689.

Werren, J.H. 1980. Sex ratio adaptations to local mate competition in a parasitic wasp. *Science* 208:1157–1159.

Werren, J.H. 1984. A model for sex ratio selection in parasitic wasps: Local mate competition and host quality effects. *Netherlands Journal of Zoology* 34:81–96.

Wilkinson, G.S. 1984. Reciprocal food sharing in the vampire bat. *Nature* 308:181–184.

Wilson, E.O. 1971. *The Insect Societies* (Cambridge, MA: Harvard University Press).

Aging and Other Life History Characters

Evolution by natural selection has engineered all organisms to perform the same single ultimate task: to reproduce. Nonetheless, organisms employ a rich variety of reproductive strategies. A few examples illustrate the diversity:

- Some mammals mature early and reproduce quickly, whereas others mature late and reproduce slowly. For example, female deer mice *(Peromyscus maniculatus)* mature at about seven weeks and have three or four litters of pups each year, whereas female black bears *(Ursus americanus)* mature at four or five years and produce cubs only once every two years (Nowak 1991).

- Plants have a wide range of reproductive life spans. Some, like the California poppy *(Eschscholtzia californica)*, live and flower for just a single season. Others, like the black cherry *(Prunus serotina)*, flower yearly for decades.

- Some bivalves produce enormous numbers of tiny eggs, whereas others produce small numbers of large eggs (Strathmann 1987). The oyster *Crassostrea gigas*, for example, releases 10 to 50 million eggs in a single spawn, each 50 to 55 micrometers in diameter. The clam *Lasaea subviridis*, in contrast, broods fewer than 100 eggs at a time, each some 300 micrometers in diameter.

The branch of evolutionary biology that attempts to make sense of the diversity in reproductive strategies is life history analysis.

An organism truly perfected for reproduction would mature at birth, continuously produce high-quality offspring in large numbers, and live forever. No such organism exists, even after 3.8 billion years of evolution by natural selection. The reason is that such an organism is impossible. Some actual organisms come close to realizing one or another of the traits of an ideal reproducer, but all such organisms fall strikingly short by one or more of the remaining measures. For example, the female thrips egg mite *(Adactylidium* sp.) is mature at birth. Furthermore, she is already inseminated, having hatched inside her mother's body and mated with her brother (Elbadry and Tawfik 1966; see also Gould 1980, essay 6). But she dies at the age of just 4 days, when her own offspring eat her alive from the inside (Figure 17.1a). Another example, the brown kiwi *(Apteryx australis mantelli)* produces high-quality offspring (Taborsky and Taborsky 1993). Female kiwis, which weigh six pounds, lay eggs that weigh one pound (Figure 17.1b). The chicks that hatch from these huge eggs become largely self-reliant within a week. However, kiwi parents cannot produce these chicks continuously, and they cannot produce them in large numbers. It takes the female over a month to make each of the eggs in a typical two-egg clutch. The male has to incubate the eggs for about three months, during which time he loses some 20% of his body weight.

Organisms face fundamental trade-offs in their use of energy and time.

As the egg mite and the kiwi suggest, the laws of physics and biology impose fundamental trade-offs. The amount of energy an organism can harvest is finite, and biological processes take time. Energy and time devoted to one activity are energy and time that cannot be devoted to another. For example, an individual can allocate energy to growth for a long time, which may enable it to reach a larger size and ultimately enable it to produce more offspring. This benefit of large size, however, is balanced by a cost. The time required to grow to a large size is time during which predators, diseases, or accidents may strike. An individual that takes the time to grow to a large size thus incurs a greater risk of dying without ever having reproduced at all. We introduced the concept of trade-offs in Chapter 8, and we discussed how trade-offs constrain the evolution of adaptations. Whenever there is a trade-off between different components of fitness, we expect natural selection to favor individuals that allocate energy and time with an optimal balance between benefits and costs, thereby maximiz-

(a)   (b)

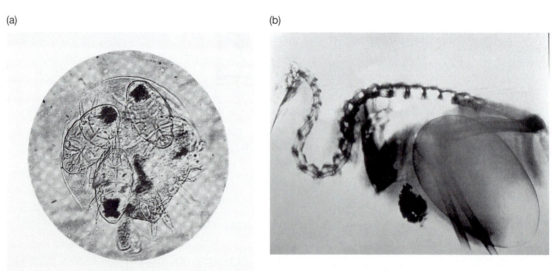

***Figure 17.1* Extreme reproductive strategies**   (a) Having devoured their mother from the inside, three thrips egg mites *(Adactylidium* sp.) prepare to depart her empty cuticle. The mother's legs are visible at lower right (180×). Reproduced by permission from E. A. Elbadry and M.S.F. Tawfik, 1966. Life cycle of the mite *Adactylidium* sp. (Acarina: Pyemotidae), a predator of thrips eggs in the United Arab Republic. *Annals of the Entomological Society of America* 59(3): 458–61, May 1966, p. 460, Fig. 6.   (b) An x-ray of a female brown kiwi *(Apteryx australis mantelli)* ready to lay an egg (Otorohanga Kiwi House, New Zealand).

ing lifetime reproductive success. Different balances are optimal in different environments.

In exploring the evolution of life histories, we analyze costs and benefits, and fitness trade-offs, as they apply to the following questions:

- Why do organisms age and die?
- How many offspring should an individual produce in a given year?
- How big should each offspring be?

In the final section, we place life history analysis in a broader evolutionary context by considering the maintenance of genetic variation and evolutionary transitions in life history.

17.1 Basic Issues in Life History Analysis

An example of a life history, one that we will return to near the end of section 17.2, appears in Figure 17.2. The figure follows the career of a hypothetical female Virginia opossum *(Didelphis virginiana)*. As a baby, this female nursed for a little more than three months, then was weaned and became independent. She continued to grow for another several months, reaching sexual maturity at an age of about $10\frac{1}{2}$ months. Shortly thereafter the female had her first litter, with eight offspring. A few months later she had a second litter, this time with seven offspring. At the age of 20 months, the female was killed by a predator.

The figure also indicates where the female opossum got her energy from at different stages of her life, and the functions to which she allocated that finite energy supply. Before she became sexually mature, the female used her energy for growth, metabolic functions like thermoregulation, and the repair of damaged tissues. After she became sexually mature the female stopped growing, thereafter using her energy for metabolism, repair, and reproduction.

Fundamentally, differences among life histories concern differences in the allocation of energy. For example, a different female opossum than the one shown in Figure 17.2 might stop allocating energy to growth at an earlier age, thereby reaching sexual maturity more quickly. This strategy involves a trade-off: The female also matures at a smaller size, which means that she has to produce smaller litters. Still another female might, after reaching sexual maturity, allocate less energy to reproduction and more to repair, thereby keeping her

Differences among life histories concern differences in the allocation of energy.

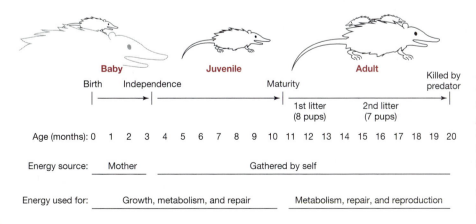

Figure 17.2 The life history of a hypothetical female Virginia opossum *(Didelphis virginiana)* This hypothetical female has a life history typical of female Virginia opossums living in the mainland United States (Austad 1988; 1993). Figure designed after Charnov and Berrigan (1993).

tissues in better condition. Again there is a trade-off: Allocating less energy to reproduction means having smaller litters. Natural selection acts on life histories to adjust energy allocation in a way that maximizes the total lifetime production of offspring.

17.2   Why Do Organisms Age and Die?

Aging, or senescence, is a late-life decline in an individual's fertility and/or probability of survival (Partridge and Barton 1993). Figure 17.3 documents aging in three animal species: a bird, a mammal, and an insect. All three show de-

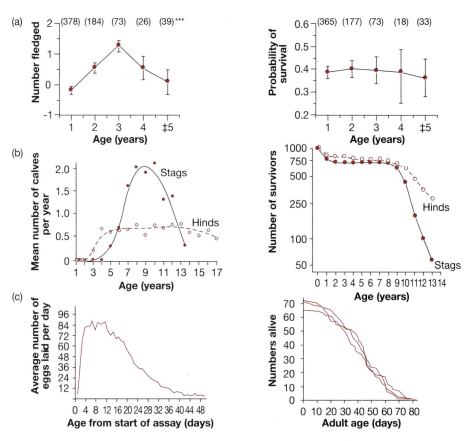

Figure 17.3   **Aging in three animals**   (a) Aging in a bird. For females in a wild population of collared flycatchers *(Ficedula albicollis),* the number of young fledged each year declines after age three (age-related differences in number fledged are significant at P < 0.001). The probability of survival from one year to the next declines slightly, but not significantly, after age two. Sample sizes in parentheses. From Gustafsson and Pärt (1990). (b) Aging in a mammal. For males in a wild population of red deer *(Cervus elaphus),* the number of calves fathered each year declines sharply after a peak at about age nine; for females, the number of calves produced each year declines gradually starting at about age 13. For both sexes, the probability of surviving from one year to the next is nearly 100% from age two to age nine. After age nine, the probability of surviving plummets. From Clutton-Brock et al. (1988).   (c) Aging in an insect. For females in a laboratory population of fruit flies *(Drosophila melanogaster),* the average number of eggs laid per day declines after the age of about 12 days. In three laboratory populations, the probability of surviving from one day to the next falls at the age of about 20 days. Modified from Rose (1984).

clines in both fertility and survival. All else being equal, aging reduces an individual's fitness. Aging should therefore be opposed by natural selection.

Aging should be opposed by natural selection.

We consider two theories on why aging persists. The first, called the rate-of-living theory, invokes an evolutionary constraint (see Chapter 8); it posits that populations lack the genetic variation to respond any further to selection against aging. The second, called the evolutionary theory, invokes, in part, a trade-off between the allocation of energy to reproduction versus repair.

The Rate-of-Living Theory of Aging

The rate-of-living theory of senescence holds that aging is caused by the accumulation of irreparable damage to cells and tissues (reviewed in Austad and Fischer 1991). Damage to cells and tissues is caused by errors during replication, transcription, and translation, and by the accumulation of poisonous metabolic by-products. Under the rate-of-living theory, all organisms have been selected to resist and repair cell and tissue damage to the maximum extent physiologically possible. They have reached the limit of biologically possible repair. In other words, populations lack the genetic variation that would enable them to evolve more effective repair mechanisms than they already have.

The rate-of-living theory makes two predictions: (1) because cell and tissue damage is caused in part by the by-products of metabolism, the aging rate should be correlated with the metabolic rate; and (2) because organisms have been selected to resist and repair damage to the maximum extent possible, species should not be able to evolve longer life spans, whether subjected to natural or artificial selection.

One theory holds that aging is a function of metabolic rate . . .

Steven Austad and Kathleen Fischer (1991) tested the first prediction, that aging rate will be correlated with metabolic rate, using comparative data on a diversity of mammals. With data from the literature, Austad and Fischer calculated the amount of energy expended per gram of tissue per lifetime for 164 mammal species in 14 orders. According to the rate-of-living theory, all species should expend about the same amount of energy per gram per lifetime, whether they burn it slowly over a long lifetime or rapidly over a short lifetime. In fact, there is wide variation among mammal species (Figure 17.4). Across all 164 species Austad and Fischer surveyed, energy expenditure ranges from 39 kilocalories per gram per lifetime in an elephant shrew to 1,102 kcal/g/lifetime in a bat. Even within orders, energy expenditure varies greatly. The range for bat species runs from 325 to 1,102 kcal/g/lifetime. As a group, bats have metabolic rates that are similar to those of other mammals of the same size, but life spans that average nearly three times longer. Energy expenditure for marsupial species ranges from 43 to 406 kcal/g/lifetime. As a group, marsupials have metabolic rates that are significantly lower than those of other mammals of the same size ($P = 0.001$), but life spans that are significantly shorter ($P < 0.001$). These patterns are not consistent with the rate-of-living theory of aging.

. . . but data on variation in metabolic rate and aging among mammals refute this theory.

Leo Luckinbill and colleagues (1984) tested the second prediction, that species cannot evolve longer life spans, by artificially selecting for longevity in laboratory populations of fruit flies *(Drosophila melanogaster)*. Luckinbill et al. collected wild flies and used them to establish four laboratory populations. In two populations, the researchers selected for early reproduction by collecting eggs from young

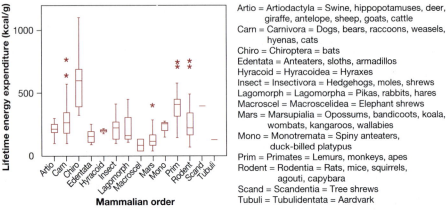

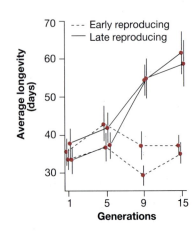

Figure 17.4   **Variation among mammals in lifetime energy expenditure**   This box plot represents the range of lifetime energy expenditure within each of 14 orders of mammals. The horizontal line dividing each box represents the median value for that order. The top and bottom of each box represent the 75th and 25th percentiles. The vertical lines extending above and below each box represent the range of values; the asterisks represent statistical outliers. From Austad and Fischer (1991).

adults (two to six days after eclosion) and using the individuals that hatched from these eggs as the next generation's breeders. Longevity in these populations did not change during 13 generations of selection (Figure 17.5). In the other two populations, Luckinbill and colleagues selected for late reproduction by collecting eggs from old adults. "Old" meant 22 days after eclosion at the beginning of the experiment, and 58 days after eclosion by the end. Longevity in these populations increased dramatically during 13 generations of selection (Figure 17.5). At the beginning of the experiment, the average life span of the flies in these populations was about 35 days; by the end, the average life span was about 60 days. Other researchers conducting similar experiments have confirmed that average life span increases in *Drosophila* populations in response to selection for late-life reproduction (Rose 1984; Partridge and Fowler 1992; Roper et al. 1993). These results are consistent with the rate-of-living theory of aging only if the long-lived populations have evolved lower metabolic rates. Phillip Service (1987) found that fruit flies selected for long life span indeed had lower metabolic rates

Selection experiments on fruitflies show that populations can evolve longer life spans.

Figure 17.5   **Artificial selection increases life span in fruit flies**   The graph shows average life span in each of four laboratory populations of *Drosophila melanogaster* over 13 generations of selection. The vertical lines show the 95% confidence intervals for the estimated population averages. From Luckinbill et al. (1984).

than controls, but only in the first 15 days of life. It is not clear that an evolved difference in metabolic rate can explain an evolved difference in life span as large as that obtained by Luckinbill and colleagues.

The Evolutionary Theory of Aging

The results discussed so far present a paradox. Fruit fly populations can evolve longer life spans, and bat species apparently have evolved longer life spans than other eutherian mammals. If natural selection can lead to longer life spans, why has it not done so in all species? The evolutionary theory of senescence offers two related mechanisms to resolve this paradox (Medawar 1952; Williams 1957; Hamilton 1966; Partridge and Barton 1993; Neese and Williams 1995). Under the evolutionary theory, aging is caused not so much by cell and tissue damage itself as by the failure of organisms to completely repair such damage. George C. Williams argues that complete repair ought to be physiologically possible (Williams 1957; Neese and Williams 1995). Given that organisms are capable of constructing themselves via a complex process of development, they should in principle be capable of the easier job of maintaining the organs and tissues thus formed. Indeed, organisms do have remarkable abilities to replace or repair damaged parts. Yet in many organisms repair is incomplete. Under the evolutionary theory of senescence, the failure to completely repair damage is ultimately caused by either (1) deleterious mutations; or (2) trade-offs between repair and reproduction.

Figure 17.6 uses a simple genetic and demographic model of a hypothetical population to show how deleterious mutations and/or trade-offs can lead to the evolution of senescence. The figure follows the life histories of individuals in the population from birth until death. Individuals in the population are always at risk of death due to accidents, predators, and diseases. Except where noted, the probability that an individual will survive from one year to the next is 0.8.

An analysis of changes in the strength of natural selection as function of age sets the stage for explaining why populations that could evolve longer life spans have failed to do so.

Figure 17.6a tracks the life histories of individuals with the wild-type genotype. The columns of the table are as follows:

- The first column lists ages.
- The second column indicates the fraction of all wild-type zygotes born that are still alive at each age. From age 1 onward, each number in the second column is simply the number immediately above it multiplied by 0.8. Few individuals survive to the age of 15. To keep the size of the table reasonable, we assume that all individuals that survive until their 16th birthday die before reproducing that year (this assumption affects only about 3% of the population).
- The third column shows that wild-type individuals reach reproductive maturity at age 3. Once the organisms start to reproduce at age 3, they each have one offspring every year (for as long as they survive).
- The fourth column shows the expected reproductive success at each age for wild-type zygotes. The expected reproductive success at age 5, for example, is simply the fraction of zygotes that will survive to age 5 multiplied by the number of offspring each survivor will have at age 5. The sum of the numbers in this column gives the expected lifetime reproductive success of wild-type zygotes.

(a)   Wild type matures at age 3 and dies at age 16; prior to age 16 annual rate of survival = 0.8

Age	Fraction of Zygotes Surviving	RS of Survivors	Expected RS for Zygotes
0	1.000	0	0.000
1	0.800	0	0.000
2	0.640	0	0.000
3	0.512	1	0.512
4	0.410	1	0.410
5	0.328	1	0.328
6	0.262	1	0.262
7	0.210	1	0.210
8	0.168	1	0.168
9	0.134	1	0.134
10	0.107	1	0.107
11	0.086	1	0.086
12	0.069	1	0.069
13	0.055	1	0.055
14	0.044	1	0.044
15	0.035	1	0.035

Expected lifetime RS: 2.419

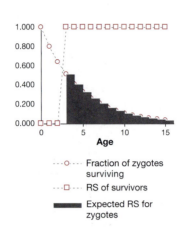

·· ○ ··· Fraction of zygotes surviving

·· □ ··· RS of survivors

▬ Expected RS for zygotes

(b)   Mutation that causes death at age 14; prior to age 14 annual rate of survival = 0.8

Age	Fraction of Zygotes Surviving	RS of Survivors	Expected RS for Zygotes
0	1.000	0	0.000
1	0.800	0	0.000
2	0.640	0	0.000
3	0.512	1	0.512
4	0.410	1	0.410
5	0.328	1	0.328
6	0.262	1	0.262
7	0.210	1	0.210
8	0.168	1	0.168
9	0.134	1	0.134
10	0.107	1	0.107
11	0.086	1	0.086
12	0.069	1	0.069
13	0.055	1	0.055
14	0.000	1	0.000
15	0.000	1	0.000

Expected lifetime RS: 2.340

·· ○ ··· Fraction of zygotes surviving

·· □ ··· RS of survivors

▬ Expected RS for zygotes

(c)   Mutation that causes maturation at age 2 and death at age 10; prior to age 10 annual rate of survival = 0.8

Age	Fraction of Zygotes Surviving	RS of Survivors	Expected RS for Zygotes
0	1.000	0	0.000
1	0.800	0	0.000
2	0.640	1	0.640
3	0.512	1	0.512
4	0.410	1	0.410
5	0.328	1	0.328
6	0.262	1	0.262
7	0.210	1	0.210
8	0.168	1	0.168
9	0.134	1	0.134
10	0.000	1	0.000
11	0.000	1	0.000
12	0.000	1	0.000
13	0.000	1	0.000
14	0.000	1	0.000
15	0.000	1	0.000

Expected lifetime RS: 2.663

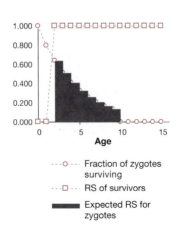

·· ○ ··· Fraction of zygotes surviving

·· □ ··· RS of survivors

▬ Expected RS for zygotes

Figure 17.6 **A simple genetic model reveals two mechanisms through which aging can evolve**   (a) The life history of individuals with the wild-type genotype. Through age 15 these individuals have a constant probability of 0.8 of surviving from one year to the next. This leads to an exponential decline over time in the fraction of individuals still alive (column 2 of table; open circles in graph). Individuals mature at age 3, from which time they have one offspring each year (column 3 of table; open squares in graph). The product of the fraction of individuals surviving to a given age and the number of offspring the survivors have at that age gives the expected reproductive success that wild-type zygotes will have when they reach that age (column 4 of table; shaded bars in graph). The sum of column 4 of the table (the area of the shaded region of the graph) gives the expected lifetime reproductive success of wild-type zygotes: 2.419.   (b) The life history of individuals with a mutation that causes death at age 14. The expected lifetime reproductive success of zygotes with this mutation is 2.340. The mutation is deleterious, but not extremely so. (c) The life history of individuals with a mutation that causes maturation at age 2 and death at age 10. The expected lifetime reproductive success of zygotes with this mutation is 2.663. The mutation is advantageous. See text for more details.

The numbers in the table are plotted in the graph; the expected lifetime reproductive success of the wild-type zygotes is equal to the area of the shaded region. The expected lifetime reproductive success of wild-type individuals is about 2.42.

We now consider two mutations that change the life histories of the individuals that carry them. If we imagine that these mutations are dominant in their effects, then our consideration covers both homozygotes and heterozygotes. If we imagine that the mutations are recessive, then our consideration covers only homozygotes.

Deleterious Mutations and Aging: The Mutation Accumulation Hypothesis

Figure 17.6b depicts a mutation that causes death at age 14. In other words, the mutation causes premature senescence. All other aspects of life history are unchanged. The mutation is obviously deleterious, but how strongly will it be selected against? As shown in the table and graph, the expected lifetime reproductive success of zygotes with the mutation is about 2.34. This is over 96% of the fitness of wild-type zygotes. Because few zygotes survive to age 14 anyway, zygotes carrying the mutation causing death at 14 do not, on average, suffer much of a penalty. The mutation is not selected against very strongly.

At first glance it is a bit surprising that a mutation causing death is only mildly deleterious. Many mutations that cause death are, in fact, highly deleterious. A mutation causing death at age 2, for example, would be selected against strongly. Zygotes carrying such a mutation would have an expected lifetime reproductive success of zero. But mutations causing death after reproduction has begun are selected against less strongly. The later in life such mutations exert their deleterious effects, the more weakly they are selected against. Mutations that are selected against only weakly can persist in mutation-selection balance (see Chapter 5). The accumulation in populations of deleterious mutations whose effects occur only late in life is one evolutionary explanation for aging (Medawar 1952).

Because selection is inherently weaker late in life, mutations that cause late-life declines in fecundity or survival are only mildly deleterious; such mutations may persist in mutation-selection balance, or even rise to high frequency by drift.

What kind of mutation could cause death, but only at an advanced age? Perhaps a mutation that reduces an organism's ability to maintain itself in good repair. Humans provide an example. Among the kinds of cellular damage that humans (and other organisms) must repair are DNA mismatch errors. Mismatched nucleotide pairs can be created by mistakes during DNA replication, or they can be induced by chemical damage to DNA (Vani and Rao 1996). Repair of these errors is performed by a suite of special enzymes. Germ-line mutations in the genes that code for these enzymes can result in the accumulation of mismatch errors, which in turn can result in cancer.

Germ-line mutations in DNA mismatch repair genes cause a form of cancer in humans called hereditary non-polyposis colon cancer (Eshleman and Markowitz 1996; Fishel and Wilson 1997). In one study, the age at which individuals were diagnosed with hereditary non-polyposis colon cancer ranged from 17 to 92. The median age of diagnosis was 48 (RodriguezBigas et al. 1996). Thus most people carrying mutations in the genes for DNA mismatch repair enzymes do not suffer the deleterious consequences of the mutations until well after the age at which reproduction begins. In an evolutionary sense, hereditary non-polyposis colon cancer is a manifestation of senescence that is caused by deleterious mutations. These deleterious mutations persist in populations because they reduce survival only late in life.

Trade-Offs and Aging: The Antagonistic Pleiotropy Hypothesis

Figure 17.6c depicts a mutation that affects two different life history characters. That is, the mutation is pleiotropic. The mutation causes reproductive maturation at age 2 instead of age 3, and the mutation causes death at age 10. In other words, the mutation involves a trade-off between reproduction early in life and survival late in life; its pleiotropic effects are antagonistic.

As shown in the table and graph, the expected lifetime reproductive success of zygotes with the mutation is about 2.66. This is 1.1 times the expected lifetime reproductive success of wild-type zygotes. Most of the zygotes born with the mutation will live long enough to reap the benefit of earlier reproduction, but few will survive long enough to pay the cost of early aging. This mutation that causes both early maturation and early senescence is therefore favored by selection. Selection for alleles with pleiotropic effects that are advantageous early in life and deleterious late in life is a second evolutionary explanation for aging (Williams 1957).

Because selection is inherently weaker late in life, mutations that enhance early-life reproduction may be favorable even if they also hasten death.

What kind of mutation could increase reproduction early in life at the same time it reduced reproduction or survival late in life? Perhaps a mutation that causes less energy to be allocated to repair early in life, and more energy to be allocated to reproduction (see Figure 17.2). Specific mutations with this kind of pleiotropic action remain to be discovered. Evolutionary biologists have, however, found evidence that there is indeed a trade-off between reproduction early in life and reproduction and/or survival late in life. In the paragraphs that follow, we review three examples.

Michael Rose (1984) studied trade-offs in the fruit fly *Drosophila melanogaster.* He conducted an artificial selection experiment similar to that of Luckinbill and colleagues (1984), described earlier. From a single ancestral laboratory population, Rose established several experimental populations. In half of these populations Rose selected for early reproduction, and in half he selected for late reproduction. Figure 17.7 shows the results. After 15 generations, the populations had diverged substantially in life span (Figure 17.7a). In addition, the populations had diverged in the age-dependence of fecundity, measured as the number of eggs laid per day by females. Flies in the populations selected for late reproduction had lower fecundity early in life and higher fecundity late in life than the flies in the populations selected for early reproduction (Figure 17.7b and c). These results indicate that there is a genetically based trade-off between reproduction early in life and reproduction and survival late in life. In other words, the flies in the early-reproduction populations enjoyed increased early reproductive success at the expense of more rapid senescence in both reproduction and survival (for alternative interpretations, see Partridge 1987; Partridge and Fowler 1992; Roper et al. 1993).

Lars Gustafsson and Tomas Pärt (1990) studied trade-offs in a bird, the collared flycatcher *(Ficedula albicollis).* Working for 10 years with a flycatcher population on the Swedish island of Gotland, Gustafsson and Pärt followed the life histories of individual birds from hatching to death. The researchers found that some female flycatchers begin breeding at age 1, whereas others wait until age 2. The females that breed at age 1 have smaller clutch sizes throughout life (Figure 17.8a), indicating that there is a cost later in life to breeding early. To further

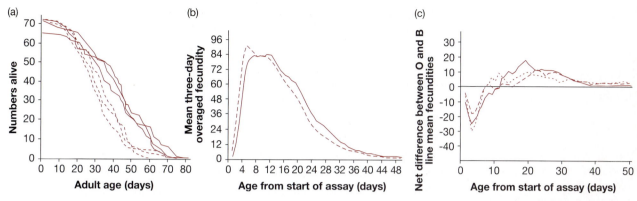

Figure 17.7   Artificial selection in fruit flies demonstrates a trade-off between early-life and late-life reproduction
(a) Number of individuals surviving as a function of age in samples from populations of *Drosophila melanogaster* selected for early reproduction (dashed lines) or late reproduction (solid lines).   (b) Fecundity as a function of age in females from populations se-lected for early reproduction (dashed lines) or late reproduction (solid lines). Fecundity is the number of eggs laid per day; the lines have been smoothed by taking a three-day running average.   (c) Difference in fecundity as a function of age in females se-lected for early versus late reproduction. Each line shows the difference between the fecundity of females in a late-reproduction population and the fecundity of females in an early-reproduction population. Values below zero indicate that the early-reproduc-tion females have higher fecundity; values above zero indicate that late-reproduction females have higher fecundity. From Rose (1984).

investigate this interpretation, Gustafsson and Pärt manipulated early reproduc-tive effort by giving some first-year breeders extra eggs. The females given extra eggs had progressively smaller clutch sizes in subsequent years, whereas control females did not begin to show reproductive senescence until age 4 (Figure 17.8b). Gustafsson and Pärt conclude that there is a trade-off in collared fly-catchers between early-life and late-life reproduction. The researchers note that despite this trade-off, first-year breeders had higher lifetime reproductive success

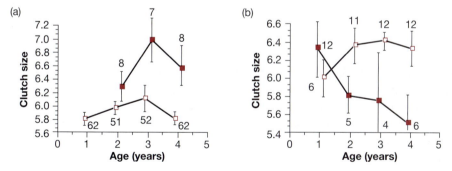

Figure 17.8   In collared flycatchers *(Ficedula albicollis)*, natural variation and ex-perimental manipulations demonstrate a trade-off between reproduction early in life versus reproduction late in life   (a) Females that first breed at age 1 (open boxes) have smaller clutches at ages 2, 3, and 4 than females that do not breed until age 2 (filled boxes). The error bars indicate the standard errors of the estimated mean clutch sizes; the numbers above the symbols give the sample sizes.   (b) Females given extra eggs at age 1 (filled boxes) have progressively smaller clutches each year at ages 2, 3, and 4. In contrast, control females (open boxes) do not begin to show a decline in clutch size until age 4. From Gustafsson and Pärt (1990).

than second-year breeders (1.24 ± 0.08 versus 0.90 ± 0.14 offspring surviving to adulthood; P < 0.05).

Truman Young (1990) studied trade-offs in plants. Young reviewed data from the literature on the energy allocated to reproduction by closely related pairs of annuals versus perennials (Table 17.1). Annuals, which reproduce once and die, always allocate more energy to their sole bout of reproduction than perennials allocate to any given bout. This indicates that there is a trade-off in plants between reproduction and survival. Annual plants enjoy enhanced reproduction in their first reproductive season at the expense of drastically accelerated senescence.

Natural Experiments on the Evolution of Aging

We now review two studies that test the evolutionary theory of aging. In both cases, researchers took advantage of natural experiments. Both studies compare populations that have been historically exposed to different rates of mortality caused by extrinsic factors such as predators, diseases, and accidents. We will call this kind of mortality "ecological mortality," in contrast to mortality caused by processes intrinsic to the organism, like the wearing out of body parts. (Mortality caused by intrinsic processes might be called physiological mortality.)

The evolutionary theory of senescence predicts that populations with lower rates of ecological mortality will evolve delayed senescence (Austad 1993). Both of the evolutionary mechanisms that lead to senescence have reduced effectiveness in populations with lower ecological mortality rates. In the case of late-acting deleterious mutations (Figure 17.6b), lower ecological mortality means that a higher fraction of zygotes will live long enough to experience the deleterious effects. Late-acting deleterious mutations are thus more strongly selected against, and will be held at lower frequency in mutation-selection balance. In

---

TABLE 17.1 Reproductive allocation in annual versus perennial plants

Each line in the table gives the amount of energy allocated to reproduction by annual plants during their only bout of reproduction, divided by the amount of energy allocated to reproduction by perennial plants during any single bout of reproduction. All the values are larger than one, indicating that annuals always allocate more per reproductive episode. The comparisons for *Oryza perennis* and *Ipomopsis aggregata* are within species; all other comparisons are between species. From Young (1990); see this source for citations to individual studies.

Species	Allocation by annuals/allocation by perennials
Oryza perennis	2.9
Oryza perennis	5.3
Gentiana spp.	2.2–3.5
Lupinus spp.	2.2–3.2
Helianthus spp.	1.7–4.0
Temperate herbs	2.8–2.9
Old field herbs	1.7
Ipomopsis aggregata	1.5–2.3
Sesbania spp.	2.1–2.3
Hypochoeris spp.	2.4–3.7

the case of mutations with pleiotropic effects (Figure 17.6c), lower ecological mortality means that a higher fraction of zygotes will live long enough to experience both the early-life benefits and the late-life costs. The change in the fraction of zygotes experiencing the benefits and costs is more pronounced, however, for the costs. Thus mutations with pleiotropic effects are less strongly favored by selection. All else being equal, individuals in populations with lower ecological mortality should therefore show later senescence. The first study we review involves populations of a mammal; the second study involves two closely related species of plants.

The mammal, studied by Steven Austad (1993), is the Virginia opossum (Figure 17.2). Austad compared a population living in the mainland southeastern United States to a population living on Sapelo Island, located off the coast of Georgia. In the mainland population, opossums have high ecological mortality rates. In one study reviewed by Austad, more than half of all naturally occurring opossum deaths were caused by predators. When identifiable, two-thirds of the predators were mammals, including bobcats and feral dogs. Mammalian predators are absent on Sapelo Island. Sapelo Island supports an opossum population that has been isolated from the mainland population for four to five thousand years. Other than the difference in mammalian predators, Sapelo Island differs little from Austad's mainland study site at Savanna River, South Carolina. The two habitats are similar in temperature, rainfall, opossum ectoparasite loads, and food available per opossum. The evolutionary theory of senescence predicts that the Sapelo Island opossums will show delayed senescence relative to the mainland opossums.

Austad put radio collars on 34 island females and 37 mainland females and followed their life histories from birth until death. By three different measures, island females indeed show delayed senescence:

- Island females show delayed senescence in month-to-month probability of survival (Figure 17.9a). The probability of surviving from one month to the next falls with age in both populations, a manifestation of senescence. Monthly survival falls more slowly, however, for island females than for mainland females. As a result, the average life span of island females is significantly longer than the average life span of mainland females (24.6 versus 20.0; P < 0.002).

- Island females show delayed senescence in reproductive performance (Figure 17.9b). Austad measured reproductive performance by monitoring the growth rates of litters of young. For mainland females, litters produced in the mother's second year of reproduction grew more slowly than litters produced in the mother's first year of reproduction. This difference indicates that second-year mothers are less efficient at nourishing their young. Island females show no such decline in performance with age.

- Island females show delayed senescence in connective tissue physiology (Figure 17.9c). As mammals age, the collagen fibers in their tendons develop cross-links between protein molecules. These cross-links reduce the flexibility of the tendons. The amount of cross-linking in a tendon can be determined by measuring how long it takes for collagen fibers from the tendon to break. In both island and mainland opossums, breaking time in tail-tendon collagen increases with age, but breaking time increases less rapidly with

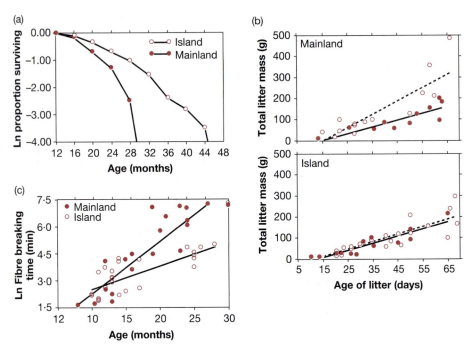

Figure 17.9   Female opossums on Sapelo Island age more slowly than female opossums on the mainland   (a) The rate of survival from one month to the next falls more rapidly with age for mainland opossums than for island opossums.   (b) Each graph shows total litter mass as a function of litter age for females in their first year of reproduction (open circles, dotted best–fit lines) and for females in their second year of reproduction (solid circles, solid best–fit lines). In mainland females (top graph), offspring in second-year litters grow more slowly than offspring in first-year litters (P < 0.001). In island females, second-year litters grow just as fast as first-year litters.   (c) Tail collagen fiber-breaking time increases more slowly with age in island females than in mainland females (P < 0.001). From Austad (1993).

Where predation risk is high, even individuals that invest heavily in repair may not survive for long. Where predators are abundant, female opossums invest more energy in early reproduction—at the cost of accelerated aging.

age in island opossums. In other words, island opossums have slower rates of physiological aging.

These results are all consistent with the evolutionary theory of senescence.

While they support the conclusion that ecological mortality is an important factor in the evolution of senescence, Austad's results do not allow us to determine which of the evolutionary theory's two hypotheses is more important. Is the more rapid aging of mainland opossums due to late-acting deleterious mutations, or to trade-offs between early reproduction and late reproduction and survival? At least part of the difference in rates of senescence appears to be due to trade-offs. Island opossums have, on average, significantly smaller litters (5.66 versus 7.61; P < 0.001). This suggests that mainland opossums are, physiologically and evolutionarily, trading increased early reproduction for decreased later reproduction and survival.

We now turn to the second natural experiment on the evolutionary theory of aging. The plant study, by Truman Young (1984; 1990), involves two closely related species of giant rosette lobelia that live on Mount Kenya (Figure 17.10). The species are called *Lobelia telekii* and *Lobelia keniensis*. Both species grow

Lobelia telekii
Semelparous

Lobelia keniensis
Iteroparous

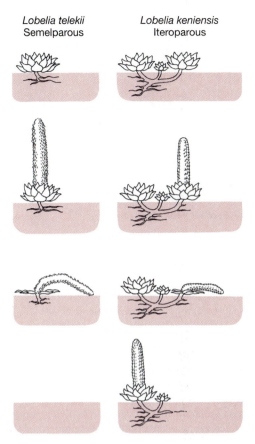

Figure 17.10 Life histories of two species of giant rosette lobelia
The circular cluster of leaves is called a rosette; the flower spike is called an inflorescence. When a rosette flowers, it transfers all of the resources stored in its roots and stem to the inflorescence. After flowering, the rosette dies. In *Lobelia telekii* (left), individual plants have only a single rosette. Thus individual plants are semelparous: They grow for 40 to 60 years, flower once, and die. In *Lobelia keniensis* (right), individual plants have several rosettes. Thus individual plants are iteroparous: They grow for 40 to 60 years, then one of their rosettes flowers and dies, and the other rosettes continue to grow. Successive rosettes flower and die every several years. After Young (1984).

slowly and require 40 to 60 years to reach reproductive maturity. *L. telekii* is semelparous, meaning that individuals flower once and die. *L. keniensis* is iteroparous, meaning that individuals flower repeatedly. After individual *L. keniensis* reach reproductive maturity, they flower once every several years.

The two species have overlapping ecological distributions (Figure 17.11, horizontal axis). *L. telekii* lives in drier, less productive environments, where resources are scarce. Individual *L. telekii* that live in relatively good environments for their species allocate their extra resources to their single bout of reproduction: They produce a larger inflorescence, with more flowers on it (Figure 17.11, filled circles).

L. keniensis lives in wetter, more productive environments, where resources are abundant (Figure 17.11, horizontal scale). Individual *L. keniensis* that live in relatively good environments for their species do not allocate their extra resources to any single bout of reproduction: Inflorescence size does not change as a function of environmental quality (Figure 17.11, open circles). Instead, *L. keniensis* living in relatively good environments invest their extra resources in a longer reproductive life span: They develop more rosettes (Figure 17.11, open triangles).

The two lobelia species suffer different rates of ecological mortality. *L. telekii,* the species that flowers once and dies, suffers higher ecological mortality than *L. keniensis.* Likewise, different populations of *L. keniensis,* the species that flowers repeatedly, suffer different rates of ecological mortality. Important among the ecological causes of mortality is the herbivorous rock hyrax *(Procavia johnstoni).*

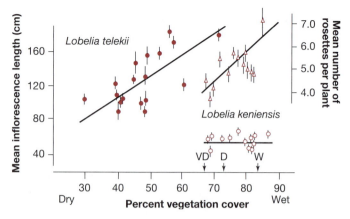

Figure 17.11   **How inflorescence size and rosette number vary with environmental quality in two species of giant rosette lobelia**   The horizontal axis represents soil moisture, from dry to wet. Soil moisture is correlated with vegetation cover and is a good indicator of environmental quality. As a result, the horizontal axis should be read as running from poor environments on the left to good environments on the right. The filled circles and left vertical axis represent *Lobelia telekii* inflorescences. The open circles and left vertical axis represent *L. keniensis* inflorescences. The open triangles and the right vertical axis represent rosette number for *L. keniensis.* The letters and arrows just above the horizontal axis indicate the environments of the three *L. keniensis* populations from which Young collected detailed demographic data: VD = very dry outlying population (this population is living in an environment more typical of *L. telekii*); D = drier (but not as dry as the very dry population); W = wetter. The error bars represent one standard error. From Young (1990).

Near the zone of overlap, individual *L. keniensis* achieve in any single bout of reproduction approximately $\frac{1}{5}$ to $\frac{1}{3}$ the reproductive success that individual *L. telekii* achieve in their only bout of reproduction. This means that *L. keniensis* has the better life history strategy if, and only if, it lives long enough to flower three to five times.

To determine whether these patterns make sense in terms of the evolutionary theory of senescence, Young (1990) sought to determine whether, according to the theory, each species has the better life history strategy in its particular environment. Young focused this research on *L. keniensis,* following the life histories of individual plants in three populations along an environmental gradient (Figure 17.11, arrows). One population lives in an environment that is, even for *L. keniensis,* relatively wet and productive. Another population lives in an environment that is drier and less productive. The third population lives in an environment that is, for *L. keniensis,* extremely dry and unproductive. This last population lives in an environment that overlaps that of *L. telekii.* If each species has the better strategy for its own environment, then the *L. keniensis* living in this last environment should exhibit lower lifetime reproductive success than the nearby *L. telekii.*

Young calculated the expected lifetime reproductive success for *L. keniensis* in the three different environments. Young's calculations were not exactly the same as the ones we made in Figure 17.6, because *L. keniensis* does not flower every year, but the logic is similar. The expected lifetime reproductive success of *L. keniensis* depends on the probability of survival from one year to the next, and also on the number of years that pass between reproductive bouts. Both of these parameters change with environmental quality. In wetter environments *L. keniensis* has a higher annual probability of survival, and it flowers more often.

Figure 17.12 shows Young's estimates of the expected lifetime reproductive success of *L. keniensis* as a function of both yearly survival probability and number of years between flowering episodes. The graph reads like a topographic map. The number of years between flowering bouts is analogous to latitude; the probability of survival from one year to the next is analogous to longitude. Reproductive success is analogous to elevation. Each curved line represents a particular level of reproductive success. As the probability of survival increases and the number of years between flowering bouts falls, expected lifetime reproductive success climbs. This climb follows a path from the lower left corner of the graph to the upper right corner. Expected lifetime reproductive success is given in multiples of the reproductive success a *Lobelia keniensis* gets in a single flowering episode. In other words, reproductive success is reported in the number of times an individual plant is expected to flower before it dies.

The three circles on the graph represent the locations in the environment of the three populations of *L. keniensis* that Young studied. At the wetter site, plants have an annual survival probability of about 0.99 and flower about once every 8 years. They are expected to flower more than 10 times. At the drier site, plants have an annual survival probability of about 0.985 and flower about once every 14 years. They are expected to flower about five times. At the very dry outlying site, plants have an annual survival probability of less than 0.98 and flower about once every 16 years. They are expected to flower less than three times.

We noted earlier that individual *L. keniensis* achieve in any single flowering episode about $\frac{1}{5}$ to $\frac{1}{3}$ the reproductive success that *L. telekii* achieve in their only flowering episode. *Lobelia keniensis* have to flower three to five times to achieve the same lifetime reproductive success as *L. telekii*. The hatched band on the

Where herbivores are abundant and resources are scarce, even plants that invest hesvily in maintenance may not survive for long. Under such conditions, Lobelias invest heavily in their first bout of reproduction — at the cost death.

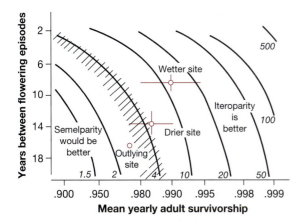

Figure 17.12 **Lifetime reproductive success in *Lobelia keniensis* varies with yearly probability of survival and number of years between flowering episodes** The horizontal axis represents the probability of survival from one year to the next (logarithmic scale). The vertical axis represents the number of years between flowering episodes (descending scale, with smaller intervals between flowering at the top). The vertical and horizontal error bars represent 95% confidence intervals. Young did not have enough data for the outlying (very dry) population to estimate the confidence intervals. From Young (1990).

graph in Figure 17.12 shows the zone in which *L. keniensis* are expected to flower three to five times. Below and to the left of that zone, *L. keniensis* would have a total lifetime reproductive success that is less than that achieved by *L. telekii* in its only bout of reproduction. The *L. keniensis* population at the outlying site is living at the extreme edge of its species' ecological range. The members of this population are achieving lower lifetime reproductive success than nearby *L. telekii*.

The data in Figure 17.12 therefore demonstrate that each species of lobelia has the better life history strategy for its particular environment. In wet environments, where survival probability is higher and the period between flowering bouts is shorter, plants get higher lifetime reproductive success by living a long time and flowering repeatedly (although with smaller inflorescences). In dry environments, where survival probability is lower and the period between flowering bouts is longer, plants get higher lifetime reproductive success by throwing all of their resources into their first flowering episode—even though the cost of this strategy is death. Young's results are consistent with the evolutionary theory of senescence.

In summary, the evolutionary theory of senescence hinges on the observation that the force of natural selection declines late in life. This is because most individuals die—due to predators, diseases, or accidents—before reaching late life. Two mechanisms can lead to the evolution of senescence: (1) deleterious mutations whose effects occur late in life can accumulate in populations; (2) when there are trade-offs between reproduction and maintenance, selection may favor investing in early reproduction even at the expense of maintaining cells and tissues in good repair. The evolutionary theory of senescence has been successful in explaining variation in life history among populations and species. Among the questions that remain are these: What is the relative importance of deleterious mutations and trade-offs in the evolution of senescence (Partridge and Barton 1993)? Can evolutionary theory explain unusual reproductive life histories, such as menopause in human females (Box 17.1)?

17.3 How Many Offspring Should an Individual Produce in a Given Year?

Section 17.2 dealt, in part, with the optimal allocation of energy to reproduction versus repair over an organism's entire life history. In this section we turn to the related issue of how much an organism should invest in any single episode of reproduction. Again we are concerned with trade-offs. Primary among them is this intuitively straightforward constraint: The more offspring a parent (or pair of parents) attempts to raise at once, the less time and energy the parent can devote to caring for each one.

Questions about the optimal number of offspring have been addressed most thoroughly by biologists studying clutch size in birds. The reason is probably that it is easy to count the eggs in a nest, and it is easy to manipulate clutch size by adding or removing eggs. Assuming that the size of individual eggs is fixed, how many eggs should a bird lay in a single clutch?

Clutch Size in Birds

The simplest hypothesis for the evolution of clutch size, first articulated by David Lack (1947), is that selection will favor the clutch size that produces the most surviving offspring. Figure 17.13 shows a simple mathematical version of this hypothesis (for a more detailed mathematical treatment, see Stearns 1992). The model assumes a fundamental trade-off in which the probability that any individual offspring will survive decreases with increasing clutch size. Many researchers have tested this assumed trade-off by adding eggs to nests; in the majority of cases they have found that adding eggs indeed reduces the survival rate for individual chicks (Stearns 1992). One explanation could be that the ability of the parents to feed any individual offspring declines as the number of offspring increases. In Figure 17.13a, we assume that the decline in offspring survival is a linear function of clutch size, but the model depends only on survival being a decreasing function. Given a function describing offspring survival, the number of surviving offspring from a clutch of a given size is just the product of the clutch size and the probability of survival (Figure 17.13b). The number of surviving offspring reaches a maximum at an intermediate clutch size. It is this most-productive clutch size that Lack's hypothesis predicts will be favored by selection.

Mark Boyce and C. M. Perrins (1987) tested Lack's hypothesis with data from a long-term study of great tits *(Parus major)* nesting in Wytham Wood, a research site near Oxford, England. Combining data for 4,489 clutches monitored over years 1960 through 1982, Boyce and Perrins plotted a histogram showing the distribution of clutch sizes in the Wytham Wood tit population (Figure 17.15, page 666). The mean clutch size was 8.53. Boyce and Perrins also determined the average number of surviving offspring from clutches of each size (Figure 17.15). This number was highest for clutches of 12 eggs. When researchers added three eggs to each of a large number of clutches, the most productive clutch size was still 12. In other words, birds that produced smaller clutches apparently could have increased their reproductive success for the year by laying 12 eggs. Taken at face value, these data seem to indicate that natural selection in Wytham Wood favors larger clutches than the birds in the population actually produce. Because the average clutch size was less than the most productive clutch size, these results are not consistent with Lack's hypothesis.

The literature on Lack's hypothesis is extensive, and many researchers have done studies similar to that of Boyce and Perrins (see reviews in Roff 1992 and Stearns 1992). The results of Boyce and Perrins are typical: The majority of studies have shown that birds lay smaller clutches than predicted. How can we explain this discrepancy? The mathematical logic of Lack's hypothesis is correct. The hypothesis must therefore make one or more implicit assumptions that often turn out to be wrong. Evolutionary biologists have identified and tested several assumptions implicit in Lack's hypothesis. We review three here.

First, Lack's hypothesis assumes that there is no trade-off between a parent's reproductive effort in one year and its survival and/or reproductive performance in future years. As we discussed in section 17.2, however, reproduction often entails exactly such costs. The data in Figure 17.8b demonstrated that when female collared flycatchers are given an extra egg in their first year, their clutch size in

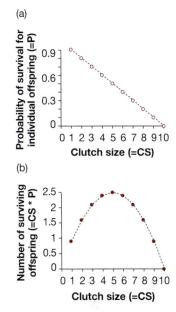

Figure 17.13 A mathematical treatment of Lack's hypothesis on the evolution of clutch size (a) The probability that any individual offspring will survive declines with increasing clutch size. Here, the probability of survival starts at 0.9 for the single offspring in a one-egg clutch, then declines by 0.1 for each one-egg increment in clutch size. (b) The number of survivung offspring per clutch is the number of eggs multiplied by the probability that any individual offspring will survive. The optimal clutch size is the one that produces the maximum number of survivors. Here, the optimal clutch size is 5.

Under simple assumptions, Lack's hypothesis predicts that birds will add eggs to their clutches until additional eggs would mean fewer surviving offspring. Many birds produce smaller clutches than predicted.

Box 17.1   Is there an evolutionary explanation for menopause?

Reproductive capacity declines earlier and more rapidly in women than in men (Figure 17.14a). The early decline in the reproductive capacity of women is puzzling, especially given that other measures of women's physiological capacity decline much more slowly (Figure 17.14b). Why should women's reproductive systems shut down by age 50, while the rest of their organs and tissues are still in good repair?

We consider two hypotheses. One hypothesis suggests that menopause is a nonadaptive artifact of our modern life-style (see Austad 1994). The other hypothesis suggests that menopause is a life history adaptation associated with the contribution grandmothers make to feeding their grandchildren (Hawkes at al. 1989).

Advocates of the artifact hypothesis point out that archaeologists reconstructing the demography of ancient peoples have often concluded that in pre-modern cultures, virtually all adults died by age 50 or 55 (see Hill and Hurtado 1991). If death by age 50 or 55 was the rule for our hunter-gatherer ancestors, then the modern situation, in which individuals often live into their 80s and 90s, is unprecedented in our evolutionary history. Menopause cannot be an adaptation, because our hunter-gatherer ancestors never lived long enough to experience it. When other mammals are kept in captivity and given modern medical care, they too live far longer than individuals do in nature. In captive mammals, females in at least some species show a decline in reproductive capacity well in advance of the decline in male reproductive capacity, and long before death (Figure 17.14c). Thus menopause may need no other explanation than that our modern life-style has extended our life span beyond that experienced by our ancestors.

Critics of the artifact hypothesis point out that in contemporary hunter-gatherer societies, many individuals live into their 60s and 70s (Figure 17.14d). These data may be more reliable indicators of the demography of our hunter-gatherer ancestors than are archaeological reconstructions (Hill and Hurtado 1991; see also Austad 1994). If a substantial fraction of our female hunter-gatherer ancestors lived long enough to experience menopause, then menopause needs an evolutionary explanation.

Advocates of the grandmother hypothesis note that human children are dependent on their mothers for food for several years after weaning. This is true in contemporary hunter-gatherer cultures, particularly when mothers harvest foods that yield a high return for adults but are difficult for children to process (Hawkes et al. 1989). Thus a woman's ability to have additional children may be substantially limited by her need to provision her older, still-dependent children. Furthermore, as a woman gets older, several relevant trends are likely to occur: (1) The probability that she will live long enough to be able to nurture another baby from birth to independence declines; (2) the risks associated with pregnancy and childbirth rise; and (3) her own daughters will themselves start to have children. The grandmother hypothesis suggests that older women may reach a point at which they can get more additional copies of their genes into future generations by ceasing to reproduce themselves, and instead helping to provision their weaned grandchildren so that their daughters can have more babies. In other words, grandmothers face a trade-off between investment in children and investment in grandchildren.

Kristen Hawkes and colleagues (1989; 1997) studied postmenopausal women in the Hadze, a contemporary hunter-gatherer society in East Africa. If the grandmother hypothesis is correct, then women in their 50s, 60s, and 70s should continue to work hard at gathering food. If the grandmother hypothesis is wrong, then we might expect older women (who no longer have dependent children) to relax. In fact, older Hadze women work harder at foraging than any other group (Figure 17.14e). Furthermore, for at least some crops at some times of year, older women are the most effective foragers (Figure 17.14f). Older women do with their extra food exactly what the grandmother hypothesis predicts: They share it with young relatives, thereby improving the childrens' nutritional status.

These data are consistent with the grandmother hypothesis, but they do not provide a definitive test. As Austad (1994) points out, the crucial issue is whether the daughters of helpful grandmothers are

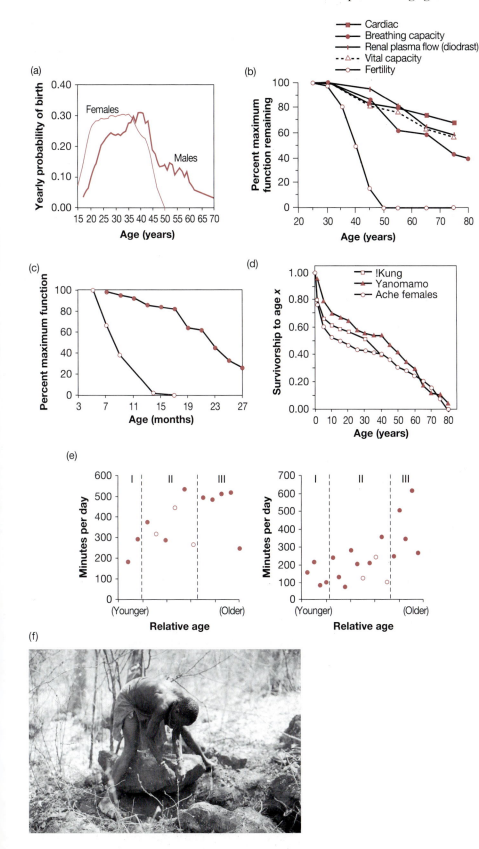

Figure 17.14 **Data on menopause** (a) Reproductive senescence in women and men. These data are from the Ache, a group of contemporary hunter-gatherers living in eastern Paraguay. The graph shows the probability that women and men will have a child born during the year that they are any given age. From Hill and Hurtado (1996). (b) This graph shows the functional capacity of various physiological systems in women as a function of age. Capacity is calculated as the fraction of youthful capacity still remaining. From Hill and Hurtado (1991); see citations therein for sources of data. (c) Reproductive capacity as a function of age in captive female (open circles) and male (filled circles) rats. From Austad (1994); see citations therein for sources of data. (d) Fraction of individuals surviving as a function of age in three contemporary hunter-gatherer cultures. From Hill and Hurtado (1991); see citations therein for sources of data. (e) Time spent foraging each day during the wet season (left) and the dry season (right) by Hadze women of different ages: I = women who have reached puberty but not yet married or begun to have children; II = women who are pregnant and/or have young children; III = women who are past child-bearing age and have no children younger than 15. Open circles represent women nursing young children. From Hawkes et al. (1989). (f) This Hadze woman is approximately 65 years old. She is using a digging stick and muscle power to dig tubers from underneath large rocks. Digging tubers requires knowledge, skill, patience, strength, and experience, making Hadze grandmothers the most productive foragers (Copyright James F. O'Connell, University of Utah).

Box 17.1   *(continued)*

actually able to have more children, and whether the grandmothers thereby achieve higher inclusive fitness (see Chapter 16) than they would by attempting to have more children of their own. Kim Hill and Magdalena Hurtado (1991; 1996) addressed this issue with data on the Ache hunter-gatherers of Paraguay. Hill and Hurtado's data show that the average 50-year-old woman has 1.7 surviving sons and 1.1 surviving daughters. The researchers calculate that by helping these children reproduce, the average Ache grandmother can gain the inclusive fitness equivalent of only 5% of an additional offspring of her own. This conclusion offers little support for the grandmother hypothesis.

More complete data on more cultures are needed for definitive evaluation of both the artifact hypothesis and the grandmother hypothesis. Steven Austad

(1994) points out that which hypothesis is correct is of more than academic interest. If menopause evolved by natural selection, then we would expect women's physiology to be adapted to adjust to the low-estrogen state their bodies enter after menopause. Thus if the grandmother hypothesis is correct, taking estrogen supplements after menopause might be counterproductive to women's overall health. If the artifact hypothesis is correct, however, then a long life span after menopause is an evolutionary novelty for women. We would not necessarily expect women's physiology to be adapted to adjust to the low-estrogen postmenopausal state. If this is the case, then treatment to alleviate unwanted consequences of menopause is less likely to cause problems by upsetting an otherwise finely tuned physiological state.

Efforts to identify which of Lack's assumptions are violated have led to the discovery of additional trade-offs, and improved estimates of lifetime fitness.

future years is lower than that of control females. In a review of the literature on reproductive costs in birds, Mats Lindén and Anders Møller (1989) found that 26 of the 60 studies that looked for trade-offs between current reproductive effort and future reproductive performance found them. In addition, Lindén and Møller found that 4 of the 16 studies that looked for trade-offs between current reproductive effort and future survival found them. When reproduction is costly and selection favors withholding some reproductive effort for the future, the optimal clutch size may be less than the most productive clutch size.

Second, Lack's hypothesis assumes that the only effect of clutch size on offspring is in determining whether the offspring survive. Being part of a large clutch may, however, impose other costs on individual offspring than just reducing their probability of survival. Dolph Schluter and Lars Gustafsson (1993) added or removed eggs from the nests of collared flycatchers, put leg bands on the chicks that hatched from the nests, and then monitored the chicks' subsequent life histories. When the female chicks matured and built nests of their

Figure 17.15   Lack's hypothesis tested with data on great tits (Parus major)   The bars and left vertical axis form a histogram showing the variation in clutch size. The average clutch size for 4,489 clutches was 8.53. The dots connected by line segments and the right vertical axis show the number of surviving young per clutch as a function of clutch size. The number of surviving young per clutch was highest for clutches of 12 eggs. From Roff (1992; redrawn from Boyce and Perrins 1987).

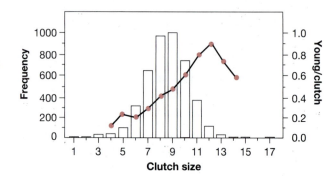

own, there was a strong relationship between the size of the clutches they produced and how much the clutch they were reared in had been manipulated (Figure 17.16). Females reared in nests from which eggs had been removed produced larger clutches, whereas females reared in nests to which eggs had been added produced smaller clutches. This indicates that clutch size affects not only offspring survival but also offspring reproductive performance. When larger clutches mean lower offspring reproductive success, the optimal clutch size will be smaller than the most numerically productive clutch size.

Third, Lack's hypothesis assumes that there is no variation from year to year in the size of the most productive clutch. When there is variation among years, the best measure of the long-term fitness of a genotype is not the arithmetic mean fitness (the average), but the geometric mean fitness. The geometric mean fitness of a list of n numbers is the nth root of the product of the numbers. Figure 17.17 uses a numerical example to demonstrate why geometric mean fitness is a better measure of the long-term fitness of a genotype. Figures 17.17a and b show Lack's hypothesis for good years, medium years, and bad years. In this example, the optimal clutch size in good years is seven. In medium years it is five; in bad years it is three. Figure 17.17c shows the arithmetic and geometric mean fitness of each clutch size over a run of years in which there are equal numbers of good, medium, and bad years. The arithmetic mean fitness indicates that the optimal clutch size is five.

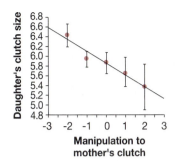

Figure 17.16 Does clutch size affect offspring reproductive performance? The graph shows the relationship between daughters' clutch sizes and the number of eggs added to or removed from the mothers' nests in which the daughters were reared. From Schluter and Gustafsson (1993).

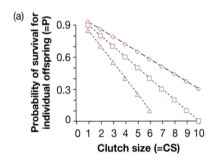

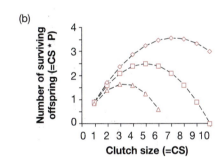

(c)

Clutch Size	Number Surviving			Mean Fitness	
	Good Year	Medium Year	Bad Year	Arithmetic	Geometric
3	2.37	2.1	1.65	2.04	2.02
4	2.88	2.4	1.6	2.29	2.23
5	3.25	2.5	1.25	2.33	2.17
6	3.48	2.4	0.6	2.16	1.71
7	3.57	2.1	0	1.89	0.00

Figure 17.17 The long-term optimal clutch when there is variation among years
(a) Probability of offspring survival as a function of clutch size in a good year (diamonds), a medium year (squares), and a bad year (triangles).   (b) The number of offspring surviving as a function of clutch size in a good year (diamonds), a medium year (squares), and a bad year (triangles).   (c) The productivity of different clutch sizes in a good year, a medium year, and a bad year. For each type of year, the productivity of the optimal clutch size is set in color. The last two columns give the arithmetic and geometric mean fitnesses across the three types of year. Judged by the arithmetic mean fitnesses, the long-term optimal clutch size is five. Judged by the geometric mean fitness, however, the long-term optimal clutch size is four. After Boyce and Perrins (1987).

Note, however, that in bad years birds laying clutches of seven eggs have zero reproductive success. If a run of bad years lasts longer than the reproductive life span of an individual bird, then birds that lay seven eggs will be lost from the population forever. As a long-term measure of fitness, the arithmetic mean across years misses this point and is therefore misleading. The geometric mean fitness captures this point, however, by giving a clutch size of seven a long-term fitness of zero. The geometric mean is thus a better way to calculate long-term fitness when there is variation among years. In our example, calculations of geometric mean fitness show that the long-term optimal clutch size is four, not five.

Boyce and Perrin's (1987) first test of Lack's hypothesis used the arithmetic mean fitness as the measure of the most productive clutch. When the researchers recalculated the long-term productivity of different clutch sizes using geometric means, they found that the optimal clutch size was nine (Figure 17.18). This value is very close to the observed most frequent clutch size (Figure 17.15).

Note that we have been assuming that clutch size is fixed for any given genotype. In fact, clutch size is often phenotypically plastic (see Chapter 8). When clutch size is plastic, and when birds can predict whether they are going to have a good year or a bad year, then we would predict that individuals will adjust their clutch size to the optimum value for each kind of year (see, for example, Sanz and Moreno 1995).

Lack's Hypothesis Applied to Parasitoid Wasps

Although it often proves to be too simple to accurately predict clutch size, the examples we have reviewed demonstrate that Lack's hypothesis is a useful null model. By explicitly specifying what we should expect under minimal assumptions, Lack's hypothesis alerts us to interesting patterns we might not otherwise have noticed. This application of Lack's hypothesis is not limited to birds.

Lack's hypothesis is a useful null model for other organisms as well.

Eric Charnov and Samuel Skinner (1985) used Lack's hypothesis to explore the evolution of clutch size in parasitoid wasps. Parasitoid wasps use a stinger-like ovipositor to inject their eggs into the eggs or body cavity of a host insect. When the larval parasitoids hatch, they eat the host alive from the inside. The larvae then pupate inside the empty cuticle of the host, finally emerging as adults to mate and repeat the life cycle.

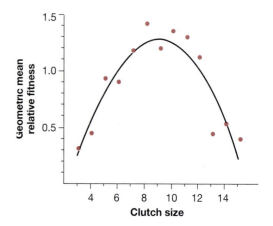

Figure 17.18 **Relative fitness of different great tit clutch sizes, calculated with geometric means** From Boyce and Perrins (1987).

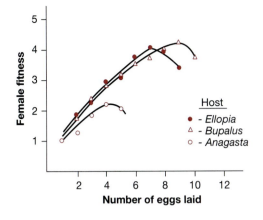

Figure 17.19 **Lack's hypothesis applied to a parasitoid wasp**
This graph shows the fitness of female parasitoid wasps *(Trichogramma embryophagum)* as a function of clutch size. Each curve represents wasp clutches deposited in a different species of host egg. From Charnov and Skinner (1985).

For a parasitoid, a host is analogous to a nest. A female parasitoid can lay a clutch of one or more eggs in a single host. The larvae compete among themselves for food, so there is a trade-off between clutch size and the survival of individual larvae. An added twist with insects is that adult size is highly flexible. In addition to reducing offspring survival, competition for food may result in larvae simply becoming smaller adults. The maternal fitness associated with a given clutch size must therefore be calculated as the product of the clutch size, the probability of survival of individual larvae, and the lifetime egg production of offspring of the size that emerge.

Charnov and Skinner used this modified version of Lack's hypothesis to analyze the oviposition behavior of female parasitoid wasps in the species *Trichogramma embryophagum*. This wasp deposits its eggs in the eggs of a variety of host insects. Using data from the literature, Charnov and Skinner calculated maternal fitness as a function of clutch size for three different host species (Figure 17.19). Table 17.2 gives the most productive clutch sizes and the actual clutch sizes female wasps lay in each species of host egg. The female wasps shift their behavior in a manner appropriate to different hosts. Females lay fewer eggs in the relatively poor hosts, and more eggs in the relatively good hosts. As with many birds, however, female wasps lay smaller clutches than those predicted by Lack's hypothesis.

Why do female wasps lay clutches smaller than the predicted sizes? Charnov and Skinner consider three reasons. Two of the three are similar to factors we discussed for birds: Larger clutch sizes may reduce offspring fitness

TABLE 17.2 Parasitoid wasp (*Trichogramma embryophagum*) clutch sizes predicted by Lack's hypothesis versus actual clutch sizes for three host species

Host species	Lack's optimal clutch size	Actual clutch size
Anagasta	4	1–2
Ellopia	7	5–8
Bupalus	9	5–8

Source: From Charnov and Skinner (1985).

in ways that Charnov and Skinner did not include in their calculations; and there may be trade-offs between a female's investment in a particular clutch and her own future survival or reproductive performance. Charnov and Skinner's third hypothesis is novel to parasitoid wasps. Unlike birds, female parasitoid wasps may produce more than one clutch in rapid succession. Soon after she has laid one clutch, a female wasp may begin looking for another host to parasitize. The appropriate measure of a wasp's fitness with regard to clutch size may not be the discrete fitness she gains from a single clutch. Instead, it may be the rate at which her fitness rises as she searches for hosts and lays eggs in them. Readers familiar with behavioral ecology may recognize this as an optimal foraging problem.

Figure 17.20 presents a graphical analysis of a female's rate of increase in fitness over time. The figure follows the female from the time she sets out to find a host egg until she leaves that host egg to look for another. While she is searching, the female gains no fitness. Once she finds a host and begins to lay eggs in it, however, her fitness begins to rise. The fitness she gets from a clutch of any given size is determined by a parabolic function, as in our original depiction of Lack's hypothesis (Figure 17.13). In this example, if a female leaves to look for a new host after laying just one egg, her total fitness gain from the first host is 0.9. Her average rate of fitness gain from the time she set out looking for the first host to the time she leaves to look for a second is given by 0.9 divided by the total elapsed time. This rate of fitness gain is equal to the slope of the diagonal line from the origin to the point representing a clutch size of one. Likewise, if the female stays to lay five eggs, her average rate of fitness gain for the whole trip is given by the slope of the upper diagonal line. In this example, the female would get the highest rate of fitness gain from this host if she left after laying four eggs. This is one egg less than the most productive clutch size. Thus if female parasitoids are selected to maximize their rate of fitness increase, they may produce smaller clutches than those predicted by Lack's hypothesis.

In summary, Lack's hypothesis is a useful starting point for the evolutionary analysis of clutch size. Assuming only that there is a trade-off between the number of offspring in a clutch and the survival of individual offspring, Lack's hypothesis predicts that parents will produce clutches of the size that maximizes

Figure 17.20 **Rate of increase in parasitoid maternal fitness with time spent searching for hosts and laying eggs** The horizontal axis represents the time spent by a female parasitoid in searching for a host egg and depositing a clutch. The vertical axis represents female fitness, in units of surviving offspring. The open circles show the relationship between number of surviving offspring and clutch size, as in Lack's hypothesis. After Charnov and Skinner (1985).

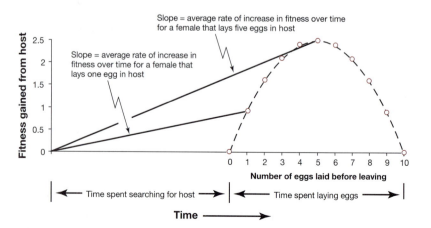

the number of surviving offspring. This prediction is often violated, with actual clutches typically smaller than expected. These violations indicate the possible presence of other trade-offs. Current parental reproductive effort may be negatively correlated with future parental survival or reproductive performance, or clutch size may be negatively correlated with offspring reproductive performance. Alternatively, a violation of predicted clutch size may indicate that we have chosen the wrong measure of parental fitness.

17.4   How Big Should Each Offspring Be?

In section 17.3 we assumed that the size of individual offspring was fixed. We now relax that assumption. Given that an organism will invest a particular amount of energy in an episode of reproduction, we can ask whether that energy should be invested in many small offspring or a few large offspring.

A trade-off between the size and number of offspring should be fundamental. A pie can be sliced into many small pieces or a few large pieces, but it cannot be sliced into many large pieces. Biologists have found empirical evidence for a size-versus-number trade-off in a variety of taxa. Mark Elgar (1990), for example, analyzed data from the literature on 26 families of fish. Elgar found a clear negative correlation between clutch size and egg size: Fish that produce larger eggs produce fewer eggs per clutch (Figure 17.21a). David Berrigan (1991) performed a similar analysis of variation in egg size and number among species in three orders of insects. In each case, Berrigan found a clear negative correlation between egg number and egg size (Figure 17.21b).

Organisms face a trade-off between making many low-quality offspring or a few high-quality offspring.

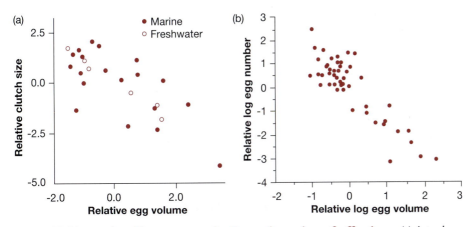

Figure 17.21   Trade-offs across taxa in size and number of offspring   (a) A trade-off across 26 fish families. Larger fish produce bigger clutches, so Elgar had to use statistical techniques to remove the effect of variation among fish families in body size. The vertical axis shows relative clutch size, or the number of eggs per clutch adjusted statistically for differences in body size among families. The horizontal axis shows relative egg volume, or egg size adjusted for differences in body size among families. The negative correlation between size and number of eggs is statistically significant ($P < 0.001$). From Elgar (1990).   (b) A trade-off across fruit fly species. Larger fruit flies produce more and larger eggs, so Berrigan also had to use statistical techniques to remove the effect of variation in body size. The vertical axis shows relative egg number; the horizontal axis shows relative egg volume. The negative correlation is statistically significant ($P < 0.001$). Berrigan found similar patterns in wasps and beetles. Provided by David Berrigan using data analyzed in Berrigan (1991).

(a)

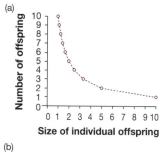

(b)

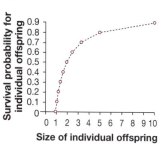

(c)

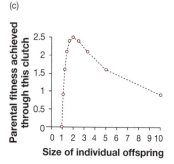

Figure 17.22   **The optimal compromise between size and number of offspring**
(a) Assumption 1: There is a trade-off between size and number of offspring. The units we have used are arbitrary. The shape of the curve may vary from species to species. Here we have used the equation: Number = 10/Size.   (b) Assumption 2: There is a minimum size below which individual offspring do not survive; above this minimum size the probability that any individual offspring will survive is an increasing function of offspring size. Again, we have used arbitrary units. The shape of the curve may vary from species to species. Here we have used the equation: Survival = 1 − (1/Size).   (c) Analysis: The parental fitness gained from a single clutch of offspring of a given size is the number of offspring in the clutch multiplied by the probability that any individual offspring will survive. For example, given the equations and units used here, if a parent makes offspring of size five it can make two of them, and each has a probability of survival of 0.8. Thus the expected fitness gained by the parent from this clutch is 2 × 0.8 = 1.6. For some (but not all) combinations of the trade-off function (a) and the survival function (b), parents achieve maximum fitness through offspring of intermediate size (as in c). After Smith and Fretwell (1974).

Selection on Offspring Size

If selection on parents is forced by a fundamental constraint to strike a balance between the size and number of offspring, what is the optimal compromise? Christopher Smith and Stephen Fretwell (1974) offered a mathematical analysis of this question. Smith and Fretwell's analysis is based on two assumptions. The first assumption is the trade-off between size and number of offspring (Figure 17.22a). The second assumption is that individual offspring will have a better chance of surviving if they are larger (Figure 17.22b). There must be a minimum size below which offspring have no chance of survival. As offspring get larger, their probability of surviving rises. If survival probability approaches one, it must do so in a saturating fashion, because survival probability cannot exceed one. Given the two assumptions, the analysis is simple: The expected fitness of a parent producing offspring of a particular size is the number of such offspring the parent can make multiplied by the probability that any individual offspring will survive. A plot of expected parental fitness versus offspring size (Figure 17.22c) reveals the size of offspring that gives the highest parental fitness. The optimal offspring size depends on the shapes of the relationships for offspring number versus size and offspring survival versus size. In many cases, as in Figure 17.22, the optimal offspring size is intermediate. In other words, selection on parental fitness often favors offspring smaller than the size favored by selection on offspring fitness. This identification of a potential conflict of interest between parents and offspring is the primary contribution of Smith and Fretwell's model.

It is possible to test Smith and Fretwell's analysis empirically only if there is substantial variation in offspring size among parents within a population. Variation in offspring size is relatively small in most species (Stearns 1992). We review two recent studies that have confirmed both the assumptions and the conclusion of Smith and Fretwell's analysis. In one study, researchers generated extra variation in egg size in a lizard by surgically manipulating females. In the other study, researchers took advantage of phenotypic plasticity in egg size in a beetle.

Selection on Offspring Size in a Lizard

Barry Sinervo and colleagues (1992) studied the side-blotched lizard *(Uta stansburiana)*, which lives in the deserts of western North America. Females lay

Selection on individual offspring favors high quality, but selection on parents favors a compromise between quality and quantity.

clutches of one to nine eggs (Sinervo and Licht 1991). Wild populations of side-blotched lizards show heritable variation in egg size (Sinervo and Doughty 1996). Sinervo and colleagues (1992) further expanded the range of egg sizes by surgically manipulating gravid females. The researchers induced near-term females to lay unusually small eggs by withdrawing a fraction of the yolk from each egg. The researchers induced early-term females to lay unusually large eggs by destroying all but two or three of the developing follicles. This caused the females to allocate to a few remaining eggs yolk that they would otherwise have allocated to several. The histograms in Figure 17.23a show the variation in egg size in two populations of lizards in each of two years.

Sinervo and colleagues (1992) tested the first assumption of Smith and Fretwell at the level of individual females. Each plot in Figure 17.23b shows an estimated relationship between clutch size and egg size for females in a population in 1989 or 1990. In both populations and both years there was a trade-off between size and number of eggs.

Sinervo and colleagues tested the second assumption of Smith and Fretwell by following the fates of individual hatchlings. The researchers marked several hundred hatchling lizards and released them into the wild. One month later, the researchers censused the hatchlings to see which had survived. Each plot in Figure 17.23c shows an estimate of the probability of hatchling survival as a function of egg mass. In 1989, selection favored larger offspring in both populations—exactly as Smith and Fretwell assumed it would. In 1990, selection favored medium-sized offspring—a surprising result.

How did the egg size versus egg number trade-off and selection on hatchlings combine to select on the mothers? Each plot in Figure 17.23d shows an estimate of the relationship between maternal fitness and egg size. In all cases selection on mothers favored eggs of intermediate size. In 1989, selection on the mothers favored smaller eggs than did selection on offspring. This is the conflict of interest that Smith and Fretwell predicted would sometimes occur. In 1990, selection on mothers favored eggs almost exactly the same size as did selection on offspring.

Phenotypic Plasticity in Egg Size in a Beetle

Charles Fox and colleagues (1997) studied the seed beetle *Stator limbatus*. The females of this small beetle lay their eggs directly onto the surface of host seeds. The larvae hatch and burrow into the seed. Inside, the larvae feed, grow, and pupate. They emerge from the seed as adults. *S. limbatus* is a generalist seed predator; it has been reared on the seeds of over 50 different host plants.

Fox and colleagues studied *S. limbatus* on two of its natural hosts: an acacia *(Acacia greggii)* and a palo verde *(Cercidium floridum)*. The acacia is a good host; most larvae living in its seeds survive to adulthood. The palo verde is a poor host; fewer than half the larvae living in its seeds survive. We can easily add hosts of different quality to the Smith and Fretwell analysis (Figure 17.24). When we do so, we get a clear prediction: Females should lay larger eggs on the poor host than on the good host. Recall from Chapter 8 that when selection favors different phenotypes at different times or in different places, organisms sometimes evolve phenotypic plasticity. The analysis in Figure 17.24 predicts that *S. limbatus* should exhibit phenotypic plasticity in egg size.

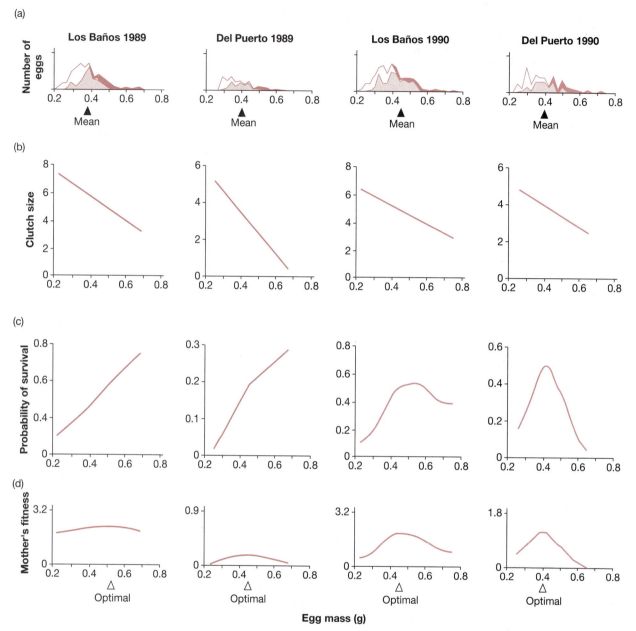

Figure 17.23   **Selection on egg size in the side-blotched lizard (*Uta stansburiana*)**   (a) These histograms show the distribution of egg sizes in the populations. The shaded region represents natural variation; the white region represents surgically induced small eggs; the black portion represents surgically induced large eggs. The mean egg size for natural eggs in each population is indicated by the black triangle.   (b) Each plot shows the estimated relationship between maternal clutch size and egg size. Clutch size was a decreasing function of egg size. This indicates a trade-off, confirming the first assumption of Smith and Fretwell's analysis.   (c) Each plot shows the estimated relationship in nature between egg size and the probability of hatchling survival to the age of one month. In 1989 offspring survival increased with egg size, confirming the second assumption of Smith and Fretwell's analysis. In 1990, offspring survival was highest for medium-sized hatchlings.   (d) Each plot shows the estimated relationship between maternal fitness (clutch size × probability that an individual offspring will survive) and egg size. In both populations in both years the optimal egg size for mothers (white triangle) was an intermediate value. In 1989 this value was smaller than the optimal egg size for offspring. From Sinervo et al. (1992).

Figure 17.24 **The optimal compromise between size and number of offspring on a good host and a poor host**   (a) As in Figure 17.22, we assume a trade-off between size and number of offspring.   (b) As in Figure 17.22, we assume that there is a minimum size below which individual offspring do not survive; above this minimum size, the probability that any individual offspring will survive is an increasing function of offspring size. The minimum size for offspring survival is smaller on the good host. Furthermore, survival is higher on the good host at all sizes above the minimum.   (c) Analysis: The parental fitness gained from a single clutch of offspring of a given size is the number of offspring in the clutch multiplied by the probability that any individual offspring will survive. The optimal offspring size (for the mother) is bigger on the poor host than on the good host. After Smith and Fretwell (1974).

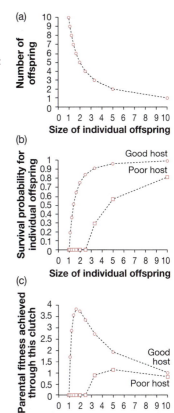

Fox and colleagues found that, as predicted, female *S. limbatus* adjust the size of the eggs that they lay to the host on which they deposit them. When the researchers took newly emerged females from the same population and gave them only one kind of seed, females given palo verde seeds (the poor host) laid significantly larger eggs than females given acacia seeds (Figure 17.25a). Confirming assumption 1 of Smith and Fretwell, these larger eggs came at the cost of fewer eggs produced over a lifetime (Figure 17.25b).

For females laying on palo verde seeds, the production of large eggs is adaptive. Fox et al. manipulated females into laying small eggs on palo verde seeds by keeping the females on acacia seeds until they laid their first egg, then moving them to palo verde seeds. Only 0.3% of the larvae hatching from small eggs on palo verde seeds survived to adulthood, whereas 24% of the larvae hatching from large eggs on palo verde seeds survived (P < 0.0001). Confirming assumption 2 of Smith and Fretwell, even among the large eggs on palo verde seeds, the probability of survival from egg to adult was positively correlated with egg size. For females laying on acacia seeds, the production of small eggs is adaptive. Given that nearly all larvae hatching on acacia seeds survive, females producing more and smaller eggs have higher lifetime reproductive success.

Fox and colleagues even showed that individual females that had started to lay size-appropriate eggs on one host could readjust their egg size when switched to the other host (Figure 17.25c). Control females left on one kind of seed consistently produced large or small seeds for life (Figure 17.25d).

In summary, selection on offspring size often involves a conflict of interest between parents and offspring. Because making larger offspring also means making fewer offspring, selection on parents can favor smaller offspring sizes than are optimal for offspring survival. The exact balance between size versus number of offspring depends on the relationship between offspring size and survival. Poor environments pose a greater obstacle to offspring survival and thus favor larger offspring.

17.5 Life Histories in a Broader Evolutionary Context

In this final section of the chapter, we place life histories in a broader evolutionary context. We briefly consider examples of research addressing two general questions: What forces maintain genetic variation in populations? How do novel traits evolve?

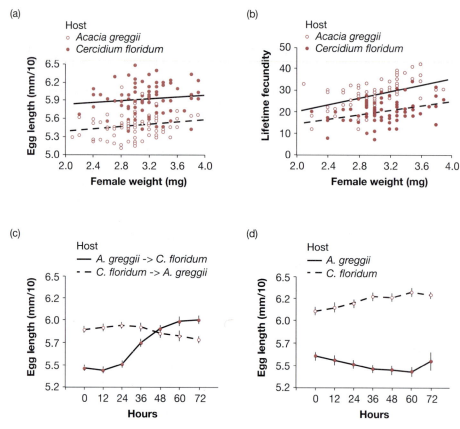

Figure 17.25   **Phenotypic plasticity in egg size in the seed beetle,** *Stator limbatus*
(a) Larger females lay slightly larger eggs, but females laying on *Cercidium floridum* lay larger
eggs for their size (P < 0.001). (Egg length is given in mm/10. A value of 6.2 on the vertical axis corresponds to an egg length of 0.62 mm.)   (b) Larger females have slightly higher
fecundity (number of eggs), but females laying on *A. greggii* have higher fecundity for their
size (P < 0.001).   (c) In a host switch experiment, Fox and colleagues placed newly
emerged females on each kind of seed, let them lay their first egg, and then switched them
to the other kind of seed. The graph shows egg size as a function of time for the subsequent
72 hours. Females that laid their first egg on acacia (*A. greggi,* the good host), produced small
eggs at first, but gradually switched to large eggs. Females that laid their first egg on palo
verde (*C. floridum,* the poor host) produced large eggs at first, but gradually switched to
smaller eggs.   (d) In a control experiment for (c), Fox and colleagues left females on each
kind of seed for life. Each group laid large or small eggs for life. From Fox et al. (1997).

The Maintenance of Genetic Variation

Natural selection on a trait should reduce the genetic variation for the trait
(Fisher 1930). A simple example illustrates why (Roff 1992). Imagine a series
of loci, each with two alleles, that affect a single trait correlated with fitness. At
each locus one allele ("0") contributes to the trait in such a way as to add zero
units to the fitness of individuals, whereas the other allele ("1") contributes
one unit to individual fitness. The genotype with the highest fitness is homozygous for allele "1" at all loci. Over time, selection should lead to the fixa-

tion of the "1" allele at each locus, and there will no longer be genetic variation in the trait.

Life history traits, because of their intimate connection with reproduction, should be more closely correlated with fitness than other kinds of traits, including behavioral, physiological, or morphological traits (Mousseau and Roff 1987). Consequently, life history traits should show less genetic variation—lower heritability—than other kinds of traits (for a discussion of heritability, see Chapter 6). From the literature, Mousseau and Roff (1987) assembled a sample of 1,120 estimates of the heritability of various traits. They found that life history traits indeed tend to have the lowest heritabilities (Figure 17.26). This result is consistent with the expectation from our simple theoretical treatment (for an alternative interpretation, see Price and Schluter 1991).

Life history traits are closely correlated with fitness and have relatively low heritabilities.

Nonetheless, Mousseau and Roff's review documents that life history traits typically have substantial genetic variation. What evolutionary forces maintain genetic variation in populations? The list of possibilities includes mutation, heterozygote advantage, frequency-dependent selection, and genotype-by-environment interaction in which different genotypes have higher fitness in different environments or at different times (see Chapter 8).

Richard Grosberg (1988) studied the maintenance of genetic variation for life history traits in a population of the sea squirt *Botryllus schlosseri*. *Botryllus schlosseri* is a colonial animal that lives attached to hard surfaces in shallow marine waters of the temperate zone. Colonies consist of a number of identical modules. The modules in a colony are physiologically connected, and their life histories are synchronous. The population Grosberg studied contains two distinct life history morphs. One morph is semelparous: Upon reaching sexual maturity, the modules in a colony reproduce once and die. The other morph is iteroparous: Colonies have at least three episodes of sexual reproduction before they die. In a series of experiments in which he grew sea squirts in a common environment and bred the morphs with each other, Grosberg demonstrated that the two morphs are genetically determined.

What maintains genetic variation for life history morphs in this sea squirt population? Grosberg tracked the seasonal frequency of the two morphs over two years (Figure 17.27). In both years, the semelparous morph dominated

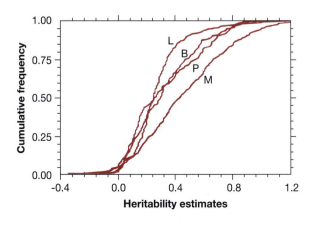

Figure 17.26   Life history traits have lower heritabilities than other kinds of traits   This plot shows four cumulative frequency distributions. A cumulative frequency distribution is a running sum, moving across a histogram, of the heights of the bars. The more rapidly the curve in a cumulative frequency distribution rises to 1, the lower the mean of the histogram. The line marked L is the cumulative frequency distribution for estimates of the heritability of life history traits; B = behavioral traits; P = physiological traits; M = morphological traits. Life history traits tend to have the lowest heritabilities. From Mousseau and Roff (1987).

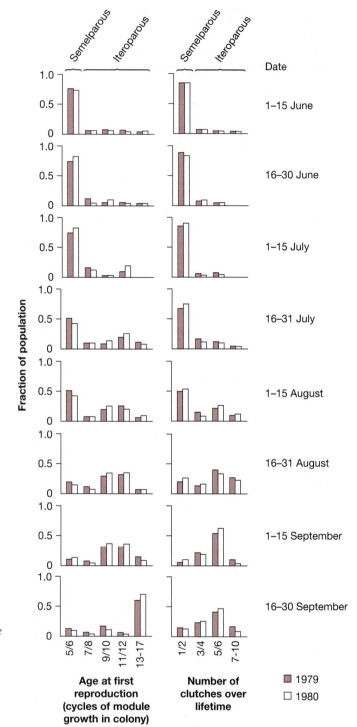

Figure 17.27 **Annual cycles in the frequencies of two life history morphs in a population of sea squirts** Each plot shows a frequency distribution for the population during a two-week period. Colonies of the semelparous morph reproduce at an early age (age = cycles of module growth in the colony) and produce only a single clutch of offspring. Colonies of the iteroparous morph reproduce at a late age and produce at least three clutches of offspring. From Grosberg (1988).

the population in the spring and early summer, whereas the iteroparous morph dominated in the late summer. This indicates that the two morphs are maintained in the population by seasonal variation in selection. One important selective factor may be competitive interactions with another sea squirt *(Botryllus leachi)*. This competitor, which becomes more abundant late in the summer, overgrows colonies of the semelparous *B. schlosseri* morph but not the iteroparous morph—a genotype-by-environment interaction.

Seasonal variation in selection maintains genetic variation for life history traits in a sea squirt.

The Evolution of Novel Traits

The evolution of novel traits represents a challenge for the theory of evolution by natural selection (see Chapter 2). Providing an example that is the focus of current research, closely related species of sea urchins can have strikingly different larval forms. Figure 17.28a shows the larvae of two urchins in the same genus: one is a pluteus; the other is a schmoo. Pluteus larvae hatch from small eggs. Before metamorphosis to the adult form, pluteus larvae live,

The repeated evolution of a novel larval form in urchins appears to have been prompted by selection on a life history trait.

(a)

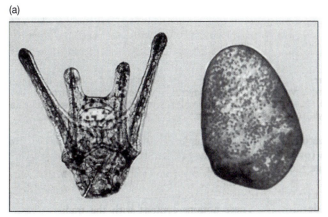

(b)

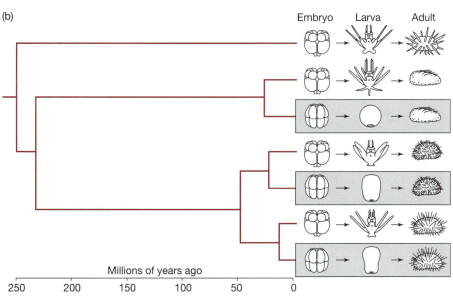

Figure 17.28 **The evolution of nonfeeding larvae in sea urchins** (a) The larva on the left is a form called a pluteus. The larva on the right is a form called a schmoo. These two larvae belong to closely related species of sea urchin: The pluteus is from *Heliocidaris tuberculata;* the schmoo is from *H. erythrogramma.* Both larvae in this photo are about three days old. Reproduced by permission from G.A. Wray, 1995. Punctuated evolution of embryos. *Science* 267: 1115–6, Feb. 24, 1995. © American Association for the Advancement of Science.   (b) The pluteus larval form is ancestral and ancient in sea urchins. In the urchin phylogeny shown here, the pluteus form is at least 250 million years old. More extensive phylogenies have shown the pluteus form to be about 500 million years old. Extant species with schmoo larvae appear to be relatively recently derived from ancestors with pluteus larvae. As the diagrams of early embryos included in the figure show, the pattern of the earliest cell divisions is different in species with pluteus versus schmoo larvae. From Wray (1995).

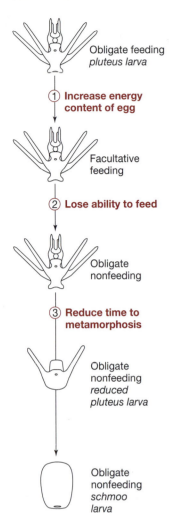

① **Increase energy content of egg**

② **Lose ability to feed**

③ **Reduce time to metamorphosis**

Obligate feeding
pluteus larva

Facultative feeding

Obligate nonfeeding

Obligate nonfeeding
reduced pluteus larva

Obligate nonfeeding
schmoo larva

Figure 17.29 **Hypothesized steps in the transition from pluteus larva to schmoo** From Wray (1996).

feed, and grow in the plankton. Schmoo larvae hatch from large eggs. Schmoo larvae undergo metamorphosis earlier than do pluteus larvae, and they do not feed.

The process of development in pluteus larvae versus schmoo larvae is as strikingly different as their morphology (Wray and Bely 1994; Wray 1995; 1996). The two larval forms differ in the pattern of the earliest cell divisions (Figure 17.28b). Likewise, the larval forms differ in the expression of a variety of genes.

Although the schmoo form is simpler than the pluteus, the pluteus form is ancestral (Figure 17.28b). Reconstructed phylogenies indicate that the pluteus form is hundreds of millions of years old. In contrast, extant species with schmoo larvae appear to have been derived within the last 50 million years from ancestors with pluteus larvae. The transition from pluteus larvae to schmoo larvae appears to be one-way. Schmoo larvae have evolved from pluteus ancestors at least 20 times in the sea urchins, whereas pluteus larvae have never evolved from schmoos (Wray and Bely 1994; Wray 1995).

Figure 17.29 shows a hypothesis for how the schmoo larval form evolves from a pluteus ancestor. In the first step in this scheme, selection favors the production of larger, more energy-rich eggs. This makes feeding optional for the derived larva. In the second step, the derived larva loses the ability to feed. Finally, selection for earlier metamorphosis causes the loss of all feeding structures, ultimately resulting in the most derived larval form, the schmoo. Michael Hart (1996) studied the larval form of one sea urchin *(Brisaster latifrons)* that appears to represent the first transitional step in this scenario. *Brisaster latifrons* larvae hatch from large eggs. Although they can feed, *B. latifrons* larvae to not have to feed in order to complete metamorphosis to the adult form. If the scenario outlined in Figure 17.29 is both correct and general, then the evolution of the novel developmental and morphological features of the schmoo form has been repeatedly initiated by selection for a simple life history trait: larger eggs.

Summary

Organisms face fundamental trade-offs. The amount of energy available is finite, and energy devoted to one function—such as growth or tissue repair—cannot be devoted to other functions—such as reproduction. Furthermore, biological processes take time. An individual growing to a large size before maturing risks dying from disease or predation before ever reproducing. Fundamental trade-offs involving energy and time mean that every organism's life history is an evolutionary compromise.

Senescence evolves because natural selection is weaker late in life. Late-acting deleterious mutations can persist in populations under mutation-selection balance. Selection may favor increased investment in reproduction early in life at the expense of repair. Both mechanisms can result in a decline in reproductive performance and survival with age.

A trade-off between the number of offspring in a clutch and the survival of individual offspring constrains the evolution of clutch size. Additional constraints may involve trade-offs between present parental reproductive effort and future reproductive performance or survival, as well as trade-offs between clutch size and offspring reproductive performance.

A trade-off between the size and number of offspring constrains the evolution of offspring size. Selection on parents may favor smaller offspring than does selection on the offspring themselves.

Because life history traits are closely related to fitness, theory predicts that life history traits should have low heritability. Life history traits do tend to have lower heritability than other kinds of traits, but nonetheless they typically show substantial genetic variation. One mechanism demonstrated to maintain genetic variation in life history traits is temporally varying selection.

Selection on life history traits can have dramatic consequences for other aspects of an organism's biology. Selection for larger eggs in echinoderms appears to have initiated, in numerous independent lineages, dramatic rearrangements in larval form and development.

Questions

1. Most domestic female rabbits will get uterine cancer if they are not spayed. The uterine cancer usually appears sometime after the age of 2 years. Describe a hypothesis for why rabbits have not evolved better defenses against uterine cancer. What do you think the average life span of a wild female rabbit might be? What do you think is a typical cause of death in wild rabbits? Why?

2. We have seen how aging can evolve due to two different phenomena: First, aging may evolve due to mutations that have deleterious effects only late in life. As a review, explain how such mutations could ever become common in a population. Second, aging may evolve due to mutations with pleiotropic effects that cause trade-offs—positive effects early and negative effects late. What would happen if a mutation arose with a reverse trade-off, that is, a mutation with negative effects early and positive effects late in life? Could such a mutation ever be selected for?

3. Look again at Figure 17.6, which shows life history trade-offs for a hypothetical species. Suppose you have a large captive population of these (hypothetical) organisms, and you notice a new mutation that causes its carriers to have two offspring per year instead of just one. The new mutation does not alter the age of maturation, which still occurs at 3 years. Your preliminary observations indicate that the new mutation may cause an early death, but you are not certain exactly how early the deaths occur. You do notice, however, that the new mutation is increasing in frequency and the wild-type allele is decreasing. Make a prediction about the minimum possible age of death of organisms that carry this mutation, and explain your reasoning.

4. Assuming for the moment that the grandmother hypothesis of menopause is correct, speculate on what aspects of a species' behavior and sociality may make menopause likely to evolve. For instance, is it important whether the species is highly social, or whether the species lives in kin groups? Might the age of independence of the young be important? Could menopause ever evolve in a species

without parental care, such as aphids or willow trees? As fuel for thought, consider the likelihood of evolution of menopause in: (1) orangutans, which are not very social (groups are small, often consisting simply of females and their dependent young); (2) lions, in which females are very social and remain with their female kin; and (3) Arabian oryx, which live in small family groups in the desert and must sometimes find distant waterholes known only to the older oryx.

5. Look again at Figure 17.9, which shows evidence for reduced aging in island opossums compared to mainland opossums. Do you think that month-to-month survival (Figure 17.9a) is a good measure of aging? How convincing are the data in Figure 17.9b, which show evidence for reproductive senescence in mainland females? If you were studying opossums, what additional observations or experiments might you perform to more clearly reveal differences in life span or reproduction between island and mainland opossums? Are your experiments practical?

6. Imagine a bird species living in an environment with good years and bad years. Further imagine that parents in this species show phenotypic plasticity for clutch size, but that they cannot tell whether a year is going to be good or bad until after they lay their eggs. What strategies do you predict these birds will use to maximize their lifetime reproductive success? What are the costs and benefits of each possible strategy?

7. The examples of the side-blotched lizards and seed beetles indicate that females probably cannot produce many large eggs—instead, they must choose between producing many small eggs or producing a few large eggs (and sometimes, in unfortunate cases, just a few small eggs). Explain, then, how it is possible for a queen honeybee to produce a very large number of relatively large eggs. (Hint: consider what the other bees are doing.) Does this suggest a general way in which a female can escape from the size-number trade-off?

Exploring the Literature

8. A restricted diet prolongs life span in mammals. One hypothesis for the mechanism behind this phenomenon is that a restricted diet reduces the metabolic rate. If this is true, then the effect of a restricted diet would be consistent with the rate-of-living theory of aging. For an experimental evaluation of the effect of diet on metabolic rate, see:

   McCarter, R., E.J. Masoro, and B.P. Yu. 1985. Does food restriction retard aging by reducing the metabolic rate? *American Journal of Physiology* 248:E488–E490.

9. Graham Bell distinguished between the rate-of-living versus evolutionary theories of aging by comparing invertebrates that lay eggs (and thus have a distinct soma and germ line) with invertebrates that reproduce by fission (and thus have no soma/germ line division). According to the rate-of-living theory, both kinds of organisms will inevitably accumulate irreparable damage. According to the evolutionary theory, only in organisms with a disposable soma would selection allow genes responsible for senescence to accumulate. See:

   Bell, G. 1984. Evolutionary and nonevolutionary theories of senescence. *American Naturalist* 124:600–603.

10. Lack's hypothesis of the optimal clutch size assumes that birds (and other organisms) are constrained only by their ability to care for offspring, not by their ability to make eggs. What happens if a bird can make only a limited number of eggs? See:

    Monaghan, P., M. Bolton, and D.C. Houston. 1995. Egg production constraints and the evolution of avian clutch size. *Proceedings of the Royal Society of London, Series B* 259:189–191.

11. For applications of Lack's hypothesis to mammals, see:

Packer, C., and A.E. Pusey. 1995. The Lack clutch in a communal breeder: Lion litter size is a mixed evolutionarily stable strategy. *American Naturalist* 145:833–841.

Risch, T.S., E.S. Dobson, and J.O. Murie. 1995. Is mean litter size the most productive? A test in Columbian ground squirrels. *Ecology* 76:1643–1654.

12. Our discussion of life history evolution assumed that life history characters evolve rapidly enough to be unconstrained by phylogenetic history. Is this actually true? See:

Lessios, H.A. 1990. Adaptation and phylogeny as determinants of egg size in echinoderms from the two sides of the isthmus of Panama. *American Naturalist* 135:1–13.

Citations

Austad, S.N. 1988. The adaptable opossum. *Scientific American* (February):98–104.

Austad, S.N. 1993. Retarded senescence in an insular population of Virginia opossums *(Didelphis virginiana). Journal of Zoology, London* 229:695–708.

Austad, S.N. 1994. Menopause: An evolutionary perspective. *Experimental Gerontology* 29:255–263.

Austad, S.N., and K.E. Fischer. 1991. Mammalian aging, metabolism, and ecology: Evidence from the bats and marsupials. *Journal of Gerontology* 46:B47–53.

Berrigan, D. 1991. The allometry of egg size and number in insects. *Oikos* 60:313–321.

Boyce, M.S., and C.M. Perrins. 1987. Optimizing great tit clutch size in a fluctuating environment. *Ecology* 68:142–153.

Charnov, E.L., and D. Berrigan. 1993. Why do female primates have such long life spans and so few babies? or Life in the slow lane. *Evolutionary Anthropology* 1:191–194.

Charnov, E.L., and S.W. Skinner. 1985. Complementary approaches to the understanding of parasitoid oviposition decisions. *Environmental Entomology* 14:383–391.

Clutton-Brock, T.H., S.D. Albon, and F.E. Guinness. 1988. Reproductive success in male and female red deer. In T.H. Clutton-Brock, ed. *Reproductive Success* (Chicago: University of Chicago Press), 325–343.

Elbadry, E.A., and M.S.F. Tawfik. 1966. Life cycle of the mite *Adactylidium* sp. (Acarina: Pyemotidae), a predator of thrips eggs in the United Arab Republic. *Annals of the Entomological Society of America* 59:458–461.

Elgar, M.A. 1990. Evolutionary compromise between a few large and many small eggs: Comparative evidence in teleost fish. *Oikos* 59:283–287.

Eshleman, J.R., and S.D. Markowitz. 1996. Mismatch repair defects in human carcinogenesis. *Human Molecular Genetics* 5:1489–1494.

Fishel, R., and T. Wilson. 1997. MutS homologs in mammalian cells. *Current Opinion in Genetics and Development* 7:105–133.

Fisher, R.A. 1930. *The Genetical Theory of Natural Selection* (Oxford: Clarendon Press).

Fox, C.W., M.S. Thakar, and T.A. Mousseau. 1997. Egg size plasticity in a seed beetle: An adaptive maternal effect. *American Naturalist* 149:149–163.

Gould, S.J. 1980. *The Panda's Thumb: More Reflections in Natural History* (New York: W.W. Norton).

Grosberg, R.K. 1988. Life-history variation within a population of the colonial ascidian *Botryllus schlosseri*. I. The genetic and environmental control of seasonal variation. *Evolution* 42:900–920.

Gustafsson, L., and T. Pärt. 1990. Acceleration of senescence in the collared flycatcher *(Ficedula albicollis)* by reproductive costs. *Nature* 347:279–281.

Hamilton, W.D. 1966. The moulding of senescence by natural selection. *Journal of Theoretical Biology* 12:12–45.

Hart, M.W. 1996. Evolutionary loss of larval feeding: Development, form, and function in a facultatively feeding larva, *Brisaster latifrons. Evolution* 50:174–187.

Hawkes, K., J.F. O'Connell, and N.G. Blurton Jones. 1989. Hardworking Hadza Grandmothers. In V. Standen and R.A. Foley, eds. *Comparative Socioecology* (Oxford: Blackwell Scientific Publications), 341–366.

Hawkes, K., J.F. O'Connell, and N.G. Blurton Jones. 1997. Hadza women's time allocation, offspring provisioning, and the evolution of long postmenopausal life spans. *Current Anthropology* 38: 551–577.

Hill, K., and A.M. Hurtado. 1991. The evolution of premature reproductive senescence and menopause in human females: An evaluation of the "grandmother hypothesis." *Human Nature* 2:313–351.

Hill, K., and A.M. Hurtado. 1996. *Ache Life History: The Ecology and Demography of a Foraging People* (New York: Aldine de Gruyter).

Lack, D. 1947. The significance of clutch size. *Ibis* 89:302–352.

Lindén, M., and A.P. Møller. 1989. Cost of reproduction and covariation of life history traits in birds. *Trends in Ecology and Evolution* 4:367–371.

Luckinbill, L.S., R. Arking, M.J. Clare, W.C. Cirocco, and S.A. Buck. 1984. Selection for delayed senescence in *Drosophila melanogaster. Evolution* 38:996–1003.

Medawar, P.B. 1952. *An Unsolved Problem in Biology* (London: H. K. Lewis).

Mousseau, T.A., and D.A. Roff. 1987. Natural selection and the heritability of fitness components. *Heredity* 1987:181–197.

Neese, R.M., and G.C. Williams. 1995. *Why We Get Sick: The New Science of Darwinian Medicine* (New York: Vintage Books).

Nowak, R.M. 1991. *Walker's Mammals of the World,* 5th ed. (Baltimore: The Johns Hopkins University Press).

Partridge, L. 1987. Is accelerated senescence a cost of reproduction? *Functional Ecology* 1:317–320.

Partridge, L., and N.H. Barton. 1993. Optimality, mutation, and the evolution of ageing. *Nature* 362:305–311.

Partridge, L., and K. Fowler. 1992. Direct and correlated responses to selection on age at reproduction in *Drosophila melanogaster. Evolution* 46:76–91.

Pough, F.H., J.B. Heiser, and W.N. McFarland. 1996. *Vertebrate Life* (Upper Saddle River, NJ: Prentice Hall).

Price, T., and D. Schluter. 1991. On the low heritability of life-history traits. *Evolution* 45:853–861.

RodriguezBigas, M.A., P.H.U. Lee, L. OMalley, T.K. Weber, O. Suh, G.R. Anderson, and N.J. Petrelli. 1996. Establishment of a hereditary nonpolyposis colorectal cancer registry. *Diseases of the Colon & Rectum* 39:649–653.

Roff, D.A. 1992. *The Evolution of Life Histories* (New York: Chapman & Hall).

Roper, C., P. Pignatelli, and L. Partridge. 1993. Evolutionary effects of selection on age at reproduction in larval and adult *Drosophila melanogaster. Evolution* 47:445–455.

Rose, M.R. 1984. Laboratory evolution of postponed senescence in *Drosophila melanogaster. Evolution* 38:1004–1010.

Sanz, J.J., and J. Moreno. 1995. Experimentally induced clutch size enlargements affect reproductive success in the Pied Flycatcher. *Oecologia* 103:358–364.

Schluter, D., and L. Gustafsson. 1993. Maternal inheritance of condition and clutch size in the collared flycatcher. *Evolution* 47:658–667.

Service, P. 1987. Physiological mechanisms of increased stress resistance in *Drosophila melanogaster* selected for postponed senescence. *Physiological Zoology* 60:321–326.

Sinervo, B., and P. Doughty. 1996. Interactive effects of offspring size and timing of reproduction on offspring reproduction: Experimental, maternal, and quantitative genetic aspects. *Evolution* 50:1314–1327.

Sinervo, B., P. Doughty, R.B. Huey, and K. Zamudio. 1992. Allometric engineering: A causal analysis of natural selection on offspring size. *Science* 258:1927–1930.

Sinervo, B., and P. Licht. 1991. Proximate constraints on the evolution of egg size, number, and total clutch mass in lizards. *Science* 252:1300–1302.

Smith, C.C., and S.D. Fretwell. 1974. The optimal balance between size and number of offspring. *American Naturalist* 108:499–506.

Stearns, S.C. 1992. *The Evolution of Life Histories* (Oxford: Oxford University Press).

Strathmann, M.F. 1987. *Reproduction and Development of Marine Invertebrates of the Northern Pacific Coast* (Seattle, WA: University of Washington Press).

Taborsky, M., and B. Taborsky. 1993. The kiwi's parental burden. *Natural History* 1993:50–56.

Vani, R.G., and M.R.S. Rao. 1996. Mismatch repair genes of eukaryotes. *Journal of Genetics* 75:181–192.

Williams, G.C. 1957. Pleiotropy, natural selection, and the evolution of senescence. *Evolution* 11:398–411.

Wray, G.A. 1995. Punctuated evolution of embryos. *Science* 267:1115–1116.

Wray, G.A. 1996. Parallel evolution of nonfeeding larvae in echinoids. *Systematic Biology* 45:308–322.

Wray, G.A., and A.E. Bely. 1994. The evolution of echinoderm development is driven by several distinct factors. *Development 1994 Supplement:* 97–106.

Young, T.P. 1984. The comparative demography of semelparous *Lobelia telekii* and iteroparous *Lobelia keniensis* on Mount Kenya. *Journal of Ecology* 72:637–650.

Young, T.P. 1990. Evolution of semelparity in Mount Kenya lobelias. *Evolutionary Ecology* 4:157–171.

Evolution and the Human Condition

Among the reasons for studying evolution is its relevance to human affairs. Throughout the book we have pointed out human applications of evolutionary analysis. Here we devote an entire chapter to human applications we have not elsewhere had the opportunity to address.

We begin with a consideration of evolution and human behavior. Chapter 16 demonstrated that selection thinking can be used to understand the behavior of animals. Attempts to apply selection thinking to human behavior are complicated by the influence of culture. However, careful research cautiously interpreted yields valuable insights.

Next we turn to evolution and human health. In Chapter 1 we discussed the evolution of drug resistance in populations of pathogens. Here we use ideas and tools from evolutionary biology to reconstruct the course of disease in individual patients, to understand the symptoms of disease, and to decipher puzzling features of human physiology.

Finally, we consider whether we humans should take responsibility for managing our own evolutionary future. There are compelling reasons to practice such mangement, and yet past efforts to do so led to disturbing social policies and disastrous consequences. New knowledge and new technologies continue to raise difficult questions.

Our discussion in this chapter demonstrates that human applications of evolutionary biology are highly controversial, even within—perhaps especially within—the scientific community. Evolutionary biologists are especially critical of each other's work when that work involves human subjects. This is as it should be, given the inherent difficulties of doing research on people, and the checkered history of efforts to combine evolutionary analysis with governmental policy. Critical review and vigorous debate have led to better evolutionary science.

18.1 Evolution and Human Behavior

This first section of the chapter explores who we are and why. What can we learn about human nature within the theoretical framework of evolution by natural selection? In considering this question, the scientists whose work we discuss follow the adaptationist program illustrated in Chapters 3, 8, 16, and elsewhere. In Chapter 3, for example, we reviewed research by Erick Greene and colleagues (1987) on the Tephritid fly *Zonosemata vittigera*. When Greene et al. saw this fly's unusual wing markings, and watched its even more unusual wing-waving behavior, they assumed that the fly's markings and behavior had been shaped by natural selection. Starting from this assumption, Greene et al. formulated hypotheses about how the fly's markings and behavior improve its prospects for survival and reproduction. The scientists then conducted experiments to distinguish among their hypotheses. They concluded that the fly's wing markings and behavior mimic the territorial displays of its jumping-spider predators, thereby protecting the fly from attack. Researchers following the adaptationist program have been successful in explaining a wide range of animal behaviors (see Chapter 16).

In using this approach to understand the behavior of humans, evolutionary psychologists view the brain as an organ whose properties as a regulator of behavior have been shaped by natural selection. As a regulator of behavior the brain is a flexible machine, not a computer slavishly converting input to output according to some fixed program. The human brain runs on a complex mix of conscious and unconscious perception, emotion, experience, and calculation, in pursuit of a variety of goals. But in the view of evolutionary psychologists:

> The ultimate objective of our conspicuously purposive physiology and psychology is not longevity or pleasure or self-actualization or health or wealth or peace of mind. It is fitness. Our appetites and ambitions and intellects and revulsions exist because of their historical contributions to this end. Our perceptions of self-interest have evolved as proximal tokens of expected gains and losses of fitness, "expected" being used here in its statistical sense of what would be anticipated on average from the cumulative evidence of the past. (Daly and Wilson 1988a, page 10)

When selection thinking is applied to human behavior, caution is required. Human behavior is influenced by culture, and most modern humans live in environments different from the one to which we are adapted.

This adaptationist approach to human behavior requires caution. In its capacity as a regulator of behavior, the human brain is influenced by culture as well as by evolutionary history. Culture evolves by its own set of rules (see Box 18.1). Furthermore, culture can manifestly induce individuals to behave in ways contrary to the interests of their genetic fitness. The mass suicide of 39 members of the Heaven's Gate cult in March of 1997 defies adaptationist explanation.

The influence of culture on human behavior means that studies on the behavior of people within a single society cannot disentangle the effects of culture from those of evolutionary history. To make a plausible claim that a psychological trait or pattern of behavior is a product of natural selection, evolutionary psychologists must show that the trait or behavior pattern is broadly cross-cultural. Cross-cultural diversity has fallen dramatically during the last century. All but the most remote and isolated traditional societies have been contacted, and western ideas and artifacts have spread virtually everywhere (see Diamond

BOX 18.1 Is cultural evolution Darwinian?

A treatment of the mechanisms of cultural evolution is beyond the scope of this chapter. In fact, the mechanisms of cultural evolution are probably beyond the scope of evolutionary biology altogether.

Richard Dawkins, in his 1976 book *The Selfish Gene,* suggested that we might develop a theory of cultural evolution by natural selection that works exactly like our theory of biological evolution. Central to this suggestion is the idea that natural selection is a generalizable process. Natural selection works on organisms because organisms have four key features: mutation, reproduction, inheritance, and differential reproductive success. In principle, natural selection should operate on any class of entities that have the same four properties.

Dawkins noted that elements of culture have these four properties, and thus should evolve by natural selection. A new word, song, idea, or style is analogous to a new allele created by mutation. The Shakers' austere and beautiful style of furniture design, for example, is an element of culture. A new piece of culture reproduces when other people adopt it and pass it on, as when a woodworker admires a Shaker table, goes home to her shop, and imitates the design. Some pieces of culture are more successful than others at getting themselves transmitted from person to person. For example, Shaker-style furniture has achieved much wider adoption than Shaker-style lifelong celibacy. Culture evolves as the relative frequencies of styles and ideas change.

Dawkins coined the term *meme* for the fundamental unit of cultural evolution. He saw the meme as analogous to the gene, the fundamental unit of biological evolution. Dawkins envisioned a detailed theory of population memetics that would be similar to the theory of population genetics we covered in Chapters 4, 5, and 6. For a recent exposition on the potential explanatory power of this idea, see Dennett (1995).

The trouble with Dawkins' suggestion, noted by Dawkins himself (see also the 1989 edition of his book), is that the effectiveness of natural selection as a mechanism of evolution depends not just on the property of inheritance, but on the details of how inheritance works. This fact was first recognized by Fleeming Jenkin, one of Darwin's critics, in 1867 (see section 2.7). In Darwin's time, the prevailing model of inheritance involved the blending in the offspring of infinitely divisible particles contributed by the parents. Jenkin pointed out that blending inheritance undermines evolution by natural selection because of the fate it implies for new variations. In a sexual population with blending inheritance, any new variation would quickly vanish, like a single drop of black paint dissolving into a bucket of white. Mendelian genetics rescues Darwin's theory, because Mendelian inheritance is particulate. Genes do not blend. A new recessive mutation can remain hidden in a population for generations. Eventually the mutant allele may reach a high enough frequency that heterozygotes start to mate with each other, producing among their offspring a few homozygous recessives.

In its correct form, then, the generalizable theory of evolution by natural selection applies to any class of entities that has the properties of mutation, reproduction, *particulate* inheritance, and differential reproductive success. The crucial question for the theory of cultural evolution by natural selection is whether memes are transmitted via particulate or blending inheritance. As Allen Orr (1996) puts it: "Do street fashion and high fashion segregate like good genes, or do they first mix before replicating in magazines or storefronts?" Nobody knows. If memes are transmitted by blending inheritance, then natural selection is at best a weak mechanism of cultural evolution. We need other mechanisms to explain cultural evolution, perhaps mechanisms entirely different from those responsible for biological evolution.

Although biological evolution and cultural evolution may proceed by different mechanisms, this does not mean that either is irrelevant to the other. Cultural evolution can set the stage for biological evolution. Most humans, for example, stop producing the enzyme lactase in childhood, but the cultural practice of dairy farming led to the evolution of lifelong lactase production in many human populations (Durham 1991). Likewise, biological evolution can influence cultural evolution. For example, the division of the visible light spectrum into verbally distinguished colors follows cross-culturally universal patterns (Durham 1991). These patterns are determined by the way our eyes and brains encode visual information, indicating that the structure of our nervous systems has constrained cultural variation in color terminology. Cultural and biological evolution are distinct but interdependent.

1992). Some evolutionary biologists feel that it is no longer possible to conduct a genuine cross-cultural study. Others feel that cross-cultural studies are still worth pursuing, particularly when new findings are combined with information extracted from databases of earlier anthropological research.

Another caveat for the study of human behavior involves recent dramatic change in the human environment. The environments most humans live in today are strikingly different from the environments all humans lived in for most of our evolutionary history. From the time of the earliest members of the genus *Homo,* over 2 million years ago (see Chapter 14), until the advent of agriculture at about 8000 B.C., all humans lived in small groups and made their living by hunting and gathering. The rapid pace of cultural change in the last ten thousand years has been largely too fast for genetic evolution to keep up. Even the most modern urbanites among us still come equipped with minds and bodies adapted to the stone age. As a result, it is of little use to ask why natural selection would have produced human behaviors, such as a willingness to ski down a mountainside at 75 miles per hour, that can only occur in a modern context. So long as we are careful to allow for our incomplete understanding of the hunter-gatherer life-style, however, it may make sense to ask why natural selection would have built into us a desire for the social rewards that we can attain, under the right circumstances, through dramatic demonstrations of superior athleticism and bravery. Scientists pursuing such an inquiry would formulate and test hypotheses about how a recipient of these social rewards, living in a hunter-gatherer society, might convert them into reproductive success.

Evolution and Parenthood

We now explore evolutionary psychology by considering aspects of mate choice and parenthood. We begin with a prediction about parenthood. On the assumption that the psychology of parenting has been shaped by natural selection, evolutionary psychologists predict that human adults should direct more of their parental caregiving to their own genetic offspring than to the genetic offspring of others. We would, in fact, make the same prediction about any organism that provides parental care. Parental care is expensive to the caregiver, and caregivers who reserve their efforts for their own genetic young should enjoy higher lifetime reproductive success than caregivers who are indiscriminate. The generality of this prediction gives us confidence that it is legitimate for human hunter-gatherers. And the prediction has hidden subtleties, as an animal example will show.

Reed buntings *(Emberiza shoeniclus)* are small ground-nesting birds in which both males and females provide parental care. Most nesting pairs are socially monogamous, meaning that each member of the pair tends no other nest than the one they tend together. Genetic testing by Andrew Dixon and colleagues (1994) revealed that there is more to the reed bunting mating system than meets the eye. The researchers found that 55% of all chicks were sired by males other than their mother's social mate, and that 86% of all nests included at least one such chick. Dixon and colleagues predicted that males, if they could tell what fraction of the chicks they had sired in any given nest, would adjust their parental effort accordingly. Dixon et al. tested this prediction by looking at the chick-

As with other species, selection thinking yields the prediction that human parents will discriminate between their own versus others' genetic offspring.

feeding behavior of 13 pairs of buntings that raised two clutches of chicks in a single season. The males fed the chicks more often in the nest in which they had sired a higher proportion of the chicks (Figure 18.1a). The females, who were the genetic mothers of all the chicks in both nests, showed no such pattern (Figure 18.1b).

We have presented the reed bunting example because evolutionary biologists using Darwinism to understand human behavior are often accused of genetic determinism (see, for example, Lewontin 1980). Genetic determinism is the notion that fundamental characteristics of human societies are unchangeably programmed into our genes. Note, however, the sense in which genes do and do not determine the parental behavior of male reed buntings. A male's genotype does not specify a particular level of parental care that the male will provide at all times, no matter what. Instead, each male's genotype specifies a range of phenotypic plasticity in parental care (see Chapter 8). That is, the bird's brain has a mechanism that adjusts the effort the male expends in caring for a brood based on cues that indicate his probable level of paternity in that brood. If a male's

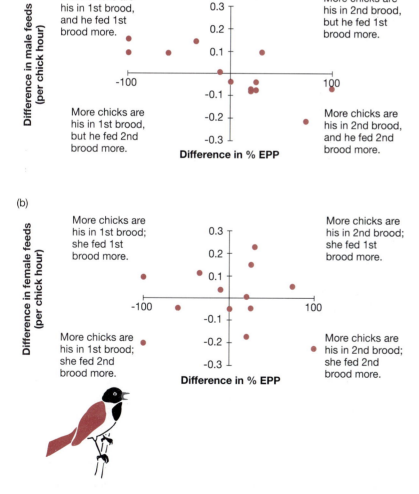

Figure 18.1 Male reed buntings adjust their parental effort depending on who they are feeding (a) Each dot represents a single male who raised two broods of chicks in a season. The horizontal axis represents the difference between the two broods in the percentage of extra-pair paternity (% EPP), or fraction of chicks sired by some other male. The vertical axis represents the difference between how frequently the male fed chicks in the first brood and how frequently he fed chicks in the second brood. Most males fed chicks more often in the nest in which they had sired a higher fraction of chicks. This association was significant at P = 0.0064. (b) Each dot represents a single female who raised two broods of chicks in a season. The horizontal axis represents the difference between the percentage of chicks in the first brood sired by extra-pair males and the percentage of chicks in the second brood sired by extra-pair males (% EPP). The vertical axis represents the difference between how frequently the female fed chicks in the first brood and how frequently she fed chicks in the second brood. Females showed no relationship between the number of feedings they provided and the relative number of chicks sired by extra-pair males. With modifications from Dixon et al. (1994).

social or biological environment changes, he alters his level of parental care accordingly, as Figure 18.1a shows. The pattern of phenotypic plasticity in a trait is called the trait's reaction norm. Reaction norms for reed bunting parental care presumably vary from male to male. It is this genetic variation in reaction norms that provides the raw material for the evolution of parental behavior. The average reaction norm of reed buntings might be described as "reed bunting nature."

Evolutionary psychologists studying human behavior are likewise interested in phenotypic plasticity—that is, reaction norms. They recognize that human reaction norms allow wide latitude for social and environmental circumstances to modify human behavior, and they recognize that reaction norms vary from person to person. What evolutionary psychologists do is formulate and test hypotheses about average human reaction norms.

Are humans as discriminating as male reed buntings in adjusting their provision of parental care? The question is difficult to study directly, at least in modern western cultures where most of the interactions between parents and children take place in private. Martin Daly and Margo Wilson therefore approach the issue by analyzing what may at first seem an odd source of data: case files of homicides in which parents killed their children (Daly and Wilson 1988a; see also Daly and Wilson 1988b; 1994a; 1994b). Daly and Wilson's rationale is that (1) the flip side of parental care is conflict of interest between parents and children; and (2) homicides are an unambiguous indication of conflict. Daly and Wilson predicted that children would be killed at a higher rate by stepparents than by biological parents.

Daly and Wilson do not suggest that killing stepchildren, in itself, is or ever was adaptive for humans. Anyone who kills someone else's child, even in a traditional hunter-gatherer society, is likely to suffer social penalties that outweigh any potential benefits of eliminating an unwanted demand for stepparental investment. Instead, what Daly and Wilson suggest is adaptive is the combination of two traits: (1) an intellectual and psychological apparatus that perceives a personal interest in the distinction between own and others' genetic offspring; and (2) the emotional motivation to turn this perception into active discrimination between the two kinds of children. Whenever such an apparatus exists, individuals will, rarely, commit errors of excess. These errors of excess become Daly and Wilson's data.

Observational data on child homicides are consistent with a selectionist prediction: Stepparents kill stepchildren at a much higher rate than biological parents kill biological children . . .

Data on murders of children in Canada dramatically confirm Daly and Wilson's prediction: Stepparents kill stepchildren at a much higher rate than biological parents kill biological children (Figure 18.2). It is worth dissecting this result a bit. In absolute numbers (that is, simply counting up homicides) more children are killed by biological parents than by stepparents (341 versus 67 in Daly and Wilson's study). But this is because only a small minority of children have stepparents. This is especially true for young children, the most common victims of parental homicide. In 1984 only 0.4% of Canadian children 1 to 4 years old lived with a stepparent. To adjust for the fact that few young children live with stepparents, Daly and Wilson report the data in Figure 18.2 as rates: the number of homicides per million child-years that parents or stepparents and children spend living together. Epidemiologists often summarize the results of such a study by reporting a relative risk. Here, the relative risk of homicide in stepchildren versus biological children is the rate at which stepparents kill stepchildren

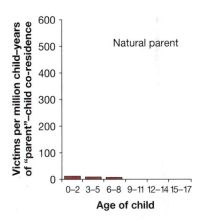
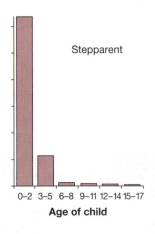

Figure 18.2 **The risk to children of being killed by a biological parent versus a stepparent** The graphs show, for biological parents (left) and stepparents (right), the rate at which parents killed children (number of homicides per million child-years that parents and children spent living in the same house). Children aged two or younger are killed by stepparents at a rate about 70 times higher than such children are killed by biological parents. The data are for homicides in Canada, 1974 through 1983. From Daly and Wilson (1988a; 1988b).

divided by the rate at which biological parents kill biological children. For children 0 to 2 years old, the relative risk of parental homicide for stepchildren versus biological children is about 70. This is an extraordinarily high relative risk. For comparison, the relative risk of lung cancer in smokers versus nonsmokers is about 11.

Daly and Wilson's discovery is clearly consistent with their prediction that human parents discriminate between their own genetic offspring and the genetic offspring of others. But before we accept the result as evidence for Daly and Wilson's Darwinian hypothesis, we need to consider that the result comes from a purely observational study. Is it really stepparent status, per se, that leads to a higher risk of killing children? Or is it some other factor that is correlated with being a stepparent? For example, people who are prone to violence might have higher divorce rates. If so, then violent people might tend to become stepparents at a higher rate than nonviolent people. Under this scenario, killing children is not a risk of stepparenthood. Rather, killing children and stepparenthood are both risks of having a violent personality.

. . . but this observation does not establish cause and effect.

If we were studying animals, we could disentangle cause and effect by performing a randomized experiment. Without regard to personality type, we would flip a coin to assign our subjects to be either stepparents or biological parents, and rearrange their families accordingly. If the randomly assigned stepparents killed children at a higher rate than the randomly assigned biological parents, then we could be confident that stepparenthood was indeed the cause. Obvious ethical and practical considerations preclude such an experiment.

Because Daly and Wilson cannot do experiments to test their hypothesis, they have been accused of simply making up adaptive stories to fit their data. Stephen George (1989), for example, seizes on an aside in one of Daly and Wilson's papers (1988). After noting that the costs associated with stepparenting may make it seem surprising that step-relationships are ever successful, Daly and Wilson point out that "people thrive by the maintenance of networks of social reciprocity and by establishing reputations for fairness and generosity that will make them attractive exchange partners." Daly and Wilson are thus using a Darwinian framework to explain both why most stepparent-stepoffspring relationships succeed, and why step-relationships fail more often than biological relationships. George speculates:

> If the frequency of killing [biological] children had turned out to be greater than the frequency of killing stepchildren, instead of the reverse, Daly and Wilson would have had no trouble "explaining" it by guessing that past selection for "maintenance of networks of social reciprocity" had been stronger than selection for "parental solicitude." (George 1989)

In other words, George is accusing Daly and Wilson of collecting their data first and framing their predictions after, like a sharpshooter who fires at the wall and paints a target around the bullet hole.

Daly and Wilson's (1989) response is that they did no such thing. A discovery that biological offspring are killed at a higher rate than stepoffspring "would be very surprizing in any animal species." Such an outcome would warrant further investigation, not the pretense of having expected it all along. Furthermore, Daly and Wilson point out:

> Risks to stepchildren . . . had never been assessed before an evolutionary model of parental inclinations inspired us to make the relevant comparisons. Our discovery that such children are several dozen times more likely to be slain than genetic offspring seems to us to warrant serious concern rather than dismissal as "storytelling."

We still must contend with the possibility that stepparenthood and homicide of children are both risks associated with some third factor, like a violent personality. Even when we cannot do experiments, additional observational studies can sometimes eliminate at least some of the alternative cause-and-effect scenarios. Mark Flinn (1988) conducted an extensive and detailed observational study of the interactions between parents and offspring in a small village in rural Trinidad. Flinn's observations on these families eliminate most of the alternative cause-and-effect scenarios that might explain Daly and Wilson's results, because they allow Flinn to compare the behavior of fathers toward their genetic offspring with the behavior of the same fathers toward their stepoffspring. Flinn interviewed all the residents of the village to determine which people were genetically related and which people were living together. Then, once or twice a day for six months, Flinn walked a standard route through the village that took him within 20 meters of every house and public building. He started at a different, randomly determined point each day, so as not to regularly pass particular places at particular times. Each time he saw any one of the village's 342 residents, Flinn recorded what the person was doing, who he or she was with, and the nature of the interaction they were having. The houses and buildings in the village are all rather open, so Flinn was able to see much that went on indoors as well as out.

Fourteen of the village's 112 households included mothers that were the genetic mothers of all the resident children, and fathers that were the genetic fathers of some of the resident children and the stepfathers of others. These 14 families included 28 genetic offspring of the fathers, and 26 stepoffspring of the fathers. There can be no hidden differences between the genetic fathers and the stepfathers, because the genetic fathers and stepfathers are the same men.

Flinn calculated the amount of time the fathers spent with their children, and the fraction of their interactions with their children that were agonistic. An ago-

Observing interactions between men and their stepchildren, and between the same men and their biological children, allows the men to be used as their own controls.

nistic interaction was one that "involved physical or verbal combat (e.g., spanking or arguing) or expressions of injury inflicted by another individual (e.g., screaming in pain or anguish or crying)" (Flinn 1988). Note that overall, only 6% of the parent/offspring interactions Flinn saw were agonistic, and 94% of these involved only verbal exchanges. During his study, Flinn was not aware of any interactions between parents and children that would be considered physical child abuse. ("Screaming in pain or anguish" may sound like evidence of physical child abuse, but anyone who has spent time with a 2-year-old knows that this is not necessarily so.) In other words, Flinn's research concerns parent-offspring interactions that most anyone would consider normal, not the errors of excess that Daly and Wilson study.

Flinn found that the 14 fathers with both genetic offspring and stepoffspring spent more of their time with their genetic offspring (Figure 18.3a). Furthermore, a smaller fraction of the father/genetic offspring interactions were agonistic (Figure 18.3b). These results are consistent with the prediction that parents will discriminate among children on the basis of their genetic relationship with those children. Furthermore, Flinn's results provide evidence that degree of genetic relationship is actually the ultimate cause of parental discrimination.

Consistent with a selectionist prediction, men who have both stepchildren and biological children spend more time—and get along better—with their biological children.

There is, however, one final potentially confounding variable. The pattern in Flinn's data could be explained by the late arrival of stepfathers in the lives of their stepchildren. Men might feel less affection and concern for their stepchildren simply because they joined the family when the stepchildren were older, whereas the men were already in the family when all of their own genetic children were born.

Flinn's dataset for this village, remarkably, includes 23 stepchildren who were born when their mothers and stepfathers were already living together, as well as

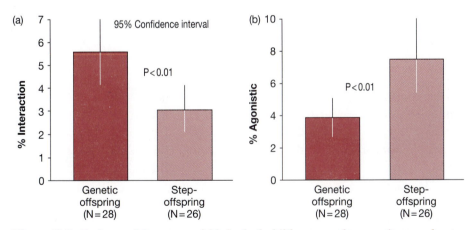

Figure 18.3 **Fathers with step- and biological children spend more time, and get along better, with their biological children**   (a) The bars show the fraction of their time (% of all observed interactions) that 14 fathers spent with their genetic children versus their stepchildren. The 95% confidence interval is an estimate of the certainty that Flinn's estimated percentage is close to the truth. Roughly speaking, we can be 95% certain that the true number is within the 95% confidence interval.   (b) The bars show the fraction of interactions between 14 fathers and their children that were agonistic (see text for definition). From Flinn (1988).

11 stepchildren born before their mothers and stepfathers moved in together. (This sample includes all the stepfathers in the village, not just those that also have genetic children living in the same house.) If a man's parental affection is simply a function of the fraction of the child's life during which the man has lived with the child, then the stepfathers who had lived with their stepchildren from the children's births should be more affectionate. In fact, the opposite appears to be true. The stepfathers in Flinn's study spent more of their time, and had a lower fraction of agonistic interactions, with stepchildren born before the stepfathers joined the family (Figure 18.4).

We noted earlier that studies within a single society offer no means of disentangling the influences of culture and evolutionary history. Both of the studies we have reviewed, Daly and Wilson's and Flinn's, concern modern western societies. We could argue that the patterns of discrimination revealed in these studies are simply a product of western culture, and have nothing to do with our species' adaptive history. Evidence is accumulating, however, that parental discrimination between own and others' genetic offspring is a cross-cultural phenomenon. One kind of evidence comes from single-culture studies of non-western societies. For example, Kim Hill and Hillard Kaplan (1988) studied the survival of biological children versus stepchildren in the Ache Indians, a traditional foraging culture in Paraguay. Hill and Kaplan found that 81% of children raised by both biological parents survived to their 15th birthday, whereas only 57% of children raised by one biological parent and one stepparent survived. Likewise, Napoleon Chagnon (1992; see also 1988) studied the Yąnomamö Indians, a traditional hunting, gathering, and gardening culture in Venezuela and Brazil. The Yąnomamö are polygynous, which means that women have little trouble finding husbands, but men often have difficulty finding wives. Chagnon reports that men work harder to find wives for their biological sons than for their stepsons.

The second kind of evidence that parental discrimination is broadly cross-cultural comes from anthropological databases. Daly and Wilson (1984), for ex-

Parental discrimination between own versus others' genetic offspring appears to be broadly cross-cultural.

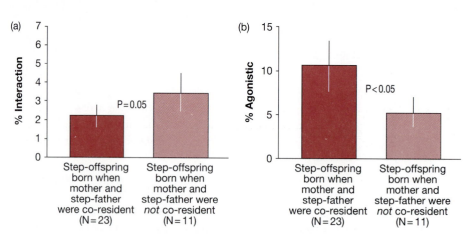

Figure 18.4 Stepfathers spend more time with their stepchildren, and get along better with them, when the stepchildren are born before the stepfather joins the family   (a) Fraction of time (% of all observed interactions) that stepfathers spent with their stepchildren, with the data separated by whether the stepfather was living in the house when the stepchild was born.   (b) Fraction of interactions between stepfathers and their stepchildren that were agonistic (see text for definition). From Flinn (1988).

ample, examined a set of 60 cultures called the "probability sample" from a database called the Human Relations Area Files. The probability sample was assembled by cultural anthropologists, and the societies it includes are independent of each other and representative of the world's cultural diversity. The files on 35 of the 60 cultures contain both records of infanticide and descriptions of the circumstances in which infanticide occurs. In 20 of these 35 cultures (57%), lack of genetic relationship between the child and the infanticidal parent is among the circumstances mentioned.

John Hartung (1981) also examined a representative subset of the Human Relations Area Files, but he took a different approach to measuring parental discrimination—one that is reminiscent of the reed bunting study. The crux of Hartung's test is the recognition of this fact: A mother is always certain that she is genetically related to her birth children, but a father's certainty that he is genetically related to his children (that is, his partner's birth children) is a function of whether and how often his partner has extra-pair sex. Hartung looked at a set of 186 societies. The files on 22 of these contained information on (1) the frequency with which women have extra-pair sex, and (2) which relatives inherit a man's wealth when he dies. In the 11 cultures in which women rarely have extra-pair sex, a man can be confident that he is genetically related to his own children. In 10 of these 11 cultures, men pass their wealth to their own children. In contrast, in the 11 cultures in which women frequently have extra-pair sex, a man's "own" children might not actually be genetic relatives at all. That is, his "own" children might actually be stepchildren. But a man always has genetic kinship with his mother's other sons and daughters, and with his sister's sons and daughters. In 6 of these 11 cultures, men pass their wealth not to their own children, but to their brothers, or to their sister's sons. The association between level of extra-pair sex and heir is statistically significant ($P = 0.0298$).

More detailed studies, like the one by Flinn, need to be done on a broader diversity of cultures. From the information currently available, however, it appears that parental discrimination between own versus others' genetic offspring is a part of human nature.

We noted earlier that from an evolutionary perspective it may seem puzzling that step-relationships are often successful, despite the inherent difficulties, and are often characterized by lasting ties of genuine mutual affection. Why should a stepparent invest at all in someone else's genetic offspring? A general Darwinian answer is that during our species' evolutionary history, such investors received returns, in the form of social or material rewards, that ultimately could be translated into reproductive success. Flinn's (1988) study gives a hint about what one such reward might be. Flinn divided his data on the rate of agonism in father/child interactions according to whether the children's mothers were present or absent at the time of the interaction (Figure 18.5). The mother's presence or absence had no effect on the rate of agonism in interactions between fathers and their genetic offspring. The mother's presence or absence did, however, have a significant effect on interactions between stepfathers and their stepoffspring. Relations were more harmonious when the mothers were present. This result suggests a hypothesis for further investigation: Men are solicitous toward actual or potential stepchildren as a courtship tactic directed at the children's mothers.

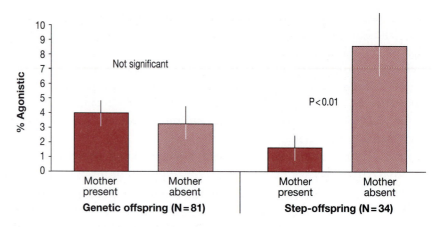

Figure 18.5 **The effect of mother's presence on how well fathers and stepfathers get along with children** (Left) Genetic fathers have the same rate of agonistic interactions with their genetic children whether the mother is present or absent. (Right) Stepfathers have fewer agonistic interactions with their stepchildren when the mother is present than when the mother is absent. From Flinn (1988).

Evolution and Human Mate Choice

We now turn from parenting to mate choice, again asking whether an evolutionary analysis can give us insight into human behavior. Among the criteria thought to be important to female mammals in choosing a mate is the mate's genotype at the loci of the major histocompatibility complex (MHC). The loci of the MHC code for cell-surface proteins that are important in the immune response. The MHC gene products are involved in distinguishing between self versus non-self proteins. The immune systems of individuals heterozygous at MHC loci may be able to recognize and respond to a greater diversity of non-self proteins, giving them an advantage in fighting disease. Thus MHC heterozygotes are thought to have higher fitness. If this scenario is correct, then a female mammal would enjoy higher fitness, via higher rates of offspring survival, if she chose a mate whose MHC genotype is different than her own. The similarity between a pair's MHC genotypes may also indicate the degree to which they are genetically related. If so, then a female chosing an MHC-different male might also enjoy higher fitness via reduced inbreeding depression (at non-MHC loci) in her offspring (Potts and Wakeland 1993). In Chapter 5 we discussed a study by Wayne Potts and colleagues (1991) showing that female mice behave as predicted, preferring under semi-natural conditions to mate with MHC-different males (see also Pomiankowski et al. 1992).

Research by Carole Ober and colleagues on the Hutterites of South Dakota suggests that human women, like other female mammals, would benefit by being choosy about their mates' MHC genotypes (Ober et al. 1992; Ober 1995). The Hutterites are a small religious sect that tend to marry within the group. As a result, the Hutterite population is more inbred and less genetically variable than most other human populations. This elevated inbreeding, in turn, means that Hutterite couples share MHC genotypes more often than couples in other populations. Ober and colleagues found that when couples share alleles at two MHC loci, called HLA-DR and HLA-B, they had lower fertility and higher rates of spontaneous abortion than couples that did not share MHC alleles.

As with other species, selection thinking yields the prediction that women should prefer as mates men whose MHC genotypes are different than their own.

Claus Wedekind and coworkers (1995) set out to determine experimentally whether women in a modern western society actually might discriminate among potential mates on the basis of MHC genotypes. How could a woman possibly tell whether a potential mate has a different MHC genotype than her own? One way is by smell. Using only their noses, rats and mice can distinguish rodents with different MHC genotypes (Beauchamp et al. 1985; Singh et al. 1987, 1988; Brown et al. 1987, 1989). A.N. Gilbert and colleagues (1986) showed that humans can, by smell alone, distinguish strains of mice that differ only in their MHC genotypes. Thus it is reasonable to expect that women can distinguish MHC-different men by smell.

Wedekind et al. enrolled 31 women and 44 men in their study. All of the subjects were student volunteers. Wedekind and colleagues determined the genotypes of all subjects at the MHC loci HLA-A, HLA-B, and HLA-DR. Then the researchers had each man sleep in a new cotton T-shirt for two nights in a row. When he returned the T-shirt, the researchers put it in a cardboard box with a hole in the lid through which the shirt could be sniffed. Each woman, alone in a room, sniffed T-shirts worn by six different men. Although she was not told this, she sniffed three T-shirts chosen at random from among those worn by men with MHC genotypes similar to her own, and three T-shirts chosen at random from among those worn by men with MHC genotypes dissimilar to her own. She rated each shirt for intensity and pleasantness of odor.

The results show that the women preferred, by a small but significant margin, the odors of men whose MHC genotypes are dissimilar to their own (Figure 18.6a). Wedekind and colleagues computed the average score each woman gave to men with similar MHC genotypes, and the average score each woman gave to men with dissimilar MHC genotypes. The women gave both categories of men equal scores for odor intensity. When scoring the men's odors for pleasantness, however, the women gave higher marks to men with MHC genotypes dissimilar to their own. Furthermore, the women more often reported that the T-shirt they were sniffing reminded them of one of their own past or present mates when that shirt had been worn by a man with a dissimilar MHC genotype (Figure 18.6b).

At least under ideal conditions, women find the odors of MHC-different men more pleasant.

Phil Hedrick and Volker Loeschcke (1996) are skeptical of the significance of these results. Wedekind and colleagues conducted a legitimate blind experiment, in that the women knew neither the identities of the men whose shirts they sniffed nor the results of the men's genetic tests. However, Hedrick and Loeschcke object to the other measures that the Wedekind team took to improve their chances of getting a positive result. Wedekind et al. describe these measures in their original paper. Before the men slept in the T-shirts, the researchers asked the men to wash their sheets and clothes with perfume-free detergent, and to bathe with perfume-free soap. During the two days on which the men slept in the shirts, the researchers asked them to live an odor-neutral lifestyle by avoiding smelly foods, rooms, and people, and by abstaining from smoking, drinking alcohol, sharing their beds with anyone else, or having sex. For two weeks prior to sniffing these men's shirts, the women used a nose spray that encourages the regeneration of the nasal mucous membrane. Wedekind and coworkers also asked the women to sensitize their odor perception by reading a novel called *Das Parfum*. Given these preparations, Hedrick and Loeschcke question whether Wedekind et al.'s results establish any meaningful role for MHC-based odor discrimination in human mate choice.

Wedekind and Thomas Seebeck (1996) reply that they were not trying to measure the actual degree of influence MHC-related odor differences have on women's mating preferences. Instead, they were simply trying to establish

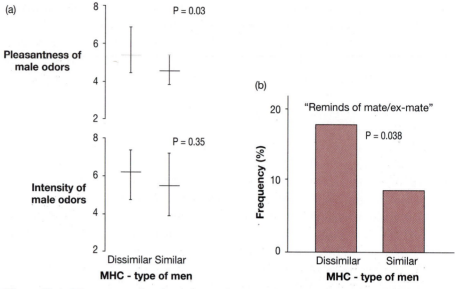

Figure 18.6 **Women prefer the odors of men with MHC genotypes different than their own** (a) Each of 31 women rated three T-shirts worn by men, whose MHC genotypes were dissimilar to her own. If we call the average of each woman's three scores her "composite" score, then the left side of each graph represents the distribution of the 31 composite scores. The horizontal line shows the median score (50th percentile), and the vertical line shows the range from the 25th percentile to the 75th percentile. The right side of each graph represents the distribution of the same 31 women's composite scores when they rated three T-shirts worn by men whose MHC genotypes were similar to their own. Women rated the odors of the two kinds of men significantly different in pleasantness, but equal in intensity.   (b) Women sniffing T-shirts reported that the odors reminded them of one of their own past or present mates more often when the shirts had been worn by men with MHC genotypes dissimilar to the women's own. Redrawn from Wedekind et al. (1995).

whether MHC-related odor differences could have any influence at all. Therefore they sought to eliminate as many confounding variables as possible, such as environmental odors that would contaminate the personal odors of their male subjects. Likewise, the researchers made sure that the women on their sniffing panel were as attentive as possible to small differences in odor. After making these preparations, Wedekind and colleagues showed that MHC-related odor differences can, in principle, influence a woman's assessment of a man's attractiveness. Much additional research will be needed to establish how important this influence is in the world outside the lab.

Research into the role of MHC compatibility in mate choice and reproductive success may seem frivolous, but it has practical implications. For example, H.N. Ho and colleagues (1994) considered MHC compatibility as a factor in whether in vitro fertilization can help couples get pregnant. Ho et al. studied 76 couples with unexplained infertility. Each of the couples had failed to get pregnant in at least three attempts with other assistive technologies (superovulation and intrauterine insemination). The researchers determined each couple's genotypes at a series of MHC loci, then treated the couples with in vitro fertilization. Of the 76 couples, 34 had successful pregnancies, 36 failed to get pregnant, and 6 got pregnant but spontaneously aborted shortly thereafter. The couples that failed to get pregnant or miscarried shared significantly more MHC alleles than the couples with successful pregnancies ($P = 0.015$). This result indicates that MHC-genotype similarity is one reason why some couples cannot get pregnant, even with in vitro fertilization. Other similar studies have produced similar results (Weckstein et al. 1991; Balasch et al. 1993). Researchers are working on immunotherapies that, combined with in vitro fertilization, may help some MHC-similar couples achieve successful pregnancies (Matsuyama et al. 1992).

Evolution and Sexual Orientation

In considering parental care and MHC-related mate choice, we simply assumed that sufficient genetic variation exists, or existed in the past, to allow the human behavior in question to evolve by natural selection. Some aspects of human behavior, however, seem to demand an analysis of standing genetic variation. One such case is sexual orientation.

Before proceeding with an evolutionary analysis of sexual orientation, we note that there is tremendous variation in sexual practice from culture to culture. For example, our own culture considers homosexual relationships between men and boys inherently abusive, and strongly prohibits them. However, in many traditional cultures in New Guinea, such relationships are an institutionalized rite of passage (see Creed 1984). Some practice of homosexual behaviors, by at least some individuals, is a broadly cross-cultural phenomenon. Furthermore, homosexuality is not confined to our species. Homosexual behaviors are common in the everyday interactions among our species' closest relatives, the common chimpanzee and the bonobo. Adult male chimpanzees, for example, mount each other as if copulating (Trivers 1985). This behavior is used for reassurance, for example when the males hear the calls of a neighboring chimpanzee group. In bonobos, adult males mount each other as an appeasement behavior, after one of them gives an agonistic display (Mori 1984; Blount 1990; De Waal 1997). These observations suggest that a capacity for at least some kinds of homosexual behaviors may be an ancestral feature of our species' sexuality.

Nonetheless, many people feel that homosexuality presents an evolutionary puzzle. This view is based on two assumptions: (1) that there is genetic variation for sexual orientation; and (2) that homosexuals have lower lifetime reproductive success than heterosexuals. If both of these assumptions are true, then we are challenged to explain the persistence of genes that reduce the evolutionary fitness of the individuals that carry them. We will consider each of the assumptions in turn.

Homosexuality presents an evolutionary puzzle, if there is genetic variation for sexual orientation and if homosexuals have lower fitness.

Dean Hamer and colleagues (1993) sought evidence of genetic variation for male sexual orientation. The researchers started by analyzing the family pedigrees of homosexual men. Some of the pedigrees suggested that there may be a gene on the X chromosome that influences whether a man is homosexual or heterosexual. In both of the pedigrees in Figure 18.7a, for example, the homosexual men in the third generation have homosexual male relatives on their mother's side of the family, but not on their father's side of the family. The men's mothers and maternal grandmothers have homosexual brothers. These pedigrees are consistent with X-linked inheritance of a gene influencing sexual orientation, but they are by no means conclusive.

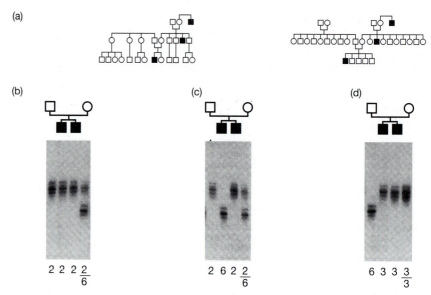

Figure 18.7   The hunt for a gene that influences male sexual orientation
(a) Pedigrees of two families suggesting X-linked inheritance of male sexual orientation. Circles represent women; open squares represent heterosexual men; solid squares represent homosexual men.   (b) A family with genotypes at marker locus DXS1108 that are consistent with a gene influencing male sexual orientation.   (c) A family with genotypes at marker locus DXS1108 that are apparently inconsistent with a gene influencing male sexual orientation.   (d) A family with genotypes at marker locus DXS1108 that provide no information about a gene influencing male sexual orientation. Reproduced by permission from D.H. Hamer et al., 1993. A linkage between DNA markers on the X chromosome and male sexual orientation. *Science* 261:324, Fig. 4, July 16, 1993. © American Association for the Advancement of Science.

Hamer and colleagues adopted as a working hypothesis a model of an X-linked gene for male sexual orientation. It will be helpful to state the model explicitly. Imagine an X-linked locus with two alleles. One allele (X^{ho}) tends to cause its male carriers to be homosexual, whereas the other allele, (X^{he}), tends to cause its male carriers to be heterosexual. Now imagine a heterozygous woman ($X^{ho}X^{he}$). If she has two sons, there is a 25% chance that she will pass X^{ho} to both sons, a 25% chance that she will pass X^{he} to both sons, and a 50% chance that she will pass X^{ho} to one son and X^{he} to the other. If both sons are $X^{ho}Y$, they will both tend to be homosexual. If both sons are $X^{he}Y$, they will both tend to be heterosexual. If one son is $X^{ho}Y$ and the other son is $X^{he}Y$, then one will tend to be homosexual and the other heterosexual.

If it exists, how could we find this hypothetical X-linked gene for male sexual orientation? Hamer and colleagues used a strategy of trial and error. The researchers investigated each of a series of polymorphic marker loci on the X chromosome. For each marker, Hamer and colleagues checked to see whether the pedigrees and individual genotypes were consistent with the model in 40 families of homosexual brothers.

Figure 18.7b illustrates the process of checking a single marker, called DXS1108, in a single family. The family pedigree is at the top of the figure. From the left, there is the father, two homosexual sons, and the mother. DXS1108 is a tandem repeat locus. Different alleles have different numbers of repeats of the same short nucleotide sequence. Hamer et al. used polymerase chain reaction (PCR) to amplify the DNA from the tandem repeat, and then ran the amplified sequences on a gel. A photograph of the gel appears under the pedigree. The dark bands on the gel represent the fragments of amplified DNA. Note first that the mother (right lane) has two bands corresponding to two sizes of amplified DNA sequence. This shows that the mother is a heterozygote; the numbers at the bottom of the gel identify the two alleles that she has. They are called 2 and 6. The sons and the father show only one band each, of course, because they are haploid for this locus. Now note that both sons inherited the same allele, allele 2, from their mother. The genotypes in this family are consistent with the possibility that DXS1108, or much more likely a gene located near DXS1108 on the X chromosome, is the hypothetical gene for male sexual orientation.

The genotypes in the family are consistent with the model, but by itself one family is not very persuasive. A pattern like that in Figure 18.7b is what we would expect half the time anyway, by chance, even if there is no gene at all on the X chromosome that has anything to do with sexual orientation.

Figure 18.7c shows a family that appears to be inconsistent with the possibility that DXS1108 is the hypothetical sexual orientation gene or is linked to it. The mother is a heterozygote, but the sons, both homosexual, inherited different alleles. One way to make this family fit the model is to posit that not every male who carries the X^{ho} allele of the sexual orientation gene is homosexual (that is, the gene has incomplete penetrance). This would almost certainly be true of any gene that influences sexual orientation. Another way to make the family fit is to argue that while the mother is heterozygous for the marker DXS1108, she must be homozygous for the actual sexual orientation gene to which the marker is linked. This is certainly possible. However, the more families we find like the one in Figure 18.7c, the less seriously we will entertain the notion that DXS1108 sits near a sexual orientation gene.

Figure 18.7d shows a family that provides no useful information at all. The mother is homozygous for DXS1108 allele 3. Thus the fact that her sons both inherited the same allele tells us nothing. They could not have inherited anything else.

Hamer and colleagues were looking for a marker that was consistent with their model of a sexual orientation gene in significantly more than 50% of the families. Most of the markers the team checked led nowhere. However, in one particular region of the X chromosome, Hamer et al. found what they were looking for. This region, called Xq28, has five markers that fit the model for a sexual orientation gene in 80% to 85% of the 40 families (combined $P < 0.00001$). Hamer and colleagues concluded that region Xq28 contains a gene, or genes, that influence male sexual orientation.

Evidence suggests that there is a gene on the X chromosome with alleles that influence sexual orientation in men . . .

Anne Fausto-Sterling and Evan Balaban (1993) were skeptical of Hamer et al.'s conclusion. They challenged Hamer and colleagues to perform a more definitive test of their hypothesis. Instead of just showing that pairs of homosexual brothers share marker alleles more often than expected by chance, Fausto-Sterling and Balaban suggested that Hamer et al. also attempt to show that homosexual/heterosexual brothers share marker alleles less often than expected by chance. If they could do so, Hamer and colleagues would make a much stronger case that there is an X-linked gene that influences male sexual orientation.

Stella Hu and colleagues (1995), working in Dean Hamer's lab, did exactly that. Looking at families that include two homosexual sons and at least one heterosexual son, Hu et al. showed that homosexual brothers share Xq28 marker alleles significantly more often than expected by chance, and that homosexual/heterosexual brothers share Xq28 markers significantly less often than expected by chance (Figure 18.8).

Hu and colleagues also looked at the sharing of Xq28 marker alleles between pairs of sisters. There is no evidence of a gene in region Xq28 that influences female sexual orientation (Figure 18.8). Elsewhere Angela Pattatucci and Dean Hamer (1995) showed that female sexual orientation runs in families, but they had insufficient evidence to determine whether this is because families share genes, environments, or both.

Assumption 1 of the homosexuality puzzle appears, then, to be at least partially true. For the men in the population that Hamer, Hu, and colleagues studied, in the environment in which the men live, there is genetic variation that influences variation in sexual orientation. However, we still have no idea what the gene in region Xq28 codes for, nor do we understand the mechanism through which the gene influences men's sexuality (see also LeVay and Hamer 1994; Byne 1994).

What about assumption 2, that homosexuals have lower lifetime reproductive success than heterosexuals? This obviously must be true for lifelong exclusive homosexuals. Many people in many cultures, however, engage in both homosexual and heterosexual behavior. In our own culture, many self-described homosexuals have children. As far as we know there are no solid data on the relative lifetime reproductive success of homosexuals versus heterosexuals. Imagine what it would take to collect such data. Simply asking people to report their sexual orientation and number of children would not be good enough. Both women and men might lie about the first, and the men might themselves be misled about the second (see Smith 1984). To get accurate data on both sexual orientation and reproductive success would require a protocol for making observations and performing genetic paternity tests that would be impractical at best, and unethical and dangerous at worst. Assumption 2 seems likely to remain a matter of speculation.

. . . but there are no reliable data on the relative fitness of homosexual versus heterosexual men.

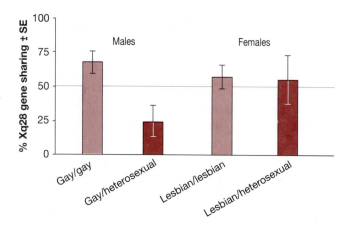

Figure 18.8 **Evidence that region Xq28 on the X chromosome contains a gene that influences sexual orientation in men but not women**   The bars show the proportion of sibling pairs that share marker alleles in region Xq28. The error bars represent the standard errors of the estimated percentages. Among men, homosexual brothers share alleles significantly more often than expected by chance, whereas homosexual/heterosexual brothers share alleles significantly less often than expected by chance (P < 0.005). Among women, both kinds of sibling pairs share alleles neither more nor less often than expected by chance. From Hu et al. (1995).

We will grant, for the sake of argument only, that both assumptions are true, and that the persistence of genes for homosexuality is a genuine evolutionary puzzle. Is there a selective scenario we can offer to solve it? The most reasonable hypothesis is probably heterozygous advantage (for a review, see Futuyma and Risch 1984).

In its original form (Hutchinson 1959), the heterozygous advantage scenario of homosexuality is analogous to the genetics of sickle cell anemia (see Chapter 5). It postulates a single autosomal locus with two alleles, one for homosexual orientation, the other for heterosexual orientation. The heterozygotes might be heterosexual or bisexual, but they would have higher survival or reproductive success than either kind of homozygote. This would maintain the gene for homosexuality in the population, even if individuals with two copies of the allele for homosexuality have reduced fitness. What might be the mechanism behind the higher fitness of the heterozygotes is anyone's guess.

In the case of an X-linked gene for male sexual orientation, like that inferred by Hamer, Hu, and colleagues, the heterozygous advantage scenario requires modification. The heterozygotes that would reap the postulated advantage would be females, while the individuals whose sexual orientation would be affected are males. An analogy for this situation is the genetics of color vision in New World primates. Squirrel monkeys, marmosets, and other New World monkeys have just one locus controlling red and green vision (Shyue et al. 1995). The locus is X-linked, and has an allele that confers red vision and an allele that confers green vision. (Humans and other Old World primates have two separate X-linked loci for color vision, one for green and one for red.) Heterozygous females can see both red and green, and are presumably at an advantage over homozygous females that can see only red or only green. Males get only one allele, so they too can see only red or green. Thus we have a case in which a male polymorphism appears to be maintained by an X-linked gene with heterozygous advantage in females.

Of course it is a rank speculation that the X-linked gene for male sexual orientation might work the same way. What function the gene might have in females, and why heterozygotes would be at an advantage, are again anyone's guesses. However, identification of the gene in region Xq28 and the discovery of its function will provide an opportunity to test the heterozygous advantage hypothesis.

18.2 Evolution and Human Health

In this section we turn from human nature to human health. In Chapter 1, we discussed one way in which an understanding of evolution is relevant to human health: Pathogens like HIV and the tuberculosis bacterium rapidly evolve resistance to the drugs used to fight them. Undertanding how this process works leads to recommendations for medical practice. Patients who start a course of treatment must finish it, so that their pathogen population is driven to extinction rather than selected for resistance. The examples we discuss here include uses of evolutionary analysis to reconstruct the course of disease in individual patients, to interpret the symptoms of disease, and to understand puzzling features of human physiology.

Tissues as Evolving Populations of Cells: Explaining a Patient's Spontaneous Recovery

All the cells in an individual's body are descended from a common ancestor, the zygote. If during the development of a tissue, a mutation occurs in a cell still capable of continued division, then we can think of the tissue as a population of reproducing cells with heritable genetic variation. If one of the genetic variants leads to increased cell survival or faster reproduction, then the tissue will evolve by natural selection, just like a population of free-living organisms.

Rochelle Hirschhorn and colleagues (1996) documented a case in which tissue evolution saved the life of a boy with a serious genetic disease. The disease is called adenosine deaminase deficiency. Adenosine deaminase (ADA), encoded by a locus on chromosome 20, is a housekeeping enzyme normally made in all cells of the body. ADA's job is to recycle purines. Cells lacking ADA accumulate two poisonous metabolites, deoxyadenosine and deoxyadenosine triphosphate. The cells in the body most susceptible to these poisons are the lymphocytes, including the T cells and B cells vital to the immune system (Youssoufian 1996). Individuals who inherit loss-of-function mutations in both copies of the ADA gene have no T cells, and B cells that are nonfunctional or absent (Klug and Cummings 1997). In consequence, these individuals suffer from severe combined immunodeficiency. Without treatment, they usually die of opportunistic infections at an early age.

Both parents of the boy that Hirschhorn and colleagues studied carry in heterozygous state a recessive loss-of-function allele for ADA. One of the boy's older siblings inherited two loss-of-function alleles, made no ADA, and died of severe combined immunodeficiency at age 2. The boy himself also inherited both his parents' mutant alleles, and during his first five years he suffered the recurrent bacterial and fungal infections characteristic of severe combined immunodeficiency. Between the ages of 5 and 8, however, the boy spontaneously and mysteriously recovered. He was 12 years old when Hirschhorn and colleagues published their paper, and he had been clinically healthy for four years.

With a careful genetic analysis of the mother, father, and son, Hirschhorn and colleagues were able to reconstruct a plausible explanation for the boy's recovery. Although the boys' parents are both carriers for ADA deficiency, they are carriers for different loss-of-function alleles. Hirschhorn and colleagues showed that the son's blood cells are a genetic mosaic (Box 18.2). The father's mutation is present in all of the boy's peripheral leukocytes (white blood cells) and lymphoid B cells. The mother's mutation is present in all of the boy's peripheral leukocytes, but absent in most of his B cells.

How could this happen? Hirschhorn and colleagues found evidence that the ancestral cell to most of the boy's existing lymphoid B cells had sustained a lucky back-mutation in the allele the boy inherited from his mother, thus spontaneously reverting to wild type (Box 18.2). Over time, the descendants of this reverted cell apparently became more and more abundant in the boy's B-cell population. Eventually reverted B cells became abundant enough, and made and released enough ADA, that the boy's clinical symptoms of ADA deficiency vanished.

Hirschhorn and colleagues believe that the increase in frequency of reverted B cells in the boy's B-cell population happened by natural selection. It is possible that the increase happened by drift alone. It is likely, however, that reverted cells are at a distinct selective advantage. Because they make their own supply of a crucial housekeeping enzyme, they should live longer than cells that have to pick up the enzyme after it has been released by the cells that make it.

The boy's story may have implications for the treatment of other individuals with ADA deficiency. The outlook for patients with ADA deficiency has improved dramatically in recent years. In 1987 researchers developed a form of injectable ADA that is an effective enzyme replacement treatment for many ADA patients (Hershfield et al. 1987). In the early 1990s, researchers began the first clinical experiments with somatic-cell gene therapy (Bordignon et al. 1995; Blaese et al. 1995). Somatic-cell gene therapy involves removing lymphocytes and/or bone marrow cells from the patient, inserting into the cells' chromosomes a functioning version of the ADA gene with its own promoter, and returning the cells to the patient's body. In other words, gene therapy is an attempt to accomplish by design the reversion mutation that happened spontaneously in the boy studied by Hirschhorn and colleagues. The early trials have shown that the engineered cells can survive for years, and that they can grow and divide. In some early trials, gene therapy appears to have been responsible for dramatic improvements in the patients' clinical health.

As a precaution against the failure of gene therapy, researchers conducting gene therapy trials have kept their patients on enzyme replacement therapy. Hirschhorn et al. suggest, however, that enzyme replacement may reduce the effectiveness of gene therapy. If enzyme replacement reduces the selective advantage the engineered cells have over ADA-deficient cells, it will slow the rate of evolution by natural selection in the patients' blood cell population. If further research supports this suggestion, then in the future doctors will have to balance the benefits of encouraging rapid selective fixation of the engineered cells against the risks of depriving patients of the insurance provided by continued enzyme replacement therapy.

Tissues as Evolving Populations of Cells: Reconstructing the History of a Cancer

Another medical context in which it is productive to view tissues as evolving populations of cells is cancer (Shibata et al. 1996). A cancer starts with a cell that has accumulated mutations that free it from the normal controls on cell division. The cell divides, and its offspring divide, and so on, to produce a large population of descendants—a tumor. The cells in some kinds of cancer have extremely high mutation rates, allowing tumors to accumulate measurable genetic diversity. If we assume that the mutation rate per cell division is constant, and that the mutations are neutral, then the genetic diversity within a particular tumor is a measure of the tumor's age.

Occasionally a cell may leave the tumor it was born in and migrate elsewhere in the body to initiate a new tumor. (This process is called metastasis.) This new tumor represents a new population of cells. Because it was founded by a single individual, this new population will have low genetic diversity. As it grows, however, the population will evolve. Like the population from which its founder came, the new tumor will accumulate genetic diversity as a result of mutation and genetic drift.

Analyses of genetic variations among tumor cells can be used to reconstruct the history of a patient's cancer.

Darryl Shibata and colleagues (1996) use the amount of genetic variation in tumors to reconstruct the history of cancers within individual patients. Their approach is similar to that used by Sarah Tishkoff and colleagues to reconstruct the history of populations in the species *Homo sapiens* (Chapter 14). Non-

---

BOX 18.2   Genetic sleuthing solves a medical mystery

Figure 18.9 shows the nucleotide sequence of a short piece of the ADA gene. Both parents have point mutations substituting an A for a G. The father's point mutation is in an intron/exon splice site, resulting in the deletion of an entire exon from the processed ADA transcript. The mother's point mutation results in the substitution of one amino acid for another in the primary structure of the ADA enzyme. Both mutations are known from other individuals who have them to virtually destroy the enzymatic activity of ADA.

Notice that the wild-type sequence contains a cutting site for the restriction enzyme *Hin*p1I, and a cutting site for the restriction enzyme *Bsr*I. The father's mutation eliminates the *Bsr*I cutting site, whereas the mother's mutation eliminates the *Hin*p1I cutting site. Hirschhorn and colleagues amplified a 254-bp long sequence from the ADA genes in both parents and in different populations of the boy's blood cells. When digested with *Bsr*I, the wild-type 254-bp fragment yields a 182-bp fragment and a 72-bp fragment. Figure 18.10a illustrates a *Bsr*I restriction fragment analysis of the family. Look first at the mother's lane on the gel. Both her wild-type allele and her mutant allele have the *Bsr*I cutting site, so her gel shows only two bands: the 182-bp fragment and the 72-bp fragment. Now look at the father's lane. His wild-type allele has the *Bsr*I cutting

site, so his lane shows the 180-bp and 72-bp bands. His mutant allele, however, lacks the *Bsr*I cutting site, so his lane also shows a 254-bp fragment. Finally, look at the boy's lanes. They show that the father's mutant allele is present in both the boy's peripheral leukocytes (white blood cells) and his lymphoid B cells.

Figure 18.10b illustrates a *Hin*p1I restriction fragment analysis. This time, the mother's mutant allele shows up as an uncut 254-bp fragment. The son's lanes show something unexpected: His mother's mutant allele is present in his peripheral leukocytes, but absent in most of his lymphoid B cells. We say most, because there is a faint band from a 254-bp fragment in the boy's B-cell lane. Closer examination of 15 different B-cell lines revealed that all carry the father's mutant allele, but only two carry the mother's mutant allele.

Hirschhorn and colleagues discovered that the mother's mutant allele also contains a unique neutral marker. When they checked the boy's lymphoid B cell lines that appear to lack the mother's mutant allele, the researchers found that these cells in fact carry the mother's neutral marker. Hirschhorn and colleagues concluded that the ancestral cell to most of the boy's existing lymphoid B cells had sustained a lucky back-mutation in the allele from the mother, thus spontaneously reverting to wild type.

Figure 18.9 A short piece of the gene for adenosine deaminase Point mutations are shown in bold; the bars indicate sites recognized by the restriction enzymes *Hin*p1I and *Bsr*I. From Hirschhorn et al. (1996).

	*Hin*p1I		*Bsr*I
Wild type:	...G C G C C A C C A G	C C C A G T...
Father's mutation:	...G C G C C A C C A G	C C C A **A** T...
Mother's mutation:	...G C **A** C C A C C A G	C C C A G T...

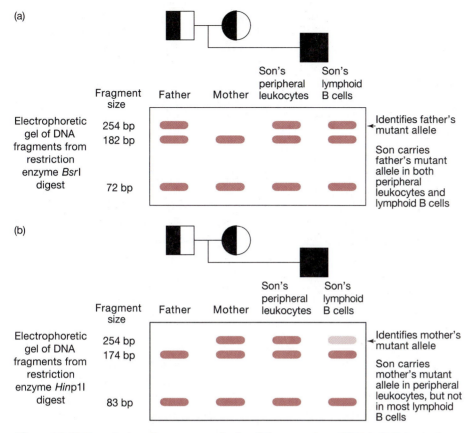

(a)

Electrophoretic gel of DNA fragments from restriction enzyme *Bsr*I digest

| Fragment size | Father | Mother | Son's peripheral leukocytes | Son's lymphoid B cells | |

254 bp — Identifies father's mutant allele

182 bp

72 bp — Son carries father's mutant allele in both peripheral leukocytes and lymphoid B cells

(b)

Electrophoretic gel of DNA fragments from restriction enzyme *Hin*p1I digest

254 bp — Identifies mother's mutant allele

174 bp

83 bp — Son carries mother's mutant allele in peripheral leukocytes, but not in most lymphoid B cells

Figure 18.10 **Restriction fragment analysis of the genetics of ADA deficiency in a family** (a) A *Bsr*I restriction fragment assay for the father's mutant allele. The 254–bp band at the top of the gel is a marker for the presence of the father's mutation, which eliminates a cutting site recognized by *Bsr*I. (b) A *Hin*p1I restriction fragment assay for the mother's mutant allele. The 254–bp band at the top of the gel is a marker for the presence of the mother's mutation, which eliminates a cutting site recognized by *Hin*p1I. After Hirschhorn et al. (1996).

African populations have lower genetic diversity than African populations. With other evidence, this pattern indicates that non-African populations were founded by emigrants from Africa.

Figure 18.11a shows the allelic diversity at a tandem repeat locus in subpopulations of two adjacent tumors in a 43-year-old patient. The patient has a cancer called a colorectal adenocarcinoma. One tumor, called an adenoma, is in glandular tissue; the other tumor, called a carcinoma, is in colorectal tissue. Three of the four adenoma subpopulations show substantially greater genetic diversity than the carcinoma subpopulations. Figure 18.11b shows estimates of the ages of the eight subpopulations, based on the estimated mutation rate for this kind of cancer and the assumption that each cancer cell divides once per day. It appears that the adenoma grew benignly for up to 10 years before it spawned a cell that migrated into the patient's colon to initiate the malignant carcinoma.

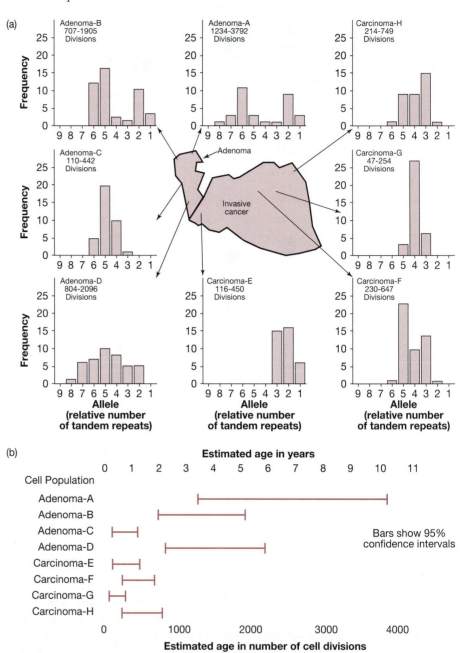

***Figure 18.11* An evolutionary reconstruction of the history of a cancer within an individual patient** (a) The diagram at the center is a map of a pair of neighboring tumors in a 43-year-old patient. At left is an adenoma, a tumor of glandular tissue that is usually benign. At right is a colorectal carcinoma, an invasive malignant tumor. The histograms that surround the map show the genetic diversity within samples of cells taken from the locations indicated by the arrows. Each histogram shows the frequencies of various alleles at a tandem repeat locus. Different alleles are identified by the number of tandem repeats they carry. Some cell samples, such as Adenoma-D, show high genetic diversity (many different alleles present). Other cell samples, such as Carcinoma-E, show low genetic diversity. From Shibata et al. (1996).   (b) Estimates of the age of the different tumor subpopulations shown in (a). The higher genetic diversity of the samples from the adenoma indicates that the adenoma is older than the carcinoma. Redrawn from Shibata et al. (1996).

Selection thinking suggests two hypotheses about the adaptive significance of fever: (1) Fever is adaptive for the pathogens that induce it, and (2) fever is adaptive for the host that manifests it.

Symptoms of Disease as Adaptive Defenses

Many people consider the symptoms that accompany illness to be a nuisance. A common response to the fever associated with a cold or flu, for example, is to take aspirin, acetaminophen, or ibuprofen. These drugs reduce the fever, but they do not combat the virus that is causing the cold or flu. Here we ask whether taking drugs to reduce fever is a good idea. To answer the question, we need to know why people get a fever when they are sick.

An evolutionary perspective suggests two interpretations of fever. One interpretation is that fever may represent manipulation of the host by the pathogen. Viruses or bacteria may release chemicals that cause the host to elevate its body temperature so as to increase the pathogen's growth rate and/or reproductive rate. If this hypothesis is correct, then reducing the fever would probably help the host combat the infection. The second interpretation is that fever may be an adaptive defense against the pathogens. The pathogens may grow and reproduce more slowly at higher temperatures, or the host's immune response may be more effective. If this hypothesis is correct, then taking drugs that alleviate fever might be counterproductive to recovery.

Matthew Kluger has for over 20 years advocated the second hypothesis—that fever is an adaptive defense against disease. In 1974, Linda Vaughn, Harry Bernheim, and Kluger discovered that the desert iguana *(Dipsosaurus dorsalis)* develops a behavioral fever in response to infection with a bacterium called *Aeromonas hydrophila*. Recall from Chapter 8 that lizards, being ectotherms, use behavior instead of physiology to regulate their body temperature. They move to hot spots to warm themselves, and cold spots to cool themselves. Vaughn et al. found that when they injected desert iguanas with dead bacteria, the lizards chose body temperatures about 2°C higher than they normally do (Figure 18.12a and b).

Is behavioral fever an adaptive response to infection, or are the bacteria manipulating the lizards? To distinguish between these hypotheses, Daniel Ringler and Miriam Anver (1975) infected desert iguanas with live bacteria, then prevented the lizards from thermoregulating by keeping them in fixed-temperature incubators. Most of the lizards kept at temperatures mimicking behavioral fever survived, whereas most of the lizards kept at lower temperatures died (Figure 18.12c). This result suggests that behavioral fever is, in fact, adaptive for desert iguanas infected with *A. hydrophila*.

Research demonstrates that when lizards are infected with a bacterium, fever is adaptive for the host.

If fever is an adaptive defense against *A. hydrophila,* then it would probably be a bad idea for infected lizards to take aspirin. This is not as silly a statement as it sounds, at least in one sense: The researchers found that the aspirin-like drug sodium salicylate reduces behavioral fever in lizards just as it reduces physiological fever in mammals. Apparently thermoregulation is controlled by similar neurological mechanisms in both groups of animals. Bernheim and Kluger (1976) infected 24 desert iguanas with bacteria, then gave half of the infected lizards sodium salicylate. The researchers allowed all the lizards to behaviorally thermoregulate. All of the control lizards developed behavioral fever, and all but one of them survived the infection. Five of the medicated lizards developed behavioral fever in spite of the medication, and all of them survived. The other seven medicated lizards failed to develop behavioral fever, and all of them died (Figure 18.12d).

Since the mid-1970s, researchers have documented behavioral fever in a wide variety of reptiles, amphibians, fishes, and invertebrates. In several animal studies researchers have shown that fever increases survival (see Kluger 1992 for a review). These results broadly support the hypothesis that fever is an adaptive response to infection. However, it is unclear whether these results are generalizable to humans. Fewer clinical studies have been done on this question than one might expect (Kluger 1992; Green and Vermeulen 1994). We review two studies, neither of which is conclusive.

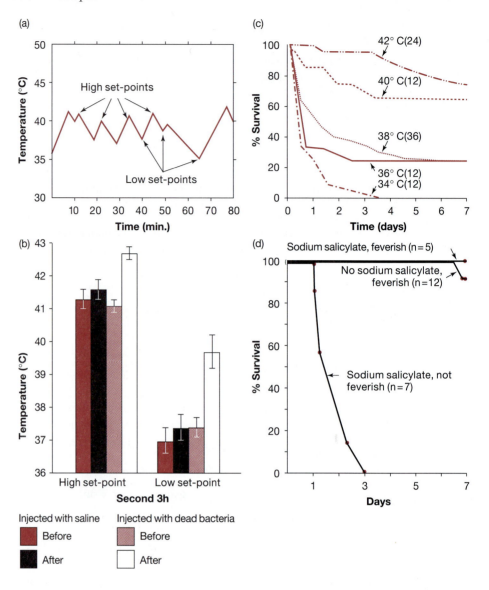

Injected with saline
■ (red) Before
■ (black) After

Injected with dead bacteria
■ (pink) Before
□ (white) After

Fever and Chickenpox

Timothy Doran and colleagues (1989) studied 68 children with chickenpox. After getting the informed consent of the children's parents, the researchers divided the children at random into two groups. The experimental group took acetaminophen, a common over-the-counter antifever medicine. The control group took a placebo (pills that looked like acetaminophen, but contained no medicine). The assignment of children to study groups was double-blind: Neither the researchers nor the parents (nor the children) knew which children were in which group until after the study was over.

By most measures of the duration and severity of illness, there was no difference between the acetaminophen and placebo groups. Where the results hint at a difference, the children taking the placebo recovered faster. It took less time, on average, for all of the vesicles to scab over in the placebo children (5.6 ± 2.5 days) than in the acetaminophen children (6.7 ± 2.3 days). By itself, this result

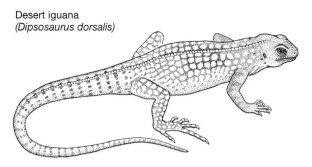

Desert iguana
(Dipsosaurus dorsalis)

Figure 18.12   Behavioral fever in the desert iguana *(Dipsosaurus dorsalis)* (a) Temperature trace for a thermoregulating desert iguana. This lizard was in a terrarium with a temperature of 30°C on one side and 50°C on the other. The lizard stayed at the warm end until its temperature climbed to about 41°C, then moved to the cool end until its temperature fell to about 37.5°C, then moved back to the warm end again. The high and low setpoints are the temperatures at which the lizard moved from one end to the other. From Vaughn et al. (1974).   (b) Linda Vaughn and colleagues injected saline into nine control lizards. The dark-filled and black bars show the average set-points (± standard error) for these lizards before and after the injection. There was no significant change in either setpoint. Vaughn and colleagues injected killed bacteria *(Aeromonas hydrophila)* into 10 experimental lizards. The light-filled and white bars show the average set-points for these lizards before and after the injection. The increase in both set-points after injection with killed bacteria was statistically significant (P < 0.001). From Vaughn et al. (1974).   (c) The graph shows the fraction of infected lizards surviving at each temperature over time (numbers of lizards in parentheses). From Ringler and Anver (1975).   (d) Twelve lizards infected with *A. hydrophila* and given no antifever medication developed a behavioral fever, and all but one survived. Five of 12 infected lizards given antifever medication developed behavioral fever anyway, and all survived. The other seven infected lizards given antifever medication failed to develop behavioral fever, and all died within three days. From Bernheim and Kluger (1976).

is statistically significant at P = 0.048; but given that several other measures failed to turn up any difference, it is not persuasive. (Remember that if we do 20 statistical tests, one of them is likely to be significant at P < 0.05 simply due to chance.) The children's itching seemed to subside faster in the placebo group (Figure 18.13), but this pattern is not statistically significant.

The simple interpretation is that the antifever medicine acetaminophen has little or no effect on the course of chickenpox, and therefore that fever is

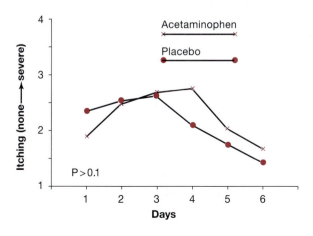

Figure 18.13   Does the antifever medicine acetaminophen have any effect on the course of chickenpox? The graph shows the severity of itching as a function of time (as judged by the children's parents) in 37 children taking acetaminophen versus 31 children taking a placebo. The placebo group appears to have recovered faster, but the difference is not significant. From Doran et al. (1989).

Research on humans has not established whether fever is an adaptive host response to chicken pox . . .

neither adaptive nor maladaptive as a defense against the disease. Kluger (1992) points out, however, that only slightly over half the children in the study ran a fever (defined by Doran et al. as a temperature of 100.4°F or higher), and that the fraction of children who ran a fever was the same in the acetaminophen group (57%) and the placebo group (55%). Kluger's interpretation is that the acetaminophen children were not given enough medicine to reduce fever, and thus that the study did not test the hypothesis that fever is adaptive. Kluger's analysis illustrates that it is important to critically examine a study's methods and results before accepting the conclusions.

Fever and the Common Cold

Neil Graham and colleagues (1990) intentionally infected 56 consenting adult volunteers with rhinovirus type 2, one of the viruses that can cause the common cold. Assigned to groups double-blind and at random, 14 subjects took a placebo. The rest took common over-the-counter antifever medications: 13 took ibuprofen, 15 took aspirin, and 14 took acetaminophen. The volunteers taking the placebo suffered less nose stuffiness (Figure 18.14a), and made more antibodies against the rhinovirus (Figure 18.14b) than did the volunteers taking antifever medicines. The reason for the reduced antibody response in volunteers taking medicine may be that the medicines prevented monocytes, a class of white blood cells, from moving from the blood to the infected tissues (Figure 18.14c). Once in the infected tissues, monocytes differentiate into macrophages, which help mount an immune response against the virus (Graham et al. 1990).

. . . nor whether fever is an adaptive host response to the common cold . . .

This time, the simple interpretation is that the antifever medications interfered with the immune response to the common cold, and therefore that fever is an adaptive defense against the disease. Kluger (1992) points out, however, that few of the subjects in the study ran a fever; and the fraction of subjects taking the placebo who ran a fever (14%) was not significantly higher than the fraction of subjects taking medicine who did so (7%). Kluger's interpretation is that few people infected with rhinovirus type 2 run a fever, and thus that the study did not test the hypothesis that fever is adaptive.

. . . and because antifever medications can alter other aspects of the host's immune response, definitive studies will be difficult.

The study by Graham and colleagues did show, however, that antifever medicines interfered with the immune response to the virus. This demonstrates that antifever medicines have multiple physiological effects. This fact will make it extremely difficult to design studies on the adaptive significance of fever in mammals. Studies using traditional antifever medicines cannot separate fever from other aspects of the immune response.

Fever and Medical Practice

Even if researchers find proof that fever in humans is, indeed, an adaptive response to some infections, no responsible doctor (or evolutionary biologist) would suggest that it is always a bad idea to suppress fever. There are several reasons:

- It may be that fever is an adaptive response against some pathogens, but not others. Some bacteria or viruses may grow and reproduce faster at fever temperatures than at normal temperatures. In other words, the adaptive-response and pathogen-manipulation-of-host hypotheses may be mutually exclusive for any particular pathogen, but they are not mutually exclusive across all pathogens.

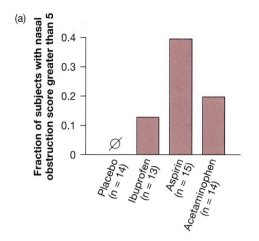

(a)

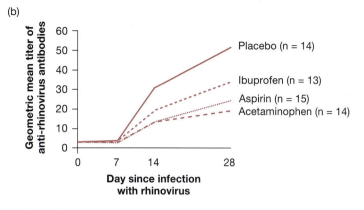

(b)

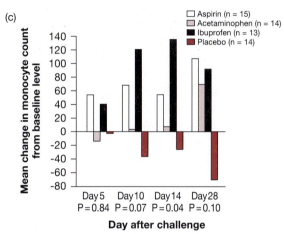

(c)

Figure 18.14 **Do antifever medicines have any effect on the course of the common cold?**   (a) The graph shows the fraction of volunteers in each group with "nasal obstruction scores" (a measure of stuffiness) over 5. Volunteers taking a placebo had less-stuffy noses than volunteers taking one of three antifever medicines. The difference among groups is significant at P = 0.02.   (b) Volunteers taking a placebo made more antibodies against the rhinovirus that caused their cold than did volunteers taking one of three antifever medicines. On day 28, the difference between the placebo group and the other three groups combined was significant at P = 0.03. (c) Monocytes are white blood cells that circulate to infected tissues, then leave the blood and differentiate into macrophages. Volunteers taking a placebo showed a drop in the concentration of monocytes in their blood over time (indicating that the cells had moved into the tissues), whereas volunteers taking one of three antifever medicines showed an increase in the concentration of monocytes. From Graham et al. (1990).

- Even when fever is beneficial, it carries costs as well (Neese and Williams 1995). In the case of mild illness and low fever, sometimes the benefits of antifever medicines in alleviating symptoms and allowing people to continue to their normal activities outweigh the costs of a somewhat diminished immune response. In the case of serious illness and high fever, fever can itself deplete nutrient reserves and can even cause temporary or permanent tissue damage.

- There are circumstances in which fever may cause damage directly, unconnected with its role in infections. For example, experiments with animals and observational studies of humans suggest that fever following a stroke causes neurological damage and reduces the likelihood of survival (Azzimondi et al. 1995).

More research is needed on the adaptive signifance of fever in humans, and on the costs and benefits of using various antifever medications.

Physiological Puzzles

Some features of human physiology are puzzling, because they seem to be maladaptive. Myopia, or nearsightedness, is an example. Most readers probably know people who are so myopic that it is difficult to imagine how they could have survived, let alone reproduced, if they had lived in a hunter-gatherer society.

How could natural selection have allowed the persistence of such poor genes for eyesight? The answer is that it did not, or at least not exactly. Recall that our species spent most of its evolutionary history living a hunter-gatherer life-style. Our hunter-gatherer ancestors spent most of their time outdoors, and they did not read. Myopia is induced, in the 25% or so of people who inherit a susceptibility to it, by close visual work done during childhood (see Neese and Williams, 1995, for a review). Very few contemporary people living as hunter-gatherers develop myopia, even while a quarter of the urbanites of the same ethnicity do. Natural selection tolerated the genes for a susceptibility to myopia because those genes did not cause myopia in the environment to which we are adapted.

Some apparently maladaptive human traits result from our exposure to novel environments.

Other features of our physiology that seem to be deleterious to evolutionary fitness may turn out, on closer examination, to be adaptive. Pregnancy sickness is an example. Pregnancy sickness is characterized by a cluster of symptoms that often occur during the first trimester of pregnancy, including food aversions, nausea, and/or vomiting. Among other things, pregnancy sickness has been interpreted as a nonfunctional by-product of the hormonal changes associated with pregnancy, or a subconscious attempt to terminate the pregnancy (see Deuchar 1995 for a review). Superficially, pregnancy sickness appears maladaptive. A pregnant women needs to eat enough to satisfy not only her own needs for energy and nutrients, but also the needs of her developing child. How could natural selection allow the persistence of food aversions and nausea, which logically would reduce a woman's food intake? Margie Profet argues that once we recognize the salient features of pregnancy sickness, it becomes clear that pregnancy sickness is adaptive (Profet 1988; 1992; 1995).

Other apparently maladaptive traits may turn out, on closer inspection, to be adaptive.

An Adaptive Hypothesis for Pregnancy Sickness

Selection thinking suggests that pregnancy sickness is an adaptation that limits the embryo's exposure to chemicals that can disrupt development.

Profet's hypothesis is that pregnancy sickness serves to prevent women from eating foods containing chemicals that could disrupt embryonic development and cause embryonic death or birth defects. All plants produce toxins (Ames et al. 1990a; 1990b). The adaptive value of these toxins for the plants is precisely that they are poisonous. They deter attack by fungi, insects and other herbivores and omnivores. Organisms that eat plants, including humans, have evolved specific and elaborate countermeasures against the natural pesticides made by plants. These countermeasures include the detection of poisons before they are eaten (by smell, taste, and memory of bad experiences), and the enzymatic detoxification of poisons after they are eaten.

Developing embryos are more vulnerable to plant toxins than are their mothers, because toxins can interfere with limb and organ formation. For example, among the poisons in potato sprouts are the molecules α-chaconine and α-solanine. When fed to pregnant Syrian hamsters, even at doses nontoxic to the mothers, these poisons can cause serious neural tube defects in the pups, such as anencephaly and spina bifida (Renwick et al. 1984). Potato tubers (the part of the plant we eat) also contain α-chaconine and α-solanine. The poisons are at low concentration immediately after harvest, higher concentration after storage, and still higher concentration after the tubers turn green or sprout. These toxins may be the reason why cultures heavily dependent on potatoes have the highest rates of anencephaly and spina bifida in the world. Potatoes are more toxic in years of bad potato blight, and the severity of blight is strongly correlated with the incidence of anencephaly and spina bifida (Renwick et al. 1984).

Profet reviews several salient features of pregnancy sickness that suggest it has adaptive value in protecting the developing embryo. Pregnancy sickness is a nearly universal feature of the first trimester of pregnancy. Estimates of the fraction of pregnant women who experience nausea range from 75% to 89%, and estimates of the fraction who vomit are usually near 55%. If food aversions were included in the estimates, and if women were surveyed during the first trimester, when most actually have pregnancy sickness, the fraction of women reporting symptoms would be even higher. Pregnancy sickness is broadly cross-cultural. In particular, to the extent that anthropological sources discuss the issue, pregnancy sickness appears to be common in modern hunter-gatherer women. Pregnancy sickness coincides with the period of limb and organ formation, during which time the developing embryo is most vulnerable to plant toxins. Limb and organ formation occur between about day 20 and day 56. Pregnancy sickness usually starts between day 14 and day 24, peaks between day 42 and day 56, and usually ends completely by day 98. In many women the symptoms of pregnancy sickness are triggered by bitter or pungent odors and tastes, which are cues of plant and bacterial toxins in food. Finally, the selective pressure to protect the developing embryo from plant toxins would have been particularly strong for our hunter-gatherer ancestors, who almost certainly ate a much broader array of plants than most modern humans.

How might we test Profet's hypothesis? It would be unethical to conduct experiments, so we have to rely on epidemiological evidence. If pregnancy sickness protects the developing embryo, then we can make two predictions: (1) women who experience pregnancy sickness should have lower rates of miscarriage, and fewer infants with birth defects, than women who do not experience pregnancy sickness; and (2) if taking drugs that suppress pregnancy sickness causes pregnant women to eat toxic foods in larger quantities than they would without medication, then women using such drugs should have more miscarriages, and more infants with birth defects.

The embryo-protection hypothesis of pregnancy sickness makes specific predictions . . .

Pregnancy Sickness, Miscarriage, and Birth Defects

Some of the epidemiological evidence on pregnancy sickness appears consistent with prediction 1. For example, Ronald Weigel and Margaret Weigel (1989) combined the results from a number of studies and found that women who experienced nausea and vomiting in pregnancy had a risk of miscarriage that was less than half that for women who did not have these symptoms. Asher Bashiri and colleagues (1995) found that women hospitalized for severe pregnancy sickness had miscarriages at a rate of 3.1%, compared to the rate of 15% reported for the general population. As predicted, both of these studies show that women who experience pregnancy sickness have lower rates of miscarriage.

However, both of these studies were observational, not experimental. It is possible that both pregnancy sickness and miscarriage risk are associated with some other factor. A good candidate is the hormone human chorionic gonadotropin (HCG). HCG is responsible both for maintaining pregnancy and for inducing pregnancy sickness. In some women, low levels of HCG may result in both the absence of pregnancy sickness and a failure to

. . . but these predictions are either complicated by potentially confounding variables . . .

maintain the pregnancy. In such cases, absence of pregnancy sickness would be associated with miscarriage, but it would not be the cause of miscarriage (Profet 1988; 1995).

Other epidemiological studies have yielded results inconsistent with prediction 1. Mark Klebanoff and James Mills (1986) followed the pregnancies of over 16,000 women, and found no significant differences in the rate of birth defects between women who vomited during pregnancy and women who did not. However, many women who experience pregnancy sickness do not vomit, so this result is not definitive.

Antinausea Drugs, Miscarriage, and Birth Defects

The epidemiological evidence does not appear to be consistent with prediction 2. For example, in a study by Stig Kullander and Bengt Källén (1976), 4,954 women had normal, live-born infants, and 423 women miscarried. If we divide these women into a group that used antinausea drugs and a group that did not, 3.6% of the women who used drugs miscarried versus 8.4% of the women who did not use drugs (Table 18.1a). This pattern contradicts the prediction that women who use antinausea drugs will have more miscarriages.

Is it possible that antinausea drugs actually help *prevent* miscarriage? Probably not. The evidence on prediction 1 showed that women who have pregnancy sickness have a lower rate of miscarriage, and it is the women who have pregnancy sickness who take antinausea drugs. If we divide the women in Kullander and Källén's study according to whether they reported experiencing pregnancy sickness, 5.5% of the women reporting sickness miscarried versus 13.9% of the women not reporting sickness (Table 18.1b). We can attempt to control for pregnancy sickness by looking only at the women who reported experiencing sickness, and dividing them into a group that used drugs and a group that did not. This procedure reveals that 3.6% of the women who used drugs miscarried versus 5.9% of the women who did not use drugs (Table 18.1c). This difference is statistically significant, but caution is in order. It is still possible that the women who used antinausea drugs had more severe pregnancy sickness than the women who did not use drugs.

So Kullander and Källén's study does not establish that antinausea drugs are protective. However, our discussion of the study has gotten us turned around. The hypothesis we set out to evaluate was that antinausea drugs are dangerous. Kullander and Källén's study certainly provides no evidence of that.

Finally, we turn to the prediction that women using antinausea drugs will have more infants with birth defects. Studies of the once-widely-used antinausea drug Bendectin have failed to establish that its use was associated with a significantly increased risk of birth defects (McKeigue et al. 1994; Brent 1995; Neutel and Johansen 1995). Reviewing studies from the literature, McKeigue and colleagues calculated the relative risk of having a malformed infant for mothers who took Bendectin versus mothers who did not. Their estimate of the relative risk was 0.95. In other words, mothers who took Bendectin appear to have had a slightly *reduced* risk of having infants with birth defects. We cannot conclude that Bendectin is dangerous. However, such studies have also failed to establish that Bendectin is absolutely safe. The 95% confidence interval of McKeigue and colleagues' estimate of the relative risk runs from 0.88 to 1.04.

TABLE 18.1 Antinausea drugs and miscarriage

(a)   All women, by use of antinausea drugs

	Used drugs	No drugs
Normal birth	591 (96.4%)	4363 (91.6%)
Miscarriage	22 (3.6%)	401 (8.4%)
Total	613 (100%)	4764 (100%)

$\chi^2 = 17.5; P < 0.001.$

(b)   All women, by report of pregnancy sickness

	Reported sickness	No sickness
Normal birth	3645 (94.5%)	1309 (86.1%)
Miscarriage	212 (5.5%)	211 (13.9%)
Total	3857 (100%)	1520 (100%)

$\chi^2 = 105.8; P \ll 0.001.$

(c)   Women who reported pregnancy sickness, by use of antinausea drugs

	Used drugs	No drugs
Normal birth	591 (96.4%)	3054 (94.1%)
Miscarriage	22 (3.6%)	190 (5.9%)
Total	613 (100%)	3244 (100%)

$\chi^2 = 5.1; 0.02 < P < 0.05.$

Source: Calculated from Kullander and Källén (1976).

In other words, McKeigue and colleagues' results do not exclude the possibility that infants whose mothers took Bendectin had an increased risk of birth defects as high as 4%.

There appears to be no definitive evidence to confirm the prediction that antinausea drugs will increase the risk of miscarriages and birth defects. However, we made this prediction on the untested assumption that women using such drugs will eat toxic foods in larger quantities than they would without medication. Drugs like Bendectin may primarily serve to bring women with moderate-to-severe pregnancy sickness down into the low-to-moderate range, and thus not actually eliminate the food aversions of their users (Profet, personal communication). We conclude that Profet's hypothesis has not been adequately tested.

. . . or have not been adequately tested.

To fairly evaluate Profet's hypothesis, we need a large study that correlates women's symptoms of pregnancy sickness, the level and kind of plant toxins in the food they eat before and during pregnancy, and their rates of spontaneous abortion and birth defects. Such a study remains to be done.

Profet's Hypothesis and Medical Practice

If Profet's hypothesis proves to be correct, then it will have implications for the management of pregnancy sickness. Profet (1995) asserts that "[b]ecause pregnancy sickness serves an important function, women should not try to suppress it with medicine unless it is so severe that they can't hold down any food." Profet also recommends against natural and herbal remedies for pregnancy sickness, such as ginger and herbs. While these treatments may alleviate nausea, they may themselves contain toxins dangerous to the developing embryo. Profet suggests that women who do not experience pregnancy sickness nonetheless behave as though they have it, avoiding plant foods that contain high concentrations of toxins. At the same time, she strongly emphasizes that pregnant women should be sure to maintain an adequate intake of vitamins, such as folic acid, that are essential to proper embryonic development.

18.3 Evolution and Future Human Populations: Eugenics

In section 18.2 we briefly discussed the genes associated with myopia. Some 25% of the people in our present-day population have inherited these genes, which predispose their owners to become nearsighted in a modern industrial environment. Given the array of technological options available for correcting myopia—including glasses, contact lenses, and even surgery—the genes for myopia are probably not under selection. Instead, the myopia genes are probably free to drift. Imagine that several thousand years from now the myopia genes have drifted to fixation. Everyone in the population has to have the lenses in their eyes replaced with artificial implants. When they look back on our own time, what will the historians say about us? Will they consider us to have been irresponsible? Perhaps they will write, "Even before the year 2000, our ancestors understood the rules of genetics and the mechanisms of evolution. They could see that the myopia genes were drifting and might well rise to fixation one day. And yet they did nothing to prevent it!"

Our knowledge of the mechanisms of evolution give us the potential to direct the evolution of human populations.

Biologists began to ponder scenarios like these almost as soon as the ink was dry on the first edition of *On the Origin of Species.* In 1865 Charles Darwin's cousin, Francis Galton, published an article entitled *Hereditary Talent and Character.* In it, Galton attempted to show that the intellectual and emotional traits associated with social and economic success are heritable. He then advocated social policies intended to guide human evolution so as to increase the frequency of these traits in the population. In 1883 Galton coined the term *eugenics;* it is Greek for "good birth" or "well born." Early volumes of the *Journal of Heredity* contain a mission statement quoting Galton's definition of eugenics as "the study of agencies under social control that may improve or impair the racial qualities of future generations, either physically or mentally." Eugenics quickly gathered enthusiastic ranks of students, researchers, and practitioners. Indeed, many of the foundations of modern population genetics, quantitative genetics, and statistics were developed by scientists whose primary interest was eugenics (for a history, see Kevles 1995).

Eugenecists advocate social control over the evolution of human populations.

Classical Eugenics: An Example

In the late 19th and early 20th centuries, eugenics attracted many prominent supporters . . .

Among the staunch advocates of eugenics in the United States was Alexander Graham Bell, the inventor of the telephone. Bell had a long-standing personal and professional interest in deafness. His mother was hard of hearing, his wife was deaf, and Bell and his father were both professional speech teachers (Lane 1984). In 1883, at a meeting of the National Academy of Sciences, Bell read a paper entitled *Upon the Formation of a Deaf Variety of the Human Race*. The paper was later published by the Academy as a memoir.

Most of Bell's memoir is concerned with a detailed presentation of data. Bell documents two phenomena. First, he shows that deaf people preferentially marry other deaf people. Second, he shows that deaf parents tend to have deaf offspring. Figure 18.15, for example, shows a pedigree of three generations in a family. Four of the women who married into this family were hearing; everyone else was deaf. Mendelian genetics, of course, remained to be rediscovered by evolutionary biologists, and the development of quantitative genetics was in its infancy. We know now that in Bell's day more cases of deafness were caused by rubella (German measles) than by genetics (Lane 1984). However, there are some genes that have alleles associated with deafness. The first English settlers of Martha's Vineyard, an island off the coast of Massachusetts, brought with them an autosomal recessive allele for deafness (see Groce 1985; Sacks 1989). The allele was at a high enough frequency that in some towns on the Vineyard, nearly a quarter of the people were deaf. Given the limited analytical tools that were available to him, Bell makes a reasonably persuasive case that at least some forms of deafness are heritable.

Given the phenomena he has documented, Bell infers the future evolutionary consequences. Because the deaf tend to intermarry, and because deaf parents tend to have deaf offspring, the deaf are on their way toward creating a deaf variety of the human race. In the final chapter of his memoir, Bell reveals his eugenic agenda:

> Those who believe as I do, that the production of a defective race of human beings would be a great calamity to the world, will examine carefully the causes that lead to the intermarriages of the deaf with the object of applying a remedy.

Members of today's Deaf community do not view themselves as defective, but rather as members of a linguistic minority (Lane and Philip 1984; Lane 1992). In his own day, however, Bell was probably safe in assuming that many members of his audience would share his view of deafness.

Bell argues that there are two primary reasons why deaf people tend to marry each other. First, deaf children are sent at an early age to be educated in residential schools for the deaf, where their social interactions are almost exclusively with other deaf people. Second, deaf people use a signed language that is different from the spoken language used by most hearing people. (The language of the deaf in North America is now known as American Sign Language, or ASL; see Wolkomir 1992.) Bell concludes that if we want to discourage deaf people

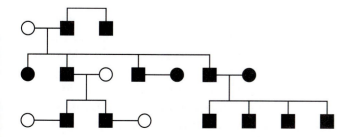

Figure 18.15 **Pedigree of a deaf family**   Open symbols indicate hearing people; filled symbols indicate deaf people. Redrawn from Bell (1883).

from marrying each other, we should close the residential schools for the deaf, instead educating deaf children at day schools where they have regular contact with hearing children. And we should discourage the use of sign language, instead teaching deaf students to speak and to speech read (that is, to read lips).

Bell's conclusions on the heritability of deafness drew vigorous debate from his contemporaries (Fay 1884). His educational policies were even more hotly contested. Edward Gallaudet (1884), for example, pointed out that even with extensive training, only a small fraction of the deaf ever learn to speak intelligibly enough to be understood by strangers. Of those that do learn to speak intelligibly, almost all are either hard of hearing rather than fully deaf, or became deaf only after they had already learned to speak. Rather than waste time on a doomed effort to teach all deaf students to speak, Gallaudet believed that it was better to encourage deaf students to develop fluency in sign language, and to use sign language as the mode of instruction for teaching other subjects.

In the debate over deaf education, Bell and other opponents of sign language won the day (Lane and Phillip 1984; Lane 1984; 1992). American Sign Language survived largely because deaf students took it underground, teaching it to each other in dorm rooms and on playgrounds, out of sight of their teachers. Even today, Donald Moores (1994) notes, "a smaller percentage of deaf children attend residential schools, are taught through Sign Language, or are instructed by deaf teachers than in Bell's time."

Other Examples of Classical Eugenics

The policies Bell advocated in his memoir amount to efforts to change patterns of nonrandom mating. Other eugenicists advocated policies of artificial selection. For example, in 1891 Victoria Woodhull wrote: "The best minds of today have accepted the fact that if superior people are desired, they must be bred; and if imbeciles, criminals, paupers, and [the] otherwise unfit are undesirable citizens they must not be bred" (quoted in Kevles 1995). Such policies increased in popularity after the rediscovery of Mendelian genetics, and as understanding of the mechanisms of evolution improved.

Efforts to increase the reproductive success of people with traits deemed desirable are known as positive eugenics. Starting in 1920, the American Eugenics Society sponsored a series of Fitter Families contests at state fairs (Kevles 1995). Many eugenicists opposed contraception on the grounds that it was more likely to be used by members of the upper classes, the desirable group whose reproduction they most wanted to encourage.

Efforts to decrease the reproductive success of people with traits deemed undesirable are known as negative eugenics. Many states passed laws mandating compulsory sterilization (Kevles 1995). Those covered included the feebleminded, the insane, alcoholics, people infected with sexually transmitted diseases, epileptics, and various sorts of criminals. By 1930, 24 states had compulsory sterilization laws (in 22 states, such laws are still on the books). By the mid 1930s some 20,000 people had been sterilized.

In just a few decades, the eugenics movement had progressed from social engineering to direct governmental control over the reproduction of individuals. Inspired in part by the eugenics movement in the United States and Britain, the

Nazis embraced eugenics as a central tenet of their ideology. The Nazis took eugenics to its logical extreme, moving beyond sterilization to genocide. The results were horrific. Today, eugenics as governmental policy carries strongly negative connotations.

. . . but the policies they advocated are now widely rejected.

Modern Eugenics

Although eugenics as explicit governmental policy went out of favor after World War II, recent advances in genetics and genetic technology have placed both individuals and governments in a position to make new kinds of eugenic decisions. Among the developments responsible are prenatal genetic screening, germ-line genetic engineering, and in vitro embryo selection.

Modern genetic and reproductive technologies are forcing us to confront new eugenic dilemmas.

Genetic Knowledge and Prenatal Genetic Screening

Extensive research has led to the discovery of the genes responsible for a great variety of traits and diseases. This knowledge gives people a great deal of information about the probable genotypes and phenotypes of their future offspring. Both members of a couple, for example, may be carriers for the same recessive allele. Should they reproduce together if each of their future children has a one-in-four chance of having cystic fibrosis? What about phenylketonuria, deafness, or albinism?

Many genetic conditions can now be diagnosed in utero by amniocentesis or chorionic villus biopsy. Prenatal genetic screening gives couples knowledge of the actual genotypes of their future offspring. If a couple discovers that their fetus has a genotype associated with a disease, they may choose to terminate the pregnancy.

Are genetic research and prenatal screening really manifestations of eugenics? Many would argue that they are not. Eugenics as defined by Galton involves social control. Neither a genetic counselor advising a couple about probabilities nor a health care worker reporting the results of a fetal diagnosis will dictate the couple's decision. They simply give the couple information. On the other hand, a recent editorial in *Nature Genetics* (January 1997) points out that "implicit in every research grant written for the study of a genetic disorder is the suggestion that . . . identification of a 'causative' gene or genes will help in population screening or fetal diagnosis." A consequence of population screening and fetal diagnosis is that some individuals will choose not to reproduce, and some couples will choose to terminate pregnancies. Thus government funding policy leads to individual action that results in changes in allele frequencies. Has this crossed the line to social control?

Germ-Line Genetic Engineering

In section 18.2 we discussed somatic-cell gene therapy for ADA deficiency. Somatic-cell gene therapy is not eugenic. As John Wagner (1997) points out, "the aim of gene therapy, like other treatments of inherited disease, is to allow patients with genetic diseases to live longer, rather than to eliminate certain genetic traits from the human gene pool."

However, genetic techniques now exist that allow direct manipulation of the germ line in mammals. Richard Palmiter and colleagues (1983) injected genes for human growth hormone into mouse zygotes, let the zygotes divide a few

(a)

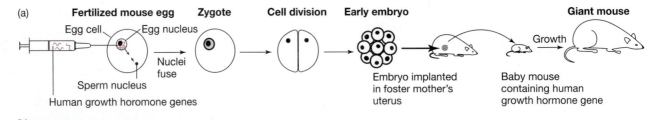

Fertilized mouse egg   Zygote   Cell division   Early embryo   Giant mouse

Egg cell   Egg nucleus

Nuclei fuse

Sperm nucleus

Human growth horomone genes

Embryo implanted
in foster mother's
uterus

Baby mouse
containing human
growth hormone gene

Growth

(b)

Figure 18.16   Germ-line genetic engineering produces giant mice
(a) The procedure used by Palmiter and colleagues (1983) to introduce genes for
human growth hormone into mouse zygotes. After Postlethwait and Hopson
(1992).   (b) A normal mouse, right, and a giant mouse, left. The giant mouse has
genes for human growth hormone in its germ line, and thus will pass those genes
to its offspring. (James Brinster, University of Pennsylvania)

times, and then inserted the resulting embryos into the uterus of a foster-
mother mouse (Figure 18.16a). The result was mice that make large quantities
of human growth hormone, and grow to twice the size of normal mice (Figure
18.16b). Note that this experiment represents an instant evolutionary change in
the mouse population in the lab. The human growth hormone genes are part of
the germ line of the giant mice, and will thus be passed on to future generations. Palmiter and colleagues were not motivated by a desire to improve their
population of lab mice by making the mice bigger. (Had they wanted bigger
rodents, they could have just bought rats.) Instead the researchers picked mice
and human growth hormone as a convenient system in which to demonstrate a
research technique with a variety of applications in genetics, developmental biology, and endocrinology. It is obvious, however, that germ-line genetic engineering also has potential applications in medicine and eugenics.

When and if germ-line genetic engineering is available for humans, should
we allow it? Imagine a couple who are both carriers for Tay-Sachs disease. Tay-
Sachs is a lipid storage disorder that results in death by age 3. It is caused by an
autosomal recessive loss-of-function allele in the gene for the enzyme hexosaminidase. Germ-line genetic engineering would offer this couple the opportunity to reproduce by in vitro fertilization, including the insertion into the zygote of the wild-type gene for hexosaminidase. This would mean that not only
would the couple's own children be freed of their one-in-four chance of Tay-
Sachs, all of the couples' descendants would be freed as well. This seems like a
good thing. How could we deny the couple the use of such a technology?

Now imagine a couple who want their descendants to be a little taller than
average, or smarter, or better looking. Would we allow them to use germ-line
genetic engineering as well? If we allow germ-line genetic engineering for Tay-
Sachs disease, how can we justify banning it for other conditions? What about
couples that are carriers for conditions, like deafness, that some people consider
defects and other people consider part of the natural variation among humans
that makes human culture diverse and interesting?

In Vitro Embryo Selection

Although germ-line genetic engineering is not yet available for humans, many of the questions it raises are with us now, thanks to in vitro embryo selection. When a couple undergoes in vitro fertilization, the first step is for the woman to take a drug that causes her to ovulate several eggs at once. The eggs are then harvested surgically and fertilized in a dish with the man's sperm. The embryos are allowed to undergo several cell divisions, then stored in a freezer for later implantantion in the woman's uterus. Before an embryo is implanted, researchers can pluck from it a single cell. The embryo still develops normally, and the single cell can be subjected to genetic screening. Each embryo is genetically unique, giving the couple their pick of which to try to implant.

In vitro embryo selection can be used for either positive or negative eugenics (Testart and Sele 1995). The couple may choose to avoid using embryos that will be carriers or sufferers of genetic diseases. This has the potential to reduce the frequencies of deleterious alleles (negative eugenics). The couple may also choose to preferentially select embryos with desirable genotypes (positive eugenics). They might base their choice on eye color, height, or sex. In essence, the couple is conducting artificial selection on their own population of offspring. Again we are confronted with difficult questions. Is it justifiable for a society to allow the use of such a technology? Is it justifiable not to allow it? These questions will only become more pressing in the future.

Summary

The insights and techniques of evolutionary biology can be applied to a variety of human issues, including human behavior, human health, and the evolutionary future of human populations.

Our species' behavioral repertoire allows a great deal of flexibility, and behavioral variation is strongly influenced by both environmental variation and cultural variation. Nonetheless, some patterns of behavior are broadly cross-cultural and can profitably be interpreted as legacies of natural selection. For example, an evolutionary perspective on parental care led to the discovery that stepparenthood is an important risk factor for infanticide.

An evolutionary perspective can be valuable in medicine. Viewing tissues as evolving populations of cells may help doctors to design effective strategies for somatic-cell gene therapy, and to reconstruct the history of tumors. Treating symptoms of disease, like fever, may be counterproductive if those symptoms are in fact adaptive defenses. Some apparently maladaptive aspects of our physiology, like myopia, are responses of certain genotypes to completely novel modern environments. Others, like pregnancy sickness, may prove on closer examination to be adaptive.

Following the discovery of evolution by natural selection, eugenicists sought to apply evolutionary principles to the management of our species' future. This effort, initially popular, ultimately led to ghastly governmental policies. Although eugenics as governmental policy went out of favor after World War II, individuals continue to face hard eugenic choices. These choices have become more difficult with advances in genetic screening, assisted reproduction, and genetic manipulation.

Questions

1. The male reed buntings in Dixon et al.'s study almost seem to be consciously aware of genetic relationships and "trying" to increase their reproductive success. Can evolution cause reed buntings (and other animals) to behave as if they are aware of the evolutionary consequences of their actions, without actually being aware of those consequences? Do you think the same could be true of some human behaviors?

2. Recall that Daly and Wilson's data on infanticide risks might be explained by stepfathers having, on average, more violent personalities than biological fathers. Could this "violent personality" explanation also apply to Flinn's data from the Trinidad village? Why or why not? Daly and Wilson's study involved general data about a large number of families, whereas Flinn's study involved detailed data on a small number of families. What are the advantages and disadvantages of each kind of study?

3. An evolutionary biologist once hypothesized that if evolution has affected human social behavior, then a mother's brothers should take a particular interest in her children—more so than the father's brothers, and perhaps even more so than the father himself. Why did he hypothesize this? (As it turned out, the biologist was right—many anthropologists have noted a "maternal uncle" phenomenon in various cultures.)

4. Look again at the family trees of homosexual men in Figure 18.7a. We have described one possible genetic explanation: an X-linked gene that predisposes men to homosexuality. Can you think of a non-genetic explanation (something relating to human psychology, learning, modern culture, etc.) for why men might be more likely to be gay if their mother's relatives are gay than if their father's relatives are gay?

5. If the X-linked gene theory is correct, how might this explain the phenomenon that exclusively homosexual women are less common in the general population than are exclusively homosexual men? How does Figure 18.8 address this issue?

6. Is something wrong with Figure 18.7b? (Why is Dad not gay?)

7. In Figure 18.12b, why was it important to give one group of lizards a saline injection?

8. We have seen how the genetic diversity of a tumor can be used to estimate the tumor's age. This analysis depends on mutation rate per cell division being constant. If a cancerous tumor has evolved a high mutation rate, how will this bias the results? Do the genetic markers we use to estimate a tumor's age need to be selectively neutral? Why or why not?

9. As we have seen, some biologists regard our bodies as small ecosystems that exert enormous selective pressure for the evolution of invasive metastatic cancer. If this is true, why don't we all get cancer? (Hint: Consider the speed of evolution vs. human life span.) However, these same biologists believe that humans have certain genes that have evolved specifically to prevent cancer. How is it possible to have both strong selection for cancer, and strong selection for anticancer genes? (Hint: Consider evolution within one person's body vs. evolution within a population of many people).

10. Suppose that fever is conclusively shown to be an effective defense against infections. Now suppose that your 2-year-old child is suffering a serious infection and has just started running a dangerously high fever of 106°F. You and your doctor decide to put your child on an antifever medication. Would you want your child's temperature to return all the way to normal, or to stabilize at a "mild fever" level (say, 101°F)? If you feel that you need more information to make this important decision, what further information would you want? (That is, what further studies should be done?)

11. Young children often develop strong food preferences that sometimes seem bizarre to their parents—such as wanting to eat only a certain bland food prepared in a familiar way, or being reluctant to eat new foods or foods with strong or bitter tastes (like coffee and some vegetables). Develop an evolutionary explanation for such preferences. What evidence would you gather to investigate your hypothesis and rule out alternative explanations? (Hint: Review the discussion of pregnancy sickness in this chapter.)

12. The literature of the Institute for Creation Research (ICR), a creationist organization, often implies that anyone who accepts evolutionary theory must also endorse any and all consequences of evolution. As an example, the ICR commonly describes evolutionists as being in favor of extreme eugenics policies (including genocide) and as believing that any behaviors that evolved by natural selection must be "good." Comment on this depiction of evolutionists. If human behavior has been shaped by evolution, does that necessarily mean that all evolved behaviors are "good"?

Exploring the Literature

13. We analyzed cross-cultural patterns in the inheritance of wealth from the evolutionary perspective of men. The cross-cultural data may make even more sense when analyzed from the evolutionary perspective of women. For more detailed theory and more extensive data, along with peer commentaries, see:

Hartung, J. 1985. Matrilineal inheritance: New theory and analysis. *Behavioral and Brain Sciences* 8:661–688.

14. For another study on the effects of antifever medication on human illness, see:

Sugimura, T., T. Fujimoto, H. Motoyama, T. Maruoka, and S. Korematu, et al. 1994. Risks of antipyretics in young children with fever due to infectious disease. *Acta Paediatrica Japonica* 36:375–378.

How strong is the evidence in this paper that fever is an adaptive response to bacterial respiratory infection? Consider that acetaminophen affects aspects of the immune response other than fever ("Fever and the Common Cold" in section 18.2). Also consider that Sugimura and colleagues conducted an observational study, not an experimental one. That is, the researchers did not assign their subjects at random to acetaminophen versus placebo groups. Instead, the researchers asked parents to keep a diary of the number of doses of acetaminophen they gave their children.

15. If your library has the earliest volumes of the *Journal of Heredity,* read:

Bell, Alexander Graham. 1914. How to improve the race. *Journal of Heredity* 5:1–7.

Keep in mind that population genetics was in its infancy; Mendelism had yet to be integrated with natural selection. What was accurate and inaccurate in Bell's understanding of the mechanisms of evolution? Would the policy Bell advocated actually have accomplished his aims? Why or why not? If so, would it have done so for the reasons Bell thought it would?

Citations

Ames, B.N., M. Profet, and L.S. Gold. 1990a. Dietary pesticides (99.99% all natural). *Proceedings of the National Academy of Sciences, USA* 87:7777–7781.

Ames, B.N., M. Profet, and L.S. Gold. 1990b. Nature's chemicals and synthetic chemicals: Comparative toxicology. *Proceedings of the National Academy of Sciences, USA* 87:7782–7786.

Azzimondi, G., L. Bassein, F. Nonino, L. Fiorani, L. Vignatelli, et al. 1995. Fever in acute stroke worsens prognosis: A prospective study. *Stroke* 26:2040–2043.

Balasch, J., I. Jov'e, J. Martorell, A. Gay'e, and J.A. Vanrell. 1993. Histocompatibility in in vitro fertilization couples. *Fertility and Sterility* 59:456–458.

Bashiri, A., L. Neumann, E. Maymon, and M. Katz. 1995. Hyperemesis gravidarum: Epidemiologic features, complications, and outcome. *European Journal of Obstetrics and Gynecology and Reproductive Biology* 63:135–138.

Beauchamp, G.K., K. Yamazaki, C.J. Wysocki, B.M. Slotnick, L. Thomas, and E.A. Boyse. 1985. Chemosensory recognition of mouse major histocompatibility types by another species. *Proceedings of the National Academy of Sciences, USA* 82:4186–4188.

Bell, A.G. 1883. *Upon the Formation of a Deaf Variety of the Human Race* (New Haven: National Academy of Sciences).

Bernheim, H.A., and M.J. Kluger. 1976. Fever: Effect of drug-induced antipyresis on survival. *Science* 193:237–239.

Blaese, R.M., K.W. Culver, A.D. Miller, C.S. Carter, T. Fleisher, et al. 1995. T lymphocyte-directed gene therapy for ADA-SCID: Initial trial results after 4 years. *Science* 270:475–480.

Blount, B.G. 1990. Issues in Bonobo *(Pan paniscus)* sexual behavior. *American Anthropologist* 92:702–714.

Bordignon, C., L.D. Notarangelo, N. Nobili, G. Ferrari, and G. Casorati, et al. 1995. Gene therapy in peripheral blood lymphocytes and bone marrow for ADA-immunodeficient patients. *Science* 270:470–575.

Brent, R.L. 1995. Bendectin: Review of the medical literature of a comprehensively studied human nonteratogen and the most prevalent tortogen-litigen. *Reproductive Toxicology* 9:337–349.

Brown, R.E., B. Roser, and P.B. Singh. 1989. Class I and class II regions of the major histocompatibility complex both contribute to individual odors in congenic inbred strains of rats. *Behavior Genetics* 19:659–674.

Brown, R.E., P.B. Singh, and B. Roser. 1987. The major histocompatibility complex and the chemosensory recognition of individuality in rats. *Physiology and Behavior* 40:65–73.

Byne, W. 1994. The biological evidence challenged. *Scientific American* (May):50–55.

Chagnon, N.A. 1988. Male manipulations of kinship classifications of female kin for reproductive advantage. In L. Betzig, M. Borgerhoff Mulder, and P. Turke, eds. *Human Reproductive Behavior: A Darwinian Perspective* (Cambridge: Cambridge University Press), 23–48.

Chagnon, N.A. 1992. *Yąnomamö* (Fort Worth: Harcourt Brace College Publishers).

Creed, G.W. 1984. Sexual subordination: Institutionalized homosexuality and social control in Melanesia. *Ethnology* 23:157–176.

Daly, M., and M. Wilson. 1984. A sociobiological analysis of human infanticide. In G. Hausfater and S.B. Hrdy, eds. *Infanticide: Comparative and Evolutionary Perspectives* (New York: Aldine de Gruyter), 487–502.

Daly, M., and M. Wilson. 1988a. *Homicide* (New York: Aldine De Gruyter).

Daly, M., and M. Wilson. 1988b. Evolutionary social psychology and family homicide. *Science* 242:519–524.

Daly, M., and M. Wilson. 1989. Evolution and family homicide, reply. *Science* 243:463–464.

Daly, M., and M.I. Wilson. 1994a. Some differential attributes of lethal assaults on small children by stepfathers versus genetic fathers. *Ethology and Sociobiology* 15:207–217.

Daly, M., and M. Wilson. 1994b. Stepparenthood and the evolved psychology of discriminative parental solicitude. In S. Parmigiani and F.S. vom Saal, eds. *Infanticide and Parental Care* (London: Harwood Academic Publishers), 121–133.

Dawkins, R. 1976, 1989. *The Selfish Gene,* 1st and 2nd ed. (Oxford: Oxford University Press).

Dennett, D.C. 1995. *Darwin's Dangerous Idea* (New York: Simon and Schuster).

Deuchar, N. 1995. Nausea and vomiting in pregnancy: A review of the problem with particular regard to psychological and social aspects. *British Journal of Obstetrics and Gynaecology* 102:6–8.

De Waal, Frans. 1997. *Bonobo: The Forgotten Ape* (University of California Press).

Diamond, J. 1992. *The Third Chimpanzee* (New York: HarperCollins).

Dixon, A., D. Ross, S.L.C. O'Malley, and T. Burke. 1994. Paternal investment inversely related to degree of extra-pair paternity in the reed bunting. *Nature* 371:698–700.

Doran, T.F., C. De Angelis, R. Baumgardner, and E.D. Mellits. 1989. Acetaminophen: More harm than good for chickenpox? *Journal of Pediatrics* 114:1045–1048.

Durham, W.H. 1991. *Coevolution: Genes, Culture, and Human Diversity* (Stanford: Stanford University Press).

Fausto-Sterling, A., and E. Balaban. 1993. Genetics and male sexual orientation. *Science* 261:1257.

Fay, E.A. 1884. Discussion by the National Academy of Sciences concerning the formation of a deaf variety of the human race. *American Annals of the Deaf* 29:70–77.

Flinn, M.V. 1988. Step- and genetic parent/offspring relationships in a Caribbean village. *Ethology and Sociobiology* 9:335–369.

Futuyma, D.J., and S.J. Risch. 1984. Sexual orientation, sociobiology, and evolution. *Journal of Homosexuality* 9:157–168.

Galton, F. 1865. Hereditary talent and character. *Macmillan's Magazine* 12:157–166, 318–327.

Gallaudet, E.M. 1884. Reply to Dr. Bell. In Bell, A.G., 1884. Fallacies concerning the deaf. *American Annals of the Deaf* 29:32–69.

George, S.A. 1989. Evolution and family homicide. *Science* 243:462.

Gilbert, A.N., K. Yamazaki, G.K. Beauchamp, and L. Thomas. 1986. Olfactory discrimination of mouse strains *(Mus musculus)* and major histocompatibility types by humans *(Homo sapiens)*. *Journal of Comparative Psychology* 100:262–265.

Graham, N.M.H., C.J. Burrell, R.M. Douglas, P. Debelle, and L. Davies. 1990. Adverse effects of aspirin, acetaminophen, and ibuprofen on immune function, viral shedding, and clinical status in rhinovirus-infected volunteers. *Journal of Infectious Diseases* 162:1277–1282.

Green, M.H., and C.W. Vermeulen. 1994. Fever and the control of gram-negative bacteria. *Research in Microbiology* 145:269–272.

Greene, E., L.J. Orsak, and D.W. Whitman. 1987. A Tephritid fly mimics the territorial displays of its jumping spider predators. *Science* 236:310–312.

Groce, N.E. 1985. *Everyone Here Spoke Sign Language: Hereditary Deafness on Martha's Vineyard* (Cambridge: Harvard University Press).

Hamer, D.H., S. Hu, V.L. Magnuson, N. Hu, and A.M.L. Pattatucci. 1993. A linkage between DNA markers on the X chromosome and male sexual orientation. *Science* 261:321–327

Hartung, J. 1981. Paternity and inheritance of wealth. *Nature* 291:652–654.

Hedrick, P., and V. Loeschcke. 1996. MHC and mate selection in humans? *Trends in Ecology and Evolution* 11:24.

Hershfield, M.S., R.H. Buckley, M.L. Greenberg, A.L. Melton, L. Schiff, et al. 1987. Treatment of adenosine deaminase deficiency with polyethylene glycol-modified adenosine deaminase. *New England Journal of Medicine* 316:589–596.

Hill, K., and H. Kaplan. 1988. Tradeoffs in male and female reproductive strategies among the Ache, part 2. In L. Betzig, M. Borgerhoff Mulder, and P. Turke, eds. *Human Reproductive Behavior: A Darwinian Perspective* (Cambridge: Cambridge University Press), 291–305.

Hirschhorn, R., D.R. Yang, J.M. Puck, M.L. Huie, C.-K. Jiang, and L.E. Kurlandsky. 1996. Spontaneous *in vivo* reversion to normal of an inherited mutation in a patient with adenosine deaminase deficiency. *Nature Genetics* 13:290–295.

Ho, H.N., Y.S. Yang, R.P. Hsieh, H.R. Lin, S.U. Chen, et al. 1994. Sharing of human leukocyte antigens in couples with unexplained infertility affects the success of *in vitro* fertilization and tubal embryo transfer. *American Journal of Obstetrics and Gynecology* 170:63–71.

Hu, S., A.M. Pattatucci, C. Patterson, L. Li, D.W. Fulker, S.S. Cherny, L. Kruglyak, and D.H. Hamer. 1995. Linkage between sexual orientation and chromosome Xq28 in males but not in females. *Nature Genetics* 11:248–256.

Hutchinson, G.E. 1959. A speculative consideration of certain possible forms of sexual selection in man. *American Naturalist* 93:81–91.

Kevles, D.J. 1995. *In the Name of Eugenics: Genetics and the Uses of Human Heredity* (Cambridge: Harvard University Press).

Klebanoff, M.A., and J.L. Mills. 1986. Is vomiting during pregnancy teratogenic? *British Medical Journal* 292:724–726.

Klug, W.S., and M.R. Cummings. 1997. *Concepts of Genetics,* 5th ed. (Upper Saddle River, NJ: Prentice Hall).

Kluger, M.J. 1992. Fever revisited. *Pediatrics* 90:846–850.

Kullander, S., and B. Källén. 1976. A prospective study of drugs and pregnancy II. Anti-emetic drugs. *Acta Obstetricia et Gynecologica Scandinavica* 55:105–111.

Lane, H. 1984. *When the Mind Hears: A History of the Deaf* (New York: Random House).

Lane, H. 1992. *The Mask of Benevolence: Disabling the Deaf Community* (New York: Vintage Books).

Lane, H., and F. Philip, eds. 1984. *The Deaf Experience: Classics in Language and Education* (Cambridge: Harvard University Press).

LeVay, S., and D.H. Hamer. 1994. Evidence for a biological influence in male homosexuality. *Scientific American* (May):44–49.

Lewontin, R.C. 1980. Sociobiology: Another biological determinism. *International Journal of Health Services* 10:347–363.

Matsuyama, T., Y. Kobayashi, I. Honda, M. Inoue, and A. Fujii. 1992. Analysis of HLA antigens and immunotherapy for infertile couples who failed to conceive after *in vitro* fertilization-embryo tubal replacement (IVF-ETR). *Acta Obstetrica et Gynaecologica Japonica* 44:1241–1247.

McKeigue, P.M., S.H. Lamm, S. Linn, and J.S. Kutcher. 1994. Bendectin and birth defects: I. A meta-analysis of the epidemiologic studies. *Teratology* 50:27–37.

Moores, D.F. 1994. Eugenics revisited: Hereditary deafness and genetic technology. *American Annals of the Deaf* 139:393.

Mori, A. 1984. An ethological study of pygmy chimpanzees in Wamba, Zaire: A comparison with chimpanzees. *Primates* 25:255–278.

Neese, R.M., and G.C. Williams. 1995. *Why We Get Sick: The New Science of Darwinian Medicine* (New York: Vintage Books).

Neutel, C.I., and H. Johansen. 1995. Measuring drug effectiveness by default: The case of Bendectin. *Canadian Journal of Public Health* 86:66–70.

Ober, C. 1995. Current topic: HLA and reproduction: Lessons from studies in the Hutterites. *Placenta* 16:569–577.

Ober, C., S. Elias, D.D. Kostyu, and W.W. Hauck. 1992. Decreased fecundability in Hutterite couples sharing HLA-DR. *American Journal of Human Genetics* 50:6–14.

Orr, H.A. 1996. Dennett's dangerous idea. *Evolution* 50:467–472.

Palmiter, R.D., G. Norstedt, R.E. Gelinas, R.E. Hammer, and R.L. Brinster. 1983. Metallothionein–Human GH fusion genes stimulate growth of mice. *Science* 222:809–814.

Pattatucci, A.M.L., and D.H. Hamer. 1995. Development and familiality of sexual orientation in females. *Behavior Genetics* 25:407–420.

Pomiankowski, A., M. Pagel, W.K. Potts, C.J. Manning, and E.K. Wakeland. 1992. Sexual selection and MHC genes. *Nature* 356:293–294.

Postlethwait, J.H., and J.L. Hopson. 1992. *The Nature of Life,* 2nd ed. (New York: McGraw-Hill).

Potts, W.K., C.J. Manning, and E.K. Wakeland. 1991. Mating patterns in seminatural populations of mice influenced by MHC genotype. *Nature* 352:619–621.

Potts, W.K., and E.K. Wakeland. 1993. Evolution of MHC genetic diversity: A tale of incest, pestilence and sexual preference. *Trends in Genetics* 9:408–412.

Profet, M. 1988. The evolution of pregnancy sickness as protection to the embryo against pleistocene teratogens. *Evolutionary Theory* 8:177–190.

Profet, M. 1992. Pregnancy sickness as adaptation: A deterrent to maternal ingestion of teratogens. In J.H. Barkow, L. Cosmides, and J. Tooby, eds. *The Adapted Mind: Evolutionary Psychology and the Generation of Culture* (Oxford: Oxford University Press), 327–365.

Profet, M. 1995. *Protecting your baby-to-be: Preventing birth defects in the first trimester* (Reading, MA: Addison-Wesley Publishing Company).

Renwick, J.H., W.D.B. Claringbold, M.E. Earthy, J.D. Few, and A.C.S. McLean. 1984. Neural-tube defects produced in Syrian hamsters by potato glycoalkaloids. *Teratology* 30:371–381.

Ringler, D.H., and M.R. Anver. 1975. Fever and survival. *Science* 188:166–168.

Sacks, O. 1989. *Seeing Voices: A Journey into the World of the Deaf* (Berkeley: University of California Press).

Shibata, D., W. Navidi, R. Salovaara, Z.-H. Li, and L.A. Aaltonen. 1996. Somatic microsatellite mutations as molecular tumor clocks. *Nature Medicine* 2:676–681.

Shyue, S.-K., E.D. Hewett, H.G. Sperling, D.M. Hunt, J.K. Bowmaker, J.D. Mollon, and W.H. Li. 1995. Adaptive evolution of color vision genes in higher primates. *Science* 269:1265–1267.

Singh, P.B., R.E. Brown, and B. Roser. 1987. MHC antigens in urine as olfactory recognition cues. *Nature* 327:161–164.

Singh, P.B., R.E. Brown, and B. Roser. 1988. Class I transplantation antigens in solution in body fluids and in the urine. Individuality signals to the environment. *Journal of Experimental Medicine* 168:195–211.

Smith, R.L. 1984. Human Sperm Competition. In R.L. Smith, ed. *Sperm Competition and the Evolution of Animal Mating Systems* (Orlando, FL: Academic Press), 601–659.

Testart, J., and B. Sele. 1995. Towards an efficient medical eugenics: Is the desirable always the feasible? *Human Reproduction* 10:3086–3090.

Trivers, R. 1985. *Social Evolution* (Menlo Park, CA: Benjamin/Cummings).

Vaughn, L.K., H.A. Bernheim, and M. Kluger. 1974. Fever in the lizard *Dipsosaurus dorsalis. Nature* 252:473–474.

Wagner, J.A. 1997. Gene therapy is not eugenics. *Nature Genetics* 15:234.

Weckstein, L.N., P. Patrizio, J.P. Balmaceda, R.H. Asch, and D.W. Branch. 1991. Human leukocyte antigen compatibility and failure to achieve a viable pregnancy with assisted reproductive technology. *Acta Europaea Fertilitatis* 22:103–107.

Wedekind, C., T. Seebeck, F. Bettens, and A. Paepke. 1995. MHC dependent mate preferences in humans. *Proceedings of the Royal Society of London,* Series B 260:245–249.

Wedekind, C., and T. Seebeck. 1996. MHC and mate selection in humans?, reply. *Trends in Ecology and Evolution* 11:24–25.

Weigel, R.M., and M.M. Weigel. 1989. Nausea and vomiting of early pregnancy and pregnancy outcome. A meta-analytical review. *British Journal of Obstetrics and Gynaecology* 96:1312–1318.

Wolkomir, R. 1992. American Sign Language: "It's not mouth stuff—it's brain stuff." *Smithsonian* 23:30–41.

Youssoufian, H. 1996. Natural gene therapy and the Darwinian legacy. *Nature Genetics* 13:255–256.

Biodiversity and Conservation

The world's human population has reached an alarming size. Between the appearance of modern *Homo sapiens* some 100,000 years ago (see Chapter 14) and the advent of agriculture about 10,000 years ago, the population grew to about 10 million people. By 2,000 years ago our numbers had reached some 100 million. The population hit 1 billion sometime around the year 1830. By 1995, there were 5.7 billion people alive. Our population's growth rate reached its all-time high of 2.1% per year between 1965 and 1970. The rate of increase has since dropped to about 1.6% per year, and is eventually expected to fall to zero. Nonetheless, the United Nations predicts that the human population will soar to between 7.5 and more than 20 billion before it peaks (Figure 19.1).

All these people have to live somewhere, and they have to eat. As their population grew, modern humans spread from Africa into Asia, Europe, Australia, and the Americas. In Chapter 13, we reviewed evidence of human complicity in the extinctions of numerous large mammals in the Americas, Eurasia, and Australia between 100,000 and 10,000 years ago. We also discussed research by David Steadman and others showing that human expansion across the islands of the South Pacific resulted in the extinction of some 2,000 species of birds. These early waves of anthropogenic extinction were the result of overexploitation, the introduction of nonnative predators and diseases, and habitat destruction. All of these mechanisms continue to cause extinctions today. However, as the human population has swelled in the last few centuries, and as people have converted land to agricultural fields, pastures, and living space, habitat destruction has become the most prominent threat to biological diversity (see also Table 19.1). If present rates of habitat destruction continue, then over the coming centuries or millenia humans will cause a mass extinction to rival the Big Five.

The social, political, and economic dimensions of the biodiversity crisis are beyond the scope of evolutionary biology. We mention them only briefly in

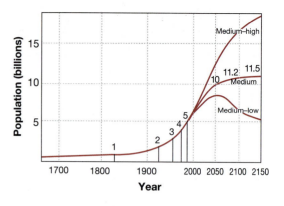

Figure 19.1 **The size of the human population from 1700 to the present, with projections to 2150** The three projections are based on different assumptions about how rapidly the population growth rate will fall to zero from its present value of 1.6% per year. From McMichael (1993).

this chapter. The same is true for many of the ecological issues involved. Instead, we focus on aspects of evolutionary biology that offer crucial insights into biodiversity and its conservation. Population genetics allows us to understand the evolutionary consequences of the reduction and fragmentation of populations. Systematics gives us tools for identifying species and populations in need of conservation and offers criteria for setting priorities.

This chapter should be read as a progress report. Research on biodiversity and conservation is accelerating rapidly, and the situation on the ground changes monthly. Scientists from every branch of evolutionary biology—from paleontologists to theoretical population geneticists to botanists and zoologists—are working on these issues, frequently in collaborations that cross disciplines and national boundaries. The work is new, dynamic, and urgent.

TABLE 19.1 Resource use varies dramatically by life-style

The ever-expanding use of energy-intensive technologies is compounding our species' impact on the environment. People living a modern western life-style in economically and technologically developed countries consume resources at a much higher rate than people living a simpler life-style in less developed countries. This difference is evident in a recent comparison between resource use in the United States and India. This table shows, for 1991, the total amount of each resource consumed by each country. The column on the right shows the ratio of per capita use in the U.S. to per capita use in India. For example, the average American used about 43 times more petroleum in 1991 than the average Indian. Although the U.S. population, at 263 million in 1995, is less than a third of India's 931 million, the United States puts a bigger strain on the global environment.

Resource (unit of measure)	Total U.S. consumption	Total Indian consumption	U.S./India per capita ratio
Petroleum (1,000 metric tons)	666,032	53,294	42.7
Natural gas (1,000 metric tons)	21,387,719	397,250	183.9
Aluminum (1,000 metric tons)	4,137	420	33.7
Iron ore (1,000 metric tons)	64,810	25,384	8.7
Pulpwood (1,000 cm³)	136,377	1,208	385.7
Beef and veal (head)	35,989	11,758	10.5

Source: World Resources Institute (1994).

19.1 How Many Species Are There?

Aside from declines during mass extinctions, the diversity of organisms has steadily increased over evolutionary time (Figure 19.2). How many species are alive today? Some 1.7 million species have been identified and named (Heywood and Watson 1995). Table 19.2a gives an accounting of protozoa and animals; Table 19.2b gives an accounting of plants. There is considerable variation among taxa in species richness. For example, about half of all the named species of organisms are insects, whereas the entire animal phylum Placozoa has only one known species, *Trichoplax*. Biologists agree that the named species represent only a small fraction of the diversity of living organisms. They also agree that we can only guess at the actual total (for example, May 1988).

Attempts to assess life's diversity are hobbled by two substantial problems. First, there are not enough trained systematists capable of recognizing and describing new species, and funding for such work is inadequate (Blackmore 1996). Second, we lack general, practical criteria for recognizing and classifying species. The number of species we identify is obviously influenced by whether we use the biological species concept, the phylogenetic species concept, or the morphospecies concept (see Chapter 9). Systematists working on different groups of organisms apply different rules in distinguishing species. For example, Robert Selander (1985) observed that different strains of bacteria identified as belonging to the single species *Legionella pneumophila* may have DNA sequence similarities of less than 50%. These strains of bacteria thus are separated by genetic distances equivalent to that between mammals and fishes. Our criteria for dividing bacteria and vertebrates into species are clearly not consistent.

Problems notwithstanding, various approaches have been used to estimate the total number of extant species. All approaches involve extrapolating in some way from relatively well-known groups of organisms to poorly known groups. Here are two examples:

- Birds and mammals are relatively well known. They are large and charismatic, and thus well studied. We can antipicate the discovery of few new species. Of the birds and mammals known, about one-third live in the temperate zone, and two-thirds live in the tropics. In contrast to birds and mammals, insects

Only a fraction of the extant species have been catalogued. Estimates of the total number of species are based on extrapolation from well-known groups to poorly known groups.

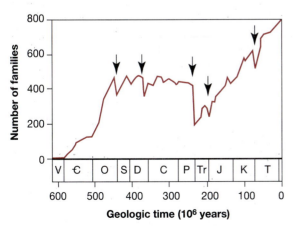

*Figure 19.2 **Biological diversity has steadily increased over evolutionary time** This graph plots the number of families of marine organisms over time from the Cambrian explosion (see Chapter 12) to the present. Arrows indicate the Big Five mass extinctions (see Chapter 13). From May et al. (1995).*

TABLE 19.2 Named species of animals and plants

(a)   This listing of animal taxa, most of which are broken down by phylum or subphylum, totals over 1.2 million named species (Brusca and Brusca 1990). There is general agreement among specialists that especially large numbers of nematodes, insects, and mites are yet to be discovered and named. Note that a few phyla are extremely species-rich, while most have few or very few species.

Taxon	Examples or description	Known species
Protozoa	Amoebae, flagellates, ciliates, others (7 phyla)	35,000
Placozoa	A marine organism called *Trichoplax*	1
Mesozoa	Microscopic, many symbiotic or parasitic (3 phyla)	100
Porifera	Sponges	9,000
Cnidaria	Jellyfish, sea anemones, corals, sea fans	9,000
Ctenophora	Comb jellies	100
Platyhelminthes	Flatworms: flukes, tapeworms, others	20,000
Nemertea	Ribbonworms	900
Gnathostomulida	Tiny, wormlike, occupy marine sands	80
Rotifera	Rotifers: microscopic, primarily in freshwater	1,800
Gastrotricha	Microscopic, aquatic organisms	450
Kinorhyncha	Microscopic; primarily in marine sands, muds	150
Nematoda	Roundworms	12,000
Nematomorpha	Horsehair worms: most occupy freshwater	230
Priapula	Wormlike, burrow in marine sediments	15
Acanthocephala	wormlike, most are parasites of fish	700
Entoprocta	Aquatic filter feeders	150
Loricifera	Microscopic; live in marine sediments	9
Annelida	Segmented worms: earthworms, leeches, others	15,000
Echiura	Unsegmented marine worms	135
Sipuncula	Peanut worms: all marine, most burrow	250
Pogonophora	Beard worms: marine, tube-dwelling	135
Vestimentifera	Marine forms, also called beard worms	8
Tardigrada	"Water bears": microscopic, in semiaquatic habitats	400
Onychophora	Resemble soft-bodied, unsegmented centipedes	80
Arthropoda: Cheliceriformes	Horseshoe crabs, spiders, mites, ticks, others	65,000
Arthropoda: Crustacea	Crabs, shrimp, barnacles, lobsters, others	32,000
Arthropoda: Uniramia	Insects and myriapods	860,000
Mollusca	Clams, mussels, chitons, snails, slugs, octopuses	100,000
Brachiopoda	Lamp shells: marine, benthic, solitary organisms	335
Ectoprocta	Moss animals: colonial, aquatic, filter feeders	4,500
Phoronida	Tube dwelling, marine, filter feeders	15
Chaetognatha	Arrow worms: planktonic, marine, worm-like	100
Echinodermata	Sea stars, brittle stars, sea urchins, sand dollars	7,000
Hemichordata	Acornworms	85
Chordata: Urochordata	Sea squirts and others	3,000
Chordata: Cephalochordata	Lancelets, amphioxus	23
Chordata: Vertebrata	Vertebrates: fish, amphibians, reptiles, mammals	47,000

(b) This listing of plant taxa by phylum or subphylum totals about 250,000 species. Two flowering plant families are far and away the most species-rich: the orchids, with over 23,000 species, and sunflowers (Asteraceae), with over 21,000 species.

Taxon	Examples or description	Known species
Bryophyta	Mosses, liverworts, hornworts	16,600
Psilophyta	Psilopsids (*Psilotum*)	9
Lycopodiophyta	Lycophytes	1,275
Equisetophyta	Horsetails	15
Filicophyta	Ferns	10,000
Gymnosperma	Gymnosperms: conifers (pines, cedars, firs, spruces), gingkoes, yews, junipers, others	529
Angiosperma: Dicotolydonae	Dicots: maples, oaks, sunflowers, cactus, figs, legumes, roses, others	170,000
Angiosperma: Monocotolydonae	Monocots: lilies, iris, grasses, orchids, palms, cattails, sedges, rushes, others	50,000

are poorly known. Nearly 1 million species have been described, but new ones are being discovered all the time. Most of the described insect species are from the temperate zone. If the ratio of temperate to tropical species is the same for insects as it is for birds and mammals, then there must be something like 2 million species of tropical insects. This pushes the total number of insect species close to 3 million, and the total number of all organisms into the range of 3 to 5 million (May 1988).

- Terry Erwin and J.C. Scott (1980) used an insecticidal fog to collect insects from a species of tropical tree called *Luehea seemannii*. Among the insects that fell from the canopy, Erwin (1982) identified over 900 different species of beetles (the majority of them undescribed). Erwin estimated that 160 of these beetles are specialists that live only on *L. seemannii*. Thanks to Erwin's work, the canopy beetles of *L. seemannii* are a relatively well known group of arthropods. The rest of the arthopods, those that live on *L. seemannii* and elsewhere, are relatively poorly known. Beetles represent 40% of the known arthropods. Assuming that the fraction of beetles is the same among arthropods that are *L. seemannii* specialists, Erwin calculated that there are 400 species of arthropods living exclusively on the canopy of this one species of tree. Assuming that the community of canopy arthropods is twice as rich as the community of trunk, root, and forest-floor arthropods, Erwin calculated that a total of 600 species of arthropods are living in specialized association with *L. seemannii*. If each of the estimated 50,000 species of tropical tree also harbors 600 specialized arthropod species, then there is a grand total of some 30 million terrestrial arthropod species. Thus the total number of species of organisms is between 30 and 35 million.

As Robert May (1988) points out, it is easy to push either of these estimates substantially higher. For example, parasites have been studied intensively only in economically important hosts. If we assume that each species of animal harbors just one specialized species of bacterial, protozoan, or helminth worm parasite, then we can roughly double the estimated total number of extant species. Furthermore, the mites—both temperate and tropical—have received less attention than the insects and may prove to be extremely diverse. Even less well known than terrestrial organisms are the communities of the ocean floor. When the ocean floor is included, estimates of the number of species of marine invertebrates range from 1/2 million to 10 million (Grassle and Maciolek 1992; May 1992; Poore et al. 1993). Finally, recent developments in molecular systematics have revealed that microorganisms, including eukaryotes, bacteria, and archaea, are much more diverse than previously suspected (see Chapter 11).

A recent estimate puts the total number of extant species at 10 to 100 million.

What is the best current estimate for global species diversity? Using extrapolations from well-known to poorly known groups, for insects as well as other species-rich taxa, Stuart Pimm and colleagues (1995; see also Hammond 1992; Stork 1996) estimate that the total number of living species is somewhere between 10 and 100 million (Figure 19.3).

Better data and more precise estimates may be forthcoming. Proposals to catalog the species present at several representative sites in the tropics are now pending (Janzen and Hallwachs 1994; Janzen 1996). If these intensive samples are completed, they will provide a stronger basis for extrapolation and a better estimate for the world total of species.

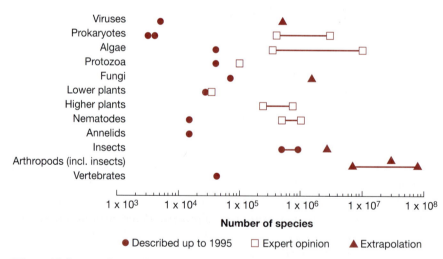

Figure 19.3 Numbers of described species and estimates of actual totals
This figure plots the number of named species in some prominent taxonomic groups, along with best-guess figures for the grand total (including yet-to-be-discovered species). The entries labeled "expert opinion" represent judgments made by taxonomic specialists in the relevant groups. Entries labeled "extrapolation" are based on small, intensive samples. The data point for fungi, for example, was calculated as follows: An average of 6 fungal species are found on each species of vascular plant in Britain; multiplying 6 by the estimated 250,000 plant species yields a value of 1.5 million fungal species worldwide. Note that species number is plotted on a logarithmic scale. From Pimm et al. (1995).

How Rapidly Are Species Being Lost?

In the absence of accurate estimates of how many species there are, it is difficult to estimate the rate at which species are going extinct. Robert May and colleagues (1995) have reviewed three approaches used in estimating the current extinction rates:

- Multiply the number of species found per hectare in different environments by rates of habitat loss measured from satellite photos.
- Quantify the rate that well-known species are moving from threatened to endangered to extinct status in the lists maintained by conservation groups.
- Estimate the probability that all species currently listed as threatened or endangered will actually go extinct over the next 100 or 200 years.

We provided more detail on these calculations in Chapter 13, when we examined evidence that a mass extinction is now underway. All three approaches suggest that extinctions are now occurring at a rate of 100 to 1,000 times the normal, or background, rate of extinction (see also Smith et al. 1993a; 1993b; Pimm et al. 1995). If current rates of habitat destruction continue, the coming centuries or millenia will see a mass extinction on the same scale as the Big Five documented in the fossil record.

Humans have increased the extinction rate by a factor of 100 to 1,000.

Where Is the Problem Most Acute?

Between 1600 A.D. and 1993, biologists observed the extinction of 486 animal species and 600 plant species (Smith et al. 1993b). Most of these extinctions occurred in North America, the Caribbean, Australasia, and the islands of the Pacific Ocean. Habitat destruction is rampant world-wide (Figure 19.4), putting many more species at risk. For example, a 1996 report by the Nature Conservancy (Stein and Chipley 1996) listed roughly one-third of all American plant and animal species as threatened (Figure 19.5). The taxa at greatest risk in the United States are the freshwater mussels, crayfish, amphibians, and freshwater fish.

Humans have caused the extinctions in many regions. Tropical rainforests are attracting attention at present because they still harbor tremendous diversity and are under threat.

Tropical rainforests are a focus of particular concern among those working to preserve global biodiversity. There are two reasons for this:

- Tropical rainforests are extraordinarily species-rich. E. O. Wilson (1988) recounts that he once collected 43 species of ants belonging to 26 genera—numbers roughly equivalent to the entire ant fauna of the British Isles—from a single tree in a Peruvian rainforest. Peter Ashton identified 700 different species of trees—the same number found in all of North America—in just ten 1-hectare sample plots from a rainforest in Borneo. With the exception of conifers, salamanders, and aphids, nearly every well-studied lineage on the tree of life shows a latitudinal gradient in diversity, with the largest number of species residing in the tropics. Why this pattern occurs is not clear (Ehrlich and Wilson 1991), but the results are striking. Tropical forests occupy less than 7% of the Earth's land area but contain at least half of all plant and animal species.
- Tropical rainforests are presently under acute threat. Many nontropical habitats in the northern hemisphere, and most oceanic islands, have been under

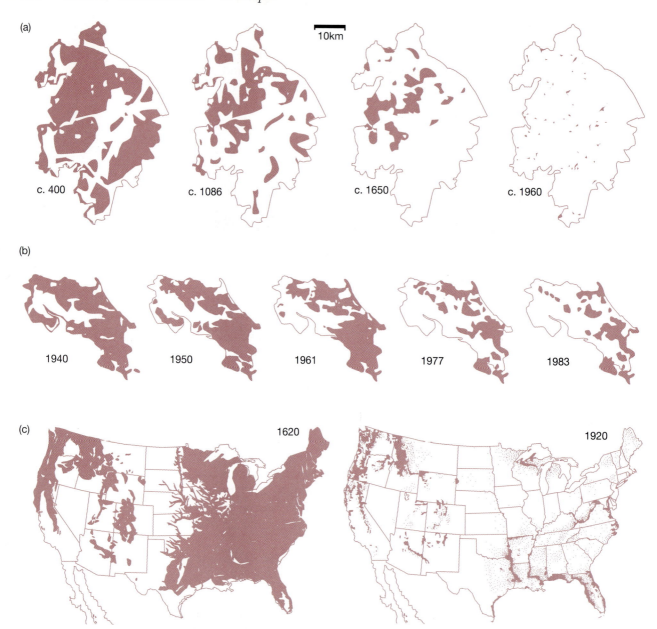

Figure 19.4   Deforestation in England, Costa Rica, and the United States   Each set of maps shows the change over time in the fraction of land area covered by forests.   (a) Warwickshire, England, 400 A.D. to 1960.   (b) Costa Rica, 1940 to 1983   (c) United States, 1620 to 1920. Each dot on the U.S. map for 1920 represents 25,000 acres. From Dobson (1996).

continuous occupation by high densities of humans for several hundred years. As a result, the flora and fauna of these nontropical habitats have already sustained numerous extinctions. Andrew Balmford (1996) suggests that the long history of dense human occupation, combined with extinctions caused by radical climate change during the ice ages of the Pliocene and Pleistocene, have put nontropical-rainforest biomes through an "extinction filter." The plant and animal communities now living in these regions are

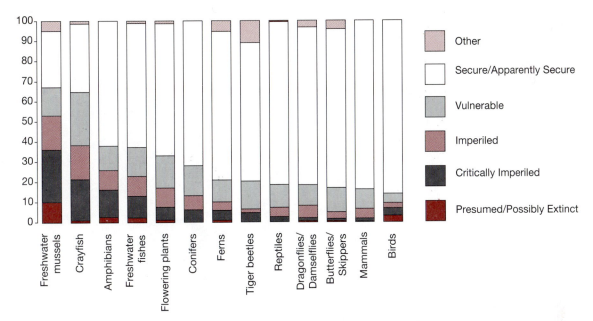

Figure 19.5 **The status of species in the United States**   Each bar shows, for a group of organisms, the fraction of species in each of six status catetories ranging from secure to presumed or possibly extinct. The Nature Conservancy report (Stein and Chipley 1996) considered a species critically imperiled if it had five or fewer populations or 1,000 or fewer individuals. Imperiled species had 6 to 20 populations or 1,001 to 3,000 individuals. Vulnerable species were rare, but with larger populations. After Stein and Chipley (1996); Dicke (1996).

expected to be relatively resilient in the face of continued human impact. In contrast, many areas of the tropics have been relatively unaffected by humans in recent history and were less affected by glaciation and sea-level changes in the Pleistocene. The tropics are now experiencing the highest rates of growth in human populations and the highest rates of habitat loss. Tropical rainforests are a habitat in which it is not too late to intervene to preserve biological diversity.

The Brazilian Amazon is the largest continuous tropical forest in the world, and a special target of concern. David Skole and Compton Tucker (1993) used photographs from the Landsat satellite system to quantify rates of deforestation in the Brazilian Amazon (Figure 19.6). Their study indicates that an average of about 15,000 square kilometers of forest were lost each year during the interval between 1978 and 1988. Although this figure is much smaller than some earlier estimates of forest destruction rates, it still represents an annual loss of forested area roughly equal in size to the state of Connecticut. Skole and Tucker also maintain that more than double this amount of land was adversely affected each year because of edge effects. Forested areas adjacent to clearings undergo dramatic changes: Light levels increase, soils dry, daily temperature fluctuations increase markedly, domestic livestock encroach, and hunting pressure by humans heightens. Skole and Tucker (1993, page 1909) note that the "Implications for biological diversity are not encouraging."

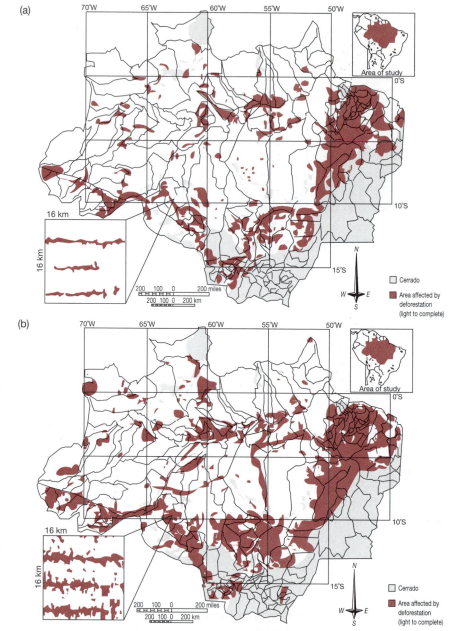

Figure 19.6 **Deforestation in the Brazilian Amazon** Skole and Tucker (1993) used satellite imagery to quantify the extent of forest, cerrado, and aquatic habitats in the Brazilian Amazon (*cerrado* is the term for tropical savanna, meaning grasslands with scattered trees). The top map was generated from photographs taken in 1978, and the bottom map from images generated in 1988. Note that deforestation is most intense along the southern and eastern edges of the Amazon, and along major transportation corridors such as the Amazon River. From Skole and Tucker (1993).

19.2 Why Is Species Loss a Problem?

We have discussed attempts to estimate the total number of extant species, and reviewed evidence that a mass extinction may now be underway. Before going on to consider how evolutionary biology can lend insight to the biodiversity crisis and inform our response, we need to ask a basic question: Why should we care? Why should human beings spend time, effort, and money to save species? In many cases, species conservation efforts require short-term economic losses. Where is the return?

Reasons for conserving biodiversity range from the practical to the aesthetic.

Biologists involved in conservation efforts point to three ways that humans benefit from species preservation (see Ehrlich and Wilson 1991; Perrings et al. 1994; Mooney et al. 1995). We consider each in turn.

Ecosystem Services

The ecosystems of planet Earth provide our life support system. Organisms regulate the composition of the atmosphere: photosynthesizers by releasing oxygen, all organisms by both absorbing and releasing carbon dioxide. Via the greenhouse effect, the composition of the atmosphere has an important influence on climate. Organisms are also responsible for soil formation, and they help control soil erosion. Organisms break down wastes, recycle nutrients, and provide us with food. Collectively, the activities of living organisms in maintaining our life support system are referred to as ecosystem services.

Costanza and colleagues (1997) estimated the total economic value of the renewable ecosystem services that are important to human health and welfare. Their figure was 16 to 54 trillion U.S. dollars per year. For comparison, the sum of the gross national products of all the countries on Earth comes to about 18 trillion U.S. dollars per year.

The loss of biodiversity may threaten ecosystem services. A case in point is the impact that deforestation is having on the global carbon budget. When tropical forests are burned and replaced with rangeland to support domestic livestock, less carbon dioxide is sequestered in terrestrial ecosystems and more is released to the atmosphere. This transfer of carbon may be exacerbating the greenhouse effect induced by the burning of fossil fuels, with repercussions for global warming. In another example, David Tilman and colleagues (1997) have shown that the functional diversity of savanna-grassland species is positively correlated with ecosystem productivity in experimental plots. Civilizations have fallen when their own activities undermined the ecosystem services on which they relied (Diamond 1992; Box 19.1). Erhlich and Wilson (1991) contend that preserving species diversity is one of the best ways we have of ensuring the long-term health of the Earth's life support systems.

Economic Benefits

Most ecosystem services, such as regulating the composition of the atmosphere, cannot be directly exploited for profit. Strictly utilitarian arguments for species preservation are based on the value that species have as exploitable economic resources. The value of wild fish and shellfish species is well documented. Futhermore, all domesticated plants and animals are derived from wild stocks. Plant breeders regularly use the wild relatives of domesticated species as a genetic reservoir—a source of alleles for artificial selection programs designed to increase yields or respond to disease and pest outbreaks (for example, Rick 1974; Iltis 1988). Developing alternative sources of domesticated food and fiber will be impossible in the future if extinctions wipe out the lineages capable of producing new varieties.

Wild species are also an important source of fungicides, insecticides, and medicines (for example, Plotkin 1988). About 25% of the prescription medications dispensed in the United States contain active ingredients derived from

BOX 19.1 Ecological disaster on Easter Island

Located 2,300 miles west of Chile, 2,400 miles southwest of the Galápagos, and 1,400 miles east of Pitcairn, Easter Island is the most isolated human habitation on Earth (Figure 19.7a). Dutch explorer Jakob Roggeveen and his crew became the first Eu-

ropean visitors in 1722. As they found it, the tiny volcanic island—just 50 square miles in area—was covered in treeless grassland and adorned with hundreds of giant stone statues (Figure 19.7b). Weighing up to 85 tons and standing up to 37 feet tall, the

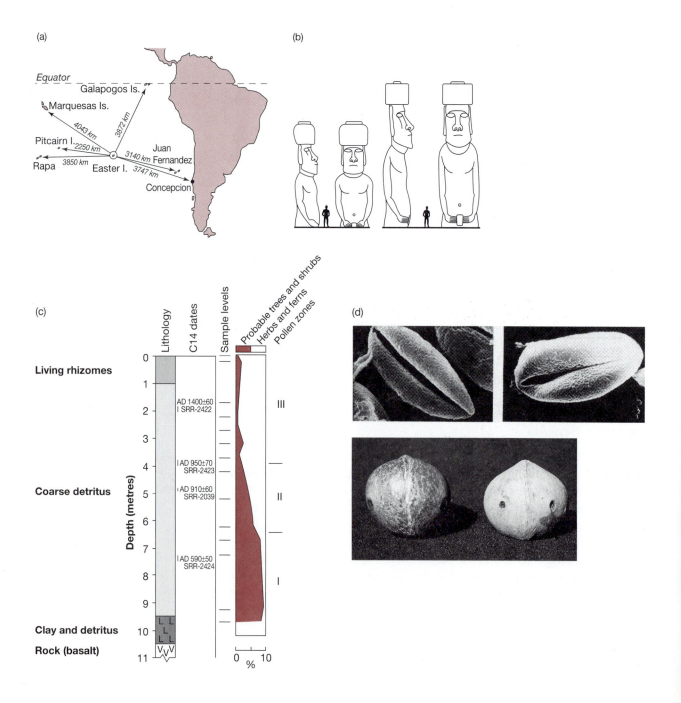

statues had been carved from lava in the island's quarries and transported up to several miles—by people using stone-age technology. Lacking trees from which to make rollers, and ropes to control a large load for transport, the remaining Easter islanders were no longer building and moving statues. With a population estimated by early European visitors at 600 to 2,000 people, the islanders subsisted on cultivated sweet potatoes and bananas, small numbers of domestic chickens, and a limited inshore fishery (Ayers 1985).

Archaeological, linguistic, and genetic data demonstrate that the native Easter islanders are Polynesians (Bahn and Flenley 1992; Hagelberg et al. 1994). Easter Island thus marks the extreme eastern edge of the Polynesians' great expansion across the South Pacific (Dye and Steadman 1990).

Analyzing pollen in sediment cores from Easter Island's three crater swamps, John Flenley and Sarah King (Flenley 1979; Flenley and King 1984) discovered that when the first Polynesian settlers arrived in 400 A.D., the island was covered with a lush forest of trees and shrubs (Figure 19.7c). The frequency of tree and shrub pollen in the sediment cores steadily declines after about 600 A.D. By 1400 A.D., tree and shrub pollen has been almost entirely replaced by pollen from herbs and ferns. Abundant in the samples from the forest period is pollen from a palm tree. Based on the morphology of this pollen and on remains of seeds found in a cave on Easter Island, J. Dransfield and colleagues (1984) established that this palm was a relative of the Chilean wine palm, *Jubaea chilensis* (Figure 19.7d). The now-extinct

Easter Island palm was probably a tall tree with a smooth trunk, suitable for use in making both ocean-going canoes and rollers for the transport of giant statues.

Excavations on Easter Island by David Steadman and colleagues (1994) and other researchers reveal that the diet of the first islanders contained an abundance of dolphins and birds. About when the deforestation of the island was completed, dolphin bones disappear from the excavated garbage piles. Of the 25 species of sea birds found in the Easter Island archaeological record, one is globally extinct, and 12 to 15 are locally extinct on Easter Island and all offshore islets (Steadman 1995). The excavations also turned up bones of six species of land birds unique to Easter Island. All of them are extinct.

Based on the pollen cores and the archaeological record, the history of Easter Island is the story of a people who taxed their environment's ecosystem services to the breaking point (Diamond 1992; 1995). The first Polynesian settlers found in their new home an abundance of natural resources: They cleared the forest to make gardens for sweet potatoes and bananas; they felled still more palm trees to make ocean-going canoes for hunting dolphins; they harvested the local birds and their eggs. Thriving amid this abundance, the Easter Island population swelled.

One thousand years after the original settlers arrived, their descendants numbered some 7,000. The island's economy provided a large enough surplus to support the stone carvers, sculptors, and moving crews that erected the giant statues. But disaster was

Figure 19.7 Easter Island (a) Easter Island is the most isolated human habitation on Earth. From Bahn and Flenley (1992). (b) Giant stone sculptures from Easter Island. From Bahn and Flenley (1992). (c) Data from a pollen core taken from one of Easter Island's crater swamps. The column labeled "Lithology" describes the physical nature of the sediments in the core as a function of depth. The column "C14 dates" gives the ages, established by carbon dating, of the sediments at four depths. The ticks in the "Sample levels" column indicate sediment layers examined for pollen analysis. The graph that forms the next column shows the changes over time in the type of plant community implied by the pollen found in the core. The shaded region represents the frequency of pollen from trees and shrubs; the white region represents the frequency of pollen from herbs and ferns. The pollen samples indicate that during the period labeled Pollen Zone I, Easter Island was dominated by a palm forest with an abundance of trees and shrubs. During Pollen Zone II, the palm forest was replaced by grassland. During Pollen Zone III, Easter Island was, as it is today, dominated by grassland. From Flenley and King (1984). (d) Pollen grains (top) and seeds (bottom) from the extinct Easter Island palm (left) and the Chilean wine palm (right). Pollen photo reprinted by permission from J. Dransfield et al. A recently extinct palm from Easter Island. *Nature* 312:750−2, Fig. 1 a & b, Dec. 20−27, 1984; seed photo reproduced by permission from P. Bahn and J. Flenley, *Easter Island; Earth Island,* page 87, Fig. 61. New York: Thames & Hudson, 1992.

Box 19.1 *(continued)*

on the way. The forest had been cleared from the entire island. Not a single palm was left standing. Consumption of the palm's seeds by the islanders, and by the rats they brought with them, prevented the palm population from regenerating itself. Because the islanders had no more palm trunks to make canoes, the dolphin fishery collapsed. With no rollers to transport statues, production ground to a halt. With the extirpation of the bird populations, the islanders were forced to rely heavily on their crops and the limited inshore fishery. The human population crashed to less than one-third its size at the height of prosperity. Warfare broke out among rival clans. The stone statues were pulled down.

vascular plants (Farnsworth and Morris 1976). The prominent anticancer drug vincristine/vinblastine was first synthesized from a plant native to Madagascar called the rosy periwinkle *(Catharanthus roseus)*. The anticancer drug taxol was discovered in extracts from the Pacific yew *(Taxus brevifolia),* which lives only in old-growth forests of the Pacific Northwest. Based on facts like these, the entomologist Thomas Eisner (1989; see also Eisner and Beiring 1994) has proposed a research program he calls chemical prospecting. Eisner's proposal is for systematists to collaborate with agricultural and pharmaceutical researchers in cataloging the diversity of species in the tropics, and in testing newly discovered plants, insects, and microbes for antifungal, antibacterial, or other economically important properties. In 1991 Costa Rica's Instituto Nacional de Biodiversidad turned the promise of profits from bioprospecting into reality by signing an agreement with the drug company Merck (Gershon 1992; Blum 1993; Meyer 1996). The agreement allows Merck to look for new drugs in Costa Rican plants and animals; in return Merck provides support for Costa Rican conservation efforts.

Moral, Esthetic, and Cultural Benefits

Humans are now a dominant life form in energy use, geographic distribution, and impact on the global environment. That one species could threaten the existence of half or more of all other species is unprecedented in the history of life. Other mass extinctions have been caused by bolide impacts or massive climate and geological change, not environmental changes wrought by a single species (see Chapter 13). Our ability to foresee the disastrous consequences our present activities will have for other species and for future generations gives us the option to take preventative action. Failure to do so would be seen as immoral by people from a great diversity of cultures, religions, and political persuasions.

These ethical reasons for preserving species are closely tied to aesthetic ones. The aesthetic argument for preserving species is simple: Most people consider their quality of life to be better when they live in a clean and healthy environment. The presence of wildlife and wild places adds values like beauty, tranquillity, and adventure to human existence. Most people perceive environments that have been traumatized by deforestation, soil erosion, and species loss as desolate and undesirable.

Culturally, certain species and landscapes play prominent roles in the identities of indigenous people and contemporary nations. Birds of paradise are protected in Papua New Guinea because they are a critical cultural resource. The sense of shame induced by the endangered status of the bald eagle, a national symbol of the United States, helped to galvanize species protection efforts there in the 1970s. Cranes are revered in many Asian cultures as symbols of long life, marital fidelity, and happiness, and are afforded special protection as a result. Loss of any of these species would have a negative effect on the health and well-being of these cultures.

19.3   The Genetics and Demography of Small Populations

We now consider ways in which evolutionary biology can improve our understanding of the biodiversity crisis and help us plan a rational response. In this section we use population genetics to predict the evolutionary consequences of habitat destruction. Habitat destruction reduces population size, which in turn leads to reduced genetic diversity and increased risk of extinction.

Habitat Destruction and Reduced Population Size

The most obvious consequence of habitat destruction is reduction in the sizes of the populations that occupy the habitat. When a population is reduced in size, genetic drift immediately becomes a more potent mechanism of evolution (see Chapter 5). Under genetic drift, allele frequencies rise and fall due to chance events. Eventually any given allele may drift to a frequency of 0 or 1, and be forever lost or fixed in the population. Because alleles drift to loss or fixation more rapidly in small populations, we can predict that a reduction in population size will result in a reduction in genetic diversity. Figure 19.8 demonstrates that there is indeed a clear relationship between population size and genetic diversity. Andrew Young and co-workers (1996) compiled data from the literature and plotted two measures of overall genetic diversity against population size in four species of plants. In almost every case, smaller populations show less genetic diversity.

Habitat destruction leads to reduced population size, which leads to reduced genetic diversity.

Genetic variation is the raw material for evolution by natural selection. As alleles drift to loss or fixation in a small population, the population loses its potential for adaptive evolution. This can be especially problematic because habitat destruction is often accompanied by more than simply a reduction in population size. The physical environment may change, and biological interactions may shift as the relative abundance of other species in the habitat changes. These changes may impose new selective pressures on populations.

A Hawaiian honeycreeper called the i'iwi *(Vestiaria coccinea)* provides an example (Smith et al. 1995; Freed et al. 1996; Winker 1996). Historically, the i'iwi fed on the nectar of lobelia flowers (Figure 19.9a). Lobelia flowers have long, curved corollas, suggesting that the long, curved bill of the i'iwi is an adaptation for feeding. Habitat destruction in Hawaii has decimated or extirpated lobelia populations, depriving the i'iwi of its favored food source. I'iwi now feed primarily on flowers of the ohia tree *(Metrosideros polymorpha)*. Ohia flowers lack corollas (Figure 19.9b). Thomas Smith and colleagues (1995) found that present-day i'iwi have bills that are significantly shorter than those of museum specimens collected before 1902 (Figure 19.9c and d). The researchers' interpretation is that i'iwi

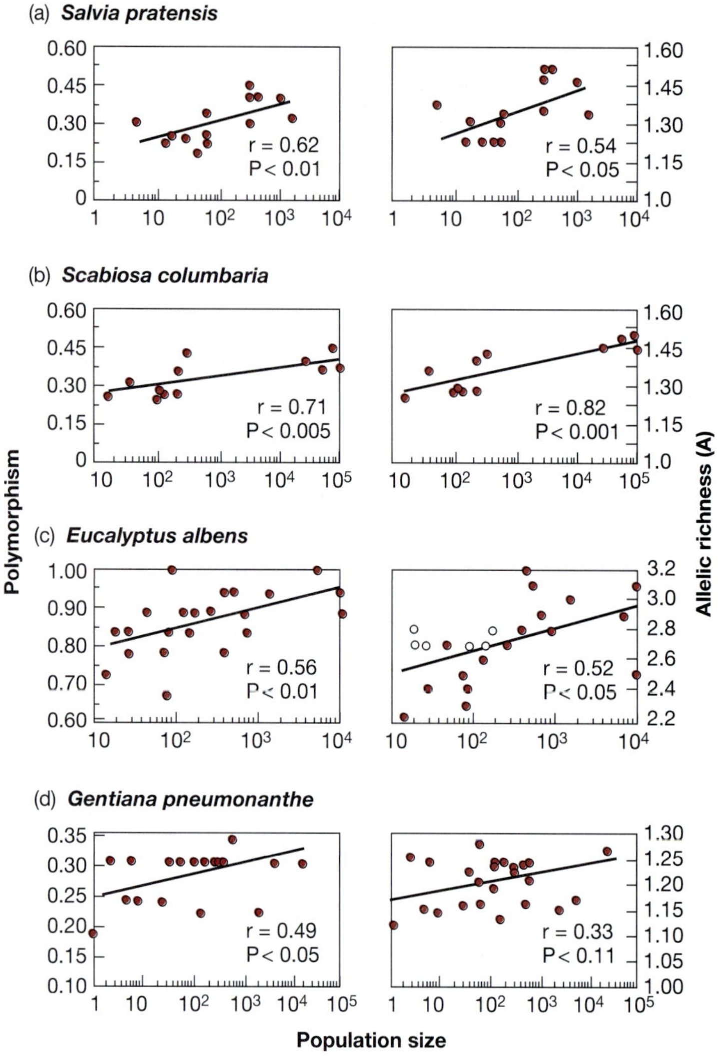

Figure 19.8   Population size and genetic diversity   Each data point on these scatterplots represents a population of flowering plants. Polymorphism, plotted on the vertical axis of the graphs at left, is the proportion of allozyme loci at which the frequency of the most common allele in the population is less than 0.99. In other words, polymorphism is the fraction of alleles that are substantially polymorphic. Allelic richness, plotted on the vertical axis of the graphs at right, is the average number of alleles per locus. The statistic r is a measure of association, called the Pearson correlation coefficient, which varies from 0 (no association between variables) to 1 (perfect correlation). P specifies the probability that the correlation coefficient is significantly different from zero. *Salvia pratensis, Scabiosa columbaria,* and *Gentiana pneumonanthe* are all flowering herbs; *Eucalyptus albens* is a tree. From Young et al. (1996).

populations have evolved in response to selection pressures associated with the switch in primary food source (Smith et al. 1995; Freed et al. 1996; Winker 1996).

Habitat destruction may thus lead to new selection pressures at the same time it contributes to loss of genetic variation in populations. The result is that populations are challenged to evolve at the same time their potential to evolve is eroding. These combined effects can increase the probability that populations will go extinct.

Habitat Destruction and Population Fragmentation

Habitat destruction typically leads to population fragmentation as well as reduced population size. Individual subpopulations lose genetic diversity even faster, although different subpopulations may retain different alleles.

As shown in Figure 19.4, habitat destruction typically leads not only to a reduction in the total amount of habitat but also to the fragmentation of the habitat into small, isolated patches. These patches are, in effect, islands. A species occupying a fragmented habitat thus not only has a smaller total population size, but its total population is also broken up into subpopulations whose size is even

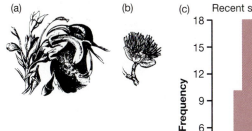

(a) (b)

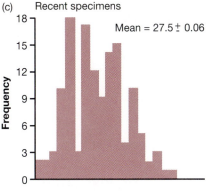

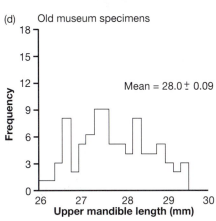

(c) Recent specimens

Mean = 27.5 ± 0.06

(d) Old museum specimens

Mean = 28.0 ± 0.09

Upper mandible length (mm)

Figure 19.9 **Natural selection on mandible length in a honeycreeper as a consequence of habitat destruction**
(a) A 19th-century illustration of the i'iwi, with a lobelia flower, the honeycreeper's historical food source. Lobelia flowers have long corolla tubes, making their nectar accessible only to birds with long bills. (b) An ohia flower, presently the i'iwi's primary nectar source. This flower lacks a corolla. (c) This histogram shows the distribution of upper mandible lengths in living i'iwi that were netted, measured, and released between 1988 and 1991. (d) This histogram shows the distribution of upper mandible lengths in museum specimens collected prior to 1902. The mean mandible length is significantly shorter in the recent specimens ($P < 0.001$). From Smith et al. (1995).

smaller still. If migration between habitat islands is sufficiently limited, then each subpopulation will evolve independently. Within any individual population, alleles will drift rapidly to fixation or loss. However, different alleles will become fixed in different populations.

Alan Templeton and colleagues (1990) tested these predictions by documenting the results of a natural experiment in Missouri's Ozark Mountains. Although now covered in oak-hickory forest, the Ozarks were part of a desert during an extended period of hot, dry climate that lasted from 8,000 to 4,000 years ago. This desert was contiguous with the desert of the American Southwest. Many southwestern desert species expanded their ranges eastward into the Ozarks. Among them was the collared lizard *(Crotaphytus collaris)*. When the warm period ended, the collared lizard's range retracted westward again. The oak-hickory forest reinvaded the Ozarks. Within this forest, on exposed rocky outcrops, are small remnants of desert habitat called glades. Living in some of these glades are relict populations of collared lizards. The relict populations are tiny; most harbor no more than a few dozen individuals.

Templeton and colleagues assayed several glade populations for genetic variation. The researchers screened lizards for their genotypes at a variety of enzyme loci, for their ribosomal DNA genotypes, and for their mitochondrial DNA genotypes. The researchers identified among the lizards seven distinct multilocus genotypes. Confirming the predicted consequences of isolation and small population size, most glade populations are fixed for a single multilocus genotype, with different genotypes fixed in different populations (Figure 19.10).

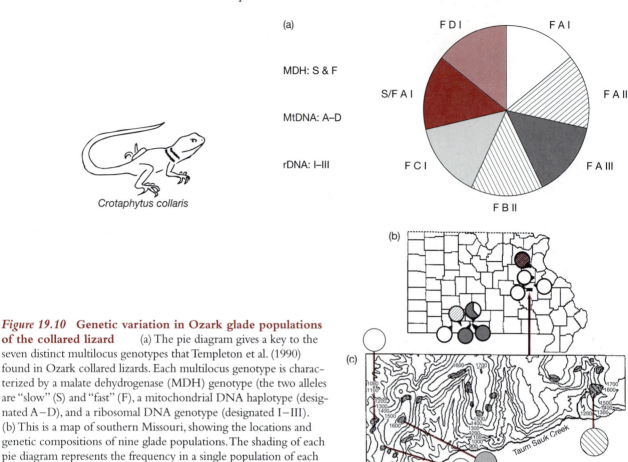

Crotaphytus collaris

Figure 19.10   Genetic variation in Ozark glade populations of the collared lizard   (a) The pie diagram gives a key to the seven distinct multilocus genotypes that Templeton et al. (1990) found in Ozark collared lizards. Each multilocus genotype is characterized by a malate dehydrogenase (MDH) genotype (the two alleles are "slow" (S) and "fast" (F), a mitochondrial DNA haplotype (designated A–D), and a ribosomal DNA genotype (designated I–III). (b) This is a map of southern Missouri, showing the locations and genetic compositions of nine glade populations. The shading of each pie diagram represents the frequency in a single population of each multilocus genotype present.   (c) This is an expanded map of a small piece of the map in (b). It gives the locations and genetic compositions of 5 more glade populations. From Templeton et al. (1990).

Fragmentation can interact with reduced population size to influence evolutionary potential. Figure 19.11 illustrates some differences between the evolutionary consequences of population reduction that does, versus does not, result in fragmentation. The key fact is that for any remaining individual population, the smaller the population's size, the faster alleles will drift to loss or fixation.

- A fragmented set of subpopulations may preserve more of the total initial genetic variation than an unfragmented but reduced population (Figure 19.11a, b, and c). This is because different alleles are likely to become rapidly fixed in different subpopulations, whereas in a reduced-but-continuous population a single allele may eventually drift to fixation.

- However, in a fragmented set of subpopulations, each subpopulation will rapidly lose genetic variation, and with it the potential to evolve (Figure 19.11c). Because it is larger than any individual subpopulation, a reduced-but-continuous population will retain genetic variation, and the potential to evolve, for a longer time (Figure 19.11b).

- In a fragmented set of subpopulations, further habitat destruction is likely to eliminate a greater proportion of the remaining total genetic variation. In

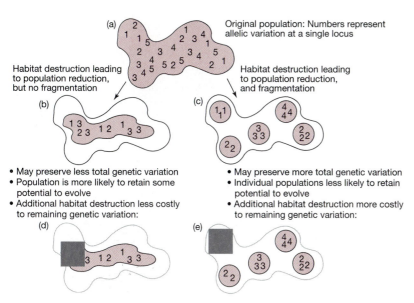

Figure 19.11 **The evolutionary consequences of population fragmentation**
(a) This illustration represents a single continuous population prior to habitat destruction. The shaded region represents the population's geographic range. The numbers represent different alleles of a single locus.   (b) and (c) These two illustrations show possible evolutionary consequences of habitat destruction. Each population contracts to a smaller geographic range (shown by the shaded regions; the old range is indicated by the outer line). In (b), the population is reduced, but remains a single continuous population. In (c), the population is reduced, and fragmented into five very small subpopulations.   (d) and (e) These two figures show an additional bout of habitat destruction, with the eliminated habitat (and its resident organisms) indicated by the gray square. See text for more details.

Figure 19.11e the extirpation of just one subpopulation eliminates 25% of the remaining total genetic variation. In contrast, in an unfragmented population the same amount of additional habitat destruction may not immediately reduce the remaining total genetic variation at all (Figure 19.11d).

An additional difference is that when the habitat is fragmented, the tiny size of the subpopulations makes them highly prone to extinction due to demographic factors that we will discuss shortly. If there is no migration between habitat patches, then once a subpopulation goes extinct it can never be reestablished. Thus a fragmented population is more prone to total extinction by the incremental loss of its subpopulations. Templeton and colleagues (1990) note that there is good evidence that many relict populations of collared lizards have gone extinct: Many apparently suitable Ozark glades are unoccupied by lizards.

Inbreeding in Small Populations

As allele frequencies in a small population drift toward loss or fixation, the frequency of heterozygotes in the population decreases. Once an allele is fixed, all individuals are homozyotes at that locus. Another way to think about the decline in the frequency of heterozygotes is to note that even with random mating, all individuals in a small population will eventually be related to each other. Inbreeding is inevitable. As we showed in Chapter 5, inbreeding decreases the frequency of heterozygotes.

Sewall Wright (1931) showed that the decrease in the frequency of heterozygotes from one generation to the next can be calculated as follows:

$$\text{Het}^* = \text{Het}\left[1 - \frac{1}{2N}\right]$$

Small populations become inbred, which may increase their vulnerability to extinction.

where Het* is the frequency of heterozygotes (or the population's heterozygosity) in this generation, Het is the heterozygosity in the previous generation, and

N is the number of individuals in the population. The value of $\left[1 - \dfrac{1}{2N} \right]$ is always between $\frac{1}{2}$ and 1, so the frequency of heterozygotes inevitably decreases. To appreciate the implications of this conclusion, suppose we were managing a captive population of an endangered species, with perhaps 50 breeding adults in zoos worldwide. Even if we could arrange the shipment of adults or semen to accomplish random mating, we would still see a loss in heterozygosity of 1% every generation due to drift.

Peter Buri's (1956) experiment on genetic drift in fruit flies (also discussed in Chapter 5—see Figure 5.24) provides an example of the loss of heterozygosity in small populations. Buri established 107 populations of flies, each consisting of eight females and eight males. Buri was interested in evolution at a single locus with two alleles. The genetic properties of the locus were such that all three genotypes were identifiable from their phenotypes. Thus Buri was able to assess directly the frequency of heterozygotes in each population. The frequency of heterozygotes in generation zero was one, so the heterozygosity in generation one was 50%. Thereafter, the 107 populations evolved by genetic drift. After 19 generations, more than half the populations had become fixed for one allele or the other.

Every generation Buri noted the frequency of heterozygotes in each population, then took the average heterozygosity across all 107 populations. Figure 19.12 plots these values for average heterozygosity over the 19 generations of the experiment. Look first at the dots, which represent the actual data. Consistent with our theoretical prediction, the average frequency of heterozygotes steadily declined.

The fit between theory and results is not perfect, however. The dashed curve in the figure shows the predicted decline in heterozygosity, using the formula we just introduced and a population size of 16. The actual decline in heterozygosity was more rapid than expected. The solid curve shows the predicted decline for a population size of 9; it fits the data well. Buri's populations lost heterozygosity as though they contained only 9 individuals instead of 16. In other words, the effective population size in Buri's experiment was 9.

The effective population size is the size of an ideal theoretical population that would lose heterozygosity at the same rate as the actual population. The ef-

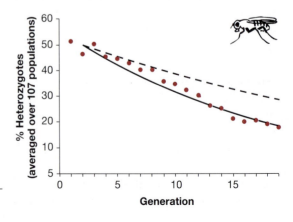

Figure 19.12 **Heterozygosity declines each generation in small populations**   The dots show the actual data. The dashed curve shows the theoretical prediction for a population of 16 flies. The solid curve shows the theoretical prediction for a population of 9 flies. The graph demonstrates that (1) heterozygosity decreases across generations in small populations; and (2) although all the populations had an actual population of 16 flies, their effective population size was roughly 9. Replotted from data in Buri (1956), after Hartl (1981).

fective population size is virtually always smaller than the actual population size. In Buri's experiment, two possible reasons for the difference in effective versus actual population size are (1) some of the flies in each bottle died (by accident) before reproducing; and (2) fruit flies exhibit sexual selection by both male-male combat and female choice—either of which could have prevented some males from reproducing.

The effective population size is particularly sensitive to differences in the number of reproductively active females versus males. When there are different numbers of each sex in a population, the effective population size (N_e) can be estimated as follows:

$$N_e = 4 N_m N_f / (N_m + N_f)$$

where N_m is the number of males and N_f is the number of females. To see how strongly an imbalanced sex ratio can reduce the effective population size, use the formula to show that: when there are 5 males and 5 females, $N_e = 10$; when there is 1 male and 9 females, $N_e = 3.6$; and when there is 1 male and 1,000 females, $N_e = 4$. Consider the logistical problems involved in maintaining a captive breeding program for a species in which the males are extremely aggressive and will not tolerate each other's presence.

The inevitable loss of heterozygosity (or the inevitable increase in homozygosity) is of concern to conservation biologists because increased homozygosity often leads to decreased fitness in experimental populations (see, for example, Polans and Allard 1989; Barrett and Charlesworth 1991). Presumably this involves the same mechanism as inbreeding depression: It exposes deleterious alleles to selection.

Michael Lynch and Wilfried Gabriel (1990) have proposed that an accumulation of deleterious recessives (a phenomenon known as genetic load) can actually lead to the extinction of small populations. They noted that when exposure of deleterious mutations produces a reduction in population size, the effectiveness of drift is increased. The speed and proportion of deleterious mutations going to fixation subsequently increases, which further decreases population size. Lynch and Gabriel termed this synergistic interaction between mutation, population size, and drift a "mutational meltdown."

Alternatively, some theoreticians have proposed that the exposure of deleterious recessives in small populations can lead to selection purging these alleles from the population. If this occurs, it would lead to an *increase* in the mean fitness of the population (for a review, see Young et al. 1996). As yet, however, neither mutational meltdowns nor purging events have been observed in laboratory or field populations.

Demography in Small Populations

In addition to the genetic issues we have discussed, a number of demographic issues affect small populations and can increase the probability of extinction due to random causes or altered population dynamics (see Lande 1988; Schemske et al. 1994; Wahlberg et al. 1996). These include:

Small populations are more vulnerable to chance catastrophes.

- **Catastrophic losses due to unusual storms, weather conditions, or other infrequent events like a predator, herbivore, or disease outbreak.**

The population of gray wolves *(Canis lupus)* on Isle Royale, in Lake Superior, recently declined from 50 individuals to 12 (Wayne et al. 1991). The decrease coincided with an outbreak of canine parvovirus in nearby wolf and dog populations, and may have been caused by the disease. In another example, 1987 was a dry year on the Costa Rican mountain top that was home to the only population of golden toads *(Bufo periglenes)*. The breeding pools dried up before the tadpoles completed metamorphosis, and as a result the species went extinct (Pounds and Crump 1994; Phillips 1994).

- **Changes in the availability of pollinators or mates that result in a reduced population growth rate.** The last five dusky seaside sparrows *(Ammodrumus maritimus nigrescens)* were all males (Hedrick et al. 1996). This extraordinarily bad luck explains why the species went extinct shortly thereafter.

- **Lower colonization rates by dispersing seeds or juveniles, because appropriate habitats have become rare or widely separated.** Spotted owls have low rates of survival during juvenile dispersal in fragmented habitats because they are vulnerable to predators when passing through clearcuts.

Demography Versus Genetics

Deciding whether demographic or genetic factors are the more important issue for conservation biologists to track is a source of continuing controversy. Although theoretical and laboratory studies provide abundant evidence for deleterious genetic effects in small populations, the situation in nature appears somewhat different. In vertebrates, at least, a variety of species have little or no variation in allozyme loci, but seemingly normal fitness. Examples include populations of bandicoots, cheetahs, members of the weasel family, Pere David's deer, and elephant seals. Partly as a result of this observation, Robert May (1995) articulated what may be an emerging consensus: Demographic factors like recruitment, survivorship, local extinctions, and colonizing ability are probably the most important parameters for conservation managers to track in the short term (see also Lande 1988). Genetic issues may be most important when managers need to design captive breeding and reintroduction programs for species in crisis, and for the management of fragmented populations in the long term.

19.4 Systematics, Conservation, and Hard Choices

Few scientists believe that the present diversity of life will emerge intact if current patterns of habitat destruction continue. Instead, most research on biodiversity, and most debate over conservation practices, centers on strategies for establishing conservation priorities. In this section we focus on how systematics informs decisions about which species are most important to conserve.

Identifying Species

Conservation priorities can be influenced by which taxa are recognized as legitimate species.

Perhaps the most straightforward contribution of evolutionary biology to conservation is in the identification of species. A species has to be recognized before it can be protected.

Research on Darwin's fox by Christopher Yahnke and colleagues (1996) illustrates this point. Darwin's fox was first described by Charles Darwin himself, from a specimen he collected in 1834 during his voyage on the Beagle. Although the fox carries its own latin name *(Dusicyon fulvipes)* and is morphologically distinct from other Chilean foxes, the systematic status of Darwin's fox has been unclear. It is relatively poorly known, with a population of fewer than 500 individuals long thought to be confined to Chiloé Island (Figure 19.13a). Many mammologists considered Darwin's fox to be a geographic subspecies of the mainland chilla fox *(D. griseus)*. Darwin's fox was listed as threatened by the Chilean governement in 1987, but because most biologists did not consider it a distinct species the fox received little attention from conservationists.

In 1990 mammologists made a discovery that suggested Darwin's fox might be a distinct species after all. Researchers found several Darwin's foxes at Nahuelbuta, 600 km north of Chiloé Island. This small population coexists with the chilla fox and with the culpeo fox *(D. culpaeus)*, apparently without interbreeding.

Following up on this discovery, Yanhke and colleagues (1996) estimated the phylogeny of the three foxes (Figure 19.13b). Darwin's fox is clearly a distinct evolutionary lineage; it is the sister taxon to the chilla and culpeo foxes (which themselves prove to be unresolvable as separate species). Yanhke and colleagues proclaim the rediscovery of Darwin's fox as a distinct species, and advocate greater attention to the protection of the fox and its habitat.

Species Concepts and Conservation

If species are the entities we seek to protect, then our conservation decisions will be influenced by which species concept we employ. In Chapter 9 we introduced three species concepts: the biological species concept, the phylogenetic species concept, and the morphospecies concept. David Hibbett and Michael Donoghue (1996) use the shiitake mushroom to demonstrate the different implications that each concept may have for the conservation of a particular taxon.

The recognition of species is influenced by the species concept employed.

The shiitake mushroom is a wood-decaying fungus (Figure 19.14a). Its natural geographic range inclues Japan, Thailand, Borneo, Papua New Guinea, Tasmania, and New Zealand (Figure 19.14b). Because it is used in northeast Asian cooking, the shiitake is also widely cultivated. Two trends threaten the naturally occuring genetic diversity of the shiitake. First, native shiitake habitats are being destroyed. Second, shiitake cultivation is rapidly expanding. Exotic mushrooms imported for cultivation may escape, and the resulting gene flow may disrupt local adapation in natural shiitake populations.

Decisions concerning conservation measures will depend on the recognition of natural variation among geographic strains. By at least one form of the biological species concept, there is only one species of shiitake. All strains of shiitake are reproductively compatible in controlled crosses in the lab; consequently, many mycologists consider all forms to belong to the single species *Lentinula edodes.* Workers employing the morphospecies concept recognize three species: *L. edodes, L. novaezelandieae,* and *L. lateritia* (Figure 19.14c). Hibbett and Donoghue analyzed ribosomal RNA sequences to estimate the phylogeny of shiitake strains from a wide range of locations (Figure 19.14c). Their phylogeny identifies four distinct clades (groups I–IV); by the phylogenetic species concept, each is a

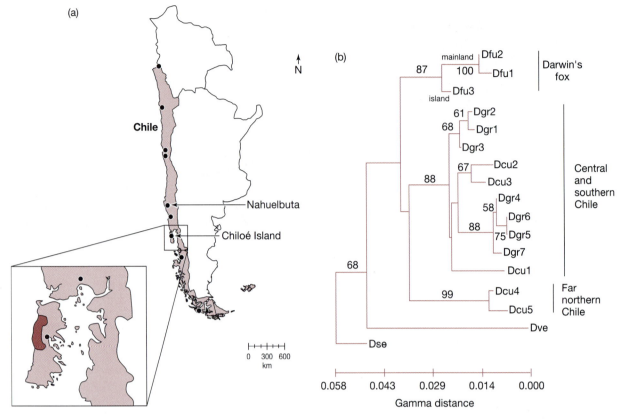

Figure 19.13   Systematics of Darwin's fox   (a) This map of Chile shows the locations (black dots) from which Yahnke et al. (1996) collected fox samples. Chiloé Island constitutes the primary range of Darwin's fox. The darker area in the enlarged map of Chiloé is a national park.   (b) A phylogeny of several species of fox, based on the neighbor-joining method applied to sequence data for mitochondrial DNA. Dfu = *Dusicyon fulvipes* (Darwin's fox); Dgr = *D. griseus* (chilla fox); Dcu = *D. culpaeus* (culpeo fox); Dve = *D. vetulus* (hoary fox, an out-group); Dse = *D. sechurae* (sechurae fox, an out-group). Numbers next to the abbreviations designate distinct mtDNA haplotypes. The scale at the bottom represents a measure of genetic distance. The numbers on branches represent the number of times that the node below and to the right of each branch appeared on the phylogeny in 100 bootstrap replicates. From Yahnke et al. (1996).

Many biologists argue that we should attempt to conserve phylogenetic species.

different species. These phylogenetic species do not correspond to the recognized morphospecies. Donoghue and Hibbett argue that if our goal is to preserve naturally occurring genetic variation, then we should seek to identify and conserve unique evolutionary lineages—that is, phylogenetic species.

Many other biologists are advocating a conservation strategy that recognizes and seeks to preserve distinct evolutionary lineages (for example, Moritz 1994; Avise 1996). Perhaps in an effort to avoid the contentiousness that has characterized the debate over species concepts (see Chapter 9), some conservation biologists have adopted the term *evolutionarily significant unit* to describe the phylogenetic entity that they feel should be the basis of conservation planning. Evolutionarily significant units are populations or groups of populations that have long been isolated from one another, and thus collectively contain a substantial fraction of the genetic diversity within a lineage. As an

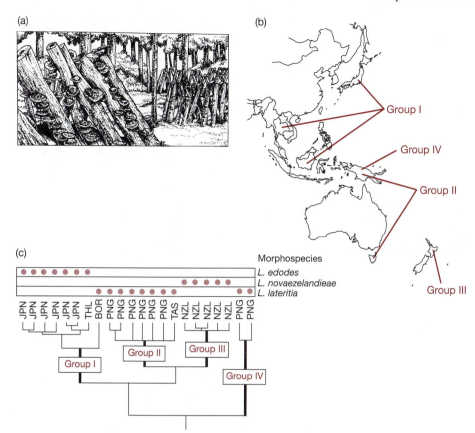

Figure 19.14 Classification of shiitake mushrooms by three different species concepts
(a) The shiitake mushroom, shown here under cultivation on cut logs.   (b) The geographic distribution of the shiitake mushroom.   (c) A phylogeny of the shiitake mushroom estimated from ribosomal RNA sequences, showing four resolvable phylogenetic species (Groups I–IV). The dots indicate species defined on the basis of morphology (names at right). The morphospecies called *Lentinula lateritia* is paraphyletic. By the biological species concept, all forms of shiitake mushroom belong to a single species, *L. edodes*. From Hibbett and Donoghue (1996).

operational definition, Craig Moritz (1994) suggests that evolutionarily significant units should be resolvable in phylogenies based on mitochondrial DNA sequences, and should show substantial differences in allele frequencies at nuclear loci.

Phylogenetic Priorities

Here we will assume that we have decided to use phylogenies as our guide in setting conservation policy: The things that we want to save are branches on phylogenies. Given that money, time, and political will are limited, we cannot save all branches on all phylogenies. Are all branches on the tree of life equally worthy of protection? Or should some kinds of branches receive higher priority than others?

The case of the tuatara illuminates this dilemma (Daugherty et al. 1990; May 1990). The tuatara are reptiles that constitute the sole surviving genus of an order that flourished during the Triassic, Jurassic, and Cretaceous periods. Superficially the tuatara resemble lizards (Figure 19.15a). But they are not lizards. They are the sister group of the lineage that contains the lizards and the snakes (Pough et al. 1996). The tuatara occur only on islands off New Zealand (Figure 19.15b). They have been protected since 1895 by a New Zealand conservation law that recognizes only one species: *Sphenodon punctatus.* However, based on their estimate of the phylogeny of the genus (Figure 19.15c), C.H. Daugherty and colleagues (1990) advocate a view that was advanced as early as 1877: The tuatara of Brothers Island (Island 12) constitute a distinct species, called *Sphenodon guntheri*.

We will assume that we are managers committed to the conservation of reptile biodiversity, but that we cannot save all lineages of all reptiles. Among the questions we face are these:

- How much importance should we place on saving the single surviving population of *S. guntheri* versus the individual populations of *S. punctatus?*

- How much importance should we place on saving *Sphenodon* versus other genera of reptiles?

The simplest approach, a sort of null conservation strategy, would be to place equal importance on all branches of all phylogenies. Concerning the relative value of different lineages of tuatara, this null strategy would assign equal worth to each of the nine resolved branches on the tuatara phylogeny (Figure 19.15c).

Figure 19.15 **A phylogeny of the tuatara**
(a) The tuatara, *Sphenodon punctatus* (Tom McHugh/Photo Researchers, Inc.).   (b) The geographic range of *Sphenodon*. Tuatara occur only on islands off New Zealand. The symbols designate islands or island groups; their names are given in (c). From Daugherty et al. (1990).

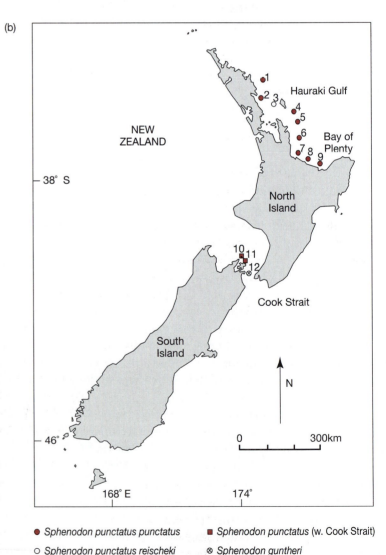

- ● *Sphenodon punctatus punctatus*
- ○ *Sphenodon punctatus reischeki*
- ■ *Sphenodon punctatus* (w. Cook Strait)
- ⊗ *Sphenodon guntheri*

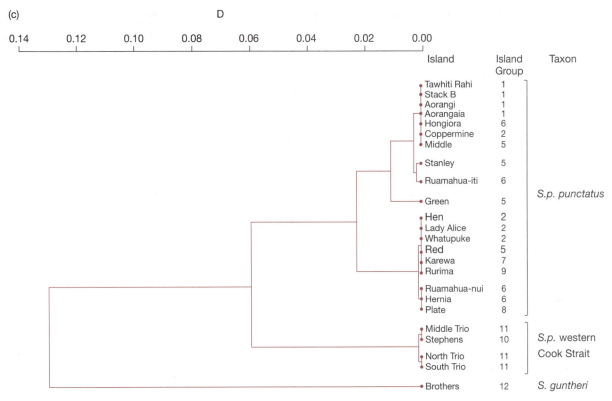

Figure 19.15 (c) A phylogeny of tuatara populations based on genetic distances calculated from 25 allozyme loci. From Daugherty et al. (1990).

The branch holding the *S. punctatus* populations on Tawhiti Rahi through Middle islands, the branch containing the *S. punctatus* populations on Middle Trio and Stephens islands, and the branch containing the sole *S. guntheri* population (on Brothers Island) are all of equal value. If we can save only four of the nine resolvable lineages, we would choose the ones to save by drawing numbers out of a hat. Concerning the relative importance of *Sphenodon* versus other genera of reptiles, the null strategy would view the conservation of *Sphenodon* as no more and no less important than the conservation of any other reptile genus.

It is probably safe to say that the null conservation strategy has few adherents. There is debate, however, over what kind of alternative system we should use in setting conservation priorities. On one side of the issue are those who feel that we should preserve as much of the evolutionary past as possible by placing a high value on phylogenetic distinctiveness. On the other side are those who feel that we should preserve as much of the evolutionary future as possible by placing a high value on within-lineage diversity.

Preserving the Evolutionary Past

Following up on a suggestion by Atkinson (1989), R.I. Vane-Wright and colleagues (1991) advocate placing high value on phylogenetic distinctiveness in setting conservation priorities. From this perspective, the protection of *S. guntheri* should be given high priority among the lineages of *Sphenodon*. Although in its entirety *S. guntheri* consists of one small population, it is the most distinct

One approach to setting priorities is to try to conserve deep branches on the tree of life.

of the extant *Sphenodon* lineages. That is, it represents one of the two deepest branches on the *Sphenodon* phylogeny; the other branch holds all the other populations in the genus. Likewise, the preservation of the genus *Sphenodon* should be given high priority among the reptiles. *Sphenodon* is the sole representative of one of the two deepest branches on the phylogeny of the lepidosaurs; the other branch holds the lizards, amphisbaenians, and snakes.

In this spirit of preserving evolutionary history, a series of algorithms have been proposed for using the phylogeny of a group to assign priority rankings for protection (see Vane-Wright et al. 1991; Holsinger 1992). Daniel Faith's (1992) system is based on a quantity he calls phylogenetic diversity (PD). Given a cladogram for a group of species or taxa with synapomorphies mapped as hash marks on the branches (see Chapter 10), Faith's algorithm works as follows: The PD of each possible subset of taxa is calculated by summing the number of synapomorphies found on branches leading to those taxa. In the example given in Figure 19.16, Faith supposes that a reserve system can be established that would preserve four of the ten taxa pictured. To maximize PD, the reserve system should be designed to include the taxa numbered 2, 6, 8, and 10 as its highest priority. This subset would preserve more evolutionary features of the clade than any other combination of four taxa.

Ross Crozier (1992) points out that a similar strategy can be used when molecular phylogenies are available. In this case the published branch lengths reflect genetic distance. The goal of a ranking system is to preserve the subset of species that maximizes the total amount of branch length on the phylogeny.

The unifying idea in this family of phylogeny-based algorithms is to preserve as much evolutionary history as possible. In practice these "taxic diversity" strategies tend to give much higher weightings to taxa that branched off early in the history of a group. As a result, they tend to promote the preservation of branches on the tree of life—at the expense of twigs.

Preserving the Evolutionary Future

Another approach to setting priorities is to try to conserve branches on the tree of life that have many twigs.

A contrasting point of view exists, however. Inspired by studies on the recovery phases of previous mass extinctions, some workers are suggesting that preservation efforts should target the youngest and most rapidly diversifying lineages living today, rather than older, relictual clades with long branch lengths but few species (Erwin 1991; Brooks et al. 1992; Fjeldså 1994; see also Moritz 1996).

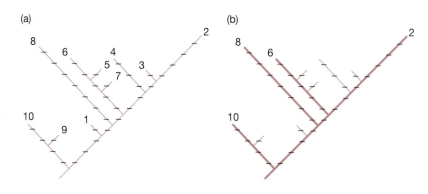

Figure 19.16 **Phylogenetic criteria for conservation priorities**   This is a hypothetical cladogram of 10 species, with unique derived features mapped as hash marks across branches. A reserve system that protected populations of species 2, 6, 8, and 10 would conserve the majority of unique evolutionary features found in this clade. From Faith (1992).

This approach is futuristic, in a sense. It seeks to identify and protect "evolutionary fronts," or clades that could rapidly radiate subsequent to what is seen as an inevitable mass extinction. From this perspective, the preservation of *Sphenodon guntheri* gets low priority among the tuatara lineages. Its single small population likely harbors little genetic diversity, and thus has limited evolutionary potential. Over the long term, this population is likely to go extinct, and the species *S. guntheri* with it. In attempting to save *S. guntheri* we would squander precious effort. Likewise, genus *Sphenodon* gets low priority among the genera of reptiles. The last relict of an otherwise extinct order, *Sphenodon* is almost certainly an evolutionary dead end. We are better off investing in the evolutionary future of more recently derived genera with large, genetically diverse populations.

Beyond Individual Taxa: Conservation Hotspots

In seeking ways to invest conservation effort wisely, a number of researchers are developing strategies for the integrated preservation of diverse assemblages of taxa (for example, Avise 1996). One such strategy involves the identification of hotspots.

Hotspots are geographic areas where species diversity is extraordinarily high. When conservation areas have to be set up quickly, and in situations where information on the diversity and distribution of affected species is limited or nonexistent, one approach is to pick the best-known taxonomic group—often birds—and give highest priority to regions where the ranges of many species overlap.

The hotspot approach is thought to be especially valuable when centers of endemism exist. These are regions where many species are found that exist nowhere else. Hotspots like these tend to be found in areas like Hawaii, the African Rift Lakes, or the Indo-Malayan Archipelago, where speciation rates have been intense because of repeated and dramatic geological changes (Brooks et al. 1992).

This hotspot strategy makes an important assumption, however: that the distribution of species in the well-known group accurately reflects the ranges of other, often more species-rich, lineages. Empirical tests of the hotspot approach have produced mixed results. J.R. Prendergast and co-workers (1993) found that hotspots identified independently for British birds, butterflies, dragonflies, aquatic plants, and liverworts showed little overlap. If conservationists in Britain were to use one of these groups as a proxy and reserve conservation areas to protect its hotspots, few benefits would extend to the other lineages in the study. In contrast, Andy Dobson and colleagues (1997) mapped the ranges of 924 endangered species of plants, fish, birds, and molluscs in 2,858 counties of the United States, and found a small subset of counties where over 50% of the endangered forms from all taxonomic groups were found (Figure 19.17). This result suggests that reserving critical areas in these counties—concentrated in Hawaii, southern California, the Edwards Plateau of Texas, south Florida, and the central Appalachians—would be an extremely efficient way of protecting large numbers of endangered species.

The report by Dobson and colleagues ignited a controversy, however (Dunn et al. 1997). Numerous conservation managers objected to one of the study's messages: that large numbers of species can be saved at little cost in terms of the

Some biologists are shifting their focus from conserving individual species to conserving species assemblages and habitats.

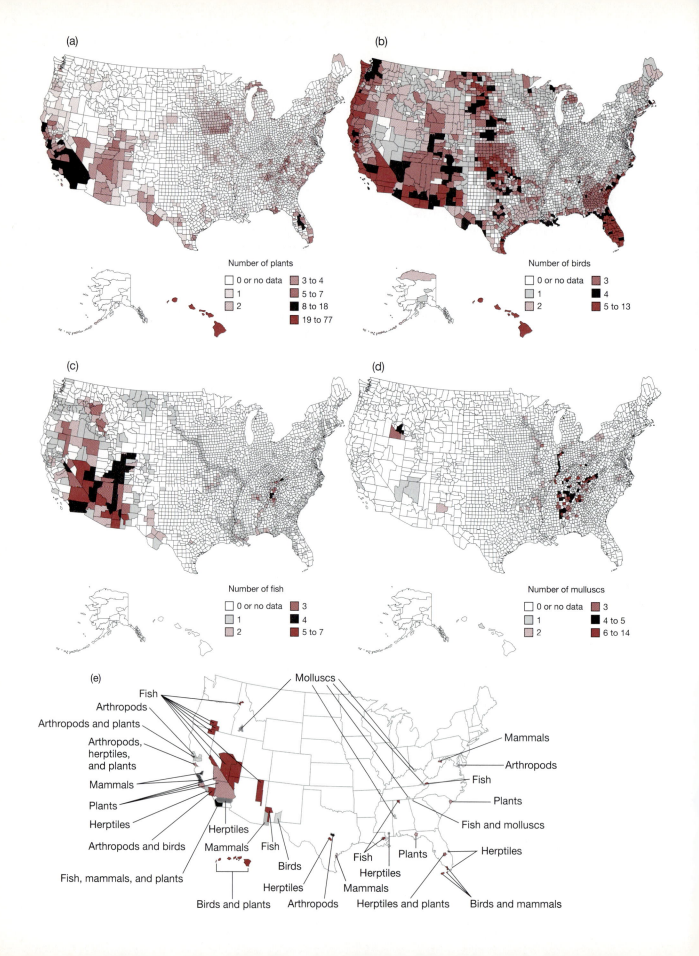

(a)

Number of plants
- 0 or no data
- 1
- 2
- 3 to 4
- 5 to 7
- 8 to 18
- 19 to 77

(b)

Number of birds
- 0 or no data
- 1
- 2
- 3
- 4
- 5 to 13

(c)

Number of fish
- 0 or no data
- 1
- 2
- 3
- 4
- 5 to 7

(d)

Number of mulluscs
- 0 or no data
- 1
- 2
- 3
- 4 to 5
- 6 to 14

(e)

Molluscs

Fish

Arthropods

Arthropods and plants

Arthropods,
herptiles,
and plants

Mammals

Plants

Herptiles

Arthropods and birds

Fish, mammals, and plants

Herptiles

Mammals

Fish

Birds

Herptiles

Fish

Herptiles

Mammals

Arthropods

Plants

Mammals

Arthropods

Fish

Plants

Fish and molluscs

Herptiles

Herptiles and plants

Birds and mammals

Birds and plants

total land area that needs to be protected from development. This debate turns in part on a more fundamental issue in conservation biology: whether we should concentrate our efforts solely on protecting individual species, or shift our focus to emphasize the preservation of biological communities.

Because of difficulties in implementing the United States' Endangered Species Act, the high cost of captive breeding programs, the continued loss of critical habitats, and an increase in species being listed as endangered, there is a growing trend among conservation managers toward landscape- or ecosystem-based conservation strategies. Instead of concentrating primarily on saving endangered species, the goal of these programs is to prevent endangered status by preserving large tracts of habitat under a mix of public and private ownership, and implement management schemes that attempt to include sustainable grazing, logging, fishing, or other types of development.

Perspective

Debates over the best way to establish conservation priorities continue. There is an important undercurrent to these scientific issues that is purely pragmatic, however: Conservation biology is by nature a science of the possible, and time is often of the essence. If a relictual population is important culturally and might inspire a successful public relations campaign in support of a new biological reserve, even ardent advocates of the evolutionary-front approach would probably support it. In some cases research programs designed only to catalog the diversity of species in a tropical reserve have not won the support of local governments, who prefer chemical prospecting campaigns because they hold the promise of economic development. Conservation biology is an arena where academic scientists, wildlife managers, microeconomists, international bankers, and cultural, political, and business leaders have to work together on the common goal of preserving biodiversity.

Finally, it is important to recognize that many of the conservation issues we have discussed are short-term, or stopgap, measures. If human population growth does not slow in the upcoming century and if the population does not eventually begin to decline, a mass extinction may be traumatic enough to make moot many of the issues discussed in this chapter. Fortunately, the evolutionary hallmarks of our species are adaptability in the face of environmental change, intelligence, and the ability to cooperate on common endeavors. The biodiversity crisis presents a formidable challenge to these traits. Over the past 500 years, we have been extremely clever about altering the environment to maximize our population growth. Over the next 100, we will need to be equally clever about preserving the biodiversity on which our future well-being depends.

Figure 19.17 **Endangered species hotspots in the United States**   (a–d) These maps show the numbers of endangered plants, birds, fish, and molluscs by county in the United States. Alaska and Hawaii are shown at the lower left of each map, but not to scale. (e) This map shows the set of counties that contain 50% of the endangered species from each taxonomic group in (a–d). Hawaii is not to scale. From Dobson et al. (1997).

Summary

Some 1.7 million species have been identified and named, but extrapolations made from limited datasets suggest that between 10 and 100 million species may actually be living today. Among both animals and plants a few lineages tend to be extremely species-rich, while others are relatively species-poor. Extrapolations made from small datasets suggest that current extinction rates are 100 to 1,000 times as intense as the background rates typical of most of Earth's history. Declining biodiversity is a concern for a variety of reasons: It may threaten the ability of the global ecosystems to provide such services as regulating the climate and the chemical composition of the atmosphere; it will destroy valuable but undiscovered biological molecules; it will be a moral, esthetic, and cultural loss.

The primary cause of extinction today is habitat destruction. Habitat destruction leads to a reduction in the size of populations. Small populations have genetic and demographic problems that can lead to extinction. Genetic problems in small populations include a loss of population-wide allelic diversity due to genetic drift, and the concomitant loss of the capacity to evolve. Genetic diversity within individuals also decreases, as drift leads to inbreeding and a steady loss of heterozygosity. It is unclear, however, how serious these genetic consequences are for natural populations. Several vertebrate species have extremely limited genetic diversity without obvious effects on their mean fitness. Demographic problems in small populations include increased vulnerability to environmental calamities, diseases, and random fluctuations in the sex ratio. At least in the short term, demographic factors like the ability to colonize new habitats and the availability of pollinators or mates may be of greater concern in managing small populations than are genetic issues.

The threat of mass extinction is so acute that saving all species is impractical. Phylogenetic studies can inform decisions about conservation priorities. Some conservation biologists argue that we should attempt to preserve evolutionary history by giving high priority to phylogenetically distinct taxa representing deep branches on the tree of life. Others argue that we should attempt to preserve the evolutionary future by giving high priority to recently derived taxa with high genetic diversity and the potential to undergo adaptive radiation. Other new approaches to setting conservation priorities include identifying biodiversity hotspots, and preserving unique habitats or ecosystems irrespective of the status of individual species.

Questions

1. Should much effort be expended trying to answer the question "How many species are there?" If there are 5 million instead of 100 million species, how would it affect our understanding of evolutionary processes and conservation needs?

2. Species- and landscape-based approaches to conservation are not mutually exclusive, but differ in emphasis. To grasp this point, write outline versions of laws titled "The Endangered Species Act" and "The Endangered Ecosystems Act." How would endangered status be defined under each law? How would recovery programs be implemented?

3. Suppose you were giving a talk on rainforest conservation to an assembly of frontier farmers in the Brazilian Amazon. Which, if any, of the three reasons for preserving

biodiversity listed in this chapter would be most persuasive to this community? Comment on the propriety of an American urging Brazilians to make short-term economic sacrifices in the interest of preserving biodiversity.

4. Conservation managers often try to purchase corridors of undeveloped habitat so that larger preserves are linked into networks. Why?

5. Loss of heterozygosity may be especially detrimental at MHC loci, because allelic variability at these loci increases disease resistance. Surveys of microsatellite loci show that the gray wolves on Isle Royale are highly inbred (Wayne et al. 1991). Recall that the wolf population crashed during an outbreak of canine parvovirus. How might these disparate facts be linked?

6. If you were a manager charged with conserving the collared lizards of the Ozarks, one of your tasks might be to reintroduce the lizards into glades in which they have gone extinct. When reintroducing lizards to a glade, you will have a choice between using only individuals from a single extant glade population, or from several extant glade populations. What would be the evolutionary and demographic consequences of each choice, for both the donor and recipient populations? Which strategy will you follow, and why?

7. Some evolutionary biologists reject the idea that priorities can be established in species conservation. They maintain that every taxon is a unique and irreplaceable resource. Comment on this point of view (which happens to be consistent with the spirit of the United States' Endangered Species Act).

8. Political leaders in developing countries are concerned about intellectual property issues raised by chemical prospecting. They maintain that developing nations should be compensated for their role as custodians of much of the world's biodiversity. Representatives from industrialized nations counter that their businesses need to recoup the huge capital investments required to find and develop products extracted from tropical species. Suppose a researcher, funded by an agricultural company based in Canada, is on a chemical prospecting trip in Indonesia. She discovers an insecticide produced by a vine that lives in tropical rainforests there. The molecule can be patented and could produce tens of millions of dollars in profits annually. Who should hold the patent? Who should reap the profits?

9. Suppose you are using phylogenetic information to establish conservation priorities. How will each of the motivations for preserving biodiversity inform your decisions? For example, if you are attempting to protect ecosystem services, will you favor the protection of distinctive and deep-branching lineages, or recently derived lineages with large, genetically diverse populations? What if you are attempting to protect untapped reserves of economically useful biological molecules? What if you are motivated by moral, esthetic, and cultural values?

Exploring the Literature

10. Determining the minimum population size necessary to make the extinction of a species unlikely over the long term is an active area of research in conservation genetics. The following papers provide recent developments on this question.

Lande, R. 1995. Mutation and conservation. *Conservation Biology* 9:782–791.

Lynch, M. 1996. A quantitative genetic perspective on conservation issues. In J.C. Avise and J. Hamrick, eds. *Conservation Genetics: Case Histories from Nature* (New York: Chapman and Hall), 471–501.

11. Why are the tropics so species-rich? A number of hypotheses have been advanced over the years, but a robust result has been elusive. Check the following papers for recent work on this question:

Fjeldså, J. 1994. Geographical patterns for relict and young species of birds in Africa and South America and implications for conservation priorities. *Biodiversity and Conservation* 3:207–226.

Gaston, K.J. 1996. Biodiversity—latitudinal gradients. *Progress in Physical Geography* 20:466–476.

Gaston, K.J., and T.M. Blackburn. 1996. The tropics as a museum of biological diversity: an analysis of the New World avifauna. *Proceedings of the Royal Society of London,* Series B 263:63–68.

Blackburn, T.M., and K.J. Gaston. 1997. The relationship between geographic area and the latitudinal gradient in species richness in New World birds. *Evolutionary Ecology* 11:195–204.

12. In a novel and dramatic application of molecular systematics to conservation, C.S. Baker and S.R. Palumbi went shopping in Japan and bought sixteen samples of meat sold under the generic description "whale." In their hotel room, the researchers extracted mitochondrial DNA from the samples and used PCR to amplify a short sequence. Back at home, they sequenced the DNA samples, and placed them on a phylogeny of the whales. This allowed the researchers to identify fourteen of the samples to species, and to determine whether the whale meat had been sold in accord with international whaling treaties.

Baker, C.S., and S.R. Palumbi. 1994. Which whales are hunted? A molecular genetic approach to monitoring whaling. *Science* 265:1538–1539.

13. Cheetahs have long been used as a classic example of a species whose low genetic diversity put it at increased risk of extinction. Recently researchers have debated the validity of this view. For a start on the literature, see:

Menotti-Raymond, M., and S.J. O'Brien. 1993. Dating the genetic bottleneck of the African cheetah. *Proceedings of the National Academy of Sciences, USA* 90:3172–3176.

Merola, M. 1994. A reassessment of homozygosity and the case for inbreeding depression in the cheetah, *Acinonyx jubatus:* Implications for conservation. *Conservation Biology* 8:961–971.

14. For other recent attempts to determine whether low genetic diversity threatens the survival of populations, see:

Ledberg, P.L. 1993. Strategies for population reintroduction: Effects of genetic variability on population growth and size. *Conservation Biology* 7:194–199.

Jimenez, J.A., K.A. Hughes, G. Alaks, L. Graham, and R.C. Lacy. 1994. An experimental study of inbreeding depression in a natural habitat. *Science* 266:271–273.

Sanjayan, M.A., K. Crooks, G. Zegers, and D. Foran. 1996. Genetic variation and the immune response in natural populations of pocket gophers. *Conservation Biology* 10:1519–1527.

Citations

Atkinson, I. 1989. Introduced animals and extinctions. In D. Western and M. Pearl, eds. *Conservation for the Twenty-first Century* (New York: Oxford University Press), 54–69.

Avise, J.C. 1996. Toward a regional conservation genetics perspective: Phylogeography of faunas in the southeastern United States. In J.C. Avise and J.L. Hamrick, eds. *Conservation Genetics: Case Histories from Nature* (New York: Chapman and Hall), 431–470.

Ayres, W.S. 1985. Easter Island Subsistence. *Journal de la Societe des Oceanistes* 80:103–124.

Bahn, P., and J. Flenley. 1992. *Easter Island; Earth Island* (New York: Thames and Hudson).

Balmford, A. 1996. Extinction filters and current resilience: The significance of past selection pressures for conservation biology. *Trends in Ecology and Evolution* 11:193–196.

Barrett, S.C.H., and D. Charlesworth. 1991. Effects of a change in the level of inbreeding on the genetic load. *Nature* 352:522–524.

Blackmore, S. 1996. Knowing the earth's biodiversity: Challenges for the infrastructure of systematic biology. *Science* 274:63–64.

Blum, E. 1993. Making biodiversity conservation profitable: A case study of the Merck/INBio agreement. *Environment* 35:16–28.

Brooks, D.R., R.L. Mayden, and D.A. McLennan. 1992. Phylogeny and biodiversity: Conserving our evolutionary legacy. *Trends in Ecology and Evolution* 7:55–59.

Brusca R.C., and G.J. Brusca. 1990. *Invertebrates* (Sunderland, MA: Sinauer).

Buri, P. 1956. Gene frequency in small populations of mutant *Drosophila*. *Evolution* 10:367–402.

Costanza, R., R. d'Arge, R. de Groot, S. Farber, and M. Grasso, et al. 1997. The value of the world's ecosystem services and natural capital. *Nature* 387:253–260.

Crozier, R.H. 1992. Genetic diversity and the agony of choice. *Biological Conservation* 61:11–15.

Daugherty, C.H., A. Cree, J.M. Hay, and M.B. Thompson. 1990. Neglected taxonomy and continuing extinctions of tuatara *(Sphenodon)*. *Nature* 347:177–179.

Diamond, J. 1992. *The Third Chimpanzee* (New York: HarperCollins).

Diamond, J. 1995. Easter's End. *Discover* (August):62–69.

Dicke, W. 1996. Numerous U.S. plant and freshwater species found in peril. *The New York Times,* January 2: B8.

Dobson, A.P. 1996. *Conservation and Biodiversity* (Scientific American Library, distributed by W. H. Freeman and Company, New York).

Dobson, A.P., J.P. Rodriguez, W.M. Roberts, and D.S. Wilcove. 1997. Geographic distribution of endangered species in the United States. *Science* 275:550–553.

Dransfield, J., J.R. Flenley, S.M. King, D.D. Harkness, and S. Rapu. 1984. A recently extinct palm from Easter Island. *Nature* 312:750–752.

Dunn, C.P., M.L. Bowles, G.B. Rabb, K.S. Jarantoski, D. Ehrenfeld, R.F. Noss, G.K. Meffe, A. Dodson, J.P. Rodriguez, W.M. Roberts, and D.S. Wilcove. 1997. Endangered species "hot spots." *Science* 275:513–517.

Dye, T., and D.W. Steadman. 1990. Polynesian ancestors and their animal world. *American Scientist* 78:207–215.

Ehrlich, P.R., and E.O. Wilson. 1991. Biodiversity studies: Science and policy. *Science* 253:758–762.

Eisner, T. 1989. Prospecting for nature's chemical riches. *Issues in Science and Technology* 6:31–34.

Eisner, T., and E.A. Beiring. 1994. Biotic exploration fund—protecting biodiversity through chemical prospecting. *BioScience* 44:95–98.

Erwin, T.L. 1982. Tropical forests: Their richness in coleoptera and other arthropod species. *Coleopterist's Bulletin* 36:74–75.

Erwin, T.L. 1991. An evolutionary basis for conservation strategies. *Science* 253:750–752.

Erwin, T.L., and J.C. Scott. 1980. Seasonal and size patterns, trophic structure, and richness of coleoptera in the tropical arboreal ecosystem: The fauna of the tree *Luehea seemannii* Triana and Planch in the Canal Zone of Panama. *Coleopterist's Bulletin* 34:305–322.

Faith, D.P. 1992. Conservation evaluation and phylogenetic diversity. *Biological Conservation* 61:1–10.

Farnsworth, N.R., and R.W. Morris. 1976. Higher plants—the sleeping giant of drug development. *American Journal of Pharmaceutical Education* 148:46–52.

Fjeldsa, J. 1994. Geographical patterns for relict and young species of birds in Africa and South America and implications for conservation priorities. *Biodiversity and Conservation* 3:207–226.

Flenley, J.R. 1979. Stratigraphic Evidence of Environmental Change on Easter Island. *Asian Perspectives* 22:33–40.

Flenley, J.R., and S.M. King. 1984. Late Quarternary Pollen Records from Easter Island. *Nature* 307:47–50.

Freed, L.A., T.B. Smith, J.H. Carothers, and J.K. Lepson. 1996. Shrinkage is not the most likely cause of bill change in i'iwi: A rejoinder to Winker. *Conservation Biology* 10:659–660.

Gershon, D. 1992. If biological diversity has a price, who sets it and who should benefit (Merck & Co's agreement with Costa Rican scientists)? *Nature* 359:565.

Grassle, J.F., and N.J. Maciolek. 1992. Deep sea species richness: Regional and local diversity estimates from quantitative bottom samples. *American Naturalist* 139:313–341.

Hagelberg, E., S. Quevedo, D. Turbon, and J.B. Clegg. 1994. DNA from ancient Easter Islanders. *Nature* 369:25–26.

Hammond, P.M. 1992. Species inventory. In B. Groombridge, ed. *Global Biodiversity* (London: Chapman and Hall), 17–39.

Hartl, D.L. 1981. *A Primer of Population Genetics* (Sunderland, MA: Sinauer).

Hedrick, P.W., R.C. Lacy, F.W. Allendorf, and M.E. Soulé. 1996. Directions in conservation biology: Comments on Caughley. *Conservation Biology* 10:1312–1320.

Heywood, V.H., and R.T. Watson, eds. 1995. *Global Biodiversity Assessment* (Cambridge: Cambridge University Press).

Hibbett, D.S., and M.J. Donoghue. 1996. Implications of phylogenetic studies for conservation of genetic diversity in shiitake mushrooms. *Conservation Biology* 10:1321–1327.

Holsinger, K.E. 1992. Setting priorities for regional plant conservation programs. *Rhodora* 94:243–257.

Iltis, H.H. 1988. Serendipity in the exploration of biodiversity. In E.O. Wilson, ed. *Biodiversity* (Washington DC: National Academy Press), 98–105.

Janzen, D.H. 1996. Wildland biodiversity management in the tropics. In M.L. Reaka-Kudla, D.E. Wilson, and E.O. Wilson, eds. *Biodiversity II* (Washington DC: National Academy Press), 411–431.

Janzen, D.H., and W. Hallwachs. 1994. All Taxa Biodiversity Inventory (ATBI) of terrestrial systems: A generic protocol for preparing wildland biodiversity of non-damaging use. Report of National Science Foundation Workshop (Washington DC: National Science Foundation).

Lande, R. 1988. Genetics and demography in biological conservation. *Science* 241:1455–1460.

Lynch, M., and W. Gabriel. 1990. Mutation load and the survival of small populations. *Evolution* 44:1725–1737.

May, R.M. 1988. How many species are there on earth? *Science* 241:1441–1449.

May, R.M. 1990. Taxonomy as destiny. *Nature* 347:129–130.

May, R.M. 1992. Bottoms up for the oceans. *Nature* 357:278–279.

May, R.M. 1995. The cheetah controversy. *Nature* 374:309–310.

May, R.M., J.H. Lawton, and N.E. Stork. 1995. Assessing extinction rates. In J.H. Lawton and R.M. May, eds. *Extinction Rates* (Oxford: Oxford University Press), 1–24.

McMichael, A.J. 1993. *Planetary Overload: Global Environmental Change and the Health of the Human Species* (Cambridge: Cambridge University Press).

Meyer, C.A. 1996. NGOs and environmental public goods: Institutional alternatives to property rights. *Development and Change* 27:453–474.

Mooney, H.A., J. Lubchenco, R. Dirzo, and O. Sala, eds. 1995. Biodiversity and ecosystem functioning. Sections 5 and 6 in V.H. Heywood and R.T. Watson, eds, *Global Biodiversity Assessment* (Cambridge: Cambridge University Press).

Moritz, C. 1994. Defining "evolutionarily significant units" for conservation. *Trends in Ecology and Evolution* 9:373–375.

Moritz, C. 1996. Uses of molecular phylogenies for conservation. In P.H. Harvey, A.J.L. Brown, J. Maynard Smith, and S. Nee, eds. *New Uses for New Phylogenies* (Oxford: Oxford University Press), 203–214.

Perrings, C., C. Folke, K-G. Maler, C.S. Holling, and B-O. Jansson, eds. 1994. *Biodiversity Loss: Ecological and Economic Issues* (Cambridge: Cambridge University Press).

Phillips, K. 1994. *Tracking the Vanishing Frogs: An Ecological Mystery* (New York: Penguin Books).

Pimm, S.L., G.J. Russell, J.L. Gittleman, and T.M. Brooks. 1995. The future of biodiversity. *Science* 269:347–350.

Plotkin, M.J. 1988. The outlook for new agricultural and industrial products from the tropics. In E.O. Wilson, ed. *Biodiversity* (Washington DC: National Academy Press), 106–116.

Polans, N.O., and R.W. Allard. 1989. An experimental evaluation of the recovery potential of ryegrass populations from genetic stress resulting from restriction of population size. *Evolution* 43:1320–1324.

Poore, G.C.B., G.D.F. Wilson, and R.M. May. 1993. Marine species richness. *Nature* 361:597–598.

Pough, F.H., J.B. Heiser, and W.N. McFarland. 1996. *Vertebrate Life* (Upper Saddle River, NJ: Prentice Hall).

Pounds, J.L., and M.L. Crump. 1994. Amphibian declines and climate disturbance: The case of the golden toad and the harlequin frog. *Conservation Biology* 8:72–85.

Prendergast, J.R., R.M. Quinn, J.H. Lawton, B.C. Eversham, and D.W. Gibbons. 1993. Rare species, the coincidence of diversity hotspots and conservation strategies. *Nature* 365:335–337.

Rick, C.M. 1974. High soluble-solids content in large-fruited tomato lines derived from a wild gree-fruited species. *Hilgardia* 42:492–510.

Schemske, D.W., B.C. Husband, M.H. Ruckelshaus, C. Goodwillie, I.M. Parker, and J.G. Bishop. 1994. Evaluating approaches to the conservation of rare and endangered plants. *Ecology* 75:584–606.

Selander, R.K. 1985. Protein polymorphism and the genetic structure of natural populations of bacteria. In T. Ohta and K. Aoki, eds. *Population Genetics and Molecular Evolution* (Berlin: Springer-Verlag), 85–106.

Skole, D., and C. Tucker. 1993. Tropical deforestation and habitat fragmentation in the Amazon: Satellite data from 1978 to 1988. *Science* 260:1905–1920.

Smith, F.D.M., R.M. May, R. Pellew, T.H. Johnson, and K.S. Walter. 1993a. Estimating extinction rates. *Nature* 364:494–496.

Smith, F.D.M., R.M. May, R. Pellew, T.H. Johnson, and K.R. Walter. 1993b. How much do we know about the current extinction rate? *Trends in Ecology and Evolution* 8:375–378.

Smith, T.B., L.A. Freed, J.K. Lepson, and J.H. Carothers. 1995. Evolutionary consequences of extinctions in populations of a Hawaiian honeycreeper. *Conservation Biology* 9:107–113.

Steadman, D.W. 1995. Prehistoric extinctions of Pacific island birds: Biodiversity meets zooarchaeology. *Science* 267:1123–1131.

Steadman, D.W., P.V. Casanova, and C.C. Ferrando. 1994. Stratigraphy, chronology, and cultural context of an early faunal assemblage from Easter Island. *Asian Perspectives* 33:80–96.

Stein, B.A., and R.M. Chipley, eds. 1996. *Priorities for Conservation: 1996 Annual Report Card for U.S. Plant and Animal Species* (Arlington, VA: The Nature Conservancy).

Stork, N.E. 1996. Measuring global biodiversity and its decline. In M.L. Reaka-Kudla, D.E. Wilson, and E.O. Wilson, eds. *Biodiversity II* (Washington DC: National Academy Press), 41–68.

Templeton, A.R., K. Shaw, E. Routman, and S.K. Davis. 1990. The genetic consequences of habitat fragmentation. *Annals of the Missouri Botanical Garden* 77:13–27.

Tilman, D., J. Knops, D. Wedin, P. Reich, M. Ritchie, and E. Siemann. 1997. The influence of functional diversity and composition on ecosystem processes. *Science* 277: 1300–1302.

Vane-Wright, R.I., C.J. Humphries, and P.H. Williams. 1991. What to protect?—Systematics and the agony of choice. *Biological Conservation* 55:235–254.

Wahlberg, N., A. Moilanen, and I. Hanski. 1996. Predicting the occurrence of endangered species in fragmented landscapes. *Science* 273:1536–1538.

Wayne, R.K., N. Lehman, D. Girman, P.J.P. Gogan, D.A. Gilbert, K. Hansen, R.O. Peterson, U.S. Seal, A. Eisenhawer, L.D. Mech, and R.J. Krumenaker. 1991. Conservation genetics of the endangered Isle Royale gray wolf. *Conservation Biology* 5:41–51.

Wilson, E.O. 1988. The current state of biodiversity. In E.O. Wilson, ed. *Biodiversity* (Washington DC: National Academy Press), 3–18.

Winker, K. 1996. Specimen shrinkage versus evolution: I'iwi morphology. *Conservation Biology* 10:657–658.

World Resources Institute. 1994. *World Resources 1994–95* (New York: Oxford University Press), 17.

Wright, S. 1931. Evolution in Mendelian populations. *Genetics* 16:97–159.

Yahnke, C.J., W.E. Johnson, E. Geffen, D. Smith, F. Hertel, et al. 1996. Darwin's fox: A distinct endangered species in a vanishing habitat. *Conservation Biology* 10:366–375.

Young, A., T. Boyle, and T. Brown. 1996. The population genetic consequences of habitat fragmentation for plants. *Trends in Ecology and Evolution* 11:413–418.

Glossary

acclimation A form of phenotypic plasticity in which an individual's physiology changes in response to the conditions under which it is living; not to be confused with *evolution* or *adaptation*.

adaptation A trait that increases the ability of an individual to survive or reproduce compared to individuals without the trait.

adaptationist program The research effort directed at identifying traits that are adaptive and understanding how those traits increase fitness.

adaptive radiation The divergence of a clade into populations adapted to many different ecological niches.

additive effect The contribution an allele makes to the phenotype that is independent of the identity of the other alleles at the same or different loci.

additive genetic variation Differences among individuals in a population that are due to the additive effects of genes.

agent of selection Any factor that causes individuals with certain phenotypes to have, on average, higher fitness than individuals with other phenotypes.

allopatry Living in different geographic areas.

allozymes Distinct forms of an enzyme, encoded by different alleles at the same locus.

altruism Behavior which decreases the fitness of the actor and increases the fitness of the recipient.

ancestral Describes a trait that was possessed by the common ancestor of the species on a branch of an evolutionary tree; used in contrast with *derived*.

assortative mating Occurs when individuals tend to mate with other individuals with the same genotype or phenotype.

back mutation A mutation that reverses the effect of a previous mutation; typically a mutation that restores function after a loss-of-function mutation.

background extinction Extinctions that occur during "normal" times, as opposed to during mass extinction events.

best-fit line The line that most accurately represents the trend of the data in a scatterplot; typically, best-fit lines are calculated by least-squares linear regression.

blending inheritance The hypothesis that heritable factors blend to produce a phenotype and are passed on to offspring in this blended form.

bootstrap value In phylogeny reconstruction, an estimate of the strength of the evidence that a particular node in a tree exists; ranges from 0% to 100%, with higher values indicating stronger support.

bottleneck A large-scale but short-term reduction in population size followed by an increase in population size.

cenancestor The most recent common ancestor of all living organisms (excluding viruses).

character displacement A pattern observed in some closely related populations, in which certain traits are more divergent where the ranges of the populations overlap.

chromosome inversion A region of DNA that has been flipped, so that the genes are in reverse order; results in lower rates of crossing-over and thus greater linkage among loci within the inversion.

clade The set of species descended from a particular common ancestor; synonymous with *monophyletic group*.

cladistics A classification scheme based on the historical sequence of divergence events (phylogeny); also used to identify a method of inferring phylogenies based on the presence of shared derived characters (synapomorphies).

cladogram An evolutionary tree reflecting the results of a cladistic analysis.

cline A systematic change along a geographic transect in the frequency of a genotype or phenotype.

clone An individual that is genetically identical to its parent, or a group of individuals that are genetically identical to each other.

codon bias A nonrandom distribution of codons in a DNA sequence.

coefficient of inbreeding (F) The probability that the alleles at any particular locus in the same individual are identical by descent.

coefficient of linkage disequilibrium (D) A calculated value that quantifies the degree to which genotypes at one locus are nonrandomly associated with genotypes at another locus.

coefficient of relatedness (r) The probability that the alleles at any particular locus in two different individuals are identical by descent.

coevolution A pattern of reciprocal adaptation, caused by two species evolving in close association.

common garden experiment An experiment in which individuals from different populations or treatments are reared together under identical conditions.

comparative method A research program that compares traits and environments across taxa and looks for correlations that test hypotheses about adaptation.

concerted evolution A pattern of little or no variation observed among loci that code for functionally identical proteins or RNAs.

confidence interval An indication of the statistical certainty of an estimate; if a study yielding an estimate is done repeatedly, and a 95% confidence interval is calculated for each estimate, the confidence interval will include the true value 95% of the time.

constraint Any factor that tends to slow the rate of adaptive evolution or prevent a population from evolving the optimal value of a trait.

control group A reference group that provides a basis for comparison; in an experiment, the control group is exposed to all conditions affecting the experimental group except one—the potential causative agent of interest.

convergent evolution Similarity between species that is caused by a similar, but evolutionarily independent, response to a common environmental problem.

cryptic species Species which are indistinguishable morphologically, but divergent in songs, calls, odor, or other traits.

degree of genetic determination Synonym for broad-sense heritability (see *heritability*).

derived Describes a trait that was not possessed by the common ancestor of the species on a branch of an evolutionary tree; an evolutionary novelty; used in contrast with *ancestral*.

differential success A difference between the average survival, fecundity, or number of matings achieved by individuals with certain phenotypes versus individuals with other phenotypes.

dioecious Describes a species in which male and female reproductive function occurs in separate individuals; usually used with plants.

directional selection Occurs when individual fitness tends to increase or decrease with the value of phenotypic trait; can result in steady evolutionary change in the mean value of the trait in the population.

disruptive selection Occurs when individuals with more extreme values of a trait have higher fitness; can result in increased phenotypic variation in a population.

dominance genetic variation Differences among individuals in a population that are due to the nonadditive effects of genes, such as dominance; typically means the genetic variation left over after the additive genetic variation has been taken into account.

drift Synonym for *genetic drift*.

effective population size (N_e) The size of a random mating population (with no selection, mutation, or migration) that would lose genetic variation via drift at the same rate as is observed in an actual population.

endosymbiosis theory The hypothesis that eukaryotic organelles such as mitochondria and chloroplasts originated when bacteria took up residence inside early eukaryotic cells.

environmental variation Differences among individuals in a population that are due to differences in the environments they have experienced.

eugenics The study and practice of social control over the evolution of human populations; positive eugenics seeks to increase the frequency of desirable traits, whereas negative eugenics seeks to decrease the frequency of undesirable traits.

eusocial A social system characterized by overlapping generations, cooperative brood care, and specialized reproductive and non-reproductive castes.

evolutionarily stable strategy (ESS) In game theory, a strategy or set of strategies that cannot be invaded by a new, alternative strategy.

evolutionary arms race Occurs when an adaptation in one species (a parasite, for example) reduces the fitness of individuals in a second species (such as a host), thereby selecting in favor of counter-adaptations in the second species. These counter-adaptations, in turn, select in favor of new adaptations in the first species, and so on.

evolutionary tree A diagram (typically an estimate) of the relationships of ancestry and descent among a group of species; in paleontological studies the ancestors may be known from fossils, whereas in studies of extant species the ancestors may be hypothetical constructs.

exon A nucleotide sequence that occurs between introns, and that remains in the messenger RNA after the introns have been spliced out.

extant Living today.

fecundity The number of gametes produced by an individual; usually used in reference to the number of eggs produced by a female.

fitness The extent to which an individual contributes genes to future generations, or an individual's score on a measure of performance expected to correlate with genetic contribution to future generations (such as lifetime reproductive success).

fixation The elimination from a population of all the alleles at a locus but one; the one remaining allele, now at a frequency of 1.0, is said to have achieved fixation, or to be fixed.

forward mutation A mutation that incapacitates a gene, so that no functional product is produced; also called a loss-of-function, knock-out, or null mutation.

founder event The establishment of a new population, usually by a small number of individuals.

frequency The proportional representation of a phenotype, genotype, gamete, or allele in a population; if six out of ten individuals have brown eyes, the frequency of brown eyes is 60%, or 0.6.

frequency-dependent selection Occurs when the fitness of a phenotype depends on the frequency of the phenotype in the population; typically occurs when a phenotype has higher fitness when rare and lower fitness when common.

gamete pool The set of all copies of all gamete genotypes in a population that could potentially be contributed by the members of one generation to the members of the next generation.

gene family A group of loci related by common descent, and sharing identical or similar function.

gene flow The movement of alleles from one population to another population, typically via the movement of individuals or via the transport of gametes by wind, water, or pollinators.

gene pool The set of all copies of all alleles in a population that could potentially be contributed by the members of one generation to the members of the next generation.

gene tree Describes the relationships of ancestry and descent among the alleles of a gene in one or more populations or species; the ancestors on the tree may be either known or hypothetical.

genetic drift Change in the frequencies of alleles in a population resulting from chance variation in the survival and/or reproductive success of individuals; results in nonadaptive evolution.

genetic linkage The tendency for alleles at different loci to be inherited together.

genetic load Reduction in the mean fitness of a population due to the presence of deleterious alleles.

genetic recombination The placement of allele copies into multilocus genotypes (on chromosomes or within gametes) that are different from the multilocus genotypes they belonged to in the previous generation; results from meiosis with crossing-over and sexual reproduction with outcrossing.

genetic variation Differences among individuals in a population that are due to differences in genotype.

genotype-by-environment interaction Differences in the effect of the environment on the phenotype displayed by different genotypes; for example, among people living in the same location some change their skin color with the seasons and others do not.

germ-line genetic engineering An alteration, through direct manipulation of the genome, of the alleles that will be passed from parent to offspring; results in an instantaneous evolutionary change in the (laboratory) population.

h² Symbol for the narrow-sense heritability (see *heritability*).

half-life The time required for half of the atoms of a radioactive material, present at any time, to decay into a daughter isotope.

haplodiploidy A reproductive system in which males are haploid and develop from unfertilized eggs, while females are diploid and develop from fertilized eggs.

haplotype Genotype for a suite of linked loci on a chromosome; typically used for mitochondrial genotypes, because mitochondria are haploid and all loci are linked.

Hardy-Weinberg equilibrium A situation in which allele and genotype frequencies in an ideal population do not change from one generation to the next, because the population experiences no selection, no mutation, no migration, no genetic drift, and random mating.

heritability In the broad sense, that fraction of the total phenotypic variation in a population that is caused by genetic differences among individuals; in the narrow sense, that fraction of the total variation that is due to the additive effects of genes.

hermaphroditic In general, describes a species in which male and female reproductive function occur in the same individual; with plants, describes a species with perfect flowers.

heterozygosity That fraction of the individuals in a population that are heterozygotes.

heterozygote superiority (heterosis) Describes a situation in which heterozygotes at a particular locus tend to have higher fitness than homozygotes.

histogram A bar chart that represents the variation among individuals in a sample; each bar represents the number of individuals, or the frequency of individuals, with a particular value (or within a particular range of values) for the measurement in question.

hitchhiking Change in the frequency of an allele due to selection on a closely linked locus.

homology Similarity between species that results from inheritance of traits from a common ancestor.

homoplasy Similarity in the characters found in different species that is due to convergent evolution, parallelism, or reversal—not common descent.

horizontal transfer Transfer of genetic material across species barriers.

hybrid zone A geographic region where differentiated populations interbreed.

identical by descent Describes alleles, within a single individual or different individuals, that have been inherited from the same ancestral copy of the allele.

inbreeding Mating among kin.

inbreeding depression Reduced fitness in individuals or populations resulting from kin matings; often due to the decrease in heterozygosity associated with kin matings, either because heterozygotes are superior or because homozygotes for deleterious alleles become more common.

inclusive fitness An individual's total fitness; the sum of its indirect fitness, due to reproduction by relatives made possible by its actions, and direct fitness, due to its own reproduction.

independence (statistical) Lack of association among data points, such that the value of a data point does not affect the value of any other data point.

inheritance of acquired characters The hypothesis that phenotypic changes in the parental generation can be passed on, intact, to the next generation.

interaction In genetics, occurs when the effect of an allele on the phenotype depends on the other alleles present at the same or different loci; in statistics, occurs when the effect of a treatment depends on the value of other treatments.

intersexual selection Differential mating success among individuals of one sex due to interactions with members of the other sex; for example, variation in mating success among males due to female choosiness.

intrasexual selection Differential mating success among individuals of one sex due to interactions with members of the same sex; for example, differences in mating success among males due to male-male competition over access to females.

intron (intervening sequence) A noncoding stretch of DNA nucleotides that occurs between the coding regions of a gene and must be spliced out after transcription to produce a functional messenger RNA.

iteroparous Describes a species or population in which individuals experience more than one bout of reproduction over the course of typical lifetime; humans provide an example.

kin recognition The ability to discern the degree of genetic relatedness of other individuals.

kin selection Natural selection based on indirect fitness gains.

life history An individual's pattern of allocation, throughout life, of time and energy to various fundamental activities, such as growth, repair of cell and tissue damage, and reproduction.

linkage (dis)equilibrium If, within a population, genotypes at one locus are randomly distributed with respect to genotypes at another locus, then the population is in linkage equilibrium for the two loci; otherwise, the population is in linkage disequilibrium.

macroevolution   Large evolutionary change, usually in morphology; typically refers to the evolution of differences among populations that would warrant their placement in different genera or higher-level taxa.

maladaptive   Describes a trait that reduces the fitness of individuals possessing it, relative to the fitness of individuals without it; the opposite of adaptive.

mass extinction   A large-scale, sudden extinction event that is geographically and taxonomically widespread.

maternal effect   Variation among individuals due to variation in non-genetic influences exerted by their mothers; for example, chicks whose mothers feed them more may grow to larger sizes, and thus be able to feed their own chicks more, even when size is not heritable.

Mendelian gene   A locus whose alleles obey Mendel's laws of segregation and independent assortment.

microevolution   Changes in gene frequencies and trait distributions that occur within populations and species.

microsatellite sequences   Di-, tri-, tetra- or pentanucleotide tandem repeats, found in the centromere and scattered throughout some genomes.

microtektites   Tiny glass particles created when minerals are melted by the heat generated in a meteorite or asteroid impact.

midoffspring value   The mean phenotype of the offspring within a family.

midparent value   The mean phenotype of an individual's two parents.

migration   In evolution, the movement of alleles from one population to another, typically via the movement of individuals or via the transport of gametes by wind, water, or pollinators.

minisatellite sequences   30- to 60-base-pair tandem repeats, widely distributed in eukaryotic genomes.

Modern Synthesis   The broad-based effort, accomplished during the 1930's and 1940's, to unite Mendelian genetics with the theory of evolution by natural selection; also called the Evolutionary Synthesis.

molecular clock   The hypothesis that base substitutions accumulate in populations in a clock-like fashion; that is, as a linear function of time.

monoecious   Typically used for plants, to describe either: (1) a species in which male and female reproductive function occur in the same individual; or (2) a species in which separate male and female flowers are present on the same individual (see also *hermaphroditic*).

monophyletic group   The set of species (or populations) descended from a common ancestor.

morph   Form; may refer to a structural, behavioral, or physiological phenotype, or to an allele or DNA sequence.

morphology   Structural form, or physical phenotype; also the study of structural form.

morphospecies   Populations that are designated as separate species based on morphological differences.

mutagen   Any agent that induces mutations; typically radiation or a chemical.

mutation-selection balance   Describes an equilibrium in the frequency of an allele that occurs because new copies of the allele are created by mutation at exactly the same rate that old copies of the allele are eliminated by natural selection.

mutualism   An interaction between two individuals, typically of different species, in which both individuals benefit.

natural selection   A difference, on average, between the survival or fecundity of individuals with certain phenoypes compared to individuals with other phenotypes.

negative selection   Selection against deleterious mutations.

neutral evolution   A theory that models the rate of fixation of alleles with no effect on fitness; also associated with the claim that the vast majority of observed base substitutions are neutral with respect to fitness.

node   A point on an evolutionary tree at which a branch splits into two or more sub-branches.

null hypothesis   The predicted outcome, under the simplest possible assumptions, of an experiment or observation; in a test of whether populations are different, the null hypothesis is typically that they are not different, and that apparent differences are due to chance.

null model   The set of simple and explicit assumptions that allows a researcher to state a null hypothesis.

outbreeding   Mating among unrelated individuals.

outgroup   A taxonomic group that diverged prior to the rest of the taxa in a phylogenetic analysis.

P value   An estimate of the statistical support for a claim about a pattern in data, with smaller values indicating stronger support; an estimate of the probability that apparent violations of the null hypothesis are due to chance (see *statistically significant*).

paired-choice test   A measurement of preference in which an experimental subject is given a choice between two alternatives.

paleontology   The study of fossil organisms.

parallelism   Similarity in closely related species that is not due to traits inherited from a common ancestor.

paraphyletic group   A set of species that includes a common ancestor and some, but not all, of its descendants.

parental investment   Expenditure of time and energy on the provision, protection, and care of an offspring; more specifically, investment by a parent that increases the fitness of a particular offspring, and reduces the fitness the parent can gain by investing in other offspring.

parsimony   A criterion for selecting among alternative patterns or explanations based on minimizing the total amount of change or complexity.

parthenogenesis   A mating system in which females produce offspring without fertilization.

perfect flower   A flower with both female and male reproductive function.

phenetics   A classification scheme based on grouping populations according to their similarities.

phenotypic plasticity   Variation, under environmental influence, in the phenotype associated with a genotype.

phylogeny   The evolutionary history of a group, usually represented graphically by a phylogenetic tree.

polyandry   A mating system in which at least some females mate with more than one male.

polygyny   A mating system in which at least some males mate with more than one female.

polymorphism   The existence within a population of more than one variant for a phenotypic trait, or of more than one allele.

polyphyletic group   A set of species that are grouped by similarity, but not descended from a common ancestor.

polyploid   Having more than two haploid sets of chromosomes.

polytomy   A node, or branch point, on a phylogeny with more than two descendent lineages emerging.

population   For sexual species, a group of interbreeding individuals and their offspring; for asexual species, a group of individuals living in the same area.

population tree   A diagram that describes the relationships of ancestry and descent among populations.

positive selection   Selection in favor of advantageous mutations.

postzygotic isolation   Reproductive isolation between populations caused by dysfunctional development or sterility in hybrid forms.

pre-biotic soup   A hypothesized solution of organic molecules, synthesized via non-biological processes, in which the primordial form appeared by spontaneous generation.

preadaptation   A trait that changes due to natural selection and acquires a new function.

prezygotic isolation   Reproductive isolation between populations caused by differences in mate choice or timing of breeding, so that no hybrid zygotes are formed.

primordial form   The first organism; the first entity capable of: (1) replicating itself through the directed chemical transformation of its environment, and (2) evolving by natural selection.

proximate causation   Explanations for how, in terms of physiological or molecular mechanisms, traits function.

pseudogene   DNA sequences that are homologous to functioning genes, but are not transcribed.

qualitative trait   A trait for which phenotypes fall into discrete categories (such as red versus black breeding colors in stickleback fish).

quantitative trait   A trait for which phenotypes do not fall into discrete categories, but instead show continuous variation among individuals; a trait determined by the combined influence of the environment and many loci of small effect.

radiometric dating   Techniques for assigning absolute ages to rock samples, based on the ratio of parent to daughter radioactive isotopes present.

reaction norm   The pattern of phenotypic plasticity exhibited by a genotype.

reciprocal altruism   An exchange of fitness benefits, separated in time, between two individuals.

recombination rate (r)   The frequency, during meiosis, of crossing over between two linked loci; ranges from 0 to 0.5.

reinforcement   Natural selection that results in assortative mating in recently diverged populations in secondary contact; also known as reproductive character displacement.

relative dating   Techniques for assigning relative ages to rock strata, based on assumptions about the relationships between newer and older rocks.

relative fitness   The fitness of an individual, phenotype, or genotype compared to others in the population; can be calculated by dividing the individual's fitness by either (1) the mean fitness of the individuals in the population, or (2) the highest individual fitness found in the population; method (1) must be used when calculating the selection gradient.

replacement substitution   A DNA substitution that changes the amino acid or RNA sequence specified by a gene.

reproductive success (RS)   The number of viable, fertile offspring produced by an individual.

response to selection (R)   In quantitative genetics, the difference between the mean phenotype of the offspring of the selected individuals in a population and the mean phenotype of the offspring of all the individuals.

retrosequences   Transposable elements that move via an RNA intermediate but do not contain their own reverse transcriptase.

retrotransposons   Transposable elements that move via an RNA intermediate and contain the coding sequence for reverse transcriptase; closely related to retroviruses.

retrovirus   An RNA virus whose genome is reverse transcribed, to DNA, by reverse transcriptase.

ribozyme   An RNA molecule that has the ability to catalyze a chemical reaction.

root   The location on a phylogeny of the common ancestor of a clade.

selection coefficient (s)   A variable used in population genetics to represent the difference in fitness between one genotype and another.

selection differential (S)   A measure of the strength of selection used in quantitative genetics; equal to the difference between the mean phenotype of the selected individuals (for example, those that survive to reproduce) and the mean phenotype of the entire population.

selection gradient   A measure of the strength of selection used in quantitative genetics; for selection on a single trait it is equal to the slope of the best-fit line in a scatterplot showing relative fitness as a function of phenotype.

self-compatible   A hermaphroditic individual capable of fertilizing its own eggs and producing viable offspring.

selfish gene   An allele that is transmitted in non-Mendelian ratios, and thus spreads at the expense of competing alleles at the same locus (irrespective of its effect on fitness).

semelparous   Describes a species or population in which individuals experience only one bout of reproduction over the course of typical lifetime; salmon provide an example.

senescence   A decline with age in reproductive performance, physiological function, or probability of survival.

sexual dimorphism   A difference between the phenotypes of females and males within a species.

sexual selection   A difference, among members of the same sex, between the average mating success of individuals with a particular phenotype versus individuals with other phenotypes.

significant   In scientific discussions, typically a synonym for *statistically significant*.

silent substitution   A DNA substitution that does not change the amino acid or RNA sequence specified by the gene.

sister species   The species that diverged from the same ancestral node on a phylogenetic tree.

somatic-cell gene therapy   Treatment of disease via the direct manipulation of the genotypes of a population of non-germline cells within the patient's body.

species   Groups of interbreeding populations that are evolutionarily independent of other populations.

species richness   The number of species in a higher-level taxon.

species tree   Describes the relationships of ancestry and descent among species; the ancestors on the tree may be either hypothesized or known.

stabilizing selection   Occurs when individuals with intermediate values of a trait have higher fitness; can result in reduced phenotypic variation in a population, and can prevent evolution in the mean value of the trait.

standard error   The likely size of the error, due to chance effects, in an estimated value, such as the average phenotype for a population.

statistically significant   Describes a claim for which there is a degree of evidence in the data; by convention, a result is considered statistically significant if the probability is less than or equal to 0.05 that the observed violation of the null hypothesis is due to chance effects.

sympatric   Living in the same geographic area.

synapomorphy   A shared, derived character; in a phylogenetic analysis, synapomorphies are used to define clades and distinguish them from outgroups.

systematics   A scientific field devoted to the classification of organisms.

taxon   Any named group of organisms, (the plural form is taxa).

trade-off   An inescapable compromise between one trait and another.

transition   In DNA, a mutation that substitutes a purine for a purine, or a pyrimidine for a pyrimidine.

transitional form   A species that exhibits traits common to ancestral and derived groups, especially when the groups are sharply differentiated.

transmission distortion   Transmission of an allele in non-Mendelian proportions, due to mechanisms which eliminate gametes not carrying the allele (also known as meiotic drive).

transposable elements   Any DNA sequence capable of transmitting itself or a copy of itself to a new location in the genome.

transposons   Transposable elements that move via a DNA intermediate, and contain insertion sequences along with a transposase enzyme and possibly other coding sequences.

transversion   In DNA, a mutation that substitutes a purine for a pyrimidine, or a pyrimidine for a purine.

ultimate causation   Explanations for why, in terms of fitness benefits, traits evolved.

unequal cross-over   A crossing-over event between mispaired DNA strands that results in the duplication of sequences in some daughter strands and deletions in others.

uniformitarianism   The assumption (sometimes called a "law") that processes identical to those at work today are responsible for events that occurred in the past; first articulated by James Hutton, the founder of modern geology.

variance   A measure of the variation among the numbers in a list; to calculate the variance of a list of numbers, first square the difference between each number and the mean of the list, then take the sum of the squared differences and divide it by the number of items in the list. (For technical reasons, when researchers calculate the variance for a sample of individuals, they usually divide the sum of the squared differences by the sample size minus one).

vestigial traits   Rudimentary traits that are homologous to fully functional traits in closely related species.

vicariance   Splitting of a population's former range into two or more isolated patches.

wild type   A phenotype or allele common in nature.

Index

Note: Italicized page numbers followed by an *f* indicate figures; *t* indicates tables